Lecture Notes in Computer Science 14072

Founding Editors

Gerhard Goos
Juris Hartmanis

T0180358

The series Lecture Notes in Computer Science (LNCS), including its subseries Lecture Notes in Artificial Intelligence (LNAI) and Lecture Notes in Bioinformatics (LNBI), has established itself as a medium for the publication of new developments in computer science and information technology research, teaching, and education.

LNCS enjoys close cooperation with the computer science R & D community, the series counts many renowned academics among its volume editors and paper authors, and collaborates with prestigious societies. Its mission is to serve this international community by providing an invaluable service, mainly focused on the publication of conference and workshop proceedings and postproceedings. LNCS commenced publication in 1973.

Frank Nielsen · Frédéric Barbaresco
Editors

Geometric Science of Information

6th International Conference, GSI 2023
St. Malo, France, August 30 – September 1, 2023
Proceedings, Part II

 Springer

Editors
Frank Nielsen ⓘ
Sony Computer Science Laboratories Inc.
Tokyo, Japan

Frédéric Barbaresco ⓘ
THALES Land and Air Systems
Meudon, France

ISSN 0302-9743   ISSN 1611-3349 (electronic)
Lecture Notes in Computer Science
ISBN 978-3-031-38298-7   ISBN 978-3-031-38299-4 (eBook)
https://doi.org/10.1007/978-3-031-38299-4

This Springer imprint is published by the registered company Springer Nature Switzerland AG
The registered company address is: Gewerbestrasse 11, 6330 Cham, Switzerland

6th Geometric Science of Information Conference (GSI'23): From Classical To Quantum Information Geometry

Saint-Malo, France, Pierre Louis Moreau de Maupertuis' Birthplace

We are celebrating the 10th anniversary of the launch of the GSI conferences cycle, which were initiated in 2013. As for GSI'13, GSI'15, GSI'17, GSI'19 and GSI'21 (https://franknielsen.github.io/GSI/), the objective of this 6th edition of the SEE GSI conference, hosted in Saint-Malo, birthplace of Pierre Louis Moreau de Maupertuis, is to bring together pure and applied mathematicians and engineers with a common interest in geometric tools and their applications for information analysis. GSI emphasizes the active participation of young researchers to discuss emerging areas of collaborative research on the topic of "Geometric Science of Information and its Applications". In 2023, GSI's main theme was "FROM CLASSICAL TO QUANTUM INFORMATION GEOMETRY", and the conference took place at the Palais du Grand Large, in Saint-Malo, France.

The GSI conference cycle was initiated by the Brillouin Seminar Team as early as 2009 (http://repmus.ircam.fr/brillouin/home). The GSI'21 event was motivated by the continuity of the first initiative, launched in 2013 (https://web2.see.asso.fr/gsi2013), at Mines ParisTech, consolidated in 2015 (https://web2.see.asso.fr/gsi2015) at Ecole Polytechnique, and opened to new communities in 2017 (https://web2.see.asso.fr/gsi2017) at Mines ParisTech, 2019 (https://web2.see.asso.fr/gsi2019) at ENAC Toulouse and 2021 (https://web2.see.asso.fr/gsi2021) at Sorbonne University. We mention that in 2011, we organized an Indo-French workshop on the topic of "Matrix Information Geometry" (https://www.lix.polytechnique.fr/~nielsen/MIG/) that yielded an edited book in 2013, and in 2017, collaborated at a CIRM seminar in Luminy on the event TGSI'17 "Topological & Geometrical Structures of Information" (https://fconferences.cirm-math.fr/1680.html).

GSI satellite events were organized in 2019 and 2020 as FGSI'19 "Foundation of Geometric Structures of Information" in Montpellier (https://fgsi2019.sciencesconf.org/) and Les Houches Seminar SPIGL'20 "Joint Structures and Common Foundations of Statistical Physics, Information Geometry and Inference for Learning" (https://franknielsen.github.io/SPIG-LesHouches2020/).

The technical program of GSI'23 covered all the main topics and highlights in the domain of the "Geometric Science of Information" including information geometry manifolds of structured data/information and their advanced applications. These Springer LNCS proceedings consist solely of original research papers that have been carefully single-blind peer-reviewed by at least two or three experts. 125 of 161 submissions were accepted for this volume. Accepted contributions were revised before acceptance.

Like GSI'13, GSI'15, GSI'17, GSI'19, and GSI'21, GSI'23 addresses inter-relations between different mathematical domains such as shape spaces (geometric statistics on manifolds and Lie groups, deformations in shape space, ...), probability/optimization and algorithms on manifolds (structured matrix manifolds, structured data/information, ...), relational and discrete metric spaces (graph metrics, distance geometry, relational analysis,...), computational and Hessian information geometry, geometric structures in thermodynamics and statistical physics, algebraic/infinite-dimensional/Banach information manifolds, divergence geometry, tensor-valued morphology, optimal transport theory, manifold and topology learning, ... and applications such as geometries of audio-processing, inverse problems and signal/image processing. GSI'23 topics were enriched with contributions from Lie Group Machine Learning, Harmonic Analysis on Lie Groups, Geometric Deep Learning, Geometry of Hamiltonian Monte Carlo, Geometric & (Poly)Symplectic Integrators, Contact Geometry & Hamiltonian Control, Geometric and structure-preserving discretizations, Probability Density Estimation & Sampling in High Dimension, Geometry of Graphs and Networks and Geometry in Neuroscience & Cognitive Sciences.

At the turn of the century, new and fruitful interactions were discovered between several branches of science: Information Sciences (information theory, digital communications, statistical signal processing), Mathematics (group theory, geometry and topology, probability, statistics, sheaf theory, ...) and Physics (geometric mechanics, thermodynamics, statistical physics, quantum mechanics, ...). The GSI biannual international conference cycle is an effort to discover joint mathematical structures to all these disciplines by elaboration of a "General Theory of Information" embracing physics science, information science, and cognitive science in a global scheme.

The GSI'23 conference was structured in 25 sessions of more than 120 papers and a poster session:

- **Geometry and Machine Learning**

  - **Geometric Green Learning** - Alice Barbara Tumpach, Diarra Fall & Guillaume Charpiat
  - **Neurogeometry Meets Geometric Deep Learning** - Remco Duits & Erik Bekkers, Alessandro Sarti
  - **Divergences in Statistics & Machine Learning** - Michel Broniatowski & Wolfgang Stummer

- **Divergences and Computational Information Geometry**

  - **Computational Information Geometry and Divergences** - Frank Nielsen & Olivier Rioul
  - **Statistical Manifolds and Hessian Information Geometry** - Michel Nguiffo Boyom

- **Statistics, Topology and Shape Spaces**

  - **Statistics, Information and Topology** - Pierre Baudot & Grégoire Seargeant-Perthuis
  - **Information Theory and Statistics** - Olivier Rioul
  - **Statistical Shape Analysis and more Non-Euclidean Statistics** - Stephan Huckemann & Xavier Pennec

- **Probability and Statistics on Manifolds** - Cyrus Mostajeran
- **Computing Geometry & Algebraic Statistics** - Eliana Duarte & Elias Tsigaridas

- **Geometry & Mechanics**

  - **Geometric and Analytical Aspects of Quantization and Non-Commutative Harmonic Analysis on Lie Groups** - Pierre Bieliavsky & Jean-Pierre Gazeau
  - **Deep Learning: Methods, Analysis and Applications to Mechanical Systems** - Elena Celledoni, James Jackaman, Davide Murari and Brynjulf Owren
  - **Stochastic Geometric Mechanics** - Ana Bela Cruzeiro & Jean-Claude Zambrini
  - **Geometric Mechanics** - Géry de Saxcé & Zdravko Terze
  - **New trends in Nonholonomic Systems** - Manuel de Leon & Leonardo Colombo

- **Geometry, Learning Dynamics & Thermodynamics**

  - **Symplectic Structures of Heat & Information Geometry** - Frédéric Barbaresco & Pierre Bieliavsky
  - **Geometric Methods in Mechanics and Thermodynamics** - François Gay-Balmaz & Hiroaki Yoshimura
  - **Fluid Mechanics and Symmetry** - François Gay-Balmaz & Cesare Tronci
  - **Learning of Dynamic Processes** - Lyudmila Grigoryeva

- **Quantum Information Geometry**

  - **The Geometry of Quantum States** - Florio M. Ciaglia
  - **Integrable Systems and Information Geometry (From Classical to Quantum)** - Jean-Pierre Francoise, Daisuke Tarama

- **Geometry & Biological Structures**

  - **Neurogeometry** - Alessandro Sarti, Giovanna Citti & Giovanni Petri
  - **Bio-Molecular Structure Determination by Geometric Approaches** - Antonio Mucherino
  - **Geometric Features Extraction in Medical Imaging** - Stéphanie Jehan-Besson & Patrick Clarysse

- **Geometry & Applications**

  - **Applied Geometric Learning** - Pierre-Yves Lagrave, Santiago Velasco-Forero & Teodora Petrisor

June 2023   Frank Nielsen
Frédéric Barbaresco

The original version of the book was revised: the book was inadvertently published with a typo in the frontmatter. This has been corrected. The correction to the book is available at https://doi.org/10.1007/978-3-031-38299-4_66

Organization

Conference Co-chairs

Frank Nielsen   Sony Computer Science Laboratories Inc., Japan
Frédéric Barbaresco   Thales Land & Air Systems, France

Local Organizing Committee

SEE Groupe Régional GRAND OUEST
Christophe Laot, SEE & IMT Atlantique, France
Alain Alcaras, SEE & THALES SIX, France
Jacques Claverie, SEE & CREC St-Cyr Cöetquidan, France
Palais du Grand Large Team, Saint-Malo

Secretariat

Imene Ahmed   SEE, France

Scientific Committee

Bijan Afsari   Johns Hopkins University, USA
Pierre-Antoine Absil   Université Catholique de Louvain, Belgium
Jesus Angulo   Mines ParisTech, France
Nihat Ay   Max Planck Institute, Germany
Simone Azeglio   ENS Paris, France
Frédéric Barbaresco   Thales Land & Air Systems, France
Pierre Baudot   Median Technologies, France
Daniel Bennequin   Paris-Diderot University, France
Pierre Bieliavsky   Université Catholique de Louvain, Belgium
Michel Boyom   Montpellier University, France
Goffredo Chirco   University of Naples Federico II, Italy
Florio M. Ciaglia   Max Planck Institute, Germany
Nicolas Couellan   ENAC, France
Ana Bela Ferreira Cruzeiro   Universidade de Lisboa, Portugal
Ariana Di Bernardo   ENS Paris, France

GSI'23 Keynote Speakers

Information Theory with Kernel Methods

Francis Bach

Inria, Ecole Normale Supérieure

Abstract. Estimating and computing entropies of probability distributions are key computational tasks throughout data science. In many situations, the underlying distributions are only known through the expectation of some feature vectors, which has led to a series of works within kernel methods. In this talk, I will explore the particular situation where the feature vector is a rank-one positive definite matrix, and show how the associated expectations (a covariance matrix) can be used with information divergences from quantum information theory to draw direct links with the classical notions of Shannon entropies.

Reference

1. Francis, B.: Information theory with kernel methods. To appear IEEE Trans. Inf. Theor (2022). https://arxiv.org/pdf/2202.08545

From Alan Turing to Contact Geometry: Towards a "Fluid Computer"

Eva Miranda

Universitat Politècnica de Catalunya and Centre de Recerca Matemàtica

Abstract. Is hydrodynamics capable of performing computations? (Moore 1991) Can a mechanical system (including a fluid flow) simulate a universal Turing machine? (Tao, 2016)

Etnyre and Ghrist unveiled a mirror between contact geometry and fluid dynamics reflecting Reeb vector fields as Beltrami vector fields. With the aid of this mirror, we can answer in the positive the questions raised by Moore and Tao. This is a recent result that mixes up techniques from Alan Turing with modern Geometry (contact geometry) to construct a "Fluid computer" in dimension 3. This construction shows, in particular, the existence of undecidable fluid paths. I will also explain applications of this mirror to the detection of escape trajectories in Celestial Mechanics (for which I'll need to extend the mirror to a singular set-up). This mirror allows us to construct a tunnel connecting problems in Celestial Mechanics and Fluid Dynamics.

References

1. Robert, C., Eva, M., Daniel, P.-S., Francisco, P.: Constructing turing complete euler flows in dimension 3. Proc. Natl. Acad. Sci. USA **118**(19), 9. Paper No. e2026818118 (2021)
2. Etnyre, J., Ghrist, R.: Contact topology and hydrodynamics: I. Beltrami fields and the Seifert conjecture. Nonlinearity **13**, 441 (2000)
3. Miranda, E., Oms, C., Peralta-Salas, D.: On the singular Weinstein conjecture and the existence of escape orbits for b-Beltrami fields. Commun. Contemp. Math. **24**(7), 25. Paper No. 2150076 (2022)
4. Tao, T.: Finite time blowup for an averaged three-dimensional Navier–Stokes equation. J. Am. Math. Soc. **29**, 601–674 (2016)
5. Turing, A.: On computable numbers, with an application to the entscheidungsproblem. Proc. London Math. Soc. **s2–42**(1), 230–265 (1937). DOI:10.1112/plms/s2-42.1.230 ISSN 0024-6115

Transverse Poisson Structures to Adjoint Orbits in a Complex Semi-simple Lie Algebra

Hervé Sabourin

Director for Strategic projects of the Réseau Figure® (network of 31 universities)
Former Regional Director of the A.U.F. (Agence Universitaire de la Francophonie)
for the Middle East
Former Vice-President of the University of Poitiers, France

Abstract. The notion of transverse Poisson structure has been introduced by Alan Weinstein stating in his famous splitting theorem that any Poisson Manifold M is, in the neighbourhood of each point m, the product of a symplectic manifold, the symplectic leaf S at m, and a submanifold N which can be endowed with a structure of Poisson manifold of rank 0 at m. N is called a transverse slice at M of S. When M is the dual of a complex Lie algebra g equipped with its standard Lie-Poisson structure, we know that the symplectic leaf through x is the coadjoint G. x of the adjoint Lie group G of g. Moreover, there is a natural way to describe the transverse slice to the coadjoint orbit and, using a canonical system of linear coordinates $(q1, \ldots, qk)$, it follows that the coefficients of the transverse Poisson structure are rational in $(q1, \ldots, qk)$. Then, one can wonder for which cases that structure is polynomial. Nice answers have been given when g is semi-simple, taking advantage of the explicit machinery of semi-simple Lie algebras. One shows that a general adjoint orbit can be reduced to the case of a nilpotent orbit where the transverse Poisson structure can be expressed in terms of quasihomogeneous polynomials. In particular, in the case of the subregular nilpotent orbit the Poisson structure is given by a determinantal formula and is entirely determined by the singular variety of nilpotent elements of the slice.

References

1. Sabourin, H.: Sur la structure transverse à une orbite nilpotente adjointe. Canad. J. Math. **57**(4), 750–770 (2005)
2. Sabourin, H.: Orbites nilpotentes sphériques et représentations unipotentes associées : Le cas SL(n). Represent. Theor. **9**, 468–506 (2005)
3. Sabourin, H.: Mémoire d'HDR, Quelques aspects de la méthode des orbites en théorie de Lie, Décembre (2005)
4. Damianou, P., Sabourin, H., Vanhaecke, P.: Transverse poisson structures to adjoint orbits in semi-simple Lie algebras, Pacific J. Math. **232**, 111–139 (2007)

5. Sabourin, H., Damianou, P., Vanhaecke, P.: Transverse poisson structures: the subregular and the minimal orbits, differential geometry and its applications. Proc. Conf. Honour Leonhard Euler, Olomouc, August (2007)
6. Sabourin, H., Damianou, P., Vanhaecke, P.: Nilpotent orbits in simple Lie algebras and their transverse poisson structures. Am. Inst. Phys. Conf. Proc. Ser. **1023**, 148–152 (2008)

Statistics Methods for Medical Image Processing and Reconstruction

Diarra Fall

Institut Denis Poisson, UMR CNRS, Université d'Orléans & Université de Tours, France

Abstract. In this talk we will see how statistical methods, from the simplest to the most advanced ones, can be used to address various problems in medical image processing and reconstruction for different imaging modalities. Image reconstruction allows the images in question to be obtained, while image processing (on the already reconstructed images) aims at extracting some information of interest. We will review several statistical methods (mainly Bayesian) to address various problems of this type.

Keywords: Image processing · Image reconstruction · Statistics · Frequentist · Bayesian · Parametrics · Nonparametrics

References

1. Fall, M.D., Dobigeon, N., Auzou, P.: A bayesian estimation formulation to voxel-based lesion symptom mapping. In: Proceedings of European Signal Processing Conference (EUSIPCO), Belgrade, Serbia, September (2022)
2. Fall, M.D.: Bayesian nonparametrics and biostatistics: the case of PET imaging. Int. J. Biostat. (2019)
3. Fall, M.D., Lavau, E., Auzou, P.: Voxel-based lesion-symptom mapping: a nonparametric bayesian approach. In: Proceedings of IEEE International Conference on Acoustics, Speech and Signl Processing (ICASSP) (2018)

Algebraic Statistics and Gibbs Manifolds

Bernd Sturmfels

MPI-MiS Leipzig, Germany

Abstract. Gibbs manifolds are images of affine spaces of symmetric matrices under the exponential map. They arise in applications such as optimization, statistics and quantum physics, where they extend the ubiquitous role of toric geometry. The Gibbs variety is the zero locus of all polynomials that vanish on the Gibbs manifold. This lecture gives an introduction to these objects from the perspective of Algebraic Statistics.

References

1. Pavlov, D., Sturmfels, B., Telen, S.: Gibbs manifolds. arXiv:2211.15490
2. Sturmfels, B., Telen, S., Vialard, F.-X., von Renesse, M.: Toric geometry of entropic regularization. arXiv:2202.01571
3. Sullivant, S.: Algebraic Statistics. graduate studies in mathematics, Am. Math. Soc. Providence, RI, 194 (2018)
4. Huh, J., Sturmfels, B.: Likelihood geometry, in combinatorial algebraic geometry. In: Conca, A., et al. Lecture Notes in Mathematics, vol. 2108, Springer, pp. 63–117 (2014)
5. Geiger, D., Meek, C., Sturmfels, B.: On the toric algebra of graphical models, Annal. Stat. **34**, 1463–1492 (2006)

Learning of Dynamic Processes

Juan-Pablo Ortega

Head, Division of Mathematical Sciences, Associate Chair (Faculty), School of
Physical and Mathematical Sciences, Nanyang Technological University, Singapore

Abstract. The last decade has seen the emergence of learning techniques
that use the computational power of dynamical systems for information
processing. Some of those paradigms are based on architectures that are
partially randomly generated and require a relatively cheap training effort,
which makes them ideal in many applications. The need for a mathematical understanding of the working principles underlying this approach,
collectively known as Reservoir Computing, has led to the construction of new techniques that put together well-known results in systems
theory and dynamics with others coming from approximation and statistical learning theory. In recent times, this combination has allowed
Reservoir Computing to be elevated to the realm of provable machine
learning paradigms and, as we will see in this talk, it also hints at various connections with kernel maps, structure-preserving algorithms, and
physics-inspired learning.

References

1. Gonon, L., Grigoryeva, L., Ortega, J.-P.: Approximation bounds for random neural
   networks and reservoir systems. To appear in The Annals of Applied Probability.
   Paper (2022)
2. Cuchiero, C., Gonon, L., Grigoryeva, L., Ortega, J.-P., Teichmann, J.: Expressive
   power of randomized signature. NeurIPS. Paper (2021)
3. Cuchiero, C., Gonon, L., Grigoryeva, L., Ortega, J.-P., Teichmann, J.: Discrete-time
   signatures and randomness in reservoir computing. IEEE Trans. Neural Netw. Learn.
   Syst. **33**(11), 6321–6330. Paper (2021)
4. Gonon, L., Ortega, J.-P.: Fading memory echo state networks are universal. Neural
   Netw. **138**, 10–13. Paper (2021)
5. Gonon, L., Grigoryeva, L., Ortega, J.-P.: Risk bounds for reservoir computing. J.
   Mach. Learn. Res. **21**(240), 1–61. Paper (2020)
6. Gonon, L., Ortega, J.-P.: Reservoir computing universality with stochastic inputs.
   IEEE Trans. Neural Netw. Learn. Syst. **31**(1), 100–112. Paper (2020)
7. Grigoryeva, L., Ortega, J.-P.: Differentiable reservoir computing. J. Mach. Learn.
   Res. **20**(179), 1–62. Paper (2019)

8. Grigoryeva, L., Ortega, J.-P.: Echo state networks are universal. Neural Netw. **108**, 495–508. Paper (2018)
9. Grigoryeva, L., Ortega, J.-P.: Universal discrete-time reservoir computers with stochastic inputs and linear readouts using non-homogeneous state-affine systems. J. Mach. Learn. Res. **19**(24), 1–40. Paper (2018)

Pierre Louis Moreau de Maupertuis, King's Musketeer Lieutenant of Science and Son of a Saint-Malo Corsaire

« *Héros de la physique, Argonautes nouveaux/Qui franchissez les monts, qui traversez les eaux/Dont le travail immense et l'exacte mesure/De la Terre étonnée ont fixé la figure./Dévoilez ces ressorts, qui font la pesanteur./Vous connaissez les lois qu'établit son auteur.* » … [Heroes of physics, new Argonauts/Who cross the mountains, who cross the waters/Whose immense work and the exact measure/Of the astonished Earth fixed the figure./Reveal these springs, which make gravity./You know the laws established by its author.] - Voltaire on Pierre Louis Moreau de Maupertuis

Son of René Moreau de Maupertuis (1664–1746) a corsair and ship owner from Saint-Malo, director of the Compagnie des Indes and knighted by Louis XIV, Maupertuis was offered a cavalry regiment at the age of twenty. His father, with whom he had a very close relationship, thus opened the doors of the gray musketeers to him, of which he became lieutenant. Between 1718 and 1721, Maupertuis devoted himself to a military career, first joining the company of gray musketeers, then a cavalry regiment in Lille, without abandoning his studies. In 1718, Maupertuis entered the gray musketeers, writes Formey in his Éloge (1760), but he carried there the love of study, and above all the taste for geometry. However, his profession as a soldier was not to last long and at the end of 1721, the learned Malouin finally and permanently went to Paris, as he could not last long in the idleness of the state of a former military officer in time of peace, and soon he took leave of it. This moment marks the official entry of Maupertuis into Parisian intellectual life, halfway between the literary cafés and the benches of the Academy. He nevertheless preferred to abandon this military career to devote himself to the study of mathematics, an orientation crowned in 1723 by his appointment as a member of the Academy of Sciences.

He then published various works of mechanics and astronomy. In 1728, Maupertuis visited London, a trip which marked a decisive turning point in his career. Elected associate member of the Royal Society, he discovered Newton's ideas, in particular universal attraction, of which he was to become an ardent propagandist in France, which D'Alembert, in the Discourse preliminary to the Encyclopedia, did not miss. Academician at 25, Pierre-Louis Moreau de Maupertuis led a perilous expedition to Lapland to verify Newton's theory and became famous as "the man who flattened the earth". Called by Frederick II to direct the Berlin Academy of Sciences, he was as comfortable in the royal courts as in the Parisian salons.

The rejection of the Newtonian approach, as well as the distrust of the Cartesian approach, led Maupertuis to the elaboration of a cosmology different from both the finalism of some and the anti-finalism of others. It is a cosmology that cannot be attributed to any particular tradition, and that must rather be read as an independent and creative elaboration. All of Maupertuis' cosmology is based on a physical principle which he was the first to formulate, namely the principle of least action, the novelty and generality of which he underlines on several occasions.

His "principle of least action" constitutes an essential contribution to physics to this day, a fundamental principle in classical mechanics. It states that the motion of a particle between two points in a conservative system is such that the action integral, defined as the integral of the Lagrangian over the time interval of motion, is minimized. Maupertuis' principle was renewed by the Cartan-Poincaré Integral Invariant in the field of geometric mechanics. In geometric mechanics, the motion of a mechanical system is described in terms of differential forms on a configuration manifold and the Cartan-Poincaré integral invariant is associated with a particular differential form called the symplectic form, which encodes the dynamics of the system. The integral invariant is defined as the integral of the symplectic form over a closed loop in the configuration manifold. More recently, Maupertuis' principle has been extended more recently by Jean-Marie Souriau through Maxwell's principle with the hypothesis that the exterior derivative of the Lagrange 2-form of a general dynamical system vanishes. For systems of material points, Maxwell's principle allows us, under certain conditions, to define a Lagrangian and to show that the Lagrange form is nothing else than the exterior derivative of the Cartan form, in the study of the calculus of variations. Without denying the importance of the principle of least action nor the usefulness of these formalisms, Jean-Marie Souriau declares that Maupertuis' principle and least action principle seem to him less fundamental than Maxwell's principle. His viewpoint seems to him justified because the existence of a Lagrangian is ensured only locally, and because there exist important systems, such as those made of particles with spin, to which Maxwell's principle applies while they have not a globally defined Lagrangian. Jean-Marie Souriau has also geometrized Noether's theorem (algebraic theorem proving that we can associate invariants to symmetries) with "moment map" (components of moment map are Noether's invariants).

« *La lumière ne pouvant aller tout-à-la fois par le chemin le plus court, et par celui du temps le plus prompt ... ne suit-elle aucun des deux, elle prend une route qui a un avantage plus réel : le chemin qu'elle tient est celui par lequel la quantité d'action est la moindre.* » *[Since light cannot go both by the shortest path and by that of the quickest time... if it does not follow either of the two, it takes a route which has a more real advantage: the path that it holds is that by which the quantity of action is least.]* - Maupertuis 1744.

ALEAE GEOMETRIA – BLAISE PASCAL's 400th Birthday

We celebrate in 2023 Blaise Pascal's 400th birthday. GSI'23 motto is "ALEA GEOMETRIA".

In 1654, Blaise Pascal submitted a paper to « Celeberrimae matheseos Academiae Parisiensi » entitled « ALEAE GEOMETRIA : De compositione aleae in ludis ipsi subjectis »

- « … et sic matheseos demonstrationes cum aleae incertitudine jugendo, et quae contraria videntur conciliando, ab utraque nominationem suam accipiens, stupendum hunc titulum jure sibi arrogat: **Aleae Geometria** »
- « … par l'union ainsi réalisée entre les démonstrations des mathématiques et l'incertitude du hasard, et par la conciliation entre les contraires apparents, elle peut tirer son nom de part et d'autre et s'arroger à bon droit ce titre étonnant: **Géométrie du Hasard** »
- « … by the union thus achieved between the demonstrations of mathematics and the uncertainty of chance, and by the conciliation between apparent opposites, it can take its name from both sides and arrogate to right this amazing title: **Geometry of Chance** »

Blaise Pascal had a multi-disciplinary approach of Science, and has developed 4 topics directly related to GSI'23:*

- **Blaise Pascal and COMPUTER**: Pascaline marks the beginning of the development of mechanical calculus in Europe, followed by Charles Babbage analytical machine from 1834 to 1837, a programmable calculating machine combining the inventions of Blaise Pascal and Jacquard's machine, with instructions written on perforated cards.
- **Blaise Pascal and PROBABILITY:** The "calculation of probabilities" began in a correspondence between Blaise Pascal and Pierre Fermat. In 1654, Blaise Pascal submitted a short paper to "Celeberrimae matheseos Academiae Parisiensi" with the title "Aleae Geometria" (Geometry of Chance), that was the seminal paper founding Probability as a new discipline in Science.
- **Blaise Pascal and THERMODYNAMICS:** Pascal's Experiment in the Puy de Dôme to Test the Relation between Atmospheric Pressure and Altitude. In 1647, Blaise Pascal suggests to raise Torricelli's mercury barometer at the top of the Puy de Dome Mountain (France) in order to test the "weight of air" assumption.
- **Blaise Pascal and DUALITY:** Pascal's Hexagrammum Mysticum Theorem, and its dual Brianchon's Theorem. In 1639 Blaise Pascal discovered, at age sixteen, the famous hexagon theorem, also developed in "Essay pour les Coniques", printed in 1640, declaring his intention of writing a treatise on conics in which he would derive the major theorems of Apollonius from his new theorem.

The GSI'23 Conference is Dedicated to the Memory of Mademoiselle Paulette Libermann, Geometer Student of Elie Cartan and André Lichnerowicz, PhD Student of Charles Ehresmann and Familiar with the Emerald Coast of French Brittany

Paulette Libermann died on July 10, 2007 in Montrouge near Paris. Admitted to the entrance examination to the Ecole Normale Supérieure de Sèvres in 1938, she was a pupil of Elie Cartan and André Lichnerowicz. Paulette Libermann was able to learn about mathematical research under the direction of Elie Cartan, and was a faithful friend of the Cartan family. After her aggregation, she was appointed to Strasbourg and rubbed shoulders with Georges Reeb, René Thom and Jean-Louis Koszul. She prepared a thesis under the direction of Charles Ehresmann, defended in 1953. She was the first ENS Sèvres woman to hold a doctorate in mathematics. She was then appointed professor at the University of Rennes and after at the Faculty of Sciences of the University of Paris in 1966. She began to collaborate with Charles-Michel Marle in 1967. She led a seminar with Charles Ehresmann until his death in 1979, and then alone until 1990. In her thesis, entitled "On the problem of equivalence of regular infinitesimal structures", she studied the symplectic manifolds provided with two transverse Lagrangian foliations and showed the existence, on the leaves of these foliations, of a canonical flat connection. Later, Dazord and Molino, in the South-Rhodanian geometry seminar, introduced the notion of Libermann foliation, linked to Stefan foliations and Haefliger Γ-structures. Paulette Libermann also deepened the importance of the foliations of a symplectic manifold which she called "simplectically complete", such as the Poisson bracket of two functions, locally defined, constant on each leaf, that is also constant on each leaf. She proved that this property is equivalent to the existence of a Poisson structure on the space of leaves, such that the canonical projection is a Poisson map, and also equivalent to the complete integrability of the subbundle symplectically orthogonal to the bundle tangent to the leaves. She wrote a famous book with Professor Charles-Michel Marle, "Symplectic Geometry and Analytical Mechanics". Professor Charles-Michel Marle told us that Miss Paulette Libermann had bought an apartment in Dinard and spent her summers just in front of Saint-Malo, and so was familiar with the emerald coast of French Brittany.

GSI'23 Sponsors

THALES (https://www.thalesgroup.com/en) and European Horizon CaLIGOLA (https://site.unibo.it/caligola/en) were both PLATINIUM SPONSORS of the SEE GSI'23 conference.

Contents – Part II

Symplectic Structures of Heat and Information Geometry

Geometric Methods in Mechanics and Thermodynamics

Fluid Mechanics and Symmetry

Learning of Dynamic Processes

The Geometry of Quantum States

Bio-Molecular Structure Determination by Geometric Approaches

Geometric Features Extraction in Medical Imaging

Applied Geometric Learning

Contents – Part I

Divergences in Statistics and Machine Learning

Computational Information Geometry and Divergences

Statistical Manifolds and Hessian Information Geometry

Statistics, Information and Topology

Information Theory and Statistics

Statistical Shape Analysis and more Non-Euclidean Statistics

Probability and Statistics on Manifolds

Computing Geometry and Algebraic Statistics

Geometric and Analytical Aspects of Quantization and Non-Commutative Harmonic Analysis on Lie Groups

Deep Learning: Methods, Analysis and Applications to Mechanical Systems

Stochastic Geometric Mechanics

Geometric Mechanics

Lie Group Quaternion Attitude-Reconstruction of Quadrotor UAV

Zdravko Terze[(⊠)], Dario Zlatar, Marko Kasalo, and Marijan Andrić

Department of Aeronautical Engineering, Faculty of Mechanical Engineering
and Naval Architecture, University of Zagreb, Ivana Lučića 5, 10002 Zagreb, Croatia
{zdravko.terze,dario.zlatar,marko.kasalo,marijan.andric}@fsb.hr

Abstract. The quadrotor unmanned aerial vehicles (UAVs) have already gained enormous popularity, both for commercial and hobby applications. They are utilized for a wide range of practical tasks, including fire protection, search and rescue, border surveillance, etc. Therefore, there is an ever increasing need for better controllers and dynamics simulators. The conventional approach to modeling UAVs is to use rotational quaternions for attitude determination, together with position vector for tracking the UAV's center of mass. The attitude is then usually updated by integrating linearized quaternion differential equations and subsequently enforcing unitary norm of the quaternion, through additional algebraic equation. The paper presents utilization of the recently introduced attitude and position update algorithms - based on Lie groups - for modeling UAV dynamics, which exhibit better computational characteristics.

Keywords: Quadrotor · UAV · Quaternion · Lie groups

1 Introduction

The quadrotor unmanned aerial vehicles (UAVs) have already gained an enormous popularity, due to their versatility, ease of use and relatively low cost. However, the high degree of nonlinearity inherent for quadrotor dynamics presents a significant challenge for quadrotor control design [1]. The quadrotor attitude is conventionally described by using rotational quaternions, in order to overcome singularities that arise when using any global three-parameter description (for example often used Euler angles [2]) of the 3D rotation.

The quadrotor UAV attitude reconstruction problem during maneuvering can be divided in two separate tasks - determination of the attitude change due to actuation and due to unknown phenomena, such as gusts of wind for example. The rotation due to the unknown phenomena is usually considered as noise and a type of error minimization technique is utilized (see for example [3]). The conventional approach for performing kinematic reconstruction of attitude described by quaternion is to numerically solve four differential equations with

F. Nielsen and F. Barbaresco (Eds.): GSI 2023, LNCS 14072, pp. 3–11, 2023.
https://doi.org/10.1007/978-3-031-38299-4_1

a chosen scheme without any consideration for the unitary norm constraint, and then after each step perform stabilization of the quaternion, by dividing it by its norm. A different approach, based on Lie group formulations, is adopted in this paper and the benefits of that approach are presented by numerical examples. Instead of solving four linear kinematic differential equations and explicitly ('brute force') enforcing algebraic unitary constraint in each time step, the attitude reconstruction is performed by solving three nonlinear kinematic differential equations in the manifold tangent plane (Lie algebra) to calculate incremental rotation vector and then reconstructing the quaternion by exponential mapping, inherently preserving unitary norm constraint, as introduced in [4].

The quaternions are used as parameters for describing rotations of a formation of quadrotor UAVs in [6]. Another controller discussed in [7] is also based on the similar quaternion-based algorithm for simulating rotations. All these papers use algorithms which are equivalent to the conventional method used in this paper (Method I) as a baseline for comparison and evaluation of the recently introduced geometric algorithms (Methods II and III) [4,5].

2   Quadrotor UAV Computational Model

The X configuration quadrotor is used for developing a computational model as shown in Fig. 1. It is important to emphasize that the reason why the rotors 2 and 4 are rotating in opposite direction from rotors 1 and 3 is to be able to control the yaw of the UAV.

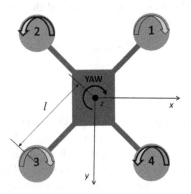

Fig. 1. Drawing of the X configuration quadrotor

2.1   Quadrotor Dynamics Modeling

The main goal of the paper is to present a novel scheme for forward dynamics simulation of the quadrotor, focusing on the better kinematical reconstruction of the UAV's position and attitude. Therefore, the complex aerodynamic phenomena are neglected, and it is assumed that the thrust force in the rotor axis direction F_T and drag torque $\boldsymbol{T}_D$ can be defined as a function depending only on rotor rotational speed and written as $F_{Ti} = f(\omega_i)$ and $\boldsymbol{T}_{Di} = f(\omega_i)$, where ω_i represent the angular velocity of the i-th rotor. To this end, rotational dynamics of the quadrotor UAV can be formulated as

$$\boldsymbol{J}\dot{\boldsymbol{\omega}} = \sum_{i=1}^{4} \boldsymbol{T}_{Ti} + \sum_{i=1}^{4} \boldsymbol{T}_{Di} - \tilde{\boldsymbol{\omega}}\boldsymbol{J}\boldsymbol{\omega} - \tilde{\boldsymbol{\omega}}\sum_{i=1}^{4} \boldsymbol{J}_{mi}\boldsymbol{\omega}_{mi}, \tag{1}$$

where $\boldsymbol{J}$ represents UAV's inertia matrix, while $\boldsymbol{J}_{mi}$ and $\boldsymbol{\omega}_{mi}$ represent inertia matrix and relative angular velocity of the i-th rotor with respect to the UAV. The thrust torque $\boldsymbol{T}_{Ti}$ is calculated as $\boldsymbol{T}_{Ti} = \sum_{i=1}^{4} \tilde{r}_i F_{Ti}$, where r_i represents origin of the reference frame fixed to the i-th rotor. Also, it is important to emphasize that the term $\sum_{i=1}^{4} \boldsymbol{J}_{mi}\dot{\boldsymbol{\omega}}_{mi}$ is neglected because the angular acceleration of the i-th rotor with respect to the UAV is assumed as sufficiently small. All terms are expressed in the coordinate system attached to the body frame of the UAV.

The translational and rotational dynamics equations can be solved by any ODE numerical scheme to obtain field of translational and angular velocities. The part that deserves a more careful treatment is the kinematical reconstruction step, to obtain position and - specially - attitude of the UAV at any given point in time.

2.2   Kinematic Reconstruction

The position of the UAV is easily reconstructed from the translational velocity by using any ODE integration scheme. On the other hand, there are multiple ways that can be used to represent UAV's attitude. Probably the most common approach taken is to use rotational quaternions due to the fact that - by using quaternions - rotational singularities of 3D parameters (such as Euler angles) are avoided.

The quaternion $\boldsymbol{Q}$ is commonly considered as a four-elements vector consisting of real part q_0 and imaginary part $\boldsymbol{q}$, such that $\boldsymbol{Q} = \begin{bmatrix} q_0 & \boldsymbol{q} \end{bmatrix}^\top$. The parameters of rotational quaternion are often called Euler parameters, and must inherently satisfy unitary norm condition $||\boldsymbol{Q}|| = q_0^2 + \boldsymbol{q}^T\boldsymbol{q} = 1$.

Method I Conventional Approach to Kinematic Reconstruction of Quaternions. The conventional approach (used for example in [6, 7]) for reconstruction of rotational quaternions from the angular velocity field is to use the expression

$$\dot{\boldsymbol{Q}} = \frac{1}{2}\boldsymbol{Q} \circ \boldsymbol{\omega}. \tag{2}$$

These four linear differential equations, together with the unitary norm condition equation, form a DAE problem, that is usually solved by integrating (2) with a chosen ODE solver, and subsequently enforcing unitary norm with equation $\boldsymbol{Q}_{out} = \boldsymbol{Q} \setminus ||\boldsymbol{Q}||$. However, as it will be presented in a numerical case later in the paper, this approach suffers at larger time steps due to the linearization of the 3D rotation, which is inherently a nonlinear phenomenon.

Method II The Geometric (Lie Group) Approach to Kinematic Reconstruction of Quaternions. The different approach is introduced in [4], that mitigates above-mentioned limitations. The approach is based on isomorphism between unit quaternion group, unit sphere in $\mathcal{R}^4$, symplectic group $Sp(1)$ and special unitary group $SU(2)$. The unit quaternions can be described as elements of the sphere in $\mathcal{R}^4$ as $\mathcal{S}^3 = \{Q \in \mathcal{R}^4 \mid ||Q|| = 1\}$. Therefore, the rotational motion of the UAV can be described by the path defined on unit sphere $Q(t) \in \mathcal{S}^3$. The velocities, on the other hand, belong to the space of skew symplectic quaternions $sp(1) = \{w \in \mathcal{R}^4 \mid w + \bar{w} = 0\}$ where $\bar{w}$ represents conjugate of the quaternion w. The $sp(1)$ is the Lie algebra of the symplectic group $Sp(1)$. The $sp(1)$ is the set of pure quaternions, which is isomorphic to $\mathcal{R}^3$ and therefore isomorphic to the $so(3)$ - Lie algebra of special orthogonal group, the group of 3D rotations. In order to ensure isomorphism between Lie algebras, the element $u \in \mathcal{R}^3$ is associated to the element $w = \frac{1}{2}\begin{bmatrix} 0 & u^T \end{bmatrix}^T \in sp(1)$ [4] and to the element $\tilde{u} \in so(3)$, where $\tilde{}$ represents the skew symmetric operator [8].

Due to the presented isomorphisms, similar to the already well known exponential mapping on $SO(3)$ group, used for example in [9], the closed form exponential mapping from vector u to $\mathcal{S}^3$, i.e. $\exp_{\mathcal{S}^3} : \mathcal{R}^3 \cong sp(1) \to Sp(1) \cong \mathcal{S}^3$ can be found as [4]

$$\exp_{\mathcal{S}^3}(w) = \cos\left(\frac{1}{2}||u||\right)\begin{bmatrix} 1 \\ 0 \end{bmatrix} + \frac{\sin\left(\frac{1}{2}||u||\right)}{||u||}\begin{bmatrix} 0 \\ u \end{bmatrix}. \tag{3}$$

The Eq. (3) yields quaternion that represents rotation around vector u by angle $||u||$. This can be used to construct an algorithm for kinematic reconstruction of rotational quaternions Q from angular velocity field ω, by first seeking for an incremental rotation vector u, which can be found by using well-known differential exponential operator [10]

$$\dot{u} = \text{dexp}_{-\tilde{u}}^{-1}(\omega) = \omega + \frac{1}{2}\tilde{u}\omega - \frac{||u||\cot\left(\frac{||u||}{2}\right) - 2}{2||u||^2}\tilde{u}^2\omega, \tag{4}$$

and then reconstructing the attitude by using Eq. (3).

Method III Dual Quaternions as a Lie Group Description of the Complete UAV Rigid Body Motion. Dual quaternions are a natural extensions of rotational quaternions that can be used for description of the full rigid body motion. Similar to the definition of the dual number, dual quaternion $\hat{Q}$ can be defined as [11]

$$\hat{Q} = Q + \varepsilon Q_\varepsilon = \begin{bmatrix} q_0 + \varepsilon q_{\varepsilon 0} \\ q + \varepsilon q_\varepsilon \end{bmatrix}, \tag{5}$$

where Q and Q_ε are two quaternions, while ε is the dual unit satisfying condition $\varepsilon^2 = 0$. In order to describe rigid body motion, dual quaternions need to have unit length, which leads to the expression $\left|\left|\hat{Q}\right|\right| = ||Q|| + \varepsilon 2Q^T Q_\varepsilon = 1 + \varepsilon 0$ [12]. In other words unit dual quaternion has to satisfy two constraints: unitary norm of the quaternion $||\boldsymbol{Q}|| = 1$ and orthogonality of two quaternions $\boldsymbol{Q}^T\boldsymbol{Q}_\varepsilon = 0$, also known as Plücker condition. Similar to unit quaternions forming an $Sp(1)$ Lie group, unit dual quaternions form the $\widehat{Sp}(1)$ 6-dimensional Lie group, by replacing quaternions in $Sp(1)$ by dual quaternions. The group $\widehat{Sp}(1)$ is isomorphic to $SE(3)$ group and can therefore be used to describe rigid body motions. The dual pure quaternions form the $\widehat{sp}(1)$ Lie algebra, which is isomorphic to $se(3)$ by mapping

$$\boldsymbol{X} = \begin{bmatrix} \boldsymbol{u} \\ \boldsymbol{\eta} \end{bmatrix} \in se(3) \rightarrow \widehat{\boldsymbol{X}} = \frac{1}{2} \begin{bmatrix} 0 \\ \boldsymbol{u} + \varepsilon\boldsymbol{\eta} \end{bmatrix} \in \widehat{sp}(1). \tag{6}$$

Due to the presented isomorphism, the exponential map for the unit dual quaternion describing rigid body motion can be written as [12]

$$\exp_{\widehat{\mathcal{S}}^3} \widehat{\boldsymbol{X}} = \exp_{\mathcal{S}^3} (\boldsymbol{w}) + \varepsilon \begin{bmatrix} -\frac{d}{2}\sin\left(\frac{1}{2}||\boldsymbol{u}||\right) \\ \frac{d}{2}\cos\left(\frac{1}{2}||\boldsymbol{u}||\right)\boldsymbol{n} + \sin\left(\frac{1}{2}||\boldsymbol{u}||\right)\boldsymbol{m} \end{bmatrix}, \tag{7}$$

where $\boldsymbol{n} = \boldsymbol{u}/||\boldsymbol{u}||$ is the unit vector along the screw axis, $\boldsymbol{m}$ is the moment vector defined as $\boldsymbol{m} = -\boldsymbol{n} \times (-\boldsymbol{n} \times \boldsymbol{\eta})/||\boldsymbol{u}||$, while d is displacement along the screw axis $d = \boldsymbol{u}\cdot\boldsymbol{\eta}/||\boldsymbol{u}|| = \boldsymbol{n}\cdot\boldsymbol{m}$. The non-dual part of (7) is the unit quaternion describing rotation in the same way as unit quaternions in the previous paragraph, while the displacement is now encoded in the dual part. The incremental screw coordinates can be found by solving differential equation $\dot{\boldsymbol{X}} = \mathrm{dexp}_{-\tilde{\boldsymbol{u}}}^{-1}(\boldsymbol{V})$ where $\boldsymbol{V} = \begin{bmatrix} \boldsymbol{\omega}^T & \boldsymbol{v}^T \end{bmatrix}^T$ stands for vector of both angular and translational velocities, expressed in the body-fixed reference frame.

To summarize, in Methods I and II completely the same dynamical equations are solved: Newton equation for translational dynamics, expressed in the global (fixed) reference frame, and Euler equations for computing angular velocity field expressed in the body-fixed reference frame (and we pursue separate kinematic reconstruction of position and attitude of the vehicle). On the other hand, in Method III the translational dynamics equations are expressed in the body-fixed reference frame, just as the rotational dynamics equations (and we treat translation and rotation as a coupled rigid body motion).

3 Numerical Experiments

In this section, numerical experiments are conducted in order to evaluate the methods described in Sect. 2. It is important to emphasize that the all statements in the paper are made neglecting the values in the order of the round-off errors for decimal numbers stored in double precision. Additionally, for each of the examples, the correct attitude actually represents the converged solution obtained by all described methods. The Euclidian norm of the difference between obtained and correct quaternion is adopted as a metric for describing error in attitude.

The chosen quadrotor example is in X configuration, as previously described, with the following properties: mass $m = 0.5$, principle moments of inertia $J_x = 0.00365$, $J_y = 0.00368$ and $J_z = 0.00703$, arm length $l = 0.17$, rotor-motor assembly inertia $J_{mzi} = 0.0002271$, thrust force $|F_{Ti}(\omega_{mi})| = 5.57 \cdot 10^{-6} \cdot \omega_{mi}^2$ and drag torque $|\boldsymbol{T}_{Di}(\omega_{mi})| = \begin{bmatrix} 0 & 0 & -1.37 \cdot 10^{-7} \cdot |\omega_{mi}| \cdot \omega_{mi} \end{bmatrix}^T$ where arm length corresponds to the horizontal distance from the UAV center of mass to the rotor axis, while rotor-motor assembly inertia represents inertia of the whole rotating assembly about the rotating axis. All involved ODE differential equations are solved by the Runge-Kutta fourth order method.

3.1   Test Case I - Yaw Rotation Coupled with Climbing

This test case involves yaw rotation coupled with climbing of the UAV. This motion is achieved by rotating rotors 1 and 3 with different angular velocities than rotors 2 and 4 making the sum of the thrust forces larger than the weight of the UAV. All tested methods converged to the same solution for the step size equal to 10^{-3}. For the evaluation purposes, the methods are tested for the increasing step sizes and the results are compared to the converged solution. The norm of the difference between converged and resulting quaternion for increasing step size is shown in Fig. 2.

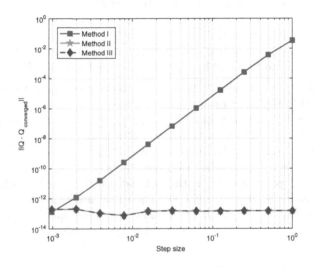

Fig. 2. Norm of the difference between converged and resulting quaternion for increasing step size

It can be seen that Method I exhibits usual convergence to the correct solution, both in position (not shown here) and attitude, as expected [4]. On the other hand, Methods II and III give correct solutions for any tested time step. It is important to emphasize that Methods I and II solve translational dynamics and kinematics in completely the same way, and the difference exists only in the kinematic reconstruction of rotation. Therefore, it can be concluded that the errors pertinent to Method I, visible in Fig. 2, arise solely from the less accurate reconstruction of the attitude from the angular velocity field.

3.2 Test Case II - Roll Maneuver

In this section, a bit more complex maneuver will be simulated. Initially the UAV is at rest at an altitude of 20 m, and then the rotors 1 and 2 are accelerated, while decelerating rotors 3 and 4 to articulate roll maneuver - the rotation of the UAV about x axis. The length of the simulation is set to 1, and all three presented methods have been used for simulating UAV motion.

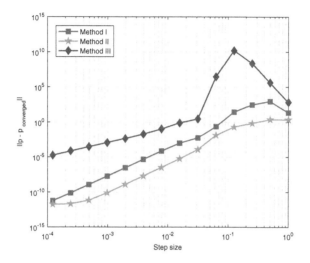

Fig. 3. Norm of the position error vector for increasing step size

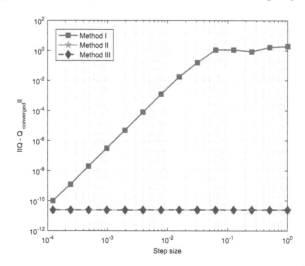

Fig. 4. Norm of the difference between converged and resulting quaternion for increasing step size

The Methods I and II converged to the same solution for the step size of 10^{-4} for both position and attitude, while Method III required step size smaller than 10^{-6} for full convergence of the position coordinates. The results for the simulation with increasing time steps with all three methods are again compared to the converged solution in terms of norm of the position error vector and norm of the difference between converged and resulting quaternion. The comparison is given in Fig. 3 and Fig. 4.

The Methods II and III (based on Lie group formulations) are again superior in terms of kinematic reconstruction of the attitude from the angular velocity field, as they exhibit exact solutions for any tested step size. On the other hand Method I again exhibits convergence to the correct solution with reducing step size, while yielding a completely useless results for larger step sizes.

For position integration none of the methods yielded accurate solutions for very large time steps, and all of the methods converged to the correct solution with reduced step sizes. However, there is a significant difference in the accuracy between methods. Method II clearly outperforms other tested methods, since it is the most accurate for any tested step size. Even more, it is at least two orders of magnitude better than the other methods for most of the tested step sizes. As already emphasized, Methods I and II update positions in exactly the same way and the only difference in the results is due to the fact that Lie group Method II outperforms standard Method I in terms of attitude reconstruction from the angular velocity field.

On the other hand, dual quaternion Method III suffers the most from the instabilities at larger step sizes and exhibits the slowest convergence to the correct solution. This is due to the fact that Method III involves solving the Newton equation in the body-fixed reference frame (coupled rigid body motion), which is - due to its complexity and additional transformations - more prone to numerical errors than the one expressed in the global reference frame. Again, although Methods I and II involve solving the same Newton equation for translational dynamics, the Lie group Method II yields better results, since the equation depends on the rotational quaternion which is better reconstructed with Method II.

References

1. Hoffman, D., Rehan, M., MacKunis, W., Reyhanoglu, M.: Quaternion-based robust trajectory tracking control of a quadrotor hover system. Int. J. Control Autom. Syst. **16**, 1–10 (2018)
2. Etkin, B., Reid, L.D.: Dynamics of Flight: Stability and Control, 3rd edn. Wiley, New York (1996)
3. Hartley, R., Trumpf, J., Dai, Y., Li, H.: Rotation averaging. Int. J. Comput. Vision **103**, 267–305 (2013)
4. Terze, Z., Müller, A., Zlatar, D.: Singularity-free time integration of rotational quaternions using non-redundant ordinary differential equations. Multibody Sys.Dyn. **38**(3), 201–225 (2016). https://doi.org/10.1007/s11044-016-9518-7
5. Terze, Z., Zlatar, D., Pandža, V.: Aircraft attitude reconstruction via novel quaternion-integration procedure. Aerosp. Sci. Technol. **97**, 105617 (2020)

6. Du, H., Zhu, W., Wen, G., Duan, Z., Lü, J.: Distributed formation control of multiple quadrotor aircraft based on nonsmooth consensus algorithms. IEEE Trans. Cybern. **49**, 342–353 (2019)
7. Dou, J., King, X., Chen, X., Wen, B.: Output feedback observer-based dynamic surface controller for quadrotor UAV using quaternion representation. Proc. Inst. Mech. Eng. Part G J. Aerosp. Eng. **231**, 2537–2548 (2017)
8. Celledoni, E., Owren, B.: Lie group methods for rigid body dynamics and time integration on manifolds. Comput. Methods Appl. Mech. Eng. **192**, 421–438 (2003)
9. Terze, Z., Müller, A., Zlatar, D.: Lie-group integration method for constrained multibody systems in state space. Multibody Syst. Dyn. **34**, 275–305 (2015)
10. Iserles, A., Munthe-Kaas, H.Z., Norsett, S.P., Zanna, A.: Lie-group methods. Acta Numer. **9**, 215–365 (2000)
11. Chevallier, D.P.: Lie algebras, modules, dual quaternions and algebraic methods in kinematics. Mech. Mach. Theory **26**, 613–627 (1991)
12. Müller, A., Terze, Z., Pandža, V.: A non-redundant formulation for the dynamics simulation of multibody systems in terms of unit dual quaternions. In: Proceedings of the ASME 2016 International Design Engineering Technical Conferences and Computers and Information in Engineering Conference, 21–24 August 2016, Charlotte, North Carolina (2016)

A Variational Principle of Minimum for Navier-Stokes Equation Based on the Symplectic Formalism

Géry de Saxcé[✉] [iD]

Univ. Lille, CNRS, Centrale Lille, UMR 9013 - LaMcube - Laboratoire de mécanique multiphysique multiéchelle, 59000 Lille, France
gery.de-saxce@univ-lille.fr
http://lamcube.univ-lille.fr

Abstract. The object of this work is to apply the formalisms of symplectic inclusions and symplectic Brezis-Ekeland-Nayroles principle to dissipative media in spatial representation. In the spirit of Newton-Cartan theory, our approach is covariant in the sense that it includes the gravitation and satisfies Galileo's principle of relativity. This aim is reached in three steps. Firstly, we develop a Lagrangian formalism for the reversible media based on the calculus of variation by jet theory. Next, we propose a corresponding Hamiltonian formalism for such media. Finally, we deduce from it a symplectic minimum principle for dissipative media and we show how to obtain a minimum principle for unstationary compressible and incompressible Navier-Stokes equation. The weak regularity of the potential of dissipation allows to encompass nonsmooth dissipative constitutive laws such as the one of Bingham fluids.

Keywords: Navier-Stokes equation · Dynamical dissipative systems · Symplectic geometry · Convex analysis · Galilean relativity

1 Introduction

This paper was written out to mark the occasion of the 200 th birthday of Navier's works that spearheaded the Navier-Stokes equation.

We are working with the phase space of which the elements are of the form

$$z = \begin{bmatrix} \xi \\ \eta \end{bmatrix}$$

where ξ are the degrees of freedom and η are the corresponding momenta. It is equipped with the symplectic form $\omega(dz, dz') = dz^T J\, dz'$ where the operator J is skew-symmetric. Introducing the Hamiltonian vector field (or symplectic gradient) $\dot{z} = X_H = J \cdot \nabla_z H(t, z)$ restitues the canonical equations. For dissipative systems, several authors proposed unified frameworks. We quickly review now these theoretical frameworks:

F. Nielsen and F. Barbaresco (Eds.): GSI 2023, LNCS 14072, pp. 12–21, 2023.
https://doi.org/10.1007/978-3-031-38299-4_2

- The metriplectic systems were introduced by Morrison [1,6,14,18] and developed further by Grmla and Öttinger [12,19] as the GENERIC systems (General Equation for Non-Equilibrium Reversible-Irreversible Coupling). They combine the Hamiltonian formulation and the Onsager one, according to the evolution law

$$\dot{z} = J \cdot \nabla_z H(z) + K \cdot \nabla_z S(z)$$

  where Onsager term is built from a symmetric and positive-definite operator K and an entropy-like function S. A variational formulation of GENERIC can be found in [13].
- The Port-Hamiltonian systems were introduced by Brockett [2] and van der Schaft [22]

$$\dot{z} = (J - R) \cdot \nabla_z H(z)$$

  where the symmetric and positive-definite operator R modelizes the resistive effects (because of the minus sign).
- The rate-independent systems proposed by Mielke and Theil [15], Mielke [16] and developed with applications in Mielke and Roubíček [17], are based on two fundamental conditions:
  - The stability condition: $\nabla_\xi E(\xi) \cdot w + \Phi(w) \geq 0, \ \forall w$
  - The power balance: $\nabla_\xi E(\xi) \cdot \dot{\xi} + \Phi(\dot{\xi}) = 0$
  where E in the energy functional and Φ is a 1-homogeneous dissipation potential depending on the velocity.
- The Hamiltonian inclusions were proposed by Buliga [3],

$$\dot{z} = J \cdot \nabla_z H(z) + J \cdot \nabla_{\dot{z}} \Phi(\dot{z}) \tag{1}$$

  where Φ is a convex dissipation potential. Because of the first term, It is clearly related to the two former formalisms (GENERIC and Port-Hamiltonian systems) but it is also inspired from the rate-independent systems because of the dissipation potential. Nevertheless it is important to remark that Φ is not necessarily 1-homogeneous or even homogeneous.

The problem of finding a variational formulation for the Navier-Stokes equations has been debated for a long time since Helmholtz and Rayleigh. A comprehensive survey can be found in [21]. Hamilton variational principle is limited to dynamical reversible systems. To modelize the dynamics of dissipative systems in a thermodynamical framework, this principle was modified by Fukagawa and Fujitani [9] by introducing a nonholonomic constraint and Lagrange multipliers. Likewise, Gay-Balmaz and Yoshimura [10] proposed a Lagrangian variational formulation of the Navier-Stokes-Fourier system by introducing new dual variables of which the time rates are the production of entropy and the temperature. The nearest approach to the present one is Ghoussoub work [11] on the application of anti-selfdual Lagrangian to the resolution of the Navier-Stokes equation. However, our approach is different because it is developed from a geometric point of view, as in Gay-Balmaz and Yoshimura work, but also, while the last term of Ghoussoub's functional is depending on specific boundary value problems, ours is general and does not depend on the boundary conditions that are only considered as constraints in the minimization.

2   Lagrangian Formalism for a Reversible Continuum

We are working in the space-time, set of the events $X = (t, x)$ where x is the position at time t. In contrast to the usual representation of the motion of a continuum by a map $x = f(t, x_0)$ where x_0 are the material coordinates of the particle, our point of view is to represent it by $x_0 = \kappa(t, x) = \kappa_t(x) = \kappa(X)$. As usual, the deformation gradient is denoted $F = \nabla_{x_0} x$. Let Ω be a bounded open subset of the space-time corresponding to the motion of the continuum. The equations of balance of the linear momentum and the energy of a reversible continuum are deduced from a space-time action

$$\alpha\,[x_0] = \int_\Omega \mathcal{L}\,(X, x_0, \nabla_X x_0)\,\mathrm{d}^4 X$$

of Hamilton's principle, using a special form of the calculus of variation by replacing the original field x_0 by its first jet prolongation $j^1 x_0$, that leads to perform variations not only on the field and its derivatives but also on the variable X. To explicit them, we consider a new parameterization given by a regular map $X = \psi\,(Y)$ of class C^1 and we perform the variation of the function ψ, the new variable being Y. After calculating the variation of the action, we will consider the particular case where the function ψ is the identity of Ω. Hence we start with

$$\alpha\,[X, x_0] = \int_{\Omega'} \mathcal{L}\,(\psi(Y), x_0, \nabla_Y x_0 \cdot \nabla_X Y)\,\det\,(\nabla_Y X)\,d^4 Y\;,$$

where $\Omega' = \psi^{-1}(\Omega)$ and the variables of the functional are now both X and x_0. For more details, the reader is referred to [7]. In order to satisfy Galileo's principle of relativity, we consider the Lagrangian

$$\mathcal{L} = \rho\,\left(\frac{1}{2}\,\|\,v\,\|^2 + A \cdot v - \phi - e_{int}(x_0, C)\right) \tag{2}$$

where v is the spatial velocity, A, ϕ are the vector and scalar potentials of the Galilean gravitation, e_{int} is the specific internal energy depending on the right Cauchy strains $C = F^T \cdot F$, and $\rho = \rho_0(x_0)/\det(F)$ is the mass density satisfying the equation of balance of mass. Introducing the linear momentum

$$\pi = \nabla_v \mathcal{L} = \rho\,(v + A)\;, \tag{3}$$

the Hamiltonian density

$$\mathcal{H} = \pi \cdot v - \mathcal{L} = \frac{1}{2\,\rho}\,\|\,\pi - \rho\,A\,\|^2 + \rho\,(\phi + e_{int})\;. \tag{4}$$

and the reversible stresses

$$\sigma_R = 2\rho\,F \cdot \nabla_C e_{int} \cdot F^T$$

by variation of the action with respect to $X = (t, x)$, we recover the balance equations of the energy

$$\frac{\partial \mathcal{H}}{\partial t} + \nabla \cdot (\mathcal{H} v - \sigma_R \cdot v) = \rho \left(\frac{\partial \phi}{\partial t} - \frac{\partial A}{\partial t} \cdot v \right)$$

and the linear momentum

$$-\frac{\partial \pi}{\partial t} + \nabla \cdot (\sigma_R - v \otimes \pi) + \rho \left((\nabla A) \cdot v - \nabla \phi \right) = 0 \tag{5}$$

After classical simplifications owing to the balance of mass, it is reduced to

$$-\rho \frac{Dv}{Dt} + \nabla \cdot \sigma_R + \rho \left(g - 2\,\Omega \times v \right) = 0$$

where occurs the material derivative $D/Dt = \partial/\partial t + v \cdot \nabla$, the gravity $g = -\nabla \phi - \partial A/\partial t$ and Coriolis' vector $\Omega = 1/2\,(\nabla \times A)$. For a barotropic fluid, $\sigma_R = -p\,I$ where p is the pressure and I is the identity matrix, then the balance of linear momentum takes the form of Euler's equations

$$-\rho \frac{Dv}{Dt} - \nabla p + \rho \left(g - 2\,\Omega \times v \right) = 0 \tag{6}$$

It is worth to remark these two equations of balance are covariant with respect to Galileo's principle of relativity.

3 Hamiltonian Formalism and Canonical Equations for a Reversible Continuum

Let Ω_t the set of positions occupied by the material particles of the continuum at time t. As v are the components of a 1-contravariant tensor and $\mathcal{H} = \pi \cdot v - \mathcal{L}$ is a density, π are the components of a 1-covariant and antisymmetric 3-contravariant tensor. Then the field $x \mapsto \pi(t, x)$ defined on Ω_t is a section of the fiber bundle $\bigwedge^3 (T\,\Omega_t) \otimes T^* \Omega_t$ of which the coordinates (x, π) in local charts are taken as canonical variables. The total energy at time t is

$$H\,[x_0, \pi] = \int_{\Omega_t} \mathcal{H}\,(x, x_0, \nabla x_0, \pi)\,d^3 x$$

where, for sake of easiness, the dependence with respect to time t is no longer explicitly expressed in the sequel.

We claim that the motion of the continuum is described by the canonical equations

$$\zeta = \left(\frac{dx}{dt}, \frac{\partial \pi}{\partial t} \right) = \left(v, \frac{\partial \pi}{\partial t} \right) = X_H$$

obtained by the calculus of the Hamiltonian vector field X_H for the canonical symplectic form

$$\omega(\zeta, \zeta') = \int_{\Omega_t} \left(\frac{dx}{dt} \cdot \frac{\partial \pi'}{\partial t} - \frac{\partial \pi}{\partial t} \cdot \frac{dx'}{dt} \right) d^3 x \tag{7}$$

As H is a functional, X_H is a variational derivative that can be calculated by the jet theory as in the previous section. Hence we consider a new parameterization $x = \psi(y)$ of class C^1 and we perform the variation of the function ψ, the new variable being y. After calculating the variation of the functional, we will consider the particular case where the function ψ is the identity of Ω. Hence we start with

$$H[x, x_0, \pi'] = \int_{\Omega_t'} \mathcal{H}\left(\psi(y), x_0, \nabla_y x_0 \cdot \nabla y, \det(\nabla y) (\nabla y)^T \cdot \pi'\right) \det(\nabla_y x)\, d^3 y \;,$$

where $\Omega_t' = \psi^{-1}(\Omega_t)$ and π' are the components of the linear momentum in the coordinates y. A calculus similar to the one of the previous section leads to the following expression of the canonical equations

$$\frac{dx}{dt} = \nabla_\pi \mathcal{H}$$

$$\frac{\partial \pi}{\partial t} = -\nabla \mathcal{H}$$

$$-\nabla \cdot \left[\nabla_{\nabla x_0} \mathcal{H} \cdot \nabla x_0 - (\mathcal{H} - \nabla_\pi \mathcal{H} \cdot \pi) I + \nabla_\pi \mathcal{H} \otimes \pi\right]$$

where the extra terms of the jet theory are given by the last line. For the Lagrangian (2), we obtain

$$\frac{dx}{dt} = \frac{\pi}{\rho} - A, \qquad -\frac{\partial \pi}{\partial t} + \nabla \cdot (\sigma_R - v \otimes \pi) + \rho\left((\nabla A)\, v - \nabla \phi\right) = 0 \qquad (8)$$

where we can recognize the definition (3) of the linear momentum and the equation (5) of balance of the linear momentum.

4 Symplectic Brezis-Ekeland-Nayroles Principle for Dissipative Continua

To build a minimum principle for such continua as a viscous fluid, we adapt the scheme proposed in [4]. The key-idea is a decomposition of ζ into reversible and irreversible parts

$$\zeta = \zeta_R + \zeta_I, \qquad \zeta_R = X_H, \qquad \zeta_I = \zeta - X_H$$

We are interested in the material continua that are modelized by a potential of dissipation Φ, convex and lower semicontinuous. We define the symplectic subdifferential of Φ at ζ as the (possibly empty) set [3]

$$\partial^\omega \Phi(\zeta) = \{\zeta_I \text{ such that } \forall \zeta', \quad \Phi(\zeta + \zeta') - \Phi(\zeta) \geq \omega(\zeta_I, \zeta')\}$$

Then the law of dissipative yielding is given by the Hamiltonian inclusion

$$\zeta_I \in \partial^\omega \Phi(\zeta)$$

In particular, if Φ is in addition differentiable, the symplectic subdifferential is reduced to the single element X_Φ and the law becomes an equality $\zeta_I = X_\Phi$ but we follow now considering the general non differentiable case.

Next, we define the symplectic Fenchel polar (or conjugate) function $\Phi^{*\omega}$ by

$$\Phi^{*\omega}(\zeta_I) = \sup_\zeta \left(\omega(\zeta_I, \zeta) - \Phi(\zeta) \right)$$

As superior envelop of affine functions, it satifies the symplectic Fenchel inequality

$$\forall \zeta', \forall \zeta_I', \qquad \Phi(\zeta') + \Phi^{*\omega}(\zeta_I') - \omega(\zeta_I', \zeta') \geq 0 \tag{9}$$

In particular, when the dynamical dissipative constitutive law is satisfied, it can be proved the equality is reached [4]

$$\zeta_I \in \partial^\omega \Phi(\zeta) \qquad \Leftrightarrow \qquad \Phi(\zeta) + \Phi^{*\omega}(\zeta_I) - \omega(\zeta_I, \zeta) = 0 \tag{10}$$

We are interested during the interval $[0, T]$ by the evolution paths $t \mapsto (\kappa_t, \zeta)$ which are admissible in the sense that they satisfy the boundary conditions and the initial conditions of the considered problem. Following pioneering works by Brezis, Ekeland and Nayroles in 1976, Buliga and the author proposed in [4] a symplectic version of the Brezis-Ekeland-Nayroles principle (in short SBEN) of which the functional is built from three functions, the symplectic form (for the dynamics), the dissipation potential (for the irreversible behavior) and the Hamiltonian (for the reversible behavior) through the Hamiltonian vector field:

SBEN principle: the natural evolution path $t \mapsto (\kappa_t, \zeta)$ minimizes the functional

$$\Pi[\kappa, \zeta] = \int_0^T \left\{ \Phi(\zeta) + \Phi^{*\omega}(\zeta - X_H) - \omega(\zeta - X_H, \zeta) \right\} dt \tag{11}$$

among all the admissible evolution paths, and the minimum is zero.

The idea is that the functional is non negative because of (9) and vanishes if and only if the dynamical dissipative constitutive law is satisfied almost everywhere because of (10).

5   SBEN Principle for Compressible Navier-Stokes Equation

With the notations of the previous section, the canonical equations (8) lead to

$$\zeta_I = \zeta - X_H = (v_I, \pi_I)$$

with

$$v_I = v - \frac{\pi}{\rho} + A, \quad \pi_I = \frac{\partial \pi}{\partial t} - \nabla \cdot (\sigma_R - v \otimes \pi) - \rho \left((\nabla A) v - \nabla \phi \right)$$

With the same classical simplifications used to transform (5) into (6), we have for a barotropic fluid

$$\pi_I = \rho \frac{Dv}{Dt} + \nabla p - \rho \left(g - 2\Omega \times v \right) \tag{12}$$

To apply the very general formalism of the previous section to Navier-Stokes equation, we need two additional hypotheses. The first one claims that the convex smooth potential Φ depends explicitly only on v ($\partial \pi / \partial t$ is ignorable):

$$\Phi(\zeta) = \varphi(v)$$

Then the symplectic Fenchel polar function has a finite value

$$\Phi^{*\omega}(\zeta_I) = \Phi^{*\omega}(v_I, \pi_I) = \varphi^*(-\pi_I)$$

if $v_I = 0$, that is (3), where φ^* is the classical Fenchel polar function of φ. As in the SBEN principle the minimum value of the functional is zero then finite, we suppose in the sequel that (3) is *a priori* satisfied. Moreover, owing to (7), the last term in the functional (11) becomes

$$-\omega(\zeta - X_H, \zeta) = \int_{\Omega_t} \left(\pi_I \cdot v - v_I \cdot \frac{\partial \pi}{\partial t} \right) d^3x = \int_{\Omega_t} \pi_I \cdot v \, d^3x$$

Then the SBEN functional becomes

$$\Pi[\kappa, \zeta] = \int_0^T \left\{ \varphi(v) + \varphi^*(-\pi_I) + \int_{\Omega_t} \pi_I \cdot v \, d^3x \right\} dt$$

Owing to (12), the functional (11) can be recast, that leads to
    SBEN principle for compressible Navier-Stokes equation:
the natural evolution path $t \mapsto (\kappa_t, v)$ minimizes the functional

$$\Pi[\kappa, v] = \int_0^T \left\{ \varphi(v) + \varphi^*\left(-\rho \frac{Dv}{Dt} - \nabla p + \rho \left(g - 2\, \Omega \times v \right) \right) \right.$$
$$\left. + \int_{\Omega_t} \left[\rho \frac{Dv}{Dt} + \nabla p - \rho g \right] \cdot v \, d^3x \right\} dt \tag{13}$$

among all the admissible evolution paths, and the minimum is zero.

It is worth to remark that $\Phi(\zeta) + \Phi^{*\omega}(\zeta_I) = \varphi(v) + \varphi^*(-\pi_I)$ is an anti-selfdual Lagrangian, a tool proposed by Ghoussoub [11] to study among others the solutions of Navier-Stokes equation. This reveals the symplectic origin of the structure of such self-antidual Lagrangians. On the other hand, while the last term of Ghoussoub's functional is depending on specific boundary value problems, ours is general and does not depend on the boundary conditions that are only considered as constraints in the minimization.

The second additional hypothesis claims that φ depends on v through its symmetric gradient $D = \mathcal{D}(v) = \nabla_s v = 1/2 \left(\nabla v + (\nabla v)^T \right)$ and is quadratic with respect to v of the form

$$\varphi(v) = \int_{\Omega_t} W(\mathcal{D}(v)) \, d^3x = \int_{\Omega_t} \mu \left[Tr(D^2) - \frac{1}{3} (Tr(D))^2 \right] d^3x$$

then the viscous part of the stress tensor is traceless (Stokes hypothesis)

$$\sigma_I = \nabla_D W(\mathcal{D}(v)) = 2\mu \left(D - \frac{1}{3} Tr(D)\, I \right) \tag{14}$$

and, as φ is a differentiable functional, its subdifferential is reduced to its variational derivative

$$\nabla_v \varphi(v) = -\nabla \cdot \sigma_I$$

Now, let us prove that the variational principle of minimum restitues Navier-Stokes equation. Indeed, if the minimum equal to zero is reached, we have almost everywhere in $[0, T]$

$$\varphi(v) + \varphi^*(-\pi_I) + \int_{\Omega_t} \pi_I \cdot v \, \mathrm{d}^3 x = 0$$

that is equivalent to the dynamical dissipative law

$$-\pi_I = \nabla_v \varphi(v) = -\nabla \cdot \sigma_I$$

Owing to (12) and (14), we recover Navier-Stokes equation

$$\rho \frac{Dv}{Dt} = -\nabla p + \mu \triangle v + \frac{\mu}{3} \nabla (\nabla \cdot v) + \rho \, (g - 2\,\Omega \times v) \ .$$

For a flow, the term of the last line in the functional (13) is the sum of the velocity head, pressure head and elevation head losses due to dissipation during the interval from 0 to T. For the limit case of inviscid flows, the potential of dissipation φ vanishes and its polar function φ^* has a finite value equal to zero if $\pi_I = 0$, *i.e.* Euler's equations (6), then the SBEN principle claims that the total head loss is zero, that is the expression of Bernoulli's principle.

6   SBEN Principle for Incompressible Navier-Stokes Equation

For this limit case, $\nabla \cdot v = 0$ and the pressure p becomes a free variable independent of κ. Navier-Stokes equation is reduced to:

$$\rho \frac{Dv}{Dt} = -\nabla p + \mu \triangle v + \rho \, (g - 2\,\Omega \times v) \ .$$

To obtain the corresponding SBEN principle, we proceed as follows. The internal energy is cancelled in (2) and (4). The incompressibility condition is introduced as a constraint in the minimization. The pressure disappears of the functional and reappears as a Lagrange multiplier of this constraint in the equation characterizing the minimizers. Then we state

SBEN principle for incompressible Navier-Stokes equation:
the natural evolution path $t \mapsto (\kappa_t, v)$ minimizes the functional

$$\Pi[\kappa, v] = \int_0^T \{\varphi(v) + \varphi^*(-\rho \frac{Dv}{Dt} + \rho \, (g - 2\,\Omega \times v))$$
$$+ \int_{\Omega_t} \rho \left[\frac{Dv}{Dt} - g\right] \cdot v \, \mathrm{d}^3 x\} \, \mathrm{d}t$$

among all the admissible evolution paths such that $\nabla \cdot v = 0$, and the minimum is zero.

If the fluid is homogeneous, the mass density ρ is constant and the functional no longer depends on the motion map κ.

7 Conclusions and Perspectives

The weak regularity of the potential of dissipation adopted in [4] allows to encompass nonsmooth dissipative constitutive laws such as the one of Bingham fluids, plasticity and viscoplasticity [5], brittle fracture [8]. Also, it is worth to observe that the expression of the functional is independent of the boundary conditions that appear only as constraints of the minimization. Moreover, it must be emphasized that the functional is not convex with respect to the unknown fields but there is (at least partial) convexity, that is favourable for the convergence of the minimization procedure. Another idea is to develop symplectic integrators [20] to respect the structure of the canonical equation of Sect. 3 and variational schemes based on the Lagrangian of the SBEN principles proposed in Sects. 5 and 6. Finally a minimum variational principle may constitute an interesting means to study the existence and smoothness of the solutions of Navier-Stokes equation in the spirit of [11].

Aknowledgements. This work was performed thanks to the project *BIpotentiels Généralisés pour le principe variationnel de Brezis-Ekeland-Nayroles en mécanique* (BigBen) supported by the *Agence Nationale de la Recherche* (ANR).

References

1. Barbaresco, F.: Symplectic foliation structures of non-equilibrium thermodynamics as dissipation model: application to metriplectic nonlinear lindblad quantum master equation. Entropy **24**, 1626 (2022)
2. Brockett, R.W.: Control theory and analytical mechanics. In: Martin, C., Hermann, R. (eds.) Geometric Control Theory. Lie Groups: History, Frontiers and Applications VII, pp. 1–46. Mathematical Science Press, Brookline (1977)
3. Buliga, M.: Hamiltonian inclusions with convex dissipation with a view towards applications. Math. Appl. **1**(2), 225–228 (2009)
4. Buliga, M., de Saxcé, G.: A symplectic Brezis-Ekeland-Nayroles principle. Math. Mech. Solids 1–15 (2016). https://doi.org/10.1177/1081286516629532
5. Cao, X., Oueslati, A., Nguyen, A.D., de Saxcé, G.: Numerical simulation of elastoplastic problems by Brezis-Ekeland-Nayroles non-incremental variational principle. Comput. Mech. **65**(4), 1006–1018 (2020)
6. Coquinot, B., Morrison, P.J.: A general metriplectic framework with application to dissipative extended magnetohydrodynamics. J. Plasma Phys. **86**, 835860302 (2020)
7. de Saxcé, G., Vallée, C.: Galilean Mechanics and Thermodynamics of Continua. Wiley-ISTE (2016)
8. de Saxcé, G.: A non incremental variational principle for brittle fracture. Int. J. Solids Struct. **252**, 111761 (2022)
9. Fukagawa, H., Fujitani, Y.: A variational principle for dissipative fluid dynamics. Progr. Theoret. Phys. **127**(5), 921–935 (2012)
10. Gay-Balmaz, F., Yoshimura, H.: A Lagrangian variational formulation for nonequilibrium thermodynamics. Part II: Continuum systems. J. Geom. Phys. **111**, 194–212 (2017)

11. Ghoussoub, N.: Anti-self-dual Lagrangians: variational resolutions of non-self-adjoint equations and dissipative evolutions. Annales de l'Institut Henri Poincaré C Analyse non linéaire **24**, 171–205 (2007)
12. Grmela, M., Öttinger, H.C.: Dynamics and thermodynamics of complex fluids. I. Development of a general formalism. Phys. Rev. E **56**(6), 6620–6632 (1997)
13. Duong, M.H., Peletier, M.A., Zimmer, J.: GENERIC formalism of a Vlasov-Fokker-Planck equation and connection to large-deviation principles. Nonlinearity **26**, 2951–2971 (2013)
14. Materassi, M., Morrison, P.J.: Metriplectic formalism: friction and much more, arXiv:1706.01455 (2017)
15. Mielke, A., Theil, F.: A mathematical model for rate-independent phase transformations with hysteresis. In: Alber, H.D., Balean, R., Farwig, R. (eds.) Workshop on Models of Continuum Mechanics in Analysis and Engineering, pp. 117–129. Shaker-Verlag (1999)
16. Mielke, A.: Evolution in rate-independent systems (Ch. 6). In: Dafermos, C., Feireisl, E. (eds.) Handbook of Differential Equations, Evolutionary Equations, vol. 2, pp. 461–559. Elsevier (2005)
17. Mielke, A., Roubíček, T.: Rate-independent damage processes in nonlinear elasticity. Math. Models Methods Appl. Sci. (M3AS) **16**(2), 177–209 (2006)
18. Morrison, P.J.: A paradigm for joined Hamiltonian and dissipative systems. Physica D **18**, 410 (1986)
19. Öttinger, H.C., Grmela, M.: Dynamics and thermodynamics of complex fluids. II. Illustrations of a general formalism. Phys. Rev. E **56**(6), 6633–6655 (1997)
20. Razafindralandy, D., Hamdouni, A., Chhay, M.: A review of some geometric integrators. Adv. Model. Simul. Eng. Sci. **5**(1), 1–67 (2018). https://doi.org/10.1186/s40323-018-0110-y
21. Sciubba, E.: Do the Navier-Stokes equations admit of a variational formulation? In: Sieniutycz, S., Farkas, H. (eds.) Variational and Extremum Principles in Macroscopic Systems. Elsevier (2005)
22. van der Schaft, A.J., System theoretic properties of physical systems. CWITract3, Centre for Mathematics and Informatics, Amsterdam (1984)

A Variational Symplectic Scheme Based on Simpson's Quadrature

François Dubois[1,2]([⊠]) and Juan Antonio Rojas-Quintero[3]

[1] Laboratoire de Mathématiques d'Orsay, Faculté des Sciences d'Orsay, Université Paris-Saclay, Gif-sur-Yvette, France
[2] Conservatoire National des Arts et Métiers, LMSSC laboratory, Paris, France
`francois.dubois@universite-paris-saclay.fr`
[3] CONACYT/Tecnológico Nacional de México/I.T. Ensenada, 22780 Ensenada, BC, Mexico
`jarojas@conacyt.mx`

Abstract. We propose a variational symplectic numerical method for the time integration of dynamical systems issued from the least action principle. We assume a quadratic internal interpolation of the state and we approximate the action in a small time step by the Simpson's quadrature formula. The resulting scheme is explicited for an elementary harmonic oscillator. It is a stable, explicit, and symplectic scheme satisfying the conservation of an approximate energy. Numerical tests illustrate our theoretical study.   [11 May 2023, GSI 2023.]

Keywords: ordinary differential equations · harmonic oscillator · numerical analysis

1 Introduction

The principle of least action is a key point for establishing evolution equations or partial differential equations, from classical to quantum mechanics and electromagnetisms [1,3,12]. An important application of this principle is proposed with the finite element method [2] and it is used for engineering applications since the 1950's. For dynamics equations and dynamical systems, a synthesis of the state of the art is proposed in [5,11].

In this contribution, we first recall the classical variational approach. It is founded on a midpoint quadrature formula for the approximate calculation of an integral. We essentially follow the contribution [6] in this Sect. 2. Then we recall in Sect. 3 the interpolation of functions with quadratic finite elements. Once this prerequisite is in place, we develop in Sect. 4 the approximation of discrete Lagrangians with Simpson's quadrature formula. The result is a numerical scheme that can be considered as a variant of the classical approach presented in Sect. 2 and we derive in Sect. 5 the discrete Euler-Lagrange equations. We notice in Sect. 6 that the scheme admits a symplectic structure and in Sect. 7 that an approximation of the energy is conserved along the discrete time integration. First numerical results are presented in Sect. 8 before some words of conclusion.

F. Nielsen and F. Barbaresco (Eds.): GSI 2023, LNCS 14072, pp. 22–31, 2023.
https://doi.org/10.1007/978-3-031-38299-4_3

2   A Classical Variational Symplectic Numerical Scheme

We consider a dynamical system described by a state $q(t)$ composed by a simple real variable to fix the ideas, and for $0 \leq t \leq T$. The continuous action S_c introduces a Lagrangian L and we have

$$S_c = \int_0^T L\left(\frac{dq}{dt}, q(t)\right) dt. \tag{1}$$

We use in this contribution a very classical Lagrangian

$$L\left(\frac{dq}{dt}, q\right) = \frac{m}{2}\left(\frac{dq}{dt}\right)^2 - V(q). \tag{2}$$

A discretization of the relation (1) is obtained by splitting the interval $[0, T]$ into N elements and we set $h = \frac{T}{N}$. At the discrete time $t_j = j\,h$, an approximation q_j of $q(t_j)$ is introduced and a discrete form S_d of the continuous action S_c can be defined according to

$$S_d = \sum_{j=1}^{N-1} L_d(q_j,\, q_{j+1}).$$

The discrete Lagrangian $L_d(q_\ell, q_r)$ is derived from the relation (2) with a centered finite difference approximation

$$\frac{dq}{dt} \simeq \frac{q_r - q_\ell}{h}$$

and a midpoint quadrature formula

$$\int_0^h V\left(q(t)\right) dt \simeq h\,V\left(\frac{q_\ell + q_r}{2}\right) :$$

$$L_d(q_\ell,\, q_r) = \frac{m\,h}{2}\left(\frac{q_r - q_\ell}{h}\right)^2 - h\,V\left(\frac{q_\ell + q_r}{2}\right). \tag{3}$$

We observe that

$$S_d = \cdots + L_d(q_{j-1},\, q_j) + L_d(q_j,\, q_{j+1}) + \cdots .$$

Then the discrete Euler Lagrange equation $\delta S_d = 0$ for an arbitrary variation δq_j of the discrete variable q_j can be written

$$\frac{\partial L_d}{\partial q_r}(q_{j-1},\, q_j) + \frac{\partial L_d}{\partial q_\ell}(q_j,\, q_{j+1}) = 0. \tag{4}$$

Taking into account the relation (3), we obtain

$$m\,\frac{q_{j+1} - 2\,q_j + q_{j-1}}{h^2} + \frac{1}{2}\left[\frac{dV}{dq}\left(\frac{q_j + q_{j+1}}{2}\right) + \frac{dV}{dq}\left(\frac{q_{j-1} + q_j}{2}\right)\right] = 0. \tag{5}$$

This numerical scheme is clearly consistent with the second order differential equation

$$m \frac{d^2 q}{dt^2} + \frac{dV}{dq} = 0 \tag{6}$$

associated with the Lagrangian proposed in (2). It is easy to verify that when

$$V(q) = \frac{1}{2} m \omega^2 q^2, \tag{7}$$

the scheme (5) is linearly stable. We suppose that the assumption (7) is satisfied until the end of this paragraph. The momentum p_r is defined by

$$p_r = \frac{\partial L_d}{\partial q_r} (q_\ell, q_r). \tag{8}$$

We have

$$p_{j+1} = m \frac{q_{j+1} - q_j}{h} - m \frac{\omega^2 h}{4} (q_{j+1} + q_j)$$

and an analogous relation for p_j. Then after some lines of algebra, we obtain a discrete system involving the momentum and the state:

$$p_{j+1} - p_j = -m \frac{\omega^2 h}{2} (q_{j+1} + q_j), \quad \frac{q_{j+1} - q_j}{h} = \frac{1}{2m} (p_{j+1} + p_j). \tag{9}$$

These relations are consistent with the first order Hamilton version

$$\frac{dp}{dt} + m \omega^2 q = 0, \quad \frac{dq}{dt} = \frac{p}{m}$$

of the equations of an harmonic oscillator. Moreover, we can write the system (9) under the form

$$\begin{pmatrix} p_{j+1} \\ q_{j+1} \end{pmatrix} = \mathbf{\Phi} \begin{pmatrix} p_j \\ q_j \end{pmatrix} \tag{10}$$

with

$$\mathbf{\Phi} = \frac{1}{1 + \frac{\omega^2 h^2}{4}} \begin{pmatrix} 1 - \frac{\omega^2 h^2}{4} & -m \omega^2 h \\ \frac{h}{m} & 1 - \frac{\omega^2 h^2}{4} \end{pmatrix}. \tag{11}$$

Because $\det \mathbf{\Phi} = 1$, the discrete flow (10) is symplectic as observed by Sanz-Serna [11]. Moreover, Kane et al. [6] have remarked that the numerical scheme (10) is one particular inconditionally stable version of the Newmark scheme [7]. Last but not least, the discrete Hamiltonian H_j defined by

$$H_j \equiv \frac{1}{2m} p_j^2 + \frac{m \omega^2}{2} q_j^2 \tag{12}$$

is conserved: we have

$$H_{j+1} = H_j \text{ for } 0 \leq j \leq N - 1.$$

We consider now a more elaborate interpolation in each interval, updating affine functions by polynomials of degree two.

3   Quadratic Interpolation

Internal interpolation between 0 and h can be written in terms of quadratic finite elements [8]. For $0 \le \theta \le 1$, we first set

$$\varphi_0(\theta) = (1 - \theta)(1 - 2\theta), \quad \varphi_{1/2}(\theta) = 4\theta(1 - \theta), \quad \varphi_1(\theta) = \theta(2\theta - 1). \quad (13)$$

With $t = h\theta$, we consider the polynomial function

$$q(t) = q_\ell \varphi_0(\theta) + q_m \varphi_{1/2}(\theta) + q_r \varphi_1(\theta). \quad (14)$$

Then $q(0) = q_\ell$, $q(\frac{h}{2}) = q_m$ and $q(h) = q_r$ and the basis functions (13) are well adapted to these degrees of freedom. We have also

$$\frac{dq}{dt} = \frac{1}{h} \left[q_\ell \frac{d\varphi_0}{d\theta} + q_m \frac{d\varphi_{1/2}}{d\theta} + q_r \frac{d\varphi_1}{d\theta} \right]$$

$$= \frac{1}{h} \left[q_\ell (4\theta - 3) + 4 q_m (1 - 2\theta) + q_r (4\theta - 1) \right]$$

$$= g_\ell (1 - \theta) + g_r \theta$$

with the derivatives g_ℓ and g_r given by a Gear scheme [4], *id est*

$$g_\ell = \frac{dq}{dt}(0) = \frac{1}{h}\left(-3 q_\ell + 4 q_m - q_r \right), \quad g_r = \frac{dq}{dt}(h) = \frac{1}{h}\left(q_\ell - 4 q_m + 3 q_r \right). \quad (15)$$

We remark also that

$$g_m = \frac{dq}{dt}\left(\frac{h}{2}\right) = \frac{1}{2}(g_\ell + g_r) = \frac{q_r - q_\ell}{h}. \quad (16)$$

Once the interpolation is defined in an interval of length h, we use it by splitting the range $[0, T]$ into N pieces, and $h = \frac{T}{N}$. With $t_j = jh$, we set

$$\begin{cases} q_j \simeq q(t_j) \text{ for } 0 \le j \le N \\ q_{j+1/2} \simeq q(t_j + \frac{h}{2}) \text{ with } 0 \le j \le N - 1. \end{cases}$$

In the interval $[t_j, t_{j+1}]$, the function $q(t)$ is a polynomial of degree 2, represented by the relation (14) with

$$t = t_j + \theta h, \quad q_\ell = q_j, \quad q_m = q_{j+1/2}, \quad q_r = q_{j+1}.$$

4   Simpson's Quadrature for a Discrete Lagrangian

For the numerical integration of a regular function ψ on the interval $[0, 1]$, the midpoint method studied previously $\int_0^1 \psi(\theta) \, d\theta \simeq \psi(\frac{1}{2})$ is exact for a polynomial ψ of degree smaller or equal to 1. To obtain a better precision, a very popular method has been proposed by Thomas Simpson (1710-1761):

$$\int_0^1 \psi(\theta) \, d\theta \simeq \frac{1}{6}\left[\psi(0) + 4\psi\left(\frac{1}{2}\right) + \psi(1) \right]. \quad (17)$$

The quadrature formula (17) is accurate up to polynomials of degree three. Then a discrete Lagrangian

$$L_h(q_\ell, q_m, q_r) \simeq \int_0^h \left[\frac{m}{2} \left(\frac{dq}{dt} \right)^2 - V(q) \right] dt$$

can be defined with the Simpson quadrature formula (17) associated with an internal polynomial approximation $q(t)$ of degree 2 presented in (14):

$$L_h(q_\ell, q_m, q_r) = \frac{m\,h}{12} \left(g_\ell^2 + 4\,g_m^2 + g_r^2 \right) - \frac{h}{6} \left(V(q_\ell) + 4\,V(q_m) + V(q_r) \right). \quad (18)$$

The discrete action Σ_d for a motion $t \longmapsto q(t)$ between the initial time and a given time $T > 0$ is discretized with N regular intervals and take the form

$$\Sigma_d = \sum_{j=1}^{N-1} L_h(q_j, q_{j+1/2}, q_{j+1}). \quad (19)$$

5   Discrete Euler-Lagrange Equations

We first write the Maupertuis's stationary-action principle $\delta \Sigma_d = 0$ with a variation $\delta q_{j+1/2}$ of the internal degree of freedom in the interval $[t_j, t_{j+1}]$. Due to the relations (15) (16), we first observe that

$$\frac{\partial g_\ell}{\partial q_m} = \frac{4}{h}, \quad \frac{\partial g_m}{\partial q_m} = 0, \quad \frac{\partial g_r}{\partial q_m} = -\frac{4}{h}.$$

Then, due to the expression (18) of the discrete Lagrangian, we have

$$\frac{\partial L_h}{\partial q_m} = \frac{h}{12} \left[\frac{8\,m}{h} g_\ell - \frac{8\,m}{h} g_r - 8 \frac{dV}{dq}(q_m) \right].$$

This partial derivative is equal to zero when

$$\delta \Sigma_d = 0.$$

Then

$$m \frac{g_r - g_\ell}{h} + \frac{dV}{dq}(q_m) = 0.$$

We observe that

$$g_r - g_\ell = \frac{4}{h} \left(q_\ell - 2\,q_m + q_r \right)$$

and the condition $\frac{\partial L_h}{\partial q_m} = 0$ is finally written

$$m \frac{4}{h^2} \left(q_\ell - 2\,q_m + q_r \right) + \frac{dV}{dq}(q_m) = 0. \quad (20)$$

We have put in evidence a second order discretization of the continuous Euler-Lagrange equation (6) of this problem. When the hypothesis (7) of an harmonic oscillator is satisfied, we can easily solve this equation and explicit the middle value q_m as a function of the extremities:

$$q_m = \frac{1}{1 - \frac{\omega^2 h^2}{8}} \frac{q_\ell + q_r}{2}. \tag{21}$$

This interpolation is not linear if $h > 0$. This property illustrates the underlying polynomial interpolation of degree two. Moreover, a stability condition is naturally emerging:

$$0 < \omega\, h < 2\sqrt{2}. \tag{22}$$

We now incorporate the relation (21) inside the expression (18) of the discrete Lagrangian. After a successful formal calculation with the help of the free software "SageMath" [10], we obtain a reduced Lagrangian

$$\left\{ \begin{aligned} L_h^r(q_\ell,\, q_r) &= \frac{1}{1 - \frac{\omega^2 h^2}{8}} \Big[\frac{1}{2}\, m\, h \left(\frac{q_r - q_\ell}{h} \right)^2 \\ &\quad - \frac{h}{2}\, m\, \omega^2 \left(\frac{22 - h^2\, \omega^2}{48} \left(q_\ell^2 + q_r^2 \right) + \frac{1}{12}\, q_\ell\, q_r \right) \Big]. \end{aligned} \right. \tag{23}$$

The discrete Euler-Lagrange (4) can now be written for this reduced Lagrangien (23). Instead of the relations (5), we obtain now the following numerical scheme:

$$\frac{1}{h^2} \left(q_{j+1} - 2\, q_j + q_{j-1} \right) + \frac{\omega^2}{24} \left(q_{j+1} + 22\, q_j + q_{j-1} \right) - \frac{\omega^4 h^2}{24}\, q_j = 0. \tag{24}$$

The scheme (24) is consistent with the ordinary differential Eq. (6) (7)

$$\frac{d^2 q}{dt^2} + \omega^2\, q(t) = 0.$$

Secondly, following the definition recalled in [9], the order of truncation of the scheme (24) is obtained by replacing the discrete variables q_{j+1}, q_j and q_{j-1} by the solution of the differential equation at the precise points $t_j + h$, t_j and $t_j - h$. Then

$$\left\{ \begin{aligned} q_{j+1} &= q_j + h\, \frac{dq}{dt} + \frac{h^2}{2}\, \frac{d^2 q}{dt^2} + \frac{h^3}{6}\, \frac{d^3 q}{dt^3} + \frac{h^4}{24}\, \frac{d^4 q}{dt^4} + \frac{h^5}{120}\, \frac{d^5 q}{dt^5} + \frac{h^6}{720}\, \frac{d^6 q}{dt^5} + O(h^7) \\ q_{j-1} &= q_j - h\, \frac{dq}{dt} + \frac{h^2}{2}\, \frac{d^2 q}{dt^2} - \frac{h^3}{6}\, \frac{d^3 q}{dt^3} + \frac{h^4}{24}\, \frac{d^4 q}{dt^4} - \frac{h^5}{120}\, \frac{d^5 q}{dt^5} + \frac{h^6}{720}\, \frac{d^6 q}{dt^5} + O(h^7). \end{aligned} \right.$$

In these conditions, the left hand side of the relation (24) is no longer equal to zero and defines the truncation error $\mathcal{T}_h(q_j)$. With the help of SageMath [10], one obtains without difficulty the relation

$$\mathcal{T}_h(q_j) = \frac{1}{1440}\, \omega^6\, h^4\, q_j + O(h^6).$$

The numerical scheme (24) is fourth order accurate in the sense of the truncation error.

A fundamental question concerns stability. With the linear structure of the finite difference equation (24), we consider the equation of degree two obtained by taking

$$q_{j-1} = 1, \ q_j = r, \ q_{j+1} = r^2.$$

The scheme is stable when the roots of the corresponding equation are of modulus smaller than 1. This equation can we written

$$a\,r^2 + b\,r + c = 0$$

with

$$a = c = 1 + \frac{h^2\,\omega^2}{24}, \ b = -\frac{1}{24}\left(48 - 22\,h^2\,\omega^2 + h^4\,\omega^4\right).$$

The discriminant $\Delta \equiv b^2 - 4\,a\,c$ can be factorized:

$$\Delta = \frac{\omega^2\,h^2}{576}\,(\omega^2\,h^2 - 24)\,(\omega^2\,h^2 - 12)\,(\omega^2\,h^2 - 8).$$

Under the stability condition (22), all the factors in the expression of the discriminant are negative and $\Delta < 0$. Then the equation $a\,r^2 + b\,r + c = 0$ has two conjugate complex roots r and $\bar{r}$. Their product $r\bar{r} = |r|^2$ is equal to 1 and the scheme (24) is stable.

6   Symplectic Structure

From the reduced Lagrangian (23), we define the momentum p_r with the analogue of the relation (8). It comes

$$p_r = m\,\frac{q_r - q_\ell}{h} - h\,\frac{m\,\omega^2}{6}\,\frac{q_\ell + 2\,q_r}{1 - \frac{\omega^2\,h^2}{8}} + h^3\,\frac{m\,\omega^4}{48}\,\frac{q_r}{1 - \frac{\omega^2\,h^2}{8}}. \tag{25}$$

This relation (25) can be explicited in the context of grid points. We have

$$p_{j+1} = m\,\frac{q_{j+1} - q_j}{h} - h\,\frac{m\,\omega^2}{6}\,\frac{q_j + 2\,q_{j+1}}{1 - \frac{\omega^2\,h^2}{8}} + h^3\,\frac{m\,\omega^4}{48}\,\frac{q_{j+1}}{1 - \frac{\omega^2\,h^2}{8}}$$

$$p_j = m\,\frac{q_j - q_{j-1}}{h} - h\,\frac{m\,\omega^2}{6}\,\frac{q_{j-1} + 2\,q_j}{1 - \frac{\omega^2\,h^2}{8}} + h^3\,\frac{m\,\omega^4}{48}\,\frac{q_j}{1 - \frac{\omega^2\,h^2}{8}}.$$

We eliminate the variable q_{j-1} from these two relations with the help of the difference scheme (24). We find a recurrence relation for the state $y_j \equiv (p_j, q_j)^{\mathrm{t}}$, similar to the equation (10), but the matrix Φ is replaced by a matrix Φ_3 that can be explicited:

$$\Phi_3 = \frac{1}{1 + \frac{\omega^2 h^2}{24}}\begin{pmatrix} 1 - \frac{11}{24}\omega^2 h^2 + \frac{\omega^4 h^4}{48} & -m\,\omega^2\,h\left(1 - \frac{\omega^2 h^2}{12}\right)\left(1 - \frac{\omega^2 h^2}{24}\right) \\ \frac{h}{m}\left(1 - \frac{\omega^2 h^2}{8}\right) & 1 - \frac{11}{24}\omega^2 h^2 + \frac{\omega^4 h^4}{48} \end{pmatrix}. \tag{26}$$

We observe that the "symplectic Simpson" numerical scheme defined by (10) (26) is an explicit scheme. It is easy with SageMath to verify that

$$\det \mathbf{\Phi}_3 = 1$$

and in consequence the scheme is symplectic.

7   Conservation of a Discrete Energy

To explicit a discrete energy that is conserved is not *a priori* obvious. For the harmonic oscillator, we search a conserved quadratic form of the type

$$Q(p, q) = \frac{1}{2} \xi p^2 + \eta p q + \frac{1}{2} \zeta q^2. \tag{27}$$

If we require that

$$Q(p_{j+1}, q_{j+1}) = Q(p_j, q_j)$$

with the variables p_{j+1}, q_{j+1}, p_j and q_j satisfying a linear dynamics such as (10) with a matrix

$$\mathbf{\Phi} = \begin{pmatrix} \alpha & \beta \\ \gamma & \delta \end{pmatrix}$$

of unit determinant, that is

$$\alpha \delta - \beta \gamma = 1,$$

then the coefficients ξ, η and ζ of the quadratic form (27) must satisfy the following homogeneous linear system

$$\Gamma \begin{pmatrix} \xi \\ \eta \\ \zeta \end{pmatrix} \equiv \begin{pmatrix} \frac{\alpha^2-1}{2} & \alpha\gamma & \frac{\gamma^2}{2} \\ \alpha\beta & \alpha\delta+\beta\gamma-1 & \gamma\delta \\ \frac{\beta^2}{2} & \beta\delta & \frac{\gamma^2-1}{2} \end{pmatrix} \begin{pmatrix} \xi \\ \eta \\ \zeta \end{pmatrix} = \begin{pmatrix} 0 \\ 0 \\ 0 \end{pmatrix}.$$

We have

$$\det \Gamma = (\alpha\delta - \beta\gamma - 1)(1 - \alpha - \delta + \alpha\delta - \beta\gamma)(1 + \alpha + \delta + \alpha\delta - \beta\gamma)$$

and this expression vanishes when $\alpha\delta - \beta\gamma = 1$. Moreover, when $\alpha = \delta$, we obtain $\eta = 0$ and a conserved quadratic form Q_c can be written

$$Q_c(p, q) = \frac{1}{2} \gamma p^2 - \frac{1}{2} \beta q^2$$

up to a multiplicative constant. Finally, if we set

$$H_d(p, q) \equiv \frac{1}{2m} p^2 + m \frac{\omega^2}{2} \left(1 - \frac{\omega^2 h^2}{8}\right) \left(1 - \frac{\omega^2 h^2}{12}\right) \left(1 - \frac{\omega^2 h^2}{24}\right) q^2, \tag{28}$$

the symplectic Simpson scheme

$$\begin{pmatrix} p_{j+1} \\ q_{j+1} \end{pmatrix} = \mathbf{\Phi}_3 \begin{pmatrix} p_j \\ q_j \end{pmatrix}$$

with $\mathbf{\Phi}_3$ explicited at the relation (26), satisfies the following conservation of energy:

$$H_d(p_{j+1}, q_{j+1}) = H_d(p_j, q_j).$$

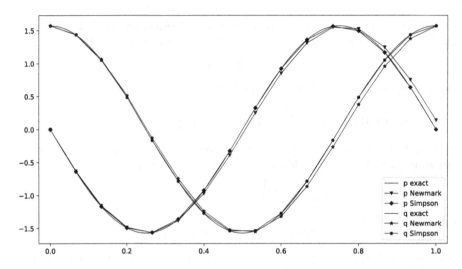

Fig. 1. Typical evolution of an harmonic oscillator. The momentum p follows a cosine curve and the state q a sine shape. Comparison of the exact solution and the Newmark and symplectic Simpson schemes for $N = 15$ meshes. Both schemes give very satisfactory results. Observe that the momentum p is very close to the exact solution with the symplectic Simpson scheme. Observe that the momentum data have been rescaled.

8   First Numerical Experiments

We have implemented the Simpson symplectic scheme and have compared it with the Newmark scheme (10) (11). Typical results for $N = 15$ meshes and one period are displayed on Fig. 1. We have chosen $q(t) = \sin \omega t$ and $p(t) = m\omega \cos \omega t$, with a period $T = 1$. Quantitative errors with the maximum norm are presented in Table 1 below. An asymptotic order of convergence can be estimated for the momentum, the state and various energies.

Table 1. Errors in the maximum norm. We observe again that the momentum is very well approximated with the symplectic Simpson scheme. The estimated order of convergence is the closest integer α measuring the ratio of successive errors in a given line by a negative power of 2 of the type $2^{-\alpha}$.

	number of meshes	10	20	40	order
Newmark	momentum	$1.91\ 10^{0}$	$5.02\ 10^{-1}$	$1.27\ 10^{-1}$	2
Symplectic Simpson	momentum	$3.41\ 10^{-3}$	$2.17\ 10^{-4}$	$1.35\ 10^{-5}$	4
Newmark	state	$2.38\ 10^{-1}$	$6.12\ 10^{-2}$	$1.55\ 10^{-2}$	2
Symplectic Simpson	state	$4.11\ 10^{-4}$	$2.55\ 10^{-5}$	$1.60\ 10^{-6}$	4
Newmark	energy (12)	$1.70\ 10^{-13}$	$4.97\ 10^{-14}$	$1.49\ 10^{-13}$	exact
Symplectic Simpson	energy (12)	$2.51\ 10^{-2}$	$1.67\ 10^{-3}$	$1.03\ 10^{-4}$	4
Newmark	energy (28)	$2.70\ 10^{-2}$	$1.67\ 10^{-3}$	$1.03\ 10^{-4}$	4
Symplectic Simpson	energy (28)	$2.13\ 10^{-14}$	$2.49\ 10^{-13}$	$1.63\ 10^{-13}$	exact

9   Conclusion and Perspectives

The symplectic Simpson numerical scheme has been developed in this contribution. It has been tested for an harmonic oscillator. The method is symplectic, conditionally stable and is fourth order accurate for state and momentum.

An important question is still open concerning the nonlinear case. The elimination of the internal degree of freedom is not possible in that case. Once this question has a satisfactory answer, the extension to systems with mutiple degrees of freedom is a natural objective for future studies.

References

1. Arnold, V.: Mathematical Methods of Classical Mechanics. Springer, New York (1974)
2. Courant, R.: Variational methods for the solution of problems of equilibrium and vibrations. Bull. Am. Math. Soc. **49**, 1–23 (1943)
3. Feynman, R.P., Hibbs, A.R.: Quantum Mechanics and Path Integrals. McGraw-Hill, New York (1965)
4. Gear, C.: Simultaneous numerical solution of differential-algebraic equations. IEEE Trans. Circuit Theory **18**, 89–95 (1971)
5. Hairer, E., Lubich, C., Wanner, G.: Geometric Numerical Integration, Structure-Preserving Algorithms for Ordinary Differential Equations. Springer, Heidelberg (2006). https://doi.org/10.1007/3-540-30666-8
6. Kane, C., Marsden, J.E., Ortiz, M., West, M.: Variational integrators and the Newmark algorithm for conservative and dissipative mechanical systems. Int. J. Numer. Meth. Eng. **49**, 1295–1325 (2000)
7. Newmark, N.M.: A method of computation for structural dynamics. J. Eng. Mech. Div. **85**(EM3), 67–94 (1959)
8. Raviart, P.A., Thomas, J.M.: Introduction à l'analyse numérique des équations aux dérivées partielles. Masson, Paris (1983)
9. Richtmyer, R.D., Morton, K.W.: Difference Methods for Initial-Value Problems. Wiley, Hoboken (1967)
10. SageMath, the Sage mathematics software system (Version 7.5.1). The Sage developers (2017). www.sagemath.org
11. Sanz-Serna, J.M.: Symplectic integrators for Hamiltonian problems: an overview. Acta Numer. **1**, 243–286 (1992)
12. Souriau, J.M.: Structure des systèmes dynamiques. Dunod, Paris (1970)
13. Wendlandt, J.M., Marsden, J.E.: Mechanical integrators derived from a discrete variational principle. Physica D **106**, 223–246 (1997)

Generalized Galilean Geometries

Eric Bergshoeff[(✉)] [ID]

Van Swinderen Institute for Particle Physics and Gravity, Nijenborgh 4,
9747 AG Groningen, The Netherlands
E.A.Bergshoeff@rug.nl

Abstract. Motivated by non-relativistic string theory, we give a classification of D-dimensional generalized Galilean geometries. They are an extension of the Galilean geometry in the sense that the two nondegenerate metrics of Galilean geometry (one to measure time intervals and another one to measure spatial distances) are replaced by two nondegenerate metrics of rank $p + 1$ and rank $D - p - 1$, respectively, with $p = 0, 1, \cdots, D - 1$. To classify these generalized geometries an important role is played by the so-called intrinsic torsion tensor indicating that this particular torsion is independent of the spin-connection. We show that there is a finite way of setting some of these intrinsic torsion tensors equal to zero and that this leads to a classification of the generalized Galilean geometries. Moreover, we show how some (but not all) of the generalized Galilean geometries that we find can be obtained by taking a special limit of general relativity.

Keywords: Galilean Geometry · Intrinsic Torsion · Branes

1 Introduction

One of the cornerstones of Einstein's description of general relativity is its underlying semi-Riemannian geometry giving a geometrical interpretation to the gravitational force. What is less known is that Newtonian gravity can be given a geometrical interpretation using a degenerate foliated geometry. Its proper formulation was given eight years after Einstein's formulation by Élie Cartan [1]. This generalization of Newtonian gravity is valid in any coordinate system and is called Newton-Cartan (NC) gravity with an underlying geometry that is called NC geometry. A characteristic feature of the NC geometry is that it contains, beyond the frame fields, an Abelian one-form that is needed to describe how massive particles move in a NC background.

Recently, there has been a growing interest in other non-Lorentzian gravity models and corresponding geometries. For some recent reviews, see [2,3]. In this work we wish to focus on generalizations of Galilean geometry in which the Abelian one-form of NC geometry is absent. This geometry can serve as the gravitational background of a so-called massless Galilean particle discussed by Souriau [4]. Recently, this massless Galilean particle has occurred as the source term of the so-called unwound string solution [5,6] of non-relativistic string theory [5,7].

F. Nielsen and F. Barbaresco (Eds.): GSI 2023, LNCS 14072, pp. 32–40, 2023.
https://doi.org/10.1007/978-3-031-38299-4_4

Motivated by non-relativistic string theory, we wish to generalize Galilean geometry to go beyond particles and consider the gravitational coupling to extended objects, also called p-branes, where $p = 0$ corresponds to a Galilean particle, $p = 1$ corresponds to a Galilean string, etc. Whereas any extended object can be coupled to general relativity, in the Galilean case each extended object requires a different Galilean geometry with a foliation that is determined by the spatial extension of the object: particles require a foliation with leaves of codimension one, but strings require a foliated geometry where the leaves are submanifolds of codimension two, that describe the dimensions transversal to the string. To distinguish between the different generalized Galilean geometries, we will denominate them as Galilean p-brane geometries such that the usual Galilean geometry corresponds to a Galilean particle geometry.

When classifying Galilean p-brane geometries, an important role is played by the so-called intrinsic torsion. When discussing torsionful geometries it is important to distinguish between the relativistic and Galilean case. In the relativistic case, the torsion tensor of a metric-compatible affine connection can be arbitrarily specified without imposing any constraints on the metric structure; in particular one may always consider a torsion-free affine connection (the Levi-Civita connection). This is no longer the case in Galilean geometry. There, part of the torsion tensor consists of so-called intrinsic torsion tensor components that form an obstruction to defining a metric compatible and torsionless connection, without imposing differential constraints on the metric structure [8]. For the purpose of classifying the Galilean p-brane geometries, we will in this work first collect all intrinsic torsion tensor components. Next, we will demonstrate which of these intrinsic torsion tensor components can be set to zero consistently with the symmetries of the structure group leading to constraints on the geometry. In this way we obtain a classification of the different geometries. Once, we have classified all Galilean p-brane geometries, we will show which ones can be obtained by taking a special limit of general relativity.

2 Galilean p-Brane Geometries

Our starting point for defining a Galilean p-brane geometry is a D-dimensional manifold with a degenerate metric structure that reduces the local structure group to [9, 10]

$$(\mathrm{SO}(p,1) \times \mathrm{SO}(D-p-1)) \ltimes \mathbb{R}^{(p+1)(D-p-1)} \tag{1}$$

for integer $0 \leq p \leq D - 2$. The Minkowskian worldvolume of a non-relativistic p-brane at rest divides up the tangent space directions of this manifold in $p + 1$ 'longitudinal' directions and $D - p - 1$ 'transversal' ones. The $\mathrm{SO}(p,1)$ and $\mathrm{SO}(D-p-1)$ factors of the structure group then correspond to Lorentz transformations of the $p + 1$ longitudinal directions and rotations of the $D - p - 1$ transversal directions, respectively. The $\mathbb{R}^{(p+1)(D-p-1)}$ factor represents boost transformations that can transform transversal directions into longitudinal ones, but not vice versa. We will refer to these as 'p-brane Galilean boosts'.

In analogy to particle or 0-brane Galilean geometry, the Cartan formulation of p-brane Galilean geometry includes two different types of one-forms (also called soldering forms): a 'longitudinal Vielbein' $\tau_\mu{}^A$ ($A = 0, 1, \cdots, p$) and a 'transversal Vielbein' $e_\mu{}^a$ ($a = p+1, \cdots, D-1$). The flat longitudinal index A can be freely raised and lowered with a $(p+1)$-dimensional Minkowski metric $\eta_{AB} = \mathrm{diag}(-1, 1, \cdots, 1)$, whereas for the flat transversal index a this is done using a $(D-p-1)$-dimensional Euclidean metric δ_{ab}. These one-forms transform under the structure group (1) in a reducible, indecomposable manner according to the following local transformation rules:

$$\delta \tau_\mu{}^A = \lambda^A{}_B \tau_\mu{}^B , \qquad\qquad \delta e_\mu{}^a = \lambda^a{}_b e_\mu{}^b - \lambda_A{}^a \tau_\mu{}^A . \qquad (2)$$

Here, $\lambda^{AB} = -\lambda^{BA}$ corresponds to the parameters of longitudinal $\mathrm{SO}(p, 1)$ Lorentz transformations, $\lambda^{ab} = -\lambda^{ba}$ to that of transversal $\mathrm{SO}(D-p-1)$ rotations, while the λ^{Aa} are the $(p+1)(D-p-1)$ p-brane Galilean boost parameters. Similar to the particle case, one introduces an 'inverse longitudinal Vielbein' $\tau_A{}^\mu$ and an 'inverse transversal Vielbein' $e_a{}^\mu$ (both are also called frame fields) such that the matrices

$$\left(\tau_\mu{}^A \ e_\mu{}^a \right) \qquad \text{and} \qquad \begin{pmatrix} \tau_A{}^\mu \\ e_a{}^\mu \end{pmatrix} \qquad (3)$$

are each other's inverse.[1]

The longitudinal and inverse transversal Vielbeine can be 'squared' to obtain two degenerate symmetric (covariant and contravariant) two-tensors that are invariant under local $\mathrm{SO}(p, 1)$, $\mathrm{SO}(D-p-1)$ and p-brane Galilean boost transformations:

$$\tau_{\mu\nu} \equiv \tau_\mu{}^A \tau_\nu{}^B \eta_{AB} , \qquad\qquad h^{\mu\nu} \equiv e_a{}^\mu e_b{}^\nu \delta^{ab} . \qquad (4)$$

These two tensors constitute a degenerate metric structure on the manifold. The covariant metric $\tau_{\mu\nu}$ is referred to as the 'longitudinal metric'. Its kernel is spanned by the $D-p-1$ vectors $e_a{}^\mu$ and it thus has rank $p+1$. The contravariant metric $h^{\mu\nu}$ is called the 'transversal metric' and has rank $D-p-1$, since its kernel is spanned by the $p+1$ one-forms $\tau_\mu{}^A$.

To define a metric compatible affine connection in p-brane Galilean geometry, we first introduce a structure group connection Ω_μ that takes values in the Lie algebra of (1)

$$\Omega_\mu = \frac{1}{2} \omega_\mu{}^{AB} J_{AB} + \frac{1}{2} \omega_\mu{}^{ab} J_{ab} + \omega_\mu{}^{Aa} G_{Aa} , \qquad (5)$$

where $J_{AB} = -J_{BA}$, $J_{ab} = -J_{ba}$ and G_{Aa} are generators of the Lie algebras of $\mathrm{SO}(p, 1)$, $\mathrm{SO}(D-p-1)$ and $\mathbb{R}^{(p+1)(D-p-1)}$, respectively. We will refer to $\omega_\mu{}^{AB} = -\omega_\mu{}^{BA}$, $\omega_\mu{}^{ab} = -\omega_\mu{}^{ba}$ and $\omega_\mu{}^{Aa}$ as spin connections for longitudinal Lorentz transformations, transversal rotations and p-brane Galilean boosts, respectively.

---

[1] In the mathematics literature one usually first introduces the frame fields and next the soldering forms.

We next introduce an affine connection $\Gamma^{\rho}_{\mu\nu}$ by imposing the following 'Vielbein postulates':

$$\partial_{\mu}\tau_{\nu}{}^{A} - \omega_{\mu}{}^{A}{}_{B}\tau_{\nu}{}^{B} - \Gamma^{\rho}_{\mu\nu}\tau_{\rho}{}^{A} = 0\,,$$
$$\partial_{\mu}e_{\nu}{}^{a} - \omega_{\mu}{}^{ab}e_{\nu b} + \omega_{\mu}{}^{Aa}\tau_{\nu A} - \Gamma^{\rho}_{\mu\nu}e_{\rho}{}^{a} = 0\,. \tag{6}$$

These postulates imply that the connection $\Gamma^{\rho}_{\mu\nu}$ is compatible with the metric structure (4). Using (6), one can express $\Gamma^{\rho}_{\mu\nu}$ in terms of the Vielbeine $\tau_{\mu}{}^{A}$, $e_{\mu}{}^{a}$, their inverses and the spin connections $\omega_{\mu}{}^{AB}$, $\omega_{\mu}{}^{ab}$, $\omega_{\mu}{}^{Aa}$ as follows:

$$\Gamma^{\rho}_{\mu\nu} = \tau_{A}{}^{\rho}\partial_{\mu}\tau_{\nu}{}^{A} + e_{a}{}^{\rho}\partial_{\mu}e_{\nu}{}^{a} - \omega_{\mu}{}^{A}{}_{B}\tau_{\nu}{}^{B}\tau_{A}{}^{\rho} - \omega_{\mu}{}^{a}{}_{b}e_{\nu}{}^{b}e_{a}{}^{\rho} + \omega_{\mu}{}^{Aa}\tau_{\nu A}e_{a}{}^{\rho}\,. \tag{7}$$

We will view the torsion $2\Gamma^{\rho}_{[\mu\nu]}$ of the affine connection as an independent and a priori arbitrary geometric ingredient. We will split it into 'longitudinal torsion' components $T_{\mu\nu}{}^{A}$ along $\tau_{A}{}^{\rho}$ and 'transversal torsion' components $E_{\mu\nu}{}^{a}$ along $e_{a}{}^{\rho}$:

$$2\Gamma^{\rho}_{[\mu\nu]} = \tau_{A}{}^{\rho}T_{\mu\nu}{}^{A} + e_{a}{}^{\rho}E_{\mu\nu}{}^{a}\,. \tag{8}$$

These equations imply that under local $SO(p,1)$, $SO(D-p-1)$ and p-brane Galilean boosts, $T_{\mu\nu}{}^{A}$ and $E_{\mu\nu}{}^{a}$ transform as follows:

$$\delta T_{\mu\nu}{}^{A} = \lambda^{A}{}_{B}T_{\mu\nu}{}^{B}\,, \qquad \delta E_{\mu\nu}{}^{a} = \lambda^{a}{}_{b}E_{\mu\nu}^{b} - \lambda_{A}{}^{a}T_{\mu\nu}{}^{A}\,. \tag{9}$$

By antisymmetrizing the Vielbein postulates (6), one obtains the following equations that are covariant with respect to the local structure group transformations:

$$T_{\mu\nu}{}^{A} = 2\partial_{[\mu}\tau_{\nu]}{}^{A} - 2\omega_{[\mu}{}^{A}{}_{B}\tau_{\nu]}{}^{B}\,, \tag{10}$$
$$E_{\mu\nu}{}^{a} = 2\partial_{[\mu}e_{\nu]}{}^{a} - 2\omega_{[\mu}{}^{ab}e_{\nu]b} + 2\omega_{[\mu}{}^{Aa}\tau_{\nu]A}\,. \tag{11}$$

We now wish to investigate which components of the above torsion two-forms T^{A} and E^{a} are independent of a spin-connection. We will call these the 'intrinsic torsion' tensor components. For this purpose, we first decompose the curved indices μ of the torsion two-forms into longitudinal and transversal indices A and a according to the following decomposition rule for any one-form V_{μ}:

$$V_{\mu} = \tau_{\mu}{}^{A}V_{A} + e_{\mu}{}^{a}V_{a} \qquad \text{or} \qquad V_{A} = \tau_{A}{}^{\mu}V_{\mu} \text{ and } V_{a} = e_{a}{}^{\mu}V_{\mu}\,. \tag{12}$$

Decomposing the 2-forms (10) and (11) in this way, we find that the following tensor components correspond to intrinsic torsion:

$$T_{a}{}^{\{AB\}} \ (p \neq 0)\,, \qquad T_{a}{}^{A}{}_{A}\,, \qquad T_{ab}{}^{A} \ (p \neq D-2)\,. \tag{13}$$

For instance, $T_{a}{}^{\{AB\}}$ denotes the following projection of the torsion two-forms $T_{\mu\nu}{}^{A}$:

$$T_{a}{}^{\{AB\}} \equiv e_{a}{}^{\mu}\tau^{\nu\{A}T_{\mu\nu}{}^{B\}}\,. \tag{14}$$

We use here a notation where $\{AB\}$ indicates the symmetric traceless part of AB. With $p \neq 0$ and $p \neq D - 2$ we indicate that the corresponding intrinsic torsion tensor components vanish for that value of p.

To classify the different constraints that one may impose on the intrinsic torsion tensor components given in (13), it is important to realize that under Galilean boosts, some components of the intrinsic torsion tensors transform to other components, and hence, those torsion tensors cannot be set to zero independently from other torsion components. One can only impose a set of constraints that is invariant under the boost transformations and does not lead to new constraints. The way that these boost transformations act on the torsion tensor components, for $p \neq 0$ and $p \neq D - 2$, can be summarized by the following sequence:

$$T_a{}^{\{AB\}}, T_a{}^A{}_A \quad \longrightarrow \quad T_{ab}{}^A \quad \longrightarrow \quad 0. \tag{15}$$

The arrows indicate the direction in which the p-brane Galilean boost transformations act. For instance, the boost transformation of $T_a{}^{\{AB\}}$ or $T_a{}^A{}_A$ gives $T_{ab}{}^A$ but not the other way around. The special cases of particles ($p = 0$) and domain-walls ($p = D - 2$) need to be discussed separately, see below.

The sequence (15) shows that, besides zero intrinsic torsion, there also exist other boost-invariant sets of constraints on the geometry. One may systematically derive these sets of constraints by starting from the right in (15) with no constraints (generic intrinsic torsion) and, next, by adding more constraints starting from the right by first setting the boost-invariant constraint $T_{ab}{}^A = 0$. A next possibility is that one adds to this constraint a second constraint by setting one of the two tensor components to the left of $T_{ab}{}^A$ to zero such that one again obtains a boost-invariant set of constraints. The fifth possibility is that, besides $T_{ab}{}^A$, one sets both $T_a{}^{\{AB\}}$ and $T_a{}^A{}_A$ to zero which leads to zero intrinsic torsion. These five distinct Galilean p-brane geometries that one obtains in this way are shortly discussed below [10].

1. The intrinsic torsion is unconstrained.
2. $\mathbf{T_{ab}{}^A = 0}$: According to the Frobenius theorem this constraint implies that the foliation by transverse submanifolds of dimension $D - p - 1$ is integrable.
3. $\mathbf{T_{ab}{}^A = T_a{}^{\{AB\}} = 0}$: the foliation is integrable and one can show that the vectors $e^\mu{}_a$ are conformal Killing vectors with respect to the longitudinal metric $\tau_{\mu\nu}$.
4. $\mathbf{T_{ab}{}^A = T_a{}^A{}_A = 0}$: the foliation is integrable and the worldvolume $(p+1)$-form Ω is closed.
5. $\mathbf{T_{\mu\nu}{}^A = 0}$: all constraints mentioned above are now valid.

The cases of particles ($p = 0$) and domain walls ($p = D - 2$) are special. Details about the domain wall case can be found in [10]. In the particle case the symmetric traceless intrinsic torsion components vanish and one recovers the standard three Galilean geometries:

1. The intrinsic torsion is unconstrained.

2. $\mathbf{T_{ab} = 0}$: the foliation by transverse submanifolds is integrable. Such a geometry is called *twistless torsional* [11]. Alternatively, the foliation is called *hypersurface orthogonal*.

3. $\mathbf{T}_{\mu\nu} = \mathbf{0}$: all intrinsic tensor components are zero, i.e. the foliation is integrable
and time is absolute.

This finishes our classification of the Galilean p-brane geometries. In the next
section we will show which ones of these geometries can be obtained by taking
a special limit of general relativity.

3   Generalized Galilei Gravity

Our starting point is the D-dimensional Einstein-Hilbert action

$$S_{\mathrm{EH}} = -\frac{1}{16\pi G_N} \int EE^{\mu}{}_{\hat{A}}E^{\nu}{}_{\hat{B}}R_{\mu\nu}{}^{\hat{A}\hat{B}}(\Omega). \tag{16}$$

Here, G_N is Newton's constant, $E_{\mu}{}^{\hat{A}}$ ($\mu, \hat{A} = 0, 1, \cdots, D-1$) is the relativistic
Vierbein and we have defined the inverse Vierbein $E^{\mu}{}_{\hat{A}}$ by

$$E_{\mu}{}^{\hat{A}}E^{\mu}{}_{\hat{B}} = \delta^{\hat{A}}{}_{\hat{B}}, \qquad\qquad E_{\mu}{}^{\hat{A}}E^{\nu}{}_{\hat{B}} = \delta_{\mu}{}^{\nu}. \tag{17}$$

Furthermore, the 2-form $R_{\mu\nu}{}^{\hat{A}\hat{B}}(\Omega)$ is defined in terms of the relativistic spin-
connection field $\Omega_{\mu}{}^{\hat{A}\hat{B}}$ as follows:

$$R_{\mu\nu}{}^{\hat{A}\hat{B}}(\Omega) = \partial_{[\mu}\Omega_{\nu]}{}^{\hat{A}\hat{B}} - 2\Omega_{[\mu}{}^{\hat{B}\hat{C}}\Omega_{\nu]\hat{C}}{}^{\hat{A}}. \tag{18}$$

To define the non-Lorentzian limit, that generalizes the limit for $p = 0$ con-
sidered in [12], we decompose the index $\hat{A}$ into longitudinal indices A ($A = 0, 1, \cdots, p$) and transverse indices a ($a = p+1, \cdots, D-1$) as $\hat{A} = (A, a)$ and
redefine the Vierbein and spin-connections with a contraction parameter ω as
follows

$$\begin{aligned}
E_{\mu}{}^{A} &= \omega\tau_{\mu}{}^{A}, & \Omega_{\mu}{}^{Aa} &= \omega^{-1}\omega_{\mu}{}^{Aa}, \\
E_{\mu}{}^{a} &= e_{\mu}{}^{a}, & \Omega_{\mu}{}^{ab} &= \omega_{\mu}{}^{ab}, & \Omega_{\mu}{}^{AB} &= \omega_{\mu}{}^{AB}.
\end{aligned} \tag{19}$$

The limit is performed by taking $\omega \to \infty$. Performing the redefinitions (19) in
the action (16) and redefining Newton's constant in such a way that the term
of leading power in ω is independent of ω, we obtain for general values of p and
finite ω three terms that are proportional to one of the following three 2-form
curvatures:

1. $R_{\mu\nu}{}^{AB}(M)$: the curvature 2-form corresponding to Lorentz transformations
in the longitudinal directions.
2. $R_{\mu\nu}{}^{Aa}(G)$: the curvature 2-form corresponding to p-brane Galilei boost transformations.

3. $R_{\mu\nu}{}^{ab}(J)$: the curvature 2-form corresponding rotations in the Euclidean transverse directions.

One may verify that the three terms in the action containing one of the curvatures $R(M), R(G)$ or $R(J)$ scale with relative powers ω^{-2}, ω^{-2} and ω^0, respectively.[2] An exception is formed by domain walls, i.e. $p = D - 2$, in which case the 2-form $R_{\mu\nu}{}^{ab}(J)$ vanishes since there are no rotations in the single transverse direction. In this special case we end up with the two other 2-form terms: $R_{\mu\nu}{}^{AB}(M)$ and $R_{\mu\nu}{}^{Aa}(G)$ that, combined with an adapted redefinition of Newton's constant, scale like ω^0. We will now discuss these two cases separately.

$\mathbf{p \neq D - 2}$. In this case, after taking the limit $\omega \to \infty$, the action takes the form

$$S_{\text{Galilei}} = -\frac{1}{16\pi G_{\text{NL}}} \int e\, e^\mu{}_a e^\nu{}_b R_{\mu\nu}{}^{ab}(J), \tag{20}$$

where, before taking the limit, we have redefined $G_{\text{N}} = \omega^{p+1} G_{\text{NL}}$ and where $e = \det(\tau_\mu{}^A, e_\mu{}^a)$. The 2-form $R_{\mu\nu}{}^{ab}(J)$ is given by

$$R_{\mu\nu}{}^{ab}(J) = 2\partial_{[\mu}\omega_{\nu]}{}^{ab} - \omega_{[\mu}{}^{ac}\omega_{\nu]c}{}^b. \tag{21}$$

Not all components of the spin connection $\omega_\mu{}^{ab}$ that occur in the action (20) are determined by the equations of motion corresponding to this action. Any component that does not occur in the quadratic spin-connection term given in the curvature (21) becomes a Lagrange multiplier imposing an intrinsic torsion constraint. To see this, we decompose the components of the spin-connection as follows:

$$\omega_\mu{}^{ab} = \tau_\mu{}^A \omega_A{}^{ab} + e_\mu{}^c \omega_c{}^{ab}. \tag{22}$$

Substituting this decomposition into the action (20), it is easy to see that the only surviving term quadratic in $\omega_\mu{}^{ab}$ is a term quadratic in the $\omega_c{}^{ab}$ components. Hence, the $\omega_A{}^{ab}$ component of the spin-connection has become a Lagrange multiplier imposing the intrinsic torsion constraint $T_{ab}{}^A = 0$. This leads to a Galilean p-brane $(p \neq D - 2)$ geometry with integrable foliation.

$\mathbf{p = D - 2}$. This case is special because there is only one transverse direction $a = z$, i.e. a domain wall, and hence $\omega_\mu{}^{ab} = 0$. Writing $e_\mu{}^z = e_\mu$ and $\omega_\mu{}^{zA} = \omega_\mu{}^A$ we obtain, after taking the limit $\omega \to \infty$, the following action:

$$S_{\text{DW}} = -\frac{1}{16\pi G_{\text{NL}}} \int e\left(\tau^\mu{}_A \tau^\nu{}_B R_{\mu\nu}{}^{AB}(M) + 2e^\mu \tau^\nu{}_A R_{\mu\nu}{}^A(G)\right), \tag{23}$$

where, before taking the limit, we have redefined $G_{\text{N}} = \omega^{p-1} G_{\text{NL}}$ and where the 2-forms $R(M)$ and $R(G)$ are given by

$$R_{\mu\nu}{}^{AB}(M) = 2\,\partial_{[\mu}\omega_{\nu]}{}^{AB} - \omega_{[\mu}{}^{AC}\omega_{\nu]C}{}^B, \tag{24}$$

$$R_{\mu\nu}{}^A(G) = 2\,\partial_{[\mu}\omega_{\nu]}{}^A - 2\omega_{[\mu}{}^{AC}\omega_{\nu]C}. \tag{25}$$

---

[2] To obtain these scalings one should also rescale the Vielbeine that occur in the action (16).

To see which flat spin-connection components become Lagrange multipliers we first make the following decompositions:

$$\omega_\mu{}^{AB} = \tau_\mu{}^C \omega_C{}^{AB} + e_\mu \bar{\omega}^{AB} \,,$$
$$\omega_\mu{}^A = \tau_{\mu C} \omega^{[AC]} + \tau_{\mu C} \omega^{\{AC\}} + \tau_\mu{}^A \bar{\omega} + e_\mu \omega^A \,, \tag{26}$$

where $\bar{\omega}^{AB} = -\bar{\omega}^{BA}$. Substituting these decompositions into the action (23), we find that both the components $\omega^{\{AB\}}$ and $\bar{\omega}$ are Lagrange multipliers. From the action (23), we derive that they impose the intrinsic torsion constraints $T^{\{AB\}} = T^A{}_A = 0$. This corresponds to a Galilean domain-wall geometry with zero torsion.

4 Discussion

The generalized Galilean geometries discussed here describe part of the geometries underlying non-relativistic string theory. There are, however, three important differences:

1. The geometry underlying non-relativistic string theory contains an additional p-form whose mathematical description is not quite clear although some of the literature seems to suggest that the introduction of the notion of gerbes is required.
2. It turns out that in non-relativistic string theory the frame fields transform under an additional local scale symmetry [13] which requires an additional dilatation gauge field beyond the spin-connections.
3. In the case of supersymmetric non-relativistic theory the geometry needs to be embedded into a so-called supergeometry which also contains fermionic intrinsic tensor components.

Irrespective of these differences, the present work sheds interesting light on some aspects of the full geometry underlying non-relativistic string theory.

Acknowledgements. The work described here is based upon work done in collaboration with Jose Figueroa-O'Farrill, Johannes Lahnsteiner, Luca Romano, Jan Rosseel, Iisakki Rotko, Tonnis ter Veldhuis and Kevin van Helden [9,10].

References

1. Cartan, E.: Sur les variétés à connexion affine et la théorie de la relativité généralisée. Annales scientifiques de l'École Normale Supérieure Série 3 Tome **40** (1923)
2. Bergshoeff, E., Figueroa-O'Farrill, J., Gomis, J.: A non-lorentzian primer arXiv:2206.12177
3. Hartong, J., Obers, N.A., Oling, G.: Review on Non-Relativistic Gravity arXiv:2212.11309

4. Souriau, J.M.: A Symplectic View of Physics, Editors Cushman, R.H. and Tuynman, G.M., Birkhäuser (1997)
5. Danielsson, U.H., Guijosa, A., Kruczenski, M.: IIA/B, wound and wrapped. JHEP **10**, 020 (2000)
6. Bergshoeff, E., Lahnsteiner, J., Romano, L., Rosseel, J.: The supersymmetric Neveu-Schwarz branes of non-relativistic string theory. JHEP **08**, 218 (2022)
7. Gomis, J., Ooguri, H.: Nonrelativistic closed string theory. J. Math. Phys. **42**, 3127–3151 (2001)
8. Figueroa-O'Farrill, J.: On the intrinsic torsion of spacetime structures arXiv:2009.01948
9. Bergshoeff, E., van Helden, K., Lahnsteiner, J., Romano, L., Rosseel, J.: Generalized Newton-Cartan geometries for particles and strings. Class. Quant. Grav. **40**(7), 075010 (2023)
10. Bergshoeff, E., Figueroa-O'Farrill, J., van Helden, K., Rosseel, J., Rotko, I., ter Veldhuis, T., work in progress
11. Christensen, M.H., Hartong, J., Obers, N.A., Rollier, B.: Torsional Newton-Cartan geometry and Lifshitz holography. Phys. Rev. D **89**, 061901 (2014)
12. Bergshoeff, E., Gomis, J., Rollier, B., Rosseel, J., ter Veldhuis, T.: Carroll versus Galilei gravity. JHEP **03**, 165 (2017)
13. Bergshoeff, E.A., Lahnsteiner, J., Romano, L., Rosseel, J., Şimşek, C.: A nonrelativistic limit of NS-NS gravity. J. High Energy Phys. **2021**(6), 1–34 (2021). https://doi.org/10.1007/JHEP06(2021)021

Continuum Mechanics of Defective Media: An Approach Using Fiber Bundles

Mewen Crespo$^{(\boxtimes)}$ ⓘ, Guy Casale ⓘ, and Loïc Le Marrec ⓘ

Univ Rennes, CNRS, IRMAR-UMR 6625, 35000 Rennes, France
{mewen.crespo,guy.casale,loic.lemarrec}@univ-rennes.fr
https://irmar.univ-rennes.fr

Abstract. The kinematics of a micro-structured material is geometrically modeled through the framework of fiber bundle geometry. The material continuum is a fiber bundle $\mathcal{M} \to \mathbb{B}$ where $\mathbb{B}$ is compact and orientable. It is commonly agreed that connections with curvature and torsion can describe defect densities in micro-structured materials. The aim of this work is to introduce a method to derive these objects from the kinematics in an intrinsic way. The material bundle $\mathcal{M}$ is therefore placed in the Euclidean fiber bundle $\mathcal{E} \equiv \mathrm{T}\mathbb{E} \to \mathbb{E}$ using a placement map $\varphi : \mathcal{M} \to \mathcal{E}$. A first-order transformation $\mathbf{F} : \mathrm{T}\mathcal{M} \to \mathrm{T}\mathcal{E}$ generalizing $\mathrm{T}\varphi$ is then introduced. Finally, using $\mathbf{F}$, a metric on $\mathbb{B}$, a connection on $\mathcal{M}$ and a solder form on $\mathcal{M}$ are inferred from the Euclidean structure on $\mathcal{E}$. These new objects are grouped into a single one, called a pseudo-metric, which allows us to describe the current state of matter through, among other things, the curvature (disclinations) and torsion (dislocations) tensors. On one hand, we see that the torsion tensor can be non-zero even in the holonomic $\mathbf{F} = \mathrm{T}\varphi$ case. On the other-hand, in order for the material to have a non-zero curvature tensor, we see that one must have a non-holonomic first-order transformation: that is, $\mathbf{F} \neq \mathrm{T}\varphi$.

Keywords: Generalized continuum · Fiber bundles · Connection · mechanics

1 Introduction

We are interested in the study of micro-structured materials. That is, materials in which the microscopic state at a point plays an import role in its macroscopic behavior and therefore have to be included in its geometric description. Mathematically, an n-dimensional micro-structured material with a k-dimensional micro-structure (or a $n \times k$ material for short) is therefore a fiber bundle manifold $\mathcal{M} \xrightarrow{\pi_{\mathcal{M}}} \mathbb{B}$ of rank k over a compact orientable n-dimensional manifold $\mathbb{B}$, which is the classical macroscopic body.

At a point b in the macroscopic body $\mathbb{B}$, the micro-structure of a micro-element is represented by the fiber $\mathcal{M}_b$ of $\mathcal{M}$ over b. Those elements are interpreted as first order infinitesimal neighborhoods of geometrical macroscopic

Supported by the ANR-11-LABX-0020-0 program Henri Lebesgue Center.

points (eventually embedded in a bigger space of higher dimension), as it is commonly considered for crystal modeling. For this reason, $\mathcal{M}$ is often closely related to a tangent space (not necessarily $T\mathbb{B}$). In our case, the material is embedded in the Euclidean space. Therefore, $n \in \{1, 2, 3\}$ and $k = 3$.

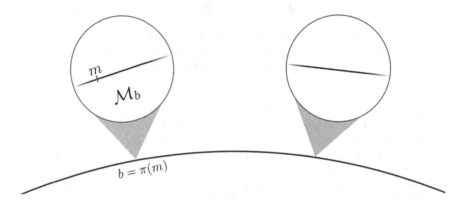

Fig. 1. The standard horizontal representation of a generic material bundle. The macroscopic space is represented horizontally and micro-elements are shown to be inside the macroscopic points. Note that neither $\mathbb{B}$ nor $\mathcal{M}_b$ is assumed to be 1-dimensional, this is only done for illustration purpose.

Examples of micro-structured materials include the 1-dimensional beams (1×3 materials), the 2-dimensional plates (2×3 materials) and the 3-dimensional Cosserat continuum (a 3×3 material). Some authors [5, p138-141] choose to use the principal bundle (see Sect. 3.2). It is a fiber bundle over the same base manifold $\mathbb{B}$ whose fibers are not the micro-elements but their possible configurations (a sub-group of GL(3)). This is why, at first, those examples may seem to have different micro-structure as, in practice, one often ends up manipulating the principal bundle. Those could for example, depending on assumptions made on the material, be isomorphic to the 1-dimensional circle SO(2) $\simeq$ S^1 (for planar inextensible beams), the 2-dimensional space SO(2) $\times \mathbb{R}^*$ (for planar extensible beams) or the 3-dimensional space SO(3) (for some plates and Cosserat continuum).

1.1 The Standard Interpretation

Figure 1 shows the standard representation of a generic material bundle. This interpretation of the mathematical model is central to our problem. It states that micro-elements are inside macroscopic points and are therefore imperceptible to an experimenter. In general, this interpretation states that the projection, to the macroscopic spaces $\mathbb{B}$ and $\mathbb{E}$, of any kinematic object of our model can be seen as a kinematic object of the classical continuum mechanics. For example, this implies that $\overline{\varphi} : \mathbb{B} \to \mathbb{E}$ (defined later on) corresponds to an usual macroscopic placement map that can be observed and measured by an experimenter.

That the macroscopic kinematics is classical does not mean that the macroscopic dynamics will be. Macroscopic kinematic object are independent of the micro-structure, but microscopic or mixed objects can depend on the macroscopic states. This means that a coupling between the macroscopic and microscopic variables can append in the dynamics' equations. In other words, in general, $\overline{\varphi}$ will not behave classically.

2   First Formalism

2.1   Placement Map

By analogy with $\mathcal{M}$, the ambient space is modeled as the Euclidean fiber bundle $\mathcal{E} \equiv \mathbb{TE} \xrightarrow{\pi_{\mathcal{E}}} \mathbb{E}$ of rank 3 over the usual ambient space $\mathbb{E}$, the 3-dimensional Euclidean space. We chose a Lagrangian approach. Accordingly, one defines a placement map. This latter is a fiber bundle embedding $\varphi : \mathcal{M} \to \mathcal{E}$ which maps the micro-element over $b \in \mathbb{B}$ to the micro-place over $\overline{\varphi}(b) \in \mathbb{E}$ where $\overline{\varphi} : \mathbb{B} \to \mathbb{E}$ is called the shadow of φ (see the commutative diagram bellow). This placement map φ defines a macroscopic material placement map, via its shadow $\overline{\varphi}$ and the differential $T\overline{\varphi}$.

$$
\begin{array}{ccccccccc}
T\mathcal{E} & \xrightarrow{\ T\pi_{\mathcal{E}}\ } & \mathbb{TE} & \xrightarrow{\ \pi_{\mathbb{TE}}\ } & \mathbb{E} & \xleftarrow{\ \pi_{\mathcal{E}}\ } & \mathcal{E} & \xleftarrow{\ \pi_{T\mathcal{E}}\ } & T\mathcal{E} \\
\big\uparrow{\scriptstyle \mathbf{F}} & & \big\uparrow{\scriptstyle T\overline{\varphi}} & & \big\uparrow{\scriptstyle \overline{\varphi}} & & \big\uparrow{\scriptstyle \varphi} & & \big\uparrow{\scriptstyle \mathbf{F}} \\
T\mathcal{M} & \xrightarrow{\ T\pi_{\mathcal{M}}\ } & \mathbb{TB} & \xrightarrow{\ \pi_{\mathbb{TB}}\ } & \mathbb{B} & \xleftarrow{\ \pi_{\mathcal{M}}\ } & \mathcal{E} & \xleftarrow{\ \pi_{T\mathcal{M}}\ } & T\mathcal{M}
\end{array}
$$

In order to also have a description of the material's defects, one must have a generalized material placement map. To complete the description of a generalized material placement map, one additionally has to introduce a generalized first order placement map $\mathbf{F} : T\mathcal{M} \to T\mathcal{E}$. The standard interpretation then requires $\mathbf{F}(V\mathcal{M}) \subset V\mathcal{E}$. Physically, this means that a microscopic change in the material causes a microscopic change in the ambient image. In particular, this implies that $\mathbf{F}$ can be seen as a fiber bundle morphism between $T\mathcal{M} \xrightarrow{T\pi_{\mathcal{M}}} \mathbb{TB}$ and $T\mathcal{E} \xrightarrow{T\pi_{\mathcal{E}}} \mathbb{TE}$ (notice the change of base spaces and projection maps, see the commutative diagram above).

The standard interpretation then also requires that the shadows of $\mathbf{F}$ with respect to the usual structures and these new structures are, respectively, $\varphi : \mathcal{M} \to \mathcal{E}$ and $T\overline{\varphi} : \mathbb{TB} \to \mathbb{TE}$. Those two requirements reduces the degrees of freedom of $\mathbf{F}$ to $k \cdot (k+n) \in \{12, 15, 18\}$ real coefficients, which are scalar fields on $\mathcal{M}$ (i.e. $\mathbf{F}$ lives in a $(k+n) \times (k \cdot (k+n))$ vectorial fiber bundle). In the classical case $\mathbf{F}$ ought to be the differential of φ – i.e. $\mathbf{F} = T\varphi$ – but as it turns out, this assumption is too restrictive and corresponds to the absence of disclinations in the micro-structure (see [7]).

2.2 Metric, Connection and Solder Form

The existence of a first order placement map allows one to superimpose the tangent of the body over the tangent of the ambient space. Doing so, first order functions of the ambient space such as tensors, can be seen in the material space. Mathematically, this correspond to a pull-back by the first order placement map.

First, there exists a metric $\mathbf{g}$ on $\mathbb{E}$: the Euclidean metric. This allows the measure of angles and lengths. Using the first order macroscopic placement $T\overline{\varphi}$, one can pull it back as a metric $\mathbf{G}$ on $\mathcal{M}$, which gives the Cauchy-Green tensor (see[1] the equation and diagram below). In the case of generalized continua, the macroscopic metric is accompanied by two micro-structured linear applications: the connection and the solder form.

$$\forall b \in \mathbb{B} \qquad \mathbf{G}_b := T^*_{\overline{\varphi}(b)}\overline{\varphi} \cdot \mathbf{g}_{\overline{\varphi}(b)} \cdot T_b\overline{\varphi}$$

$$
\begin{array}{ccc}
T^*\mathbb{B} & \xleftarrow{\ T^*\overline{\varphi}\ } & T^*\mathbb{E} \\[2pt]
\scriptstyle G \big\uparrow & & \big\uparrow \scriptstyle g \\[2pt]
T\mathbb{B} & \xrightarrow{\ T\overline{\varphi}\ } & T\mathbb{E}
\end{array}
$$

Figure 2 shows the vertical representation of a generic connection on a material bundle. A connection on $\mathcal{E}$ can be defined as a right-inverse $\mathbf{\Gamma} : T\mathbb{B} \to T\mathcal{M}$ of $T\pi_{\mathcal{M}} : T\mathcal{M} \to T\mathbb{B}$ (i.e. $T\pi_{\mathcal{M}} \cdot \mathbf{\Gamma} = \mathrm{Id}_{T\mathbb{B}}$ − please note that other equivalent definitions also exist [1,5,6]). While it does not induce a notion of length or angle, it still induces a notion of local geodesics and, as shown on the figure, parallel transport of micro-elements along the horizontal lift of a macroscopic path. This later, allows to connect close microscopic spaces by following the flow generated by $\mathbf{\Gamma}$. If the macroscopic path is a closed curve in $\mathbb{B}$, one may witness some lack of closure of the horizontal lift in $\mathcal{M}$. This lack of closure is measured by the curvature and torsion in a way similar to the way Frank and Burgers vectors are introduced and used in crystallography (see [3]).

On $\mathcal{E}$, one has the Levi-Civita connexion, which is defined as the unique connexion with no torsion whose geodesics are those of the Euclidean metric. Coordinate-wise, it corresponds to the inclusion $T\mathbb{E} \simeq \mathcal{E} \hookrightarrow \mathcal{E} \times \{0\} \subset \mathcal{E} \times \mathbb{R}^6 \simeq T\mathcal{E}$. In a similar way to what has been done for the metric, the connection on $\mathcal{M}$, which is not a tensor, is then obtained by pulling back this Euclidean Levi-Civita connection γ of $\mathcal{E}$ into a connection $\mathbf{\Gamma}$ on $\mathcal{M}$. This involves the use of the fiber bundle morphism $\mathbf{F} : T\mathcal{M} \to T\mathcal{E}$ to superimpose $T\mathcal{M}$ over $T\mathcal{E}$. The formula and associated diagram are as follows:

$$\forall b \in \mathbb{B},\ \forall m \in \mathcal{M}_b, \quad \mathbf{\Gamma}_m := \mathbf{F}_b^{-1} \cdot \gamma_{\varphi(m)} \cdot T_b\overline{\varphi}$$

$$
\begin{array}{ccc}
T\mathcal{M} & \xrightarrow{\ \mathbf{F}\ } & T\mathcal{E} \\[2pt]
\scriptstyle\Gamma \big\uparrow \big\downarrow \scriptstyle T\pi_{\mathcal{M}} & & \scriptstyle T\pi_{\mathcal{E}} \big\downarrow \big\uparrow \scriptstyle\gamma \\[2pt]
T\mathbb{B} & \xrightarrow{\ T\overline{\varphi}\ } & T\mathbb{E}
\end{array}
$$

A solder form on $\mathcal{E}$ is, in an analog fashion, an injective[2] map $\vartheta : T_x\mathbb{E} \hookrightarrow V_p\mathcal{E}$ for any $x \in \mathbb{E}, p \in \mathcal{E}_x$. While a connection glues microscopic grains together,

---

[1] Here $T^*\overline{\varphi}: T^*\mathbb{E} \to T^*\mathbb{B}$ is the transpose of the differential $T\overline{\varphi}: T\mathbb{B} \to T\mathbb{E}$ of $\overline{\varphi}: \mathbb{B} \to \mathbb{E}$.

[2] A quick comparison of dimensions gives $\dim(T_x\mathbb{E}) = 2\dim(\mathbb{E}) = 6 = 2\,\mathrm{rank}(\mathcal{E}) = \dim(V_p\mathcal{E})$. Hence, in the Euclidean case, ϑ is a bijection.

a solder form glues a grain to the macroscopic space. From a vector at a macroscopic point, it generates a vector field inside the microscopic grain at that macroscopic point. As an ambient solder form, we choose the canonical Euclidean solder form ϑ which comes from the identification $\mathcal{E} \simeq T\mathbb{E}$. Just like for γ, ϑ can be pulled back onto $\mathcal{M}$ as $\Theta : T\mathbb{B} \hookrightarrow V\mathcal{M}$. The formula and associated diagram are as follows:

$$\forall b \in \mathbb{B}, \ \forall m \in \mathcal{M}_b, \quad \Theta_m := \mathbf{F}_b^{-1} \cdot \vartheta_{\varphi(m)} \cdot T_b\overline{\varphi}$$

$$
\begin{array}{ccc}
V\mathcal{M} & \xrightarrow{\ \mathbf{F}\ } & V\mathcal{E} \\
{\scriptstyle \Theta}\big\uparrow & & {\scriptstyle \vartheta^{-1}}\big\updownarrow{\scriptstyle \vartheta} \\
T\mathbb{B} & \xrightarrow[\ T\varphi\ ]{} & T\mathbb{E}
\end{array}
$$

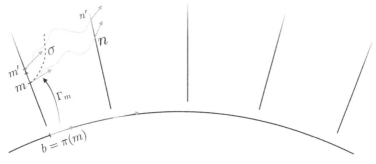

Fig. 2. The vertical representation of a generic material bundle. The macroscopic space is represented horizontally and micro-elements are unfolded vertically above it. Note that neither $\mathbb{B}$ nor $\mathcal{M}_b$ is assumed to be 1-dimensional, this is only done for illustration purpose.

2.3   Pseudo-metric

Another way to interpret the connection and solder form is as embeddings. The connection can be seen as an application which takes a macroscopic vector $\mathbf{v} \in T\mathbb{E}$ and embeds it in the total tangent space as $\gamma(\mathbf{v}) \in T\mathcal{E}$ which can be interpreted as the corresponding vector with the same macroscopic part $(T\pi_{\mathcal{E}} \cdot \gamma \cdot \mathbf{v} = \mathbf{v})$ and no vertical[3] part $(v_\gamma \cdot \gamma \cdot \mathbf{v} = 0)$.

Similarly, $\vartheta(\mathbf{v}) \in V\mathcal{E} \subset T\mathcal{E}$ can be interpreted as the corresponding vector with the same vertical part $(\vartheta^{-1} \cdot v_\gamma \cdot \vartheta \cdot \mathbf{v} = \mathbf{v})$ and no horizontal part $(h_\gamma \cdot \vartheta \cdot \mathbf{v} = 0)$. In this interpretation, $\vartheta^{-1} \cdot v_\gamma : T\mathcal{E} \to T\mathbb{E}$ can be seen as a "microscopic" projection. In particular,

$$T\varpi = T\pi + \vartheta^{-1} \cdot v_\gamma : T\mathcal{E} \to T\mathbb{E}$$

is also a projection which will be called **the projection of interpretation** as it gives the sum of the macroscopic part and the microscopic part, both seen in the macroscopic space $T\mathbb{B}$, which is how one interprets $\mathbf{v}$.

---

[3] We use here the standard notation $h_\gamma = \gamma \cdot T\pi_{\mathcal{E}}$ and $v_\gamma = \mathrm{Id}_{T\mathcal{E}} - h_\gamma$ such that $T\pi_{\mathcal{E}} \cdot h_\gamma = T\pi_{\mathcal{E}}$ and $T\pi_{\mathcal{E}} \cdot v_\gamma = \mathbf{0}$.

This projection of interpretation gives an alternative approach for the generalization of the metric. Indeed, one can simply pull-back the Euclidean metric $\mathbf{g} : T\mathbb{E} \to T^*\mathbb{E}$ onto $T\mathcal{E}$ using $T\varpi$. One then obtains

$$\mathfrak{g} : T\mathcal{E} \longrightarrow T^*\mathcal{E}$$
$$:= T^*\varpi \cdot \mathbf{g} \cdot T\varpi$$

Its kernel is the kernel of $T\varpi$. Those are the vectors whose macroscopic part is the opposite of their microscopic part. In particular their microscopic and macroscopic part are of the same magnitude, which mean these are not physically acceptable vectors. The kernel of $\mathfrak{g}$, as it turns out, is the horizontal space $H_{\gamma-\vartheta} = \mathrm{Im}\,(\gamma - \vartheta)$ of the connection $\gamma - \vartheta$. The solder form ϑ is, by construction, isometric with respect to $\mathfrak{g}_{|V\mathcal{E}} : V\mathcal{E} \to V^*\mathcal{E}$ and $\mathbf{g}$. This means that, from $\mathfrak{g}$, one can extract $\gamma - \vartheta$ and ϑ up to a "rotation" in $\mathcal{O}\left(\mathbf{g}, \mathfrak{g}_{|V\mathcal{E}}\right)$.

The tensor $\mathfrak{g}$ is called **the ambient pseudo-metric** as it is symmetric and positive but its kernel is non-trivial (otherwise it would be a metric). As $\mathbf{g}$ with $T\overline{\varphi}$, $\mathfrak{g}$ can be pulled-back on $T\mathcal{M}$ using $\mathbf{F}$. One obtains the material pseudo-metric

$$\mathfrak{G} : T\mathcal{M} \longrightarrow T^*\mathcal{M}$$
$$:= \mathbf{F}^* \cdot \mathfrak{g} \cdot \mathbf{F}$$

Similarly to the ambient case, one has that $\ker\,(\mathfrak{G}) = H_{\Gamma-\Theta}$ and that Θ is an isometry for $\mathbf{G}$ and $\mathfrak{G}_{|V\mathcal{E}}$. This means that, if $\mathbf{G}$ is known, one can extract the connection $\Gamma - \Theta$ and the orbit of Θ under $\mathcal{O}\left(\mathbf{G}, \mathfrak{G}_{|V\mathcal{E}}\right)$ from $\mathfrak{G}$. Nevertheless, $\mathbf{G} : T\mathbb{B} \to T^*\mathbb{B}$ cannot be extracted from $\mathfrak{G} : T\mathcal{M} \to T^*\mathcal{M}$ directly, unless $\mathbf{F} : T\mathcal{M} \to T\mathcal{E}$ is entirely known (in which case $\mathbf{G}$ is also known). The pseudo-metric $\mathfrak{G}$ is a canonical extension of the right Cauchy-Green tensor $\mathbf{G}$. As $\mathbf{G}$ plays a central role in the classical case, one expect $\mathfrak{G}$ to also play a key role in the generalized case. These observations therefore justify the use of $\mathbf{G}$, Γ and Θ as our primary tools.

3   Constitutive Equations and Structure Group

3.1   Stress and Constitutive Equations

By formulating continuum mechanics using differential geometry (as in [1,2,4]) one obtains a geometric formulation of static equilibrium that can be generalized to the micro-structured case. For example, the macroscopic Cauchy stress $\overline{\mathbf{S}}$ and the macroscopic second Piola tensor $\overline{\Sigma}$ are:

$$\overline{\mathbf{S}} : T^*\mathbb{E} \longrightarrow T\mathbb{E} \qquad\qquad \overline{\Sigma} : T^*\mathbb{B} \longrightarrow T\mathbb{B}$$

In the generalized version, tensors on $\mathbb{E}$ (resp. $\mathbb{B}$) naturally become tensors on $\mathcal{E}$ (resp. $\mathcal{M}$). This gives

$$\mathbf{S} : T^*\mathcal{E} \longrightarrow T\mathcal{E} \qquad\qquad \Sigma : T^*\mathcal{M} \longrightarrow T\mathcal{M}$$

Which, for consistency, are required to be morphisms for both structures $T\mathcal{E} \xrightarrow{\pi_{T\mathcal{E}}} \mathcal{E}$ and $T\mathcal{E} \xrightarrow{T\pi_{\mathcal{E}}} TE$ $\left(\text{resp. } T\mathcal{M} \xrightarrow{\pi_{T\mathcal{M}}} \mathcal{M}\right)$ and $T\mathcal{M} \xrightarrow{T\pi_{\mathcal{M}}} T\mathbb{B}$. This is a manifestation of the standard interpretation, just like it was for $\mathbf{F}$, which also requires that the shadows be $\mathrm{Id}_{\mathcal{E}}$ and $\overline{\mathbf{S}}$ (resp. $\mathrm{Id}_{\mathcal{M}}$) and $\overline{\mathbf{\Sigma}}$.

The notion of elasticity and hyper-elasticity can be generalized to the microstructured case. The elasticity simply states that $\mathbf{S}$ (resp. $\mathbf{\Sigma}$) is a linear function $\widehat{\mathbf{S}}$ $\left(\text{resp. } \widehat{\mathbf{\Sigma}}\right)$ of $\mathbf{F}$, or a linear function of $\mathfrak{G}$ in the isotropic case. The hyperelasticity on the other side states that this function $\widehat{\mathbf{\Sigma}}$ is obtainable as the differential $T\mathbf{W}$ of a functional $\mathbf{W}$ of $\mathbf{F}$ (resp. $\mathfrak{G}$ in the isotropic case) called the energy density. Such an energy, in the isotropic case, can depend on invariants of $\mathbf{G}$, $\mathbf{\Gamma}$ and $\mathbf{\Theta}$. Already existing formulae for the macroscopic case can be extended by replacing $\mathbf{G}$ with $\mathfrak{G}$. For example, identifying $T\mathcal{M}$ and $T^*\mathcal{M}$ for simplicity (this is roughly equivalent to having a reference pseudo-metric), the Saint Venant-Kirchhoff constitutive law gives, under the substitution $\mathbf{G} \to \mathfrak{G}$:

$$\mathbf{W}\left(\mathfrak{G}\right) := \frac{a}{4} \operatorname{tr}\left(\mathfrak{G} - \mathrm{Id}\right) + \frac{b}{4} \operatorname{tr}\left(\left(\mathfrak{G} - \mathrm{Id}\right)^2\right)$$
$$\mathbf{\Sigma} := a \operatorname{tr}\left(\mathfrak{G} - \mathrm{Id}\right)\mathrm{Id} + b\left(\mathfrak{G} - \mathrm{Id}\right)$$

Under a specific frame induced by $\mathbf{\Gamma}$, $\mathfrak{G}$ splits into blocks as

$$\begin{bmatrix} T\pi_{\mathcal{M}}^* \cdot \mathbf{G} \cdot T\pi_{\mathcal{M}} & 0 \\ 0 & \left(\mathrm{Id} - \mathbf{\Gamma} \cdot T\pi_{\mathcal{M}}\right)^* \cdot \mathfrak{G} \cdot \left(\mathrm{Id} - \mathbf{\Gamma} \cdot T\pi_{\mathcal{M}}\right) \end{bmatrix}$$

Hence, we see that even with a naive extension of the Saint Venant-Kirchhoff formula, there is a micro/macro coupling occurring through the term

$$\operatorname{tr}\left(\mathfrak{G} - \mathrm{Id}\right) = \operatorname{tr}\left(\mathbf{G}\right) - k + \operatorname{tr}\left(\left(\mathrm{Id} - \mathbf{\Gamma} \cdot T\pi_{\mathcal{M}}\right)^* \cdot \mathfrak{G} \cdot \left(\mathrm{Id} - \mathbf{\Gamma} \cdot T\pi_{\mathcal{M}}\right)\right) - 3$$

If we place ourselves in the holonomic case $\varphi \equiv T\overline{\varphi}$ and $\mathbf{F} = T\varphi$. That is, if the microscopic kinematics is a copy of the macroscopic one. Further computations show that one has $\operatorname{tr}(\mathfrak{G}) = 2\operatorname{tr}(\mathbf{G})$ and, up to a constant rescaling of a and b, $\mathbf{\Sigma}$ is just two (diagonal) copies of $\overline{\mathbf{\Sigma}}$. That is, the microscopic part of the stress is just an independent copy of its macroscopic part and therefore the dynamics is classical. In general, however, it is not the case.

It should be noted that, simply extending classical formulae this way does not provide all possible energies, as other invariants exist. For example, even $\operatorname{tr}(\mathbf{G})$ will not be accessible as it will be replaced with $\operatorname{tr}(\mathfrak{G})$ in the generalized case. Furthermore, this would not necessarily lead to geometrically exact laws as the micro-structure is ignored. Therefore, invariants of the pseudo-metric $\mathfrak{G}$ need to be used instead. That is, invariants of the connection $\mathbf{\Gamma}$ and the torsion $\mathbf{\Theta}$ on $\mathcal{M}$ must be used, in addition to those of the metric $\mathbf{G}$ on $\mathbb{B}$. In theory, this could allow the formulation of geometrically exact constitutive equations making use of the micro-structure. Equations which could use, for instance, the torsion and curvature tensors of the connection to quantify these generalized strain and stress.

3.2 Structure Group and Principal Formalism

Although we took care to omit this detail until now, fiber bundles are not merely spaces locally isomorphic to Cartesian products. They are actually equipped with a group, called the structure group [1], acting on its typical fiber. This group can be physically interpreted as the group of admissible change of coordinates or the set of microscopic configurations. Objects of the theory are therefore required to interact properly with it. The structure group is usually simpler than the typical fiber of $\mathcal{M}$. For this reason, using the principal bundle associated to $\mathcal{M}$ is sometimes easier. The principal bundle associated to a fiber bundle $\mathcal{M} \to \mathbb{B}$ is a fiber bundle $\mathcal{P}^{\mathcal{M}}$ whose base space, structure group and (in some sens) transition maps are the same but whose typical fiber is the structure group $\mathcal{G}_{\mathcal{M}}$ itself. For example, in the case of planar and inextensible Timoshenko-Ehrenfest beams, the structure group is SO(2) which is isomorphic to the circle S^1. Therefore, the principal material bundle is $\mathcal{P}^{\mathcal{M}} \to \mathbb{B}$ whose typical fiber is $\mathcal{G}_{\mathcal{M}} \simeq \mathrm{S}^1$.

Objects defined on $\mathcal{M}$ and $\mathcal{E}$ can canonically be defined (or transferred) on $\mathcal{P}^{\mathcal{M}}$ and $\mathcal{P}^{\mathcal{E}}$ (see [1,5]). A placement map $\varphi : \mathcal{M} \to \mathcal{E}$ can for example be decomposed as a macroscopic placement map $\overline{\varphi} : \mathbb{B} \to \mathbb{E}$ and a material section (i.e. a microscopic configuration) $\mu : \mathbb{B} \to \mathcal{G}_{\mathcal{M}}$. In the case of the planar and inextensible Timoshenko-Ehrenfest beam, we see that a material placement map is therefore described by a classical placement map $\overline{\varphi} : \mathbb{B} \to \mathbb{E}$ and an angle $\theta_b \in \mathrm{S}^1 \simeq [0, 2\pi[$ at each $b \in \mathbb{B}$ (see Fig. 3).

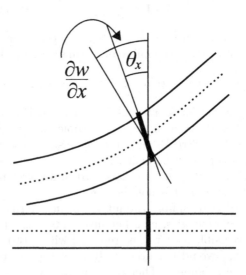

Fig. 3. Deformation of a Timoshenko beam. The normal section rotates by an amount θ_x with $\frac{\partial w}{\partial x} \neq \sin(\theta_x)$ ($\frac{\partial w}{\partial x}$ is the z-aligned part of $\mathrm{T}\overline{\varphi}$). Source: Wikipedia

Acknowledgements. We would like to thank the C.N.R.S' research group GDR GDM C.N.R.S. for stimulating the interactions between differential geometry and mechanics. An interaction which is at the heart of this work. We would also like to thank CNRS, Univ Rennes, ANR-11-LABX-0020-0 program Henri Lebesgue Center for their support.

References

1. Epstein, M.: The Geometrical Language of Continuum Mechanics. Springer, Heidelberg (2010)
2. Gonzalez, O., Stuart, A.M.: A First Course in Countinuum Mechanics. Cambridge University Press, Cambridge (2008)
3. Katanaev, M.O.: Geometric theory of defects. Phys. Usp. **48**(7), 675–701 (2005). https://doi.org/10.1070/PU2005v048n07ABEH002027. https://ufn.ru/en/articles/2005/7/b/
4. Kolev, B., Desmorat, R.: Éléments de géométrie différentielle á l'usage des mécaniciens (2020)
5. Marsden, J., Hughes, T.: Mathematical Foundations of Elasticity. Dover Civil and Mechanical Engineering Series, Dover (1994). https://books.google.fr/books?id=RjzhDL5rLSoC
6. Michor, P.W.: Topics in Differential Geometry. https://www.mat.univie.ac.at/michor/dgbook.pdf
7. Nguyen, V.H., Casale, G., Le Marrec, L.: On tangent geometry and generalized continuum with defects. Math. Mech. Solids (2021). https://doi.org/10.1177/10812865211059222

Singular Cotangent Models in Fluids with Dissipation

Baptiste Coquinot[1(✉)], Pau Mir[2], and Eva Miranda[2,3]

[1] Laboratoire de Physique de l'École Normale Supérieure, ENS, Université PSL, CNRS, Sorbonne Université, Université Paris Cité, 24 rue Lhomond, 75005 Paris, France
baptiste.coquinot@ens.fr

[2] Laboratory of Geometry and Dynamical Systems, Universitat Politècnica de Catalunya, Avinguda del Doctor Marañon 44-50, 08028 Barcelona, Spain

[3] CRM Centre de Recerca Matemàtica, Campus de Bellaterra Edifici C, 08193 Bellaterra, Barcelona, Spain

Abstract. In this article we analyze several mathematical models with singularities where the classical cotangent model is replaced by a *b*-cotangent model. We provide physical interpretations of the singular symplectic geometry underlying in *b*-cotangent bundles. The twisted cotangent model includes (for linear potentials) the case of fluids with dissipation. We also discuss more general physical interpretations of the twisted and non-twisted *b*-symplectic models. These models offer a Hamiltonian formulation for systems which are dissipative, extending the horizons of Hamiltonian dynamics and opening a new approach to study non-conservative systems.

1 Introduction

The study of fluid mechanics has a long and rich history, revealing a complex structure on both the physical and the mathematical levels. We point to recent work detailing how new geometric facets of this complexity have been revealed through several reincarnations (see [CMPSP21, CMPS23]). As it is well-known, in the Navier-Stokes equation the Reynolds number provides a measure of fluid

B. Coquinot is funded by the J.-P. Aguilar grant of the CFM Foundation. P. Mir and E. Miranda are partially supported by the AEI grant PID2019-103849GB-I00 of MCIN/AEI /10.13039/501100011033. E. Miranda is supported by the Catalan Institution for Research and Advanced Studies via an ICREA Academia Prizes 2016 and 2021 and by the Spanish State Research Agency, through the Severo Ochoa and María de Maeztu Program for Centers and Units of Excellence in R&D (project CEX2020-001084-M). E. Miranda also acknowledges partial support from the grant "Computational, dynamical and geometrical complexity in fluid dynamics", Ayudas Fundación BBVA a Proyectos de Investigación Científica 2021. B. Coquinot would like to acknowledge the (numerical) hospitality of the Laboratory of Geometry and Dynamical Systems of the Universitat Politècnica de Catalunya and especially E. Miranda for her supervision.

F. Nielsen and F. Barbaresco (Eds.): GSI 2023, LNCS 14072, pp. 50–59, 2023.
https://doi.org/10.1007/978-3-031-38299-4_6

complexity, giving rise to turbulence for high Reynolds number flows (with infinite Reynolds number corresponding to the Euler flow). The present work is not specifically about fluid mechanics, but an aligned investigation of the singular geometric nature of the case of a 0 (or very low) Reynolds number flow, corresponding to a laminar flow, expressed in terms of a finite-dimensional analogy.

Symplectic geometry provides the landscape where classical mechanics take place. The pair of position and momenta is the physical manifestation of the existence of a cotangent bundle underlying this picture. The role of the base and fibers of the cotangent bundle is an important landmark that fixes and makes precise Hamiltonian dynamics. However, this perfect symplectic picture is often insufficient to describe the complexity of physical phenomena. Poisson geometry provides a more general scenery appropriate to capture the intricacy of physical systems. Nevertheless, Poisson geometry is, in general, too involved and even the existence of appropriate local coordinates is a difficult battleground. From this perspective, singular symplectic manifolds provide a much more controlled terrain to fulfill some of these needs. In this article we explore some physical systems that can be described as singular symplectic manifolds. We focus on the class of b-symplectic manifolds and identify two models: a canonical and a twisted one. We associate relevant physical systems to these two models.

In the context of symplectic geometry, singular forms have been an important object of study in the last years. A main class of such singular forms is the class of b-symplectic forms, formally introduced in [GMP11] and [GMP14]. They provide a way to model systems with boundary and to study manifolds through compactification.

In [Mor86], Morrison introduced the metriplectic formalism as an extension of the Hamiltonian formalism so as to include dissipation while maintaining a conserved energy-like quantity. This formalism couples Poisson brackets, coming from the Hamiltonian symplectic formalism, with metric brackets, coming from out-of-equilibrium thermodynamics (see also [Mor84b, Mor84a, Mor98, MM17] and [CM20]). Thus, the formalism describes systems with both Hamiltonian and dissipative components that can model friction, electric resistivity, collisions and more, in various contexts ranging over biophysics, geophysics, and plasma physics. The construction builds in asymptotic convergence to a pre-selected equilibrium state.

Following these ideas, in this article we make use of Hamilton's equations to model a system which is dissipative in the classical sense. The original idea is that we do not rearrange the conservative Hamilton's equations but, instead, we introduce a singularity at the level of the symplectic structure of the manifold, which we equip with a twisted b-symplectic form. Thus, we give a new application of the twisted cotangent model. In particular, we present the Stokes' Law of motion for free-falling particles in fluids with viscosity as a twisted cotangent model. We prove that, in general, a one-dimensional motion with a dissipation which is proportional to the velocity can be modeled by a twisted b-symplectic form.

This paper is a short version of [CMM23]. In particular, we shorten the preliminaries on b-Symplectic geometry and skip varieties of examples and generalisations to go straight to the point. The interested reader can consult [CMM23] for more information. In Sect. 3 we introduce the new model for fluids with dissipation based on a twisted b-symplectic structure. We start with the one-dimensional case and the linear potential, which provides an analogue of the Stokes' Law, and we extend it to more general potentials. In Sect. 4 we consider time-dependent singular models in which friction arises from a re-scaling of time. Finally, in Sect. 5 we summarize the results of the paper and present our conclusions.

2 Preliminaries

2.1 b-Symplectic Geometry

A symplectic manifold is a manifold M which admits a *symplectic form* ω which is closed and non-degenerate 2-form. Given a function H over a symplectic manifold, called Hamiltonian, it is useful to consider its associated *Hamiltonian flow*, which is the flow of the vector field X defined by $\iota_X \omega = -dH$. The existence and uniqueness of X and its flow are a consequence of the non-degeneracy of the symplectic form.

In physics, the usual and more general formalism used to study dynamics is Poisson geometry [MR99]). Poisson manifolds are generalizations of symplectic manifolds in which the symplectic form ω is replaced by a bivector Π. Indeed, a symplectic form ω in a symplectic manifold (M, ω) may be seen as a smooth map from the space of vector fields $\mathfrak{X}(M)$ to the space of 1-forms $\Omega^1(M)$. Among the large class of Poisson manifolds we find b-symplectic manifolds, that can also be considered a wider class of manifolds which contains symplectic manifolds.

The basic definitions of b-symplectic geometry start with the notions of *b-manifold* (a pair (M, Z) where Z is a hypersurface in a manifold M), *b-map* (a map $f : (M_1, Z_1) \longrightarrow (M_2, Z_2)$ between b-manifolds with f transverse to Z_2 and $Z_1 = f^{-1}(Z_2)$) and *b-vector field* (a vector field on M which is tangent to Z at all points of Z).

Let (M^n, Z) be a b-manifold. If x is a local defining function for Z on an open set $U \subset M$ and $(x, y_1, \ldots, y_{n-1})$ is a chart on U, then the set of b-vector fields on U is a free $C^\infty(M)$-module with basis

$$(x\frac{\partial}{\partial x}, \frac{\partial}{\partial y_1}, \ldots, \frac{\partial}{\partial y_{n-1}}).$$

There exists a vector bundle associated to this module called *b-tangent bundle* and denoted by ^bTM. The *b-cotangent bundle* $^bT^*M$ of M is defined to be the vector bundle dual to ^bTM.

For each $k > 0$, let $^b\Omega^k(M)$ denote the space of sections of the vector bundle $\Lambda^k(^bT^*M)$, which are called *b-de Rham k-forms*. For any defining function f of Z, every b-de Rham k-form can be written as

$$\omega = \alpha \wedge \frac{df}{f} + \beta, \text{ with } \alpha \in \Omega^{k-1}(M) \text{ and } \beta \in \Omega^k(M). \tag{1}$$

A special class of closed b-de Rham 2-forms is the class of b-*symplectic forms* as defined in [GMP14]. It contains forms with singularities and can be introduced formally for b-symplectic manifolds, making it possible to extend the symplectic structure from $M\backslash Z$ to the whole manifold M.

Definition 1 (b-symplectic manifold). *Let (M^{2n}, Z) be a b-manifold and $\omega \in {}^b\Omega^2(M)$ a closed b-form. We say that ω is b-symplectic if ω_p is of maximal rank as an element of $\Lambda^2({}^bT_p^*M)$ for all $p \in M$. The triple (M, Z, ω) is called a b-symplectic manifold.*

2.2 b-Cotangent Lifts

The cotangent bundle of a smooth manifold M is naturally equipped with a symplectic structure, since there is always an intrinsic canonical linear form λ on T^*M defined by

$$\langle \lambda_p, v \rangle = \langle p, d\pi_p v \rangle, \qquad p = (m, \xi) \in T^*M, v \in T_p(T^*M),$$

where $d\pi_p : T_p(T^*M) \longrightarrow T_m M$ is the differential of the canonical projection at p. In local coordinates (q_i, p_i), the form is written as $\lambda = \sum_i p_i \, dq_i$ and is called the *Liouville 1-form*. Its differential $\omega = d\lambda = \sum_i dp_i \wedge dq_i$ is a symplectic form on T^*M.

For b-symplectic manifolds there are two natural choices for the singular Liouville form, each of them giving a different symplectic form ω:

1. Non-twisted forms: $\lambda = \frac{c}{q_1} p_1 dq_1 + \sum_{i=2}^n p_i dq_i$ and $\omega = \frac{c}{q_1} dp_1 \wedge dq_1 + \sum_{i=2}^n dp_i \wedge dq_i$,
2. Twisted forms: $\lambda = c \log(p_1) dq_1 + \sum_{i=2}^n p_i dq_i$ and $\omega = \frac{c}{p_1} dp_1 \wedge dq_1 + \sum_{i=2}^n dp_i \wedge dq_i$.

The non-twisted, or canonical, symplectic form carries the singularity at the base (the transversal hypersurface Z is given by $q_1 = 0$), while the twisted symplectic form carries the singularity at the fiber (the transversal hypersurface Z is given by $p_1 = 0$). The constant c in the expression of the forms is called the *modular weight*.

The cotangent lift of a group action is defined in the following way.

Definition 2. *Let $\rho : G \times M \longrightarrow M$ be a group action of a Lie group G on a smooth manifold M. For each $g \in G$, there is an induced diffeomorphism $\rho_g : M \longrightarrow M$. The cotangent lift of ρ_g, denoted by $\hat{\rho}_g$, is the diffeomorphism on T^*M given by*

$$\hat{\rho}_g(q, p) := (\rho_g(q), ((d\rho_g)_q^*)^{-1}(p)), \qquad with \ (q, p) \in T^*M,$$

which makes the following diagram commute:

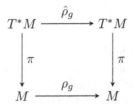

Given a diffeomorphism $\rho : M \longrightarrow M$, its cotangent lift is defined in an analogous way and it preserves the Liouville 1-form λ. As a consequence, it also preserves the symplectic form on T^*M. In the twisted case, the twisted b-cotangent lift preserves the twisted 1-form $\lambda = c \log(p_1)dq_1 + \sum_{i=2}^{n} p_i dq_i$ and the twisted b-symplectic form $\omega = \frac{c}{p_1}dp_1 \wedge dq_1 + \sum_{i=2}^{n} dp_i \wedge dq_i$.

3   The Twisted b-Symplectic Model for Dissipation

In this section, we describe how b-symplectic geometry offers a way to model, in a Hamiltonian fashion, a particle moving in a dissipative fluid with viscosity. In particular, we construct an example that uses the twisted b-symplectic form in the cotangent bundle of $\mathbb{R}$. This example gives precisely the equation of the friction drag force exerted on a small spherical particle moving through a viscous laminar fluid in one dimension, the so-called Stokes' Law. Then, we generalize this model to higher dimensions and to other configuration spaces different from $\mathbb{R}^n$.

Take $M = \mathbb{R}$ and $T^*M \cong \mathbb{R}^2$ with coordinates (q, p). Consider the Hamiltonian

$$H(q, p) = \frac{p^2}{2} + f(q), \tag{2}$$

which corresponds to the energy of a massive particle subject to a potential $f(q)$. The Hamilton's equations derived from $\iota_{X_H}\omega = -dH$ with the standard symplectic form $\omega = dp \wedge dq$ provide the following system, which models the main models in classical mechanics:

$$\begin{cases} \dot{q} = p \\ \dot{p} = -\frac{\partial f}{\partial q} \end{cases}. \tag{3}$$

But, more interestingly, the Hamilton's equations derived from $\iota_{X_H}\omega = -dH$ with the twisted b-symplectic form

$$\omega = \frac{1}{p}dp \wedge dq$$

are:

$$\begin{cases} \dot{q} = p^2 \\ \dot{p} = -p\frac{\partial f}{\partial q} \end{cases}. \tag{4}$$

At $p = 0$ there are just fixed points and system (4) gives no dynamics. Hence, we can reduce the dynamical study to $p > 0$, and for $p < 0$ it will be symmetric up to a change of sign.

Differentiating the first equation of system (4) and substituting into the second one, we find

$$\ddot{q} = -2\dot{q}\frac{\partial f}{\partial q}, \tag{5}$$

which is a second order ODE depending only on q. Notice that, although we have associated q to the position coordinate, $\dot{q}$ is not equal to the standard physical momentum p but to p^2. However, we can still think of $p = \sqrt{\dot{q}}$ as a modified physical momentum, since it is an increasing function of $\dot{q}$. Taking into account this point of view, we proceed to obtain various models of dynamics for different families of potentials $f(q)$.

A natural choice for the potential $f(q)$ is a function of linear type. This simple model already gives an original way of considering dissipation as a b-symplectic model, as the following result proves.

Theorem 1 (Dissipation as a twisted singular cotangent model). *Consider the twisted b-symplectic model in $T^*\mathbb{R}$, given by Eq. (4). The particular case $f(q) = \frac{\lambda}{2}q$ corresponds to the model of a spherical particle moving in a fluid with viscosity and suffering a friction proportional to its velocity, i.e., to the Stokes' Law.*

Proof. Consider $f(q) = \frac{\lambda}{2}q$, with $\lambda > 0$, in the case of the Hamilton's equations coming from the twisted b-symplectic form, namely, in system (4). Explicitly, Hamilton's equations are

$$\begin{cases} \dot{q} = p^2 \\ \dot{p} = -\frac{\lambda}{2}p \end{cases}. \tag{6}$$

The corresponding second order ODE becomes

$$\ddot{q} = -\lambda\dot{q}, \tag{7}$$

which corresponds exactly to the equation of a free massive particle moving in one dimension and affected by viscous friction. In fact, the Stokes' Law (8) describes precisely the same case, which appears in the study of non-ideal fluids. It states that the frictional force F is:

$$F = 6\pi\mu Rv, \tag{8}$$

where μ is the dynamic viscosity, R is the radius of the particle and v is the flow velocity relative to the object (or minus the object velocity relative to the flow). The Stokes' Law computes the magnitude of the drag force that is acting against the particle motion and slowing it. This force is proportional to the velocity of the particle with respect to the fluid and of opposite direction.

Denoting the velocity v by $\dot{q}$, assuming that the force F is proportional to the acceleration $\ddot{q}$ and combining physical constants, we deduce that Eq. (7) is equivalent to the Stokes' Law.

Remark 1. In the classical symplectic setting, the particular case $f(q) = \frac{\lambda}{2}q$ in Eq. (4), with $\lambda > 0$, gives rise to the dynamics of a rectilinear motion with constant acceleration (of $\frac{\lambda}{2}$). It is, for instance, the model for the free fall of a particle subject to a one-dimensional constant gravity field. Notice that there is no loss of energy of the system.

Considering a nonlinear potential $f(q)$, other kinds of dissipations may be described by this model. For instance, a quadratic potential models a particle crossing the interior of a box at a slow speed when it is near each edge and at a high speed in the middle. One may observe that the orbits in the twisted model "break" like in the linear twisted case, allowing an infinite number of "escape orbits" [CMM23]. This example has a bonus, as the escape orbits here correspond exactly to genuine *singular periodic orbits* as the ones described in [MO21]. These singular periodic orbits are indeed the union of 4 different trajectories: two symmetric hetero-clinic half-circles and the two fixed points on the horizontal axis at their ends.

4   Time-Dependent Singular Models

In order to generalize this friction model to multiple dimensions, the key idea is to extend the configuration space Q to $Q \times \mathbb{R}$. The $\mathbb{R}$ component in $Q \times \mathbb{R}$ describes the real time t while the dynamics inside the phase space is computed according to a curvilinear time s. This is conceptually the idea of the well-known method of characteristics in PDEs. After computing the solution, one only needs to project the trajectory on the space Q and read the time on the real axis. We require $t > 0$ to be consistent and we denote by q the position in Q and by p the associated momentum. We also denote by E the conjugated variable associated with t, since the energy is the natural conjugate of time in physics. Using the results of Sect. 3, where we have seen how to introduce dissipation in one dimension thanks to a b-symplectic form, our goal is to include the dissipation in this new energy variable. Therefore, the non-dissipative dynamics will proceed classically, with an energy which is dissipated through time.

To start, consider the Hamiltonian

$$H(p, q, t, E) = \frac{p^2}{2} + V(q, t) - E. \tag{9}$$

Assuming E is the energy of the system, one expects the preservation of the Hamiltonian (the conservation of $H = 0$) along the physical trajectory. We use the canonical symplectic form

$$\omega = \sum_i \mathrm{d}p_i \wedge \mathrm{d}q_i - \mathrm{d}E \wedge \mathrm{d}t. \tag{10}$$

The associated dynamics writes

$$\dot{q}_i = p_i \qquad \dot{p}_i = -\frac{\partial V(q, t)}{\partial q_i} \tag{11}$$

$$\dot{t} = 1 \qquad \dot{E} = \frac{\partial V(q, t)}{\partial t} \tag{12}$$

Therefore, in this case, the curvilinear coordinate is the real time: $s = t$. The particle follows the expected dynamics with a potential that may depend on time.

Now, to model friction, it is natural to consider adding to the Hamiltonian a factor depending on a friction coefficient λ. The friction will slow down the dynamics and thus t compared with s. However, the potential remains associated to the real time and thus it appears accelerated compared with the curvilinear time. In order to use this effective time, we need to re-scale the Hamiltonian to deduce the suitable time re-scaling. When considering dissipative dynamics, it is natural to expect an exponential re-scaling. Indeed, a close-to-the-equilibrium relaxation mode provides a Lyapunov coefficient to control the decay of the perturbation [GM13, GP71]. Such re-scaling ideas have already been suggested in different contexts, see for instance [FL85]. For our purpose, we consider the following Hamiltonian

$$H(p, q, t, E) = \frac{p^2}{2} + \frac{e^{2\lambda t}}{\lambda^2} V(q, t) - \frac{e^{\lambda t}}{\lambda} E, \tag{13}$$

with the same canonical symplectic form. The associated dynamics writes

$$\dot{q}_i = p_i \qquad \dot{p}_i = -\frac{e^{2\lambda t}}{\lambda^2} \frac{\partial V(q, t)}{\partial q_i} \tag{14}$$

$$\dot{t} = \frac{e^{\lambda t}}{\lambda} \qquad \dot{E} = \frac{e^{2\lambda t}}{\lambda^2} \frac{\partial V(q, t)}{\partial t} + \frac{2e^{2\lambda t}}{\lambda} V(q, t) - e^{\lambda t} E \tag{15}$$

The two first terms describe the energy linked with the time-dependence of the potential. The last term describes the loss of energy caused by the viscous dissipation. The equation for t can be solved exactly: $t(s) = -\frac{\ln(-s)}{\lambda}$. In particular, $ds = \lambda e^{-\lambda t} dt$. Let us now reconstruct the particle dynamics in real time:

$$\frac{dq_i}{dt} = \lambda e^{-\lambda t} \dot{q}_i = \lambda e^{-\lambda t} p_i \qquad \frac{dp_i}{dt} = \lambda e^{-\lambda t} \dot{p}_i = -\frac{e^{\lambda t}}{\lambda} \frac{\partial}{\partial q_i} V(q, t), \tag{16}$$

and, therefore,

$$\frac{d^2 q_i}{dt^2} = -\lambda \frac{dq_i}{dt} - \frac{\partial}{\partial q_i} V(q, t), \tag{17}$$

which is the equation of a particle in a n-dimensional space with a viscous friction of coefficient λ and in a time-dependent potential $V(q, t)$.

The friction arises from an exponential re-scaling of time. Such a re-scaling is actually the source of a singularity and, then, singular geometry arises naturally after a change of variables from t to s in the symplectic form using $s(t) = e^{-\lambda t}$ and $dt = -\frac{ds}{\lambda s}$. For convenience, we also redefine $E_s = E/\lambda$. Then, we obtain

$$\omega = \sum_i dq_i \wedge dp_i + \frac{1}{s} ds \wedge dE_s, \tag{18}$$

which is the non-twisted canonical b-symplectic form. In these coordinates, the Hamiltonian becomes

$$H(p, q, s, E_s) = \frac{p^2}{2} + \frac{V(q, t(s))}{(\lambda s)^2} - \frac{E_s}{s}, \tag{19}$$

which has a singularity of higher order. Indeed, it is a b^2-function and not a b-function. Such a discrepancy between the degree of the singularity in the symplectic form and the degree of the singularity in total energy of the system is not new (see [DKM17] for other examples).

Summing up, the Hamiltonian is simpler in these coordinates. But the main advantage is that the intrinsic time (the curvilinear coordinate) now corresponds to the coordinate s. Indeed, the equations of motion now read as follows:

$$\dot{q}_i = p_i \qquad \dot{p}_i = -\frac{1}{(\lambda s)^2}\frac{\partial V(q, t(s))}{\partial q_i} \tag{20}$$

$$\dot{s} = 1 \qquad \dot{E} = \frac{\partial}{\partial s}\left(\frac{1}{(\lambda s)^2}V(q, t(s))\right) + \frac{E_s}{s^2} \tag{21}$$

The coordinate s is now trivial and we may omit this dimension, leaving a standard Hamiltonian dynamics with a modified time-dependent potential. The dynamics then writes as:

$$\ddot{q}_i(s) = -\frac{1}{(\lambda s)^2}\frac{\partial V(q, s)}{\partial q_i} \tag{22}$$

and the real-time solution is obtained by undoing the change of variables $s(t) = e^{-\lambda t}$.

5   Conclusions

This paper aims to provide a finite-dimensional analogy of fluid mechanics using the techniques of b-symplectic geometry to model dissipation in conditions of no turbulence. The twisted b-symplectic model presented here is suited for the case of laminar viscous flows in which the Reynolds number is small enough. In general, the model is good for flows of low complexity and no turbulence and for which the Stokes' Law is a valid approximation.

We have seen that dissipation naturally emerges from a singular symplectic form in the direction of the singularity. In particular, this provides a simple model for uni-dimensional friction. This can be generalized to multiple dimensions with arbitrary external field by including an additional dimension to describe the physical time and energy. The dissipation and then the singularity must be applied on this extra-dimension, while the Hamiltonian must be re-scaled accordingly to the dissipation coefficient. Therefore, any d-dimensional system with an external potential and a global dissipation given by a fixed dissipation factor can be naturally described by an Hamiltonian dynamics on a $(d+1)$-dimensional b-symplectic manifold.

References

[CM20] Coquinot, B., Morrison, P.J.: A general metriplectic framework with application to dissipative extended magnetohydrodynamics. J. Plasma Phys. **86**(3), 835860302 (2020)

[CMM23]   Coquinot, B., Mir, P., Miranda, E.: Singular cotangent models in fluids
with dissipation. Physica D **446**, 133655 (2023)

[CMPS23]   Cardona, R., Miranda, E., Peralta-Salas, D.: Computability and Beltrami
fields in Euclidean space. J. Math. Pures Appl. **9**(169), 50–81 (2023)

[CMPSP21] Cardona, R., Miranda, E., Peralta-Salas, D., Presas, F.: Constructing
turing complete euler flows in dimension 3. Proc. Natl. Acad. Sci. USA
118(19), Paper No. e2026818118, 9 (2021)

[DKM17]   Delshams, A., Kiesenhofer, A., Miranda, E.: Examples of integrable and
non-integrable systems on singular symplectic manifolds. J. Geom. Phys.
115, 89–97 (2017)

[FL85]   Feix, M.R., Lewis, H.R.: Invariants for dissipative nonlinear systems by
using rescaling. J. Math. Phys. **26**(1), 68–73 (1985)

[GM13]   De Groot, S.R., Mazur, P.: Non-Equilibrium Thermodynamics. Courier
Corporation (2013). Google-Books-ID: mfFyG9jfaMYC

[GMP11]   Guillemin, V., Miranda, E., Pires, A.R.: Codimension one symplectic foliations and regular Poisson structures. Bull. Braz. Math. Soc. (N.S.) **42**(4),
607–623 (2011)

[GMP14]   Guillemin, V., Miranda, E., Pires, A.R.: Symplectic and Poisson geometry
on *b*-manifolds. Adv. Math. **264**, 864–896 (2014)

[GP71]   Glansdorff, P., Prigogine, I.: Thermodynamic Theory of Structure, Stability and Fluctuations. Wiley-Interscience (1971). Google-Books-ID:
vf9QAAAAMAAJ

[MM17]   Materassi, M., Morrison, P.: Metriplectic formalism: friction and much
more (2017)

[MO21]   Miranda, E., Oms, C.: The singular Weinstein conjecture. Adv. Math.
389, Paper No. 107925, 41 (2021)

[Mor84a]   Morrison, P.J.: Some observations regarding brackets and dissipation.
Technical report PAM-228, University of California at Berkeley (1984)

[Mor84b]   Morrison, P.J.: Bracket formulation for irreversible classical fields. Phys.
Lett. A **100**(8), 423–427 (1984)

[Mor86]   Morrison, P.J.: A paradigm for joined Hamiltonian and dissipative systems. Phys. D **18**(1–3), 410–419 (1986). Solitons and coherent structures
(Santa Barbara, Calif., 1985)

[Mor98]   Morrison, P.J.: Hamiltonian description of the ideal fluid. Rev. Modern
Phys. **70**(2), 467–521 (1998)

[MR99]   Marsden, J.E., Ratiu, T.S.: Introduction to Mechanics and Symmetry:
A Basic Exposition of Classical Mechanical Systems. Texts in Applied
Mathematics, vol. 17, 2nd edn. Springer, New York (1999). https://doi.
org/10.1007/978-0-387-21792-5

Multisymplectic Unscented Kalman Filter for Geometrically Exact Beams

Tianzhi Li and Jinzhi Wang$^{(\boxtimes)}$

Department of Mechanics and Engineering Science, State Key Laboratory
for Turbulence and Complex Systems, Peking University, Beijing 100871, China
tlee@stu.pku.edu.cn, jinzhiw@pku.edu.cn

Abstract. This paper introduces an unscented Kalman filter for the
dynamics of geometrically exact beams based on multisymplectic geometry and Hamel's formalism for classical field theories. The presented approach is a field-theoretic analogue of the well-known unscented Kalman filter. The discrete-time propagation equations are derived from the discrete
variational principle, rather than a direct discretization of the continuous-
time differential equations. As such, the proposed estimation scheme
respects intrinsic physical structures of mechanical systems and exhibits
excellent numerical properties such as energy conservation and structure
preservation. The approach is also applicable to a wide range of models
characterized by one-dimensional field theory. Properties of the proposed
estimation algorithm are illustrated with numerical simulations.

Keywords: Computationally efficient estimation · Geometrically
exact beam · Classical field theory · Unscented Kalman filter

1 Introduction

In the past decades, efficient estimation methods for mechanical systems have
received considerable attention in control theory and engineering applications.
One of the most well-known estimation method is the Kalman fiter (KF) [1],
which concentrates on the problem of estimating linear systems. It was further
extended to the nonlinear case with the extended Kalman filter (EKF) by directly
linearizing the system dynamics [2]. After that, the unscented Kalman filter
(UKF) [3] was proposed for estimating nonlinear systems by using the unscented
transformations. However, the above filters are based on vector spaces that are
endowed with linear structures, and there are many non-Euclidean and nonlinear
estimation problems arising in practical implementations. One important example of non-Euclidean estimation is the attitude filtering problem. Conventional
methods that are based on vector space structures suffer from problems like the
well-known gimbal lock or singularities.

Fortunately, modern differential geometry provides a natural language to
describe nonlinearity and non-Euclidean problems. Therefore, geometry based

filters have been studied extensively in recent years [4–6]. However, most geometric filters for manifolds or Lie groups rely on the filtering dynamics on the non-linear structures, and therefore this results in computational complexities and costs. In addition, intrinsic geometric structures are ignored during the direct discretization of the continuous-time equations, and this leads to the numerical dissipation in long-time simulations.

In this paper, a multisymplectic unscented Kalman filter is presented for the estimation of the dynamics of the geometrically exact beam. The formulation is derived based on multisymplectic geometry and Hamel's formalism for classical field theories. The spacetime propagation step and the measurement update step of the proposed filter are constructed on the Lie algebra, and naturally reduce the computational complexities. Furthermore, due to the variational discretization scheme of the filtering dynamics, the proposed filter preserves intrinsic structures (Lie group structure, momenta, energy) and exhibits excellent numerical performances in long-time simulations. Moreover, the proposed multisymplectic UKF can be regarded as a generalization of the conventional UKF.

2   Dynamics of the Geometrically Exact Beam

In this section, a brief overview of multisymplectic geometry and Hamel's formalism for classical field theories are given, since they apply to the geometrically exact beam dynamics. We refer the reader to [7–11, 15–20] for more information.

2.1   Classical Field Theory

In the field-theoretic description of geometrically exact beams, the configuration is characterized by the fiber bundle $\pi_{\mathbf{XY}} : \mathbf{Y} \to \mathbf{X}$, where $\mathbf{X} = S \times \mathbb{R}$ represents the base manifold with $S = [0, L]$ being the range for the single spatial variable s. Thus, it is evident that dim $\mathbf{X} = 2$. The physical fields are smooth local sections of $\pi_{\mathbf{XY}}$ denoted by $\phi : U \subseteq \mathbf{X} \to \mathbf{Y}$. In order to describe the dynamics as in classical mechanics, the field-theoretic analogue of the tangent bundle is the first jet bundle $J^1\mathbf{Y}$ of the configuration bundle $\pi_{\mathbf{XY}}$. It is the affine bundle over $\mathbf{Y}$ with fibers

$$J_y^1\mathbf{Y} = \{\psi : T_x\mathbf{X} \to T_y\mathbf{Y} | \psi \text{ is linear}, T\pi_{\mathbf{XY}} \circ \psi = Id_{T_x\mathbf{X}}, y \in \mathbf{Y}_x\}. \quad (1)$$

The field-theoretic analogues of the velocities in classical mechanics are represented by sections $j^1\phi$ of $J^1\mathbf{Y}$ which are called the first jet prolongations of the field ϕ, that is,

$$j^1\phi : x \in \mathbf{X} \longrightarrow T_x\phi \in J_{\phi(x)}^1\mathbf{Y}. \quad (2)$$

As there is only one spatial variable for the field-theoretic description of the geometrically exact beam, we mainly focus on (1+1)-D field theory. The Lagrangian density is a smooth bundle map given by

$$\mathfrak{L} : J^1\mathbf{Y} \to \Lambda^2\mathbf{X}, \quad (3)$$

where $\Lambda^2\mathbf{X}$ is the vector bundle of 2-forms on $\mathbf{X}$.

2.2 Field-Theoretic Hamel's Equations for the Beam

Hamel's formalism [8–14] provides a tool to do symmetry reduction by measuring the velocity of the system with respect to frames that are not dependent on coordinates of the configuration. This results in a simpler and more accurate representation of the system. As in [12, 13], by choosing the velocity operator

$$\Psi_g : \mathbb{R}^6 \to T_g SE(3), \tag{4}$$

with $g \in SE(3)$ and $\mathbf{Y} = SE(3) \times \mathbf{X} = SE(3) \times ([0, L] \times \mathbb{R})$, the Lagrangian (3) of the geometrically exact beam can be reduced as

$$\ell(\phi, \zeta, \gamma) = \ell(\phi, \xi) = L(\phi, \Psi\xi) = \frac{1}{2} \langle K\zeta, \zeta \rangle - \frac{1}{2} \langle P\gamma, \gamma \rangle, \tag{5}$$

where K and P are diagonal, $\xi = \Psi^{-1}d\phi$, and $d\phi = \Psi_\phi\gamma ds + \Psi_\phi\zeta dt$. The Lie algebraic elements ζ and γ are called nonmaterial velocity and nonmaterial strain, respectively. Then, by Hamilton's principle in field-theoretic Hamel's formalism, the equations of motion for the geometrically exact beam are [12]

$$\frac{\partial}{\partial t}K\zeta - \frac{\partial}{\partial s}P\gamma - [\zeta, K\zeta]^* + [\gamma + e_2, P\gamma]^* = 0, \tag{6}$$

with the kinematic equations

$$d\phi = \Psi_\phi\xi, \tag{7}$$

where $[\cdot, \cdot]$ and $[\cdot, \cdot]^*$ coincide with the Lie bracket and its dual on $\mathfrak{se}(3)$, respectively. As mentioned in [13], to ensure the integrability of (7) one has to impose the compatibility condition

$$\frac{\partial}{\partial t}\gamma - \frac{\partial}{\partial s}\zeta - [\gamma + e_2, \zeta] = 0. \tag{8}$$

3 Multisymplectic Unscented Kalman Filter

3.1 System Dynamics

This subsection describes the variational discretization of the beam dynamics, which will be served as the basis for the proposed multisymplectic unscentd Kalman filter. Inspired by [17, 18] and [13], the discretization starts with the discrete parameter space $\mathbb{P}_d = \mathbb{S}_d \times \mathbb{T}_d$, where $\mathbb{T}_d = \{0, 1, ..., N\}$ is the discrete time sequence and $\mathbb{S}_d = \{0, 1, ..., M\}$ represents the discretization of the one-dimensional space (see Fig. 1). Therefore, one gets rectangles given by

$$\square_n^j = \{(n, j), (n + 1, j), (n, j + 1), (n + 1, j + 1)\}, \tag{9}$$

where $n = 0, 1, ..., N - 1$ and $j = 0, 1, ..., M - 1$. Based on this, the discrete base $\mathbf{X}_d$ is obtained by a bijection

$$\Phi_{\mathbb{P}_d, \mathbf{X}_d}(n, j) = (t_n, s_n^j) =: x_n^j. \tag{10}$$

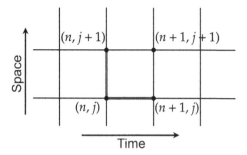

Fig. 1. Discrete spacetime $\mathbb{P}_d$.

Then, the discrete configuration bundle is represented by

$$\pi^d_{\mathbf{XY}} : \mathbf{Y}_d = \mathbf{X}_d \times SE(3) \to \mathbf{X}_d. \tag{11}$$

Hence, the discrete fields ϕ_d are sections of $\pi^d_{\mathbf{XY}}$, with values at discrete base points x^j_n denoted by $\phi^j_n = \phi_d(x^j_n)$. Let $\square_{\mathbb{P}_d}$ be the set of all rectangles determined by $\mathbb{P}_d$. The discrete first jet bundle is

$$J^1\mathbf{Y}_d = \mathbf{X}^\square_d \times SE(3) \times ... \times SE(3) \to \mathbf{X}^\square_d, \tag{12}$$

where $\mathbf{X}^\square_d = \Phi_{\mathbb{P}_d, \mathbf{X}_d}(\square_{\mathbb{P}_d})$. The first jet prolongation of a discrete field ϕ_d is defined by $j^1\phi_d(\square^j_n) = \left(\phi^j_n, \phi^j_{n+1}, \phi^{j+1}_n, \phi^{j+1}_{n+1}\right)$. Then, the discrete Lagrangian density corresponding to (5) is

$$\ell_d = \text{Vol}(\square)\left(\frac{1}{2}\left\langle K\zeta^j_{n+1/2}, \zeta^j_{n+1/2}\right\rangle - \frac{1}{2}\left\langle P\gamma^{j+1/2}_n, \gamma^{j+1/2}_n\right\rangle\right). \tag{13}$$

From the discrete Hamilton's principle, one obtains the discrete field-theoretic Hamel's equations [12]

$$\frac{1}{\Delta t}K\left(\zeta^j_{n+1/2} - \zeta^j_{n-1/2}\right) - \frac{1}{\Delta s}P\left(\gamma^{j+1/2}_n - \gamma^{j-1/2}_n\right)$$
$$-\frac{1}{2}\left[\zeta^j_{n+1/2}, K\zeta^j_{n+1/2}\right]^* - \frac{1}{2}\left[\zeta^j_{n-1/2}, K\zeta^j_{n-1/2}\right]^* \tag{14}$$
$$+\frac{1}{2}\left[\gamma^{j+1/2}_n + e_2, P\gamma^{j+1/2}_n\right]^* + \frac{1}{2}\left[\gamma^{j-1/2}_n + e_2, P\gamma^{j-1/2}_n\right]^* = 0.$$

In addition, the compatibility condition (8) is discretized as

$$\frac{1}{\Delta t}\left(\gamma^{j+1/2}_{n+1} - \gamma^{j+1/2}_n\right) - \frac{1}{\Delta s}\left(\zeta^{j+1}_{n+1/2} - \zeta^j_{n+1/2}\right)$$
$$-\left[\frac{1}{2}\left(\gamma^{j+1/2}_{n+1} + \gamma^{j+1/2}_n + 2e_2\right), \frac{1}{2}\left(\zeta^{j+1}_{n+1/2} + \zeta^j_{n+1/2}\right)\right] = 0, \tag{15}$$

with the discrete boundary condition $\gamma^{1/2}_n = \gamma^{M+1/2}_n = 0$, $n = 1, ..., N$.

3.2 Filtering Dynamics

By considering Lie algebraic noises in (14), the discrete-time dynamics with noise is

$$
\begin{aligned}
&\frac{1}{\Delta t} K \left(\zeta_{n+1/2}^{j} - \zeta_{n-1/2}^{j} - p_{n-1/2}^{j} \right) - \frac{1}{\Delta s} P \left(\gamma_n^{j+1/2} - \gamma_n^{j-1/2} + p_n^{j+1/2} - p_n^{j-1/2} \right) \\
&- \frac{1}{2} \left[\zeta_{n+1/2}^{j}, K \zeta_{n+1/2}^{j} \right]^{*} - \frac{1}{2} \left[\zeta_{n-1/2}^{j} + p_{n-1/2}^{j}, K \left(\zeta_{n-1/2}^{j} + p_{n-1/2}^{j} \right) \right]^{*} \\
&+ \frac{1}{2} \left[\gamma_n^{j+1/2} + p_n^{j+1/2} + e_2, P \left(\gamma_n^{j+1/2} + p_n^{j+1/2} \right) \right]^{*} \\
&+ \frac{1}{2} \left[\gamma_n^{j-1/2} + p_n^{j-1/2} + e_2, P \left(\gamma_n^{j-1/2} + p_n^{j-1/2} \right) \right]^{*} = 0,
\end{aligned}
\tag{16}
$$

where $p_{n-1/2}^{j} \sim \mathcal{N}_{\mathfrak{se}(3)}(0, Q_1)$ and $p_n^{j+1/2} \sim \mathcal{N}_{\mathfrak{se}(3)}(0, Q_2)$ are Gaussian process noises on the Lie algebra $\mathfrak{se}(3)$ with covariances Q_1 and Q_2.

For the measurements, it is assumed that we can measure the velocities and strains of the beam, and that the measurements are given by

$$
\begin{aligned}
Y_{n+1/2}^{j} &= \zeta_{n+1/2}^{j} + v_{n+1/2}^{j} \in \mathfrak{se}(3) \approx \mathbb{R}^6, \\
Y_n^{j+1/2} &= \zeta_n^{j+1/2} + v_n^{j+1/2} \in \mathfrak{se}(3) \approx \mathbb{R}^6,
\end{aligned}
\tag{17}
$$

where $v_{n+1/2}^{j} \sim \mathcal{N}_{\mathfrak{se}(3)}(0, N_1)$ and $v_n^{j+1/2} \sim \mathcal{N}_{\mathfrak{se}(3)}(0, N_2)$ are Gaussian measurement noises on $\mathfrak{se}(3)$ with covariances N_1 and N_2.

3.3 Spacetime Update

The spacetime update in the proposed multisymplectic unscented Kalman filter tackles the problem of uncertainty propagation in the filtering. This process is the field-theoretic analogue of the time update in the conventional UKF.

For the beam dynamics, the dimension of the spacetime is 2, hence two families of sigma points will be constructed. The first family of sigma points are constructed as

$$
{}^a \boldsymbol{\mu}_{n-1/2,n-1/2,i}^{j} = \left[\boldsymbol{\mu}_{n-1/2,n-1/2,i}^{j(\text{state})}, \boldsymbol{\mu}_{n-1/2,n-1/2,i}^{j(\text{noise})} \right],
\tag{18}
$$

where $j = 1, ..., M$, $i = 0, 1, ..., 2d$, and $d = \dim(\mathfrak{se}(3)) = 12$. The vectors $\boldsymbol{\mu}_{n-1/2,n-1/2,i}^{j(\text{state})} \in \mathbb{R}^6 \approx \mathfrak{se}(3)$ correspond to the nonmaterial velocities $\zeta_{n+1/2}^{j}$, and $\boldsymbol{\mu}_{n-1/2,n-1/2,i}^{j(\text{noise})}$ are related to the process noises. Similarly, the second family of sigma points are constructed as ${}^a \boldsymbol{\mu}_{n,n,i}^{j+1/2} = \left[\boldsymbol{\mu}_{n,n,i}^{j+1/2(\text{state})}, \boldsymbol{\mu}_{n,n,i}^{j+1/2(\text{noise})} \right]$, where $j = 1, ..., M-1$ and $i = 0, 1, ..., 2d$. The vectors $\boldsymbol{\mu}_{n,n,i}^{j+1/2(\text{state})} \in \mathbb{R}^6 \approx \mathfrak{se}(3)$ correspond to the nonmaterial strains $\gamma_n^{j+1/2}$, and $\boldsymbol{\mu}_{n,n,i}^{j+1/2(\text{noise})}$ are related to the process noises.

The first family of sigma points are propagated according to the discrete-time filtering dynamics (16), which gives

$$
\frac{1}{\Delta t} K \left(\mu_{n+\frac{1}{2}|n-\frac{1}{2},i}^{j(\text{state})} - \mu_{n-\frac{1}{2}|n-\frac{1}{2},i}^{j(\text{state})} - \mu_{n-\frac{1}{2}|n-\frac{1}{2},i}^{j(\text{noise})} \right) - \frac{1}{2} \left[\mu_{n+\frac{1}{2}|n-\frac{1}{2},i}^{j(\text{state})}, K \mu_{n+\frac{1}{2}|n-\frac{1}{2},i}^{j(\text{state})} \right]^*
$$
$$
- \frac{1}{\Delta s} P \left(\mu_{n|n,i}^{j+\frac{1}{2}(\text{state})} - \mu_{n|n,i}^{j-\frac{1}{2}(\text{state})} + \mu_{n|n,i}^{j+\frac{1}{2}(\text{noise})} - \mu_{n|n,i}^{j-\frac{1}{2}(\text{noise})} \right)
$$
$$
- \frac{1}{2} \left[\mu_{n-\frac{1}{2}|n-\frac{1}{2},i}^{j(\text{state})} + \mu_{n-\frac{1}{2}|n-\frac{1}{2},i}^{j(\text{noise})}, K \left(\mu_{n-\frac{1}{2}|n-\frac{1}{2},i}^{j(\text{state})} + \mu_{n-\frac{1}{2}|n-\frac{1}{2},i}^{j(\text{noise})} \right) \right]^*
$$
$$
+ \frac{1}{2} \left[\mu_{n|n,i}^{j+\frac{1}{2}(\text{state})} + \mu_{n|n,i}^{j+\frac{1}{2}(\text{noise})} + e_2, P \left(\mu_{n|n,i}^{j+\frac{1}{2}(\text{state})} + \mu_{n|n,i}^{j+\frac{1}{2}(\text{noise})} \right) \right]^*
$$
$$
+ \frac{1}{2} \left[\mu_{n|n,i}^{j-\frac{1}{2}(\text{state})} + \mu_{n|n,i}^{j-\frac{1}{2}(\text{noise})} + e_2, P \left(\mu_{n|n,i}^{j-\frac{1}{2}(\text{state})} + \mu_{n|n,i}^{j-\frac{1}{2}(\text{noise})} \right) \right]^* = 0.
$$

$$(19)$$

The second family of sigma points are propagated according to the discrete compatibility condition (15), that is,

$$
\frac{1}{\Delta t} \left(\mu_{n+1|n,i}^{j+\frac{1}{2}(\text{state})} - \mu_{n|n,i}^{j+\frac{1}{2}(\text{state})} \right) - \frac{1}{\Delta s} \left(\mu_{n+\frac{1}{2}|n-\frac{1}{2},i}^{j+1(\text{state})} - \mu_{n+\frac{1}{2}|n-\frac{1}{2},i}^{j(\text{state})} \right)
$$
$$
- \left[\frac{1}{2} \left(\mu_{n+1|n,i}^{j+\frac{1}{2}(\text{state})} + \mu_{n|n,i}^{j+\frac{1}{2}(\text{state})} + 2 e_2 \right), \frac{1}{2} \left(\mu_{n+\frac{1}{2}|n-\frac{1}{2},i}^{j+1(\text{state})} + \mu_{n+\frac{1}{2}|n-\frac{1}{2},i}^{j(\text{state})} \right) \right] = 0.
$$

$$(20)$$

After the propagation, the predicted means and covariances are computed as weighted sums. Then, the prior estimations $\left\{ \zeta_{n+1/2|n-1/2}^{j} \right\}_{j=1}^{M}$ and $\left\{ \gamma_{n+1|n}^{j+1/2} \right\}_{j=1}^{M+1}$ of nonmaterial velocities and strains are obtained, respectively.

3.4 Measurement Update

The measurement update is formulated on the vector space of the Lie algebra. This scheme is new compared with the conventional UKF [3] and the UKFs on Lie groups [4–6]. The sigma points of the measurement update are constructed as $^a\eta_{n+1/2|n-1/2,i}^{j} = \left[\eta_{n+1/2|n-1/2,i}^{j(\text{state})}, \eta_{n+1/2|n-1/2,i}^{j(\text{noise})} \right]$ and $^a\eta_{n+1|n,i}^{j+1/2} = \left[\eta_{n+1|n,i}^{j+\frac{1}{2}(\text{state})}, \eta_{n+1|n,i}^{j+\frac{1}{2}(\text{noise})} \right]$. They are propagated on $\mathfrak{se}(3)$ by

$$
\Omega_{n+1/2,i}^{j} = \eta_{n+1/2|n-1/2,i}^{j(\text{state})} + \eta_{n+1/2|n-1/2,i}^{j(\text{noise})} \in \mathfrak{se}(3) \approx \mathbb{R}^6,
$$
$$
\Omega_{n+1,i}^{j+1/2} = \eta_{n+1|n,i}^{j+1/2(\text{state})} + \eta_{n+1|n,i}^{j+1/2(\text{noise})} \in \mathfrak{se}(3) \approx \mathbb{R}^6,
$$

$$(21)$$

where $\Omega_{n+1/2,i}^{j}$ and $\Omega_{n+1,i}^{j+1/2}$ correspond to the i-th sigma point. Then, the posterior means and covariances are computed as weighted sums. The posterior estimations $\left\{ \zeta_{n+1/2|n+1/2}^{j} \right\}_{j=1,\ldots,M}$ and $\left\{ \gamma_{n+1|n+1}^{j+1/2} \right\}_{j=1,\ldots,M+1}$ of nonmaterial velocities and nonmaterial strains are obtained.

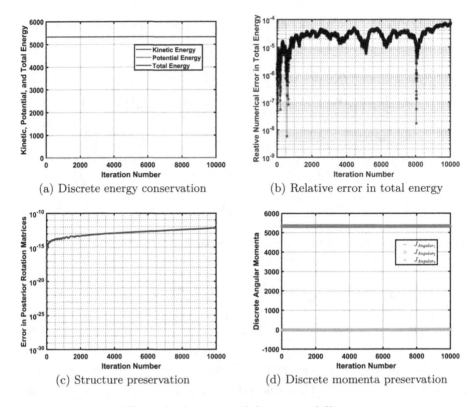

Fig. 2. Performances of the proposed filter.

3.5   Reconstruction Step

The aim of this step is to reconstruct poses of each cross section of the beam from the posterior nonmaterial velocities and strains by using the exponential map Exp : $\mathfrak{se}(3) \rightarrow SE(3)$. Specifically, the posterior estimations of poses are reconstructed as

$$g_{n+1}^j = g_n^j \mathrm{Exp}\left(\Delta t \widehat{\zeta}_{n+1/2|n+1/2}^j\right), \tag{22}$$

where $\widehat{\zeta}_{n+1/2|n+1/2}^j$ are the matrix-form nonmaterial velocities, which are related with $\zeta_{n+1/2|n+1/2}^j$ by the hat map $\widehat{\cdot}$. See [21] for more information on the hat map and its inverse.

3.6   Numerical Results

We utilize the proposed filter to estimate the pure rotation dynamics of a geometrically exact beam (see Fig. 2). The following characteristics are chosen: length $L = 10$ m, square cross section with edge length $a = 2$ m, mass density $\rho = 1000$ kg/m^3, Young's modulus $E = 10^7$ Pa, and Poisson ratio $\nu = 0.35$.

In the simulation, the temporal discretization step is $\Delta t = 5 \cdot 10^{-5}$ s. The centerline of the beam is partitioned into $M = 10$ identical elements, with spatial discretization step $\Delta s = L/M = 1$ m. The initial state of the beam is $g_0^j = I_4$, $j = 1, ..., M$. The reference velocities and strains are $\zeta_{\mathrm{Ref}} = (0, 2, 0, 0, 0, 0)^T$ and $\gamma_{\mathrm{Ref}} = (0, 0, 0, 0, 0, 0)^T$, which imply that the reference motion is a constant-speed rotation. The measurement noise matrix is chosen to be $N_1 = N_2 = \sigma_m^2 I_6$, with the deviation $\sigma_m = \frac{\pi}{180}$. The process noise matrix is chosen to be $Q_1 = Q_2 = \sigma_p^2 I_6$, with $\sigma_p = 0.1$. Figure 2 shows that the proposed algorithm preserves the discrete energy, the discrete momenta, and the Lie group structure of $SE(3)$.

4   Concluding Remarks

This paper presented a multisymplectic unscented Kalman filter for geometrically exact beams, which exhibits high accuracy in preserving key structures of the system. The obtained spacetime update step of the proposed filter includes the scheme of the conventional UKF as a particular instance. The proposed approach is derived by using multisymplectic geometry and the discretization of classical field theories. The proposed filter will serve as a basis for constructing computationally efficient filters for other field-theoretic models and models with more independent spatial variables.

Acknowledgements. The authors would like to thank the editor and the anonymous reviewers for their valuable comments. The authors also wish to thank Professor Donghua Shi of Beijing Institute of Technology, Professor Wei Chen of Peking University, and Doctors Xinchao Hu, Ganghui Cao, and Yuxi Cao for their kind help and valuable discussions. The authors acknowledge the financial support from NSFC Grant T2121002 and 61973005.

References

1. Kalman, R.E., Bucy, R.S.: New results in linear filtering and prediction theory. J. Basic Eng. **83**(1), 95–108 (1961)
2. Reif, K., Gunther, S., Yaz, E., Unbehauen, R.: Stochastic stability of the continuous-time extended Kalman filter. IEE Proc.-Control Theory Appl. **147**(1), 45–52 (2000)
3. Julier, S., Uhlmann, J.K., Durrant-Whyte, H.F.: A new method for the nonlinear transformation of means and covariances in filters and estimators. IEEE Trans. Autom. Control **45**(3), 477–482 (2000)
4. Lee, T.: Bayesian attitude estimation with the matrix Fisher distribution on SO(3). IEEE Trans. Autom. Control **63**(10), 3377–3392 (2018)
5. Sjøberg, A.M., Egeland, O.: Lie algebraic unscented Kalman filter for pose estimation. IEEE Trans. Autom. Control **67**(8), 4300–4307 (2022)
6. Ćesić, J., Marković, I., Bukal, M., Petrović, I.: Extended information filter on matrix Lie groups. Automatica **82**, 226–234 (2017)

7. Gotay, M.J., Isenberg, J., Marsden, J.E., Montgomery, R.: Momentum maps and classical relativistic fields, part I: covariant field theory (1997). http://www.cds.caltech.edu/marsden/bib/Notes.html

8. Zenkov, D.V., Leok, M., Bloch, A.M.: Hamel's formalism and variational integrators on a sphere. In: 2012 IEEE 51st IEEE Conference on Decision and Control (CDC), Maui, HI, USA, pp. 7504–7510 (2012)

9. Ball, K.R., Zenkov, D.V.: Hamel's formalism and variational integrators. In: Chang, D.E., Holm, D.D., Patrick, G., Ratiu, T. (eds.) Geometry, Mechanics, and Dynamics. FIC, vol. 73, pp. 477–506. Springer, New York (2015). https://doi.org/10.1007/978-1-4939-2441-7_20

10. Shi, D., Kogan, Y.B., Zenkov, D.V., Bloch, A.M.: Hamel's formalism for infinite-dimensional mechanical systems. J. Nonlinear Sci. **27**, 241–283 (2017)

11. Shi, D., Zenkov, D.V., Bloch, A.M.: Hamel's formalism for classical field theories. J. Nonlinear Sci. **30**, 1307–1353 (2020)

12. Wang, L., An, Z., Shi, D.: Hamel's field variational integrator for geometrically exact beam. Acta Scientiarum Naturalium Universitatis Pekinensis **52**, 692–698 (2016)

13. An, Z., Gao, S., Shi, D., Zenkov, D.V.: A variational integrator for the Chaplygin-Timoshenko sleigh. J. Nonlinear Sci. **30**, 1381–1419 (2020)

14. Gao, S., Shi, D., Zenkov, D.V.: Discrete Hamiltonian variational mechanics and Hamel's integrators. J. Nonlinear Sci. **33**, 26 (2023)

15. Demoures, F., Gay-Balmaz, F., Desbrun, M., Ratiu, T.S., Alejandro, A.: A multisymplectic integrator for elastodynamic frictionless impact problems. Comput. Methods Appl. Mech. Eng. **315**, 1025–1052 (2017)

16. Demoures, F., Gay-Balmaz, F., Kobilarov, M., Ratiu, T.S.: Multisymplectic Lie group variational integrators for a geometrically exact beam in $\mathbb{R}^3$. Commun. Nonlinear Sci. Numer. Simulat. **19**(10), 3492–3512 (2014)

17. Demoures, F., Gay-Balmaz, F.: Multisymplectic variational integrators for fluid models with constraints. In: Nielsen, F., Barbaresco, F. (eds.) GSI 2021. LNCS, vol. 12829, pp. 283–291. Springer, Cham (2021). https://doi.org/10.1007/978-3-030-80209-7_32

18. Demoures, F., Gay-Balmaz, F.: Unified discrete multisymplectic Lagrangian formulation for hyperelastic solids and barotropic fluids. J. Nonlinear Sci. **32**(94) (2022)

19. Marsden, J.E., Patrick, G.W., Shkoller, S.: Multisymplectic geometry, variational integrators and nonlinear PDEs. Comm. Math. Phys. **199**, 351–395 (1998)

20. Lew, A., Marsden, J.E., Ortiz, M., West, M.: Asynchronous variational integrators. Arch. Ration. Mech. Anal. **167**(2), 85–146 (2003)

21. Marsden, J.E., Ratiu, T.S.: Introduction to Mechanics and Symmetry, 2nd edn. Springer, New York (1999). https://doi.org/10.1007/978-0-387-21792-5

Towards Full 'Galilei General Relativity': Bargmann-Minkowski and Bargmann-Galilei Spacetimes

Christian Y. Cardall[(✉)] [ID]

Physics Division, Oak Ridge National Laboratory, Oak Ridge, TN 37831-6354, USA
cardallcy@ornl.gov

Abstract. Galilei-Newton spacetime $\mathbb{G}$ with its Galilei group can be understood as a 'degeneration' as $c \to \infty$ of Minkowski spacetime $\mathbb{M}$ with its Poincaré group. $\mathbb{G}$ does not have a spacetime metric and its Galilei symmetry transformations do not include energy; but Bargmann-Galilei spacetime $B\mathbb{G}$, a 5-dimensional extension that preserves Galilei physics, remedies these infelicities. Here an analogous Bargmann-Minkowski spacetime $B\mathbb{M}$ is described. While not necessary for Poincaré physics, it may illuminate a path towards a more extensive 'Galilei general relativity' than is presently known, which would be a useful—and conceptually and mathematically sound—approximation in astrophysical scenarios such as core-collapse supernovae.

Keywords: Relativity · Poincaré group · Galilei group · Bargmann group

1 Introduction

The terms 'relativistic physics' and 'non-relativistic physics' refer to what might be called something else—perhaps 'Poincaré relativity' and 'Galilei relativity' respectively. In terms of space, so-called 'non-relativistic physics' is in an important sense just as relativistic as 'relativistic physics'. Einstein's essential innovation in so-called 'relativistic physics' is not relativity in general, but specifically the relativity of time, or more precisely, the relativity of simultaneity: space is mixed into time in Lorentz transformations but not in homogeneous Galilei transformations. However, time is mixed into space in both Lorentz transformations and homogeneous Galilei transformations. Thus the presence of 'relativity'

This manuscript has been authored by UT-Battelle, LLC, under contract DE-AC05-00OR22725 with the US Department of Energy (DOE). The US government retains and the publisher, by accepting the article for publication, acknowledges that the US government retains a nonexclusive, paid-up, irrevocable, worldwide license to publish or reproduce the published form of this manuscript, or allow others to do so, for US government purposes. DOE will provide public access to these results of federally sponsored research in accordance with the DOE Public Access Plan (http://energy.gov/downloads/doe-public-access-plan).

F. Nielsen and F. Barbaresco (Eds.): GSI 2023, LNCS 14072, pp. 69–78, 2023.
https://doi.org/10.1007/978-3-031-38299-4_8

in both cases—albeit space only in one case, and both space and time in the other—justifies more careful reference to 'Galilei relativity' and 'Poincaré relativity' instead of 'non-relativistic physics' and 'relativistic physics'.

Now, what about Einstein circa 1905 vs. Einstein circa 1915? Having freed the term 'relativity' from specific attachment to the world according to Einstein and recognizing its relevance to the world according to Newton and Galilei, the terms 'special relativity' and 'general relativity' must be reconsidered as well. The spacetime of Einstein circa 1905 is an affine space, which can be regarded as a flat differentiable manifold. In contrast, the spacetime of Einstein circa 1915 is a more general pseudo-Riemann manifold whose curvature is determined by the energy and momentum of matter and radiation upon it.

This distinction—between flat and curved spacetime—is what ought to be meant by the terms 'special relativity' and 'general relativity', without regard for whether the physics is governed by the Poincaré group or the Galilei group [2]. In this perspective the key difference is not between 'relativistic physics'—whether 'special' or 'general'—governed by the Poincaré group on the one hand, and 'non-relativistic physics' governed by the Galilei group on the other. Instead, what distinguishes 'special relativity' from 'general relativity' is whether the group in question—whether Poincaré, or Galilei—applies to spacetime *globally*, in which case it is an affine space; or only *locally*, in which case its curvature is determined by its energy/momentum/mass content. The proper references, then, would be to 'Poincaré special relativity' and 'Poincaré general relativity', and to 'Galilei special relativity' and 'Galilei general relativity'.

One might hypothesize that these linguistic shifts, unavoidably associated also with conceptual shifts, point toward a unified perspective on Poincaré and Galilei physics that may bear fruit in a Galilei general relativity more extensive than that presently known (described for instance in [4,6]). This would play out as follows. A 5D (5-dimensional) Bargmann extension of 4D Galilei-Newton spacetime is needed to include energy in a tensor formalism on a pseudo-Riemann manifold [5,6]. A 5D Bargmann extension of 4D Minkowski spacetime is not *needed*, but it is *allowed*, and may illuminate a path to a *full* Galilei general relativity, in which *full* spacetime curvature (including possible curvature of 'space slices') is *determined* by the energy/momentum/mass content (not just given by assumption of Newton's gravitational potential as an ad hoc input). As a first step towards this goal, a 5D Bargmann extension of 4D Minkowski spacetime that limits nicely to the 5D Bargmann extension of 4D Galilei-Newton spacetime is described here, and the metric tensors of possible corresponding 5D curved spacetime generalizations are displayed.

2 Minkowski Spacetime $\mathbb{M}$

Minkowski spacetime $\mathbb{M}$ is a 4-dimensional affine space with underlying vector space $V_{\mathbb{M}}$. The invariant structure on $V_{\mathbb{M}}$ that governs causality is the null cone, embodied in a 4-metric g. With respect to a Minkowski basis (e_0, e_1, e_2, e_3) of

V_M, the metric g is represented by

$$g = \eta = \begin{bmatrix} -c^2 & 0 \\ 0 & 1 \end{bmatrix} = \begin{bmatrix} -c^2 & 0_j \\ 0_i & 1_{ij} \end{bmatrix}.$$

This 'Minkowski matrix' is invariant under Lorentz transformations represented by matrices P_M:

$$P_M^T \eta P_M = \eta.$$

It is well known that an element P_M^+ of the identity component of the Lorentz group (restricted Lorentz group) can be factored into a boost and a rotation:

$$P_M^+ = L_M R.$$

Here

$$R = \begin{bmatrix} 1 & 0 \\ 0 & R_S \end{bmatrix},$$

with $R_S \in SO(3)$ a rotation of the subspace V_S of V_M spanned by (e_1, e_2, e_3). A boost is parametrized by a 3-column $u \in \mathbb{R}^{3 \times 1}$:

$$L_M = \begin{bmatrix} \Lambda_u & \frac{1}{c^2} \Lambda_u u^T \\ \Lambda_u u & 1 + \frac{1}{\|u\|^2}(\Lambda_u - 1) u u^T \end{bmatrix},$$

where

$$\Lambda_u = \left(1 - \frac{\|u\|^2}{c^2}\right)^{-1/2}$$

is the Lorentz factor associated with u, and $\|u\|^2 = u^T u$ is the squared Euclid norm with respect to an orthonormal basis of V_S (naturally appropriate to a Minkowski basis of V_M).

The inverse metric $\overleftrightarrow{g}$ is represented by

$$\overleftrightarrow{\eta} = \begin{bmatrix} -\frac{1}{c^2} & 0 \\ 0 & 1 \end{bmatrix} = \begin{bmatrix} -\frac{1}{c^2} & 0^j \\ 0^i & 1^{ij} \end{bmatrix},$$

and is also invariant, according to

$$P_M^{-1} \overleftrightarrow{\eta} P_M^{-T} = \overleftrightarrow{\eta}.$$

Given g there is metric duality between vectors and linear forms ('raising and lowering of indices'). Let V_{M*} be the vector space of linear forms on V_M. For $a \in V_M$ and $\omega \in V_{M*}$,

$$\underline{a} = g(a, \cdot) \in V_{M*}$$
$$\overleftrightarrow{\omega} = \overleftrightarrow{g}(\omega, \cdot) \in V_M.$$

For natural contractions (here, never scalar products!), use the dot operator, for example

$$\underline{a} = g \cdot a \qquad = a \cdot g,$$
$$\overleftrightarrow{\omega} = \omega \cdot \overleftrightarrow{g} \qquad = \overleftrightarrow{g} \cdot \omega,$$

$$\overleftarrow{F} = \overleftrightarrow{g} \cdot F,$$
$$\overleftrightarrow{F} = \overleftrightarrow{g} \cdot F \cdot \overleftrightarrow{g} = \overleftarrow{F} \cdot \overleftrightarrow{g}.$$

3   Galilei-Newton Spacetime $\mathbb{G}$

Galilei-Newton spacetime $\mathbb{G}$ is a 'degeneration' of Minkowski spacetime $\mathbb{M}$ as $c \to \infty$. The metric g asymptotes (without a true limit) in a manner that suggests that the linear form $\tau = dt$, where t is the time coordinate, becomes the invariant structure governing causality, embodying absolute time. With respect to what will be called a Galilei basis, the 'time form' τ is represented by

$$\tau = \begin{bmatrix} 1 & 0 \end{bmatrix} = \begin{bmatrix} 1 & 0_i \end{bmatrix}.$$

The inverse metric $\overleftrightarrow{g}$ limits sensibly to another invariant structure, the $(2,0)$ tensor $\overleftrightarrow{\gamma}$. With respect to a Galilei basis, $\overleftrightarrow{\gamma}$ is represented by

$$\overleftrightarrow{\gamma} = \begin{bmatrix} 0 & 0 \\ 0 & 1 \end{bmatrix} = \begin{bmatrix} 0 & 0^j \\ 0^i & 1^{ij} \end{bmatrix}.$$

The covector τ is invariant according to

$$\begin{bmatrix} 1 & 0_i \end{bmatrix} \mathsf{P}_{\mathbb{G}} = \begin{bmatrix} 1 & 0_i \end{bmatrix},$$

and the tensor $\overleftrightarrow{\gamma}$ is invariant according to

$$\mathsf{P}_{\mathbb{G}}^{-1} \begin{bmatrix} 0 & 0^j \\ 0^i & 1^{ij} \end{bmatrix} \mathsf{P}_{\mathbb{G}}^{-T} = \begin{bmatrix} 0 & 0^j \\ 0^i & 1^{ij} \end{bmatrix}.$$

Here the homogeneous Galilei transformations $\mathsf{P}_{\mathbb{G}}$ are the $c \to \infty$ limit of the Lorentz transformations $\mathsf{P}_{\mathbb{M}}$. As with the restricted Lorentz group, elements $\mathsf{P}_{\mathbb{G}}^{+}$ of the identity component of the homogeneous Galilei group can be factored into a boost and a rotation:

$$\mathsf{P}_{\mathbb{G}}^{+} = \mathsf{L}_{\mathbb{G}} \, \mathsf{R}.$$

Here R is the same as before, and the Galilei boost is

$$\mathsf{L}_{\mathbb{G}} = \begin{bmatrix} 1 & 0 \\ u & 1 \end{bmatrix}.$$

The tensor $\overleftrightarrow{\gamma}$ derived from $\overleftrightarrow{g}$ does not qualify as an inverse metric tensor on $V_{\mathbb{G}}$. It has no inverse because it is degenerate:

$$\overleftrightarrow{\gamma}(\tau, \cdot) = \tau \cdot \overleftrightarrow{\gamma} = 0.$$

There is no spacetime metric on $\mathbb{G}$. Tensor algebra is more constrained: there is no metric duality—no 'raising and lowering of indices'. There are spacetime tensors but they can only be of fixed type.

4   Decomposition of $\mathbb{M}$ and $\mathbb{G}$ into Time and Space

Theories formulated in terms of tensors on spacetime can only be compared with experiments once spacetime is broken into 'time' and 'space' (and tensors are decomposed accordingly).

Affine spacetimes permit 'inertial observers' with straight worldlines and no rotation. The splitting of space and time as perceived by a single inertial observer is formally similar on $\mathbb{M}$ and $\mathbb{G}$. Select an event $\mathbf{O}$ of $\mathbb{M}$ or $\mathbb{G}$ as origin. Select a Minkowski basis of $V_{\mathbb{M}}$ or a Galilei basis of $V_{\mathbb{G}}$, designated $(e_\mu) = (e_0, e_1, e_2, e_3)$. A point $\mathbf{X} \in \mathbb{M}, \mathbb{G}$ is given in terms of coordinates $(X^\mu) = (t, x^i)$ by

$$\mathbf{X} = \mathbf{O} + e_\alpha X^\alpha.$$

The time axis $\mathbb{T}$ is the straight line

$$\mathbb{T} = \{\mathbf{O} + e_0 t \mid t \in \mathbb{R}\}.$$

Interpret $\mathbb{T}$ as as the worldline of a fiducial (and inertial) observer whose tangent vector is the constant 4-velocity $n = e_0$. Let $V_{\mathbb{S}}$ be the subspace of $V_{\mathbb{M}}$ or $V_{\mathbb{G}}$ spanned by (e_1, e_2, e_3). For a given time $t \in \mathbb{R}$, consider a one-to-one mapping

$$V_{\mathbb{S}} \to \mathbb{M} \text{ or } \mathbb{G}$$
$$x \mapsto \mathbf{O} + n\,t + x.$$

The image of this mapping is a hyperplane $\mathbb{S}_t$ through the event $\mathbf{O} + n\,t$:

$$\mathbb{S}_t = \{\mathbf{O} + n\,t + e_i\,x^i \mid (x^i) \in \mathbb{R}^3\}.$$

$\mathbb{S}_t$ is a 3-dimensional affine subspace of $\mathbb{M}$ or $\mathbb{G}$ with underlying vector space $V_{\mathbb{S}}$. Interpret $\mathbb{S}_t$ as 'space' according to the fiducial observer at her time t—a surface of 'simultaneity'. It is evident from the factorization $\mathsf{P}^+ = \mathsf{L}\,\mathsf{R}$ that $V_{\mathbb{S}}$ is rotationally invariant. Thus $V_{\mathbb{S}}$ is endowed with a Euclid metric γ defining the usual scalar product on $\mathbb{R}^3$. Each hypersurface $\mathbb{S}_t$ is a level surface of the coordinate function t. The complete collection $(\mathbb{S}_t)_{t \in \mathbb{R}}$ is said to be a foliation of $\mathbb{M}$ or $\mathbb{G}$.

5   A Material Particle on $\mathbb{M}$ and $\mathbb{G}$

A material particle is represented by a timelike curve $\mathbf{X}(\tau)$ in spacetime, parametrized by the particle's proper time τ. The tangent vector $U(\tau) = d\mathbf{X}/d\tau$, the 4-velocity, satisfies $g(U, U) = -c^2$ on $\mathbb{M}$ and $\tau(U) = 1$ on $\mathbb{G}$. Select a fiducial observer with global coordinates (t, x^i) associated with a choice of origin $\mathbf{O}$ of $\mathbb{M}$ or $\mathbb{G}$ and a Minkowski or Galilei basis (n, e_i) for $V_{\mathbb{M}}$ or $V_{\mathbb{G}}$. Decompose U into measurable pieces parallel to $\mathbb{T}$ and tangent to $\mathbb{S}_t$:

$$U = \frac{dt}{d\tau}\frac{d\mathbf{X}}{dt} = \frac{dt}{d\tau}\,(n + v).$$

This follows from the 4-column representations

$$\mathsf{X} = \begin{bmatrix} t \\ \mathsf{x}(t) \end{bmatrix}, \quad \mathsf{n} = \begin{bmatrix} 1 \\ 0 \end{bmatrix}, \quad \mathsf{v} = \begin{bmatrix} 0 \\ d\mathsf{x}/dt \end{bmatrix}.$$

The leading factor $dt/d\tau$ is determined by the fundamental structures $\boldsymbol{g}$ and $\boldsymbol{\tau}$ governing causality. Proper time increments $d\tau$ are given by

$$c\, d\tau = \sqrt{-\boldsymbol{g}\,(d\mathbf{X}, d\mathbf{X})} = c\, \Lambda_v^{-1}\, dt \quad (\text{on } \mathbb{M}),$$
$$d\tau = \boldsymbol{\tau}\,(d\mathbf{X}) \qquad\quad = dt \qquad (\text{on } \mathbb{G}),$$

so that $dt/d\tau = \Lambda_v$ on $\mathbb{M}$ and $dt/d\tau = 1$ on $\mathbb{G}$.

So far so good on both $\mathbb{M}$ and $\mathbb{G}$: a spacetime description of particle kinematics—specifying where a particle is (a point $\mathbf{X}(\tau)$ on its worldline), and how fast it is moving (the 4-velocity $\boldsymbol{U}$ tangent to the worldline)—is unproblematic in either case.

However, a spacetime formulation of particle dynamics turns out to be more problematic on $\mathbb{G}$. Because of the absence of a spacetime metric there is no equivalence between inertia and total energy. The best one can do is include kinetic energy in the time component of a 'relative energy momentum covector' $\boldsymbol{\Pi}$ [3]. But this is not fully satisfying because the notion of kinetic energy (energy of motion) inherently depends on a choice of observer (motion relative to whom?): the fiducial observer covector $\underline{\boldsymbol{n}}$ is built into the definition of the 4-covector $\boldsymbol{\Pi}$ whose time component is the kinetic energy relative to the fiducial observer. The unsatisfying result is that Lorentz or homogeneous Galilei transformations cannot transform the components of $\boldsymbol{\Pi}$ in such a way as to demonstrate the transformation rule of kinetic energy. This motivates extensions of the Lorentz and homogeneous Galilei groups that address the transformation of kinetic energy.

6 Bargmann Spacetimes $B\mathbb{M}$ and $B\mathbb{G}$

Work backwards towards Bargmann-Minkowski (or B-Minkowski) spacetime $B\mathbb{M}$ and Bargmann-Galilei (or B-Galilei) spacetime $B\mathbb{G}$ by considering a '5-velocity' $\hat{\boldsymbol{U}}$ that extends the 4-velocity $\boldsymbol{U}$ on $\mathbb{M}$ or $\mathbb{G}$. The fifth component will be the specific kinetic energy—kinetic energy per unit mass—involving only 3-velocity. With respect to a B-Minkowski or B-Galilei basis (fiducial observer):

$$\hat{\mathsf{U}} = \begin{bmatrix} \Lambda_v \\ \Lambda_v\, \mathsf{v} \\ c^2\,(\Lambda_v - 1) \end{bmatrix} \quad (\text{on } B\mathbb{M}), \qquad \hat{\mathsf{U}} = \begin{bmatrix} 1 \\ \mathsf{v} \\ \frac{1}{2}\|\mathsf{v}\|^2 \end{bmatrix} \quad (\text{on } B\mathbb{G}).$$

The additional dimension requires an additional coordinate. A point $\hat{\mathbf{X}}(\tau)$ along the particle worldline is represented by a 5-column

$$\hat{\mathsf{X}} = \begin{bmatrix} t \\ \mathsf{x}(t) \\ \eta(t) \end{bmatrix} = \begin{bmatrix} t \\ x^i(t) \\ \eta(t) \end{bmatrix}.$$

The proper time τ is governed by $\boldsymbol{g}$ or $\boldsymbol{\tau}$ as before; these are now regarded as tensors on $B\mathbb{M}$ or $B\mathbb{G}$ respectively. The fifth component $\hat{U}^\eta$ of the 5-velocity $\hat{U} = \mathrm{d}\hat{\mathbf{X}}/\mathrm{d}\tau$ must satisfy

$$\hat{U}^\eta = \frac{\mathrm{d}\eta}{\mathrm{d}\tau} = \frac{\mathrm{d}t}{\mathrm{d}\tau}\frac{\mathrm{d}\eta}{\mathrm{d}t} = c^2\,(\Lambda_v - 1) \quad \text{(on } B\mathbb{M}),$$

$$= \frac{1}{2}\|\mathsf{v}\|^2 \qquad \text{(on } B\mathbb{G}). \tag{1}$$

It is apparent that η has units of action/mass; call it the 'action coordinate'. The above 'action coordinate relation' will prove crucial to the geometry of $B\mathbb{M}$ and $B\mathbb{G}$.

Next, determine the 5×5 B-Lorentz transformation matrices $\hat{\mathsf{P}}^+_{B\mathbb{M}}$ and homogeneous B-Galilei transformation matrices $\hat{\mathsf{P}}^+_{B\mathbb{G}}$ that appear in the 5-velocity transformation

$$\hat{U} = \hat{\mathsf{P}}^+\,\hat{U}',$$

or in (4+1)-dimensional form

$$\begin{bmatrix} U \\ U^\eta \end{bmatrix} = \begin{bmatrix} \mathsf{P}^+ & 0 \\ \Phi & 1 \end{bmatrix} \begin{bmatrix} U' \\ U'^\eta \end{bmatrix}.$$

The 4-column $0 = \begin{bmatrix} 0^\mu \end{bmatrix}$ in $\hat{\mathsf{P}}^+$ ensures that the 4D relation $U = \mathsf{P}^+\,U'$ on $\mathbb{M}$ or $\mathbb{G}$ is preserved when embedded in the 5D setting of $B\mathbb{M}$ or $B\mathbb{G}$. It also ensures that the matrix representations of $\boldsymbol{g}$ and $\boldsymbol{\tau}$ do not acquire non-vanishing components in the η dimension when these are regarded as tensors on $B\mathbb{M}$ and $B\mathbb{G}$. This means that the 'timelike 4-velocity' character of U on $\mathbb{M}$ or $\mathbb{G}$ is preserved when it is extended to the 5-velocity $\hat{U}$ on $B\mathbb{M}$ or $B\mathbb{G}$. The 4-row Φ in $\hat{\mathsf{P}}^+$ is determined by the requirement that the fifth component of the above transformation of $\hat{U}$ yield the transformation rule for (specific) kinetic energy. For Poincaré physics this can be derived most easily from the time component of the 4D relation $U = \mathsf{P}^+_{\mathbb{M}}\,U'$. For Galilei physics it is derived from the transformed 3-velocity (Galilei velocity addition with rotation), the space components of the 4D relation $U = \mathsf{P}^+_{\mathbb{G}}\,U'$. The resulting expressions for Φ are

$$\Phi = \begin{bmatrix} c^2\,(\Lambda_u - 1) & \Lambda_u\,\mathsf{u}^{\mathrm{T}}\,\mathsf{R}_{\mathbb{S}} \end{bmatrix} \quad \text{(on } B\mathbb{M}),$$

$$= \begin{bmatrix} \frac{1}{2}\|\mathsf{u}\|^2 & \mathsf{u}^{\mathrm{T}}\,\mathsf{R}_{\mathbb{S}} \end{bmatrix} \quad \text{(on } B\mathbb{G}). \tag{2}$$

No new parameters beyond $\mathsf{u} \in \mathbb{R}^{3\times1}$ and $\mathsf{R}_{\mathbb{S}} \in \mathrm{SO}(3)$ already present in a Lorentz transformation $\mathsf{P}^+_{\mathbb{M}}$ or homogeneous Galilei transformation $\mathsf{P}^+_{\mathbb{G}}$ are introduced.

The set of B-Lorentz transformations $\hat{\mathsf{P}}^+_{B\mathbb{M}}$ and the set of homogeneous B-Galilei transformations $\hat{\mathsf{P}}^+_{B\mathbb{G}}$ are subgroups of $\mathrm{GL}(5)$. It is evident that these sets of matrices contain the identity ($\mathsf{u} = 0$ and $\mathsf{R}_{\mathbb{S}} = 1$). Once again there is a factorization $\hat{\mathsf{P}}^+ = \hat{\mathsf{L}}\,\hat{\mathsf{R}}$, so that inverses are given by $\hat{\mathsf{P}}^{+^{-1}} = \hat{\mathsf{R}}^{\mathrm{T}}\,\hat{\mathsf{L}}^{-1}$ with $\hat{\mathsf{L}}^{-1}$ obtained from $\hat{\mathsf{L}}$ via $\mathsf{u} \mapsto -\mathsf{u}$. Closure under matrix multiplication is shown by considering the product

$$\hat{\mathsf{P}}^{+''} = \hat{\mathsf{P}}^+\,\hat{\mathsf{P}}^{+'},$$

or

$$\begin{bmatrix} \mathsf{P}^{+\prime\prime} & 0 \\ \Phi'' & 1 \end{bmatrix} = \begin{bmatrix} \mathsf{P}^+ & 0 \\ \Phi & 1 \end{bmatrix} \begin{bmatrix} \mathsf{P}^{+\prime} & 0 \\ \Phi' & 1 \end{bmatrix} = \begin{bmatrix} \mathsf{P}^+ \, \mathsf{P}^{+\prime} & 0 \\ \Phi \, \mathsf{P}^{+\prime} + \Phi' & 1 \end{bmatrix}.$$

The 4×4 matrix relation

$$\mathsf{P}^{+\prime\prime} = \mathsf{P}^+ \, \mathsf{P}^{+\prime} \tag{3}$$

in the upper-left block is simply the known closure of the restricted Lorentz or homogeneous Galilei group. The remaining question is whether the 4-row

$$\Phi'' = \Phi \, \mathsf{P}^{+\prime} + \Phi'$$

is in the form of Eq. (2), with the relevant expressions involving u'' and R'' determined consistently from Eq. (3). Direct computation shows that the answer is yes, completing the demonstration of closure.

The existence of a 'Bargmann metric' G is suggested by the 'action coordinate relation' in Eq. (1) relating coordinate variations along a material particle worldline, and it turns out to be invariant under B-Lorentz or homogeneous B-Galilei transformations, making it a fundamental structure on $B\mathbb{M}$ or $B\mathbb{G}$. On $B\mathbb{M}$, use $\Lambda_{\mathrm{v}} = dt/d\tau$ and $c^2 \, d\tau^2 = c^2 \, dt^2 - \|dx\|^2$ in Eq. (1) to deduce

$$-2 \, d\eta \, dt + dx^a \, 1_{ab} \, dx^b + \frac{1}{c^2} \, d\eta^2 = 0 \quad (\text{on } B\mathbb{M}).$$

On $B\mathbb{G}$, use $d\tau = dt$ and $\|v\|^2 \, dt^2 = \|dx\|^2$ to deduce analogously

$$-2 \, d\eta \, dt + dx^a \, 1_{ab} \, dx^b = 0 \quad (\text{on } B\mathbb{G}).$$

In both cases the left-hand side looks like a line element, suggestive of a Bargmann metric (or B-metric) G represented by the B-Minkowski or B-Galilei matrix

$$G = \hat{\eta}_{BM} = \begin{bmatrix} 0 & 0_j & -1 \\ 0_i & 1_{ij} & 0_i \\ -1 & 0_j & \frac{1}{c^2} \end{bmatrix} \quad (\text{on } B\mathbb{M}), \qquad G = \hat{\eta}_{BG} = \begin{bmatrix} 0 & 0_j & -1 \\ 0_i & 1_{ij} & 0_i \\ -1 & 0_j & 0 \end{bmatrix} = \quad (\text{on } B\mathbb{G})$$

with respect to a B-Minkowski or B-Galilei basis. The invariance condition reads

$$\hat{\mathsf{P}}^{\mathrm{T}} \, \hat{\eta} \, \hat{\mathsf{P}} = \hat{\eta}$$

and is verified by direct computation for both $\hat{\mathsf{P}}^+_{BM}$ with $\hat{\eta}_{BM}$ and $\hat{\mathsf{P}}^+_{BG}$ with $\hat{\eta}_{BG}$. However, the 6-dimensional Lie groups of B-Lorentz and homogeneous B-Minkowski transformations are only subgroups of the 10-dimensional Lie groups that preserve G for $B\mathbb{M}$ and $B\mathbb{G}$ respectively.

The above calculation suggesting the existence of G also shows that

$$G\left(\hat{U}, \hat{U}\right) = \hat{U}^{\mathrm{T}} \, \hat{\eta} \, \hat{U} = 0,$$

that is, that $\hat{U}$ is null with respect to G. This is so even though $\hat{U}$ remains timelike with respect to g or τ as appropriate, as noted previously.

The inverse metric $\overleftrightarrow{G}$ is represented by

$$\overleftrightarrow{G} = \overleftrightarrow{\eta}_{BM} = \begin{bmatrix} -\frac{1}{c^2} & 0^j & -1 \\ 0^i & 1^{ij} & 0^i \\ -1 & 0^j & 0 \end{bmatrix} \quad \text{(on } B\mathbb{M}), \qquad \overleftrightarrow{G} = \overleftrightarrow{\eta}_{BG} = \begin{bmatrix} 0 & 0^j & -1 \\ 0^i & 1^{ij} & 0^i \\ -1 & 0^j & 0 \end{bmatrix} \quad \text{(on } B\mathbb{G})$$

with respect to a B-Minkowski or B-Galilei basis.

Note the remarkable difference in the relationship between $\mathbb{M}$ and $\mathbb{G}$ on the one hand and between $B\mathbb{M}$ and $B\mathbb{G}$ on the other, including startlingly different geometric consequences. Whereas the spacetime $\mathbb{M}$ is a pseudo-Riemann manifold with metric g and inverse $\overleftrightarrow{g}$, the spacetime $\mathbb{G}$ obtained as $c \to \infty$ is not: instead of a metric and its true inverse, one is left with an invariant time form τ and an invariant degenerate inverse 'metric' $\overleftrightarrow{\gamma}$. In contrast both $B\mathbb{M}$ and $B\mathbb{G}$ are pseudo-Riemann manifolds with a (flat) metric G and inverse $\overleftrightarrow{G}$, the versions of both of these on $B\mathbb{M}$ limiting smoothly to those on $B\mathbb{G}$ as $c \to \infty$, as is evident from the above expressions relative to B-Minkowski and B-Galilei bases.

7   Conclusion

This account of Bargmann-Minkowski spacetime $B\mathbb{M}$ with its metric G, deduced from Eq. (1), extends to Poincaré physics an ingenious elementary introduction given by de Saxcé and Vallée [6] of the Bargmann group as an extension to the Galilei group. This approach to the Bargmann group is simple and direct in comparison with its origins in the study of projective representations of Lie groups in quantum mechanics [1], but the necessity of transforming kinetic energy is a shared underlying motivation: the Hamiltonian in the 'non-relativistic' (forgive the lapse) Schrödinger equation contains kinetic energy, and this equation cannot be shown to be Galilei covariant without taking the projective phase into account [8].

The next step is to relax the assumption of an affine space, allowing instead spacetime curvature determined by its energy-momentum content. Call the 4D spacetime of standard general relativity 'Einstein spacetime' $\mathcal{E}$; in its $3+1$ formulation [7] in terms of the lapse function α, shift 3-vector $\boldsymbol{\beta}$, and 3-metric $\boldsymbol{\gamma}$, the Lorentz factor of a material particle is $\Lambda = \alpha\, \mathrm{d}t/\mathrm{d}\tau$ and proper time intervals are given by $c^2\, \mathrm{d}\tau^2 = c^2\alpha^2\, \mathrm{d}t^2 - \boldsymbol{\gamma}\,(\mathrm{d}\boldsymbol{x} + \boldsymbol{\beta}\, \mathrm{d}t, \mathrm{d}\boldsymbol{x} + \boldsymbol{\beta}\, \mathrm{d}t)$. Using these expressions in Eq. (1) yields

$$\beta_a\beta^a\, \mathrm{d}t^2 - 2\, \mathrm{d}t\, \beta_a \mathrm{d}x^a - 2\,\alpha\, \mathrm{d}\eta\, \mathrm{d}t + \mathrm{d}x^a\, \gamma_{ab}\, \mathrm{d}x^b + \frac{1}{c^2}\, \mathrm{d}\eta^2 = 0 \quad \text{(on } B\mathcal{E}),$$

suggestive of a 5D Bargmann-Einstein spacetime $B\mathcal{E}$ with metric G and inverse $\overleftrightarrow{G}$ represented by

$$\mathsf{G} = \begin{bmatrix} \beta_a\beta^a & \beta_j & -\alpha \\ \beta_i & \gamma_{ij} & 0_i \\ -\alpha & 0_j & \frac{1}{c^2} \end{bmatrix}, \qquad \overleftrightarrow{\mathsf{G}} = \begin{bmatrix} -\frac{1}{c^2\alpha^2} & \frac{1}{c^2\alpha^2}\,\beta^j & -\frac{1}{\alpha} \\ \frac{1}{c^2\alpha^2}\,\beta^i & \gamma^{ij} - \frac{1}{c^2\alpha^2}\,\beta^i\beta^j & \frac{1}{\alpha}\,\beta^i \\ -\frac{1}{\alpha} & \frac{1}{\alpha}\,\beta^j & 0 \end{bmatrix} \quad \text{(on } B\mathcal{E}).$$

As $c \to \infty$ this limits smoothly to

$$
\mathsf{G} = \begin{bmatrix} \beta_a \beta^a & \beta_j & -\alpha \\ \beta_i & \gamma_{ij} & 0_i \\ -\alpha & 0_j & 0 \end{bmatrix}, \qquad
\overleftrightarrow{\mathsf{G}} = \begin{bmatrix} 0 & 0^j & -\frac{1}{\alpha} \\ 0^i & \gamma^{ij} & \frac{1}{\alpha}\beta^i \\ -\frac{1}{\alpha} & \frac{1}{\alpha}\beta^j & 0 \end{bmatrix} \quad (\text{on } B\mathcal{G}),
$$

suggestive of a hitherto unknown 'Galilei general relativistic' spacetime $B\mathcal{G}$. (In both cases these reduce to the previous expressions on $B\mathbb{M}$ and $B\mathbb{G}$ as $\alpha \to 1$ and $\beta \to 0$.) Thus there is a reasonable prospect that recasting the 3+1 formulation of the Einstein equations on $\mathcal{E}$ as a 1+3+1 formulation on $B\mathcal{E}$ and taking the $c \to \infty$ limit could yield a Galilei gravitation of enhanced strength in which energy density and stress contribute as sources and give rise to space as well as spacetime curvature, beyond the flat space slices and spacetime curvature determined by mass density alone in Cartan's reformulation of Newtonian gravitation. This would be a useful—and conceptually and mathematically sound—approximation in astrophysical scenarios such as core-collapse supernovae, in which the energy density and pressure of the nascent neutron star contribute to enhanced gravity at the 10–20% level, but for which the computationally/numerically fraught phenomena of 'Minkowski' bulk fluid flow and back-reaction of gravitational radiation are much less significant.

Acknowledgements. Thanks to Géry de Saxcé for pointing out that preservation of the B-metric $\boldsymbol{G}$ is not sufficient to prove closure of the B-Lorentz transformations $\hat{\mathsf{P}}_{BM}^{+}$, but that closure directly follows instead from relations obtained from closure of the Lorentz group.

References

1. Bargmann, V.: On unitary ray representations of continuous groups. Annals Math. **59**, 1–46 (1954)
2. Cardall, C.Y.: Minkowski and Galilei/Newton fluid dynamics: a geometric 3+1 spacetime perspective. Fluids **4**, 1 (2019)
3. Cardall, C.Y.: Combining 3-momentum and kinetic energy on Galilei/Newton spacetime. Symmetry **12**(11), 1775 (2020)
4. de Saxcé, G.: 5-dimensional thermodynamics of dissipative continua. In: Frémond, M., Maceri, F., Vairo, G. (eds.) Models, Simulation, and Experimental Issues in Structural Mechanics. SSSSM, vol. 8, pp. 1–40. Springer, Cham (2017). https://doi.org/10.1007/978-3-319-48884-4_1
5. de Saxcé, G., Vallée, C.: Bargmann group, momentum tensor and Galilean invariance of Clausius-Duhem inequality. Int. J. Eng. Sci. **50**, 216–232 (2012)
6. de Saxcé, G., Vallée, C.: Galilean Mechanics and Thermodynamics of Continua. Wiley, Hoboken (2016)
7. Gourgoulhon, E.: 3+1 Formalism in General Relativity: Bases of Numerical Relativity. Lecture Notes in Physics, vol. 846. Springer, Heidelberg (2012). https://doi.org/10.1007/978-3-642-24525-1
8. Lévy-Leblond, J.M.: Quantum fact and classical fiction: clarifying Landé's pseudo-paradox. Am. J. Phys. **44**, 1130–1132 (1976)

New trends in Nonholonomic Systems

Existence of Global Minimizer for Elastic Variational Obstacle Avoidance Problems on Riemannian Manifolds

Leonardo Colombo[1(✉)] and Jacob Goodman[2,3]

[1] Centre for Automation and Robotics (CSIC-UPM), Ctra. M300 Campo Real, Km 0,200, Arganda del Rey, 28500 Madrid, Spain
leonardo.colombo@csic.es
[2] Instituto de Ciencias Matematicas (CSIC-UAM-UC3M-UCM), Calle Nicolas Cabrera 13-15, 28049 Madrid, Spain
jacob.goodman@icmat.es
[3] Escuela Politécnica Superior, Universidad Antonio de Nebrija, C/ Santa Cruz de Marcenado 27, 28015 Madrid, Spain

Abstract. This work is devoted to studying existence of global minimizers for optimal control problems with obstacle avoidance. We show the existence of global extrema in the general setting of Riemannian manifolds. This is a problem that consists of minimizing a suitable energy functional among a set of admissible configurations. The given energy functional depends on state variables (trajectory), control variables (velocity, covariant acceleration), and on an artificial potential function used for avoiding obstacles.

Keywords: Variational problems on Riemannian manifolds · Obstacle avoidance · Existence of global extrema

1 Introduction and Problem Formulation

Optimal control and path planning on nonlinear spaces such as Riemannian manifolds is as an active field of interest due to its numerous applications in robotics. To construct paths connecting some set of knot points interpolating some set of given positions and velocities, the use of variationally defined curves, in particular, the so-called Riemannian cubic splines [1,2] are a particularly pervasive interpolant, which themselves are composed of Riemannian cubic polynomials-curves which minimize the total squared (covariant) acceleration among all sufficiently regular curves satisfying some boundary conditions in positions and velocities-that are subsequently glued together to create a spline. Riemannian

The authors acknowledge financial support from Grant PID2019-106715GB-C21 funded by MCIN/AEI/ 10.13039/501100011033. The project that gave rise to these results received the support of a fellowship from "la Caixa" Foundation (ID 100010434). The fellowship code is LCF/BQ/DI19/11730028.

F. Nielsen and F. Barbaresco (Eds.): GSI 2023, LNCS 14072, pp. 81–88, 2023.
https://doi.org/10.1007/978-3-031-38299-4_9

cubic polynomials carry a rich geometry with them which often parallels the theory of geodesics. This has been studied extensively in the literature (see [3,4] for a detailed account of Riemannian cubics and [5] for some results with higher-order Riemannian polynomials).

Energy-optimal path planning on nonlinear spaces such as Riemannian manifolds has been an active field of interest in the last decades due to its numerous applications in manufacturing, aerospace technologies, and robotics [6,7]. It is often the case that-in addition to interpolating points-there are obstacles or regions in space which need to be avoided. In this case, a typical strategy is to augment the action functional with an artificial potential term that grows large near the obstacles and small away from them (in that sense, the minimizers are expected to avoid the obstacles) [8–10]. This strategy was used for instance in [11–15], where necessary conditions for extrema in obstacle avoidance problems on Riemannian manifolds were derived, and applications to interpolation problems on manifolds and to energy-minimum problems on Lie groups and symmetric spaces endowed with a bi-invariant metric were studied. Similar strategies have been implemented for collision avoidance problems for multiagent systems evolving on Riemannian manifolds, as in [14,16]. Existence of global minimizers and safety guarantees for the obstacle avoidance problem were studied in [16]. What is currently lacking in the literature regarding energy-optimal obstacle avoidance problems-and what this paper aims to address-is the existence of global minimizers for elastic variational obstacle avoidance problems.

In particular, the variational problem we will study in this paper is the following: For $\sigma \in \mathbb{R}$, consider a complete connected Riemannian manifold Q, and we define the set Ω of curves on Q of Sobolev class H^2 satisfying the boundary conditions:

$$q(0) = q_0, \quad \dot{q}(0) = v_0, \quad q(T) = q_T, \quad \dot{q}(T) = v_T,$$

and define the functional J on Ω as

$$J(x) = \int_0^T \frac{1}{2} \left(\left\| \frac{D^2 x}{dt^2}(t) \right\|^2 + \sigma \left\| \frac{dx}{dt}(t) \right\|^2 + V(x(t)) \right) dt, \tag{1}$$

where D denotes the covariant derivative associated with the Levi-Civita connection on Q.

Variational Obstacle Avoidance Problem: Find a curve $q \in \Omega$ minimizing the functional J, where $V : Q \to \mathbb{R}$ is a smooth, convex and non-negative function called the *artificial potential*.

The solution curves can be seen as generalizations of the classical splines in tension for the Euclidean spaces.

In order to minimize the functional J among the set Ω, we want to find curves $q \in \Omega$ such that $J(q) \leq J(\tilde{q})$ for all admissible curves $\tilde{q}$ in an H^2-neighborhood of q. The next result from [11] characterizes first-orderd necessary conditions for optimality in the variational obstacle avoidance problem.

Proposition 1. *[11] A curve $q \in \Omega$ is a critical point of the functional J if and only if it is smooth on $[0,T]$ and satisfies:*

$$\frac{D^4 q}{dt^4} + R\left(\frac{D^2 q}{dt^2}, \frac{dq}{dt}\right)\frac{dq}{dt} + \sigma\frac{Dq}{dt} = -\operatorname{grad} V(q(t)). \tag{2}$$

For the obstacle avoidance problem, the potential will most often depend on the distance between the curve q and some collection of fixed points $\{q_n\}$, which are treated as point-obstacles. As shown in [17], obstacles with "volume" can be reasonably approximated by collections of point-obstacles on its boundary, so we may assume without loss of generality that the potential is of the form $V(q) = \sum_{i=1}^{n} f_i(d(q, q_i))$, where $f_i : \mathbb{R} \to \mathbb{R}$ is a smooth, non-negative function and $d(q, q_i)$ denotes the distance (induced by the Riemannian metric) between q and the point-obstacle q_i. Note that this distance is well-defined since Q is assumed to be complete as a Riemannian manifold. For the purposes of existence of solutions to (2), the particular form of V will not be considered, it is only necessary that V be smooth and non-negative.

2   Riemannian Geometry

Let Q be an n-dimensional *Riemannian manifold* endowed with a non-degenerate symmetric covariant 2-tensor field g called the *Riemannian metric*. That is, to each point $q \in Q$ we assign an inner product $g_q : T_q Q \times T_q Q \to \mathbb{R}$, where $T_q Q$ is the *tangent space* of Q at q. The length of a tangent vector is determined by its norm, $\|v_q\| = g(v_q, v_q)^{1/2}$ with $v_q \in T_q Q$. A *Riemannian connection* ∇ on Q is a map that assigns to any two smooth vector fields X and Y on Q a new vector field, $\nabla_X Y$. For the properties of ∇, see [18]. The operator ∇_X, which assigns to every vector field Y the vector field $\nabla_X Y$, is called the *covariant derivative of Y with respect to X*.

Consider a vector field W along a curve q on Q. The kth-order covariant derivative of W along q is denoted by $\dfrac{D^k W}{dt^k}$, $k \geq 1$. We also denote by $\dfrac{D^{k+1} q}{dt^{k+1}}$ the kth-order covariant derivative of the velocity vector field of q along q, $k \geq 1$.

A vector field X along a piecewise smooth curve q in Q is said to be *parallel along q if $\dfrac{DX}{dt} \equiv 0$.* Given vector fields X, Y and Z on Q, the vector field $R(X,Y)Z$ given by $R(X,Y)Z = \nabla_X \nabla_Y Z - \nabla_Y \nabla_X Z - \nabla_{[X,Y]} Z$ is called the *curvature endomorphism* on Q. R is trilinear in X, Y and Z.

If we assume that Q is *complete*, then any two points x and y in Q can be connected by a minimal length geodesic $\gamma_{x,y}$, and the Riemannian distance $d : Q \times Q \to \mathbb{R}$ between two points in Q can be defined by $d(x,y) = \int_0^1 \left\|\dfrac{d\gamma_{x,y}}{ds}(s)\right\| ds$. The idea of a geodesic is useful because it provides a map from $T_q Q$ to Q in the following way: $v \in T_q Q \mapsto \gamma(1)$, $\gamma(0) = q$, $\dot{\gamma}(0) = v$, where γ is a geodesic. This map is called the *Riemannian exponential map* and is denoted by $\exp_q : T_q Q \to Q$. In particular, $\exp_q$ is a diffeomorphism from some star-shaped neighborhood of $0 \in T_q Q$ to a *geodesically convex* open neighborhood $\mathcal{B}$

of $q \in Q$. That is, any two points in $\mathcal{B}$ can be connected by a unique minimizing geodesic. Moreover, if $y \in \mathcal{B}$, we can express the Riemannian distance locally by means of the Riemannian exponential as $d(q, y) = \|\exp_q^{-1} y\|$.

The *Lebesgue space* $L^p([0, 1]; \mathbb{R}^n)$, $p \in (1, +\infty)$ is the space of $\mathbb{R}^n$-valued functions on $[0, 1]$ such that each of their components is p-integrable, that is, whose integral of the absolute value raised to the power of p is finite. A sequence (f_n) of functions in $L^p([0, 1]; \mathbb{R}^n)$ is said to be *weakly convergent* to f if for every $g \in L^r([0, 1]; \mathbb{R}^n)$, with $\frac{1}{p} + \frac{1}{r} = 1$, and every component i, $\lim_{n \to \infty} \int_{[0,1]} f_n^i g^i = \int_{[0,1]} f^i g^i$. A function $g: [0, 1] \to \mathbb{R}^n$ is said to be the *weak derivative* of $f: [0, 1] \to \mathbb{R}^n$ if for every component i of f and g, and for every compactly supported C^∞ real-valued function φ on $[0, 1]$, $\int_{[0,1]} f^i \varphi' = - \int_{[0,1]} g^i \varphi$. The *Sobolev space* $W^{k,p}([0, 1]; \mathbb{R}^n)$ is the space of functions $u \in L^p([0, 1]; \mathbb{R}^n)$ such that for every $\alpha \leq k$, the αth weak derivative $\frac{d^\alpha u}{dt^\alpha}$ of u exists and $\frac{d^\alpha u}{dt^\alpha} \in L^p([0, 1]; \mathbb{R}^n)$. In particular, $H^k([0, 1]; \mathbb{R}^n)$ denotes the Sobolev space $W^{k,2}([0, 1]; \mathbb{R}^n)$, and its norm may be expressed as

$$\|f\| = \left(\int_{[0,1]} \sum_{p=0}^{k} \left\| \frac{d^k}{dt^k} f(t) \right\|_{\mathbb{R}^n}^2 dt \right)^{1/2} \quad \text{for all } f \in H^k([0, 1]; \mathbb{R}^n), \text{ where } \| \cdot \|_{\mathbb{R}^n}$$

denotes the Euclidean norm on $\mathbb{R}^n$. $(f_n) \subset W^{k,p}([0, 1]; \mathbb{R}^n)$ is said to be *weakly convergent* to f in $W^{k,p}([0, 1]; \mathbb{R}^n)$ if for every $\alpha \leq k$, $\frac{d^\alpha f_n}{dt^\alpha} \rightharpoonup \frac{d^\alpha f}{dt^\alpha}$ weakly in $L^p([0, 1]; \mathbb{R}^n)$.

We denote by $H^2([0, 1]; Q)$ the set of all curves $q: [0, 1] \to Q$ such that for every chart $(\mathcal{U}, \varphi)$ of Q and every closed subinterval $I \subset [0, 1]$ such that $q(I) \subset \mathcal{U}$, the restriction of the composition $\varphi \circ q|_I$ is in $H^2([0, 1]; \mathbb{R}^m)$. Note that $H^2([0, 1]; Q)$ is an infinite-dimensional Hilbert Manifold modeled on $H^2([0, 1]; \mathbb{R}^m)$, and given $\xi = (q_0, v_0)$, $\eta = (q_T, v_T) \in TQ$, the space $\Omega_{\xi, \eta}^T$ (denoted simply by Ω unless otherwise necessary) defined as the space of all curves $\gamma \in H^2([0, 1]; Q)$ satisfying $\gamma(0) = q_0$, $\gamma(T) = q_T$, $\dot{\gamma}(0) = v_0$, $\dot{\gamma}(T) = v_T$ is a closed submanifold of $H^2([0, 1]; Q)$ (see [19]). The tangent space $T_x \Omega$ consists of vector fields along x of class H^2 which vanish at the endpoints together with their first covariant derivatives. We consider the Hilbert structure on $T_x \Omega$ induced by the inner product $\langle V, W \rangle := \int_0^T g \left(\frac{D^2}{dt^2} V, \frac{D^2}{dt^2} W \right) dt$. This inner product induces (fiberwise) a Riemannian metric on Ω, which itself induces a metric in the usual way. It is known that the completeness of Ω follows from the completeness of Q (see [20]).

3   Existence of Global Minimizer

Before beginning the proof we will introduce a lemma that simplifies the analysis considerably.

Lemma 1. *Let Q be an m-dimensional complete Riemannian manifold, and suppose that $\{q_n\} \subset \Omega$ is a sequence such that $\sup_{n \in} J(q_n) < +\infty$. Then, $\{q_n\}$ and $\{\dot{q}_n\}$ are uniformly bounded, and there exists a subsequence of $\{q_n\}$ which converges weakly to some $q \in \Omega$ with respect to the norm on H^2.*

Proof: Suppose that $\{q_n\}$ is such a sequence. Setting $G_0 := g(v_0, v_0)$ and using the Fundamental Theorem of Calculus and the Cauchy-Schwarz inequality, for all $n \in$ we have

$$g(\dot{q}_n(t), \dot{q}_n(t)) = g(\dot{q}_n(0), \dot{q}_n(0)) + \int_0^t \frac{d}{du} g(\dot{q}_n(u), \dot{q}_n(u)) \, du$$

$$\leq G_0 + 2 \int_0^T \left| g\left(\frac{D^2}{\partial u^2} q_n(u), \dot{q}_n(u) \right) \right| du$$

$$\leq G_0 + 2 \left[\int_0^T g\left(\frac{D^2}{\partial u^2} q_n(u), \frac{D^2}{\partial u^2} q_n(u) \right) dt \right]^{1/2} \left[\int_0^T g(\dot{q}_n(u), \dot{q}_n(u)) \, du \right]^{1/2}$$

$$\leq G_0 + 2T \left[\sup_{k \in} J(q_k) \right]^{1/2} \left[\sup_{t \in [0,T]} g(\dot{q}_n(t), \dot{q}_n(t)) \right]^{1/2},$$

where we have used the fact that V and $\left\| \frac{dx}{dt}(t) \right\|^2$ are non-negative in the last inequality. Let $c := \sup_{k \in} J(q_k)$ and $G := \sup_{t \in [0,T]} g(\dot{q}_n(t), \dot{q}_n(t))$. Taking the supremum of the inequality over $t \in [0, T]$, we have

$$G \leq G_0 + 2T\sqrt{cG} \implies G \leq \left(T\sqrt{c} + \sqrt{T^2 c + G_0} \right)^2 := r^2.$$

Now observe that the sequence of lengths of the curves similarly satisfies

$$L(q_n) = \int_0^T \sqrt{g(\dot{q}_n(t), \dot{q}_n(t))} dt \leq T\sqrt{G} \leq Tr.$$

Hence, the image of q_n is contained in the closed geodesic ball $\bar{B}_{Tr}(q_n)$, which is well-defined by completeness of Q, compact, and is independent of n. Therefore, the sequence $\{q_n\}$ is uniformly bounded over $[0, 1]$. Now observe that, for the Riemannian distance $d(\cdot, \cdot)$ and for all $0 \leq t < \tau \leq T$,

$$d(q_n(t), q_n(\tau)) \leq L\left(q_n|_{(t,\tau)} \right) = \int_t^\tau \sqrt{g(\dot{q}_n(u), \dot{q}_n(u))} du \leq r(\tau - t),$$

where $q_n|_{(t,\tau)}$ denotes the restriction of q_n to the interval $(t, \tau) \subset [0, T]$. Therefore $\{q_n\}$ is equicontinuous on $[0, T]$, and by the Arzela-Ascoli Theorem, there then exists a subsequence $\{q_{n_k}\} \subset \{q_n\}$ which converges uniformly to a continuous curve q_n satisfying the boundary conditions in position.

Let (U_i, φ_i) be a finite collection of charts on Q and I_i an accompanying finite partition of $[0, 1]$ such that, for sufficiently large n, there exists a compact subset $K \subset U_i$ containing $q_n(I_i)$. In local coordinates, we may consider q_n to be a curve on $\mathbb{R}^m$ (however, we will abuse this notation by continuing to call it q_n both on the chart U_i and its image in $\mathbb{R}^m$). Note that

$$\frac{d}{dt}\dot{q}_n = \frac{D}{dt}\dot{q}_n + \Gamma(q_n; \dot{q}_n, \dot{q}_n). \tag{3}$$

where $\Gamma : \mathbb{R}^{3m} \to \mathbb{R}^m$ is continuous in the first argument and bilinear in the last two-and is determined by the ordinary Christoffel Symbols induced by the connection and chart. Hence, in K, we have

$$\|q_n\|_{H^2}^2 = \|q_n\|_{L^2}^2 + \|\dot{q}_n\|_{L^2}^2 + \left\|\frac{d}{dt}\dot{q}_n\right\|_{L^2}^2$$

$$= \int_{I_i} \|q_n(t)\|_{\mathbb{R}^m}^2 \, dt + \int_{I_i} \|\dot{q}_n(t)\|_{\mathbb{R}^m}^2 \, dt + \int_{I_i} \left\|\frac{d}{dt}\dot{q}_n(t)\right\|_{\mathbb{R}^m}^2 \, dt.$$

The first integral is bounded since φ_i is a homeomorphism and K is compact-hence $\varphi_i(K) \supset (\varphi_i \circ q_n)(I_i)$ is bounded in $\mathbb{R}^m$. For the second integral, note that for some scalars α, β we have $\alpha\|X\|_{\mathbb{R}^m}^2 \leq g(X, X) \leq \beta\|X\|_{\mathbb{R}^m}^2$ for all $X \in T_x Q$ with $x \in K$. Hence, the boundedness of the second integral is equivalent to the boundedness of

$$\int_{I_i} g\left(\dot{q}_n(t), \dot{q}_n(t)\right) dt,$$

which follows immediately by the uniform boundedness of $g(\dot{q}_n, \dot{q}_n)$ on $[0, 1]$ (and hence also on the subset I_i).

Similarly, the boundedness of the final integral is equivalent to the boundedness of

$$\int_{I_i} g\left(\frac{d}{dt}\dot{q}_n(t), \frac{d}{dt}\dot{q}_n(t)\right) dt.$$

Note that Γ and V are uniformly bounded on I_i by continuity and the fact that each q_n and $\dot{q}_n$ are uniformly bounded. It then follows by (3) and the fact that $\sup_{n \in} J(q_n) < +\infty$ that the above integral is bounded. Hence, (q_n) is bounded in H^2. Since H^2 is a Hilbert space, we then get weak convergence of some subsequence of (q_n) to $q \in \Omega$ in H^2. □

Theorem 1. *The functional J attains its minimum in Ω.*

Proof: Suppose that $\{q_n\} \subset \Omega$ is a minimizing sequence. That is, $\lim_{n \in} J(q_n) = \inf_{q \in \Omega} J(q) \geq 0$. Note that, such a sequence satisfies the assumptions of Lemma (1), so that there exists a subsequence of q_n (which we also denote by $\{q_n\}$ for convenience) that converges weakly to some $q \in \Omega$ with respect to the norm on H^2. It then suffices to show that $J(q) \leq \liminf_{n \to \infty} J(q_n)$. Since $g(\dot{q}_n, \dot{q}_n)$ is uniformly bounded, there exists a finite collection of charts (U_i, φ_i) on Q and I_i an accompanying finite partition of $[0, 1]$ such that, for sufficiently large n, there exists a compact subset $K \subset U_i$ containing $q_n(I_i)$. In local coordinates, we may consider q_n to be a curve on $\mathbb{R}^m$ (as before, we will abuse this notation by continuing to call it q_n both on the chart U_i and its image in $\mathbb{R}^m$).

Observe first that $V(q_n)$ converges to $V(q)$ uniformly on $[0, T]$ since the interval is compact, V is continuous, and $q_n \to q$ uniformly. Moreover, the weak H^2 convergence of q_n to q implies that $\dot{q}_n \to \dot{q}$ uniformly, so that $g(\dot{q}_n, \dot{q}_n) \to g(\dot{q}, \dot{q})$.

Therefore, $J(q) \leq \liminf\limits_{n \to \infty} J(q_n)$ is equivalent to showing that for all intervals I_i and $n \in$, we have

$$\int_{I_i} g\left(\frac{D}{dt}\dot{q}(t), \frac{D}{dt}\dot{q}(t)\right) dt \leq \liminf_{n \to \infty} \int_{I_i} g\left(\frac{D}{dt}\dot{q}_n(t), \frac{D}{dt}\dot{q}_n(t)\right) dt. \tag{4}$$

Note that $\frac{d}{dt}\dot{q}_n = \frac{D}{dt}\dot{q}_n + \Gamma(q_n; \dot{q}_n, \dot{q}_n)$ where $\Gamma : \mathbb{R}^{3m} \to \mathbb{R}^m$ is continuous in the first argument and bilinear in the last two-and it is determined by the induced Christoffel symbols. It follows that $\Gamma(q_n; \dot{q}_n, \dot{q}_n) \to \Gamma(q; \dot{q}, \dot{q})$ uniformly on I_i, so that the above inequality is equivalent to

$$\int_{I_i} g\left(\ddot{q}(t), \ddot{q}(t)\right) dt \leq \liminf_{n \to \infty} \int_{I_i} g\left(\ddot{q}_n(t), \ddot{q}_n(t)\right) dt. \tag{5}$$

Thus far, we have suppressed the dependence of the Riemannian metric on the point at which we are evaluating the tangent vectors. Note that this is not problematic by the uniform convergence of q_n to q. That is,

$$\left| \int_{I_i} g_{q(t)}\left(\ddot{q}_n(t), \ddot{q}_n(t)\right) dt - \int_{I_i} g_{q_n(t)}\left(\ddot{q}_n(t), \ddot{q}_n(t)\right) dt \right| \to 0,$$

as $n \to \infty$, so we may assume that the metric is evaluated at $q(t)$ on both sides of inequality (5). We now consider the set

$$L_g^2(I_i, \mathbb{R}^m) := \left\{ \gamma : I_i \to \mathbb{R}^m \; : \int_{I_i} g_{q(t)}\left(\gamma(t), \gamma(t)\right) dt < +\infty \right\},$$

which can be endowed with the structure of a normed linear space, with the norm $\|\gamma\| = \int_{I_i} g_{q(t)}\left(\gamma(t), \gamma(t)\right) dt$.

Note that the Euclidean and Riemannian norms are (bi-Lipchitz) equivalent in the compact chart image of $TQ|_K$, which further implies that $L^2(I_i, \mathbb{R}^m)$ and $L_g^2(I_i, \mathbb{R}^m)$ are equivalent as normed linear spaces (indeed, L_g^2 can be thought of as a weighted L^2 space in local coordinates). Hence, they induce the same weak topology, and so the weak L^2-convergence of $\ddot{q}_n$ to $\ddot{q}$ further implies its weak L_g^2-convergence—from which (5) follows immediately.   □

References

1. Noakes, L., Heinzinger, G., Paden, B.: Cubic splines on curved spaces. IMA J. Math. Control. Inf. **6**, 465–473 (1989)
2. Crouch, P., Leite, F.S.: The dynamic interpolation problem: on Riemannian manifolds, lie groups, and symmetric spaces. J. Dyn. Control Syst. **1**, 177–202 (1995)
3. Giambò, R., Giannoni, F., Piccione, P.: An analytical theory for Riemannian cubic polynomials. IMA J. Math Control Inf. **19**, 445–460 (2002)
4. Camarinha, M., Silva Leite, F., Crouch, P.: On the geometry of Riemannian cubic polynomials. Diff. Geom. Appl. **15**(2), 107–135 (2001)
5. Giambò, R., Giannoni, F., Piccione, P.: Optimal control on Riemannian manifolds by interpolation. MCSS **16**, 278–296 (2004)
6. Bloch, A., Gupta, R., Kolmanovsky, I.: Neighboring extremal optimal control for mechanical systems on Riemannian manifolds. J. Geom. Mech. **8**(3), 257 (2016)

7. Hussein, I., Bloch, A.: Dynamic interpolation on Riemannian manifolds: an application to interferometric imaging. In: Proceedings of American Control Conference, Boston, pp. 413–418 (2004)

8. Koditschek, D.E., Rimon, E.: Robot navigation functions on man- ifolds with boundary. Adv. Appl. Math. **11**(4), 412–442 (1990)

9. Chang, D., Shadden, S., Marsden, J., Olfati-Saber, R.: Collision avoidance for multiple agent systems. In: 42nd IEEE International Conference on Decision and Control, pp. 539–543 (2003)

10. Bloch, A., Colombo, L., Gupta, R., de Diego, D.M.: A geometric approach to the optimal control of nonholonomic mechanical systems. In: Analysis and Geometry in Control Theory and its Applications 35–64 (2015)

11. Bloch, A., Camarinha, M., Colombo, L.: Variational obstacle avoidance problem on Riemannian manifolds. In: Proceedings of the IEEE International Conference on Decision and Control, pp. 146–150 (2017)

12. Bloch, A., Camarinha, M., Colombo, L.J.: Dynamic interpolation for obstacle avoidance on Riemannian manifolds. Int. J. Control **94**(3), 588–600 (2021)

13. Bloch, A., Camarinha, M., Colombo, L.: Variational point-obstacle avoidance on Riemannian manifolds. Math. Control Signals Syst. **33**(1), 109–121 (2021)

14. Chandrasekaran, R.S., Colombo, L.J. Camarinha, M., Banavar, R., Bloch, A.: Variational collision and obstacle avoidance of multi-agent systems on Riemannian manifolds. In: 2020 European Control Conference (ECC), pp. 1689–1694 (2020). https://doi.org/10.23919/ECC51009.2020.9143986

15. Assif, M., Banavar, R., Bloch, A., Camarinha, M., Colombo, L.: Variational collision avoidance problems on Riemannian manifolds. In: Proceedings of the IEEE International Conference on Decision and Control, pp. 2791–2796 (2018)

16. Goodman, J., Colombo, L.: Collision avoidance of multiagent systems on Riemannian manifolds. SIAM J. Control. Optim. **60**(1), 168–188 (2021)

17. Goodman, J., Colombo, L.: Variational obstacle avoidance with applications to interpolation problems in hybrid systems. In: Proceedings of the 7th IFAC Workshop on Lagrangian and Hamiltonian Methods in Nonlinear Control (2021)

18. Boothby, W.M.: An Introduction to Differentiable Manifolds and Riemannian Geometry. Academic Press Inc., Orlando, FL (1975)

19. Palais, R., Terng, C.-L.: Critical Point Theory and Submanifold Geometry. Springer, Berlin (1988). https://doi.org/10.1007/BFb0087442

20. Bauer, M., Maor, C., Michor, P.W.: Sobolev metrics on spaces of manifold valued curves. Annali di Scienze (2022)

Virtual Affine Nonholonomic Constraints

Efstratios Stratoglou[1], Alexandre Anahory Simoes[2(✉)], Anthony Bloch[3],
and Leonardo Colombo[4]

[1] Universidad Politécnica de Madrid (UPM), José Gutiérrez Abascal, 2,
28006 Madrid, Spain
ef.stratoglou@alumnos.upm.es
[2] School of Science and Technology, IE University, Segovia, Spain
alexandre.anahory@ie.edu
[3] Department of Mathematics, University of Michigan, Ann Arbor, MI 48109, USA
abloch@umich.edu
[4] Centre for Automation and Robotics (CSIC-UPM), Ctra. M300 Campo Real,
Km 0,200, Arganda del Rey, 28500 Madrid, Spain
leonardo.colombo@csic.es

Abstract. Virtual constraints are relations imposed on a control system via feedback control that become invariant via feedback, as opposed to physical constraints acting on the system. Nonholonomic systems are mechanical systems with non-integrable constraints on the velocities. In this work, we introduce the notion of **virtual affine nonholonomic constraints** in a geometric framework. More precisely, it is a controlled invariant affine distribution associated with an affine connection mechanical control system. We show the existence and uniqueness of a control law defining a virtual affine nonholonomic constraint.

Keywords: Nonholonomic Systems · Virtual Constraints · Nonlinear Control

1 Introduction

Virtual constraints are relations on the configuration variables of a control system which are imposed through feedback control and the action of actuators, instead of through physical connections such as gears or contact conditions with the environment. The class of virtual holonomic constraints became popular in applications to biped locomotion where it was used to express a desired walking gait, as well as for motion planning to search for periodic orbits and its employment in the technique of transverse linearization for stabilizing such orbits.

Virtual nonholonomic constraints are a class of virtual constraints that depend on velocities rather than only on the configurations of the system. Those constraints were introduced in [4] to design a velocity-based swing foot placement

The authors acknowledge financial support from Grant PID2019-106715GB-C21 funded by MCIN/AEI/ 10.13039/501100011033 and the LINC Global project from CSIC "Wildlife Monitoring Bots" INCGL20022. A.B. was partially supported by NSF grants DMS-1613819 and DMS-2103026, and AFOSR grant FA 9550-22-1-0215.

F. Nielsen and F. Barbaresco (Eds.): GSI 2023, LNCS 14072, pp. 89–96, 2023.
https://doi.org/10.1007/978-3-031-38299-4_10

in bipedal robots. In particular, this class of virtual constraints was used in [5–8] to encode velocity-dependent stable walking gaits via momenta conjugate to the unacatuated degrees of freedom of legged robots and prosthetic legs.

The recent work [9] introduces an approach to defining rigorously virtual nonholonomic constraints, but the nonlinear nature of the constraints makes a thorough mathematical analysis difficult. In the work [1], we provide a formal definition of linear virtual nonholonomic constraints, i.e., constraints that are linear on the velocities. Our definition is based on the invariance property under the closed-loop system and coincides with that of [9] in the linear case. In this paper we extend the results of [1] to the case of affine constraints on the velocities.

In particular, a virtual affine nonholonomic constraint is described by an affine non-integrable distribution on the configuration manifold of the system for which there is a feedback control making it invariant under the flow of the closed-loop system. We provide conditions for the existence and uniqueness of such a feedback law defining the virtual affine nonholonomic constraint.

The remainder of the paper is structured as follows. Section 2 introduces nonholonomic systems with affine constraints. We define virtual nonholonomic constraints in Sect. 3, where we provide conditions for the existence and uniqueness of a control law defining a virtual nonholonomic constraint. In Sect. 4, we provide an example of a marine vessel with a payload subject to a position-dependent stream to illustrate the theoretical results.

2   Nonholonomic Systems with Affine Constraints

We begin with a Lagrangian system on an n-dimensional configuration space Q and Lagrangian $L : TQ \to \mathbb{R}$, where TQ denotes the tangent bundle of Q. We assume the Lagrangian has the mechanical form

$$L = K - V \circ \pi, \tag{1}$$

where K is a function on TQ describing the kinetic energy of the system, that is, $K = \frac{1}{2}\mathscr{G}(v_q, v_q)$, where $\mathscr{G}$ is the Riemannian metric on Q, $V : Q \to \mathbb{R}$ is a function on Q representing the potential energy, and $\pi : TQ \to Q$ is the tangent bundle projection, locally given by $\pi(q, \dot{q}) = q$ with $(q, \dot{q})$ denoting local coordinates on TQ. In addition, we denote by $\mathfrak{X}(Q)$ the set of vector fields on Q and by $\Omega^1(Q)$ the set of 1-forms on Q. If $X, Y \in \mathfrak{X}(Q)$, then $[X, Y]$ denotes the standard Lie bracket of vector fields.

Consider affine nonholonomic constraints, that is, for each $q \in Q$ the velocities belong to an affine subspace $\mathscr{A}_q$ of the tangent space T_qQ. Thus, $\mathscr{A}_q$ can be written as a sum of a vector field $X \in \mathfrak{X}(Q)$ and a nonintegrable distribution $\mathscr{D}$ on Q, i.e. $\mathscr{A}_q = X(q) + \mathscr{D}_q$, where $\mathscr{D}$ is of constant rank r, with $1 < r < n$. In this case, we say that the affine space $\mathscr{A}_q$ is modelled on the vector subspace $\mathscr{D}_q$. In local coordinates $\mathscr{D}$ can be expressed as the null space of a q-dependent matrix $S(q)$ of dimension $m \times n$ and rank $S(q) = m$, with $m = n - r$ as $\mathscr{D}_q = \{\dot{q} \in T_qQ : S(q)\dot{q} = 0\}$. The rows of $S(q)$ can be represented by the coordinate functions of m independent 1-forms $\mu^b = \mu_i^b dq^i$, $1 \le i \le n$, $1 \le b \le m$. The affine distribution is $\mathscr{A}_q = \{\dot{q} \in T_q Q : S(q)(\dot{q} - X(q)) = 0\}$, hence $\mathscr{A} = \{(q, \dot{q}) \in TQ : \phi(q, \dot{q}) = 0\}$, with $\phi(q, \dot{q}) = S(q)\dot{q} + Z(q)$ and $Z(q) = -S(q)X(q) \in \mathbb{R}^m$ (see [2,3] for instance). Throughout the paper, to avoid confusion between affine and standard distributions, we will refer to the latter as linear distributions.

Definition 1. *A mechanical system with* **affine nonholonomic constraints** *on a smooth manifold Q is given by the triple $(\mathscr{G}, V, \mathscr{A})$, where $\mathscr{G}$ is a Riemannian metric on Q, representing the kinetic energy of the system, $V : Q \to \mathbb{R}$ is a smooth function representing the potential energy, and $\mathscr{A}$ is an affine distribution on Q describing the affine nonholonomic constraints.*

On any Riemannian manifold, there is a unique connection $\nabla^{\mathscr{G}} : \mathfrak{X}(Q) \times \mathfrak{X}(Q) \to \mathfrak{X}(Q)$ called the **Levi-Civita connection** satisfying the following two properties:

1. $[X, Y] = \nabla^{\mathscr{G}}_X Y - \nabla^{\mathscr{G}}_Y X$ (symmetry)
2. $X(\mathscr{G}(Y, Z)) = \mathscr{G}(\nabla^{\mathscr{G}}_X(Y, Z) + \mathscr{G}(Y, \nabla^{\mathscr{G}}_X Z)$ (compatibillity of the metric).

The Levi-Civita connection helps us describe the trajectories of a mechanical Lagrangian system. The trajectories $q : I \to Q$ of a mechanical Lagrangian determined by a Lagrangian function of the form (1) satisfy the following equations

$$\nabla^{\mathscr{G}}_{\dot{q}} \dot{q} + \mathrm{grad}_{\mathscr{G}} V(q(t)) = 0, \tag{2}$$

where the vector field $\mathrm{grad}_{\mathscr{G}} V \in \mathfrak{X}(Q)$ is characterized by $\mathscr{G}(\mathrm{grad}_{\mathscr{G}} V, X) = dV(X)$, for every $X \in \mathfrak{X}(Q)$. Observe that if the potential function vanishes, then the trajectories of the mechanical system are just the geodesics with respect to the connection $\nabla^{\mathscr{G}}$.

3   Virtual Affine Nonholonomic Constraints

In this section we present a detailed construction of virtual affine nonholonomic constraints. As will be clear from the definition, their existence is essentially linked to a controlled system, rather than to the affine distribution $\mathscr{A}$ defined by the constraints. Hence we give the necessary tools for controlled mechanical systems.

Given a Riemannian metric $\mathscr{G}$ on Q, we can use its non-degeneracy property to define the musical isomorphism $\flat : \mathfrak{X}(Q) \to \Omega^1(Q)$ defined by $\flat(X)(Y) = \mathscr{G}(X, Y)$ for any $X, Y \in \mathfrak{X}(Q)$. Also, denote by $\sharp : \Omega^1(Q) \to \mathfrak{X}(Q)$ the inverse musical isomorphism, i.e., $\sharp = \flat^{-1}$.

Given an external force $F^0 : TQ \to T^*Q$ and a control force $F : TQ \times U \to T^*Q$ of the form

$$F(q, \dot{q}, u) = \sum_{a=1}^{m} u_a f^a(q), \tag{3}$$

where $f^a \in \Omega^1(Q)$ with $m < n$, $U \subset \mathbb{R}^m$ the set of controls and $u_a \in \mathbb{R}$ with $1 \le a \le m$ the control inputs, consider the associated mechanical control system of the form

$$\nabla^{\mathscr{G}}_{\dot{q}(t)} \dot{q}(t) = Y^0(q(t), \dot{q}(t)) + u_a(t) Y^a(q(t)), \tag{4}$$

with $Y^0 = \sharp(F^0)$ and $Y^a = \sharp(f^a)$ the corresponding force vector fields.

Hence, q is the trajectory of a vector field of the form

$$\Gamma(v_q) = G(v_q) + u_a(Y^a)^V_{v_q}, \tag{5}$$

where G is the vector field determined by the unactuated forced mechanical system

$$\nabla^{\mathscr{G}}_{\dot{q}(t)}\dot{q}(t) = Y^0(q(t), \dot{q}(t))$$

and where the vertical lift of a vector field $X \in \mathfrak{X}(Q)$ to TQ is defined by

$$X^V_{v_q} = \frac{d}{dt}\Big|_{t=0} (v_q + tX(q)).$$

Definition 2. *The distribution $\mathscr{F} \subseteq TQ$ generated by the vector fields $\sharp(f_i)$ is called the* **input distribution** *associated with the mechanical control system (4).*

Now we will introduce the concept of virtual affine nonholonomic constraint.

Definition 3. *A* **virtual affine nonholonomic constraint** *associated with the mechanical control system (4) is a controlled invariant affine distribution $\mathscr{A} \subseteq TQ$ for that system, that is, there exists a control function $\hat{u} : \mathscr{A} \to \mathbb{R}^m$ such that the solution of the closed-loop system satisfies $\psi_t(\mathscr{A}) \subseteq \mathscr{A}$, where $\psi_t : TQ \to TQ$ denotes its flow.*

Before we proceed to the theorem which gives the necessary conditions for the existence and uniqueness of a control law that turns an affine distribution into a controlled invariant affine distribution (virtual affine nonholonomic constraint), we present some necessary preliminaries.

Definition 4. *If W is an affine subspace of the vector space V modelled on the vector subspace W_0, then the* **dimension** *of the affine subspace W is defined to be the dimension of the model vector subspace W_0.*

Two affine subspaces W_1 and W_2 of a vector space V are **transversal** and we write $W_1 \pitchfork W_2$ if

1. $V = W_1 + W_2$.
2. $\dim V = \dim W_1 + \dim W_2$, i.e., the dimensions of W_1 and W_2 are complementary with respect to the ambient space dimension.

Remark 1. If W_1 and W_2 are subspaces of V then the previous definition implies that $V = W_1 \oplus W_2$.

Remark 2. If W_1 and W_2 are affine subspaces of V modelled on vector subspaces W_{10} and W_{20}, respectively, then $W_1 \pitchfork W_2$ if and only if $V = W_{10} \oplus W_{20}$.

Proposition 1. *Given a linear distribution $\mathscr{F}$ on a manifold Q and an affine distribution $\mathscr{A}$, with the linear distribution $\mathscr{D}$ being the associated model distribution, and X a vector field on Q satisfying $\mathscr{A} = X + \mathscr{D}$, the tranversality condition for $\mathscr{A}$ and $\mathscr{F}$ is an inherited property from the transversality of the model distribution $\mathscr{D}$ and vice versa, namely, $\mathscr{A} \pitchfork \mathscr{F}$ if and only if $\mathscr{D} \pitchfork \mathscr{F}$.*

Proof. First suppose that $\mathscr{A} \pitchfork \mathscr{F}$ which means that for every $q \in Q$ we have

$$T_q Q = \mathscr{A}_q + \mathscr{F}_q = X(q) + \mathscr{D}_q + \mathscr{F}_q;$$

Hence, for every $v_q \in T_q Q$ there exist vectors $d_q \in \mathscr{D}_q$ and $f_q \in \mathscr{F}_q$ such that

$$v_q = X(q) + d_q + f_q \Leftrightarrow v_q - X(q) = d_q + f_q.$$

Since $v_q - X(q) \in T_q Q$ and v_q is arbitrary we have $T_q Q = \mathscr{D}_q + \mathscr{F}_q$ for every $q \in Q$. Together with the fact that $\dim \mathscr{D}_q + \dim \mathscr{F}_q = \dim T_q Q$, we have that $\mathscr{D} \pitchfork \mathscr{F}$.

Now suppose that $\mathscr{D} \pitchfork \mathscr{F}$. Note that this is the same as $TQ = \mathscr{D} \oplus \mathscr{F}$. Hence, as before, for $v_q \in T_q Q$ there are $d_q \in \mathscr{D}_q$ and $f_q \in \mathscr{F}_q$ such that

$$v_q = d_q + f_q \Leftrightarrow v_q + X(q) = X(q) + d_q + f_q.$$

By the same argument as above, $v_q + X(q) \in T_q Q$ and v_q is arbitrary. Thus, together with the dimension condition, we conclude that $\mathscr{A} \pitchfork \mathscr{F}$.

Proposition 2. *Consider two distributions $\mathscr{A}$ and $\mathscr{F}$ where the first is an affine distribution as defined previously i.e. $\mathscr{A} = X + \mathscr{D}$, with $X \in \mathfrak{X}(Q)$ and $\mathscr{D}$ its associated model distribution. Then, for $v_q \in \mathscr{A}$ we have*

$$\mathscr{A} \pitchfork \mathscr{F} \implies T_{v_q}(TQ) = T_{v_q} \mathscr{A} \oplus \mathscr{F}_{v_q}^V,$$

where $\mathscr{F}_{v_q}^V$ is the vertical lift of $\mathscr{F}_{v_q}$.

Proof. From the structure of $\mathscr{A}$, i.e., from the fact that each $v_q \in \mathscr{A}_q$ can be written as $v_q = Z(q) + d_q$ where $d_q \in \mathscr{D}_q$, we may conclude that $\mathscr{A}_q$ is a r-dimensional manifold, where r is the rank of the distribution $\mathscr{D}$. Thus $\mathscr{A}$ is a fiber bundle whose base space is the n dimensional manifold Q and whose fibers are r dimensional affine subspaces. Hence,

$$\dim(T_{v_q} \mathscr{A}) = \dim(T_{d_q} \mathscr{D}) = n + r$$

and since $\dim \mathscr{F}_{v_q}^V = n - r = m$, we have that

$$\dim T_{v_q}(TQ) = \dim T_{v_q} \mathscr{A} + \dim \mathscr{F}_{v_q}^V = n + r + m = 2n.$$

So, in order to prove that both subspaces are transversal it suffices to prove that their intersection contains only the zero tangent vector. Indeed, suppose that $v_q \in \mathscr{A}$ and $X_{v_q} \in T_{v_q} \mathscr{A}$. Since $\mathscr{A}$ is defined to be the set of vectors satisfying the equation $\phi = 0$, the tangent vector satisfies $T_{v_q} \phi(X_{v_q}) = 0$. If, in addition, $X_{v_q} \in \mathscr{F}_{v_q}^V$, then it can be written as

$$X_{v_q} = c^i \sharp(f_i)_{v_q}^V.$$

However,

$$T_{v_q} \phi(\sharp(f_i)_{v_q}^V) = (S(q) \sharp(f_i))_{v_q}^V$$

from whence it follows that if $c^i \sharp(f_i)_{v_q}^V$ was in the null space of the linear map $T_{v_q} \phi$, then $c^i \sharp(f_i)$ would be in the null space of $S(q)$ which is false, since these are vectors in $\mathscr{D}_q$ and $\mathscr{F}$ and $\mathscr{D}$ are transversal using Proposition 1. Thus $X_{v_q} = 0$.

Theorem 1. *If the affine distribution $\mathscr{A}$ and the control input distribution $\mathscr{F}$ are transversal, then there exists a unique control function making the distribution a virtual affine nonholonomic constraint associated with the mechanical control system* (4).

Proof. Suppose that $\mathscr{A} \pitchfork \mathscr{F}$ and that trajectories of the control system (4) may be written as the integral curves of the vector field Γ defined by (5). From Proposition 2 we have

$$T_{v_q}(TQ) = T_{v_q}\mathscr{A} \oplus \mathscr{F}_{v_q}^V,$$

where $v_q \in \mathscr{A}$ and $\mathscr{F}_{v_q}^V = \text{span}\{(Y^a)_{v_q}^V\}$. Using the uniqueness decomposition property arising from transversality, we define a function $\tau^* : \mathscr{D} \to \mathbb{R}^m$ where $\tau^*(v_q)$ is the unique vector $\tau^*(v_q) = (\tau_1^*(v_q), \cdots, \tau_m^*(v_q)) \in \mathbb{R}^m$ such that $\Gamma(v_q) = G(v_q) + \tau_a^*(v_q)(Y^a)_{v_q}^V \in T_{v_q}\mathscr{A}$. Next, we show that Γ depends smoothly on v_q. We will prove that τ^* is a smooth function. If $\mathscr{A}$ is defined by m constraints of the form $\phi^b(v_q) = 0$, $1 \le b \le m$, then the condition above may be rewritten as $d\phi^b(G(v_q) + \tau_a^*(v_q)(Y^a)_{v_q}^V) = 0$, which is equivalent to

$$\tau_a^*(v_q)d\phi^b((Y^a)_{v_q}^V) = -d\phi^b(G(v_q)).$$

Note that, the equation above is a linear equation of the form $P(v_q)\tau = b(v_q)$, where $b(v_q)$ is the vector $(-d\phi^1(G(v_q)), \ldots, -d\phi^m(G(v_q))) \in \mathbb{R}^m$ and $P(v_q)$ is the $m \times m$ matrix with entries $P_a^b(v_q) = d\phi^b((Y^a)_{v_q}^V) = \mu^b(q)(Y^a)$, where the last equality may be deduced by computing the expressions in local coordinates. That is, if $(q^i \dot{q}^i)$ are natural bundle coordinates for the tangent bundle, then

$$d\phi^b((Y^a)_{v_q}^V) = \left(\frac{\partial \mu_i^b}{\partial q^j}\dot{q}^i dq^j + \frac{\partial Z_i}{\partial q^j}dq^j + \mu_i^b d\dot{q}^i \right)\left(Y^{a,k}\frac{\partial}{\partial \dot{q}^k} \right)$$
$$= \mu_i^b Y^{a,i} = \mu^b(q)(Y^a).$$

In addition, $P(v_q)$ has full rank, since its columns are linearly independent. In fact suppose that

$$\begin{bmatrix} \mu^1(c_1 Y^1 + \cdots + c_m Y^m) \\ \vdots \\ \mu^m(c_1 Y^1 + \cdots + c_m Y^m) \end{bmatrix} = 0.$$

However, from Proposition 1 we have $\mathscr{D} \cap \mathscr{F} = \{0\}$ which implies that $c_1 Y^1 + \cdots + c_m Y^m = 0$. Since $\{Y_i\}$ are linearly independent we conclude that $c_1 = \cdots = c_m = 0$ and P has full rank. But, since P is an $m \times m$ matrix, and $\mathscr{D}$ is a regular distribution, it must be invertible. Therefore, there is a unique vector $\tau^*(v_q)$ satisfying the matrix equation and $\tau^* : \mathscr{D} \to \mathbb{R}^m$ is smooth since it is the solution of a matrix equation depending smoothly on v_q. Hence, Γ is a smooth vector field tangent to $\mathscr{A}$ and its flow preserves $\mathscr{A}$. $\qquad \square$

4   An Example

Consider a boat with a payload on the sea with a position-dependent stream. The position of the boat's center of mass is modeled by the configuration manifold $\mathbb{R}^2$ to which we add an orientation to obtain a complete description of its location in space, so that the system total configuration manifold is $\mathbb{R}^2 \times \mathbb{S}$ with local coordinates $q = (x, y, \theta)$. The sea's current is modeled by the vector field $C :$ $\mathbb{R}^2 \to \mathbb{R}^2$, $C = (C^1(x, y), C^2(x, y))$.

The boat is well modeled by a forced mechanical system with Lagrangian function $L = \frac{m}{2}(\dot{x}^2 + \dot{y}^2) + \frac{I}{2}\dot{\theta}^2$, where m is the boat's mass, I is the moment of inertia, and the external force is denoted by $F^{ext} = W^1 dx + W^2 dy$ accounting for the action of the current on the center of mass of the boat and to which we add a control force $F = u(\sin\theta dx - \cos\theta dy + d\theta)$.

The functions W^1 and W^2 are defined according to

$$\begin{cases} W^1 &= m\, d\left(\sin^2\theta C^1 - \sin\theta\cos\theta C^2\right)(\dot{q}), \\ W^2 &= m\, d\left(-\sin\theta\cos\theta C^1 + \cos^2\theta C^2\right)(\dot{q}), \end{cases}$$

where d represents the differential of the functions inside the parenthesis. The external force assures that in the absence of controls, the dynamics of the boat satisfies the following kinematic equations

$$\begin{cases} \dot{x} = & \sin^2\theta C^1 - \sin\theta\cos\theta C^2 \\ \dot{y} = & -\sin\theta\cos\theta C^1 + \cos^2\theta C^2, \end{cases}$$

whenever the initial velocities in the x and y direction vanish. The corresponding controlled forced Lagrangian system is

$$m\ddot{x} = u\sin\theta + W^1, \quad m\ddot{y} = -u\cos\theta + W^2, \quad I\ddot{\theta} = u,$$

and, as we will show, it has the following virtual affine nonholonomic constraint

$$\sin\theta\dot{x} - \cos\theta\dot{y} = C^2\cos\theta - C^1\sin\theta.$$

The input distribution $\mathscr{F}$ is generated just by one vector field

$$Y = \frac{\sin\theta}{m}\frac{\partial}{\partial x} - \frac{\cos\theta}{m}\frac{\partial}{\partial y} + \frac{1}{I}\frac{\partial}{\partial\theta},$$

while the virtual nonholonomic constraint is the affine space $\mathscr{A}$ modelled on the distribution $\mathscr{D}$ defined as the set of tangent vectors $v_q \in T_qQ$ where $\mu(q)(v) = 0$, with $\mu = \sin\theta dx - cos\theta dy$. Thus, we may write it as

$$\mathscr{D} = \mathrm{span}\left\{X_1 = \cos\theta\frac{\partial}{\partial x} + \sin\theta\frac{\partial}{\partial y},\, X_2 = \frac{\partial}{\partial\theta}\right\}.$$

The affine space is given as the zero set of the function $\phi(q, v) = \mu(q)(v) + Z(q)$ with $Z(q) = \cos\theta C^2(x, y) - \sin\theta C^1(x, y)$ or, equivalently, as the set of vectors v_q satisfying $v_q - C(q) \in \mathscr{D}_q$.

We may check that $\mathscr{A}$ is controlled invariant for the controlled Lagrangian system above. In fact, the control law $\hat{u}(x, y, \theta, \dot{x}, \dot{y}, \dot{\theta}) = -m\dot{\theta}(\cos\theta\dot{x} + \sin\theta\dot{y})$ makes the affine space invariant under the closed-loop system, since in this case, the dynamical vector field arising from the controlled Euler-Lagrange equations given by

$$\Gamma = \dot{x}\frac{\partial}{\partial x} + \dot{y}\frac{\partial}{\partial y} + \dot{\theta}\frac{\partial}{\partial\theta} + \left(\frac{\hat{u}\sin\theta + W^1}{m}\right)\frac{\partial}{\partial\dot{x}} + \left(-\frac{\hat{u}\cos\theta - W^2}{m}\right)\frac{\partial}{\partial\dot{y}} + \frac{\hat{u}}{I}\frac{\partial}{\partial\dot{\theta}}$$

is tangent to $\mathscr{A}$. This is deduced from the fact that

$$\Gamma(\sin\theta\dot{x} - \cos\theta\dot{y} + \cos\theta C^2(x, y) - \sin\theta C^1(x, y)) = 0.$$

References

1. Anahory Simoes, A., Stratoglou, E., Bloch, A., Colombo, L.: Virtual Nonholonomic Constraints: A Geometric Approach. arXiv e-prints. arXiv: 2207.01299 (2022)
2. Fasso, F., Sansonetto, N.: Conservation of energy and momenta in nonholonomic systems with affine constraints. Regul. Chaotic Dyn. **20**(4), 449–462 (2015)
3. Fasso, F., García-Naranjo, L., Sansonetto, N.: Moving energies as first integrals of nonholonomic systems with affine constraints. Nonlinearity **31**(3), 755 (2018)
4. Griffin, B., Grizzle, J.: Nonholonomic virtual constraints for dynamic walking. In: 54th IEEE Conference on Decision and Control, pp. 4053–4060 (2015)
5. Hamed, K., Ames, A.: Nonholonomic hybrid zero dynamics for the stabilization of periodic orbits: application to underactuated robotic walking. IEEE Trans. Control Syst. Technol. **28**(6), 2689–2696 (2019)
6. Horn, J., Mohammadi, A., Hamed, K., Gregg, R.: Nonholonomic virtual constraint design for variable-incline bipedal robotic walking. IEEE Robot. Autom. Lett. **5**(2), 3691–3698 (2020)
7. Horn, J., Mohammadi, A., Hamed, K., Gregg, R.: Hybrid zero dynamics of bipedal robots under nonholonomic virtual constraints. IEEE Control Syst. Lett. **3**(2), 386–391 (2018)
8. Horn, J., Gregg, R.: Nonholonomic virtual constraints for control of powered prostheses across walking speeds. IEEE Trans. Control Syst. Technol. (2021)
9. Moran-MacDonald, A.: Energy injection for mechanical systems through the method of Virtual Nonholonomic Constraints. University of Toronto, Thesis (2021)

Nonholonomic Systems with Inequality Constraints

Alexandre Anahory Simoes[1](✉) and Leonardo Colombo[2]

[1] School of Science and Technology, IE University, Madrid, Spain
alexandre.anahory@ie.edu
[2] Centre for Automation and Robotics (CSIC-UPM), Ctra. M300 Campo Real,
Km 0,200, Arganda del Rey, 28500 Madrid, Spain
leonardo.colombo@csic.es

Abstract. In this paper we derive the equations of motion for nonholonomic systems subject to inequality constraints, both in continuous-time and discrete-time. The last is done by discretizing the continuous time-variational principle which defines the equations of motion for a nonholonomic system subject to inequality constraints. An example is shown to illustrate the theoretical results.

Keywords: Nonholonomic Systems · Inequality constraints · Variational integrators

1 Introduction

Some mechanical systems have a restriction on the configurations or velocities that the system may assume. Systems with such restrictions are generally called constrained systems. Nonholonomic systems [2,4,8,12] are, roughly speaking, mechanical systems with constraints on their velocity that are not derivable from position constraints. They arise, for instance, in mechanical systems that have rolling contact (e.g., the rolling of wheels without slipping) or certain kinds of sliding contact.

Mechanical systems subject to inequality constraints are confined within a region of space with boundary. Collision with the boundary activates constraint forces forbiding the system to cross the boundary into a non-admissible region of space. Inequality constraints appear for instance in the problem of rigid-body collisions, mechanical grasping models and biomechanical locomotion [1,11].

Structure preserving integration of systems with inequality constraints has been addressed in many papers due to its applicability in engineering problems that require nonsmooth techniques (see [10]). In [9,10], the authors use variational techniques to deduce the equations of motion and integrators for unconstrained mechanical systems with inequality constraints. In [9], the authors extend the space of solutions to a non-autonomous space depending on time in

The authors acknowledge financial support from Grant PID2019-106715GB-C21 funded by MCIN/AEI/ 10.13039/501100011033.

F. Nielsen and F. Barbaresco (Eds.): GSI 2023, LNCS 14072, pp. 97–104, 2023.
https://doi.org/10.1007/978-3-031-38299-4_11

order to remove the non-smoothness during the collision with the boundary. However, in [10], the authors use nonsmooth analysis to deal with collisions and obtain better structure preservation: for instance, nearly energy conservation.

In this paper we consider nonholonomic systems subject to inequality constraints. The prototype example we examine is that of a wheel rolling without sliding inside a circular table. We extend the technique in [9] to derive the equations of motion via an adaptation of the Lagrange-D'Alembert principle for nonholonomic systems subject to inequality constraints, and then we use a modification of discrete Lagrange-d'Alembert principle [7] to derive variational integrators for these systems. Although dealing smoothly with the impact with the boundary, our integrator suffers from the same problems as the ones identified in [10], in particular, non-conservation of energy during the impact. However, we consider that this paper introduces a first approach to the geometric integration of nonholonomic systems with inequality constraints and motivates the search for other strategies such as DELI equations (see [10]) for nonholonomic systems.

The remainder of the paper is as follows: Sect. 2 introduces mechanical systems with inequality constraints. In Sect. 3, we review nonholonomic systems and introduce the variational principle that gives the equations of motion for nonholonomic systems with inequality constraints. In Sect. 4, we develop the discrete counterpart of the results in the preceding section. Finally, in Sect. 5, we examine the example of a disk rolling without slipping in a circular table.

2   Mechanical Systems with Inequality Constraints

In this paper, we will analyse the dynamics of nonholonomic systems evolving on the configuration manifold Q which are subjected to inequality constraints, i.e., constraints determined by a submanifold with boundary C of the manifold Q. The boundary ∂C is a smooth manifold of Q with codimension 1. Locally, the boundary ∂C is a smooth manifold of the type $\partial C = \{q \in Q \mid g(q) = 0\}$ and the manifold C is $C = \{q \in Q \mid g(q) \leqslant 0\}$ for some smooth function $g : Q \to \mathbb{R}$.

In convex geometry, given a closed convex set K of $\mathbb{R}^n$, the *polar cone* of K is the set $K^p = \{z \in \mathbb{R}^n \mid \langle z, y \rangle \leqslant 0, \forall y \in K\}$ (see [5]). The *normal cone* to K at a point $x \in K$ is given by $N_K(x) = K^p \cap \{x\}^T$, where $\{x\}^T$ is the orthogonal subspace to x with respect to the Euclidean inner product.

Based on this construction, we will only use a minimal definition of normal cone suiting the kind of inequality constraints we will be dealing with. Given a submanifold with boundary C as before, the normal cone to a point $q \in \partial C$ is the set $N_C(q) = \{\lambda dg(q) \mid \lambda \geqslant 0\}$. The two definitions match if C is a closed convex set of $\mathbb{R}^n$ with boundary being a hypersurface of dimension $n - 1$.

Given a Lagrangian function $L : TQ \to \mathbb{R}$ describing the dynamics, with local coordinates $(q^i, \dot{q}^i)$, $i = 1, \ldots, n = \dim Q$, the equations of motion under the presence of inequality constraints are given by Euler-Lagrange equations $\dfrac{d}{dt}\dfrac{\partial L}{\partial \dot{q}^i} - \dfrac{\partial L}{\partial q^i} = 0$ whenever the trajectory is in the interior of the constraint submanifold $C \setminus \partial C$. At impact times $t_i \in \mathbb{R}$ of the trajectory with the boundary $q(t_i) \in \partial C$, there is a discontinuity in the state variables of the system, often called a jump. This jump is determined by the equations:

$$\frac{\partial L}{\partial \dot{q}}|_{t=t_i^+} - \frac{\partial L}{\partial \dot{q}}|_{t=t_i^-} \in -N_C, \quad E_L|_{t=t_i^+} = E_L|_{t=t_i^-}. \tag{1}$$

Remark 1. We note that a negative sign in the previous equation appears as a consequence of the non-interpenetrability of the constraint, i.e., the mechanical system may not cross the boundary of the admissible variational constraint. We will see exactly how the negative signs appears in the following section.

Throughout the paper, L will be a regular mechanical Lagrangian, i.e., it has the form kinetic minus potential energy [2] and the Legendre transform $\mathbb{F}L : TQ \to T^*Q$ with $\mathbb{F}L(q, \dot{q}) = (q, \frac{\partial L}{\partial \dot{q}})$ is a local diffeomorphism.

3 Nonholonomic Systems with Inequality Constraints

Assume that there are velocity constraints imposed on the system. We will restrict to constraints that are linear in the velocities. Consider a distribution $\mathcal{D}$ on the configuration space Q describing these constraints, that is, $\mathcal{D}$ is a collection of linear subspaces of TQ ($\mathcal{D}_q \subset T_q Q$ for each $q \in Q$). A curve $q(t) \in Q$ will be said to satisfy the constraints if $\dot{q}(t) \in \mathcal{D}_{q(t)}$ for all t. Locally, the constraint distribution can be written as $\mathcal{D} = \{\dot{q} \in TQ | \mu_i^a(q)\dot{q}^i = 0, \quad a = 1, \dots, m\}$.

The Lagrange-d'Alembert equations of motion for the system are those determined by $\delta \int_a^b L(q, \dot{q})dt = 0$, where we choose variations $\delta q(t)$ of the curve $q(t)$ that satisfy $\delta q(a) = \delta(b) = 0$ and $\delta q(t) \in \mathcal{D}_{q(t)}$ for each $t \in [a, b]$. Note that here the curve $q(t)$ itself satisfies the constraints. Variations are taken before imposing the constraints and hence, the constraints are not imposed on the family of curves defining the variations.

The nonholonomic equations of motion are obtained from Lagrange-d'Alembert principle and its local expression is

$$\frac{d}{dt}\frac{\partial L}{\partial \dot{q}^i} - \frac{\partial L}{\partial q^i} = \lambda_a \mu_i^a, \quad \mu_i^a(q)\dot{q}^i = 0 \tag{2}$$

where λ_a is a Lagrange multiplier that might be computed using the constraints.

If C is an inequality constraint on the nonholonomic system, then Lagrange-d'Alembert equations are still valid in the interior of C. However, the jump conditions must now be changed to accommodate the constraints our system has on velocities as we will see in the following result.

Theorem 1. *Let $q : [0, h] \to Q$ be a nonholonomic trajectory of the nonholonomic system $(L, \mathcal{D})$ subjected to the inequality constraint $q(t) \in C$. Suppose that this system has an impact against the boundary ∂C at the time $t_i \in [0, h]$. Then the trajectory satisfies Lagrange-d'Alembert equations (2) in the intervals $[0, t_i^-[$ and $]t_i^+, h]$ and at the impact time t_i, the following conditions hold:*

$$\frac{\partial L}{\partial \dot{q}}|_{t=t_i^+} - \frac{\partial L}{\partial \dot{q}}|_{t=t_i^-} \in -N_C \cup \mathcal{D}^o, \quad E_L|_{t=t_i^+} = E_L|_{t=t_i^-}, \quad \dot{q}(t_i^+) \in \mathcal{D}_{q(t_i^+)}, \tag{3}$$

where $\mathcal{D}^o$ denotes the anihilator of the distribution $\mathcal{D}$.

Proof. The Lagrange-d'Alembert principle for systems with impacts is defined on the path space $\Omega = \{(c, t_i) \mid c : [0, h] \to Q$ is a smooth curve and $t_i \in \mathbb{R}\}$.

If the mapping $\mathcal{A} : \Omega \to \mathbb{R}$ is the action, then the Lagrange,d'Alembert principle states that the derivative of the action should annihilate all variations $(\delta q, \delta t_i)$ with $\delta q \in \mathcal{D}$. Since,

$$\delta \mathcal{A} = \int_0^{t_i^-} \left[\frac{\partial L}{\partial q^i} - \frac{d}{dt} \frac{\partial L}{\partial \dot{q}^i} \right] \delta q \, dt + \int_{t_i^+}^h \left[\frac{\partial L}{\partial q^i} - \frac{d}{dt} \frac{\partial L}{\partial \dot{q}^i} \right] \delta q \, dt - \left[\frac{\partial L}{\partial \dot{q}^i} \delta q + L \delta t_i \right]_{t_i^-}^{t_i^+}$$

the fact that Lagrange-d'Alembert equations hold on the intervals $[0, t_i^-[$ and $]t_i^+, h]$ follows from the application of the fundamental theorem of calculus of variations together with the fact that $\delta q \in \mathcal{D}$. The jump condition follows from the fact that $q(t_i) \in \partial C$ from where $\delta(q(t_i)) \in T(\partial C) \implies \delta q(t_i) + \dot{q}(t_i)\delta t_i \in T(\partial C)$.

The variations satisfying the previous equation are spanned by variations $\delta q(t_i) \in T(\partial C)$ and $\delta t_i = 0$ or $\delta t_i = 1$ and $\delta q(t_i) = -\dot{q}(t_i)$. From the latter we immediately deduce that $\left[\frac{\partial L}{\partial \dot{q}^i} \dot{q} - L \right]_{t_i^-}^{t_i^+} = 0$, which is the energy conservation condition in the jump equations. From $\delta t_i = 0$, we get that

$$\frac{\partial L}{\partial \dot{q}}\Big|_{t=t_i^+} - \frac{\partial L}{\partial \dot{q}}\Big|_{t=t_i^-} = \mathbb{F}L\Big|_{t=t_i^+} - \mathbb{F}L\Big|_{t=t_i^-}$$

annihilates δq if either it is on the annihilator of ∂C or it belongs to the annihilator of the distribution $\mathcal{D}$, since δq is in $T(\partial C) \cap \mathcal{D}$.

Now, in order to have $g(q(t)) \leqslant 0$ and since $g(q(t_i)) = 0$, we must have that $dg(\dot{q}(t_i^-)) \geqslant 0$ and $dg(\dot{q}(t_i^+)) \leqslant 0$, otherwise $q(t)$ would violate the inequality constraint. Noting that $(\partial C)^o$ is the union of N_C and $-N_C$, let us show that $\mathbb{F}L\big|_{t=t_i^+} - \mathbb{F}L\big|_{t=t_i^-}$ is not in N_C. Suppose it was on the normal cone then

$$\mathbb{F}L\big|_{t=t_i^+} - \mathbb{F}L\big|_{t=t_i^-} = \lambda dg(q_i), \ \lambda \geqslant 0.$$

This is equivalent to $\dot{q}(t_i^+) - \dot{q}(t_i^-) = \lambda(\mathbb{F}L)^{-1}(dg(q_i))$. Applying $dg(q_i)$ to both sides of the equation we get $dg(\dot{q}(t_i^+)) = dg(\dot{q}(t_i^-)) + \lambda dg(q_i)((\mathbb{F}L)^{-1}(dg(q_i)))$, where the right-hand side is greater or equal than 0, which is not possible. Therefore, $\mathbb{F}L\big|_{t=t_i^+} - \mathbb{F}L\big|_{t=t_i^-} \in -N_C$. This is precisely the first jump equation. The third one follows from the nonholonomic constraints. $\qquad\square$

Remark 2. The previous jump equations are in accordance with the equations obtained in [3] from Weierstrass-Erdemann conditions for impacts.

4   Nonholonomic Integrators for Systems with Inequality Constraints

We review here the formalism proposed in [7] (see also [6]) which gives rise to the discrete Lagrange-d'Alembert equations. Consider the discrete Lagrangian function $L_d : Q \times Q \times \mathbb{R} \rightarrow \mathbb{R}$ on the discrete velocity space $Q \times Q$. Let $\mathcal{D}$ be a distribution on Q and consider a discrete constraint space $\mathcal{D}_d \subseteq Q \times Q$ whose dimension agrees with that of the distribution $\mathcal{D}$ as a submanifold of TQ, $\dim\mathcal{D}_d = \dim\mathcal{D}$ and such that the diagonal set of $Q \times Q$ is contained in the discrete constraint space, $(q, q) \in \mathcal{D}_d$ for all $q \in Q$.

Then, the discrete Lagrange-d'Alembert principle asserts that the discrete flow is a critical value of the discrete action map $S_d : C_d^N(Q) \rightarrow \mathbb{R}$, which is given by $S_d(q_d) = \sum_{k=0}^{N-1} L_d(q_k, q_{k+1}, h)$, but this time we impose the restriction $\delta q_k \in \mathcal{D}_{q_k}$, that is, the infinitesimal variation of the sequence must lie in the constraint distribution. The Lagrange-d'Alembert principle states the following:

Definition 1 (Discrete Lagrange-d'Alembert principle). *The discrete flow of the discrete nonholonomic Lagrangian system determined by the discrete Lagrangian function L_d, the distribution $\mathcal{D}$ and the discrete constraint space $\mathcal{D}_d$ satisfies the constraint $(q_k, q_{k+1}) \in \mathcal{D}_d$ for all $k \in \{0, ..., N-1\}$ and is a critical value of the discrete action map S_d among all variations of sequences with fixed end-points whose infinitesimal variations satisfy $\delta q_k \in \mathcal{D}_{q_k}$.*

As it happens with its continuous counterpart, the application of the discrete Lagrange-d'Alembert principle leads to a set of equations which will be the necessary and sufficient conditions to find critical values subordinated to the imposed restrictions. Assume in the following that $\mu^a \in \Omega^1(Q)$ where $a = 1, ..., n-k$ are 1-forms on Q defining the distribution $\mathcal{D} = \{v \in TQ \mid \mu^a(v) = 0\}$ and μ_d^a are a set of $n-k$ functions on $Q \times Q$ whose zero set is the discrete constraint space $\mathcal{D}_d$.

A sequence $\{q_k\}_{k=1}^N$ of points in Q satisfies the discrete-Lagrange d'Alembert principle for the triple $(L_d, \mathcal{D}, \mathcal{D}_d)$ if and only if it satisfies the equations

$$D_2 L_d(q_{k-1}, q_k, h) + D_1 L_d(q_k, q_{k+1}, h) = \lambda_a \mu^a(q_k), \quad \mu_d^a(q_k, q_{k+1}) = 0. \tag{4}$$

4.1 Discrete Equations with Inequality Constraints

Consider a sequence of points $\{q_k\}_{k=0}^N$ contained in the inequality constraint set C. This sequence shall be considered as a discretization of a continuous smooth curve $q : [0, Nh] \rightarrow C$ satisfying $q(kh) = q_k$. In fact, since we will also need the time sequence we will use the notation $t_k = kh$. Now suppose that this curve has an impact against the boundary of C at the point $\bar{q} \in \partial C$ and that this impact occurs at time $\bar{t} := t_{i-1} + \alpha h$, for some $\alpha \in]0, 1[$ and $i \in \{1, ..., N\}$, so that $t_{i-1} < \bar{t} < t_i$. We will also use the notation $q_d : \{t_0, ..., t_{i-1}, \bar{t}, t_i, ..., t_N\} \rightarrow Q$ to denote the sequence $\{q_k\}_{k=0}^N \cup \{\bar{q}\}$ in functional notation.

In the following we will consider the discrete path space $\mathcal{M}_d$ formed by sequences such as the one described in the last paragraph:

$$\mathcal{M}_d =]0, 1[\times \{q_d : \{t_0, ..., t_{i-1}, \bar{t}, t_i, ..., t_N\} \rightarrow Q \mid \bar{q} \in \partial C \}.$$

This discrete path space is actually a manifold since it is isomorphic to $]0, 1[\times Q \times \cdots \times \partial C \times \cdots \times Q$.

To obtain the discrete equations of motion of a nonholonomic system under inequality constraints we must find the number $\alpha \in]0, 1[$ and the sequence q_d such that the differential of the action $S_d : \mathcal{M}_d \to \mathbb{R}$, given by

$$S_d(\alpha, q_d) = \sum_{k=0}^{i-2} L_d(q_k, q_{k+1}, h) + \sum_{k=i}^{N-1} L_d(q_k, q_{k+1}, h) +$$
$$L_d(q_{i-1}, \bar{q}, \alpha h) + L_d(\bar{q}, q_i, (1-\alpha)h)$$

annihilates variations $(\delta\alpha, \delta q_d) \in T\mathcal{M}_d$ satisfying $\delta q_d \in \mathcal{D}$, i.e., $\delta q_i, \delta\bar{q} \in \mathcal{D}$, where $\delta q_d = (\delta q_1, \ldots, \delta q_{i-1}, \delta\bar{q}, \delta q_i, \ldots, \delta q_N)$, $\delta q_0 = \delta q_N = 0$ and $(q_{i-1}, \bar{q})$, $(\bar{q}, q_i)$, $(q_k, q_{k+1}) \in \mathcal{D}_d$ for all $k \neq i-1$.

Theorem 2. *Let $\{q_k\}$ be a nonholonomic discrete trajectory of the nonholonomic system $(L_d, \mathcal{D}, \mathcal{D}_d)$ subjected to the inequality constraint $q_k \in C$. Suppose that this system has an impact against the boundary ∂C at the time $\bar{t} \in [0, Nh]$. Then the trajectory satisfies discrete Lagrange-d'Alembert equations (4) for $k \neq i-1, i$, and at the impact time $\bar{t}$, the following conditions hold:*

$$D_2 L_d(q_{i-2}, q_{i-1}, h) + D_1 L_d(q_{i-1}, \bar{q}, \alpha h) = \lambda_a \mu^a(q_{i-1})$$
$$D_2 L_d(\bar{q}, q_i, (1-\alpha)h) + D_1 L_d(q_i, q_{i+1}, h) = \tilde{\lambda}_a \mu^a(q_i)$$
$$D_2 L_d(q_{i-1}, \bar{q}, \alpha h) + D_1 L_d(\bar{q}, q_i, (1-\alpha)h) \in -N_C(\bar{q}) \cup \mathcal{D}_{\bar{q}}^o \qquad (5)$$
$$D_3 L_d(q_{i-1}, \bar{q}, \alpha h) - D_3 L_d(\bar{q}, q_i, (1-\alpha)h) = 0$$
$$\bar{q} \in \partial C, \ (q_{i-1}, \bar{q}), (\bar{q}, q_i), (q_i, q_{i+1}) \in \mathcal{D}_d.$$

Proof. First, we compute the variations of the action map and using $\delta q_0 = \delta q_N = 0$, we obtain that

$$\delta S_d(\alpha, q_d) \cdot (\delta\alpha, \delta q_d) = \sum_{k=1}^{i-2} [D_2 L_d(q_{k-1}, q_k, h) + D_1 L_d(q_k, q_{k+1}, h)] \delta q_k$$
$$+ \sum_{k=i+1}^{N-1} [D_2 L_d(q_{k-1}, q_k, h) + D_1 L_d(q_k, q_{k+1}, h)] \delta q_k$$
$$+ [D_2 L_d(q_{i-2}, q_{i-1}, h) + D_1 L_d(q_{i-1}, \bar{q}, \alpha h)] \delta q_{i-1}$$
$$+ [D_2 L_d(q_{i-1}, \bar{q}, \alpha h) + D_1 L_d(\bar{q}, q_i, (1-\alpha)h)] \delta\bar{q}$$
$$+ [D_2 L_d(\bar{q}, q_i, (1-\alpha)h) + D_1 L_d(q_i, q_{i+1}, h)] \delta q_i$$
$$+ h [D_3 L_d(q_{i-1}, \bar{q}, \alpha h) - D_3 L_d(\bar{q}, q_i, (1-\alpha)h)] \delta\alpha.$$

Using the facts that $\delta q_d \in \mathcal{D}$ and $(q_k, q_{k+1}) \in \mathcal{D}_d$ for all $k \neq i-1$, we immediately get Lagrange-d'Alembert equations (4) for $k \neq i-1, i$ and the first two equations in (5). From $\delta\bar{q} \in \mathcal{D} \cap T(\partial C)$ we conclude that

$$D_2 L_d(q_{i-1}, \bar{q}, \alpha h) + D_1 L_d(\bar{q}, q_i, (1-\alpha)h) \in \mathcal{D}^o \cup (-N_C),$$

where we used the fact that the jump during the impact must produce a new point $q_i \in C$. Finally, since $\delta\alpha$ is arbitrary we get the last equation in (5).   □

Remark 3. The Lagrange-d'Alembert equations in the inequality constraint setting may be used as a numerical method to integrate the equations of motion. Given two initial points q_0, q_1 satisfying the discrete constraint $\mathcal{D}_d$, we may use discrete Lagrange-d'Alembert equations to obtain the sequence $\{q_0, \ldots, q_{i-1}\}$. Then we use the first equation in (5) to obtain $\bar{q}$ from where we may use the third to obtain q_i and then the second to obtain q_{i+1}. Then, we may use again Lagrange-d'Alembert equations to integrate the remaining points.

5   Example

We will consider the motion of a vertical rolling disk without sliding in a circular table and use our previous construction to find an integrator for the impact time.

The vertical rolling disk is described by four coordinates: x, y determine the position of the center of mass in the table, θ indicates the angle that a fixed point in the disk border makes with the vertical axis and φ indicates the orientation of the disk with respect to the x-axis. Below m is the mass of the disk, I and J are its moments of inertia and R is the disk radius. The dynamics of the vertical rolling disk with unit radius is given by the Lagrangian function

$$L = \frac{m}{2}\left(\dot{x}^2 + \dot{y}^2\right) + \frac{I}{2}\dot{\theta}^2 + \frac{J}{2}\dot{\varphi}^2$$

together with the non-slipping constraints $\dot{x} = R\dot{\theta}\cos\varphi$, $\dot{y} = R\dot{\theta}\sin\varphi$ generating the distribution $\mathcal{D} = \left\langle\left\{\frac{\partial}{\partial\theta} + R\cos\varphi\frac{\partial}{\partial x} + R\sin\varphi\frac{\partial}{\partial y}, \frac{\partial}{\partial\varphi}\right\}\right\rangle$.

The non-interpenetrability condition of the circular table with radius $a > 0$ implies the inequality constraints C_+ and C_- determined by

$$C_\pm = \{(x, y, \theta, \varphi) \mid (x \pm R\cos\varphi)^2 + (y \pm R\sin\varphi)^2 \leqslant a^2\}$$

which express the fact that the disk, counting with its radius, cannot leave the table. The proposed integrator uses the discrete Lagrange-d'Alembert integrator until the first impact. When we first obtain a non-admissible solution, we switch to the impact equations to determine the exact impact point. Afterwards, we return to the DLA scheme until the next impact. The discrete Lagrangian used is

$$L_d = \frac{m}{2h}\left((x_1 - x_0)^2 + (y_1 - y_0)^2\right) + \frac{I}{2h}(\theta_1 - \theta_0)^2 + \frac{J}{2h}(\varphi_1 - \varphi_0)^2$$

and the discrete constraint was

$$x_1 - x_0 = R\cos(\frac{\varphi_1 + \varphi_0}{2})(\theta_1 - \theta_0), \quad y_1 - y_0 = R\sin(\frac{\varphi_1 + \varphi_0}{2})(\theta_1 - \theta_0).$$

Near an impact point with, for instance, ∂C_+ the first equation in (5) is used together with the constraints $\bar{q} \in \partial C_+$ and $(q_{i-1}, \bar{q}) \in \mathcal{D}_d$ to obtain $\bar{q}, \alpha$ and the multiplier λ_a. Next, using these variables, we use the third and fourth equations in (5), together with the constraint $(\bar{q}, q_i) \in \mathcal{D}_d$ to obtain q_i and the multipliers. Finally, we use the previous variables to obtain q_{i+1} and the multiplier $\tilde{\lambda}_a$ from the second equation in (5) together with the discrete constraints.

Our integrator behaves as expected before the first impact. During the impact time, it is able to deal successfully with the impact and produce an admissible trajectory. However, it has proven to be unable to preserve energy thus introducing artificial energy drift into the problem.

This fact shows that further study should be made to look for algorithms that preserve the impact structure. In particular, a promising direction is the extension of DELI integrators to nonholonomic systems.

References

1. Anahory Simoes, A., López-Gordón, A., Bloch, A., Colombo, L.:Discrete Mechanics and Optimal Control for Passive Walking with Foot Slippage (2022). arXiv preprint arXiv:2209.14255
2. Bloch, A.M.: Nonholonomic mechanics. In: Nonholonomic Mechanics and Control. IAM, vol. 24, pp. 207–276. Springer, New York (2003). https://doi.org/10.1007/b97376_5
3. Clark, W., Bloch, A.: The bouncing penny and nonholonomic impacts. In: 2019 IEEE 58th Conference on Decision and Control (CDC), pp. 2114–2119. IEEE (2019)
4. Cortés, J., de León, M., Martín de Diego, D. and Martínez, S., 2001. Mechanical systems subjected to generalized non-holonomic constraints. Proceedings of the Royal Society of London. Series A: Mathematical, Physical and Engineering Sciences, 457(2007), pp. 651–670
5. Brogliato, B., Tanwani, A.: Dynamical systems coupled with monotone set-valued operators: formalisms, applications, well-posedness, and stability. SIAM Rev. **62**(1), 3–129 (2020)
6. Cortés Monforte, J.: Geometric, Control and Numerical Aspects of Nonholonomic Systems. LNM, vol. 1793. Springer, Heidelberg (2002). https://doi.org/10.1007/b84020
7. Cortés, J., Martinez. S.: Nonholonomic integrators. Nonlinearity **14**, 1365–1392 (2001)
8. de León, M., de Diego, D.M.: On the geometry of non-holonomic Lagrangian systems. J. Math. Phys. 37(7), 3389–3414 (1996)
9. Fetecau, R.C., Marsden, J.E., Ortiz, M., West, M.: Nonsmooth Lagrangian mechanics and variational collision integrators. SIAM J. Appl. Dyn. Syst. **3**(2), 381–416 (2003)
10. Kaufman, D.M., Pai, D.K.: Geometric numerical integration of inequality constrained, nonsmooth Hamiltonian systems. SIAM J. Sci. Comput. **34**(5), A2670–A2703 (2012)
11. López-Gordón, A., Colombo, L., de León, M.: Nonsmooth Herglotz variational principle. arXiv preprint arXiv:2208.02033 (2022)
12. Nemark, J.I., Fufaev, N.A.: Dynamics of nonholonomic systems. Am. Math. Soc. **33** (2004)

Nonholonomic Brackets: Eden Revisited

Manuel de León[1,2]([envelope])[ID], Manuel Lainz[1][ID], Asier López-Gordón[1][ID],
and Juan Carlos Marrero[3][ID]

[1] Instituto de Ciencias Matemáticas (CSIC-UAM-UC3M-UCM), Madrid, Spain
mdeleon@icmat.es
[2] Real Academia de Ciencias Exactas, Físicas y Naturales, Madrid, Spain
[3] ULL-CSIC Geometría Diferencial y Mecánica Geométrica, Departamento de
Matemáticas Estadística e Investigación Operativa and Instituto de Matemáticas y
Aplicaciones (IMAULL), University of La Laguna, San Cristóbal de La Laguna, Spain

Abstract. The nonholonomic dynamics can be described by the so-
called nonholonomic bracket in the constrained submanifold, which is a
non-integrable modification of the Poisson bracket of the ambient space,
in this case, of the canonical bracket in the cotangent bundle of the configuration manifold. This bracket was defined in [2,10], although there
was already some particular and less direct definition. On the other hand,
another bracket, also called noholonomic, was defined using the description of the problem in terms of skew-symmetric algebroids. Recently,
reviewing two older papers by R. J. Eden, we have defined a new bracket
which we call Eden bracket. In the present paper, we prove that these
three brackets coincide. Moreover, the description of the nonholonomic
bracket à la Eden has allowed us to make important advances in the
study of Hamilton-Jacobi theory and the quantization of nonholonomic
systems.

Keywords: Nonholonomic systems · Nonholonomic brackets · Eden
bracket · skew-symmetric algebroids

1 Introduction

One of the most important objects in mechanics is the Poisson bracket, which
allows us to obtain the evolution of an observable by bracketing it with the Hamiltonian energy, or to obtain new conserved quantities of two given ones, using the
Jacobi identity satisfied by the bracket. Moreover, the Poisson bracket is fundamental to proceed with the quantization of the system using what Dirac called
the analogy principle, also known as the correspondence principle, according to

M. de León, M. Lainz and A. López-Gordón—Acknowledge financial support from
Grants PID2019-106715GB-C21 and CEX2019-000904-S funded by MCIN/AEI/
10.13039/501100011033. Asier López-Gordón—would also like to thank MCIN for the
predoctoral contract PRE2020-093814. J. C. Marrero—Ackowledges financial support
from the Spanish Ministry of Science and Innovation and European Union (Feder)
Grant PGC2018-098265-B-C32.

© The Author(s), under exclusive license to Springer Nature Switzerland AG 2023
F. Nielsen and F. Barbaresco (Eds.): GSI 2023, LNCS 14072, pp. 105–112, 2023.
https://doi.org/10.1007/978-3-031-38299-4_12

which the Poisson bracket becomes the commutator of the operators associated to the quantized observables.

For a long time, no similar concept has existed in the case of nonholonomic mechanical systems, until van der Schaft and Maschke [11] introduced a bracket similar to the canonical Poisson but without the benefit of integrability. In [2,3] (see also [10]), we have developed a geometric and very simple way to define nonholonomic brackets, in the time-dependent as well time-independent cases. Indeed, it is possible to decompose the tangent bundle and the cotangent bundle along the constraint submanifold in two different ways. Both result in that the nonholonomic dynamics can be obtained by projecting the free dynamics, and furthermore, if we evaluate by the canonical symplectic form the projections of the Hamiltonian fields of two functions on the configuration manifold (prior arbitrary extensions to the whole cotangent), two non-integrable brackets are obtained. The first decomposition is due to de León and Martín de Diego [4], and the second one to Bates and Sniatycki [1]. The advantage of this second decomposition is that it turns out to be symplectic, and it is the one we will use in the present paper. In any case, we proved that both brackets coincided on the submanifold of constraints [2].

On the other hand, by studying the Hamilton–Jacobi equation, we develop a description of nonholonomic mechanics in the setting of skew-symmetric (or almost Lie) algebroids. Note that the "almost" is due to the lack of integrability of the distribution determining the constraints, showing the consistency of the description. So, in [5] we define a new almost Poisson bracket that we also called nonholonomic. So far, although both nonholonomic brackets have been used in these two different contexts as coinciding, no such proof has ever been published. This paper provides this evidence for the first time.

But the issue does not end there. In 1951, R. J. Eden, who wrote his doctoral thesis on nonholonomic mechanics under the direction of P. A. M. Dirac (see his papers [7,8]). In the first paper, Eden introduced an intriguing γ operator that mapped free states to constrained states. With that operator (a kind of tensor of type $(1,1)$ that has the properties of a projector), Eden obtained the equations of motion, could calculate brackets of all observables, obtained a simple Hamilton–Jacobi equation, and even used it to construct a quantization of the nonholonomic system. These two papers by Eden have had little impact despite their relevance. Firstly, because they were written in terms of coordinates that made their understanding difficult, and secondly, because it was in the 1980s when the study of non-holonomic systems became part of the mainstream of geometric mechanics.

Recently, we have carefully studied these two papers by Eden, and realized that the operator γ is nothing else a projection defined by the orthogonal decomposition of the cotangent bundle provided by the Riemannian metric given by the kinetic energy. So, we have defined a new bracket that we call Eden bracket, and proved that coincides with the previous nonholonomic brackets.

2 Lagrangian and Hamiltonian Mechanics: A Brief Survey

2.1 Lagrangian Mechanics

Let $L : TQ \longrightarrow \mathbb{R}$ be a Lagrangian function, where Q is a configuration n-dimensional manifold. Then, $L = L(q^i, \dot{q}^i)$, where (q^i) are coordinates in Q and $(q^i, \dot{q}^i)$ are the induced bundle coordinates in TQ. We denote by $\tau_Q : TQ \longrightarrow Q$ the canonical projection such that $\tau_Q(q^i, \dot{q}^i) = (q^i)$.

We will assume that L is regular, that is, the Hessian matrix $(\partial^2 L/\partial \dot{q}^i \partial \dot{q}^j)$ is regular. Using the canonical endomorphism $S = dq^i \otimes \frac{\partial}{\partial \dot{q}^i}$ on TQ one can construct a 1-form $\lambda_L = S^*(dL)$, and the 2-form $\omega_L = -d\lambda_L$. Then, ω_L is symplectic if and only if L is regular [6].

Consider now the vector bundle isomorphism $\flat_L : T(TQ) \to T^*(TQ)$ given by $\flat_L(v) = i_v \omega_L$. We define the Hamiltonian vector field $\xi_L = X_{E_L}$ by $\flat_L(\xi_L) = dE_L$, where $E_L = \Delta(L) - L$ is the energy. Now, if $(q^i(t), \dot{q}^i(t))$ is an integral curve of ξ_L, then it satisfies the usual Euler-Lagrange equations

$$\frac{d}{dt}\left(\frac{\partial L}{\partial \dot{q}^i}\right) - \frac{\partial L}{\partial q^i} = 0. \tag{1}$$

2.2 Legendre Transformation

Let us recall that the Legendre transformation $FL : TQ \longrightarrow T^*Q$ is a fibred mapping (that is, $\pi_Q \circ FL = \tau_Q$, where $\pi_Q : T^*Q \longrightarrow Q$ denotes the canonical projection of the cotangent bundle of Q). Indeed, FL is the fiber derivative of L. In local coordinates, we have

$$FL(q^i, \dot{q}^i) = (q^i, p_i), \ p_i = \frac{\partial L}{\partial \dot{q}^i},$$

and thus L is regular if and only if FL is a local diffeomorphism.

Along this paper we will assume that FL is, in fact, a global diffeomorphism (in other words, L is hyperregular) which is the case when L is a Lagrangian of mechanical type, say $L = T - V$, where T is the kinetic energy defined by a Riemannian metric on Q, and $V : Q \longrightarrow \mathbb{R}$ is a potential energy.

2.3 Hamiltonian Description

The Hamiltonian counterpart is developed in the cotangent bundle T^*Q of Q. Denote by $\omega_Q = dq^i \wedge dp_i$ the canonical symplectic form, where (q^i, p_i) are the canonical coordinates on T^*Q.

The Hamiltonian energy is just $H = E_L \circ (FL)^{-1}$ and the Hamiltonian vector field is the solution of the symplectic equation

$$i_{X_H} \omega_Q = dH.$$

As it is well-known, the integral curves $(q^i(t), p_i(t))$ of X_H satisfy the Hamilton equations

$$\dot{q}^i = \frac{\partial H}{\partial p_i}, \quad \dot{p}_i = -\frac{\partial H}{\partial q^i}. \tag{2}$$

Since $FL^* \omega_Q = \omega_L$, we deduce that ξ_L and X_H are FL-related, and consequently FL transforms the solutions of the Euler-Lagrange equations (1) into the solutions of the Hamilton equations (2).

On the other hand, we can define a bracket of functions

$$\{\, ,\, \}_{can} : C^\infty(T^*Q) \times C^\infty(T^*Q) \longrightarrow C^\infty(T^*Q)$$

as follows

$$\{F, G\}_{can} = \omega_Q(X_F, X_G) = X_G(F) = -X_F(G).$$

This bracket is a Poisson bracket, that is, $\{\, ,\, \}_{can}$ is $\mathbb{R}$-bilinear and:

- it is skew-symmetric: $\{G, F\}_{can} = -\{F, G\}_{can}$,
- it satisfies the Leibniz rule: $\{FF', G\}_{can} = F\{F', G\}_{can} + F'\{F, G\}_{can}$, and
- it satisfies the Jacobi identity:

$$\{F, \{G, H\}_{can}\}_{can} + \{G, \{H, F\}_{can}\}_{can} + \{H, \{F, G\}_{can}\}_{can} = 0.$$

3   Nonholonomic Mechanical Systems

3.1   The Lagrangian Description

A **nonholonomic mechanical system** is a quadruple (Q, g, V, D) where Q is the configuration manifold of dimension n; g is a Riemannian metric on Q; V is a potential function, $V \in C^\infty(Q)$, and D is a non-integrable distribution of rank $k < n$ on Q.

As in Sect. 2.2, the metric g and the potential energy V define a Lagrangian function of mechanical type $L(v_q) = \frac{1}{2} g_q(v_q, v_q) - V(q)$.

The nonholonomic dynamics is provided by the Lagrangian L subject to the nonholonomic constraints given by D; that means that the permitted velocities should belong to D.

The nonholonomic problem is to find the equations of motion

$$\frac{d}{dt}\left(\frac{\partial L}{\partial \dot{q}^i}\right) - \frac{\partial L}{\partial q^i} = \lambda_\alpha \mu_i^\alpha(q), \quad \mu_i^\alpha(q)\dot{q}^i = 0, \tag{3}$$

where $\{\mu^\alpha\}$ is a local basis of D° (the annihilator of D) such that $\mu^\alpha = \mu_i^\alpha \, dq^i$. Here, λ_α are Lagrange multipliers to be determined.

A geometric description of the above equations can be obtained using the symplectic form ω_L and the vector bundle of 1-forms, F, defined by $F = \tau_Q^*(D^\circ)$. So, the above equations are equivalent to these ones

$$i_X \omega_L - dE_L \in \tau_Q^*(D^\circ), \quad X \in TD. \tag{4}$$

These equations have a unique solution, ξ_{nh}, which is called the nonholonomic vector field.

The Riemannian metric g induces a isomorphism of vector bundles between TQ and T^*Q given by $\flat_g(q)(v_q) = i_{v_q}g$ (which again induces an isomorphism between vector fields and 1-forms). The inverse of $\flat_g$ will be denoted by $\sharp_g$.

We can define the orthogonal complement, $D^{\perp_g}$, of D with respect to g, as follows:

$$D_q^{\perp_g} = \{v_q \in T_qQ \mid g(v_q, w_q) = 0, \forall w_q \in D\}.$$

$D^{\perp_g}$ is again a distribution on Q, or, if we prefer, a vector sub-bundle of TQ such that we have the Whitney sum

$$TQ = D \oplus D^{\perp_g}. \tag{5}$$

3.2   The Hamiltonian Description

We can obtain the Hamiltonian description of the nonholonomic system (Q, g, V, D) using the Legendre transformation FL, which in our case coincides with the isomorphism $\flat_g$ associated to the metric g.

So, we can define the corresponding Hamiltonian function $H : T^*Q \longrightarrow \mathbb{R}$, $H = E_L \circ (FL)^{-1}$, and constraint submanifold $M = FL(D) = \flat_g(D) = (D^{\perp_g})^\circ$. Therefore, we obtain a new orthogonal decomposition (or Whitney sum)

$$T^*Q = M \oplus D^\circ, \tag{6}$$

since $FL(D^{\perp_g}) = \flat_g(D^{\perp_g}) = D^\circ$. This decomposition is orthogonal with respect to the induced metric on tangent covectors, and it is the translation of (5) to the Hamiltonian side. Again, M and D° are vector sub-bundles of $\pi_Q : T^*Q \longrightarrow Q$ over Q. We have the following canonical inclusion $i_M : M \longrightarrow T^*Q$ and orthogonal projection $\gamma : T^*Q \longrightarrow M$, respectively.

The equations of motion for the nonholonomic system on T^*Q can now be written as follows:

$$\dot{q}^i = \frac{\partial H}{\partial p_i}, \quad \dot{p}_i = -\frac{\partial H}{\partial q^i} - \bar{\lambda}_\alpha \mu_j^\alpha g^{ij}, \tag{7}$$

together with the constraint equations $\mu_i^\alpha g^{ij} p_j = 0$. Notice that here the $\bar{\lambda}_\alpha$ are Lagrange multipliers to be determined.

Now the vector bundle of constrained forces generated by the 1-forms $\tau_Q^*(\mu^\alpha)$, can be translated to the cotangent side and we obtain the vector bundle generated by the 1-forms $\pi_Q^*(\mu^\alpha)$, say $\pi_Q^*(D^\circ)$. Therefore, the nonholonomic Hamilton equations for the nonholonomic system can be then rewritten in an intrinsic form as

$$(i_X \omega_Q - dH)_{|M} \in \pi_Q^*(D^\circ), \quad X_{|M} \in TM. \tag{8}$$

These equations have a unique solution, X_{nh}, which is called the nonholonomic vector field. Of course, X_{nh} and ξ_{nh} are related by the Legendre transformation restricted to D, say, $T(FL)_{|D}(\xi_{nh}) = X_{nh} \circ (FL)_D$.

3.3 The Skew-Symmetric Algebroid Approach

In [5] (see also [9]) we have developed an approach to nonholonomic mechanics based on the skew-symmetric algebroid setting.

We denote by $i_D : D \longrightarrow TQ$ the canonical inclusion. The canonical projection given by the decomposition $TQ = D \oplus D^\perp$ on D is denoted by $P : TQ \longrightarrow D$.

Then, the vector bundle $(\tau_Q)|_D : D \longrightarrow Q$ is a skew-symmetric algebroid. The anchor map is just the canonical inclusion $i_D \longrightarrow TQ$, and the skew-symmetric bracket $\|\,,\,\|$ on the space of sections $\Gamma(D)$ is given by

$$\|X, Y\| = P([X, Y]), \quad \text{for } X, Y \in \Gamma(D).$$

Here, $[\,,\,]$ is the standard Lie bracket of vector fields.

We also have the vector bundle morphisms provided by the adjoint operators:

$$i_D^* : T^*Q \longrightarrow D^*, P^* : D^* \longrightarrow T^*Q, \tag{9}$$

where D^* is the dual vector bundle of D.

We define now an almost Poisson bracket on D^* as follows (see [5]):

$$\{\,,\,\}_{D^*} : C^\infty(D^*) \times C^\infty(D^*) \longrightarrow C^\infty(D^*)$$
$$\{f, g\}_{D^*} = \{f \circ i_D^*, g \circ i_D^*\}_{can} \circ P^*.$$

The bracket $\{\,,\,\}_{D^*}$ has the same properties as a Poisson bracket except maybe the Jacobi identity, that is, $\{\,,\,\}_{D^*}$ is $\mathbb{R}$-bilinear, skew symmetric and satisfies the Leibniz rule.

Moreover, if $FL_{nh} : D \longrightarrow D^*$ is the nonholonomic Legendre transformation given by $FL_{nh} = i_{D^*} \circ FL \circ i_D$ and Y_{nh} is the nonholonomic dynamics in D^*, then

$$T(FL_{nh})(\xi_{nh}) = Y_{nh} \circ FL_{nh}.$$

Hence, the bracket $\{\,,\,\}_{D^*}$ may be used to give the evolution of an observable $f \in C^\infty(D^*)$. In fact, if $h : D^* \longrightarrow \mathbb{R}$ is the constrained Hamiltonian function defined by

$$h = (E_L)|_D \circ (FL_{nh})^{-1},$$

we have that

$$\dot{f} = Y_{nh}(f) = \{f, h\}_{D^*}, \forall f \in C^\infty(D^*).$$

4 The Nonholonomic Bracket

Consider the vector sub-bundle $T^D M$ over M defined by

$$T^D M = \{Z \in TM \mid T\pi_Q(Z) \in D\}.$$

As we know [1,2], $T^D M$ is a symplectic vector sub-bundle of the symplectic vector bundle $(T_M(T^*Q), \omega_Q)$, where we are denoting again by ω_Q the restriction to any fiber of $T_M(T^*Q)$. Thus, we have the following symplectic decomposition

$$T_M(T^*Q) = T^D M \oplus (T^D M)^{\perp_{\omega_Q}}, \tag{10}$$

where $(T^D M)^{\perp_{\omega_Q}}$ denotes the symplectic complement of $T^D M$. Therefore, we have associated projections

$$\mathcal{P} : T_M(T^*Q) \longrightarrow T^D M, \quad \mathcal{Q} : T_M(T^*Q) \longrightarrow (T^D M)^{\perp_{\omega_Q}}.$$

One of the most relevant applications of the above decomposition is that

$$X_{nh} = \mathcal{P}(X_H), \text{ along } M.$$

In addition, the above decomposition allows us to define the so-called nonholonomic bracket as follows. Given $f, g \in C^\infty(M)$, we set

$$\{f, g\}_{nh} = \omega_Q(\mathcal{P}(X_{\tilde{f}}), \mathcal{P}(X_{\tilde{g}})) \circ i_M, \tag{11}$$

where $i_M : M \longrightarrow T^*Q$ is the canonical inclusion, and $\tilde{f}, \tilde{g}$ are arbitrary extensions to T^*Q of f and g, respectively (see [2,10]). Since the decomposition (10) is symplectic, one can equivalently write

$$\{f, g\}_{nh} = \omega_Q(X_{\tilde{f}}, \mathcal{P}(X_{\tilde{g}})) \circ i_M. \tag{12}$$

Remark 1. Notice that $f \circ \gamma$ and $g \circ \gamma$ are natural extensions of f and g to T^*Q, so we can also define the above nonholonomic bracket as follows

$$\{f, g\}_{nh} = \omega_Q(X_{f \circ \gamma}, \mathcal{P}(X_{g \circ \gamma})) \circ i_M. \tag{13}$$

It is worth noting that $\{\,,\,\}_{nh}$ is an almost Poisson bracket. Moreover, it may be used to give the evolution of an observable, namely,

$$\dot{f} = X_{nh}(f) = \{\, f, H \circ i_M \,\}_{nh}.$$

5 Eden Bracket

Using the projector $\gamma : T^*Q \longrightarrow M$, we can define an almost Poisson bracket, called the **Eden bracket**, on M as follows:

$$\{\,,\,\}_E : C^\infty(M) \times C^\infty(M) \longrightarrow C^\infty(M)$$
$$\{f, g\}_E = \{f \circ \gamma, g \circ \gamma\}_{can} \circ i_M. \tag{14}$$

It may be used to give the evolution of an observable, namely,

$$\dot{f} = \{\, f, H \circ i_M \,\}_E.$$

6 Comparison of Brackets

First of all, one can prove that the almost Poisson brackets defined on D^* and M are isomorphic.

Theorem 1. *The vector bundle isomorphism*

$$i_{M,D^*} : M \longrightarrow D^*$$

given by the composition

$$i_{M,D^*} = i_D^* \circ i_M$$

is an almost Poisson isomorphism between the almost Poisson manifolds $(M, \{\, , \}_E)$ *and* $(D^*, \{\, , \}_{D^*})$.

Additionally, one can prove that the Eden bracket is just the nonholonomic bracket. First if follows

Proposition 1. *We have*

- $\mathcal{P}(Z) = T\gamma(Z)$ *for every* $Z \in T_M^D(T^*Q) = \{Y \in T_M(T^*Q) \mid (T\pi_Q)(Y) \in D\})$.
- *For any function* $f \in C^\infty(M)$, *we have* $T\pi_Q(x)(X_{f \circ \gamma}) \in D_{\pi_Q(x)}$ *for every* $x \in M$. *In consequence,* $X_{f \circ \gamma}(x) \in T_x^D(T^*Q), \forall x \in M$.

Theorem 2. *The nonholonomic bracket* $\{\, , \}_{nh}$ *on* M *coincides with the Eden's bracket; in other words, the identity map*

$$\mathrm{id} \colon (M, \{\, , \}_{nh}) \longrightarrow (M, \{\, , \}_E)$$

is a skew-symmetric (or almost Poisson) isomorphism.

References

1. Bates, L., Sniatycki, J.: Nonholonomic reduction. Rep. Math. Phys. **32**, 99–115 (1992)
2. Cantrijn, F., de León, M., de Diego, D.M.: On almost-Poisson structures in nonholonomic mechanics. Nonlinearity. **12**(3), 721–737 (1999)
3. Cantrijn, F., de León, M., Marrero, J.C., de Diego, D.M.: On almost-Poisson structures in nonholonomic mechanics II. The time-dependent framework. Nonlinearity. **13**(4), 1379–1409 (2000)
4. de León, M., Martín de Diego, D.: On the geometry of non-holonomic Lagrangian systems. J. Math. Phys. **37**(7), 3389–3414 (1996)
5. de León, M., Marrero, J.C., de Diego, D.M.: Linear almost Poisson structures and Hamilton-Jacobi equation. Appl. Nonholon. Mech. J. Geom. Mech. **2**(2), 159–198 (2010)
6. de León, M., Rodrigues, P.R.: Methods of differential geometry in analytical mechanics. North-Holland Mathematics Studies, 158. North-Holland Publishing Co., pp. x+483. Amsterdam (1989). ISBN: 0-444-88017-8
7. Eden, R.J.: The Hamiltonian dynamics of non-holonomic systems. Proc. Roy. Soc. London Ser. A **205**, 564–583 (1951)
8. Eden, R.J.: The quantum mechanics of non-holonomic systems. Proc. Roy. Soc. London Ser. A **205**, 583–595 (1951)
9. Grabowski, J., de León, M., Marrero, J.C., de Diego, D.M.: Nonholonomic constraints: a new viewpoint. J. Math. Phys. **50**(1), 17, 013520 (2009)
10. Ibort, A., de Leon, M., Marrero, J.C., de Diego, D.M.: Dirac brackets in constrained dynamics. Fortschr. Phys. **47**(5), 459–492 (1999)
11. van der Schaft, A.J., Maschke, B.M.: On the Hamiltonian formulation of nonholonomic mechanical systems. Rep. Math. Phys. **34**, 225–33 (1994)

Symplectic Structures of Heat and Information Geometry

The Momentum Mapping of the Affine Real Symplectic Group

Richard Cushman[(✉)] [ID]

University of Calgary, Calgary AB T2N 1N4, Canada
r.h.cushman@gmail.com

Abstract. In this paper we explain how the cocycle of the momentum map of the action of the affine symplectic group on $\mathbb{R}^{2n}$ gives rise to a coadjoint orbit of the odd real symplectic group with a modulus.

Keywords: momentum map · cocycle · modulus of coadjoint orbit

1 Basic Setup

Consider the set $\mathrm{AfSp}(\mathbb{R}^{2n}, \omega)$ of invertible affine real symplectic mappings

$$(A, a) : (\mathbb{R}^{2n}, \omega) \to (\mathbb{R}^{2n}, \omega) : v \mapsto Av + a.$$

Using the multiplication $(A, a) \cdot (B, b) = (AB, Ab + a)$, which corresponds to composition of affine linear mappings, $\mathrm{AfSp}(\mathbb{R}^{2n}, \omega)$ is a group. Identifying (A, a) with the matrix $\begin{pmatrix} A & a \\ 0 & 1 \end{pmatrix}$, $\mathrm{AfSp}(\mathbb{R}^{2n}, \omega)$ becomes a closed subgroup of $\mathrm{Sp}(\mathbb{R}^{2n}, \omega) \times \mathbb{R}^{2n}$. Thus $\mathrm{AfSp}(\mathbb{R}^{2n}, \omega)$ is a Lie group. Its Lie algebra is $\mathrm{afsp}(\mathbb{R}^{2n}, \omega) = \{(X, x) \in \mathrm{sp}(\mathbb{R}^{2n}, \omega) \times \mathbb{R}^{2n}\}$ with Lie bracket $[(X, x), (Y, y)] = ([X, Y], Xy - Yx)$.

$\mathrm{AfSp}(\mathbb{R}^{2n}, \omega)$ acts on $(\mathbb{R}^{2n}, \omega)$ by

$$\Phi : \mathrm{AfSp}(\mathbb{R}^{2n}, \omega) \times (\mathbb{R}^{2n}, \omega) \to (\mathbb{R}^{2n}, \omega) : \big((A, a), v\big) \mapsto Av + a. \qquad (1)$$

Since the symplectic form ω on $\mathbb{R}^{2n}$ is invariant under translation, for every $(A, a) \in \mathrm{AfSp}(\mathbb{R}^{2n}, \omega)$ the affine linear mapping $\Phi_{(A,a)}$ preserves ω. The infinitesimal generator $X^{(X,x)}$ of the action Φ in the direction $(X, x) \in \mathrm{afsp}(\mathbb{R}^{2n}, \omega)$ is the vector field $X^{(X,x)}(v) = Xv + x$ on $\mathbb{R}^{2n}$. We now show

Proposition 1. *The* $\mathrm{AfSp}(\mathbb{R}^{2n}, \omega)$ *action* Φ *(1) is Hamiltonian.*

Proof. For every $(Y, y) \in \mathrm{afsp}(\mathbb{R}^{2n}, \omega)$ let

$$J^{(Y,y)} : \mathbb{R}^{2n} \to \mathbb{R} : v \mapsto J^Y(v) + \omega^{\sharp}(y)v = \tfrac{1}{2}\omega(Yv, v) + \omega(y, v) \qquad (2)$$

and set $J : \mathbb{R}^{2n} \to \mathrm{afsp}(\mathbb{R}^{2n}, \omega)^*$, where $J(v)(Y, y) = J^{(Y,y)}(v)$. Then for every $(X, x) \in \mathrm{afsp}(\mathbb{R}^{2n}, \omega)$, every $v \in \mathbb{R}^{2n}$, and every $w \in T_v\mathbb{R}^{2n} = \mathbb{R}^{2n}$

$$\mathrm{d}J^{(X,x)}(v)w = \big(T_v J\,(X, x)\big)w = \omega(Xv, w) + \omega(x, w)$$
$$= \omega(Xv + x, w) = \omega(X^{(X,x)}(v), w),$$

that is, $X^{(X,x)} = X_{J(X,x)}$. Hence the action Φ is Hamiltonian. $\qquad \square$

F. Nielsen and F. Barbaresco (Eds.): GSI 2023, LNCS 14072, pp. 115–123, 2023.
https://doi.org/10.1007/978-3-031-38299-4_13

The above argument shows that the map J is the momentum map of the Hamiltonian action Φ (1). The following discussion is motivated by theorem 11.34 of Souriau [7, p.143].

Lemma 1. *The mapping*

$$\sigma : \mathrm{AfSp}(\mathbb{R}^{2n}, \omega) \to \mathrm{afsp}(\mathbb{R}^{2n}, \mathbb{R})^* : g \mapsto J\big(\Phi_g(v)\big) - \mathrm{Ad}^T_{g^{-1}} J(v) \qquad (3)$$

does not depend on $v \in \mathbb{R}^{2n}$.

Proof. For each $\eta \in \mathrm{afsp}(\mathbb{R}^{2n}, \mathbb{R})$ we have

$$\mathrm{d}\big(J{\circ}\Phi_g\big)^\eta(v) = T_v \Phi_g\, X^\eta(v) \lrcorner\, \omega(v) = X^{\mathrm{Ad}_g \eta}(v) \lrcorner\, \omega(v) = \mathrm{d}(\mathrm{Ad}^T_{g^{-1}} J)^\eta(v),$$

that is, $\mathrm{d}\big(J{\circ}\Phi_g - \mathrm{Ad}^T_{g^{-1}} J\big)(v) = 0$. Since $\mathbb{R}^{2n}$ is connected the function $J{\circ}\Phi_g - \mathrm{Ad}^T_{g^{-1}} J : \mathbb{R}^{2n} \to \mathbb{R}$ is constant. $\qquad\square$

Corollary 1. *For every* $g, g' \in \mathrm{AfSp}(\mathbb{R}^{2n}, \omega)$

$$\sigma(gg') = \sigma(g) + \mathrm{Ad}^T_{g^{-1}} \sigma(g'). \qquad (4)$$

Proof. We compute.

$$\begin{aligned}
\sigma(gg') &= J{\circ}\Phi_{gg'} - \mathrm{Ad}^T_{(gg')^{-1}} J \\
&= (J{\circ}\Phi_g - \mathrm{Ad}^T_{g^{-1}} J){\circ}\Phi_{g'} + \mathrm{Ad}^T_{g^{-1}} (J{\circ}\Phi_{g'} - \mathrm{Ad}^T_{(g')^{-1}} J) \\
&= \sigma(g) + \mathrm{Ad}^T_{g^{-1}} \sigma(g').
\end{aligned}$$

$\qquad\square$

Evaluating σ (3) at $\exp t\eta$ and then at $\zeta \in \mathrm{afsp}(\mathbb{R}^{2n}, \mathbb{R})$ gives

$$\big(\sigma(\exp t\eta)\big)\zeta = J^\zeta\big(\Phi_{\exp t\eta}(v)\big) - \big(\mathrm{Ad}^T_{\exp -t\eta} J(v)\big)\zeta. \qquad (5)$$

Differentiating (5) with respect to t and then setting t equal to 0 gives

$$\begin{aligned}
(T_e\sigma\, \eta)\zeta &= \mathrm{d}J^\zeta(v) X^\eta(v) + \big(\mathrm{ad}^T_\eta J(v)\big)\zeta \\
&= L_{X^\eta} J^\zeta(v) + J(v)\mathrm{ad}_\eta\zeta = \{J^\zeta, J^\eta\}(v) - J^{[\zeta,\eta]}(v). \qquad (6)
\end{aligned}$$

Let $\Sigma^\sharp : \mathrm{afsp}(\mathbb{R}^{2n}, \mathbb{R}) \to \mathrm{afsp}(\mathbb{R}^{2n}, \mathbb{R})^*$ be the linear mapping $\eta \mapsto \Sigma^\sharp(\eta) = -(T_e\sigma)\eta \in \mathrm{afsp}(\mathbb{R}^{2n}, \mathbb{R})^*$. Equation (6) may be written as

$$\{J^\eta, J^\zeta\}(v) = J^{[\eta,\zeta]}(v) + \Sigma(\eta, \zeta), \qquad (7)$$

where $\Sigma(\eta, \zeta) = \Sigma^\sharp(\eta)\zeta$. From Eq. (7) it follows that the bilinear map Σ is skew symmetric.

Lemma 2. Σ *is an* $\mathrm{afsp}(\mathbb{R}^{2n}, \mathbb{R})$ *cocycle, that is,*

$$\Sigma(\zeta, [\xi, \eta]) = \Sigma([\zeta, \xi], \eta) + \Sigma(\xi, [\zeta, \eta]), \qquad (8)$$

for every ξ, η, *and* $\zeta \in \mathrm{afsp}(\mathbb{R}^{2n}, \mathbb{R})$.

Proof. Since $\left(C^\infty(\mathbb{R}^{2n}), \{\ ,\ \}\right)$ is a Lie algebra

$$\{J^\varsigma, \{J^\xi, J^\eta\}\} = \{\{J^\varsigma, J^\xi\}, J^\eta\} + \{J^\xi, \{J^\varsigma, J^\eta\}\}.$$

Using Eq. (7) the above equation reads

$$\begin{aligned} \{J^\varsigma, J^{[\xi,\eta]}\} + \{J^\varsigma, \Sigma(\xi,\eta)\} &= \{J^{[\varsigma,\xi]}, J^\eta\} + \{\Sigma(\varsigma,\xi), J^\eta\} \\ &\quad + \{J^\xi, J^{[\varsigma,\eta]}\} + \{J^\varsigma, \Sigma(\varsigma,\eta)\}. \end{aligned}$$

Using (7) again gives

$$J^{[\varsigma,[\xi,\eta]]} + \Sigma(\varsigma, [\xi,\eta]) = J^{[[\varsigma,\xi],\eta]} + \Sigma([\varsigma,\xi],\eta) + J^{[\xi,[\varsigma,\eta]]} + \Sigma(\xi,[\varsigma,\eta]),$$

since $\Sigma(\xi,\eta)$, $\Sigma(\varsigma,\xi)$ and $\Sigma(\varsigma,\eta)$ are constant functions on $\mathbb{R}^{2n}$. By linearity and the Jacobi identity $J^{[\varsigma,[\xi\eta]]} = J^{[[\varsigma,\xi],\eta]} + J^{[\xi,[\varsigma,\eta]]}$ Eq. (8) holds.   □

Proposition 2. *The momentum map J has the cocycle*

$$\Sigma : \mathrm{afsp}(\mathbb{R}^{2n}, \omega) \times \mathrm{afsp}(\mathbb{R}^{2n}, \omega) \to \mathbb{R} : \big((Y,y),(Z,z)\big) \mapsto \omega(y,z). \tag{9}$$

Proof. We compute.

$$\begin{aligned} \{J^{(Y,y)}, J^{(Z,z)}\}(v) &= \left(L_{X^{(Z,z)}} J^{(Y,y)}\right)(v) = dJ^{(Y,y)}(v)X^{(Z,z)}(v) \\ &= \omega(Yv, Zv + z) + \omega(y, Zv + z) \\ &= \omega(Yv, Zv) + \omega(Yv, z) + \omega(y, Zv) + \omega(y, z). \end{aligned}$$

Now

$$\begin{aligned} \tfrac{1}{2}\omega([Y,Z]v, v) &= \tfrac{1}{2}\omega\big((YZ - ZY)v, v\big) = \tfrac{1}{2}\omega(YZv, v) - \tfrac{1}{2}\omega(ZYv, v) \\ &= -\tfrac{1}{2}\omega(Zv, Yv) + \tfrac{1}{2}\omega(Yv, Zv) = \omega(Yv, Zv). \end{aligned}$$

Thus

$$\begin{aligned} \{J^{(Y,y)}, J^{(Z,z)}\}(v) &= \tfrac{1}{2}\omega([Y,Z]v, v) + \omega(Yz - Zy, v) + \omega(y, z) \\ &= J^{[Y,Z]}(v) + \omega(Yz - Zy, v) + \omega(y, z) \\ &= J^{[(Y,y),(Z,z)]}(v) + \omega(y, z). \end{aligned}$$

□

Define the map

$$\Psi : \mathrm{AfSp}(\mathbb{R}^{2n}, \omega) \times \mathrm{afsp}(\mathbb{R}^{2n}, \omega)^* \to \mathrm{afsp}(\mathbb{R}^{2n}, \omega)^* : (g, \alpha) \longmapsto \mathrm{Ad}^T_{g^{-1}}\alpha + \sigma(g),$$

where σ is given by Eq. (3).

Proposition 3. *The map Ψ is an action of $\mathrm{AfSp}(\mathbb{R}^{2n}, \omega)$ on $\mathrm{afsp}(\mathbb{R}^{2n}, \omega)^*$.*

Proof. We compute. For g, $g' \in \mathrm{AfSp}(\mathbb{R}^{2n}, \omega)$ and $\alpha \in \mathrm{afsp}(\mathbb{R}^{2n}, \omega)^*$ we have

$$\Psi_{gg'}\alpha = \mathrm{Ad}^T_{(gg')^{-1}}\alpha + \sigma(gg') = \mathrm{Ad}^T_{g^{-1}}(\mathrm{Ad}^T_{(g')^{-1}}\alpha) + \sigma(g) + \mathrm{Ad}^T_{g^{-1}}\sigma(g')$$

$$\text{using corollary 1}$$

$$= \mathrm{Ad}^T_{g^{-1}}\big(\mathrm{Ad}^T_{(g')^{-1}}\alpha + \sigma(g')\big) + \sigma(g) = \mathrm{Ad}^T_{g^{-1}}(\Psi_{g'}\alpha) + \sigma(g)$$

$$= \Psi_g(\Psi_{g'}\alpha).$$

$\square$

Proposition 4. *The momentum mapping J is coadjoint equivariant under the action Ψ.*

Proof. We compute. For every $g \in \mathrm{AfSp}(\mathbb{R}^{2n}, \omega)$ and every $w \in \mathbb{R}^{2n}$

$$\Psi_g\big(J(w)\big) = \mathrm{Ad}^T_{g^{-1}}J(w) + \sigma(g) = \mathrm{Ad}^T_{g^{-1}}J(w) + J\big(\Phi_g(w)\big) - \mathrm{Ad}^T_{g^{-1}}J(w)$$

$$= J\big(\Phi_g(w)\big).$$

$\square$

2   Extension

Following Wallach [8] we find a central extension of Lie algebra $\mathrm{afsp}(\mathbb{R}^{2n}, \omega)$ the dual of whose adjoint map is

$$T_e\Psi(X, x)\alpha = -\mathrm{ad}^T_{(X,x)}\alpha + T_e\sigma(X, x) = -\mathrm{ad}^T_{(X,x)}\alpha + \Sigma^\sharp(X, x),$$

where $(X, x) \in \mathrm{sp}(\mathbb{R}^{2n}, \omega) \times \mathbb{R}^{2n} = \mathrm{afsp}(\mathbb{R}^{2n}, \omega)$.

Let $\widehat{\mathfrak{g}} = \{(X, v, \xi) \in \mathrm{afsp}(\mathbb{R}^{2n}, \omega) \times \mathbb{R}\}$ be the Lie algebra with Lie bracket

$$[(X, v, \xi), (Y, w, \eta)] = ([X, Y], Xw - Yv, \omega(v, w)). \tag{10}$$

From (10) it follows that $(0, 0, 1)$ lies in the center of $\widehat{\mathfrak{g}}$, that is, $[(0, 0, 1), (X, v, \xi)] = (0, 0, 0)$. Also

$$[(X, v, 0), (Y, v, 0)] = ([X, Y], Xw - Yv, 0) + \omega(v, w)(0, 0, 1).$$

Thus the Lie algebra $\widehat{\mathfrak{g}}$ is a central extension of the Lie algebra $\mathrm{afsp}(\mathbb{R}^{2n}, \omega)$ by the cocycle ω. Since we can write (10) as

$$\mathrm{ad}_{[X,x,\xi]}[Y, y, \eta] = \mathrm{ad}_{[X,x]}[Y, y] + \Sigma(\xi, \eta), \tag{11}$$

$\widehat{\mathfrak{g}}$ is the sought for Lie algebra.

3   The Odd Real Symplectic Group

We now find a connected linear Lie group $\widehat{G}$ whose Lie algebra is $\widehat{\mathfrak{g}}$. Consider the group $\widehat{G} \subseteq \mathrm{AfSp}(\mathbb{R}^{2n}, \frac{1}{2}\omega) \times \mathbb{R}$ with multiplication

$$((A,v),r) \cdot ((B,w),s) = ((AB, Aw + v), r + s + \tfrac{1}{2}\omega(A^{-1}v, w)).$$

The map $\widehat{\pi} : \widehat{G} \to \mathrm{AfSp}(\mathbb{R}^{2n}, \omega) : (A,v,r) \mapsto (A,v)$ is a surjective group homomorphism, whose kernel is the normal subgroup $\widehat{Z} = \{(I, 0, r) \in \widehat{G} \mid r \in \mathbb{R}\}$, which is the center of $\widehat{G}$. Note that $\pi_1(\widehat{G}) = \pi_1(\mathrm{Sp}(\mathbb{R}^{2n}, \omega)) = \mathbb{Z}$. $\widehat{G}$ is a Lie group with Lie algebra $\widehat{\mathfrak{g}}$, whose Lie bracket is given by (10).

Theorem 1. *The group $\widehat{G}$ is isomorphic to the odd real symplectic group.*

Proof. Consider the map

$$\rho : \widehat{G} \to \mathrm{Gl}(\mathbb{R}^{2n+2}, \mathbb{R}) : (A,v,r) \mapsto \begin{pmatrix} 1 & 0 & 0 \\ v & A & 0 \\ r & \tfrac{1}{2}\omega^{\sharp}(A^{-1}v) & 1 \end{pmatrix}. \tag{12}$$

The map ρ is an injective homomorphism. Here we have $\frac{1}{2}\omega^{\sharp}(A^{-1}v) = -\frac{1}{2}(v^T J A)$, since for every $z \in \mathbb{R}^{2n}$

$$\tfrac{1}{2}\omega^{\sharp}(A^{-1}v)z = \tfrac{1}{2}\omega(A^{-1}v, z) = \tfrac{1}{2}\omega(v, Az), \text{ because } A \in \mathrm{Sp}(\mathbb{R}^{2n}, \tfrac{1}{2}\omega)$$
$$= -\tfrac{1}{2}\omega(Az, v) = -\tfrac{1}{2}(v^T J A)z.$$

The following calculation shows that ρ is a homomorphism.

$$\rho\big((A,v,r) \cdot (B,w,s)\big) = \rho\big(AB, Aw + v, r + s + \tfrac{1}{2}\omega(A^{-1}v, w)\big)$$
$$= \begin{pmatrix} 1 & 0 & 0 \\ v + Aw & AB & 0 \\ r + s + \tfrac{1}{2}\omega^{\sharp}(A^{-1}v)w & \tfrac{1}{2}\omega^{\sharp}((AB)^{-1}(v + Aw)) & 1 \end{pmatrix}. \tag{13}$$

Now

$$\tfrac{1}{2}\omega^{\sharp}\big((AB)^{-1}(v + Aw)\big) = \tfrac{1}{2}\omega^{\sharp}\big(B^{-1}(A^{-1}v)\big) + \tfrac{1}{2}\omega^{\sharp}(B^{-1}w)$$
$$= B^T \tfrac{1}{2}\omega^{\sharp}(A^{-1}v) + \tfrac{1}{2}\omega^{\sharp}(B^{-1}w), \text{ since } B \in \mathrm{Sp}(\mathbb{R}^{2n}, \tfrac{1}{2}\omega)$$
$$= \tfrac{1}{2}\omega^{\sharp}(A^{-1}v)B + \tfrac{1}{2}\omega^{\sharp}(B^{-1}w). \tag{14}$$

But

$$\rho(A,v,r)\,\rho(B,w,s) = \begin{pmatrix} 1 & 0 & 0 \\ v & A & 0 \\ r & \tfrac{1}{2}\omega^{\sharp}(A^{-1}v) & 1 \end{pmatrix} \begin{pmatrix} 1 & 0 & 0 \\ w & B & 0 \\ s & \tfrac{1}{2}\omega^{\sharp}(B^{-1}w) & 1 \end{pmatrix}$$
$$= \begin{pmatrix} 1 & 0 & 0 \\ v + Aw & AB & 0 \\ r + s + \tfrac{1}{2}\omega^{\sharp}(A^{-1}v)w & \tfrac{1}{2}\omega^{\sharp}(A^{-1}v)B + \tfrac{1}{2}\omega^{\sharp}(B^{-1}w) & 1 \end{pmatrix}. \tag{15}$$

Using (14) we see that the right hand sides of Eqs. (13) and (15) are equal, that is, $\rho\big((A, v, r) \cdot (B, w, s)\big) = \rho(A, v, r)\,\rho(B, w, s)$. Thus the map ρ (12) is a group homomorphism. The map ρ is injective, for if $\rho(A, v, r) = (I_{2n}, 0, 0)$, then $v = 0$ and $r = 0$. So $(A, v, r) = (I_{2n}, 0, 0)$, the identity element of $\widehat{G}$.

Since $\widehat{G}$ is a subgroup of $\mathrm{AfSp}(\mathbb{R}^{2n}, \tfrac{1}{2}\omega) \times \mathbb{R}$, it follows that if $(A, v, r) \in \widehat{G}$, then $A \in \mathrm{Sp}(\mathbb{R}^{2n}, \tfrac{1}{2}\omega)$. Thus the image of the map ρ is contained in $\mathrm{Sp}(\mathbb{R}^{2n+2}, \Omega)$. Here the matrix of the symplectic form Ω with respect to the basis $\{e_0, e_1, \dots, e_n,$ $f_1, \dots, f_n, f_{n+1}\}$ of $\mathbb{R}^{2n+2}$ is $\begin{pmatrix} 0 & 0 & 1 \\ 0 & \frac{1}{2}J & 0 \\ -1 & 0 & 0 \end{pmatrix}$. and $J = \begin{pmatrix} 0 & I_n \\ -I_n & 0 \end{pmatrix}$ is the matrix of the symplectic form ω with respect to the basis $\{e_1, \dots, e_n, f_1, \dots, f_n\}$ of $\mathbb{R}^{2n}$. The image of the map ρ (12) is the *odd real symplectic group* $\mathrm{Sp}(\mathbb{R}^{2n+2}, \Omega)_{f_{n+1}} = \{A \in \mathrm{Sp}(\mathbb{R}^{2n+2}, \Omega) \,|\, A f_{n+1} = f_{n+1}\}$. Thus $\widehat{G}$ is isomorphic to $\mathrm{Sp}(\mathbb{R}^{2n+2}, \Omega)_{f_{n+1}}$. $\square$

The Lie algebra $\mathrm{sp}(\mathbb{R}^{2n+2}, \Omega)_{f_{n+1}}$ of the Lie group $\mathrm{Sp}(\mathbb{R}^{2n+2}, \Omega)_{f_{n+1}}$ is

$$\{\widehat{X} = \begin{pmatrix} 0 & 0 & 0 \\ x & X & 0 \\ \xi & \frac{1}{2}\omega^\sharp(x) & 0 \end{pmatrix} \in \mathrm{Sp}(\mathbb{R}^{2n+2}, \Omega) \,|\, x \in \mathbb{R}^{2n}, \ X \in \mathrm{sp}(\mathbb{R}^{2n}, \tfrac{1}{2}\omega), \text{ and } \xi \in \mathbb{R}\}$$

with Lie bracket

$$[\widehat{X}, \widehat{Y}] = \left[\begin{pmatrix} 0 & 0 & 0 \\ x & X & 0 \\ \xi & \frac{1}{2}\omega^\sharp(x) & 0 \end{pmatrix}, \begin{pmatrix} 0 & 0 & 0 \\ y & Y & 0 \\ \eta & \frac{1}{2}\omega^\sharp(y) & 0 \end{pmatrix} \right] = \begin{pmatrix} 0 & 0 & 0 \\ Xy - Yx & [X, Y] & 0 \\ \omega(x, y) & \frac{1}{2}\omega^\sharp(Xy - Yx) & 0 \end{pmatrix}. \quad (16)$$

Here $\frac{1}{2}\omega^\sharp(x) = -\frac{1}{2}x^T J$, since $\frac{1}{2}\omega^\sharp(x)z = -\frac{1}{2}\omega(z, x) = (-\frac{1}{2}x^T J)z$ for each $z \in \mathbb{R}^{2n}$. The map

$$\mu : \widehat{\mathfrak{g}} \to \mathrm{sp}(\mathbb{R}^{2n+2}, \Omega)_{f_{n+1}} : (X, x, \xi) \mapsto \widehat{X} = \begin{pmatrix} 0 & 0 & 0 \\ x & X & 0 \\ \xi & \frac{1}{2}\omega^\sharp(x) & 0 \end{pmatrix}$$

is a Lie algebra isomorphism, because it is a bijective linear map and

$$\mu\big([(X, x, \xi), (Y, y, \eta)]\big) = \mu\big([X, Y], Xy - Yx, \omega(x, y)\big)$$
$$= \begin{pmatrix} 0 & 0 & 0 \\ Xy - Yx & [X, Y] & 0 \\ \omega(x, y) & \frac{1}{2}\omega^\sharp(Xy - Yx) & 0 \end{pmatrix} = [\widehat{X}, \widehat{Y}] = [\mu(X, x, \xi), \mu(Y, y, \eta)].$$

This verifies that the Lie algebra of the Lie group $\widehat{G}$ has Lie bracket given by (10), because the group $\widehat{G}$ is isomorphic to $\mathrm{Sp}(\mathbb{R}^{2n+2}, \Omega)_{f_{n+1}}$.

Proposition 5. *The action*

$$\widehat{\Phi} : \mathrm{Sp}(\mathbb{R}^{2n+2}, \Omega)_{f_{n+1}} \times (\mathbb{R}^{2n}, \omega) \to (\mathbb{R}^{2n}, \omega) :$$
$$\left(\begin{pmatrix} 1 & 0 & 0 \\ v & A & 0 \\ r & \frac{1}{2}\omega^\sharp(A^{-1}v) & 1 \end{pmatrix}, w \right) \mapsto Aw + v \quad (17)$$

is Hamiltonian.

Proof. The infinitesimal generator $X^{\widehat{X}}$ of the action $\widehat{\Phi}$ in the direction $\widehat{X} \in$ $\mathrm{sp}(\mathbb{R}^{2n+2}, \Omega)_{f_{n+1}}$ is the vector field $X^{\widehat{X}}(w) = X(w) + x$ on $\mathbb{R}^{2n}$. For every $\widehat{Y} = \begin{pmatrix} 0 & 0 & 0 \\ y & Y & 0 \\ \eta & \frac{1}{2}\omega^\sharp(y) & 0 \end{pmatrix} \in \mathrm{sp}(\mathbb{R}^{2n+2}, \Omega)_{f_{n+1}}$ let

$$J^{\widehat{Y}} : \mathbb{R}^{2n} \to \mathbb{R} : w \mapsto \tfrac{1}{2}\omega(Yw, w) + \omega(y, w) + \eta. \tag{18}$$

Then

$$\mathrm{d}\widehat{J}^{\widehat{Y}}(v)w = T_v \widehat{J}(\widehat{Y})w = \omega(Yv, w) + \omega(y, w) = \omega\big(X^{\widehat{Y}}(v), w\big).$$

Thus $X^{\widehat{Y}} = X_{J^{\widehat{Y}}}$. So the action $\widehat{\Phi}$ (17) is Hamiltonian. $\qquad\square$

Since the mapping $\widehat{Y} \mapsto J^{\widehat{Y}}(w)$ is linear for every $w \in \mathbb{R}^{2n}$, the action $\widehat{\Phi}$ (17) has a momentum mapping $\widehat{J} : \mathbb{R}^{2n} \to \mathrm{sp}(\mathbb{R}^{2n+2}, \Omega)^*_{f_{n+1}}$ with $\widehat{J}(w)\widehat{Y} = J^{\widehat{Y}}(w)$.

Theorem 2. *The momentum mapping $\widehat{J}$ of the $\mathrm{Sp}(\mathbb{R}^{2n+2}, \Omega)_{f_{n+1}}$ action $\widehat{\Phi}$ (17) is coadjoint equivariant, that is, $\widehat{J}\big(\widehat{\Phi}_g(w)\big) = \mathrm{Ad}^T_{g^{-1}}\widehat{J}(w)$ for every $w \in \mathbb{R}^{2n}$ and every $g \in \mathrm{Sp}(\mathbb{R}^{2n+2}, \Omega)_{f_{n+1}}$.*

Proof. It is enough to show that

$$\{\widehat{J}^{\widehat{Y}}, \widehat{J}^{\widehat{Z}}\} = \widehat{J}^{[\widehat{Y}, \widehat{Z}]}, \text{ for every } \widehat{Y}, \widehat{Z} \in \widehat{\mathfrak{g}}, \tag{19}$$

because (19) is the infinitesimalization of the coadjoint equivariance condition and $\mathrm{Sp}(\mathbb{R}^{2n+2}, \Omega)_{f_{n+1}}$ is generated by elements which lie in the image of the exponential mapping, since it is connected. We compute

$$\widehat{J}^{[\widehat{Y}, \widehat{Z}]}(w) = \tfrac{1}{2}\omega([Y, Z]w, w) + \omega(Yz - Zy, w) + \omega(y, z),$$
$$\text{using Eqs. (16) and (18)}$$
$$= \omega(Yw, Zw) + \omega(Yw, z) + \omega(y, Zw) + \omega(y, z)$$
$$= \omega(Yw, Zw + z) + \omega(y, Zw + z) = \mathrm{d}\widehat{J}^{\widehat{Y}}(w)X^{\widehat{Z}}(w)$$
$$= \big(L_{X_{\widehat{J}\widehat{Z}}}\widehat{J}^{\widehat{Y}}\big)(w) = \{\widehat{J}^{\widehat{Y}}, \widehat{J}^{\widehat{Z}}\}(w).$$

$\qquad\square$

4   Coadjoint Orbit

In this section using results of [5] we algebraically classify the coadjoint orbit $\mathcal{O}\big(\widehat{J}(e_1)\big)$ of the odd real symplectic group $\mathrm{Sp}(\mathbb{R}^{2n+2}, \Omega)_{f_{n+1}}$ through $\widehat{J}(e_1) \in$ $\mathrm{sp}(\mathbb{R}^{2n+2}, \Omega)^*_{f_{n+1}}$. We show that this coadjoint orbit has a modulus.

First we note that the action $\widehat{\Phi}$ (17) of $\mathrm{Sp}(\mathbb{R}^{2n+2}, \Omega)_{f_{n+1}}$ on $\mathbb{R}^{2n}$ is transitive. Thus to determine the $\mathrm{Sp}(\mathbb{R}^{2n+2}, \Omega)_{f_{n+1}}$ coadjoint orbit through $\widehat{J}(w)$ for a fixed $w \in \mathbb{R}^{2n}$, it suffices to determine the coadjoint orbit $\mathcal{O}(\widehat{J}(e_1))$ through $\widehat{J}(e_1) \in \mathrm{sp}(\mathbb{R}^{2n+2}, \Omega)^*_{f_{n+1}}$. We have

$$\widehat{Y} = \left(\begin{array}{c|cc|c} 0 & 0 & 0 & 0 \\ \hline y^1 & A & B & 0 \\ y^2 & C & -A^T & 0 \\ \hline \eta & \frac{1}{2}(y^2)^T & -\frac{1}{2}(y^1)^T & 0 \end{array} \right),$$

where $Y = (Y_{ij}) = \left(\begin{smallmatrix} A & B \\ C & -A^T \end{smallmatrix} \right) \in \mathrm{sp}(\mathbb{R}^{2n}, \frac{1}{2}\omega)$ with A, B, and $C \in \mathrm{gl}(\mathbb{R}^n, \mathbb{R})$, where $B = B^T$ and $C = C^T$; $y^T = ((y^1)^T \,|\, (y^2)^T) = (y_1, \dots, y_n \,|\, y_{n+1}, \dots, y_{2n}) \in \mathbb{R}^{2n}$; and $\eta \in \mathbb{R}$. Then using (18) we get

$$\widehat{J}(e_1)\widehat{Y} = \tfrac{1}{2}e_1^T \left(\begin{array}{cc} C & -A^T \\ -A & -B \end{array} \right) e_1 + e_1^T \left(\begin{array}{c} y^2 \\ -y^1 \end{array} \right) + \eta = \tfrac{1}{2}Y_{n+1,1} + y_{n+1} + \eta.$$

Therefore

$$\widehat{J}(e_1) = \tfrac{1}{2}E^*_{n+1,1} + \tfrac{1}{2}E^*_{n+1,0} + E^*_{2n+1,1} + E^*_{2n+1,0}.$$

Here $\{E^*_{ij}\}^{2n+1}_{i,j=0}$ is a basis of $\mathrm{gl}(\mathbb{R}^{2n+2}, \mathbb{R})^*$, which is dual to the standard basis $\{E_{ij}\}^{2n+1}_{i,j=0}$ of $\mathrm{gl}(\mathbb{R}^{2n+2}, \mathbb{R})$ and $E_{ij} = (\delta_{ik}\delta_{j\ell})$.

We now use results of [5] to algebraically characterize the coadjoint orbit $\mathcal{O}(\widehat{J}(e_1))$. The *affine cotype* ∇ represented by the tuple $(\mathbb{R}^{2n+2}, Z, f_{n+1}; \Omega)$ corresponds to the coadjoint orbit $\mathcal{O}(\widehat{J}(e_1))$, see proposition 4 of [5]. Here

$$Z = \widehat{J}(e_1)^T = \left(\begin{array}{cccc} 0 & 0 & r^T & 1 \\ 0 & 0 & D & 2r \\ 0 & 0 & 0 & 0 \\ 0 & 0 & 0 & 0 \end{array} \right) \in \mathrm{sp}(\mathbb{R}^{2n+2}, \Omega)$$

with $r^T = (1/2, 0, \dots, 0) \in \mathbb{R}^n$ and $D = \mathrm{diag}(1/2, 0, \dots, 0) \in \mathrm{gl}(\mathbb{R}^n, \mathbb{R})$. Since $Z^2 = 0$, the cotype ∇ is nilpotent and has height 1. The parameter of ∇ is 1. Thus by proposition 7 of [5] the affine cotype $\nabla = \nabla_1(0), 1 + \Delta$. Here $\nabla_1(0), 1$ is the nilpotent indecomposable cotype of height 1 and *modulus* 1, which is represented by the tuple $\left(\mathbb{R}^2, \left(\begin{smallmatrix} 0 & 1 \\ 0 & 0 \end{smallmatrix} \right), f_1; \left(\begin{smallmatrix} 0 & 1 \\ -1 & 0 \end{smallmatrix} \right) \right)$, and Δ is a semisimple type, see [3]. Since Δ is a nilpotent of height 0, it equals 0.

We give another argument which proves the above assertion. The tuples $(\mathbb{R}^{2n+2}, Z, f_{n+1}; \Omega)$ and $(\mathbb{R}^{2n+2}, PZP^{-1}, f_{n+1}; \Omega)$ are equivalent when P lies in $\mathrm{Sp}(\mathbb{R}^{2n+2}, \Omega)_{f_{n+1}}$ and thus represent the same cotype ∇. Let $P = \left(\begin{smallmatrix} 1 & 0 & 0 \\ d & I_{2n} & 0 \\ f & -\frac{1}{2}d^T J & 1 \end{smallmatrix} \right)$ $\in \mathrm{Sp}(\mathbb{R}^{2n+2}, \Omega)_{f_{n+1}}$. We compute.

$$PZP^{-1} = \left(\begin{array}{ccc} 1 & 0 & 0 \\ d & I_{2n} & 0 \\ f & -\frac{1}{2}d^T J & 1 \end{array} \right) \left(\begin{array}{ccc} 0 & \widetilde{r}^T & 1 \\ 0 & \widetilde{D} & \widetilde{s} \\ 0 & 0 & 0 \end{array} \right) \left(\begin{array}{ccc} 1 & 0 & 0 \\ -d & I_{2n} & 0 \\ -f & \frac{1}{2}d^T J & 1 \end{array} \right),$$

where $\widetilde{r}^T = (0 \,|\, r^T)$, $\widetilde{s}^T = (2J\widetilde{r})^T = (2r \,|\, 0)$, and $\widetilde{D} = \left(\begin{smallmatrix} 0 & D \\ 0 & 0 \end{smallmatrix} \right)$. Set $d = -\widetilde{s}$ and $f = 0$.

Then

$$PZP^{-1} = \begin{pmatrix} 1 & 0 & 0 \\ -\widetilde{s} & I_{2n} & 0 \\ 0 & \frac{1}{2}\widetilde{s}^T J & 1 \end{pmatrix} \begin{pmatrix} 0 & 0 & 1 \\ 0 & 0 & \widetilde{s} \\ 0 & 0 & 0 \end{pmatrix} = \begin{pmatrix} 0 & 0 & 1 \\ 0 & 0 & 0 \\ 0 & 0 & 0 \end{pmatrix}.$$

Acknowledgment. I would like to thank one of the referees for pointing out references [2] and [6].

References

1. Bates, L., Cushman, R.: Removing the cocycle in a momentum map. JP J. Geom. Topol. **5**, 103–107 (2005)
2. Beckett, A., Figueroa-O'Farrill, J.: Symplectic actions and central extensions. arxiv:2203.07405
3. Burgoyne, N., Cushman, R.: Conjugacy classes in linear groups. J. Alg. **44**, 483–529 (1977)
4. Cushman, R.H., Bates, L.M.: Global Aspects of Classical Integrable Systems, 2nd edn. Birkhäuser, Basel (2015)
5. Cushman, R.: Coadjoint orbits of the odd real symplectic group. arxiv:2212.0067v2
6. Saxcé, G., Vallée, C.: Construction of a central extension of a Lie group from its class of symplectic cohomology. J. Geom. Phys. **60**, 165–174 (2010)
7. Souriau, J.-M., : Structure of dynamical systems: a symplectic point of view of physics Birkhäuser, Boston (1997)
8. Wallach, N.: Symplectic Geometry and Fourier Analysis. Math Sci Press, Brookline, MA (1977)

Polysymplectic Souriau Lie Group Thermodynamics and the Geometric Structure of Its Coadjoint Orbits

Mohamed El Morsalani$^{(\boxtimes)}$

Landesbank Baden-Wuerttemberg, Stuttgart, Germany
morsalan@hotmail.de

Abstract. In 1969, Jean-Marie Souriau introduced a "Lie Groups Thermodynamics" in Statistical Mechanics in the framework of Geometric Mechanics [19]. Frederic Barbaresco and his collaborators have proved in many papers how the Souriau's model can be applied within information geometry and geometric deep learning. In this paper we will focus on the extension of Souriau's symplectic model to the polysymplectic case. We will describe the polysymplectic model and the fact that coadjoint orbits have a polysymplectic and poly-Poisson structure.

Keywords: higher order thermodynamics · Lie groups thermodynamics · poly-symplectic and poly-Poisson manifold · coadjoint orbits · KKS theorem · polysymplectic foliation

1 Introduction

Polysymplectic geometry, as introduced in the seminal work of C. Günther [13] and further developed in [18], arises as a special case of multisymplectic geometry which is the natural geometric setting of classical field theories, see, e.g., [17]. The first extension of Souriau ideas to the polysymplectic setting has been initiated by F. Barbaresco in his paper [2]. We will focus on the polysymplectic, poly-Poisson structures and results developed in [14,18].

It is worth to mention that the polysymplectic model is motivated by higher-order model of statistical physics. For instance, for small data analytics (rarified gases, sparse statistical surveys), the density of maximum entropy should consider higher order moments constraints, so that the Gibbs density is not only defined by first moment but fluctuations request 2nd order and higher moments (see [1,2] for further references).

2 Polysymplectic Manifolds: Basic Facts

2.1 Polysymplectic Manifolds, Actions and Momentum Maps

Definition 1. *[18] A polysymplectic manifold (M,ω) is a manifold M endowed with a closed nondegenerate $\mathbb{R}^k$-valued 2-form. We can identify ω with a collection $(\omega^1, \cdots, \omega^k)$, of closed 2-forms with*

F. Nielsen and F. Barbaresco (Eds.): GSI 2023, LNCS 14072, pp. 124–133, 2023.
https://doi.org/10.1007/978-3-031-38299-4_14

$$\bigcap_{i=1}^{k} ker\omega^i = \{0\} \tag{1}$$

Throughout this paper we use this characterization of a polysymplectic structure.

A Lie group action $\phi : G \times M \to M$ of a Lie group G on a polysymplectic manifold $(M, \omega^1, \cdots, \omega^k)$ is said to be a polysymplectic action, if for each $g \in G$, the diffeomorphism $\phi_g : M \to M$; $m \mapsto \phi(g, m)$ is polysymplectic; that is for $1 \leq j \leq k$

$$\phi_g^* \omega^j = \omega^j.$$

As in the symplectic case, the notion of a momentum map for polysymplectic actions can be introduced in a natural way:

Definition 2. *Let $(M, \omega^1, \cdots, \omega^k)$ be a polysymplectic manifold and $\phi : G \times M \to M$ a polysymplectic action. let $\mathfrak{g}$ be the Lie algebra of G and $\mathfrak{g}^*$ its dual. A mapping $\mathfrak{J} \equiv (\mathfrak{J}^1, \cdots, \mathfrak{J}^k) : M \to \mathfrak{g}^* \times \overset{k}{\cdots} \times \mathfrak{g}^*$ is said to be a momentum mapping for the action ϕ if for each $\xi \in G$,*

$$i_{\xi_M} \omega^k = d\hat{\mathfrak{J}}_\xi^k$$

where $\hat{\mathfrak{J}}_\xi^j : M \to \mathbb{R}$ is defined by

$$\hat{\mathfrak{J}}_\xi^j(m) = \mathfrak{J}^j(m).\xi, \ m \in M$$

and ξ_M is the infinitesimal generator of the action corresponding to ξ.

In the polysymplectic case the actions of $(Ad_g)^k$ and $(Ad_g^*)^k$ are defined in a natural way. Denote $\mathfrak{g}_k := \mathfrak{g} \times \overset{k}{\cdots} \times \mathfrak{g}$ then

- $(Ad_g)_k$ acts on $\beta = (\beta_1, \cdots, \beta_k) \in \mathfrak{g}_k$ by $(Ad_g)^k \beta = (Ad_g \beta_1, \cdots, Ad_g \beta_k)$.
- The coadjoint representation $(Ad_g^*)^k$ acts on $\nu \in \mathfrak{g}_k^* = \mathfrak{g}^* \times \overset{k}{\cdots} \times \mathfrak{g}^*$ by $(Ad_g^*)^k (\nu_1, \cdots, \nu_k) = (Ad_g^* \nu_1, \cdots, Ad_g^* \nu_k)$ where $Ad_g^* \nu_j = \nu_j \circ Ad_{g^{-1}}$.

2.2   Equivariance with Respect to Lie Group Actions

Definition 3. *A momentum mapping $\mathfrak{J} \equiv (\mathfrak{J}^1, \cdots, \mathfrak{J}^k) : M \to \mathfrak{g}^* \times \overset{k}{\cdots} \times \mathfrak{g}^*$ for the action ϕ is said to be G-equivariant if, for every $g \in G$ and $m \in M$,*

$$\mathfrak{J}(\phi_g(m)) = Ad_{g^{-1}}^* (\mathfrak{J}(m)).$$

This equivariance condition can be written as follows: $\mathfrak{J}^j(\phi_g(m)) = Ad_{g^{-1}}^* (\mathfrak{J}^j(m))$; for every $j = 1, \cdots, k$.
A polysymplectic manifold endowed with a polysymplectic action of a Lie group and a G-equivariant momentum map, $(M, \omega^1, \cdots, \omega^k, \phi, \mathfrak{J})$, is said to be a polysymplectic Hamiltonian G-space.

2.3 Polysymplectic Cocycles

In a similar way with the symplectic case, if M is connected, there is a group one-cocycle $\theta \in C^\infty(G, \mathfrak{g}_k^*)$, $\theta = (\theta^1, \cdots, \theta^k)$ defined by

$$\theta^i(g) = \mathfrak{J}^i(\phi(m)) - Ad^*_{g^{-1}}\left(\mathfrak{J}^i(m)\right).$$

where $Ad^*_{g^{-1}}$ is the coadjoint action.
Furthermore one can define the following map:

$$\Theta : \mathfrak{g} \times \mathfrak{g} \to \mathbb{R}^k \tag{2}$$

$$\Theta(\xi, \eta) := \left.\frac{d}{dt}\right|_{t=0} \theta\left(\exp(t\xi)\right)(\eta) \in \mathbb{R}^k.$$

Taking the derivative of the relation above, we get

$$\Theta^i(\xi, \eta) = \mathfrak{J}^i_{[\xi, \eta]} - \omega^i(\xi_M, \eta_M), \tag{3}$$

ξ_M and η_M are the infinitesimal generators of the action corresponding to ξ and η. From [13] we can infer that Θ is skew-symmetric, and satisfies the two-cocycle identity

$$\Theta([\xi, \eta], \zeta) + \Theta([\eta, \zeta], \xi) + \Theta([\zeta, \xi], \eta) = 0. \tag{4}$$

3 Souriau Lie Group Thermodynamics Based on Polysymplectic Model

Let $(M, \omega^1, \cdots, \omega^k, \phi, \mathfrak{J})$ be a polysymplectic Hamiltonian G-space and μ a volume form.

3.1 Notations and Definitions

The elements of the Lie algebra $\mathfrak{g}_k$ will be denoted β as they are generalisations of the inverse temperature. A duality pairing between elements ν of the dual space $\mathfrak{g}_k^*$ and the elements $\beta \in \mathfrak{g}_k$ is denoted as $\langle \nu, \beta \rangle$.
Denote by $\Omega \subset \mathfrak{g}_k$ the largest open set such that for all $\beta \in \Omega$ the two integrals

$$\int_M e^{-\langle \mathfrak{J}(m), \beta \rangle} d\mu \in \mathbb{R}$$

and

$$\int_M \mathfrak{J}(m) e^{-\langle \mathfrak{J}(m), \beta \rangle} d\mu \in \mathfrak{g}_k^*$$

converge. The Gibbs densities are not defined on the whole $\mathfrak{g}_k$ but only on the open subset Ω. From now on we assume that Ω is not empty. Its elements are called geometric temperatures.

On Ω one can define the so-called partition function (or characteristic function) $\psi : \Omega \to \mathbb{R}$, given by

$$\psi(\beta) = \int_M e^{-\langle \mathfrak{I}(m), \beta \rangle} d\mu.$$

This permits us to define the generalized Gibbs probability densities associated to each $\beta \in \Omega$:

$$p_\beta(m) = \frac{1}{\psi(\beta)} e^{-\langle \mathfrak{I}(m), \beta \rangle}.$$

The generalized probability density p_β can be written using the Massieu potential $\Phi : \Omega \to \mathbb{R}$

$$p_\beta(m) = e^{\Phi(\beta) - \langle \mathfrak{I}(m), \beta \rangle}, \quad \forall \beta \in \Omega \tag{5}$$

and

$$\Phi(\beta) = -\log(\psi(\beta)).$$

Another interesting quantity in this model is the thermodynamic heat $Q : \Omega \to \mathfrak{g}_k^*$ which is the first derivative of the Massieu potential, i.e.,

$$Q(\beta) := D\Phi(\beta) = \int_M \mathfrak{I}(m) p_\beta(m) d\mu = \mathbb{E}_\beta[\mathfrak{I}] \in \mathfrak{g}_k^*$$

where $\mathbb{E}_\beta$ denotes the expectation with respect to p_β.

Let Ω^* be the image of the Ω by the thermodynamic heat Q and assume that $Q : \Omega \to \Omega^*$ is a diffeomorphism. In this case, the entropy $s : \Omega^* \to \mathbb{R}$ can be defined as the Legendre transform of the Massieu potential $\Phi : \Omega \to \mathbb{R}$, namely

$$s(\nu) := \langle \nu, \beta \rangle - \Phi(\beta), \text{ where } \beta = Q^{-1}(\nu) \tag{6}$$

It has been proved in [1] that for every $\beta \in \Omega$,

$$s(Q(\beta)) = S(p_\beta), \text{ where } S(p) = -\int_M p \log p \, d\mu$$

where S is the entropy of the probability density p and $Q(\beta)$ is the thermodynamic heat.

4   Kirillov-Kostant-Souriau (KKS) Theorem for k-Coadjoint Orbits

In this section we will use the results of [18] that the coadjoint orbits are an example of polysymplectic structure.

Let G be a Lie group, $\mathfrak{g}$ be the Lie algebra of G and $\mathfrak{g}^*$ its dual. Denote by $(\tau_1, \cdots, \tau_k)$ an element of $\mathfrak{g}_k^* = \mathfrak{g}^* \times \overset{k}{\cdots} \times \mathfrak{g}^*$. The k-coadjoint orbit is defined as

$$\mathfrak{O}_{(\tau_1, \cdots, \tau_k)} = \{(Ad_g^*)^k(\tau_1, \cdots, \tau_k) \mid g \in G\}.$$

It is well known that for $k = 1$ and if the action is G-equivariant (KKS theorem) the coadjoint orbit $\mathfrak{D}_\tau$ has a symplectic structure ω_τ defined by the expression

$$\omega_\tau(\nu)\left(\xi_{\mathfrak{g}^*}(\nu), \eta_{\mathfrak{g}^*}(\nu)\right) = \pm\langle \nu, [\xi, \eta]\rangle$$

where $\nu \in \mathfrak{D}_\tau$ and $\xi_{\mathfrak{g}^*}(\nu), \eta_{\mathfrak{g}^*}(\nu) \in T_\nu \mathfrak{D}_\tau$.
For every $(\nu_1, \cdots, \nu_k) \in \mathfrak{D}_{(\tau_1, \cdots, \tau_k)}$ we have that

$$T_{(\nu_1, \cdots, \nu_k)}\mathfrak{D}_{(\tau_1, \cdots, \tau_k)} = \{\xi_{\mathfrak{g}^* \times \overset{k}{\cdots} \times \mathfrak{g}^*}(\nu_1, \cdots, \nu_k)\}$$

where $\xi_{\mathfrak{g}^* \times \overset{k}{\cdots} \times \mathfrak{g}^*}$ is the infinitesimal generator of the k-coadjoint action corresponding to ξ.
Define the canonical projection

$$pr_j : \mathfrak{D}_{(\tau_1, \cdots, \tau_k)} \to \mathfrak{D}_{\tau_j}$$
$$(\nu_1, \cdots, \nu_k) \mapsto \nu_j.$$

Then for $j \in \{1, \cdots, k\}$ and each $(\nu_1, \cdots, \nu_k) \in \mathfrak{D}_{(\tau_1, \cdots, \tau_k)}$ we obtain that

$$(pr_j)_*(\nu_1, \cdots, \nu_k)\left(\xi_{\mathfrak{g}^* \times \overset{k}{\cdots} \times \mathfrak{g}^*}(\nu_1, \cdots, \nu_k)\right) = \xi_{\mathfrak{g}^*}(\nu_j)$$

using the definition of the projection and the coadjoint representation. As a consequence of the above facts we have the following relations: $T_{(\nu_1, \cdots, \nu_k)}\mathfrak{D}_{(\tau_1, \cdots, \tau_k)} \subseteq T_{\nu_1}\mathfrak{D}_{\tau_1} \times \cdots \times T_{\nu_k}\mathfrak{D}_{\tau_k}$ and $\xi_{\mathfrak{g}^* \times \overset{k}{\cdots} \times \mathfrak{g}^*}(\nu_1, \cdots, \nu_k) \equiv (\xi_{\mathfrak{g}^*}(\nu_1), \cdots, \xi_{\mathfrak{g}^*}(\nu_k))$.

Proposition 1. *[18] Let ω_{τ_j} be the symplectic structure of the coadjoint orbit $\mathfrak{D}_{\tau_j}$ then the family $(\omega_\tau^1, \cdots, \omega_\tau^k)$ given by: $\omega_\tau^i := (pr_j)^*\omega_{\tau_j}$ is a k-polysymplectic structure on the k-coadjoint orbit $\mathfrak{D}_{(\tau_1, \cdots, \tau_k)}$ at $(\tau_1, \cdots, \tau_k)$.*

By definition, every ω_{τ_j} is a closed 2-form on $\mathfrak{D}_{(\tau_1, \cdots, \tau_k)}$. To prove the polysymplectic structure we need to verify that $\bigcap_{i=j}^k ker\omega_\tau^j = \{0\}$. Indeed using the expressions above we have:

$$\omega_\tau^j(\nu_1, \cdots, \nu_k)\left(\xi_{\mathfrak{g}^* \times \overset{k}{\cdots} \times \mathfrak{g}^*}(\nu_1, \cdots, \nu_k), \eta_{\mathfrak{g}^* \times \overset{k}{\cdots} \times \mathfrak{g}^*}(\nu_1, \cdots, \nu_k)\right) =$$

$$[(pr_j)^*\omega_\tau^j](\nu_1, \cdots, \nu_k)\left(\xi_{\mathfrak{g}^* \times \overset{k}{\cdots} \times \mathfrak{g}^*}(\nu_1, \cdots, \nu_k), \eta_{\mathfrak{g}^* \times \overset{k}{\cdots} \times \mathfrak{g}^*}(\nu_1, \cdots, \nu_k)\right) =$$

$\omega_{\tau_i}(\nu_j)\left(\xi_{\mathfrak{g}^*}(\nu_j), \eta_{\mathfrak{g}^*}(\nu_j)\right) = \pm\langle \nu_j, [\xi, \eta]\rangle$ depending whether it is left or right action.
Let $\xi_{\mathfrak{g}^* \times \overset{k}{\cdots} \times \mathfrak{g}^*}(\nu_1, \cdots, \nu_k)$ be an element of $\bigcap_{j=1}^k ker\omega_\tau^j$. As a consequence of last expressions we obtain that $\nu_j[\xi, \eta] = 0, \forall \eta \in \mathfrak{g}$ which is equivalent to $\xi_{\mathfrak{g}^*}(\nu_j) = 0$.
Therefore using the identification

$$\xi_{\mathfrak{g}^* \times \overset{k}{\cdots} \times \mathfrak{g}^*}(\nu_1, \cdots, \nu_k) \equiv (\xi_{\mathfrak{g}^*}(\nu_1), \cdots, \xi_{\mathfrak{g}^*}(\nu_k))$$

we obtain that

$$\xi_{\mathfrak{g}^* \times \overset{k}{\cdots} \times \mathfrak{g}^*}(\nu_1, \cdots, \nu_k) = 0 \text{ and thus } \bigcap_{i=j}^k ker\omega_\tau^j = \{0\}.$$

As a generalization of the polysymplectic KKS theorem we can use the same ideas developed before in the case of non-Null cohomology where the coadjoint orbit equation is $\mathfrak{O}_\tau = \{(Ad_g^*)^k\tau + \theta(g) \mid g \in G\}$ and $\tau \in \mathfrak{g}^* \times \overset{k}{\cdots} \times \mathfrak{g}^*$, of the affine left action of G on $\mathfrak{g}^* \times \overset{k}{\cdots} \times \mathfrak{g}^*$ given by $\tau \to (Ad_g^*)^k\tau + \theta(g)$ with the group one-cocycle $\theta = (\theta^1, \cdots, \theta^k) \in C^\infty(G, \mathfrak{g}_k^*)$ defined by

$$\theta^i(g) = \mathfrak{J}^i(\phi(m)) - Ad_{g^{-1}}^* \left(\mathfrak{J}^i(m)\right)$$

where $Ad_{g^{-1}}^*$ is the coadjoint action. The coadjoint orbit is endowed with a natural polysymplectic $\omega = (\omega^1, \cdots, \omega^k)$ with ω^j defined by

$$\omega^j(\tau)\left(ad_\xi^*\tau - \Theta(\xi,.), ad_\eta^*\tau - \Theta(\eta,.)\right) = \langle \tau_j, [\xi, \eta]\rangle - \Theta^j(\xi, \eta)$$

where the Lie algebra two-cocycle $\Theta : \mathfrak{g} \times \mathfrak{g} \to \mathbb{R}^k$ is a map defined by

$$\Theta^j(\xi, \eta) = \mathfrak{J}_{[\xi,\eta]}^j - \omega^j(\xi_M, \eta_M).$$

Recall that $\mathfrak{J}$ is the polysymplectic momentum map and ξ_M, η_M are the infinitesimal generators.

5   Poly-Poisson Structure of the Coadjoint Orbits

In [14] the authors have introduced poly-Poisson as a higher-order extension of Poisson structures. They showed that any poly-Poisson structure is endowed with a polysymplectic foliation. It is also proved that if a Lie group acts polysymplectically on a polysymplectic manifold then, under certain regularity conditions, the reduced space is a poly-Poisson manifold. Here we are interested in their results about coadjoint orbits within the polysymplectic Souriau's model.

5.1   Definitions k-poly-Poisson Structures

We recall in the following some definitions and results from [14] about k-poly-Poisson structures.

Definition 4. *A k-poly-Poisson structure on a manifold M is a couple $(S, \bar{\Lambda}^\#)$, where S is a vector subbundle of $(T_k^1 M)^* = T^*M \oplus \overset{k}{\cdots} \oplus T^*M$ and $\bar{\Lambda}^\# : S \to TM$ is a vector bundle morphism which satisfies the following conditions:*

i) $\bar{\alpha}\left(\bar{\Lambda}^\#(\bar{\alpha})\right) = 0$ for $\bar{\alpha} \in S$
ii) If $\bar{\alpha}\left(\bar{\Lambda}^\#(\bar{\beta})\right) = 0$ for every $\bar{\beta} \in S$ then $\bar{\Lambda}^\#(\bar{\alpha}) = 0$
iii) $\bar{\alpha}, \bar{\beta} \in \Gamma(S)$ are sections of S then we have the following integrability condition holds:

$$[\bar{\Lambda}^\#(\bar{\alpha}), \bar{\Lambda}^\#(\bar{\beta})] = \left(\bar{\Lambda}^\# \mathcal{L}_{\bar{\Lambda}^\#(\bar{\alpha})}\bar{\beta} - \mathcal{L}_{\bar{\Lambda}^\#(\bar{\beta})}\bar{\alpha} - d(\bar{\beta}(\bar{\Lambda}^\#(\bar{\alpha})))\right) \quad (7)$$

The triple $(M, S, \bar{\Lambda}^\#)$ will be called a k-poly-Poisson manifold or simply a poly-Poisson manifold.

A polysymplectic manifold is a poly-Poisson manifold. Indeed in the same mentioned paper above the authors proved that polysymplectic structures can be thought of as vector bundle morphisms, obtaining an equivalent definition of polysymplectic manifolds.

Proposition 2. *[14] A k-polysymplectic structure on a manifold M is a vector bundle morphism $\bar{\omega}^b : TM \to (T_k^1 M)^*$ of the tangent bundle TM to M on the cotangent bundle of k-covelocities of M satisfying the conditions:*

i) Skew-symmetry: $\bar{\omega}^b(X)(X) = (0, \cdots, 0)$, $X \in T_p M$, $p \in M$.
ii) Non degeneracy ($\bar{\omega}^b$ is a monomorphism): $Ker\left(\bar{\omega}^b\right) = \{0\}$
iii) Integrability condition holds:

$$\bar{\omega}^b\left([X, Y]\right) = \mathcal{L}_X \bar{\omega}^b(Y) - \mathcal{L}_Y \bar{\omega}^b(X) + d(\bar{\omega}^b(X)(Y))$$

As one would expect, polysymplectic manifolds are a particular example of poly-Poisson manifolds. Let $(M, \bar{\omega})$ be a polysymplectic manifold. Since $\bar{\omega}^b$ is a monomorphism, $S := Im(\bar{\omega}^b)$ is a vector subbundle of $(T_k^1 M)^*$. Moreover, we consider $\bar{\Lambda}^\# := (\bar{\omega}^b)^{-1}_{|S} : S \to TM$. One can verify that the conditions of definition (4) are satisfied.

It is clear that Poisson manifolds are 1-poly-Poisson manifolds. Indeed, if Π is a Poisson structure on M then $(S = T^*M, \Pi^\#)$ is poly-Poisson, where $\Pi^\# : T^*M \to TM$ is the vector bundle morphism induced by the Poisson 2-vector Π (see [16] page 13). Moreover, it is well known that the generalized distribution $p \in M \to \Pi^\#(T_p^* M) \subseteq T_p M$ is a symplectic generalized foliation. Indeed let us define $C = \Pi^\#(T^*M)$ to be the characteristic field of the Poisson manifold $(M, S = T^*M, \Pi^\#)$. Then Alan Weinstein [20] proves that, loosely speaking, a Poisson manifold is the disjoint union of symplectic manifolds, arranged in such a way that the union is endowed with a differentiable structure.

In [14] the authors have been able to prove that a k-poly-Poisson manifold admits a k-polysymplectic generalized foliation.

Theorem 1. *[14] Let $(S, \bar{\Lambda}^\#)$ be a k-poly-Poisson structure on a manifold M. Then M admits a k-polysymplectic generalized foliation F whose characteristic space at the point $p \in M$ is $\mathfrak{F}(p) = \bar{\Lambda}^\#(S_p) \subseteq T_p M$. $\mathfrak{F}$ is said to be the canonical k-polysymplectic foliation of M.*

There is a converse of this theorem namely:

Theorem 2. *[14] Let S be a vector subbundle of $(T_k^1)^* M$ and $\bar{\Lambda}^\# : S \subseteq (T_k^1)^* M \to TM$ a vector bundle morphism such that:*

i) $\bar{\alpha}\left(\bar{\Lambda}^\#(\bar{\alpha})\right) = 0$ for $\bar{\alpha} \in S$ and
ii) If $\bar{\alpha}\left(\bar{\Lambda}^\#(\bar{\beta})\right) = 0$ for every $\bar{\beta} \in S$ then $\bar{\Lambda}^\#(\bar{\alpha}) = 0$.

Assume also that the generalized distribution

$$p \in M \to \mathfrak{F}(p) = \bar{\Lambda}^\#(S_p) \subseteq T_p M$$

is involutive and for every leaf L the corresponding $\mathbb{R}^k$-valued 2-form $\bar{\omega}$ on L (induced by the morphism $\bar{\Lambda}^\#$) is closed. Then, the couple $(S, \bar{\Lambda}^\#)$ is a k-poly-Poisson structure on M and $\mathfrak{F}$ is the canonical k-polysymplectic foliation of M.

5.2 Poly-Poisson Structures and Reduction of Polysymplectic Manifolds

As it is well known, one can obtain Poisson manifolds from symplectic manifolds through a reduction procedure. Let us recall some details of the process. Let (M, ω) be a symplectic manifold and G be a Lie group acting freely and properly on M. If the action $\Phi : G \times M \to M$ preserves the symplectic structure (i.e., $\Phi_g^* \omega = \omega$, for all $g \in G$) then the orbit space M/G inherits a Poisson structure Λ in such a way that the bundle projection $\pi : M \to M/G$ is a Poisson map, that is,

$$\Lambda^\#(\pi(x)) = T_x \pi \circ (\omega(x))^{-1} \circ T_x^* \pi, \text{ for } x \in M.$$

The authors in [14] used the introduced poly-Poisson structures to play the same role with respect to polysymplectic manifolds and the next theorem asserts that under certain regularity conditions the reduction of a polysymplectic manifold is a poly-Poisson manifold. One application of this theorem will show that the coadjoint orbits inherit a poly-Poisson structure.

Theorem 3. *[14] Let $(M, \omega_1, \cdots, \omega_k)$ be a polysymplectic manifold and G be a Lie group that acts over M $(\Phi : G \times M \to M)$ freely, properly and satisfies $\Phi_g^* \omega_i = \omega_i$, for every $g \in G$ and $1 \leq i \leq k$. Suppose that $V\pi$ is the vertical bundle of the principal bundle projection $\pi : M \to M/G$ and that the following conditions hold:*

1. *$Im(\bar{\omega}^b) \cap (V\pi)^{\circ k}$ is a vector subbundle, where $(V\pi)^{\circ k} = (V\pi)^\circ \times \overset{k}{\cdots} \times (V\pi)^\circ$ (where $\circ$ denotes the annihilator)*
2. *$\left((\bar{\omega}^b)(p)\right)^{-1} \left(((V_p\pi)^\perp)^{\circ k} \cap (V\pi)^{\circ k} \cap S(p)\right) \subseteq V_p\pi$, where $(V\pi)^\perp$ is the polysymplectic orthogonal, that is, $(V\pi)^\perp = \cap_{i=1}^k (\omega_i^b)^{-1}((V\pi)^\circ)$ and S is the image of $\bar{\omega}^b$ that is $S = Im(\bar{\omega}^b)$.*

Then, there is a natural k-poly-Poisson structure on M/G.

Let G be a Lie group. The cotangent bundle of k-covelocities of G, $(T_k^1)^* G$ has a canonical k-polysymplectic structure. In addition, the action of G on itself by left translations

$$L : G \times G \to G$$
$$(g, h) \mapsto L(g, h) = g.h$$

can be lifted to an action $L^{(T_k^1)^*}$ on $(T_k^1)^* G$ by

$$L^{(T_k^1)^*} : G \times (T_k^1)^* G \to (T_k^1)^* G$$
$$(g, (\alpha_1, \cdots, \alpha_k)) \mapsto (L_{g^{-1}}^* \alpha_1, \cdots, L_{g^{-1}}^* \alpha_k).$$

If we consider the left trivialization of the tangent bundle to G, $TG \cong G \times \mathfrak{g}$, and the corresponding trivialization $(T_k^1)^* G \cong G \times \mathfrak{g}_k^*$, where $\mathfrak{g}_k^* = \mathfrak{g}^* \times \overset{k}{\cdots} \times \mathfrak{g}^*$. The lift action can be written as

$$L^{(T_k^1)^*} : G \times (G \times \mathfrak{g}_k^*) \to G \times \mathfrak{g}_k^*$$
$$(g, (h, \tau_1, \cdots, \tau_k)) \mapsto (g.h, \tau_1, \cdots, \tau_k),$$

and the associated principal bundle $\pi : (T_k^1)^*G \to ((T_k^1)^*G)/G$ can be identified with the trivial principal bundle $\pi : G \times \mathfrak{g}_k^* \to \mathfrak{g}_k^*$. Thus, the vertical bundle takes the expression

$$V\pi_{(g,\tau_1,\cdots,\tau_k)} = \{(X,0,\cdots,0) \text{ such that } X \in T_pG\}.$$

Therefore, its annihilator is

$$(V\pi)^\circ_{(g,\tau_1,\cdots,\tau_k)} = \{(0,\xi_1,\xi_2,\cdots,\xi_k) \mid \xi_1,\cdots,\xi_k \in \mathfrak{g}\}.$$

The authors in [14] proved that this action of G on $(T_k^1)^*G \cong G \times \mathfrak{g}_k^*$, endowed with the canonical polysymplectic structure, satisfies the hypotheses of Theorem (3).

As a consequence, this leads to the fact that the polysymplectic leaves are just the orbits of the k-coadjoint action and there is a natural k-poly-Poisson structure on them.

References

1. Barbaresco, F., Gay-Balmaz, F.: Lie group cohomology and (Multi)symplectic integrators: new geometric tools for lie group machine learning based on Souriau geometric statistical mechanics. Entropy **22**, 498 (2020)
2. Barbaresco, F.: Higher order geometric theory of information and heat based on polysymplectic geometry of Souriau Lie groups thermodynamics and their contextures: The bedrock for Lie Group machine learning. Entropy **20**, 840 (2018)
3. Barbaresco, F.: Symplectic foliation structures of non equilibrium thermodynamics as dissipation model: application to Metriplectic nonlinear lindblad quantum master equation. Entropy **24**, 1626 (2022)
4. Barbaresco, F.: Lie group statistics and lie group machine learning based on Souriau Lie groups thermodynamics & Koszul-Souriau-Fisher metric: new entropy definition as generalized casimir invariant function in coadjoint representation. Entropy **22**, 642 (2020)
5. Barbaresco, F.: Symplectic foliation model of information geometry for statistics and learning on lie groups. SEE MaxEnt'22 conference, Institut Henri Poincaré, July 18th 2022; video: https://www.carmin.tv/fr/video/symplectic-foliation-model-of-information-geometry-for-statistics-and-learning-on-lie-groups
6. Barbaresco, F.: Symplectic theory of heat and information geometry. In: Nielsen, F., Rao, A.S., Rao, C.R., (Eds) Handbook of statistics n°46 "Geometry and Statistics", 1st edn., Elsevier, Amsterdam (2022). ISBN: 9780323913454
7. Barbaresco, F.: Jean-Marie Souriau's Symplectic Model of Statistical Physics: Seminal Papers on Lie Groups Thermodynamics - Quod Erat Demonstrandum. In: Barbaresco, F., Nielsen, F. (eds.) SPIGL 2020. SPMS, vol. 361, pp. 12–50. Springer, Cham (2021). https://doi.org/10.1007/978-3-030-77957-3_2
8. Barbaresco, F.: Souriau-Casimir Lie groups thermodynamics and machine learning. In: Barbaresco, F., Nielsen, F. (eds.) SPIGL 2020. SPMS, vol. 361, pp. 53–83. Springer, Cham (2021). https://doi.org/10.1007/978-3-030-77957-3_3
9. Barbaresco, F.: Koszul lecture related to geometric and analytic mechanics, Souriau's Lie group thermodynamics and information geometry. Inf. Geom. **4**, 245–262 (2021)

10. Barbaresco, F.: Densité de probabilité gaussienne à maximum d'entropy pour les groupes de Lie basée sur le modèle symplectique de Jean-Marie Souriau. In: Proceedings of the GRETSI'22 Conference, Nancy, France, 6–9 September 2022

11. Barbaresco, F.: Théorie symplectique de l'Information et de la chaleur: Thermodynamique des groupes de Lie et définition de l'entropy comme fonction de Casimir. In: Proceedings of the GRETSI'22 Conference, Nancy, France, 6–9 September 2022

12. Barbaresco, F.: Entropy geometric structure as casimir invariant function in coadjoint representation: geometric theory of heat & information geometry based on Souriau Lie groups thermodynamics and Lie Algebra Cohomology. In: Frontiers in Entropy Across the Disciplines; World Scientific: Singapore, pp. 133–158, Chapter 5 (2022)

13. Günther, C.: The polysymplectic Hamiltonian formalism in field theory and calculus of variations I: the local case. J. Diff. Geom. **25**, 23–53 (1987)

14. Iglesias, D., Marrero, J.C., Vaquero, M.: Poly-Poisson structures. Lett. Math. Phys. **103**, 1103–1133 (2013)

15. Koszul, J.L.; Zou, Y.M.: Introduction to Symplectic Geometry. Springer NatureSingapre and Science Press, Cham (2019)

16. Marle, C.M.: Symmetries of Hamiltonian systems on symplectic and Poisson manifolds. arXiv:1401.8157v2 (2014)

17. Gotay, M.J.; Isenberg, J.; Marsden, J.E.; Montgomery, R.; Sniatycki, J.; Yasskin, P.B.: Momentum maps and classical fields. Part I: Covariant field theory. arXiv 1997, arXiv:physics/9801019v2

18. Marrero, J. C.; Narciso R. R,; Salgado, M.; Vilarino, S.: Reduction of polysymplectic manifolds, J. Phys. A: Math. Theor. **48**, 055206 (2015)

19. Souriau, J.-M.: Structure des Systèmes dynamiques. Dunod, Paris, France (1969)

20. Weinstein, A.: The local structure of Poisson manifolds, J. Diff. Geom. **18** (1983) 523?557 and 22 (1985) 255

Polysymplectic Souriau Lie Group Thermodynamics and Entropy Geometric Structure as Casimir Invariant Function

Mohamed El Morsalani[✉]

Landesbank Baden -Wuerttemberg, Stuttgart, Germany
morsalan@hotmail.de

Abstract. In this paper we will continue our work on the extension of Souriau's symplectic model to the polysymplectic case that has been started in [14]. This paper contains a summary of some original results we will publish in a coming paper in preparation [15]. Here we will show that the entropy is still a Casimir Function as in the Souriau standard model. One of the original ideas is the introduction of an extended Lie-Poisson bracket. With its help we could recover many of the properties and results about the entropy as well as dissipative and production dynamics known from the Souriau standard model.

Keywords: higher order thermodynamics · Lie groups thermodynamics · Poly-symplectic and poly-Poisson manifold · coadjoint orbits · KKS theorem · Polysymplectic foliation · Extended Lie-Poisson bracket · entropy · casimir function

1 Introduction

The first extension of Souriau ideas to the polysymplectic setting has been initiated by F. Barbaresco in his paper [2]. F. Barbaresco and F. Gay-Balmaz used a general framework in [1] to study Souriau Lie group thermodynamics. Among other things they have been able to show many interesting results in the polysymplectic context we will recall in the following sections and apply. For the poly-Poisson structure ideas we will use the results proved in the paper [16].

It is worth to mention that the polysymplectic model is motivated by higher-order model of statistical physics [2]. For instance, for small data analytics (rarified gases, sparse statistical surveys, . . .), the density of maximum entropy should consider higher order moments constraints, so that the Gibbs density is not only defined by first moment but fluctuations request 2nd order and higher moments (see [1,2] for references).

In this summary we will highlight the fact that the entropy keeps its property of Casimir invariance using two methods: the first one using the poly-Poisson structure introduced in [16] and the second one using the extended Lie-Poisson bracket we introduced in [15].

© The Author(s), under exclusive license to Springer Nature Switzerland AG 2023
F. Nielsen and F. Barbaresco (Eds.): GSI 2023, LNCS 14072, pp. 134–143, 2023.
https://doi.org/10.1007/978-3-031-38299-4_15

2   Polysymplectic Souriau Lie Group Thermodynamics Model

In this section we will present the basic facts and known results especially from [1].

2.1   Notations and Definitions

For our setting we will need:

- A polysymplectic manifold (M, ω) which is a manifold endowed with a closed nondegenerate $\mathbb{R}^k$-valued 2-form. We can identify ω with a collection $(\omega^1, \cdots, \omega^k)$, of closed 2-forms with $\cap_{i=1}^k ker\omega^i = \{0\}$.
- A polysymplectic Lie group action $\phi : G \times M \to M$ of G on M is polysymplectic, meaning $\phi_g^* \omega^i = \omega^i \quad \forall g \in G, 1 \leq i \leq k$
- Let $\mathfrak{g}$ be the Lie algebra of G and $\mathfrak{g}^*$ its dual Lie algebra. The action is Hamiltonian and admits a polysymplectic momentum map $\mathfrak{J} : M \to \mathfrak{g}_k^* = \mathfrak{g}^* \times \overset{k}{\cdots} \times \mathfrak{g}^*$, which satisfies

$$i_{\xi_M} \omega^j = d\mathfrak{J}_\xi^j, \ \forall j$$

  where $\mathfrak{J}_\xi : M \to \mathbb{R}^k$ is defined by

$$\mathfrak{J}_\xi(m) = \mathfrak{J}(m).\xi$$

  and ξ_M is the infinitesimal generator of the action corresponding to ξ.
- if M is connected, there is group one-cocycle $\theta \in C^\infty(G, \mathfrak{g}_k^*)$, $\theta = (\theta^1, \cdots, \theta^k)$ defined by

$$\theta^i(g) = \mathfrak{J}^i(\Phi(m)) - Ad_{g^{-1}}^* \left(\mathfrak{J}^i(m) \right)$$

  where $Ad_{g^{-1}}^*$ is the coadjoint action.

- $\mathfrak{g}_k = \mathfrak{g} \times \overset{k}{\cdots} \times \mathfrak{g}$ a vector space, whose elements will be denoted $\beta = (\beta_1, \cdots, \beta_k)$ the generalisations of the inverse temperature.
- A duality pairing between elements ν of the dual space $\mathfrak{g}_k^*$ and elements $\beta \in \mathfrak{g}_k$ denoted as $\langle \nu, \beta \rangle$.
- The manifold M is endowed with a volume form $d\mu$.

Denote by $\Omega \subset \mathfrak{g}_k$ the largest open set such that for all $\beta \in \Omega$ the two integrals

$$\int_M e^{-\langle \mathfrak{J}(m), \beta \rangle} d\mu \in \mathbb{R} \text{ and } \int_M \mathfrak{J}(m) e^{-\langle \mathfrak{J}(m), \beta \rangle} d\mu \in \mathfrak{g}_k^* \tag{1}$$

converge.

On Ω one can define the so-called partition function (or characteristic function) $\psi : \Omega \to \mathbb{R}$, given by

$$\psi(\beta) = \int_M e^{-\langle \mathfrak{J}(m), \beta \rangle} d\mu \tag{2}$$

This permits us to define the generalized Gibbs probability densities associated to each $\beta \in \Omega$:

$$p_\beta(m) = \frac{1}{\psi(\beta)} e^{-\langle \mathfrak{I}(m), \beta \rangle} \tag{3}$$

There are two facts which are very important to mention: For applications in information geometry it is required that $\beta \to p_\beta$ is injective. Furthermore the Gibbs densities are not defined on the whole vector space $\mathfrak{g}_k$ but only on the open subset Ω. From now on we assume that Ω is not empty. Its elements are called geometric temperatures.

If the integrals defining Ω are normally convergent then for example Ω is convex. The generalized probability density ψ can be written using the Massieu potential $\Phi : \Omega \to \mathbb{R}$

$$p_\beta(m) = e^{\Phi(\beta) - \langle U(m), \beta \rangle}, \quad \forall \beta \in \Omega \tag{4}$$

and $\Phi(\beta) = -\log(\psi(\beta))$. Another interesting quantity in this framework is the thermodynamic heat $Q : \Omega \to \mathfrak{g}_k^*$ which is the first derivative of the Massieu potential, i.e.,

$$Q(\beta) := D\Phi(\beta) = \int_M \mathfrak{I}(m) p_\beta(m) d\mu = \mathbb{E}_\beta[\mathfrak{I}] \in \mathfrak{g}_k^* \tag{5}$$

where $\mathbb{E}_\beta$ denotes the expectation with respect to p_β.

2.2 Entropy and Some Associated Results

Let Ω^* be the image of the Ω by the thermodynamic heat Q and assume that $Q : \Omega \to \Omega^*$ which is a diffeomorphism. In this case, the entropy $s : \Omega^* \to \mathbb{R}$ can be defined as the Legendre transform of the Massieu potential $\Phi : \Omega \to \mathbb{R}$, namely

$$s(\nu) := \langle \nu, \beta \rangle - \Phi(\beta), \text{ where } \beta = Q^{-1}(\nu) \tag{6}$$

The next result has been proved in [1].

Lemma 1. *For every $\beta \in \Omega$, the following equality holds true:*

$$s(Q(\beta)) = S(p_\beta), \text{ where } S(p) = \int_M p \log p \, d\mu \tag{7}$$

S is the entropy of the probability density p and $Q(\beta)$ is the thermodynamic heat.

The following equivariance properties have been proved in a general framework in [1] and applied there also for the polysymplectic case:

$$(Ad_g)^k \Omega = \Omega, \quad \psi\left((Ad_g)^k \beta\right) = \psi(\beta) e^{\langle \theta(g^{-1}), \beta \rangle}, \quad p_\beta \circ \phi_\beta = p_{(Ad_{g^{-1}})^k \beta}$$

and

$$\Phi\left((Ad_g)^k \beta\right) = \Phi(\beta) - \langle \theta(g^{-1}), \beta \rangle \tag{8}$$

$$Q\left((Ad_g)^k \beta\right) = (Ad_{g^{-1}}^*)^k(Q(\beta)) + \theta(g) \tag{9}$$

$$s\left((Ad^*_{g^{-1}})^k(\nu) + \theta(g)\right) = s(\nu) \tag{10}$$

$$K\left((Ad_g)^k\beta\right)\left((Ad_g)^k(\delta\beta_1), (Ad_g)^k(\delta\beta_2)\right) = K(\beta)\left(\delta\beta_1, \delta\beta_2\right) \tag{11}$$

for every $g \in G$.

Furthermore Ω^* is invariant under the affine action $\nu \in \mathfrak{g}^*_k \mapsto (Ad^*_{g^{-1}})^k(\nu) + \theta(g) \in \mathfrak{g}^*_k$

3 Casimir Property of the Entropy Within the Poly-Poisson Structure

In this section we want to show the Casimir property of the entropy within the poly-Poisson structure.

3.1 Poly-Poisson Structure on $\mathfrak{g}^*_k$

In [16] the authors have introduced poly-Poisson as a higher-order extension of Poisson structures. They showed that any poly-Poisson structure is endowed with a polysymplectic foliation. It is also proved that if a Lie group acts polysymplectically on a polysymplectic manifold then, under certain regularity conditions, the reduced space is a poly-Poisson manifold. In our coming paper [15], we use their results to show that coadjoint orbits within the polysymplectic Souriau's model possess a poly-Poisson structure.

Definition 1. *A k-poly-Poisson structure on a manifold M is a couple $(S, \bar{\Lambda}^\#)$, where S is a vector subbundle of $(T^1_k M)^* = T^*M \oplus \overset{k}{\cdots} \oplus T^*M$ and $\bar{\Lambda}^\# : S \to TM$ is a vector bundle morphism which satisfies the following conditions:*

1. *$\bar{\alpha}\left(\bar{\Lambda}^\#(\bar{\alpha})\right) = 0$ for $\bar{\alpha} \in S$*
2. *If $\bar{\alpha}\left(\bar{\Lambda}^\#(\bar{\beta})\right) = 0$ for every $\bar{\beta} \in S$ then $\bar{\Lambda}^\#(\bar{\alpha}) = 0$*
3. *$\bar{\alpha}, \bar{\beta} \in \Gamma(S)$ are sections of S then we have the following integrability condition holds:*

$$[\bar{\Lambda}^\#(\bar{\alpha}), \bar{\Lambda}^\#(\bar{\beta})] = \left(\bar{\Lambda}^\#\mathcal{L}_{\bar{\Lambda}^\#(\bar{\alpha})}\bar{\beta} - \mathcal{L}_{\bar{\Lambda}^\#(\bar{\beta})}\bar{\alpha} - d(\bar{\beta}(\bar{\Lambda}^\#(\bar{\alpha})))\right) \tag{12}$$

The triple $(M, S, \bar{\Lambda}^\#)$ will be called a k-poly-Poisson manifold or simply a poly-Poisson manifold.

Let $\mathfrak{g}$ be the Lie algebra associated to the Lie group G and recall that $\mathfrak{g}_k = \mathfrak{g} \times \overset{k}{\cdots} \times \mathfrak{g}$; $\mathfrak{g}^*_k = \mathfrak{g}^* \times \overset{k}{\cdots} \times \mathfrak{g}^*$. In the case of non-null cohomology as explained in [15] the coadjoint orbit is endowed with a natural polysymplectic structure infered from the affine KKS theorem where the infinitesimal generator of the affine-coadjoint action associated is $ad^*_\xi \tau - \Theta(\tau, .)$ where Θ is the 2-cocycle. Using the machinery in [16] we extend in [15] their construction to the case of non-null cohomology to exhibit the natural poly-Poisson structure on $\mathfrak{g}^*_k$ foliated by the coadjoint orbits as polysymplectic leaves.

Let $\tau = (\tau_1, \cdots, \tau_k) \in \mathfrak{g}_k^*$, then S is defined by

$$S(\tau) = \{(\tau, ((\xi, 0, \cdots, 0), (0, \xi, \cdots, 0), \cdots, (0, 0, \cdots, \xi))) \mid \xi \in \mathfrak{g}\}$$

and $\bar{\Lambda}^{\#}$ is characterized by the following expression

$$\bar{\Lambda}_\tau^{\#} (\tau, ((\xi, 0, \cdots, 0), (0, \xi, \cdots, 0), \cdots, (0, 0, \cdots, \xi))) =$$
$$(ad_\xi^* \tau_1 - \Theta(\tau_1, .), \cdots, ad_\xi^* \tau_k - \Theta(\tau_k, .))$$

3.2   Entropy as Casimir Function of the Poly-Poisson Structure

Recall that the entropy $s : \Omega^* \to \mathbb{R}$ is the Legendre transform of the Massieu potential $\Phi : \Omega \to \mathbb{R}$, defined in (6)

$$s(\nu) := \langle \nu, \beta \rangle - \Phi(\beta), \text{ where } \beta = Q^{-1}(\nu). \tag{13}$$

One of the important results in [1] is:

$$s\left((Ad_{g^{-1}}^*)^k \tau_0 + \theta(g)\right) = s(\nu); \quad \forall g \in G \tag{14}$$

which means that the entropy s is constant on the affine coadjoint orbits defined by

$$\mathfrak{O}_{\tau_0} = \{(Ad_{g^{-1}}^*)^k \tau_0 + \theta(g) \mid g \in G\}$$

for $\tau_0 \in \mathfrak{g}_k^*$. This means that affine coadjoint orbits in the polysymplectic Souriau's model are foliated into level sets of the entropy.
In the previous sections we exhibited a poly-Poisson structure on $\mathfrak{g}_k^*$ which admits a polysymplectic foliation whose leaves are the k-coadjoint orbits. It is known that a Poisson manifold naturally decomposes into symplectic manifolds. This holds true for the Lie-Poisson structure on $\mathfrak{g}^*$ and these symplectic manifolds are the coadjoint orbits of G in $\mathfrak{g}^*$ and furthermore the Casimir functions are the Ad^*-invariant functions ([18], Proposition 7.7).
If we denote by $An(\mathfrak{g}_k^*)$ the space of analytic function on the dual space of the Lie algebra $\mathfrak{g}_k$, a function $F \in An(\mathfrak{g}_k^*)$ is a Casimir invariant if for any $g \in G$, $X \in \mathfrak{g}_k^*$, we have $F(Ad_g^* X) = F(X)$.
Using the same arguments as in ([4], 8.3) we can observe that Souriau Entropy is an extended Casimir invariant function in case of non-null cohomology in the polysymplectic context. Indeed we have seen that one of the equivariance equations is:

$$s\left((Ad_{g^{-1}}^*)^k(\nu) + \theta(g)\right) = s(\nu) \quad \forall \nu \in \mathfrak{g}_k^*$$

and that is the equivalent equation in the polysymplectic equation that F. Barbaresco used in [4] to prove Souriau Entropy is invariant in coadjoint representation in general case of non-null cohomology, that we could write

$$s\left((Ad_g^{\#}(\nu))\right) = s(\nu) \quad \forall \nu \in \mathfrak{g}_k^*$$

where $Ad_g^\#(\nu) = (Ad_{g^{-1}}^*)^k(\nu) + \theta(g)$ denotes the affine coadjoint action. This is also true in case of null-cohomology when the Souriau cocycle cancels $\theta(g) = 0$, and we recover classical generalized Casimir invariant function definition on coadjoint representation for Entropy $s\left((Ad_{g^{-1}}^*)^k(\nu)\right) = s(\nu)$.

Proposition 1. *The poly-Poisson structure on $\mathfrak{g}_k^*$ exhibited previously admits a polysymplectic foliation whose leaves are the k-coadjoint orbits that admit the entropy of the polysymplectic Souriau model as a Casimir function.*

3.3 Souriau Entropy Equation from His 1974 Paper in the Polysymplectic Context

Souriau has developed the following equation in greater depth [20]. He observed that the following identity holds and it is still true in the polysymplectic context:

$$s\left[Q\left((Ad_g)^k(\beta)\right)\right] = s\left[(Ad_g^*)^k(Q) + \theta(g)\right] = s(Q) \tag{15}$$

with $\beta = (\beta_1, \cdots, \beta_k) \in \mathfrak{g}_k$, $Q : \Omega \to \mathfrak{g}_k^*$ and $\theta = (\theta_1, \cdots, \theta_k)$ is the one-cocycle. Now recall that the entropy can be written as

$$s(Q_1, \cdots, Q_k) = \sum_{j=1}^k \langle \beta_j, Q_j \rangle - \Phi(\beta_1, \cdots, \beta_k) \tag{16}$$

with higher order geometric temperature β_j and higher order heat Q_j

$$\beta_j = \frac{\partial s(Q_1, \cdots, Q_k)}{\partial Q_j} \text{ and } Q_j = \frac{\partial \Phi(\beta_1, \cdots, \beta_k)}{\partial \beta_j} \tag{17}$$

where the dual potential function Φ has the expression

$$\Phi(\beta_1, \cdots, \beta_k) = \int_M e^{-\sum_{j=1}^k \langle \beta_j, \mathfrak{J}^j(\xi) \rangle} d\mu \tag{18}$$

Consider the curve $t \mapsto (Ad)^k_{\exp(tZ)}\beta$ with $\beta \in \mathfrak{g}_k$, $t \in \mathbb{R}$ and $Z \in \mathfrak{g}$. The curve $(Ad)^k_{\exp(tZ)}\beta$ passes, for $t = 0$, through the point β, since $(Ad)^k_{\exp(0)}$ is the map for the identity of the Lie algebra $\mathfrak{g}$. This curve is in the adjoint orbit of β. So, by taking its derivative with respect to t, for $t = 0$, a tangent vector in β is computed at the adjoint orbit of this point. When Z takes arbitrary values in $\mathfrak{g}$, the vectors generate all the vector space tangent in β to the orbit of this point:

$$\left.\frac{d\Phi\left((Ad)^k_{\exp(tZ)}\beta\right)}{dt}\right|_{t=0} = \sum_{j=1}^k \left\langle \frac{d\Phi}{d\beta_j}, \left(\left.\frac{d(Ad_{\exp(tZ)}\beta_j)}{dt}\right|_{t=0}\right) \right\rangle \tag{19}$$

$$= \sum_{j=1}^k \langle Q_j, ad_Z\beta_j \rangle = \sum_{j=1}^k \langle Q_j, [Z, \beta_j] \rangle$$

We know that
$$\Phi\left((Ad_g)^k\beta\right) = \Phi(\beta) - \langle\theta(g^{-1}),\beta\rangle \tag{20}$$

If we set $g = \exp(tZ)$ in this equation, we obtain

$$\Phi\left((Ad_{\exp(tZ)})^k\beta\right) = \Phi(\beta) - \langle\theta(\exp(-tZ)),\beta\rangle$$

then we differentiate this equation with respect to t at $t = 0$, we recover the equation of Souriau in the polysymplectic framework

$$\left.\frac{d\Phi\left((Ad)^k_{\exp(tZ)}\beta\right)}{dt}\right|_{t=0} = \sum_{j=1}^{k}\langle Q_j,[Z,\beta_j]\rangle = -\sum_{j=1}^{k}\langle d\theta(-Z),\beta_j\rangle \tag{21}$$

which yields using the property $\Theta(X,Y) = -\langle d\theta(X),Y\rangle$ to the polysymplectic equation equivalent to the Souriau equation in the standard model:

$$\sum_{j=1}^{k}\langle Q_j,[Z,\beta_j]\rangle + \sum_{j=1}^{k}\Theta^j(\beta_j,Z) = 0 \tag{22}$$

Recall that $\beta_j = \frac{ds}{dQ_j}$, $ad_{\beta_j}Z = [Z,\beta_j]$ and $\Theta^j(\beta_j,Z) = \langle\Theta^j(\beta_j,.),Z\rangle$ then we can write the last equation as

$$\left\langle\sum_{j=1}^{k}ad^*_{\frac{\partial s}{\partial Q_j}}Q_j - \Theta^j(\frac{\partial s}{\partial Q_j},.),Z\right\rangle = 0, \quad \forall Z \tag{23}$$

In this way we recover in the polysymplectic context the natural extension of the Casimir property of the entropy s observed in the Souriau standard model:

$$\sum_{j=1}^{k}ad^*_{\frac{\partial s}{\partial Q_j}}Q_j - \sum_{j=1}^{k}\Theta^j(\frac{\partial s}{\partial Q_j},.) = 0 \tag{24}$$

given for the standard model by the expression $ad^*_{\frac{\partial s}{\partial Q}}Q - \Theta(\frac{\partial s}{\partial Q},.) = 0$.

4   The Extended Lie-Poisson Bracket on $\mathfrak{g}_k^*$ and the Entropy as Casimir Invariant

Following the natural extension of the Casimir property to the polysymplectic context, let us define the set $C^\infty(\mathfrak{g}_k^*)$ of the functions $f : \mathfrak{g}_k^* \to \mathbb{R}$. Let us define the extended Lie-Poisson bracket on $\mathfrak{g}_k^*$ by the following expression for $f,g \in C^\infty(\mathfrak{g}_k^*)$ and $\beta = (\beta_1,\cdots,\beta_k) \in \mathfrak{g}_k^*$:

$$\{f,g\}_\Theta(\beta) = \sum_{j=1}^{k}\left\langle\beta_j,\left[\frac{\partial f}{\partial\beta_j},\frac{\partial g}{\partial\beta_j}\right]\right\rangle - \sum_{j=1}^{k}\Theta^j\left(\frac{\partial f}{\partial\beta_j},\frac{\partial g}{\partial\beta_j}\right) \tag{25}$$

Recall that as the 2-cocycle Θ is skew-symmetric, and satisfies the two-cocycle identity, its components Θ^j possess the same properties. Adding to this fact that the first sum in the extended Lie-Poisson bracket depends on Lie-Bracket in a linear way implies that $(C^\infty(\mathfrak{g}_k^*), \{.,.\}_\Theta)$ is a Poisson manifold (see [15]). Using the extended Poisson bracket we can rewrite

$$\sum_{j=1}^{k} ad^*_{\frac{\partial s}{\partial Q_j}} Q_j - \sum_{j=1}^{k} \Theta^j \left(\frac{\partial s}{\partial Q_j}, . \right) = 0$$

by considering $h \in C^\infty(\mathfrak{g}_k^*)$ and then we can show easily that

$$\{s, h\}_\Theta (Q) = \sum_{j=1}^{k} \left\langle Q_j, \left[\frac{\partial s}{\partial Q_j}, \frac{\partial h}{\partial Q_j} \right] \right\rangle - \sum_{j=1}^{k} \Theta^j \left(\frac{\partial s}{\partial Q_j}, \frac{\partial h}{\partial Q_j} \right) = 0 \qquad (26)$$

which holds for every $h \in C^\infty(\mathfrak{g}_k^*)$ which characterizes Entropy as an invariant Casimir function in coadjoint representation in the polysymplectic context. This yields to the following proposition.

Proposition 2. *The entropy s of the polysymplectic Souriau model is a Casimir function for the extended Lie-Poisson bracket with 2-cocycle equation, i.e., it satisfies*

$$\{s, h\}_\Theta = 0$$

for every smooth function $h : \mathfrak{g}_k^* \to \mathbb{R}$.

As in the standard model we have in the polysymplectic model that the information manifold foliates into level sets of the entropy, containing a family of coadjoint orbits, that could be interpreted in Thermodynamics: motion remaining on theses level sets is non-dissipative, whereas motion transversal to these level sets is dissipative. The affine Kirillov-Kostant-Souriau form makes each orbit into a homogeneous symplectic manifold.

4.1 Hamiltonian System Associated to the Extended Lie-Poisson Bracket

In the context of the polysymplectic model, a natural generalisation of the Lie-Poisson equations with cocycle equation are (see [1])

$$\sum_{j=1}^{k} \frac{\partial}{\partial x_j} \tau_j + \sum_{j=1}^{j} ad^*_{\frac{\partial h}{\partial \tau_j}} \tau_j = \sum_{j=1}^{k} \Theta_j \left(\frac{\partial h}{\partial \tau_j}, . \right),$$

for a map $\tau : x(x_j, \cdots, x_k) \in \mathfrak{U} \subset \mathbb{R} \mapsto \tau(x) = (\tau(x_1), \cdots, \tau(x_k)) \in \mathfrak{g}_k^*$, with $h : \mathfrak{g}_k^* \to \mathbb{R}$ a given Hamiltonian.

The Hamiltonian system associated with the extended Lie-Poisson bracket to a given Hamiltonian function $h : \mathfrak{g}_k^* \to \mathbb{R}$ is given by the Lie-Poisson equations with cocycle (or affine Lie-Poisson equations). This equation can be expressed now with the help of our introduced extended Lie-Poisson bracket as

$$\sum_{j=1}^{k} \frac{\partial f}{\partial x_j} + \sum_{j=1}^{k} \left\langle \tau_j, \left[\frac{\partial f}{\partial \tau_j}, \frac{\partial h}{\partial \tau_j} \right] \right\rangle - \sum_{j=1}^{k} \Theta^j \left(\frac{\partial f}{\partial \tau_j}, \frac{\partial h}{\partial \tau_j} \right) = 0$$

which yields

$$\sum_{j=1}^{k} \frac{\partial f}{\partial x_j} = \{f, h\}_\Theta, \quad \forall f : \mathfrak{g}_k^* \to \mathbb{R} \tag{27}$$

which reduces to the known dynamical system $\frac{df}{dt} = \{f, h\}_\Theta$ in the standard Souriau model.

5   Concluding Remark

– The geometric heat Fourier equation in the standard Souriau's model [3] has the following form in the situation of non-null cohomology:

$$\frac{\partial Q}{\partial t} = ad^*_{\frac{\partial h}{\partial Q}} Q + \Theta \left(\frac{\partial h}{\partial Q} \right)$$

One of the reviewers asked if this equation has an extension in the polysymplectic context. It is still a work in progress.
– Furthermore I am extending some results about transerve Poisson structures in the polysymplectic framework which is also a work in progress.

References

1. Barbaresco, F., Gay-Balmaz, F.: Lie group cohomology and (Multi)Symplectic integrators: new geometric tools for lie group machine learning based on Souriau geometric statistical mechanics. Entropy **22**, 498 (2020)
2. Barbaresco, F.: Higher order geometric theory of information and heat based on polysymplectic geometry of Souriau Lie groups thermodynamics and their contextures: The bedrock for Lie Group machine learning. Entropy **20**, 840 (2018)
3. Barbaresco, F.: Symplectic foliation structures of non-equilibrium thermodynamics as dissipation model: application to Metriplectic nonlinear lindblad quantum master equation. Entropy **24**, 1626 (2022)
4. Barbaresco, F.: Lie group statistics and lie group machine learning based on Souriau Lie groups thermodynamics & Koszul-Souriau-fisher metric: new entropy definition as generalized Casimir invariant function in coadjoint representation. Entropy **22**, 642 (2020)
5. Barbaresco, F.: Symplectic foliation model of information geometry for statistics and learning on Lie Groups. In: SEE MaxEnt 2022 Conference, Institut Henri Poincaré, July 18th 2022; video: https://www.carmin.tv/fr/video/symplectic-foliation-model-of-information-geometry-for-statistics-and-learning-on-lie-groups

6. Barbaresco, F.: Symplectic theory of heat and information geometry. In: Nielsen, F., Rao, A.S., Rao, C.R. (eds.) Handbook of Statistics n°46 "Geometry and Statistics", 1st edn. Elsevier: Amsterdam, The Netherlands (2022). ISBN 9780323913454

7. Barbaresco, F.: Jean-Marie Souriau's Symplectic model of statistical physics: seminal papers on lie groups thermodynamics - Quod Erat demonstrandum. In: Barbaresco, F., Nielsen, F. (eds.) SPIGL 2020. SPMS, vol. 361, pp. 12–50. Springer, Cham (2021). https://doi.org/10.1007/978-3-030-77957-3_2

8. Barbaresco, F.: Souriau-Casimir Lie groups thermodynamics and machine learning. In: Barbaresco, F., Nielsen, F. (eds.) SPIGL 2020. SPMS, vol. 361, pp. 53–83. Springer, Cham (2021). https://doi.org/10.1007/978-3-030-77957-3_3

9. Barbaresco, F.: Koszul lecture related to geometric and analytic mechanics, Souriau's Lie group thermodynamics and information geometry. Inf. Geom. **4**, 245–262 (2021)

10. Barbaresco, F.: Densité de probabilité gaussienne à maximum d'Entropie pour les groupes de Lie basée sur le mod'ele symplectique de Jean-Marie Souriau. In Proceedings of the GRETSI'22 Conference, Nancy, France, 6–9 September 2022

11. Barbaresco, F.: Théorie symplectique de l'Information et de la chaleur: Thermodynamique des groupes de Lie et définition de l'Entropie comme fonction de Casimir. In: Proceedings of the GRETSI 2022 Conference, Nancy, France, 6–9 September 2022

12. Barbaresco, F.: Entropy geometric structure as casimir invariant function in coadjoint representation: geometric theory of Heat & information geometry based on Souriau Lie groups thermodynamics and Lie Algebra Cohomology. In: Frontiers in Entropy Across the Disciplines; World Scientific: Singapore, pp. 133–158 (2022), Chapter 5

13. Barbaresco, F.: Souriau entropy based on symplectic model of statistical physics: three Jean-Marie Souriau's Seminal Papers on Lie Groups Thermodynamics. In: Frontiers in Entropy Across the Disciplines; World Scientific: Singapore, pp. 55–90 (2022), Chapter 3

14. El Morsalani, M.: Polysymplectic Souriau Lie Group thermodynamics and the geometric structure of its coadjoint orbits. Accepted for publication in Proceedings of GSI'2023

15. El Morsalani, M.: Poly-Poisson structure for Souriau Lie Group thermodynamics and entropy geometric structure as casimir invariant function. In preparation

16. Iglesias, D., Marrero, J.C., Vaquero, M.: Poly-Poisson structures. Lett. Math. Phys. **103**, 1103–1133 (2013)

17. Koszul, J.L.; Zou, Y.M.: Introduction to Symplectic Geometry. Springer NatureSingapre and Science Press, Singapore (2019). https://doi.org/10.1007/978-981-13-3987-5

18. Laurent-Gengoux, C., Pichereau, A., Vanhaecke, P.: Poisson Structures. , Grundlehren der Mathematischen Wissenschaften. Springer, Berlin/Heidelberg (2013)

19. Souriau, J.-M.: Structure des Systèmes dynamiques. Dunod, Paris, France (1969)

20. Souriau, J.-M.: Mécanique statistique, groupes de Lie et cosmologie. In: Géométrie symplectique et physique mathématique; Éditions du C.N.R.S: Aix-en-Provence, pp. 59–113, France, 1974

Riemannian Geometry of Gibbs Cones Associated to Nilpotent Orbits of Simple Lie Groups

Pierre Bieliavsky[1]([✉]), Valentin Dendoncker[2], Guillaume Neuttiens[3], and Jérémie Pierard de Maujouy[4]

[1] Université Catholique de Louvain, Ottignies-Louvain-la-Neuve, Belgium
pierre.bieliavsky@gmail.com
[2] ICHEC Brussels Management School, Brussels, Belgium
[3] University of Jena, Jena, Germany
guillaume.neuttiens@uni-jena.de
[4] Université Catholique de Louvain, Ottignies-Louvain-la-Neuve, Belgium
jeremie.pierard@uclouvain.be

Abstract. In this short note, we prove that the Gibbs cone of generalized temperatures associated to a minimal coadjoint orbit of a simple Lie group G of Kähler type is not empty. We study the Fisher-Rao metric in the particular case of $G = \mathrm{SL}_2(\mathbb{R})$. We prove that, in this case, the Gibbs cone equipped with the Fisher-Rao metric is a Riemannian symmetric space.

1 Introduction

In its book *Structure des systèmes dynamiques* [7], Souriau determined the generalized temperatures for the cases of the dynamical groups of Galileo and Poincaré. Other examples have been performed, for instance by Marle and by Neuttiens. However, a systematic study of large classes of Lie groups has not been made. In particular, contrarily to what is stated in [5], it is observed in a work due to Neuttiens [6] that the nilpotent orbits of $\mathrm{SL}_2(\mathbb{R})$ do admit Gibbs states. In the present work, we prove that this phenomenon is not an artifact of the small dimension, but actually generalizes to the minimal nilpotent orbits of all simple Lie groups of Kähler type.

2 Thermodynamics of Lie Groups

Let G be a connected Lie group and M be a symplectic manifold. In the case M is a Hamiltonian symplectic G-manifold, the momentum map $\mu : M \to \mathfrak{g}^*$ generalizes the conserved energy of an isolated system. As such, it is suggested in [7] to formulate equilibrium statistical mechanics using μ instead of the energy.

F. Nielsen and F. Barbaresco (Eds.): GSI 2023, LNCS 14072, pp. 144–151, 2023.
https://doi.org/10.1007/978-3-031-38299-4_16

In statistical mechanics, the probability density of a state with energy E is described by the Boltzmann distribution

$$P_\beta(E) = \frac{1}{Z(\beta)} e^{-\beta E}$$

with $\beta \in \mathbb{R}_+^*$ a parameter which is physically identified with an inverse temperature [1,7] and $Z(\beta)$ a normalization factor called the partition function. In Souriau's *thermodynamics of Lie groups*, this is generalized to a *Gibbs distribution*

$$P_\beta(E) = \frac{1}{Z(\beta)} e^{-\langle \beta, E \rangle}$$

with $\beta \in \mathfrak{g}$. In order for this to be a well-defined probability measure, there is an integrability requirement, which motivates the following definition:

Definition 1 (Generalized temperature [5]). *Let λ_M be the Liouville measure of the symplectic manifold M. An element $\beta \in \mathfrak{g}$ is called a* generalized temperature *if there exists a function $f : M \to \mathbb{R}^+$ which is integrable and a neighbourhood $\mathcal{U}$ of β such that*

$$\forall \beta' \in \mathcal{U}, \forall x \in M, \quad e^{-\langle \beta', \mu(x) \rangle} \leqslant f(x)$$

The set of generalized temperatures, usually denoted Ω, is an open convex G-submanifold of $\mathfrak{g}$.

Definition 2 (Thermodynamic potential). *The thermodynamic potential associated with the Hamiltonian G-manifold M is the following map:*

$$z : \Omega \to \mathbb{R}$$

$$\beta \to \ln \left(\int_M e^{-\langle \beta, \mu(x) \rangle} \lambda_M \right)$$

The thermodynamic potential is a G-invariant $\mathcal{C}^\infty$ function on Ω; its iterated differentials can be computed under the integral.

As an open subset of a vector space, Ω is equipped with a natural flat affine connection, which we note D. It can be use to define the *mean value* of z:

$$Q = -Dz$$

and the *Fisher-Rao metric* which is the Hessian of z:

$$I = D^2 z$$

The Fisher-Rao metric is always a definite positive metric on Ω.

3   Simple Lie Groups of Kähler Type

Let us recall the notion of simple Lie group of Kähler type:

Definition 3 (Simple Lie group of Kähler type). *A connected simple Lie group G, assumed to have discrete center, is of* Kähler type *if its maximal compact subgroups admit a non-discrete center.*

The center of a maximal compact subgroup K of a simple Lie group of Kähler type has dimension 1 and can be used to define a G-invariant Kähler structure on the symmetric space G/K.

Let $G = KAN$ be an associated Iwasawa decomposition. Denote by $\mathfrak{g}$, $\mathfrak{k}$, $\mathfrak{a}$ and $\mathfrak{n}$ the Lie algebras of G, K, A and N respectively. Let μ be a *maximal positive restricted root* and consider a non-zero element E of the root space $\mathfrak{g}_\mu$. Denote by $\mathcal{O}$ the adjoint orbit of E. This is the Hamiltonian G-manifold we will be interested in. Since the Killing form B of $\mathfrak{g}$ identifies $\mathfrak{g}$ with $\mathfrak{g}^*$ in a G-equivariant manner, we view $\mathcal{O}$ as a homogeneous Hamiltonian symplectic G-space (i.e. a co-adjoint orbit). In particular, we work with a trivial symplectic cocycle [1,5,7].

For all $(k, e^\alpha, n) \in K \times A \times N$, we can compute

$$\mathrm{Ad}_k \, \mathrm{Ad}_{e^\alpha} \, \mathrm{Ad}_n \, E = e^{\mu(\alpha)} \, \mathrm{Ad}_k \, E$$

If we write $\mathfrak{a}_\mu \subsetneq \mathfrak{a}$ the Killing orthogonal of $\ker \mu$ and $K_\mu = \mathrm{Stab}_K(E)$ the isotropy subgroup of E in K, then there is a global diffeomorphism

$$\phi : K/K_\mu \times \mathfrak{a}_\mu \to \mathcal{O}$$
$$([k], \alpha) \mapsto e^{\mu(\alpha)} \, \mathrm{Ad}_k \, E$$

In order to keep formulas simple we will also use the notation ϕ for the map $K \times \mathfrak{a}_\mu \to K/K_\mu \times \mathfrak{a}_\mu \xrightarrow{\phi} \mathcal{O}$. We want to use ϕ in order to identify the generalized temperatures on $\mathcal{O}$ and compute the thermodynamic potential.

First, we need to identify the pullback of the KKS form to $K \times \mathfrak{a}_\mu$. For this we will use the following forms:

Definition 4 (Maurer-Cartan forms). *For $k \in K$ let us write L_k and R_k the respective left and right actions on K (we will also use this notation for A). We define the following $\mathfrak{k}$-valued 1-forms:*

$$\omega^L : X \in T_k K \mapsto D_k L_k^{-1}(X)$$
$$\omega^R : X \in T_k K \mapsto D_k R_k^{-1}(X)$$

They satisfy the following properties:

$$\forall k \in K, \forall X \in T_k K, \; \omega^R(X) = \mathrm{Ad}_k \, \omega^L(X) \tag{1}$$
$$\forall k \in K, \; R_k^* \omega^L = \mathrm{Ad}_k^{-1} \, \omega^L \tag{2}$$

as well as the Maurer-Cartan equations:

$$\mathrm{d}\omega^L + [\omega^L, \omega^L] = 0$$
$$\mathrm{d}\omega^R - [\omega^R, \omega^R] = 0$$

We can now express the KKS form on $K \times \mathfrak{a}_\mu$:

Proposition 1. *The pullback to $K \times \mathfrak{a}_\mu$ of the KKS form of $\mathcal{O}$ takes the following form:*

$$\phi^* \omega_{\mathrm{KKS}} = e^{\mu(\alpha)} \left(B\left(E, [\omega^L, \omega^L]\right) + B\left(E, \omega^L\right) \wedge \mu \right)$$
$$= e^{\mu(\alpha)} \left(-\mathrm{d}B\left(E, \omega^L\right) + B\left(E, \omega^L\right) \wedge \mu \right)$$

Proof. Let $(k, \alpha) \in K \times \mathfrak{a}_\mu$ and $\xi_1, \xi_2 \in \mathfrak{k}, \alpha' \in \mathfrak{a}$. We compute the differential $D\phi$ of ϕ:

$$D_{ka}\phi(DR_k(\xi_i)) = \mathrm{ad}_{\xi_i} \phi(ka)$$
$$D_{ka}\phi(DR_{e^\alpha}(\alpha')) = \mathrm{Ad}_k \circ \mathrm{ad}_{\alpha'} \circ \mathrm{Ad}_a E = \mathrm{ad}_{\mathrm{Ad}_k \alpha'} \phi(ka)$$

This allows us to compute the pullback of the KKS form:

$$B\left(\phi(ka), [\xi_1, \xi_2]\right) = e^{\mu(\alpha)} B\left(\mathrm{Ad}_k E, [\xi_1, \xi_2]\right)$$
$$= e^{\mu(\alpha)} B\left(E, \left[\mathrm{Ad}_k^{-1} \xi_1, \mathrm{Ad}_k^{-1} \xi_2\right]\right)$$

and

$$B\left(\phi(ka), [\xi_1, \alpha']\right) = e^{\mu(\alpha)} B\left(\mathrm{Ad}_k E, [\xi_1, \mathrm{Ad}_k \alpha']\right)$$
$$= e^{\mu(\alpha)} B\left([\alpha', E], \mathrm{Ad}_k^{-1} \xi_1\right)$$
$$= e^{\mu(\alpha)} \mu(\alpha') B\left(E, \mathrm{Ad}_k^{-1} \xi_1\right)$$

Using Equation (1) we can therefore write:

$$\phi^* \omega_{\mathrm{KKS}} = e^{\mu(\alpha)} \left(B\left(E, [\omega^L, \omega^L]\right) + B\left(E, \omega^L\right) \wedge \mu \right)$$

Due to Eq. (2), the 1-form $B\left(E, \omega^L\right)$ on K is invariant under right translation by K_μ: let us call ω_E its factorization to K/K_μ. We denote by $2n$ the dimension of $\mathcal{O}$. We can then compute the following positive integral on $K/K_\mu \times \mathfrak{a}_\mu$, for $\beta \in \mathfrak{g}$:

$$\int_{([k], e^\alpha) \in K/K_\mu \times \mathfrak{a}_\mu} e^{-e^{\mu(\alpha)} B(\beta, \mathrm{Ad}_k E)} e^{n\mu(\alpha)} |\mu \wedge \omega_E \wedge (\mathrm{d}\omega_E)^{n-1}|$$

$$= \int_{([k], e^\alpha) \in K/K_\mu \times \mathbb{R}_+^*} e^{-rB\left(\mathrm{Ad}_k^{-1} \beta, E\right)} r^{n-1} |\mathrm{d}r \wedge \omega_E \wedge (\mathrm{d}\omega_E)^{n-1}| \qquad (3)$$

$$= \int_{([k], e^\alpha) \in K/K_\mu} \left(\int_{\mathbb{R}_+^*} e^{-rB\left(\mathrm{Ad}_k^{-1} \beta, E\right)} r^{n-1} \mathrm{d}r \right) |\omega_E \wedge (\mathrm{d}\omega_E)^{n-1}|$$

Since K is compact, we obtain a condition for the convergence of the integral:

Proposition 2. *If* $\min_{k \in K} B(E, \operatorname{Ad}_k \beta) > 0$ *then the integral is convergent and has value*

$$\int_{[k] \in K/K_\mu} \frac{(n-1)!}{B\left(\operatorname{Ad}_k^{-1} \beta, E\right)^n} |\omega_E \wedge (\mathrm{d}\omega_E)^{n-1}|$$

Proof. We use the following identity which holds for any $\lambda > 0$:

$$\int_0^{+\infty} e^{-\lambda s} s^{n-1} \mathrm{d}s = \frac{1}{\lambda^n}(n-1)!$$

We can now prove that the set of generalized temperatures is non-empty and contains half of the center of $\mathfrak{k}$:

Proposition 3. *Any element* Z *of the center* $\mathfrak{z}(\mathfrak{k})$ *such that* $B(E, Z) > 0$ *is a generalized temperature associated to the Hamiltonian action of* G *on* $\mathcal{O}$.

Proof. If $Z \in \mathfrak{z}(\mathfrak{k})$ then since K is connected for all k in K, $\operatorname{Ad}_k Z = Z$. Proposition 2 then implies that if $B(E, Z) > 0$, the integral (3) converges. We need to prove that $e^{-rB(\operatorname{Ad}_k \beta, E)}$ can be bounded above by an integrable function independent of β, for β in a neighbourhood of Z.

Let C be a compact neighbourhood of Z in $\mathfrak{g}$ and consider the map

$$f : (k, \beta) \in K \times C \mapsto B(\operatorname{Ad}_k \beta, E)$$

According to the Heine-Cantor theorem, f is uniformly continuous. In particular, since $\min_{K \times \{Z\}} f > 0$ there exists a neighbourhood $\mathcal{U}$ of Z in C such that $\inf_{K \times \mathcal{U}} f = \eta > 0$. Therefore for β in $\mathcal{U}$ there is the upper bound $e^{-rB(\operatorname{Ad}_k \beta, E)} < e^{-\eta r}$ which is an integrable function on $K \times A_\mu$ along $|r^n \mathrm{d}r \wedge \omega_E \wedge (\mathrm{d}\omega_E)^{n-1}|$.

4   Thermodynamic Geometry for $G = \mathbf{SL_2(\mathbb{R})}$

Let us compute the Fisher-Rao geometry of the nilpotent orbit in the simplest example: $SL_2(\mathbb{R})$. It was remarked by Neuttiens that the nilpotent orbit does admit generalized temperatures [6], contrary to what was affirmed in [4].

4.1   The Adjoint Orbits of $SL_2(\mathbb{R})$

Let (e, f, h) be a standard $\mathfrak{sl}_2$-triple and (e^*, f^*, h^*) its dual basis. The Killing form B of $\mathfrak{sl}_2(\mathbb{R})$ takes the following form

$$B = 8\left(h^* \otimes h^* + \frac{1}{2}(e^* \otimes f^* + f^* \otimes e^*)\right)$$

$$= 2\left(4h^* \otimes h^* + (e^* + f^*) \otimes (e^* + f^*) - (e^* - f^*) \otimes (e^* - f^*)\right)$$

It is of signature $(++-)$ and the basis $\frac{1}{2\sqrt{2}}(h, e+f, e-f)$ is pseudo-orthonormal. We will use the following Iwasawa decomposition of $\mathfrak{sl}_2(\mathbb{R})$:

$$\mathfrak{sl}_2(\mathbb{R}) \simeq \underbrace{\mathbb{R}(e-f)}_{\mathfrak{k}} \oplus \underbrace{\mathbb{R}h}_{\mathfrak{a}} \oplus \underbrace{\mathbb{R}e}_{\mathfrak{n}}$$

The adjoint orbits are exactly the connected components of the pseudo-spheres of B [4]. Defining a time orientation by setting $e - f$ future-oriented, the adjoint orbits are of the following types:

1. $\mathcal{O}_0 := \{0\}$.
2. $\mathcal{O}_\lambda := \{B(x, x) = \lambda^2\}$ with $\lambda \in \mathbb{R}_+^*$, called **hyperbolic orbits**.
3. $\mathcal{O}_\lambda^\pm := \{B(x, x) = -\lambda^2$ and $-B(x, e - f) \in \mathbb{R}_\pm^*\}$ with $\lambda \in \mathbb{R}_+^*$, called **elliptic orbits**.
4. $\mathcal{O}^\pm := \{B(x, x) = 0$ and $-B(x, e - f) \in \mathbb{R}_\pm^*\}$, called **parabolic orbits** [3].

The sets of generalized temperatures are as follows [6]:

1. The case of $\mathcal{O}_0$ is degenerate and the set of generalized temperatures is $\mathfrak{sl}_2(\mathbb{R})$.
2. The hyperbolic orbits admit no generalized temperatures.
3. The elliptic orbits $\mathcal{O}_\lambda^\pm$ admit $\{\beta \in \mathfrak{sl}_2(\mathbb{R}) \mid B(\beta, \beta) < 0$ and $\mp B(\beta, e - f) < 0\}$ as set of generalized temperatures.
4. The parabolic orbits $\mathcal{O}^\pm$ also admit $\{\beta \in \mathfrak{sl}_2(\mathbb{R}) \mid B(\beta, \beta) < 0$ and $\mp B(\beta, e - f) < 0\}$ as set of generalized temperatures.

The nilpotent orbits, which are of interest to us, are $\mathcal{O}^\pm$. They can be parametrized as follows:

$$(\theta, \xi) \in \mathbb{T} \times \mathbb{R} \mapsto \psi_\pm(\theta, \xi) = \pm \mathrm{Ad}_{\exp(\frac{\theta}{2}(e - f))} \mathrm{Ad}_{\exp(\frac{\xi}{2}h)} e$$

$$= \pm \frac{e^\xi}{2} (e - f + \cos\theta(e + f) + \sin\theta h)$$

which takes the following form, in the standard matrix representation of (e, f, h):

$$\psi_\pm(\theta, \xi) = \pm \frac{e^\xi}{2} \begin{bmatrix} \sin\theta & (1 + \cos\theta) \\ -(1 - \cos\theta) & -\sin\theta \end{bmatrix}$$

4.2 The KKS Form

We compute

$$\partial_\theta \psi_\pm(\theta, \xi) = \left[\frac{e - f}{2}, \psi_\pm(\theta, \xi) \right]$$

and

$$\partial_\xi \psi_\pm(\theta, \xi) = \left[\mathrm{Ad}_{\exp(\frac{\theta}{2}(e - f))} \frac{h}{2}, \psi_\pm(\theta, \xi) \right]$$

Therefore if we write ω_{KKS} the pullback through B of the KKS form on the coadjoint orbit $B(\mathcal{O}^\pm)$, we can compute

$$\psi_\pm^* \omega_{\mathrm{KKS}}(\partial_\theta, \partial_\xi)|_{\theta, \xi} = B\left(\psi_\pm(\theta, \xi), \left[\frac{e - f}{2}, \mathrm{Ad}_{\exp(\frac{\theta}{2}(e - f))} \frac{h}{2} \right] \right)$$

$$= B\left(\pm e, \left[\mathrm{Ad}_{\exp(-\frac{\xi}{2}h)} \left(\frac{e - f}{2} \right), \frac{h}{2} \right] \right)$$

$$= \pm \frac{1}{2} \partial_\xi \left[B\left(e, \mathrm{Ad}_{\exp(-\frac{\xi}{2}h)}(e - f) \right) \right]|_{\theta, \xi}$$

with

$$B\left(e, \mathrm{Ad}_{\exp(-\frac{\xi}{2}h)}(e-f)\right) = B\left(e, \cosh\xi(e-f) - \sinh\xi(e+f)\right) = -4e^{\xi}$$

Therefore

$$\psi_{\pm}^{*}\omega_{\mathrm{KKS}} = \mp 2d\theta \wedge de^{\xi}$$

4.3   Thermodynamic Potential

From now on we consider one fixed orbit $\mathcal{O}_{+}$. We know that the set of generalized temperatures Ω_{+} (resp. Ω_{-}) is the convex cone of past-oriented (resp. future-oriented) timelike vectors of $\mathfrak{sl}_{2}(\mathbb{R})$. Let $\beta \in \Omega_{\pm}$. The thermodynamic potential $z(\beta)$ is defined as

$$e^{z(\beta)} = \int_{\mathcal{O}\pm} e^{-B(\beta,x)}|\omega_{\mathrm{KKS}}|$$

Let us define $m : \beta \in \Omega_{\pm} \mapsto \sqrt{-B(\beta,\beta)}$. Since z is invariant under the adjoint action (which is the case in general up to a cocycle term [2,5,6]) and β belongs to the orbit of $\mp\frac{m}{2\sqrt{2}}(e-f)$, we have

$$e^{z(\beta)} = \int_{\mathcal{O}\pm} e^{-B\left(\mp\frac{m}{2\sqrt{2}}(e-f),x\right)}|\omega_{\mathrm{KKS}}(x)| = \int_{\mathbb{T}\times\mathbb{R}_{+}^{*}} e^{-m\sqrt{2}u}2d\theta du = \frac{2\sqrt{2}\pi}{m}$$

and

$$z(\beta) = \ln\left(\frac{2\sqrt{2}\pi}{m}\right)$$

4.4   Fisher-Rao Metric

We want to compute the Hessian of $z(\beta)$ for the affine structure of $\mathfrak{sl}_{2}(\mathbb{R})$: this will be the Fisher-Rao metric I [2,6]. First notice that $dm = -\frac{B(\beta)}{m}$ with $B(\beta) = B(\beta, \cdot) \in \mathfrak{sl}_{2}(\mathbb{R})^{*}$. Therefore

$$Q := -dz = \frac{dm}{m} = -\frac{B(\beta)}{m^{2}}$$

and

$$\begin{aligned}
I &= Ddz \\
&= \frac{B}{m^{2}} + \left(-2\frac{1}{m^{3}}\frac{-B(\beta)}{m}\right) \otimes B(\beta) \\
&= \underbrace{\frac{1}{m^{2}}\left(B + \frac{B(\beta)\otimes B(\beta)}{m^{2}}\right)}_{B^{\perp}} + \underbrace{\frac{dm\otimes dm}{m^{2}}}_{dz\otimes dz}
\end{aligned}$$

with $B^{\perp}$ the restriction of B on the hyperboloids of B, which are orthogonal to the radial direction at every $\beta \in \Omega_{\pm}$. The KKS form on $\mathrm{Ad}(G)(Q)$ can always be expressed in terms of I, as follows:

$$Q([x, y]) = -I([x, \beta], y)$$

which takes in the present case the explicit form

$$\frac{1}{m^2} B(\beta, [x, y]) = -\frac{1}{m^2} B([x, \beta], y)$$

Let $\mathbb{H}^-$ (resp. $\mathbb{H}^+$) be the past sheet (resp. future sheet) of the unit two-sheeted hyperboloid of B. There is then a diffeomorphism

$$\Omega_{\pm} \xrightarrow{\sim} \mathbb{R} \times \mathbb{H}^{\mp}$$

$$\beta \mapsto \left(z = \ln\left(\frac{2\sqrt{2}\pi}{m}\right), \frac{1}{m}\beta\right)$$

under which I is mapped to the product metric

$$dz \otimes dz + B|_{\mathbb{H}^{\mp}}$$

This identifies the Fisher-Rao geometry of $\Omega_{\pm}$ with the Riemannian product of a line with the hyperbolic plane.

References

1. Barbaresco, F.: Jean-Marie Souriau's symplectic model of statistical physics: seminal papers on Lie groups thermodynamics - Quod Erat demonstrandum. In: Barbaresco, F., Nielsen, F. (eds.) SPIGL 2020. SPMS, vol. 361, pp. 12–50. Springer, Cham (2021). https://doi.org/10.1007/978-3-030-77957-3_2
2. Barbaresco, F.: Chapter 4 - symplectic theory of heat and information geometry. In: Nielsen, F., Srinivasa Rao, A.S., Rao, C. (eds.) Geometry and Statistics, Handbook of Statistics, vol. 46, pp. 107–143. Elsevier, Amsterdam (2022)
3. Cairns, G., Ghys, E.: The Local Linearization Problem for Smooth Sl(n)-Actions. L'Enseignement Mathématique **43** (1997)
4. Marle, C.M.: Examples of gibbs states of mechanical systems with symmetries. J. Geom. Symm. Phys. **58**, 55–79 (2020)
5. Marle, C.M.: On gibbs states of mechanical systems with symmetries. J. Geom. Symm. Phys. **57**, 45–85 (2020)
6. Neuttiens, G.: États de Gibbs d'une action hamiltonienne
7. Souriau, J.M.: Structure des systèmes dynamiques

Symplectic Foliation Transverse Structure and Libermann Foliation of Heat Theory and Information Geometry

Frédéric Barbaresco(✉)

THALES Land & Air Systems, 78140 Velizy-Villacoublay, France
frederic.barbaresco@thalesgroup.com

Abstract. We introduce a symplectic bifoliation model of Information Geometry and Heat Theory based on Jean-Marie Souriau's Lie Groups Thermodynamics to describe transverse Poisson structure of metriplectic flow for dissipative phenomena. This model gives a cohomological characterization of Entropy, as an invariant Casimir function in coadjoint representation. The dual space of the Lie algebra foliates into coadjoint orbits identified with the Entropy level sets. In the framework of Thermodynamics, we associate a symplectic bifoliation structure to describe non-dissipative dynamics on symplectic leaves (on level sets of Entropy as constant Casimir function on each leaf), and transversal dissipative dynamics, given by Poisson transverse structure (Entropy production from leaf to leaf). The symplectic foliation orthogonal to the level sets of moment map is the foliation determined by hamiltonian vector fields generated by functions on dual Lie algebra. The orbits of a Hamiltonian action and the level sets of its moment map are polar to each other. The space of Casimir functions on a neighborhood of a point is isomorphic to the space of Casimirs for the transverse Poisson structure. Souriau's model could be then interpreted by Mademoiselle Paulette Libermann's foliations, clarified as dual to Poisson Γ-structure of Haefliger, which is the maximum extension of the notion of moment in the sense of J.M. Souriau, as introduced by P. Molino, M. Condevaux and P. Dazord in papers of "Séminaire Sud-Rhodanien de Geometrie". The symplectic duality to a symplectically complete foliation, in the sense of Libermann, associates an orthogonal foliation. Paulette Libermann proved that a Legendre foliation on a contact manifold is complete if and only if the pseudo-orthogonal distribution is completely integrable, and that the contact form is locally equivalent to the Poincaré-Cartan integral invariant. Paulette Libermann proved a classical theorem relating to co-isotropic foliations, which notably gives a proof of Darboux's theorem. Finally, we explore Edmond Fédida work on the theory of foliation structures in the language of fully integrable Pfaff systems associated with the Cartan's moving frame.

Keywords: Bifoliation · Coadjoint Orbits · Moment Map · Casimir Function · Metriplectic Flow · Transverse Poisson Structure · Information Geometry · Maurer-Cartan · Moving Frame

F. Nielsen and F. Barbaresco (Eds.): GSI 2023, LNCS 14072, pp. 152–164, 2023.
https://doi.org/10.1007/978-3-031-38299-4_17

1 Seminal Works on Foliation and Caratheodory Function Groups

"Sur une durée de quarante années l'immeuble s'est édifié; des centaines d'ouvriers ont œuvré. L'édifice n'est pas achevé, mais on peut visiter. Oui, visiter est le mot." - Georges Reeb

Foliation theory is a natural generalized qualitative theory of differential "equations, initiated by H. Poincaré, and developed by C. Ehresmann and G. Reeb, with contribution by A. Haefliger, P. Molino, B.L. Reinhart [6–8, 11]. Riemannian foliations generated by metric functions were developed by Ph. Tondeur. Notion of foliation in thermodynamics appears as soon as 1900 C. Caratheódory paper [4] where horizontal curves roughly correspond to adiabatic processes, performed in the language of Carnot cycles. The properties of the couple of Poisson manifolds was also previously explored by C. Caratheódory in 1935 [5], under the name of *"function groups, polar to each other"*, where he observed that the two families of differentiable functions formed by the first integrals of *F* (a completely integrable vector subbundle of TM) and its orthogonal *orthF*, respectively, called "function groups", are "polar" of the other. This seminal work of C. Caratheodory leads to the concept of a Poisson structure which was first defined and treated in depth by A. Lichnerowicz and independently by A. Kirillov. André Haefliger observed *"Generally, for a field of planes of codimension 1 given by a Pfaff form ω, the integrability condition is equivalent to ω ∧ dω = 0. In this case there locally exists a non-zero function λ, called a integrating factor, and a function φ, called a first integral, such that ω = λdφ. The level varieties of φ are the integral varieties. Caratheódory gave in 1909 a local geometric characterization of the complete integrability of a Pfaff form ω, namely: ω is completely integrable if and only if, for any neighborhood U of any point x, there exists a point of U which cannot be linked to x by a curve in U tangent to the kernel of ω. He used this characterization to express in a remarkably concise and conceptual way the second law of thermodynamics."*

Symplectic Geometry in conjunction with Analytical Mechanics has grown considerably over the past decades; inspired by the work of S. Lie and E. Cartan, A. Lichnerowicz, G. Reeb [10], J.M. Souriau [1, 2], as well as F. Gallissot [3], who were the initiators of this revival of Analytical Mechanics. In paper [9], G. Reeb asked the following questions about foliation structures *"Why have they been studied. How were they studied? Is it "profitable" to continue these investigations?"* and proposed motivations for studying foliations. Among motivations, G. Reeb identified two key use-cases, Lie Groups action and integrable Pfaff forms of Thermodynamics, that we develops in our paper: *"(1) Lie groups action theory (theory much older than that of foliations) often leads to consider generated foliations. Similarly, the "moving frame" theory (Cartan) ("dual" in a rather vague sense of the previous one) suggests classes of foliations with a remarkable transverse structure. (2) Thermodynamics has long accustomed mathematical physics [cf. Duhem P.] to the consideration of completely integrable Pfaff forms: the elementary heat dQ [notation of thermodynamicists] representing the elementary heat yielded in an infinitesimal reversible modification is such a completely integrable form. This point does not seem to have been explored since then."*. 1st topic on Cartan's moving frame was developed by Reeb PhD student, Edmond Fédida [12].

2 Souriau's Lie Groups Thermodynamics and Casimir Entropy

We present a new symplectic model of Information Geometry based on Jean-Marie Souriau's Lie Groups Thermodynamics [1, 2, 26, 27] with seminal work of F. Gallissot [3]. Souriau model was initially described in chapter IV "Statistical Mechanics" of his book "Structure of dynamical systems" published in 1969. This model gives a purely geometric characterization of Entropy, which appears as an invariant Casimir function in coadjoint representation, characterized by Poisson cohomology. Souriau has proved that we can associate a symplectic manifold to coadjoint orbits of a Lie group by the KKS 2-form (Kirillov, Kostant, Souriau) in the affine case (affine model of coadjoint operator equivariance via Souriau's cocycle), that we have identified with Koszul-Fisher metric from Information Geometry for Lie Groups Thermodynamics. The dual space of the Lie algebra foliates into coadjoint orbits that are also the Entropy level sets: dynamics on these symplectic leaves are non-dissipative, whereas transversal dynamics are dissipative.

Jean-Marie Souriau extended Information Geometry in the framework of statistical mechanics geometry to ensure the covariance of Gibbs density with regard to the action of the Lie group. The Souriau model provides Koszul-Fisher metric $I(\beta)$:

$$I(\beta) = -\frac{\partial^2 \Phi}{\partial \beta^2} \text{ with } \Phi(\beta) = -\log \int_M e^{-\langle U(\xi), \beta \rangle} d\lambda_\omega \text{ and } U : M \to \mathfrak{g}^*$$

$$(1)$$

preserving the Legendre transform of Information Geometry:

$$S(Q) = \langle Q, \beta \rangle - \Phi(\beta) \text{ with } Q = \frac{\partial \Phi(\beta)}{\partial \beta} \in \mathfrak{g}^* \text{ and } \beta = \frac{\partial S(Q)}{\partial Q} \in \mathfrak{g}$$

$$(2)$$

β is a "geometric" (Planck) temperature, element of Lie algebra $\mathfrak{g}$ of the group, and Q is a "geometric" heat, element of the dual space of the Lie algebra $\mathfrak{g}^*$ of the group.

Souriau proved that all coadjoint orbits of a Lie group given by $O_F = \{Ad_g^* F, g \in G\}$ subset of $\mathfrak{g}^*$, $F \in \mathfrak{g}^*$, carry, by a closed G-invariant 2-form, a natural homogeneous symplectic structure. If we define $K = Ad_g^* = (Ad_{g^{-1}})^*$ and $K_*(X) = -(ad_X)^*$ with $\langle Ad_g^* F, Y \rangle = \langle F, Ad_{g^{-1}} Y \rangle, \forall g \in G, Y \in \mathfrak{g}, F \in \mathfrak{g}^*$, the Riemannian metric proposed by Souriau has been identified by us as a generalization of the Fisher metric:

$$I(\beta) = \left[g_\beta \right] \text{ with } g_\beta \left([\beta, Z_1], [\beta, Z_2] \right) = \tilde{\Theta}_\beta \left(Z_1, [\beta, Z_2] \right) \qquad (3)$$

with $\tilde{\Theta}_\beta(Z_1, Z_2) = \tilde{\Theta}(Z_1, Z_2) + \langle Q, ad_{Z_1}(Z_2) \rangle$ where $ad_{Z_1}(Z_2) = [Z_1, Z_2]$

When a Lie group acts transitively by a Hamiltonian action on a symplectic manifold, the symplectic manifold is a covering space of a coadjoint orbit. The non-equivariance induced an additional term by a symplectic cocycle, which corresponds to $\tilde{\Theta}(Z_1, [\beta, Z_2])$. To define this extended Fisher metric, the tensor $\tilde{\Theta}$ used is defined by the moment map $J(x)$, application from M (homogeneous symplectic manifold) to the dual space of the Lie algebra $\mathfrak{g}^*$, given by $\tilde{\Theta}(X, Y) = J_{[X,Y]} - \{J_X, J_Y\}$ $J(x) : M \rightarrow \mathfrak{g}^*$ such that $J_X(x) = \langle J(x), X \rangle$, $X \in \mathfrak{g}$. As the tangent space of the cocycle $\theta(g) \in \mathfrak{g}^*$, $\tilde{\Theta}$, $\tilde{\Theta}$ could be derived (the non-equivariance of the coadjoint operator Ad_g^* generates this cocycle that modifies the action of the group on the dual space of the Lie algebra, so that the moment map could recover an equivariant relative to this new affine action):

$$Q\big(Ad_g(\beta)\big) = Ad_g^*(Q) + \theta(g) \tag{4}$$

The cocycle $\theta(g) \in \mathfrak{g}^*$ characterizes the lack of equivariance of the moment map:

$$\tilde{\Theta}(X,Y) : \mathfrak{g} \times \mathfrak{g} \rightarrow \mathfrak{R} \qquad \text{with } \Theta(X) = T_e\theta(X(e))$$
$$X, Y : \quad \langle \Theta(X), Y \rangle \tag{5}$$

Souriau entropy $S(Q)$ is found to be constant on the affine coadjoint orbit of the group by observing that $S\big(Ad_g^{\#}(Q)\big) = S(Q)$ if we note the affine coadjoint operator $Ad_g^{\#}(Q) = Ad_g^*(Q) + \theta(g)$. A function on M is a Casimir function when M is a Poisson manifold if and only if this function is constant on every symplectic leaf. Entropy is traditionally defined by Shannon's axiomatic approach. **Entropy will be defined in this essay as the Casimir equation's solution for affine equivariance**. The two equations characterizing entropy as an invariant Casimir function are related by:

$$\forall H, \{S, H\}_{\tilde{\Theta}}(Q) = \left\langle ad^*_{\frac{\partial S}{\partial Q}} Q + \Theta\left(\frac{\partial S}{\partial Q}\right), \frac{\partial H}{\partial Q} \right\rangle = 0 \Rightarrow ad^*_{\frac{\partial S}{\partial Q}} Q + \Theta\left(\frac{\partial S}{\partial Q}\right) = 0 \tag{6}$$

This equation appears in the Souriau paper published in 1974, observing that geometric temperature β is a kernel of $\tilde{\Theta}_\beta$, which is written as follows:

$$\beta \in Ker\tilde{\Theta}_\beta \Rightarrow \langle Q, [\beta, Z] \rangle + \tilde{\Theta}(\beta, Z) = 0 \tag{7}$$

and can be developed to recover the Casimir equation:

$$\langle Q, ad_\beta Z \rangle + \tilde{\Theta}(\beta, Z) = 0 \Rightarrow \left\langle ad_\beta^* Q, Z \right\rangle + \tilde{\Theta}(\beta, Z) = 0$$
$$\underset{\beta = \frac{\partial S}{\partial Q}}{\Rightarrow} \left\langle ad^*_{\frac{\partial S}{\partial Q}} Q, Z \right\rangle + \tilde{\Theta}\left(\frac{\partial S}{\partial Q}, Z\right) = \left\langle ad^*_{\frac{\partial S}{\partial Q}} Q + \Theta\left(\frac{\partial S}{\partial Q}\right), Z \right\rangle \tag{8}$$
$$= 0, \forall Z \Rightarrow ad^*_{\frac{\partial S}{\partial Q}} Q + \Theta\left(\frac{\partial S}{\partial Q}\right) = 0$$

Geometric heat equation is then given by $\frac{dQ}{dt} = ad^*_{\frac{\partial H}{\partial Q}} Q + \Theta\left(\frac{\partial H}{\partial Q}\right)$

with stable equilibrium when $H = S \Rightarrow \frac{dQ}{dt} = ad^*_{\frac{\partial S}{\partial Q}} Q + \Theta\left(\frac{\partial S}{\partial Q}\right) = 0$. We can observe that

$$dS = \tilde{\Theta}_\beta\left(\frac{\partial H}{\partial Q}, \frac{\partial S}{\partial Q}\right) dt \quad \text{where} \quad \tilde{\Theta}_\beta\left(\frac{\partial H}{\partial Q}, \frac{\partial S}{\partial Q}\right) = \tilde{\Theta}\left(\frac{\partial H}{\partial Q}, \beta\right) + \left\langle Q, \left[\frac{\partial H}{\partial Q}, \beta\right]\right\rangle$$

(9)

showing that the second law of thermodynamics could be deduced from the Souriau tensor positive definiteness related to Fisher–Koszul information metric.

3 Metriplectic Flow as Symplectic Model for Dissipation

The metriplectic bracket was first introduced in 1983 by A. N. Kaufman and P. J. Morrison [28]. This bracket formalism provides both energy conservation and a non-decrease in entropy, and it reduces to the traditional Poisson bracket formalism in the limit of no dissipation. Parallel axiomatization of this model was performed by Grmela and Öttinger. There are three main types of dissipation: thermal diffusion with energy conservation and entropy production through heat transfer; viscosity, which takes energy from the system (e.g., Navier–Stokes equation); and transport equations with collision operators. These types of systems that adhere to both the first and second laws of thermodynamics are included in metriplectic dynamics. A new bracket in the metriplectic formalism provides the evolution equation $\{\{.,.\}\}$:

$$\frac{df}{dt} = \{\{f, F\}\} = \{f, F\} + (f, F) \text{ with Hamiltonian } F = H + S \tag{10}$$

The 2nd metric flow bracket has 2 constraints $(f, F) = (F, f)$ and $(f, f) \geq 0$ with the entropy S selected from the set of Casimir invariants of the non-canonical Poisson bracket. Metriplectic flow is compliant with 2 first thermodynamics principles:

- **First thermodynamics principle: energy conservation**

$$\frac{dH}{dt} = \{H, F\} + (H, F) = \{H, H\} + \{H, S\} + (H, H) + (H, S) = 0 \tag{11}$$

because $\{H, H\} = 0$ by symmetry, $\{f, S\} = 0, \forall f$ and $(H, f) = 0, \forall f$

- **Second thermodynamics principle: entropy production**

$$\frac{dS}{dt} = \{S, F\} + (S, F) = 0 + (S, H) + (S, S) = (S, S) \geq 0 \tag{12}$$

with $\{S, f\} = 0 \forall f$, $(f, H) = 0 \forall f$ and M positive semi - definite

Finally, two compatible brackets, a Poisson bracket and a symmetric bracket, determine the geometry in metriplectic systems:

$$\frac{df}{dt} = \{\{f, F\}\} = \{f, H\} + (f, S) \tag{13}$$

The energy H is a Casimir invariant of the dissipative bracket, and the entropy S is a Casimir invariant of the Poisson bracket:

$$\{S, H\} = 0 \,\forall H \text{ and } (H, S) = 0 \,\forall S \tag{14}$$

The symmetry requirement generalizes the Onsager symmetry of linear irreversible thermodynamics to nonlinear issues; however, in the traditional metriplectic model, the possibility of Casimir symmetry is not taken into account. The bracket proposed by Kaufman is more general than the metriplectic bracket.

4 Polar Foliation, Bifoliation and Bifibration

We study bifoliation as described in [24]. Consider P as a symplectic manifold equipped with a form ω and consider Ψ be foliation of P such that the quotient space N_Ψ of P is a manifold over an equivalence relation set up by Ψ. Let $\Im_\Psi$ the space of functions on P which are constant along the leaves of Ψ. Consider $T\Psi$ be a bundle of vectors tangent to the leaves of Ψ, and consider $T\Psi^\perp$ be its orthogonal complement with respect to ω. Let χ_f be the hamiltonian vector field of $f \in \Im_\Psi$. It follows as f is constant along Ψ, for any $v \in T\Psi$, that $\omega(v, \chi_f) = vf = 0$. Then χ_f lies in $T\Psi^\perp$. We observe that $T\Psi^\perp$ is spanned at each point by the hamiltonian vector fields of functions in $\Im_\Psi$. For any $f, h \in \Im_\Psi$ one has $[\chi_f, \chi_h] = \chi_{\{f,h\}}$ and the distribution of planes $T\Psi^\perp$ is integrable since $\Im_\Psi$ is a Lie subalgebra under the Poisson bracket, by the Frobenius theorem. Therefore, it **is tangent to another foliation** $\Psi^\perp$ **named polar to** Ψ. As we require that $\Im_\Psi$ is a Poisson subalgebra, we deduce the **existence of a polar foliation**. We can also consider by $\Im_{\Psi^\perp}$ a set of functions on P constant along $\Psi^\perp$. The hamiltonian vector fields of $\Im_{\Psi^\perp}$ are tangent to Ψ and, $\Im_{\Psi^\perp}$ is also a Poisson subalgebra since the latter is a foliation. **A foliation polar to** $\Psi^\perp$ **is then** Ψ **and a pair** $(\Psi, \Psi^\perp)$ **is named bi-foliation of** P. $\Im_\Psi$ is considered as functions on N_Ψ. Choose in $\Im_\Psi$ a functionally independent basis $\{\psi_i\}$, $1 \le i \le m$, $\{\psi_i, \psi_j\} = K_{ij}(\psi_1, ..., \psi_m) \equiv (\psi_i, \psi_j)_{N_\Psi}$. The result of this computation can serve as the definition of a Poisson bracket on N_Ψ with K_{ij} being the corresponding Poisson tensor, although the bracket was computed on P. The projection $\pi_\Psi : P \to N_\Psi$ is a Poisson map and N_Ψ is a Poisson manifold. By the equivalence relation set up by the foliation $\Psi^\perp$, we can define a quotient space $N_{\Psi^\perp}$ of P. For a basis $\{\varphi_{i'}\}$, $1 \le i' \le m'$, we have $\{\varphi_{i'}, \varphi_{j'}\} = K_{i'j'}(\varphi_1, ..., \varphi_{m'}) \equiv \{\varphi_{i'}, \varphi_{j'}\}_{N_{\Psi^\perp}}$.

Such that a projection $\pi_{\psi\perp} : P \to N_{\psi\perp}$ is a Poisson map, this defines the Poisson structure on $N_{\psi\perp}$. The ultimate goal of the reduction procedure consists in determining the corresponding symplectic leaves and the Poisson structures on the reduced manifolds N_ψ and $N_{\psi\perp}$ are degenerate. We have $\{\mathfrak{I}_\psi, \mathfrak{I}_{\psi\perp}\} = 0$ by definition of $\mathfrak{I}_{\psi\perp}$. Hence, the Poisson maps π_ψ and $\pi_{\psi\perp}$ form a dual pair for P. The intersection $\mathfrak{I}_\psi \cap \mathfrak{I}_{\psi\perp}$ can be characterized as a set of functions on N_ψ that Poisson commute with all functions on N_ψ or, as functions in $N_{\psi\perp}$ that Poisson commute with all functions on $N_{\psi\perp}$. Thus, $\mathfrak{I}_\psi \cap \mathfrak{I}_{\psi\perp}$ **is a set of Casimir functions to both Poisson manifolds** N_ψ and $N_{\psi\perp}$. Assuming that P carries the action $G \times P \to P$ of some Lie group G with Lie algebra $\mathfrak{g}$. The vector fields χ_X for $X \in \mathfrak{g}$ generate an integrable distribution corresponding to a foliation of P by the G-action orbits. Selecting a basis e_i in $\mathfrak{g}$ for the vector fields $\chi_i \equiv \chi_{e_i}$ we note that: $[\chi_i, \chi_j] = c_{ij}^k \chi_k$ where c_{ij}^k are the structure coefficients of $\mathfrak{g}$. We characterize the foliation of P by G-orbits with $\Psi^\perp$, so that $N_{\psi\perp}$ is interpreted as a coset $N_{\psi\perp} = P/G$. A set $\mathfrak{I}_{\psi\perp}$ is spanned by all functions invariant under the G-action. Hamiltonian vector fields of $\mathfrak{I}_\psi$ are tangent to the leaves of $\Psi^\perp$ from the point of view of a foliation Ψ polar to $\Psi^\perp$. There exists a function basis $\{\psi_i\}$ in $\mathfrak{I}_\psi$ such that $[\chi_{\psi_i}, \chi_{\psi_j}] = c_{ij}^k(x)\chi_{\psi_k}$ by the Frobenius criterion, and we have 2 integrable distributions: $[\chi_i, \chi_j] = c_{ij}^k \chi_k$ and $[\chi_{\psi_i}, \chi_{\psi_j}] = c_{ij}^k(x)\chi_{\psi_k}$ that both give rise to the same foliation $\Psi^\perp$. Thus, any hamiltonian vector field $\chi_f, f \in \mathfrak{I}_\psi$ admit an expansion over a basis χ_i: $\chi_f = A^i(f)\chi_i$ and vice versa, where $A^i(f)$ are coordinate-dependent coefficients used to define a Lie algebra valued element $A(f) = A^i(f)e_i \in \mathfrak{g}$. $A(\{f, h\}) = \chi_{A(f)}A(h) - \chi_{A(h)}A(f) + [A(f), A(h)]$ for any $f, h \in \mathfrak{I}_\psi$ is implied by $[\chi_f, \chi_h] = \chi_{\{f,h\}}$. In the basis χ_i, $\chi_f = A^i(f)\chi_i$ reads $\{\psi_i, .\} = A^j(\psi_i)\chi_j \equiv A_i^j(x)\chi_j$. The G-action is equivalent to the hamiltonian action when the map A is χ-independent, and $A(\{f, h\}) = \chi_{A(f)}A(h) - \chi_{A(h)}A(f) + [A(f), A(h)]$ is the standard hamiltonian action moment map. The dual pair associated to this hamiltonian action is $J = (A^{-1})_i^j \psi_j(x)e^i \in \mathfrak{g}^*$ where π_ψ is identified with J. A is not a constant and

$$A(\{f, h\}) = \chi_{A(f)}A(h) - \chi_{A(h)}A(f) + [A(f), A(h)] \tag{15}$$

In the following figure, the first Foliation Ψ is by levels of a function set $\mathfrak{I}_\psi$ closed under the Poisson bracket on P. The Hamiltonian vector fields of functions from $\mathfrak{I}_\psi$ span an integrable distribution $T\Psi^\perp$. The leaves tangent to this distribution form the second foliation $\Psi^\perp$ of P (Fig. 1).

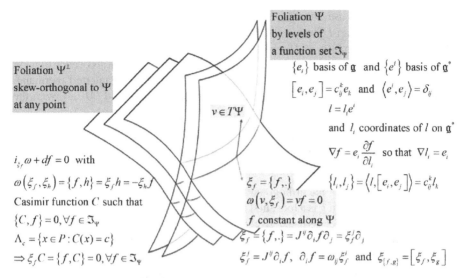

Fig. 1: Symplectic bifoliation structures

5 Libermann Foliation and Haefliger Γ-Structures

In the framework of Stefan foliations, Libermann's foliations [16, 17] are the general-ization of foliations studied in regular case by P. Libermann under the name of **symplectically complete foliations**. The study of these foliations has been clarified by the introduction of the notion of Poisson Γ-structure, which is the maximum extension of the notion of moment map in the sense of J.M. Souriau. This notion of foliation then appears as dual to that of Poisson Γ-structure. Generalized Moment has been introduced by P. Molino, M. Condevaux et P. Dazord [18–23] in papers of "Séminaire Sud-Rhodanien de Geometrie», with the translation in terms of symplectic duality between Haefliger Γ-structures and Libermann singular foliations of the notion of Souriau's moment of a Hamiltonian action. Let (M, ω) a symplectic manifold, G a Lie group with a Lie algebra $G \setminus P \leftarrow P \rightarrow \mathfrak{g}^*$, we consider a differentiable action $\phi : G \times M \rightarrow M$. Let $\phi_g(x) = \phi(g, x)$, for all $\mathfrak{g}$, we note $\xi_M = \frac{d}{dt}\phi(\exp(t\xi), x)\big|_{t=0}$, where ξ_M is the fundamental field of action associated to ξ. The action ϕ is symplectic if $\phi_g^*\omega = \omega$ for all $g \in G$, In this case, the Lie derivative $L_{\xi_M}\omega = i_{\xi_M}d\omega + d(i_{\xi_M}\omega)$ cancels for each fundamental field, and so $d(i_{\xi_M}\omega) = 0$ because $d\omega = 0$. Then, fundamental fields are locally Hamiltonian. If moreover all the fundamental fields are globally Hamiltonian, and so $i_{\xi_M}\omega$ is an exact form for all ξ, the action ϕ is said to be Hamiltonian. Given a basis $(\xi_1, \ldots, \xi_p)$ of $\xi \in \mathfrak{g}$, we can introduce functions so that $i_{\xi_i,M}\omega = df_{\xi_i}$ for $i = 1, \ldots, p$. We define a map J from M to $\mathfrak{g}$, moment of Hamiltonian action ϕ such that:

$$\left\langle J(x), \sum_{i=1}^p \lambda^i \xi_i \right\rangle = \sum_{i=1}^p \lambda^i f_{\xi_i}(x), \forall x \in M \text{ and } \lambda^1, \ldots, \lambda^p \in \mathbb{R} \tag{16}$$

In the particular case where the orbits of ϕ define a foliation $\mathfrak{I}$ of dimension r on M, the moment map J has constant rank r. The connected components of the level submanifolds of J then form on M a new foliation $\mathfrak{I}^\perp$, of codimension r. The foliations $\mathfrak{I}$ and $\mathfrak{I}^\perp$ are orthogonal in the symplectic sense, and the local first integrals of each of them define Hamiltonian fields tangent to the other. Based on work of Paulette Libermann, a **Libermann Foliation** is a foliation $\mathfrak{I}$ on the symplectic manifold (M, ω), with the following properties that are equivalent:

- (P1) The field of contact elements orthogonal to $\mathfrak{I}$ is completely integrable.
- (P2) $\mathfrak{I}$ is locally generated by Hamiltonian fields.
- (P3) The Poisson bracket of 2 local first integrals of $\mathfrak{I}$ is again a local first integral.

We will denote by $\mathfrak{I}^\perp$ the orthogonal foliation which is itself a Libermann foliation and we will observe that if f is a local first integral of $\mathfrak{I}$ (respectively of $\mathfrak{I}^\perp$), its Hamiltonian field X_f is foliated for $\mathfrak{I}$ (respectively $\mathfrak{I}^\perp$) and tangent to $\mathfrak{I}^\perp$ (respectively to $\mathfrak{I}$). A Poisson manifold (P, Λ) is a manifold endowed with an antisymmetric contravariant 2-tensor, so that the Poisson bracket operation defined on $C^\infty(P)$ by $\{l, h\}_\Lambda = \Lambda(dl \wedge dh)$ is a Lie algebra bracket. Its Hamiltonian field X_h is defined by the relation $X_h.l = \{l, h\}_\Lambda$, The Hamiltonian fields form a Lie algebra of vector fields whose orbits constitute a singular foliation S_Λ in the sense of P. Stefan. Leaves of S_Λ are endowed with a symplectic structure for which the Poisson bracket is induced by the bracket $\{.,.\}_\Lambda$. S_Λ is the symplectic foliation of (P, Λ).

if $\mathfrak{I}$ is a Libermann foliation on (M, ω), the fact that the Poisson bracket of two local first integrals is still a local first integral makes it possible to define a natural structure of Poisson manifold on the transversals to the foliation. The transverse structure of a Libermann foliation is a Poisson manifold structure, and the local projection onto a transverse is a Poisson morphism. An important Poisson structure is the Lie-Poisson structure $\mathfrak{g}^*$ on the dual of the Lie algebra $\left(\mathfrak{g}^*, \Lambda_0\right)$ of the Lie group G by setting an affine Poisson structure $\{\eta, \xi\}_{\Lambda_0}(\upsilon) = \{\eta, \xi\}(\upsilon) + \tilde{\Theta}$ where $\tilde{\Theta}$ is a 2-cocycle on the Lie algebra $\mathfrak{g}$. We can observe that the Souriau moment of a Hamiltonian action is a Poisson morphism endowed with such an affine structure. If we consider the Hamiltonian action of a Lie group on (M, ω), its orbits in general no longer form a foliation, but a singular foliation. Similarly, the Souriau moment no longer determines a foliation, but simply a Poisson morphism on a transverse Poisson manifold. Thus, the relation defined by symplectic duality between a Libermann foliation $\mathfrak{I}$ and its orthogonal $\mathfrak{I}^\perp$ loses its character of symmetry: $\mathfrak{I}$ becomes a singular foliation in the contravariant sense (the orbits of a Lie algebra), $\mathfrak{I}^\perp$ a singular foliation in the covariant meaning (defined by a family of 1-forms). Then, a general framework is given for the notion of moment, the relation established by symplectic duality between certain Stefan foliations (contravariant notion) and certain Haefliger Γ-structures (covariant notion). A Stefan foliation can be defined on a manifold M as a partition $\mathfrak{I}$ into immersed submanifolds (the leaves) such that orbits of the module of the differentiable vector fields tangent to the leaves are precisely the leaves. In the case of a symplectic manifold (M, ω), **Libermann foliation is a Stefan foliation such that any vector tangent at any point to a leaf is the value of a local Hamiltonian vector field tangent to the leaves.** Consider on (M, ω) a Poisson Γ-structure. We thus define on (M, ω) a sheaf of germs of local Hamiltonian fields. It is immediate to verify that the orbits of this bundle determine on the symplectic manifold a Libermann foliation, which we will denote by $\mathfrak{F}$. We will say that $\mathfrak{F}$ is the Libermann foliation "action" of the Poisson Γ-structure M. Conversely, M is a generalized moment for $\mathfrak{F}$. Given an arbitrary Libermann foliation on (M, ω), the existence of a generalized moment is not assured, and if there is one, there is no uniqueness.

6 Lie Foliations Associated with Elie CARtan'S Moving Frame

Based on the duality between the algebra of exterior differential forms and the Lie algebra of vector fields, Edmond Fédida [12–14] has explored the theory of foliation structures, either in the language of fully integrable Pfaff systems, or in that of involution vector fields. Within the theory of foliation structures, groups of Lie transformations, which correspond to systems of vector fields in involution, associated with a Lie algebra, has a counterpart, which is the theory of foliations associated with the Cartan's moving frame. For group of Lie transformation, a Manifold M of dimension n is equipped with vector fields X_i and structure coefficients given by C_{ij}^k by $[X_i, X_j] = C_{ij}^k X_k$, and as counterpart for moving frame, we consider a system of Pfaff form ω_i with structure coefficients where $d\omega_i = C_{ij}^k \omega_j \wedge \omega_k$. The ω_i form a constant rank system at all points. The classes form a foliation of M of whose transverse structure is modeled on that of a subspace of G. On the counterpart, The X_i form a constant rank system at all points: trajectories, the trajectories then constitute a foliation of M; leaves are homogeneous spaces. The ω_i are linearly independent at any point: the associated foliation has a transverse structure modeled on G. Such a foliation then deserves the name of Lie foliation. We can always come back to this case, a structure is given on M by a 1-form m with values in a Lie algebra of dimention q that verify Maurer-Cartan equation: $d\omega + \frac{1}{2}[\omega, \omega] = 0$. One can "desingularize" the foliation defined by ω on M by considering on the trivial principal bundle $M \times G$, the form of connection Ω induced by ω; implying that Ω is a form of flat connection. In particular Ω is a 1-form on $M \times G$ satisfying the Maurer-Cartan equation: $d\Omega + \frac{1}{2}[\Omega, \Omega] = 0$. $\mathfrak{g}$ is surjective for all $x \in M$. Under these conditions ω determines a foliation of codimension q of M, and is a Lie G-foliation of M. Molino has proved that this foliation is transversely parallelizable. We can then complete in a Riemanian structure of M by an "horizontal" metric subject to the only condition that at any point $x \in M$, the space tangent to the leaf is orthogonal, to the transverse space. Hence the foliation has a quasi-fibered metric.

We will illustrate for the Lie algebra $su(1,1)$ of Lie group $SU(1,1)$ [25]:

$$u_0 = -\frac{i}{2}\begin{pmatrix} 1 & 0 \\ 0 & -1 \end{pmatrix}, u_1 = -\frac{i}{2}\begin{pmatrix} 0 & -1 \\ 1 & 0 \end{pmatrix}, u_2 = \frac{1}{2}\begin{pmatrix} 0 & 1 \\ 1 & 0 \end{pmatrix} \tag{17}$$

Structure coefficients given by $[u_i, u_j] = C_{ij}^k u_k$: $C_{01}^2 = 1, C_{02}^1 = -1, C_{12}^0 = -1 C_{01}^2 = 1, C_{02}^1 = -1, C_{12}^0 = -1$

We introduce complex combinations $u_+ = u_1 + iu_2, u_- = u_1 - iu_2$ with

$$[u_0, u_+] = -iu_+, [u_0, u_-] = iu_-, [u_+, u_-] = 2iu_0 \tag{18}$$

The group has inverse metric : $\left[g^{ij} \right]_{ij} = \begin{bmatrix} -1 & 0 & 0 \\ 0 & 1 & 0 \\ 0 & 0 & 1 \end{bmatrix}$ (19)

We consider $H_1^1 = \{(z_1, z_2) \in \mathbb{C}^2 / |z_1|^2 - |z_2|^2 = 1\}$ a submanifold of $\mathbb{C}^2$. The matrix $h \in H_1^1$ of determinant one is parametrized by (z_1, z_2): $h = \begin{pmatrix} z_1 & z_2^* \\ z_2 & z_1^* \end{pmatrix}$ and we can compute a real left-invariant Maurer-Cartan 1-form encoding the geometry:

$$h^{-1}dh = \omega_0 u_0 + \omega_1 u_1 + \omega_2 u_2 \text{ with } d\omega_k + \frac{1}{2}C_{ij}^k \omega_i \wedge \omega_j = 0 \quad (20)$$

We define the complex combinations $\omega = \omega_1 + i\omega_2, \omega^* = \omega_1 - i\omega_2$ with
$\begin{cases} d\omega = i\omega \wedge \omega^0 \\ d\omega^0 = \frac{i}{2}\omega \wedge \omega^* \end{cases}$ where

$$\begin{cases} \omega = 2i(z_1 dz_2 - z_2 dz_1) \\ \omega^0 = i(z_1^* dz_1 - z_2^* dz_2 - z_1 dz_1^* + z_2 dz_2^*) \end{cases} \quad (21)$$

In terms of the left invariant 1-forms the metric and orientation are:

$$ds^2 = \frac{1}{4}\left(-(\omega_0)^2 + (\omega_1)^2 + (\omega_2)^2\right) \text{ and } Vol_{H_1^1} = \frac{1}{8}\omega_1 \wedge \omega_0 \wedge \omega_2 \quad (22)$$

The dual left-invariant vector fields, X_i, generate the right-action $h \to h_{u_i}$ and have the commutators $[X_i, X_j] = C_{ij}^k X_k$ and using the combinations $\begin{cases} X_+ = X_1 + iX_2 \\ X_- = X_1 - iX_2 \end{cases}$, we have

$$[X_0, X_+] = -iX_+, [X_0, X_-] = iX_-, [X_+, X_-] = 2iX_0 \quad (23)$$

where the left invariant vector fields take the form with $\partial_i = \frac{\partial}{\partial z_i}$:

$$X_0 = -\frac{i}{2}\left(z_1 \partial_1 + z_2 \partial_2 - z_1^* \partial_1^* - z_2^* \partial_2^*\right), X_- = -i\left(z_1^* \partial_2 + z_2^* \partial_1\right), X_+ = X_-^* \quad (24)$$

with the non $-$ zero pairing : $\omega_0(X_0) = 1, \omega(X_-) = \omega^*(X_+) = 2$

H_1^1 is that it is a circle fibration over a Riemann surface M characterized by a constant Gauss curvature $K = -1$. The projection is given by z a local complex coordinate on $M: \pi : H_1^1 \to M, h \mapsto z = \frac{z_2}{z_1}$ with global section $s : z \mapsto \frac{1}{\sqrt{1-|z|}}\begin{pmatrix} 1 & z^* \\ z & 1 \end{pmatrix}$.

The Maurer-Cartan 1-form is related to the Cartan connection for the Riemann surface M. We can then consider a principal H-bundle G $\to$ G/H with a natural flat connection, the Maurer-Cartan 1-form on G. A Cartan geometry is a manifold that is locally, or infinitesimally, Kleinian. For example, Cartan geometries are given by Kleinian geometries, with flat Cartan connection a homogeneous space (A Klein geometry (G, H) is a pair of a Lie group G and a closed subgroup $H \subset G$ such that the quotient G/H is smooth and connected). $H_1^1 = H^2 \times S^1$. If we consider $M = H^2 = SU(1,1)/U(1)$, the metric is $ds^2 = \frac{4}{(1+K|z|^2)^2} dz dz^*$ with the complexified co-frame $e = \frac{2dz}{1+K|z|^2}$ with

$$\begin{cases} de - ie \wedge \Gamma = 0 \\ d\Gamma = \Re = \frac{i}{2} Ke \wedge e^* \end{cases}, \quad \begin{cases} K = -1, \Re = Kw \\ w = \frac{i}{2} e \wedge e^* \end{cases} \quad \text{and } \Gamma = iK \frac{zdz^* - z^* dz}{1+K|z|^2}.$$

Previous structure and Gauss equations characterize the flatness of the Cartan connection $\rho = -\Gamma u_0 + \frac{i}{2}(eu_- - e^* u_+)$ (connection on the principal $U(1)$ bundle: $H_{-K}^1 \to M = H_{-K}^1/U(1))$, related to Maurer-Cartan 1-form on H_{-K}^1. The curvature of ρ vanishes (K: Gauss Curvature; Γ Spin Connection):

$$K_\rho = d\rho + \frac{1}{2}[\rho, \rho] = -\left(\Re - K\frac{i}{2}e \wedge e^*\right)u_0 + \frac{i}{2}(de - i\Gamma \wedge e)u_- - \frac{i}{2}(de^* + i\Gamma \wedge e*)u_+$$

Using the global section $s : z \mapsto \frac{1}{\sqrt{1-|z|}}\begin{pmatrix} 1 & z^* \\ z & 1 \end{pmatrix}$, the Cartan connection ρ on M is

the pullback of Maurer-Cartan 1-form: $\rho = s^*\left(h^{-1}dh\right)$ due to $\begin{cases} s^*(\sigma) = ie \\ s^*\sigma^0 = -\Gamma \end{cases}$.

By projection $\pi : H_1^1 \to M, h \mapsto z = \frac{z_2}{z_1}$, $h^{-1}dh$ can be expressed in terms of $\pi^*(\rho)$, up to gauge transformation $\pi^*e = -i\frac{z_1^*}{z_1}\sigma, \pi^*\Gamma = -\sigma^0 + id \ln\left(\frac{z_1}{z_1^*}\right)$.

7 Conclusion

As observed by Dazord and Delzant, Mechanics and Quantum Physics request exploration of symplectically complete foliation introduced by Libermann, where a foliation is symplecticaly complete if and only if its symplectic orthogonal is a foliation. For each transversely oriented foliation, we have the famous Godbillon-Vey characteristic class [15]. When the symplectic foliations of regular Poisson structures are transversely oriented, they have the Godbillon-Vey characteristic classes. It should be interesting to explore their Godbillon-Vey classes in terms of Poisson structure for Souriau model.

References

1. Souriau, J.M.: Structure des systèmes dynamiques. Dunod (1969)
2. Souriau, J.M.: Mécanique statistique, groupes de Lie et cosmologie. In: Colloque International du CNRS "Géométrie symplectique et physique Mathématique", 1974
3. Gallisot, F.: Les formes extérieures en mécanique. Annales de I 'Institut Fourier, Grenoble **4**, 145–297, (1952)
4. Carathéodoty, C.: Untersuchungen über die Grundlagen der Thermodynamik. Math. Ann. **67**, 355–386 (1909)
5. Caratheodory, C.: Calculus of Variations and Partial Differential Equations of the First Order. Volumes I and II. Holden Day, San Francisco (1967)
6. Haefliger, A.: Naissance des feuilletages d'Ehresmann-Reeb à Novikov. J 2(5), 99–110 (2016)
7. Reinhart, B.L.: Differential Geometry of Foliations, vol. 99, Springer Verlag, Cham (1983)

8. Reeb, G.: Structures feuilletées. Bulletin de la Société Mathématique de France, Tome **87**, 445–450 (1959)
9. Reeb, G.: Structures feuilletées, Differential Topology, Foliations and Gelfand-Fuks cohomology, Rio de Janeiro, 1976. Springer Lecture Notes in Math. **652**, 104–113 (1978)
10. Reeb, G.: Propriétés topologiques des trajectoires des systèmes dynamiques, Mém. Acad. Se. Bruxelles **27** (1952)
11. Reeb, G., Ehresmann, C., Thom, R., Libermann, P.: Structures feuilletées, CNRS édition, Colloques Internationaux Du Cnrs (CIDC) (1964)
12. Fedida, E.: Feuilletages du plan, feuilletages de Lie. University of Strasbourg, Thèse (1973)
13. Fedida, E.: Sur l'existence des feuilletages de Lie; C.R. Acad. Sci. Paris § 278 pp. 835–837 (1974)
14. Fedida, E. : Sur la théorie des feuilletages associée au repère mobile: cas des feuilletages de Lie. Lecture Notes in Mathematics, vol. 652. Springer (1978)
15. Godbillon, C., Vey, J.: Un invariant des feuilletages de codimension 1, C. R. Acad. Sci. Paris Ser. A **112**, 92–95 (1971)
16. Libermann, P.: Problèmes d'équivalence et géométrie symplectique. Astérisque, tome **107–108**, 43–68 (1983)
17. Libermann, P: Legendre foliations on contact manifolds, differential geometry and its applications, n°1, 57–76, North-Holland (1991)
18. Dazord, P., Molino, P. : Γ-Structures poissonniennes et feuilletages de Libermann, Publications du Département de Mathématiques de Lyon, fascicule 1B, « Séminaire Sud-Rhodanien 1ère partie », chapitre II , pp. 69–89 (1988)
19. Condevaux, M., Dazord, P., Molino, P. : Géométrie du moment, Publications du Département de Mathématiques de Lyon, fascicule 1B, « Séminaire Sud-Rhodanien 1ère partie », chapitre V , pp. 131–160 (1988)
20. Molino, P.: Dualité symplectique, feuilletage et géométrie du moment. Publicacions Matemátiques **33**, 533–541 (1989)
21. Dazord, P.: Sur la géométrie des sous-fibrés et des feuilletages lagrangiens. Annales scientifiques de l'École Normale Supérieure **14**(4), 465–480 (1981)
22. Dazord, P.: Feuilletages en geometrie symplectique. C. R. Acad. Sc. Paris **294**, 489–491 (1982)
23. Dazord, P. : Feuilletages et mécanique hamiltonienne. Séminaire de géométrie, université de Lyon I, fascicule 3 B (1983)
24. Arutyunov, G., Elements of Classical and Quantum Integrable Systems, Springer, Cham (2019)
25. Ross, C.D.H.: The Geometry of Integrable Vortices. Ph.D. Thesis, Heriot Watt University (2019)
26. Barbaresco, F.: Symplectic theory of heat and information geometry. In: Chapter 4, Handbook of Statistics, Vol. 46, pp. 107–143, Elsevier (2022)
27. Barbaresco, F.: Jean-Marie Souriau's symplectic model of statistical physics: seminal papers on lie groups thermodynamics - Quod Erat demonstrandum. In: Barbaresco, F., Nielsen, F. (eds.) SPIGL 2020. SPMS, vol. 361, pp. 12–50. Springer, Cham (2021). https://doi.org/10.1007/978-3-030-77957-3_2
28. Barbaresco, F.: Symplectic foliation structures of non-equilibrium thermodynamics as dissipation model: application to metriplectic nonlinear lindblad quantum master equation. Entropy **24**, 1626 (2022)

Poisson Geometry of the Statistical Frobenius Manifold

Noemie Combe$^{(\boxtimes)}$, Philippe Combe, and Hanna Nencka

Max Planck Institute for Mathematics in Sciences, Insel str. 22,
04103 Leipzig, Germany
noemie.combe@mis.mpg.de
https://www.mis.mpg.de/f-soda/index.html

Abstract. New insights on parametric families of probability distributions (exponential type) are investigated. As shown by Combe-Manin (2020), flat exponential statistical manifolds are Frobenius manifolds. Frobenius manifolds correspond to a geometrization of the PDE equation named after Witten–Dijkgraaf–Verlinde–Verlinde. In this paper, we prove that this source of Frobenius manifolds is a Poisson manifold. This is shown using Dubrovin-Novikov's approach to Frobenius structures, based on equations of hydrodynamical type. This finding makes connections with Koszul-Souriau-Vinberg's past results.

Keywords: Koszul–Souriau–Vinberg theory · equations of hydrodynamical type · Frobenius manifolds · parametric families of probability distributions · related to exponential families · Poisson manifold

1 Introduction

New insights on parametric families of probability distributions (exponential type) are investigated. As shown in [4–7], flat exponential statistical manifolds are Frobenius manifolds. Frobenius manifolds correspond to a geometrization of the PDE equation named after Witten–Dijkgraaf–Verlinde–Verlinde. The notion of Frobenius manifolds/ algebras relates tightly to Topological Field theory (TFT) [8]. This raises thus many questions regarding the relation between statistical flat exponential manifolds and (TFT).

The topic of this paper is to prove that the statistical Frobenius manifold is a Poisson manifold, i.e. symplectic with Poisson structures. By statistical Forbenius manifold we mean the manifold of probability distributions (exponential type) satisfying the WDVV equation and discussed in [4–7]. In Yu. Manin's classification, this class of statistical manifolds corresponds to the fourth Frobenius manifolds.

This result is shown using Dubrovin-Novikov's approach to Frobenius structures, based on equations of hydrodynamical type. This finding makes connections

Supported by the Max Planck Institute for Mathematics in Sciences.

with Koszul-Souriau-Vinberg's past results [2,3]. In particular, it allows to highlight the deep connection existing between those two different insights.

The paper is devoted to proving the following main theorem:

Theorem 1 (Main theorem). *The statistical Frobenius manifold is a Poisson manifold.*

We will proceed in two steps. First, we will prove in a proposition that the information geometry's Frobenius manifold is a symplectic manifold. In a second step we prove that it is equipped with Poisson structures, thus giving a Poisson manifold. It is known that a symplectic manifold is an integrable system, equipped with a Hamiltonian action (for instance see [13]), relying on the metric tensor of the manifold. The first part of the discussion is decomposed into two principal steps:

– The fourth Frobenius manifold is an integrable system.
– This integrable system is a Hamiltonian one, depending on the tensor metric of the manifold.

Smaller steps are needed to prove this statement (we choose not to mention them here, for simplicity). For recollections on paracomplex geometry we refer to [4,6,7].

2   The Statistical Frobenius Manifold is Symplectic

We prove the following statement.

Proposition 1. *The statistical Frobenius manifold is a symplectic manifold.*

Proof. **Part 1: on integrable systems**

To prove that we have an integrable system, we invoke the following theorem [12, 13]:

The class of flat, torsionless submanifolds in any Euclidean or pseudo-Euclidean space is an integrable system.

From [7,8,11] the Frobenius manifold is necessarily *flat and torsionless*. Therefore, the statistical Frobenius manifold is an integrable system.

Now we show that the statistical Frobenius manifold, being an integrable system, is a Hamiltonian one.

Part 2: on Hamiltonian system.

Note that the system has a Hamiltonian structure if and only if the *Lagrangian* $\mathcal{L}$ depends on the tensor metric, i.e. $\mathcal{L}(x,\xi) = g_{ij}(x)\xi^i(x)\xi^j(x) - U(x)$, where x are canonical coordinates on the manifold, g_{ij} is the tensor metric, U is a scalar field and ξ^i, ξ^j are vector fields.

Let us sketch the Hamiltonian construction for the statistical Frobenius manifold. From [7], it is known that the information geometry F-manifold has a *paracomplex structure*. It implies that the construction of differential forms for paracomplex geometry is necessary. We refer to [4,6,7] for some notions on paracomplex geometry. In order to obtain the Hamiltonian on the paracomplex manifold, we introduce a local paracomplex Dolbeault (1,1)-form. This is given by:

$$\omega := \partial_+ \partial_- \phi$$

where ϕ is the potential function defined uniquely modulo subspace of local functions $ker\partial_+\partial_-$ and $\partial_\pm$ are given in canonical paracomplex coordinates.

To show that the Hamiltonian relies on the metric tensor, it is sufficient to consider the scalar product on a space with paracomplex structure. By definition the scalar product is defined as:

$$\langle \xi, \eta \rangle_{\mathfrak{e}} = g_{jk}\xi^j \overline{\eta^k}.$$

This scalar product is a Hermitian one, since we have the equality:

$$\langle \xi, \eta \rangle_{\mathfrak{e}} = \overline{\langle \eta, \xi \rangle_{\mathfrak{e}}}.$$

Then, the paracomplex differentiable 2-form is given by:

$$\Omega_{\mathfrak{e}} = \frac{e}{2} g_{jk} dz^j \wedge d\bar{z}^k.$$

By definition, it is known that the Hamiltonian is a 2-form, so we have proved that a Frobenius manifold—with paracomplex structure—is an integrable system with Hamiltonian.

We apply this construction to our manifold. To show that this manifold is symplectic notice that the Hamiltonian (discussed above) is associated to the following functional:

$$\langle \mu, f \rangle = \int_{\tilde{\Omega}} f(\omega) \mu \{d\omega\}, \tag{1}$$

where f is a $\mathcal{F}$-measurable function and μ is a dominated measure on the measurable space $(\tilde{\Omega}, \mathcal{F})$. In the discrete case, this is given by:

$$\langle \mu, f \rangle = \sum_{j=1}^{m} f^j \mu_j.$$

From [5,7] statistical manifolds (satisfying the Frobenius condition) have the geometry of the Lie group of type $SO(1, n-1)$. The bilinear form for this group, is given by :

$$\kappa_1 ds^2 = \lambda_{\mu\nu} dz^\mu dz^\nu,$$

where $\{\lambda_{\mu\nu}\}$ is a diagonal matrix $(1, 1, 1, -1)$ and κ_1 is a constant.

So, the Lagrangian written in terms of the tangent vector is given by: $\mathcal{L} = \frac{1}{2}C(\xi^\mu \xi_\mu - 1) + \kappa_2 \xi^\mu A_\mu$ where A_μ corresponds to the Gauge Abelian field; C

and κ_2 are constants. The so-called "momentum" and "force" are derived from the Lagrangian, respectively as follows:

$$p_\mu = \frac{\partial \mathcal{L}}{\partial \xi^\mu} = \xi_\mu + \kappa_2 A_\mu,$$

$$f_\mu = \frac{\partial \mathcal{L}}{\partial z^\mu} = \xi^\mu \frac{\partial A_\nu}{\partial z^\mu}.$$

This allows to write the Hamiltonian using the Lagrangian view point for our case:

$$\mathcal{H} = \xi^\mu \frac{\partial \mathcal{L}}{\partial \xi^\mu} - \mathcal{L} = \xi^\mu p_\mu - \mathcal{L}.$$

This shows that our manifold has a Hamiltonian structure. Finally, as already mentioned the manifold under consideration satisfies torsionless and flatness properties and therefore, it forms an integrable system. The existence of the Hamiltonian implies that this Frobenius manifold is symplectic. $\square$

We have thus ended the proof of the main theorem. In the next part, we discuss other implications of this on the geometry.

3   Geometrical Implications: Poisson Geometry

In this section, we present geometrical implications for our F-manifold, proving the second part our main theorem. Consider a symplectic manifold. In the works of Novikov on symplectic geometry, a special type of Poisson brackets for this type of manifold was introduced. We call them the *Novikov–Poisson brackets*. These Novikov–Poisson brackets were initially introduced to construct a theory of conservative systems of hydrodynamic type. This structure strongly interferes with the existence of a Frobenius manifold structure.

From the previous statements, we have:

Proposition 2. *The statistical Frobenius manifold is a symplectic (paracomplex) manifold* $(M, \Omega_\mathfrak{e}, \Omega)$ *with the 2-form* $\Omega_\mathfrak{e} = \frac{e}{2} g_{jk} dz^j \wedge d\bar{z}^k$, *where* $e^2 = 1$ *and* g_{jk} *is a paracomplex bilinear form and* $\Omega = dz^j \wedge dp_k$.

This statement has an importance of its own. However, it can be adapted to Poisson geometry. A Poisson structure is a Lie algebra bracket $\{\cdot, \cdot\}$ on the vector space of smooth functions on a manifold M which form a Poisson algebra with the point-wise multiplication of functions. Therefore, we can say that:

Corollary 1. *The statistical Frobenius manifold carries a Poisson structure and it is thus a Poisson manifold.*

The Poisson bracket on the paracomplex space $\{\cdot, \cdot\}_\mathfrak{e}$ is given by

$$\{\xi, \eta\}_\mathfrak{e} = \frac{1}{2} Im\langle \xi, \eta \rangle_\mathfrak{e} = \frac{e}{2} \left(\frac{\langle \xi, \eta \rangle_\mathfrak{e} - \overline{\langle \xi, \eta \rangle_\mathfrak{e}}}{2} \right),$$

where e satisfies $e^2 = 1$ and Im is the imaginary part of a paracomplex number.

We consider this Frobenius manifold as a certain type of phase space, i.e. $(z^1, \cdots, z^n, p_1, \cdots, p_n, \Lambda_1, \cdots, \Lambda_m)$, where the z^μ are paracomplex numbers, $p_\mu = p_\mu(z)$ are 1-forms of the fibered cotangent bundle attached at the point z and the Λ_i are m functions on the phase space.

The set $\{\Lambda_i\}_{i=1}^m$ can have a Lie algebra structure, if the Poisson bracket of a pair of functions Λ_i and Λ_j is obtained by a linear combination of those functions, for $i, j \in \{1, \cdots, m\}$. More precisely, we have $\{\Lambda_i, \Lambda_j\} = -\Lambda_k \gamma_{ij}^k$, where γ_{ij}^k are structure constants of the algebra.

Let A, B be a pair of functions on the phase space $(z^1, \cdots, z^n, p_1 \cdots, p_n)$. These are observables and can be for example the momentum. A and B are real functions depending on z and p. Then, the Poisson bracket is given by:
$\{A, B\} = \frac{\partial A}{\partial p_\mu} \frac{\partial B}{\partial z^\mu} - \frac{\partial B}{\partial p_\mu} \frac{\partial A}{\partial z^\mu}$.

In the generalised phase space, where the Λ_i functions enter the game, $(z^1, \cdots, z^n, p_1, \cdots, p_n, \Lambda_1, \cdots, \Lambda_m)$ we have the Poisson bracket defined as:

$$\{A, B\} = \frac{\partial A}{\partial p_\mu} \frac{\partial B}{\partial z^\mu} - \frac{\partial B}{\partial p_\mu} \frac{\partial A}{\partial z^\mu} - \Lambda_k \gamma_{ij}^k \frac{\partial A}{\partial \Lambda_i} \frac{\partial B}{\partial \Lambda_j}.$$

Note that for an arbitrary Hamiltonian $\mathcal{H}(p, z, \Lambda, s)$ and a smooth phase space function $Q(p, z, \Lambda)$ the equation of motion is described by $\dot{Q} = \frac{dQ}{ds} = \{\mathcal{H}, Q\}$. This is identified to the notion of a *symplectic gradient*. Indeed, a symplectic gradient is a unique vector field $X_\mathcal{H}$ described by the condition $X_\mathcal{H}(Q) = \{\mathcal{H}, Q\}$.

For instance, one can write the following symplectic gradient: $X_\mathcal{H}(z^\mu) = \frac{dz^\mu}{ds} = \{\mathcal{H}, z^\mu\}$, where $\mathcal{H}$ is the Hamiltonian. More formally, the symplectic gradient of $\mathcal{H}$ is the vector field given by: $X_\mathcal{H} := \Omega_{\mathfrak{e}}^{-1} d_{dR} \mathcal{H}$, where the symbol d_{dR} is the *de Rham differential*, mapping smooth functions on M to 1-forms on M. On the fiber (cotangent) bundle space $T^*(M)$, we can indeed define a symplectic gradient. Since the 2-form $\Omega_{\mathfrak{e}}$ is closed, the Poisson brackets introduce on the linear space of functions a Lie algebra structure.

So, we have the important corollary, that the statistical Frobenius manifold is a Poisson manifold. This implies that it comes equipped with Poisson Geometry.

In what follows, we show a connection between our statements and works of [3]. Our last statement establishes a bridge to works of Koszul–Souriau [2,3, 10,15]. Indeed, the Frobenius structure is closely related to the Poisson bracket of hydrodynamic type (see [1,9]). By [1,9], the Novikov–Poisson bracket is a Poisson bracket of hydrodynamic type, defined by the following equation:

$$\{u^i(x), u^j(y)\} = g^{ki,\alpha}(u(x))\partial_\alpha \delta'(x - y) + b_j^{ki,\alpha}(u(x))\partial_\alpha u^j \delta(x - y), \qquad (2)$$

where u^j are fields (usually the density of momentum and energy or mass) and b_k^{ij} is a tensor.

This leads us to the following.

Lemma 1. *Let $z = (z^1, \cdots, z^n)$ and $w = (w^1, \cdots, w^n)$ be two points on a paracomplex manifold and let $u(z) = (u^1(z), \cdots, u^n(z))$, $v(w) = (v^1(w), \cdots, v^n(w))$ (with $u^i(z) = u(z^i)$) be two "observables" on the phase space, associated to the* *paracomplex Frobenius manifold. Then,* $\{u^i(z), u^j(w)\} = g^{ij}(u(z)\delta'(z-w)) + \frac{\partial u^k}{\partial z} b_k^{ij}(u(z)\delta(z-w)).$

Proof. This follows from the result in [7] showing the existence of paracomplex geometry of the statistical Frobenius manifold on one hand side and on the other from [9].

The classification of Poisson brackets has been done in [1]. Note that it depends linearly on u^k and is relative to linear changes $u^k = A_j^k w^j$. One can classify them throughout the classification of Lie algebras and Frobenius algebras. The *simplest local Lie algebra* arises from the following:

$$g^{ij} = C_k^{ij} u^k + g_0^{ij}, \quad b_k^{ij} = \text{cst}, \quad g_0^{ij} = \text{cst};$$

$$[p,q]_k(z) = b_k^{ij}(p_i(z)q_j'(z) - q_i(z)p_j'(z))$$

where

$$b_k^{ij} + b_k^{ji} = C_k^{ij} = \partial g^{ij}/\partial u^k. \tag{3}$$

The tensors b_k^{ij} (from Eq. (3)) play a key role. It defines a local, translationally invariant Lie algebra of first order if and only if the following condition is fulfilled:

Condition: *The tensor* b_k^{ij} *are the constant structures in the multiplication law of a finite dimensional algebra* $\mathcal{A}$ *defined by the identities:*

$$a,b,c \in \mathcal{A}, \quad e^i, e^j \in \mathcal{A} \quad e^i e^j = b_k^{ij} e^k;$$

$$a(bc) = b(ac), \quad (ab)c - a(bc) = (ac)b - a(cb).$$

The algebra $\mathcal{A}$ is associative and commutative if $2b_{ij} = 2b_{ji} = C_{ij}$. This algebra is Frobenius if there exists a given non-degenerate inner product satisfying the associativity condition i.e. such that: $\langle e_i \circ e_j, e_k \rangle = \langle e_i, e_j \circ e_k \rangle$. The necessary and sufficient condition for $\mathcal{A}$ (equipped with identity) to be Frobenius is that $2b_{ij} u^k = C_{ij} u^k$ is *non-degenerate*, at a generic point.

We explicit our thought. The $b_j^{ki,\alpha}$ (resp. b_{ji}^k in a less general context) plays a decisive role, regarding the Frobenius structure in the following way. Those tensors correspond to the constant structures in the multiplication law of the Frobenius algebra $\mathcal{A}_p$, associated to the Frobenius manifold M. More precisely, the Frobenius algebra is associated to the tangent space of a Frobenius manifold $T_p M$, at a given point p. This leads to the formulation of the statement below.

Lemma 2. *For the statistical Frobenius manifold, there exist Poisson brackets of hydrodynamic type, (Eq. (2)) such that the tensors* b_k^{ij} *correspond to the constant structures in the multiplication law of the Frobenius algebra, associated to the tangent space to the Frobenius manifold.*

Proof. This follows from the definition of Frobenius manifold introduced by Dubrovin in [8] and from the main statement in [1].

In particular, we have the following statement:

Lemma 3. *Let (M, Ω) be the symplectic statistical Frobenius manifold. Then, the phase space associated to it has an algebraic structure depending on the tensors $b_j^{ki,\alpha}$, defined in Eq. (2).*

Proof. Let us map T_pM to its dual space T_p^*M by a linear operator Φ. Then, there exists a map $\tilde{\phi} : \mathcal{A}_p \to \mathcal{A}^*{}_p$, where $\mathcal{A}^*{}_p$ is the algebra attached to T_p^*M. Following the commutative diagram below, we have that $\tilde{\phi} = \Psi^{-1} \circ \Phi \circ \chi$ is a homomorphism.

$$
\begin{array}{ccc}
\mathcal{A}_p & \xrightarrow{\ \tilde{\phi}\ } & \mathcal{A}^*{}_p \\
\chi \downarrow & & \downarrow \Psi \\
T_pM & \xrightarrow{\ \Phi\ } & T_p^*M
\end{array}
$$

Therefore, by duality (and only for finite dimensional cases!) the $b_j^{ki,\alpha}$ are the constant structures of the algebra associated to the cotangent space $\mathcal{A}^*{}_p$. So, the coefficients $b_j^{ki,\alpha}$ determine the underlying algebraic structure of the phase space and thus of the symplectic paracomplex manifold.

Finally, one last remark concerning the Poisson geometry of this Frobenius manifolds. The Novikov–Poisson brackets (hydrodynamic type) are tightly related to the metric of the manifold in the following way.

Proposition 3. *On the fourth Frobenius manifold, there exist flat pencils of Koszul–Souriau metrics, being related to compatible Novikov–Poisson brackets.*

Proof. Suppose we have a flat metric in some coordinate system $\{x^i\}$. If the components of the metric $g^{ij}(x)$ and the Christoffel symbols Γ_k^{ij} of the corresponding Levi–Civita connection depends linearly on the coordinate x^1, then the metrics form a *flat pencil:*

$$
g^{ij} \quad \text{and} \quad g_2^{ij} := \partial g^{ij}
$$

(it is assumed that the determinant of g_2^{ij} is not zero).

A pair of Poisson brackets (hydrodynamic type) $\{\cdot, \cdot\}$ are called *compatible* if an arbitrary linear combination with constant coefficients of those brackets defines back a Poisson bracket. Flat pencils of metrics correspond to *compatible pairs of poisson brackets* (hydrodynamic type) (see [8]). Therefore, we have the existence of flat pencils of Koszul–Souriau metrics for the fourth Frobenius manifolds.

We can end this section by remarking the following fact.

Remark 1. Orbits (that means geodesics) in the fourth Frobenius manifold are given by some trajectories given in paraholomorphic coordinates. Motions along these paraholomorphic orbits are *non dissipative* [14]. In particular, by non dissipative motion we mean that the production of entropy (during the motion) is null. In the latter case—and referring to Souriau's theory—we have an equality between the inner statistical states and the outer statistical states. Whereas the motion of transversal vectors to these surfaces is dissipative [3].

References

1. Balinskii, A.A., Novikov, S.P.: Poisson brackets of hydrodynamic type, Frobenius algebras and lie algebras. Soviet Math. Dokl. **32** (1985)
2. Barbaresco, F.: Geometric theory of heat from Souriau lie groups thermodynamics and Koszul hessian geometry: applications in information geometry for exponential families. Entropy **18**, 386 (2016)
3. Barbaresco, F.: Lie group statistics and lie group machine learning based on Souriau lie groups thermodynamics and Koszul-Souriau-fisher metric: new entropy definition as generalized Casimir invariant function in coadjoint representation. Entropy **22**(6), 642 (2020)
4. Combe, N., Combe, P., Nencka, H.: Frobenius statistical manifolds and geometric invariants. In: Nielsen, F., Barbaresco, F. (eds.) GSI 2021. LNCS, vol. 12829, pp. 565–573. Springer, Cham (2021). https://doi.org/10.1007/978-3-030-80209-7_61
5. Combe, N., Combe, P., Nencka, H.: Pseudo-elliptic geometry of a class of Frobenius-manifolds & Maurer-Cartan structures. ArXiv:2107.01985 (2021)
6. Combe, N., Combe, P., Nencka, H.: Algebraic properties of the information geometry's fourth Frobenius manifold. In: Arai, K. (eds.) Advances in Information and Communication. FICC 2022. Lecture Notes in Networks and Systems, vol. 438, pp. 356–370. Springer, Cham (2022). https://doi.org/10.1007/978-3-030-98012-2_27
7. Combe, N., Manin, Y..: F-manifolds and information geometry, Bull. London Maths Soc. **52**(5) (2020)
8. Dubrovin, B.: Geometry of 2D topological field theories. In: Francaviglia, M., Greco, S. (eds.) Integrable Systems and Quantum Groups. LNM, vol. 1620, pp. 120–348. Springer, Heidelberg (1996). https://doi.org/10.1007/BFb0094793
9. Dubrovin, B.A., Novikov, S.P.: On poisson brackets of hydrodynamic type. Dokl. Akad. Nauk SSSR, 294–297 (1984)
10. Koszul, J.L.: Introduction to Symplectic Geometry. Springer, Cham (2019)
11. Manin, Y.I.: Three constructions of Frobenius manifolds: a comparative study. Asian J. Math **3**(1), 179–220 (1999)
12. Mokhov, O.I.: Theory of submanifolds, associativity equations in $2d$ topological quantum field theories, and Frobenius manifolds. Theor. Math. Phys. **152**, 1183–1190 (2007)
13. Mokhov, O.I.: Realization of Frobenius manifolds as submanifolds in pseudo-euclidean spaces. Proc. Steklov Inst. Math. **267**(1), 217–234 (2009)
14. Nencka, H., Streater, R.F.: Information geometry for some lie algebras. Infin. Dimens. Anal. Quantum Probab. Relat. Top. **2**(3), 441–460 (1999)
15. Souriau, J.M.: Géométrie de l'espace des phases, calcul des variations et mécanique quantique

Canonical Hamiltonian Systems in Symplectic Statistical Manifolds

Michel Nguiffo Boyom[✉]

Alexander Grothendieck Research Institute, IMAG, CNRS UNIV MONTPELLIER,
Montpellier, France
nguiffo.boyom@gmail.com

Abstract. What is named Hamiltonian system in a symplectic manifold (M, ω) is an abelian Poisson subalgebra of the Poisson algebra $(C^\infty(M), \pi_\omega)$ which is defined by (M, ω). If the manifold M is compact then a Hamiltonian system may be regarded as a moment map of an action of an abelian Lie group. We aim to point out some canonical examples of Hamiltonian systems in symplectic statistical manifolds. The examples we are interested in are eigenfunctions of recursions operators of compatible symplectic statistical manifolds.

The abstract should briefly rize the contents of the paper in 15–250 words.

Keywords: Hamiltonian system · Symplectic statistical manifold · Gauge invariant · Canonical Koszul class

1 Basic Notions

Given a differentiable manifold M, $C^\infty(M)$ is the associative commutative algebra of real valued functions which are defined on M, $\mathcal{X}(M)$ is the Lie algebra of differentiable vector fields on M, $\mathcal{T}_1^2(M)$ is the sheaf of (2,1)-tensors whose sections are $C^\infty(M)$-bilinear maps of $\mathcal{X}(M)$ in itself.

1.1 Gauge Group

A gauge structure on M is a couple (M, ∇) where ∇ is a Koszul connection in M. The category of gauge structures on M is denoted by $\mathcal{G}(M)$.
A gauge isomorphism of $\mathcal{G}(M)$ is an invertible section of the vector bundle $Hom(TM, TM)$.
The group of gauge isomorphisms is denoted by $\mathcal{G}(M)$.
Every pair of gauge structures $[(M, \nabla), (M, \nabla')]$ is associated with the following differential equations,
$FE(\nabla\nabla' : \nabla' \otimes \phi - \phi \otimes \phi = O$; ϕ is a section of $Hom(TM, TM)$. Both $\nabla' \otimes \phi$ and $\phi \otimes \nabla$ are (2,1)-tensors which are defined as it follows,

$$\nabla' \otimes \phi(X, Y) = \nabla'_X \phi(Y),$$

F. Nielsen and F. Barbaresco (Eds.): GSI 2023, LNCS 14072, pp. 173–180, 2023.
https://doi.org/10.1007/978-3-031-38299-4_19

$$\phi \otimes \nabla(X, Y) = \phi(\nabla_X Y).$$

The sheaf of solutions of $FE(\nabla\nabla'$ is denoted by $\mathcal{J}_{\nabla nabla'}(M)$

1.2   Compatibility of Symplectic Structures

In a symplectic structure (M, ω) the Hamiltonian vector X_f which is associated with $f \in C^\infty(M)$ is defined as it follows,

$$\omega(X_f, Y) = df(Y).$$

The Poisson tensor π_ω which is associated with ω is defined as it follows,

$$\pi_\omega(df, df') = df'(X_f).$$

By setting

$$f, f' = \pi_\omega(df, df')$$

one has the Lie algebra $(C^\infty(M), -, -)$.

Definition 1. *[2, 4] The symplectic structures (M, ω) and $(M, \omega*)$ are compatible if $\pi_\omega + \pi_{\omega*}$ is a Poisson tensor.*

Given two sympectic structures (M, ω) and (M, ω') there is a unique gauge isomorphism Θ subject to the following identity,

$$\omega'(X, Y) = \omega(\Theta(X), Y).$$

The gauge isomorphism Θ is called recursion operator.

We recall the Nijenhuis tensor of Θ,

$$N(\Theta)(X, Y) = [X, Y] + \Theta[\Theta(X), Y] + \Theta[X, \Theta(Y)] - [\Theta(X), \Theta(Y)]$$

The following assertions are equivalent [1, 2, 4],

(1) (M, ω) and (M, ω') are compatible;
(2) $N(\Theta) = 0$.

2   Symplectic Statistical Manifolds

We aim to point out some examples of construction of compatible symplectic structures.

2.1   Canonical Koszul Classes of a Statistical Manifold

Let W be a left module of a Lie algebra h let

$$\Lambda_h = \oplus_q \Lambda^q(h)$$

be the exterior algebra of the vector space h. For convenient we remind the Chevalley-Eilenberg complex

$$\to Hom(\Lambda^{q-1}(h), W) \to Hom(\Lambda^q(h), W) \to Hom(\Lambda^{q+1}(h), W) \to$$

The coboundary operator δ is defined as it follows
$\delta f(a_0 \wedge .. \wedge a_q = \Sigma_0^q (-1)^i a_i f(.. \wedge \hat{a}_i \wedge ..) + \Sigma_{i<j}(-1)^{i+j} f([a_i, a_j] \wedge ..\hat{a}_i \wedge ..\hat{a}_j \wedge ..)$
The operator satisfies the requirement $\delta^2 = 0$. The qth cohomology space is

$$H^q_{CE}(h, W) = \frac{Ker(C^q(h, W) \to C^{q+1}(h, W)}{\delta(C^{q-1}(h, W)},$$

here

$$C^q(h, W) = Hom(\Lambda^q(h), W).$$

Henceforth we deal with statistical manifolds in the following presentation, $(M, g, \nabla, \nabla^\star)$,

$$Xg(Y, Z) - g(\nabla_X^\star Y, Z) - g(Y, \nabla_X Z) = 0.$$

The vector spaces of solutions of $FE(\nabla\nabla^\star)$ and $FE(\nabla^\star\nabla)$ are denoted by $J_{\nabla,g}$ and $J_{\nabla^\star,g}$ respectively.
Every $\phi \in J_{\nabla,g}$ gives rise to the couple

$$(\Phi, \Phi^\star) \subset J_{\nabla,g}$$

which is defined as it follows

$$g(\phi(X), Y) + g(X, \phi(Y)) = 2g(Phi(X), Y),$$

$$g(\phi(X), Y) - g(X, \phi(Y)) = 2g(\Phi^\star(X), Y).$$

Then let us put $S(\nabla) = \min\{rank(Ker(\Phi), \phi \in J_{\nabla,g}\}$,
$s^\star(\nabla) = \min\{rank(Ker(\Phi^\star), \phi \in J_{\nabla,g}\}$.
$J(\nabla) = \{\phi s t s(\nabla) = rank(Ker(\Phi))\}$,
$J^\star(\nabla) = \{phists^s tar(\nabla) = rank(Ker(\Phi^\star)\}$.
We define the following Lie sub-algebras of the Lie algebra $\mathcal{X}(M)$:
$\mathbf{G}(\nabla) = \Sigma_{\phi \in J(\nabla)} Ker(\Phi)$;
$\mathbf{G}_\star(\nabla) = \Sigma_{\phi \in J_\star(\nabla)} Ker(\Phi^\star)$.
Under the Lie derivative $T_1^2(M)$ is a left module of any Lie sub-algebra of $\mathcal{X}(M)$. Given a Koszul connection D we are concerned with the following Chevelley-Eilenberg $T_1^2(M)$-valued cocycle:

$$\mathbf{G} \ni X \to k_\infty(X) = L_X D \in T_1^2(M).$$

The cohomology class of k_∞ does not depend on the choice of the connection D. This class is denoted by

$$[k_\infty(\nabla)] \in H^1_{CE}(\mathbf{G}(\nabla), T^2_1(M)).$$

Similarly we define the cohomology class

$$[k_{\star\infty}(\nabla)] \in H^1_{CE}(\mathbf{G}_\star(\nabla), T^2_1(M)).$$

These cohomology classes are called canonical Koszul classes

Definition 2. *A symplectic statistical manifold is a datum* (M, g, ∇, ω) *where* (M, g, ∇) *is a statistical manifold,* (M, ω) *is a symplectic manifold and* ∇ *is a special symplectic connection, viz*

$$\nabla_X \omega = 0.$$

We recall that given a symplectic statistical manifold (M, g, ∇, ω) the dual (M, g, ∇) is denoted by $(M, g, \nabla^\star)$; it is defined by the following identity:

$$g(\nabla^\star_X Y, Z) = Xg(Y, Z) - g(Y, \nabla_X Z).$$

Given a symplectic statistical manifold (M, g, ∇, ω) there is a unique section of $\mathcal{G}(M)$, namely Φ such that

$$\omega(X, Y) = g(\Phi(X), Y).$$

2.2   Properties

(a) Φ and Φ^-1 are solutions of $FE(\nabla\nabla^\star)$ and $FE(\nabla^\star\nabla)$ respectively.

(b) Define the differential 2-form

$$\omega^\star(X, Y) = g(\Phi^{-1}(X), Y).$$

(c) $(M, g, \nabla^\star, \omega^\star)$ is a symplectic statistical manifold.
     Hint. Regarding the property (c), at one side Φ^{-1} is a solution of the gauge equation $FE(\nabla^\star\nabla)$, this implies the closeness of $\omega^\star$; at the other side the spectrum of ϕ^2 are negative functions; this implies the nondegeneracy of $\omega^\star$.

Proposition 1. *In a statistical manifold* (M, g, ∇) *the following assertions are equivalent:*

(a) $[k_{\star\infty}(\nabla)] = 0,$

(b) (M, g, ∇) *admits a structure symplectic statistical manifold* (M, g, ∇, ω).

Hint. If the canonical Koszul class $[k_{\star\infty}(\nabla)]$ vanishes then the Lie algebra $\mathbf{G}_\star(\nabla$ is a Lie subalgebra of some affine algebra $aff(M, D)$. This implies $\mathcal{G}_\star(\nabla)$ is zero.

3 Compatibility of Symplectic Forms

We know that every symplectic statistical manifold (M, g, ∇, ω) has its dual $(M, g, \nabla^*, \omega^*)$ where

$$\omega(X, Y) = g(\Phi(X), Y),$$

$$\omega^*(X, Y) = g(\Phi^{-1}(X, Y).$$

Proposition 2. *For all $(\lambda, \mu) \in \mathbb{R}^2 - (0, 0)$ with $\lambda\mu \leq 0$ consider the following closed differential 2-form*

$$\omega_{\lambda\mu} = \lambda\omega + \mu\omega^s tar,$$

it is nondegenerate.

Hint. One has

$$\omega_{\lambda\mu}(X, Y) = g(\Phi^{-1}(\lambda\Phi^2(X) + \mu X, Y).$$

Since the spectrum of Φ^2 is negative the mapping

$$\mathcal{X}(M) \ni X \to \lambda\Phi^2(X) + \mu X$$

is a gauge isomorphism.
For instance,

$$\omega_{1,-1} = \omega - \omega^*$$

is a symplectic form. Therefore the symplectic forms ω and $-\omega^*$ are compatible. Let Θ be their recursion operator:

$$-\omega^*(X, Y) = \omega(\Theta(X), Y);$$

then

$$\Theta = \Phi^{-2}$$

According to well known facts the spectrum of Φ^2 is a Hamiltonian system, [2], [3]
Every eigenfunction of Φ^2 is of even order.
Henceforward we assume that all of the eigenfunctions of Φ^2 is of order two. Let $2n = dim(M)$, then we put

$$spect(\Phi^2) = (f_1, .., f_n)$$

and

$$M \ni x \to F(x) = (f_1(x), .., f_n(x)) \in \mathbb{R}^{2n}$$

For all $x \in M$ each connected component of $F^{-1}(F(x))$ is a Lagrangian submanifold of $(M, \omega - \omega^*)$.

4   Symplectic Statistical Foliation

In this section we sketch the foliation versus of the notion of symplectic statistical manifold.

Definition 3. *A symplectic statistical foliation is a datum* $(M, g, \nabla), \omega)$ *where* (M, g, ∇) *is a statistical manifold,* ω *is a differential 2-form subject to the following requirement:*
$\nabla_X \omega = 0$ *for all vector field* X.

Warnings. The parallelism condition

$$\nabla \omega = 0$$

implies that ω is closed and the rank of ω) is constant.
Let Φ be a gauge homomorphism which satisfies the following requirement

$$\omega(X, Y) = g(\Phi(X), Y).$$

Since ω is ∇- parallel Φ is a solution of the gauge equation $FE(\nabla \nabla^\star)$. Therefore the tangent bundle TM is decomposed as it follows,

$$TM = Ker(\Phi) \oplus Im(\Phi),$$

Further, $Ker(\phi)$ is orthogonal to $Im(\phi)$ and both have the following parallelism properties:

$$(I): \quad \nabla_X(Ker(\Phi)) = Ker(\Phi),$$

$$(II): \quad \nabla^\star(Im(\phi)) = Im(\phi).$$

The property (II) implies that the distribution $Im(\Phi)$ satisfies the Frobenius condition of complete integrability; furthermore each leaf of $Im(\Phi)$ is a symplectic statistical manifold.

4.1   Examples Construction of Symplectic Statistical Foliation

We keep presenting a statistical manifold as a quadruplet $(M, g, \nabla, \nabla^s tar)$.
This presentation yields the differential equations $FE(\nabla \nabla^\star)$ and $FE(\nabla^\star \nabla)$. Let us focus on the space of solutions of $FE(\nabla \nabla^\star)$, namely $J_{\nabla, g}$. We have defined the mapping

$$J_{\nabla, g} \ni \phi \to (\Phi, \Phi^\star) \in J_{\nabla, g} \times J_{\nabla, g},$$

$$\frac{1}{2}(g(\phi(X), Y) + g(X, \phi(Y))) = g(\Phi(X), Y),$$

$$\frac{1}{2}(g(\phi(X), Y) - g(X, \phi(Y))) = g(\Phi^\star(X), Y).$$

We put

$$\omega(X, Y) = g(\Phi^\star(X), Y).$$

Properties of the couple $(\omega, \Phi^\star)$:

(I) we have the orthogonal decomposition

$$TM = Ker(\Phi^\star) \oplus Im(\Phi^\star);$$

(II) the parallelism properties

$$\nabla_X \omega = O;$$

$$\nabla_X^\star Im(\Phi^\star) = Im(\phi^\star).$$

(III) property (II) implies that (M, g, ∇, ω) is symplectic statistical foliations. Thus, generically each non trivial solution of the gauge equation $FE(\nabla\nabla^\star)$ yields a structure of symplectic statistical foliations.

Similarly every non trivial solution ϕ of the differential equation $FE(\nabla^\star\nabla)$ gives rise to a symplectic statistical foliation $(M, g, \nabla^\star, \omega^\star)$.

4.2   Compatibility of Symplectic Statistical Foliations

Here we restrict to attention to the samples which are constructed as we just done.

Let us consider the data:

$$J_{\nabla,g} \ni \phi \rightarrow (\Phi, \Phi^\star) \subset J_{\nabla,g}.$$

We go to involve the orthogonal decomposition

$$TM = Ker(\Phi^\star) \oplus Im(\Phi^\star).$$

A vector field X is decomposed as

$$X = X_1 + X_2$$

where X_1 is a section of $Ker(\Phi^\star$ and X_2 is a section of $Im(\Phi^\star)$ If ψ is a linear endomorphism of TM which agrees with the decomposition above then we present ψ as

$$\psi = \psi_2 + \psi_2$$

so that

$$\psi_1(Im(\Phi^\star)) = 0,$$

$$\psi_2(Ker(\Phi^\star)) = 0.$$

Therefore we present $\Phi^\star$

$$\Phi^\star = O + \Phi_2^\star;$$

consequently $\Phi_2^\star$ is an isomorphism of $Im(\Phi^\star)$. Then we put

$$\Phi^{\star-1} = 0 + (\Phi_2^\star)^{-1}.$$

Property: (a) $\Phi^{\star-1}$ is a solution of $FE(\nabla^\star\nabla)$;
(b) $Im(\Phi^{\star-1}) = Im(\Phi^\star)$;

(c) we put $\omega^\star(X,Y) = g(\Phi^{\star-1}(X),Y)$;

then $(M, g, \nabla^\star, \omega^\star)$ is a symplectic statistical foliation. Arguments similar to those involved in symplectic statistical manifolds walk in symplectic statistical manifolds. This means that for all non trivial pair of real number (λ, μ) such that $\lambda\mu \leq 0$ the differential 2-form $\lambda\omega + \mu\omega^star$ induces a symplectic structure in each leaf of $Im(\Phi^\star)$.

References

1. Brouzet, R.: About the existence of recursin operators for completely integrable hamiltonian systems near a Liouville torus. J. Math. Phys. **34**, 1309–1313 (1993)
2. Brouzet, R., Molino, P., Turiel, F.J.: Géométrie des suystèmes bihamiltoniens. Indag. Math. **4**, 269–296 (1993)
3. Gel'fand, I.M., Dorfman, I.: Ja: Hamiltoninan operator and algebraic structures related to them. Funct. Anal. Appl. **13**, 248–262 (1979)
4. Magri, F.: A simple model of the integrable hamiltonian equation. J. Math. Phys. **19**, 1156–1162 (1978)

Geometric Methods in Mechanics and Thermodynamics

A Dually Flat Geometry for Spacetime in 5d

Jan Naudts[(✉)] [iD]

Physics Department, Universiteit Antwerpen, B2610 Antwerpen, Belgium
jan.naudts@uantwerpen.be

Abstract. A model of spacetime is presented. It has an extension to five dimensions where the geometry is flat because it is by assumption the dual of the Euclidean geometry w.r.t. an arbitrary positive-definite metric. The 4-d geodesics are characterized by five conserved quantities, one of which is chosen freely.

Euler-Lagrange equations of motion are established for the 4-d coordinates. They contain external forces which originate from motion in the fifth dimension.

An example is worked out in which the motion of the fifth coordinate is that of a dynamical system with attracting and repelling states.

Keywords: Dually-flat geometry · Spacetime · Information Geometry · Induced matter theory

1 Introduction

The present work is a first exploration of a vast family of spacetime models all obtained by restriction to four dimensions of a flat geometry in five dimensions. The exploration is for the moment purely mathematical and not ready for physical interpretations.

Proponents of the space-time-matter theory [7,8] invoke the theorem of Campbell-Magaard [6] to embed the four-dimensional curved spacetime into a five-dimensional Ricci-flat Riemannian manifold. Higher-dimensional embeddings are in use in the context of string theory and its successor theories. The present work starts from a specific flat geometry for the five-dimensional spacetime and explores the curved geometries that can be obtained by selecting a subclass of geodesics and ignoring the motion in the fifth dimension with the argument that it can be observed only in an indirect manner.

The geometry of four-dimensional spacetime is described locally in good approximation by the Levi-Civita connection of the Minkowski metric which is a pseudometric with signature $(+, -, -, -)$ or $(-, +, +, +)$. The metric considered in the present work is positive-definite. This is meaningful because the geometry under study is *not* the one described by the Levi-Civita connection. The positive-definite metric and the pseudometric can both coexist with the geometry of a four-dimensional spacetime. In a recent paper [11] I show that,

© The Author(s), under exclusive license to Springer Nature Switzerland AG 2023
F. Nielsen and F. Barbaresco (Eds.): GSI 2023, LNCS 14072, pp. 183–191, 2023.
https://doi.org/10.1007/978-3-031-38299-4_20

given the positive-definite metric, one can construct a pseudometric in such a way that its metric connection has the same geodesics as those used in combination with the positive-definite metric. In this way, the present approach is reconciliated with common practices in Physics.

The flat geometry chosen in the present work for the five-dimensional space is the dual of the Euclidean geometry w.r.t. an arbitrary metric. The dually flat geometry is a corner stone of Information Geometry [5,9,10]. It derives its relevance from the importance of exponential families, known in Statistical Physics as Boltzmann-Gibbs distributions. The natural parameters of a model belonging to an exponential family are affine parameters of the so-called e-connection, which is one of two flat geometries linked by duality. The hope is that the dual of the Euclidean geometry can play a similar role in the present context.

The pair of flat connections dual w.r.t. some arbitrary positive-definite metric is a core feature of Information Geometry. Originally [2,5], the Fisher metric was adopted as the relevant metric of a statistical manifold. Eguchi [3] presented a method to derive the structure of Information Geometry, including a rather general metric tensor and a corresponding alpha-family of connections, from any suitable divergence function.

The coordinates x^μ determine positions in four dimensional spacetime. The unification of electromagnetism and gravity originally proposed by Kaluza [1] requires that the four-dimensional spacetime is replaced by a five-dimensional space, referred to as the hyperspace in what follows. The additional coordinate is denoted x^4, historically x^5. Summations in four-dimensional spacetime involve Greek indices μ, ν, ρ, σ, $\cdots$. In hyperspace Latin capitals A, B, C, $\cdots$ are used.

In Relativistic Mechanics the four components p_μ of the momentum of a free particle are conserved quantities. The original motivation for extending 4-d spacetime to five dimensions is that the particle can be charged and that the charge is a conserved quantity of the enlarged system. The fifth component P_4 of the momentum is then interpreted as e/c, with e the elementary charge and c the speed of light. More recent interpretations relate the fifth conserved quantity to the mass of a test particle. See for instance [7] or Sect. 7.6 of [4].

The next two sections introduce the flat geometry. The geodesics of this flat geometry are characterized by five conserved quantities, denoted P_A. These are calculated in Sect. 4.

Section 5 shows that fixing P_4 restores the one-to-one relation between geodesics with common starting point x and initial velocities $\dot{x}^\mu$. Section 6 considers Lagrangian equations of motion. Forces acting on a test particle in spacetime are equal to changes of momenta. Section 7 discusses an example. It exhibits geodesics for which the fifth coordinate x^4 evolves towards a constant value. The final section discusses some particular topics.

2   The Dual Geometry

In five dimensions a geodesic $t \mapsto x_t$ with affine parameter t satisfies the equations of motion

$$\ddot{x}^A + \Gamma_{BC}^A \dot{x}^B \dot{x}^C = 0. \tag{1}$$

The coefficients Γ^A_{BC} are the connection coefficients of the geometry and may still depend on x in a smooth way. The 5-d space is treated here as a differentiable manifold. The derivatives $\dot{x}^B = \mathrm{d}x^B/\mathrm{d}t$ are components of a tangent vector field.

The metric is denoted $g_{AB}(x)$. It is chosen to be the positive-definite solution of

$$\partial_C g_{DB} = g_{BA}\Gamma^A_{CD}, \tag{2}$$

if such a solution exists. This is the main assumption of the present work. The reason for this choice is explained below.

The connection coefficients Γ^{*A}_{BC} of the dual geometry, dual w.r.t. the metric tensor g_{AB}, are by definition fixed by

$$\partial_C g_{DB} = g_{BA}\Gamma^A_{CD} + g_{DA}\Gamma^{*A}_{CB}.$$

The assumption (2) is therefore equivalent with the assumption that a metric g exists so that the dual geometry has vanishing connection coefficients or, equivalently, that the coordinates x^A are affine coordinates for the dual geometry.

A well-known theorem states that a geometry has vanishing curvature if and only if its dual has vanishing curvature. See for instance Theorem 3.3 of [5]. One concludes that the assumption that a metric $g(x)$ exists for which (2) holds implies that the geometry has vanishing curvature because it is the dual of the Euclidean geometry.

The following result is what one expects intuitively. The proof is straightforward.

Proposition 1. *Let g be a solution of (2). Are equivalent*

(a) g is the Hessian of a potential $\Phi(x)$;
(b) The connection with coefficients Γ^A_{BC} is torsion-free.

Note that a curvature-free connection is said to be flat if there exist affine coordinates. The latter implies that the connection has vanishing torsion. By the proposition, the connection with coefficients Γ^A_{BC} is flat when the corresponding metric g is Hessian.

3   Inverse Formulas

From now on $g^{\mu\rho}$ denotes the components of the inverse of the 4-dimensional matrix with components $g_{\mu\nu}$. Note that the latter is positive-definite because the 5-dimensional matrix g is positive-definite.

Introduce the notations

$$\lambda^\mu = g^{\mu\rho}g_{\rho4},$$
$$\lambda_\nu = g_{\nu\mu}\lambda^\mu = g_{\nu4}$$
$$\gamma = g_{44} - \lambda_\mu g^{\mu\nu}g_{\nu4} = g_{44} - \lambda_\mu\lambda^\mu.$$

Proposition 2. *The positive-definitness of the metric tensor g implies that $0 < \gamma \leq g_{44}$.*

Proof. The scalar $\lambda_\mu \lambda^\mu$ is non-negative. This implies $\gamma \leq g_{44}$.
  For any 5-vector u^A is

$$0 \leq u^A g_{AB} u^B = u^\mu g_{\mu\nu} u^\nu + 2 u^\mu \lambda_\mu u^4 + g_{44}(u^4)^2,$$

with strict inequality if the vector u does not vanish. Choose $u^\mu = \lambda^\mu$ and $u^4 = -1$. Then it follows that $0 < \gamma$.

□

   The connection coefficients Γ_{CD}^A can be calculated starting from the 5-dimensional metric tensor g. This follows from

Proposition 3. *Expression (2) implies*

$$\gamma \Gamma_{CD}^\rho = \gamma X^{\rho\sigma} \partial_C g_{D\sigma} - \lambda^\rho \partial_C g_{D4}, \tag{a}$$
$$\gamma \Gamma_{CD}^4 = \partial_C g_{D4} - \lambda^\mu \partial_C g_{D\mu} \tag{b}$$

with $X^{\rho\sigma} = g^{\rho\sigma} + \frac{1}{\gamma}\lambda^\rho\lambda^\sigma$.

Proof. Write (2) as

$$\partial_C g_{D\mu} = g_{\mu\nu}\Gamma_{CD}^\nu + \lambda_\mu \Gamma_{CD}^4,$$
$$\partial_C g_{D4} = \lambda_\nu \Gamma_{CD}^\nu + g_{44}\Gamma_{CD}^4.$$

Multiply the former line with $g^{\rho\mu}$ to obtain

$$\Gamma_{CD}^\rho = g^{\rho\mu}\partial_C g_{D\mu} - \lambda^\rho \Gamma_{CD}^4. \tag{3}$$

Insert this result in the latter line. This gives

$$\partial_C g_{D4} = \lambda^\mu \partial_C g_{D\mu} + \gamma \Gamma_{CD}^4.$$

This is result (b). Combine (3) with (b) to obtain (a).

□

Note that $\gamma X^{\rho\sigma}\lambda_\sigma = g_{44}\lambda^\rho$.

4   Conserved Quantities

From now on assume connection coefficients Γ_{BC}^A which are such that a metric $g(x)$ exists solving (2). Multiply expression (1) with g_{DA} and use (2) to find

$$0 = g_{DA}\ddot{x}^A + g_{DA}\Gamma_{BC}^A \dot{x}^B \dot{x}^C$$
$$= g_{DA}\ddot{x}^A + [\partial_B g_{CD}]\dot{x}^B \dot{x}^C$$
$$= g_{DA}\ddot{x}^A + \dot{x}^C \frac{d}{dt}g_{CD}$$
$$= \frac{d}{dt}[g_{DA}\dot{x}^A].$$

This shows that the quantities $P_D = g_{DA}\dot{x}^A$ are constant along any geodesic. Write them as

$$P_\mu = p_\mu + \lambda_\mu \dot{x}^4 \quad \text{with} \quad p_\mu = g_{\mu\nu}\dot{x}^\nu, \tag{4}$$

$$P_4 = \lambda_\mu \dot{x}^\mu + g_{44}\dot{x}^4. \tag{5}$$

Solving these equations yields

$$\gamma \dot{x}^\nu = \gamma X^{\nu\mu} P_\mu - \lambda^\nu P_4, \tag{6}$$

$$\gamma \dot{x}^4 = P_4 - \lambda^\mu P_\mu. \tag{7}$$

The conserved quantities P_A can be used as local coordinates w.r.t. a reference point. The exponential map determines a point x in 5d spacetime by following a geodesic $t \mapsto x_t$ with initial velocities $\dot{x}^A$ starting from the reference point x_0 at $t = 0$ up to $x_1 = x$ at $t = 1$. In this way the geodesic characterized by the conserved quantities P_A is associated with a point x in 5d spacetime.

5 Geodesics in 4-D Spacetime

Distinct geodesics in five dimensions with common starting point x can have the same initial velocity $\dot{x}^\mu$ when projected on four-dimensional spacetime, i.e. when neglecting what happens in the fifth dimension. They are characterized by the following proposition.

Proposition 4. *Geodesics in five-dimensional space with common starting point x have the same initial velocity $\dot{x}^\mu$ in 4-d spacetime if and only if the corresponding conserved quantities P_A give the same value at the point x to the quantities $g_{44}P_\rho - \lambda_\rho P_4$.*

The proof is straightforward. The proposition shows that fixing the value of P_4 restores the one-to-one relation between initial velocity $\dot{x}^\nu$ and geodesic.

6 Lagrangian Forces

Introduce the notation $\bar{g}_{AB}(t) = g_{AB}(x_t)$ for the metric tensor along the geodesic. The geodesic is a solution of the Euler-Lagrange equations

$$\frac{\mathrm{d}}{\mathrm{d}t}\frac{\partial}{\partial \dot{x}^\mu}\mathscr{L} = \frac{\partial}{\partial x^\mu}\mathscr{L}$$

for the Lagrangian

$$\mathscr{L} = \mathscr{L}_0 + F_\mu(t)\, x^\mu \tag{8}$$

with $\mathscr{L}_0$ given by

$$\mathscr{L}_0 = \frac{1}{2}\bar{g}_{\mu\nu}(t)\dot{x}^\mu \dot{x}^\nu$$

and forces $F_\mu(t)$ defined by

$$F_\mu(t) = -\frac{d}{dt}\left(\lambda_\mu \dot{x}_t^4\right).$$

This follows from (4) and the fact that the P_μ are constants of motion. The conclusion is that the motion in the fifth dimension causes a force field acting in 4-d spacetime.

Note that the forces $F_\mu(t)$ satisfy $F_\mu = \dot{p}_\mu$. To verify this use that P_μ is constant along a geodesic. As a consequence, the Lagrangian $\mathscr{L}$ as given by (8) does not need reference to the fifth dimension to describe motion in 4-d spacetime. The relation $F_\mu = \dot{p}_\mu$ generalizes the second law of Newton: The forces acting on a test particle are equal to the changes in momentum. Hence, p_μ can be identified with the momentum of a test particle moving in 4-d along the geodesic $t \mapsto x_t$. Due to the motion in the fifth dimension the momentum is not conserved.

7   Example

Let be given a positive-definite metric tensor $\breve{g}_{\mu\nu}$ on $\mathbb{R}^4$, together with a non-vanishing vector field $\breve{\lambda}_\mu$. Let $x \mapsto \breve{x}$ denote the projection from $\mathbb{R}^5$ to $\mathbb{R}^4$ obtained by neglecting x^4. Introduce a 5-d metric tensor g defined by

$$g_{\mu\nu}(x) = (1 + (x^4)^2)\,\breve{g}_{\mu\nu}(\breve{x})$$
$$\lambda_\mu(x) = g_{\mu 4}(x) = x^4\,\breve{\lambda}_\mu(\breve{x}),$$
$$g_{44} = \alpha\,(1 + (x^4)^2).$$

Here, α is a strictly positive constant.

Introduce the notation $\hat{g}^{\mu\nu}$ for the inverse of $\breve{g}_{\mu\nu}$, i.e.

$$\hat{g}^{\mu\nu}\breve{g}_{\nu\rho} = g^\mu_\rho.$$

This notation is needed because raising and lowering indices is done with $g_{\mu\nu}$, not with $\breve{g}_{\mu\nu}$. Similarly, let

$$\hat{\lambda}^\mu = \hat{g}^{\mu\nu}\breve{\lambda}_\nu.$$

With these notations one obtains

$$g^{\mu\nu} = \frac{1}{1 + (x^4)^2}\,\hat{g}^{\mu\nu}$$

and

$$\lambda_\rho \lambda^\rho = \frac{(x^4)^2}{1 + (x^4)^2}\,\breve{\lambda}_\rho \hat{\lambda}^\rho.$$

Note that $\lambda_\rho \lambda^\rho = 0$ occurs only when $x^4 = 0$ because by assumption $\hat{\lambda}_\mu$ is a non-vanishing vector field.

Proposition 5. *Assume that* $\check{\lambda}_\rho\hat{\lambda}^\rho < 4\alpha$. *Then the matrix* g *is positive-definite.*

Proof. Let us first show that $\check{\lambda}_\rho\hat{\lambda}^\rho < \alpha$ implies that $\gamma > 0$. When $\check{\lambda}_\rho\hat{\lambda}^\rho \leq \alpha$ then the expression

$$\gamma = \alpha(1 + (x^4)^2) - \frac{(x^4)^2}{1 + (x^4)^2}\,\check{\lambda}_\rho\hat{\lambda}^\rho. \tag{9}$$

is minimal at $x^4 = 0$. The minimal value is α, which is strictly positive by assumption. Hence, one concludes that $\gamma \geq \alpha > 0$.

If $\check{\lambda}_\rho\hat{\lambda}^\rho \geq \alpha$ then the minimum is reached at

$$1 + (x^4)^2 = \sqrt{\frac{\check{\lambda}_\rho\hat{\lambda}^\rho}{\alpha}}.$$

The minimal value is

$$\sqrt{\check{\lambda}_\rho\hat{\lambda}^\rho}(2\sqrt{\alpha} - \sqrt{\check{\lambda}_\rho\hat{\lambda}^\rho}).$$

It is non-negative when

$$\sqrt{\check{\lambda}_\rho\hat{\lambda}^\rho} < 2\sqrt{\alpha}.$$

This condition is equivalent with $\check{\lambda}_\rho\hat{\lambda}^\rho < 4\alpha$.

Next use that $\gamma > 0$ to show that the tensor with components g_{AB} is positive-definite. For any 5-vector u is

$$u^A g_{AB} u^B = u^\mu g_{\mu\nu} u^\nu + 2\lambda_\mu u^\mu u^4 + g_{44}(u^4)^2$$
$$= (u^\mu + \lambda^\mu u^4)g_{\mu\nu}(u^\nu + \lambda^\nu u^4) + \gamma(u^4)^2$$

with $\gamma = g_{44} - \lambda_\mu\lambda^\mu$ as before. Hence, $\gamma > 0$ implies that $u^A g_{AB} u^B \geq 0$ for all u. Finally, one verifies that $u^A g_{AB} u^B = 0$ implies $u = 0$.

$\square$

Consider now a geodesic $t \mapsto x_t$ characterized by the conserved quantities P_A. For convenience, introduce the notation

$$m = \frac{1}{2}\hat{\lambda}^\mu P_\mu.$$

First assume $P_4 > 0$. If $|m| < P_4$ then $\dot{x}^4$ cannot be negative. Hence, x^4 is an increasing function of t. For large values of t the coordinate x^4 increases like $(3P_4 t/\alpha)^{1/3}$, the derivative $\dot{x}^4$ slows down like $t^{-2/3}$.

In the other case, i.e. $|m| \geq P_4$, there exists two values of x^4 for which the derivative $\dot{x}^4$ vanishes. At the thresholds $m = \pm P_4$ these values coincide and are given by $x^4 = 1$, respectively $x^4 = -1$. See Fig. 1. In the region between the two zeroes the sign of $\dot{x}^4$ is negative. As a consequence, in the region $|x^4| < 1$ the lines of vanishing $\dot{x}^4$ are attracting, in the regions $|x^4| > 1$ the lines are repelling. The points $(1, P_4)$ and $(-1, -P_4)$ are critical points.

The discussion of the case $P_4 < 0$ is similar.

Finally consider the case that $P_4 = 0$. Then the points of the line $x^4 = 0$ are stationary points. They are the only ones when $m \neq 0$. They are attractive for $m > 0$ and repelling for $m < 0$.

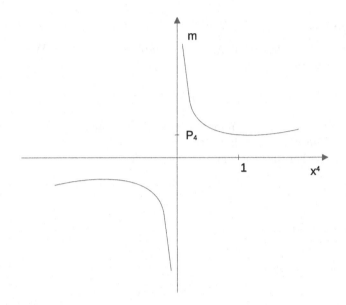

Fig. 1. Lines in the (x^4, m)-plane where $\dot{x}^4$ vanishes (case $P_4 > 0$).

8   Discussion

This paper investigates a family of models the Riemannian curvature of which vanishes when an extra parameter is added to the model. The presentation is kept concrete by considering models of General Relativity. It has been proposed, see for instance [8], that the real world is flat in 3 and in 5 dimensions while it is known to be curved in 4 dimensions. The fifth dimension was proposed by Kaluza [1] shortly after Einstein argued that the world is four-dimensional. The addition of a fifth coordinate simplifies the treatment of classical electromagnetism.

Pseudo-metrics, which are used systematically in the Physics Literature, are not essential for the present paper. The reader interested in the interplay between the positive-definite metric g of the present paper and the pseudometric G of General Relativity can consult [11].

Each geodesic in 5-d is characterized by five conserved quantities. Expressions for them are derived in Sect. 4. By fixing the fifth conserved quantity the remaining four characterize the geodesics of the 4-d spacetime. In Sect. 6 it is shown that the motion in the fifth dimension induces forces that modify geodesics in four dimensions.

The four spacetime coordinates x^μ, $\mu = 0, 1, 2, 3$, can be seen as a parameterization of a 4-dimensional submanifold of the 5-dimensional hyperspace. A statistical model belonging to an exponential model defines a submanifold of a larger space of probability distributions. Information Geometry studies a dually flat geometry on the submanifold. The present approach utilizes the dual geometry of the embedding space, rather than that of the submanifold itself, to define an embedded model with tractable properties. This point of view opens the way to further generalizations.

Notice the link with curved exponential families (Sect. 4.3 of [5]). They describe submanifolds of models belonging to an exponential family. If the 5-d coordinates are used to parameterize a family of probability distributions belonging to an exponential family then the projection to four dimensions, obtained by forcing the conserved quantity P_4 to vanish, defines a 4-d model belonging to a curved exponential family. Indeed, Eqs. (6, 7) with $P_4 = 0$ express the new coordinates P_μ, $\mu = 0, 1, 2, 3$, as a function of the five parameters x^A, $A = 0, 1, 2, 3, 4$.

The example of Sect. 7 is chosen with the intention to show that singularities of the kind occurring in General Relativity can be handled by introduction of additional parameters, *in casu* the coordinate of a fifth dimension. The example of Sect. 7 suggests that it is possible to replace the singularities of the pseudo-metric by attracting and repelling regions in the phase space of the 5-d model.

References

1. Kaluza, T.: Zum Unitätsproblem in der Physik, Sitzungsber. Preuss. Akad. Wiss. Berlin. Math. Phys. 966–972 (1921)
2. Rao, C.R.: Information and accuracy attainable in the estimation of statistical parameters. Bull. Calcutta Math. Soc. **37**, 81–91 (1945)
3. Eguchi, S.: Second order efficiency of minimum contrast estimators in a curved exponential family. Ann. Stat. **11**, 793–803 (1983)
4. Marsden, J.E., Ratiu, T.S.: Introduction to Mechanics and Symmetry, 2nd ed. Springer, New York (1999). https://doi.org/10.1007/978-0-387-21792-5
5. Amari, S., Nagaoka, H.: Methods of Information Geometry, Translations of Mathematical Monographs 191 Oxford University Press, Oxford (2000)
6. Seahra, S.S., Wesson, P.S.: Application of the Campbell-Magaard theorem to higher-dimensional physics. Class. Quantum Gravity **20**, 1321 (2003)
7. Wesson, P.S.: Five-Dimensional Physics. World Scientific, Singapore (2006)
8. Wesson, P.: The status of modern five-dimensional gravity. Int. J. Mod. Phys. D **24**, 1530001 (2015)
9. Amari, S.: Information Geometry and Its Applications. AMS, vol. 194. Springer, Tokyo (2016). https://doi.org/10.1007/978-4-431-55978-8
10. Ay, N., Jost, J., Lê, H.V., Schwachhöfer, L.: Information Geometry. Ergebnisse der Mathematik und ihrer Grenzgebiete. 3. Folge / A Series of Modern Surveys in Mathematics, vol. 64. Springer, Cham (2017). https://doi.org/10.1007/978-3-319-56478-4
11. Naudts, J.: A dually flat embedding of spacetime. Entropy **25**, 651 (2023)

Structure-preserving Discretization of the Cahn-Hilliard Equations Recast as a Port-Hamiltonian System

Antoine Bendimerad-Hohl, Ghislain Haine(✉)⬤, and Denis Matignon⬤

ISAE-SUPAERO, Université de Toulouse, Toulouse, France
antoine.bendimerad-hohl@student.isae-supaero.fr,
{ghislain.haine,denis.matignon}@isae.fr

Abstract. The structure-preserving discretization of the Cahn-Hillard equation, a phase field model describing phase separation with diffuse interface, is proposed using the Partitioned Finite Element Method. The discrete counter-part of the power balance is proved and a sufficient condition for the phase preservation is provided.

Keywords: Phase field · port-Hamiltonian system · Structure-preserving discretization

1 Introduction

Eutectic freeze crystallization is a promising process for desalinizing water for it requires fewer energy than other methods [1,11]. Due to the thermodynamic nature of this process, the port-Hamiltonian (pH) framework is an interesting approach for modelling this control system [9,10]. Indeed, pH systems are especially well-suited for modelling energy exchange through boundaries [13,14].

In this article, we will focus on the structure-preserving discretization of the phase separation problem using the Cahn-Hilliard equation [5,7]. The Partitioned Finite Element Method (PFEM) [6] is applied to the pH formulation of this problem, with the aim to achieve simulations of a separation process [7].

The PFEM has already been successfully applied to the Allen-Cahn equation [2,3], a model of solidification process. The Cahn-Hilliard equation is more challenging since a second order differential operator is to be found in its pH formulation [16]. Furthermore, in addition to the power balance satisfied by the Hamiltonian, the preservation of the phase at the discrete level would clearly provide more physically meaningful simulations.

The paper is organized as follows: in Sect. 2, the modelling of the phase field system is presented. The PFEM is applied in Sect. 3, and it is shown that the method is able to mimic the free energy balance and the phase preservation at the discrete level. Some perspectives are discussed in Sect. 5.

Supported by the AID from the French Ministry of the Armed Forces.

2   Cahn-Hilliard Model as a Port-Hamiltonian System

2.1   Phase Field Modelling

The modelling of phase separation or transition can be tackle by considering a *phase field* representation with *diffuse interface*. Let us consider $\phi(x,t) : \Omega \times \mathbb{R} \to [0,1]$ the phase function, that represents *e.g.* the state of the phase field at a given point or the concentration of a solute in a solution. The dynamics of the system is then given by minimizing the Landau-Ginzburg free energy:

$$\mathcal{G}(\phi) := \int_\Omega g(\phi) + \frac{1}{2}\kappa \, \mathbf{grad}(\phi) \cdot \mathbf{grad}(\phi),$$

where g is the bulk free energy (often a double-well potential) and κ the coefficient corresponding to the interface energy, that prevents the system from having an infinitely thin interface between the two phases.

2.2   Cahn-Hilliard as a Port-Hamiltonian System

This section was made after the work of Benjamin Vincent on the Allen-Cahn and Cahn-Hilliard equations [5,16].

The Cahn-Hilliard model proposes the following dynamics for the phase:

$$\begin{cases} \partial_t \phi = -\mathrm{div}(\boldsymbol{j}_\phi), \\ \boldsymbol{j}_\phi = -\Gamma \, \mathbf{grad}(\frac{\delta\mathcal{G}}{\delta\phi}), \end{cases} \tag{1}$$

where $\Gamma > 0$ represents the interface mobility. The phase field ϕ is transported by the flux $\boldsymbol{j}_\phi$: thus, it is a conserved quantity.

In order to recast (1) in the pH formalism, the state is augmented by considering $\boldsymbol{\psi} := \mathbf{grad}(\phi)$. The potential is rewritten as:

$$\tilde{\mathcal{G}}(\phi, \boldsymbol{\psi}) = \int_\Omega g(\phi) + \frac{1}{2}\kappa\|\boldsymbol{\psi}\|^2.$$

Let us introduce the *flow* variable:

$$\boldsymbol{F}_\phi := -\mathbf{grad}\left(\frac{\delta\tilde{\mathcal{G}}}{\delta\phi} - \mathrm{div}(\frac{\delta\tilde{\mathcal{G}}}{\delta\boldsymbol{\psi}})\right), \tag{2}$$

so that the system reads [16]:

$$\begin{pmatrix} \partial_t \phi \\ \partial_t \boldsymbol{\psi} \\ \boldsymbol{F}_\phi \end{pmatrix} = \begin{pmatrix} 0 & 0 & -\mathrm{div} \\ 0 & 0 & -\mathbf{grad}(\mathrm{div}(\cdot)) \\ -\mathbf{grad}\,\mathbf{grad}(\mathrm{div}(\cdot)) & 0 \end{pmatrix} \begin{pmatrix} \frac{\delta\tilde{\mathcal{G}}}{\delta\phi} \\ \frac{\delta\tilde{\mathcal{G}}}{\delta\boldsymbol{\psi}} \\ \boldsymbol{j}_\phi \end{pmatrix},$$

together with the constitutive relations:

$$\frac{\delta\tilde{\mathcal{G}}}{\delta\phi} = g'(\phi), \quad \frac{\delta\tilde{\mathcal{G}}}{\delta\boldsymbol{\psi}} = \kappa\boldsymbol{\psi}, \quad \boldsymbol{j}_\phi = \Gamma\boldsymbol{F}_\phi.$$

Let us compute $\frac{d\tilde{\mathcal{G}}(\phi,\psi)}{dt}(t)$ in order to obtain the free energy balance equation, and identify physically relevant boundary terms.

Proposition 1 ([16]). *The Landau-Ginzburg free energy satisfies:*

$$\frac{d\tilde{\mathcal{G}}(\phi,\psi)}{dt} = - \int_{\Omega} \boldsymbol{F}_{\phi} \cdot \boldsymbol{j}_{\phi} + \int_{\partial\Omega} \left[\frac{\delta\tilde{\mathcal{G}}}{\delta\phi} \boldsymbol{j}_{\phi}.\boldsymbol{n} + div(\boldsymbol{j}_{\phi}) \frac{\delta\tilde{\mathcal{G}}}{\delta\psi}.\boldsymbol{n} - div\left(\frac{\delta\tilde{\mathcal{G}}}{\delta\psi}\right) \boldsymbol{j}_{\phi}.\boldsymbol{n} \right].$$
(3)

The boundary ports $(f^{\partial}, e^{\partial})$ are then defined as follows:

$$f^{\partial} := \begin{pmatrix} \frac{\delta\tilde{\mathcal{G}}}{\delta\phi}_{|\partial\Omega} \\ \frac{\delta\tilde{\mathcal{G}}}{\delta\psi}_{|\Omega} \cdot \boldsymbol{n} \\ -\boldsymbol{j}_{\phi|\partial\Omega} \cdot \boldsymbol{n} \end{pmatrix}, \quad e^{\partial} := \begin{pmatrix} \boldsymbol{j}_{\phi|\partial\Omega} \cdot \boldsymbol{n} \\ div(\boldsymbol{j}_{\phi})_{|\partial\Omega} \\ div\left(\frac{\delta\tilde{\mathcal{G}}}{\delta\psi}_{|\partial\Omega}\right) \end{pmatrix}.$$

Thanks to $\boldsymbol{j}_{\phi} = \Gamma\boldsymbol{F}_{\phi}$, the pH system is *lossy*: $\frac{d\tilde{\mathcal{G}}(\phi,\psi)}{dt}(t) \leq \int_{\partial\Omega} f^{\partial} \cdot e^{\partial}$.

3   Structure-preserving Discretization of Cahn-Hilliard Model

In order to spatially discretize the system, PFEM is used, which has already allowed for discretizing a various amount of pHs (see *e.g.* [2,6,15]). The method consists of 3 steps: (1) write a weak formulation of the problem, (2) select a *partition* of the variables, use Stokes theorem to perfom an integration by parts which makes appear the useful control in the boundary term (3) Choose a set of finite element families for the state and control variables. Thanks to this method, a finite-dimensional power balance is then satisfied at the discrete level. Note that a structure-preserving method has already been proposed in [8] for the Allen-Cahn equations.

Variational Problem. Let $\lambda \in C^{\infty}(\Omega, \mathbb{R})$ and $\boldsymbol{\mu}, \boldsymbol{\xi} \in C^{\infty}(\Omega, \mathbb{R}^3)$ be three test functions corresponding to the flow variables: $\frac{\delta\tilde{\mathcal{G}}}{\delta\phi}, \frac{\delta\tilde{\mathcal{G}}}{\delta\psi}$ and $\boldsymbol{F}_{\phi}$. Then the variational problem is:

$$\begin{cases} \int_{\Omega} \lambda \partial_t \phi = - \int_{\Omega} \lambda \, div(\boldsymbol{j}_{\phi}), \\ \int_{\Omega} \boldsymbol{\mu} \cdot \partial_t \psi = - \int_{\Omega} \boldsymbol{\mu} \cdot \mathbf{grad}(div(\boldsymbol{j}_{\phi})), \\ \int_{\Omega} \boldsymbol{\xi} \cdot \boldsymbol{F}_{\phi} = - \int_{\Omega} \boldsymbol{\xi} \cdot \mathbf{grad}(\frac{\delta\tilde{\mathcal{G}}}{\delta\phi}) + \int_{\Omega} \boldsymbol{\xi} \cdot \mathbf{grad}(div(\frac{\delta\tilde{\mathcal{G}}}{\delta\psi})). \end{cases}$$
(4)

And for the constitutive relations:

$$\begin{cases} \int_{\Omega} \lambda \frac{\delta\tilde{\mathcal{G}}}{\delta\phi} = \int_{\Omega} \lambda \, g'(\phi), \\ \int_{\Omega} \boldsymbol{\mu} \cdot \frac{\delta\tilde{\mathcal{G}}}{\delta\psi} = \int_{\Omega} \boldsymbol{\mu} \cdot (\kappa\psi), \\ \int_{\Omega} \boldsymbol{\xi} \cdot \boldsymbol{j}_{\phi} = \int_{\Omega} \boldsymbol{\xi} \cdot (\Gamma\boldsymbol{F}_{\phi}). \end{cases}$$
(5)

Choice of Causality. In order to make the boundary control (i.e. choose a causality), one needs to integrate by parts the previous equations. Here, choosing the integration by parts have to be made carefully, indeed the formal adjoint of $\mathrm{div}(\cdot)$ is $-\mathbf{grad}(\cdot)$, and only one integration by parts on the first or third line is required. But $\mathbf{grad}(\mathrm{div}(\cdot))$ is formally *symmetric*, and we will need to integrate by parts twice to make a skew symmetric matrix appear. Therefore we can choose between integrating the second or third line two times or both lines one time. In our case the idea is the following: integrating by parts on the first line makes the $\mathbf{j}_\phi$ (the flux of ϕ at the boundary) control appear, which is physically meaningful. Finally integrating by parts the second term of the third line as well as the second line allows us to use divergence conforming first order Finite Elements (FE), e.g. Raviart-Thomas elements, instead of second order ones.

Let us integrate by parts on the first, second and third line:

$$-\int_\Omega \lambda \mathrm{div}(\mathbf{j}_\phi) = \int_\Omega \mathbf{grad}(\lambda) \cdot \mathbf{j}_\phi - \int_{\partial\Omega} \lambda \mathbf{j}_\phi \cdot \mathbf{n},$$

$$-\int_\Omega \boldsymbol{\mu} \cdot \mathbf{grad}(\mathrm{div}(\mathbf{j}_\phi)) = \int_\Omega \mathrm{div}(\boldsymbol{\mu}) \mathrm{div}(\mathbf{j}_\phi) - \int_{\partial\Omega} \mathrm{div}(\mathbf{j}_\phi) \boldsymbol{\mu} \cdot \mathbf{n},$$

$$\int_\Omega \boldsymbol{\xi} \cdot \mathbf{grad}(\mathrm{div}(\frac{\delta\tilde{\mathcal{G}}}{\delta\psi})) = -\int_\Omega \mathrm{div}(\boldsymbol{\xi}) \mathrm{div}(\frac{\delta\tilde{\mathcal{G}}}{\delta\psi}) + \int_{\partial\Omega} \mathrm{div}(\frac{\delta\tilde{\mathcal{G}}}{\delta\psi}) \boldsymbol{\xi} \cdot \mathbf{n}.$$

From this result, we deduce that 3 different scalar boundary controls are required: $e_{\mathbf{j}_\phi} := \mathbf{j}_\phi \cdot \mathbf{n}_{|\partial\Omega}$, $e_d := \mathrm{div}(\mathbf{j}_\phi)_{|\partial\Omega}$ and $e_\psi := \mathrm{div}(\frac{\delta\tilde{\mathcal{G}}}{\delta\psi})_{|\partial\Omega}$.

Finite Elements Families. Now let us choose, 6 finite elements families: 3 for the flow and effort variables, and 3 for the control. Let $(\lambda_i)_{i\in[1,n]} \in L^2(\Omega, \mathbb{R})$, $(\boldsymbol{\mu}_i)_{i\in[1,m]} \in L^2(\Omega, \mathbb{R}^3)$ and $(\boldsymbol{\xi}_i)_{i\in[1,k]} \in H^2(\Omega, \mathbb{R}^3)$ be the families corresponding to the flow variables ϕ, ψ and $\mathbf{F}_\phi$ of cardinal $n, m, k \in \mathbb{N}$ respectively . And let $(\gamma_i)_{i\in[1,n_\phi]}, (\eta_i)_{i\in[1,n_d]}, (\nu_i)_{i\in[1,n_\psi]} \in L^2(\partial\Omega, \mathbb{R})$ be the families corresponding to the control variables $e_{\mathbf{j}_\phi} := \mathbf{j}_\phi \cdot \mathbf{n}_{|\partial\Omega}$, $e_d := \mathrm{div}(\mathbf{j}_\phi)_{|\partial\Omega}$ and $e_\psi := \mathrm{div}(\frac{\delta\tilde{\mathcal{G}}}{\delta\psi})_{|\partial\Omega}$., of cardinal n_ϕ, n_d and n_ψ respectively. Then by decomposing the flow, effort and control variables over these families, it yields:

$$\begin{cases} \text{for the flow variables:} \\ \phi^d(x,t) = \sum_1^n \phi_i(t)\lambda_i(x) \\ \psi^d(x,t) = \sum_1^m \psi_i(t)\boldsymbol{\mu}_i(x) \\ \mathbf{F}_\phi^d(x,t) = \sum_1^k \mathbf{F}_\phi^i(t)\boldsymbol{\xi}_i(x) \end{cases} \quad \begin{cases} \text{for the effort variables:} \\ \frac{\delta\tilde{\mathcal{G}}}{\delta\phi}^d(x,t) = \sum_1^n \partial_\phi\mathcal{G}^i(t)\lambda_i(x) \\ \frac{\delta\tilde{\mathcal{G}}}{\delta\psi}^d(x,t) = \sum_1^m \partial_\phi\mathcal{G}^i(t)\boldsymbol{\mu}_i(x) \\ \mathbf{j}_\phi^d(x,t) = \sum_1^k \mathbf{j}_\phi^i(t)\boldsymbol{\xi}_i(x) \end{cases}$$

$$\begin{cases} \text{and for the boundary control variables:} \\ e_{\mathbf{j}_\phi} = \sum_1^{n_{j_\phi}} e_{\mathbf{j}_\phi}^i(t)\gamma_i(x) \\ e_d = \sum_1^{n_d} e_d^i(t)\eta_i(x) \\ e_\psi = \sum_1^{n_\psi} e_\psi^i(t)\nu_i(x) \end{cases} \tag{6}$$

Let us note:

$$\begin{cases} \overline{\phi}(t) := (\phi_i)_{i\in[1,n]} \\ \overline{\psi}(t) := (\psi_i)_{i\in[1,m]} \\ \overline{F_\phi}(t) := (F_\phi^i)_{i\in[1,k]} \end{cases} , \begin{cases} \overline{\partial_\phi \mathcal{G}}(t) := (\partial_\phi \mathcal{G}^i)_{i\in[1,n]} \\ \overline{\partial_\psi \mathcal{G}}(t) := (\partial_\psi \mathcal{G}^i)_{i\in[1,m]} \\ \overline{j_\phi}(t) := (j_\phi^i)_{i\in[1,k]} \end{cases} , \begin{cases} \overline{e_{j_\phi}} := (e_{j_\phi}^i)_{i\in[1,n_\phi]} \\ \overline{e_d} := (e_d^i)_{i\in[1,n_d]} \\ \overline{e_\psi} := (e_\psi^i)_{i\in[1,n_\psi]} \\ u_\partial := \begin{bmatrix} \overline{e_{j_\phi}}^\mathsf{T} & \overline{e_d}^\mathsf{T} & \overline{e_\psi}^\mathsf{T} \end{bmatrix}^\mathsf{T} \end{cases} \tag{7}$$

The variational problem then becomes:

$$\begin{cases} \forall j \in [1,n], \sum_i^n \phi_i(t) \int_\Omega \lambda_j \lambda_i = \sum_i^k j_\phi^i(t) \int_\Omega \mathbf{grad}(\lambda_j) \cdot \boldsymbol{\xi}_i - \sum_i^{n_{j_\phi}} e_{j_\phi}^i(t) \int_{\partial\Omega} \lambda_j \gamma_i \,, \\ \forall j \in [1,m], \sum_i^m \psi_i(t) \int_\Omega \boldsymbol{\mu}_j \cdot \boldsymbol{\mu}_i = \sum_i^k j_\phi^i(t) \int_\Omega \mathrm{div}(\boldsymbol{\mu}_j) \mathrm{div}(\boldsymbol{\xi}_i) - \sum_i^{n_d} e_d^i \int_{\partial\Omega} \lambda_j \eta_i \,, \\ \forall j \in [1,k], \sum_i^k F_\phi^i(y) \int_\Omega \boldsymbol{\xi}_i \cdot \boldsymbol{\xi}_j = -\sum_i^n \partial_\phi \mathcal{G}^i \int_\Omega \mathbf{grad}(\lambda_i) \cdot \boldsymbol{\xi}_j \\ \qquad\qquad\qquad - \sum_i^m \partial_\psi \mathcal{G}^i(t) \int_\Omega \mathrm{div}(\boldsymbol{\mu}_i) \mathrm{div}(\boldsymbol{\xi}_j) \\ \qquad\qquad\qquad + \sum_i^{n_\psi} e_\psi^i \int_{\partial\Omega} \mathrm{div}(\boldsymbol{\xi}_j) \nu_i \,. \end{cases} \tag{8}$$

And for the constitutive relations:

$$\begin{cases} \forall j \in [1,n], \sum_i^n \partial_\phi \mathcal{G}^i(t) \int_\Omega \lambda_i \lambda_j = \int_\Omega \lambda_j \, g'(\sum_i^n \phi_i \lambda_i) \,, \\ \forall j \in [1,m], \sum_i^m \partial_\psi \mathcal{G}^i(t) \int_\Omega \boldsymbol{\mu}_i \cdot \boldsymbol{\mu}_j = \sum_i^m \psi_i(t) \int \boldsymbol{\mu}_j \cdot (\kappa \boldsymbol{\mu}_i) \,, \\ \forall j \in [1,k], \sum_i^k j_\phi^i(t) \int_\Omega \boldsymbol{\xi}_i \cdot \boldsymbol{\xi}_j = \sum_i^k F_\phi^i(t) \int_\Omega \Gamma \boldsymbol{\xi}_i \cdot \boldsymbol{\xi}_j \,. \end{cases}$$

Note that g' is *non linear*, this means that the integral in the first constitutive relation has to be computed at each time step, which increases the computational time. However, if g' is *polynomial*, which happens to be the case for the double-well potential [4], one can take advantage of off-line computations if required, as proposed in [6], allowing for a possible trade-off between computation time and memory usage.

Matrices Definition. Finally let us define the matrices of the finite-dimensional system. Let $M_\lambda := (\int_\Omega \lambda_i \lambda_j)_{i,j\in[1,n]}$, $M_\mu := (\int_\Omega \boldsymbol{\mu}_i \cdot \boldsymbol{\mu}_j)_{i,j\in[1,m]}$ and $M_\xi := (\int_\Omega \boldsymbol{\xi}_i \boldsymbol{\xi}_j)_{i,j\in[1,k]}$ be the mass matrices. Let us define two rectangular matrices $D_\nabla := (\int_\Omega \mathbf{grad}(\lambda_i) \cdot \boldsymbol{\xi}_j)_{i,j}$ of size $n \times k$ and $D_{\mathrm{divdiv}} := (\int_\Omega \mathrm{div}(\boldsymbol{\mu}_i) \mathrm{div}(\boldsymbol{\xi}_j))_{i,j}$ of size $m \times k$ as the structure matrices. Let $C_\kappa := (\int_\Omega \boldsymbol{\mu}_i \cdot (\kappa \boldsymbol{\mu}_j))_{i,j\in[1,m]}$ and $C_\Gamma := (\int_\Omega \boldsymbol{\xi}_i \cdot (\Gamma \boldsymbol{\xi}_j))_{i,j\in[1,k]}$ be the two constitutive relations matrices, and let $\overline{h}(\overline{\phi}) := (h_i(\overline{\phi}))_{i\in[1,n]}$ be the $n \times 1$ vector corresponding to the non-linear constitutive relation:

$$h_i(\overline{\phi}) = \int_\Omega \lambda_i(x) \, g'(\sum_j^n \phi_j(t) \lambda_j(x)) \,.$$

Finally let $B_{j_\phi} := (\int_{\partial\Omega} \gamma_j \boldsymbol{\xi}_i \cdot \boldsymbol{n})_{(i,j)\in[1,k],[1,n_{j_\phi}]}$, $B_d := (\int_{\partial\Omega} \eta_j \boldsymbol{\xi}_i \cdot \boldsymbol{n})_{(i,j)\in[1,k],[1,n_d]}$ and $B_\psi := (\int_{\partial\Omega} \nu_j \mathrm{div}(\boldsymbol{\xi}_i))_{(i,j)\in[1,k],[1,n_\phi]}$ be the partial control matrices. Let us note $B := [-B_{j_\phi} \;\; B_d \;\; B_\psi]^\mathsf{T}$ the control matrix.

Fully Spatially Discretized System. With finite-dimensional vectors, the fully discretized system then reads:

$$
\begin{cases}
\begin{bmatrix} M_\lambda & 0 & 0 \\ 0 & M_\mu & 0 \\ 0 & 0 & M_\xi \end{bmatrix} \begin{pmatrix} \partial_t\overline{\phi} \\ \partial_t\overline{\psi} \\ \overline{F_\phi} \end{pmatrix} = \begin{bmatrix} 0 & 0 & D_\nabla \\ 0 & 0 & D_{\text{divdiv}} \\ -D_\nabla^T & -D_{\text{divdiv}}^T & 0 \end{bmatrix} \begin{pmatrix} \overline{\partial_\phi\mathcal{G}} \\ \overline{\partial_\psi\mathcal{G}} \\ \overline{j_\phi} \end{pmatrix} + Bu_\partial \,, \\[6pt]
\text{With the collocated observations:} \\
M_\partial y_\partial = B^\top \begin{pmatrix} \overline{\partial_\phi\mathcal{G}} \\ \overline{\partial_\psi\mathcal{G}} \\ \overline{j_\phi} \end{pmatrix} , \\[6pt]
\text{and the constitutive relations:} \\
M_\lambda\overline{\partial_\phi\mathcal{G}} = \overline{h}(\overline{\phi}) \,, \\
M_\mu\overline{\partial_\psi\mathcal{G}} = M_\kappa\overline{\psi} \,, \\
M_\xi\overline{j_\phi} = M_\Gamma\overline{F_\phi} \,.
\end{cases} \tag{9}
$$

Denote J the $(n+m+k) \times (n+m+k)$ skew-symmetric structure matrix and M the $(n+m+k) \times (n+m+k)$ symmetric mass matrix on the first line of (9).

Discrete Free Energy Balance Equation: Let us define the discrete free energy functional:

$$
\overline{G}(\overline{\phi},\overline{\psi}) := \int_\Omega g(\phi^d) + \frac{1}{2}\psi^d \cdot \kappa\psi^d = \int_\Omega g(\sum \phi_i\lambda_i) + \frac{1}{2}\overline{\psi} \cdot C_\kappa\overline{\psi} \,.
$$

The following theorem can then be proved:

Theorem 1 *The discrete free energy balance is given by:*

$$
\boxed{\frac{d}{dt}\overline{G}(\overline{\phi},\overline{\psi}) = -\overline{F_\phi}M_\gamma\overline{F_\phi} + y_\partial^T M_\partial u_\partial \quad \le y_\partial^T M_\partial u_\partial \,.}
$$

Proof: Let us compute the discrete free energy balance:

$$
\left(\overline{\partial_\phi\mathcal{G}}^T \ \overline{\partial_\psi\mathcal{G}}^T \ \overline{j_\phi}^T\right) M \begin{pmatrix} \partial_t\overline{\phi} \\ \partial_t\overline{\psi} \\ \overline{F_\phi} \end{pmatrix} = \left(\overline{\partial_\phi\mathcal{G}}^T \ \overline{\partial_\psi\mathcal{G}}^T \ \overline{j_\phi}^T\right) \left(J \begin{pmatrix} \overline{\partial_\phi\mathcal{G}} \\ \overline{\partial_\psi\mathcal{G}} \\ \overline{j_\phi} \end{pmatrix} + B \begin{bmatrix} \overline{e_{j_\phi}} \\ e_d \\ \overline{e_\psi} \end{bmatrix} \right) \,,
$$

$$
= 0 + \left(\overline{\partial_\phi\mathcal{G}}^T \ \overline{\partial_\psi\mathcal{G}}^T \ \overline{j_\phi}^T\right) B \begin{pmatrix} \overline{e_{j_\phi}} \\ e_d \\ \overline{e_\psi} \end{pmatrix} \,;
$$

$$
\tag{10}
$$

and therefore:

$$
y_\partial^T M_\partial u_\partial = \overline{\partial_\phi\mathcal{G}}^T M_\lambda\partial_t\overline{\phi} + \overline{\partial_\psi\mathcal{G}}^T M_\mu\partial_t\overline{\psi} + \overline{j_\phi}^T M_\xi\overline{F_\phi} \,. \tag{11}
$$

Also we can compute more precisely:

$$\frac{d}{dt}\int_\Omega g(\sum \phi_i\lambda_i) = \sum_j \int_\Omega g'(\sum \phi_i\lambda_i)\lambda_j\phi_j'(t) = \overline{h}(\overline{\phi}) \cdot \frac{d}{dt}\overline{\phi} = \overline{\partial_\phi \mathcal{G}}^T M_\lambda \partial_t \overline{\phi},$$

which gives:

$$\frac{d}{dt}\overline{G}(\overline{\phi}, \overline{\psi}) = \overline{\partial_\phi \mathcal{G}}^T M_\lambda \partial_t \overline{\phi} + \overline{\partial_\psi \mathcal{G}}^T M_\mu \partial_t \overline{\psi},$$

which yields the *exact* free energy balance:

$$\frac{d}{dt}\overline{G}(\overline{\phi}, \overline{\psi}) = -\overline{j_\phi}^T M_\xi \overline{F_\phi} + y_\partial^T M_\partial u_\partial,$$
$$= -\overline{F_\phi} M_\gamma \overline{F_\phi} + y_\partial^T M_\partial u_\partial,$$
$$\leq y_\partial^T M_\partial u_\partial.$$

Note that this discrete free energy balance mimics the previous one (3) obtained in the continuous setting.

Discrete Phase Balance Equation: Let us first recall the phase balance equation in the *continuous* setting:

$$\frac{d}{dt}\int_\Omega \phi = -\int_{\partial\Omega} \boldsymbol{j_\phi} \cdot \boldsymbol{n} = -\int_{\partial\Omega} e_{j_\phi}.$$

Let us note $c : x \mapsto 1$ the constant function equal to 1 over Ω, and let us note $V^h = Vect(\lambda_1, ..., \lambda_n)$. Let us note p_λ the orthogonal projection of $H^1(\Omega)$ on V^h, and for any $f \in H^1(\Omega)$, $\overline{p_\lambda}(f)$ the vector of size n corresponding to the coefficients of this projection over $\lambda_1, ..., \lambda_n$.

Theorem 2. *Let us assume* $c \in V^h$, *then:*

$$\boxed{\frac{d}{dt}\int_\Omega \phi^d = -1_\lambda^T B_{j_\phi} \overline{e_{j_\phi}}.} \tag{12}$$

Proof: Firstly, let us compute the following :

$$0 = \mathbf{grad}(c) = \mathbf{grad}(p_\lambda(c)) = \mathbf{grad}(\sum_i \alpha_i \lambda_i) = \sum_i \alpha_i \mathbf{grad}(\lambda_i)$$

Then,

$$\forall i \in [\![1, k]\!], \quad (D_\nabla^T \overline{p_\lambda}(c))_i = \sum_j^n \int_\Omega \xi_i \cdot \mathbf{grad}(\lambda_j)\alpha_j$$
$$= \int_\Omega \xi_i \cdot (\sum_j \alpha_j \mathbf{grad}(\lambda_j)) \tag{13}$$
$$= 0$$

Therefore, $\mathbf{1}_\lambda := \overline{p_\lambda}(c) \in \ker(D_\nabla^T)$. Then, let us compute $\frac{d}{dt} \int_\Omega \phi^d$:

$$\frac{d}{dt} \int_\Omega \phi^d = \frac{d}{dt} \int_\Omega 1 \sum_i \lambda_i \, \phi_i = \mathbf{1}_\lambda^T M_\lambda \frac{d}{dt} \overline{\phi} = \mathbf{1}_\lambda^T (D_\nabla \overline{j_\phi} - B_{j_\phi} \overline{e_{j_\phi}}) = -\mathbf{1}_\lambda^T B_{j_\phi} \overline{e_{j_\phi}}.$$

Remark: Note that the hypothesis of Theorem 2 is really a very weak hypothesis. Indeed, the constant function 1 does belong to most finite element approximation spaces. Thus, the phase balance equation is preserved at the discrete level.

4   Numerical Experiment

In order to show the efficiency of the approach, we consider a square of length π and a given smooth distribution of ϕ at initial time. Parameters are chosen as follows: $\kappa = 0.0004$, $\Gamma = 10$, and $g(\phi) = 0.25 \, \phi^2 (1 - \phi)^2$. Note that κ is taken to correspond to the square of the parameter ε ($= 0.02$) appearing in the *classical* statement of the Cahn-Hilliard equations [7]. Futhermore, Γ is taken large enough to observe a displacement of the interface in a relatively small time. For the sake of simplicity, controls are taken equal to zero in this first example.

Regarding the discretization in space, the mesh size is $\Delta x = \Delta y = 0.1$, and we use Lagrange finite element of order 1 for all variables (both scalar and vector fields). The final system has about $13{,}000 \, {}^\circ C$ of freedom. The SCRIMP[1] simulation environment has been used.

For the time discretization of the obtained port-Hamiltonian Differential Algebraic Equation (pH-DAE), following e.g. [12] and references therein, a Backward Differentiation Formula (BDF) of order 4 has been chosen, and the nonlinearity induced by $h(\phi) = g'(\phi)$ has been treated explicitly as a right-hand side, making use of the previous time step.

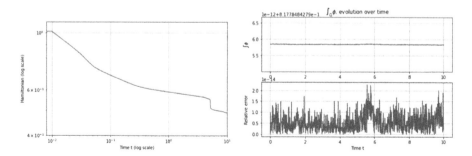

Fig. 1. Evolution of the Hamiltonian over time (left), visualisation of the phase preservation with relative error at machine precision (right).

---

[1] https://github.com/g-haine/scrimp.

On the left of Fig. 1, one may appreciate the evolution of the Hamiltonian, which indeed shows the expected decaying behavior of Theorem 1. On the right of the same figure, the phase preservation is verified, in accordance with Theorem 2. More precisely, the relative error is at machine precision.

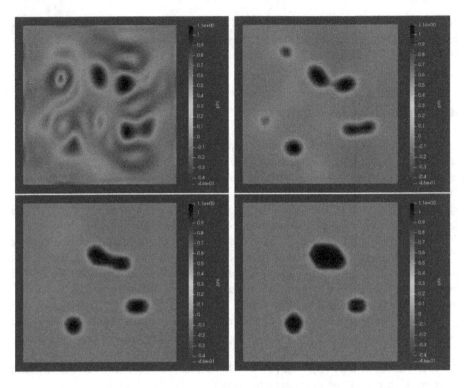

Fig. 2. Evolution of the phase at different times: $t = 0$, $t = 0.005$, $t = 0.05$ and $t = 0.3$.

On Fig. 2, four snapshots of the evolution of the phase distribution are presented, and one may observe the phase separation process as described by the Cahn-Hilliard equations. These results have been obtained at a relatively low numerical cost, showing that the PFEM is indeed able to capture the relevant physical properties of the continuous model, even at low resolution.

5   Conclusion

At the discrete level, the PFEM is able to mimic both the free energy balance, and the conservation of the phase of the Cahn-Hilliard equations. Moreover, making use of the SCRIMP software, simulation results have been obtained, showing the relevance of the port-Hamiltonian framework together with structure-preserving discretization.

References

1. Beier, N., Sego, D., Donahue, R., Biggar, K.: Laboratory investigation on freeze separation of saline mine waste water. Cold Reg. Sci. Technol. **48**(3), 239–247 (2007)
2. Bendimerad-Hohl, A., Haine, G., Matignon, D., Maschke, B.: Structure-preserving discretization of a coupled Allen-Cahn and heat equation system. IFAC-PapersOnLine **55**(18), 99–104 (2022)
3. Bendimerad-Hohl, A., Matignon, D., Haine, G.: Spatial discretization and simulation of the Allen-Cahn and Cahn-Hilliard equations as port-Hamiltonian systems. Technical Report, ISAE-Supaero (2022). https://oatao.univ-toulouse.fr/29098/
4. Boettinger, W.J., Warren, J.A., Beckermann, C., Karma, A.: Phase-field simulation of solidification. Ann. Rev. Mater. Res. **32**(1), 163–194 (2002)
5. Cahn, J.W., Hilliard, J.E.: Free energy of a nonuniform system. I. interfacial free energy. J. Chem. Phys. **28**(2), 258–267 (1958)
6. Cardoso-Ribeiro, F.L., Matignon, D., Lefèvre, L.: A partitioned finite element method for power-preserving discretization of open systems of conservation laws. IMA J. Math. Control Inf. **38**(2), 493–533 (2021)
7. Church, J.M., et al.: High accuracy benchmark problems for Allen-Cahn and Cahn-Hilliard dynamics. Commun. Comput. Phys. **26**(4) (2019)
8. Egger, H., Habrich, O., Shashkov, V.: On the energy stable approximation of Hamiltonian and gradient systems. Comput. Methods Appl. Math. **21**(2), 335–349 (2021)
9. Gay-Balmaz, F., Yoshimura, H.: A Lagrangian variational formulation for nonequilibrium thermodynamics. Part I: discrete systems. J. Geom. Phys. **111**, 169–193 (2017)
10. Gay-Balmaz, F., Yoshimura, H.: A Lagrangian variational formulation for nonequilibrium thermodynamics. Part II: Continuum systems. J. Geom. Phys. **111**, 194–212 (2017)
11. van der Ham, F., Witkamp, G.J., De Graauw, J., Van Rosmalen, G.: Eutectic freeze crystallization: application to process streams and waste water purification. Chem. Eng. Process. Process Intensification **37**(2), 207–213 (1998)
12. Mehrmann, V., Unger, B.: Control of port-Hamiltonian differential-algebraic systems and applications. Acta Numerica **32**, 395–515 (2023)
13. van der Schaft, A.J.: Port-Hamiltonian systems: an introductory survey. In: Proceedings of the International Congress of Mathematicians, vol. 3, pp. 1339–1365 (2006)
14. van der Schaft, A.J., Maschke, B.M.: Hamiltonian formulation of distributed-parameter systems with boundary energy flow. J. Geom. Phys. **42**, 166–194 (2002)
15. Serhani, A., Haine, G., Matignon, D.: Anisotropic heterogeneous nD heat equation with boundary control and observation: II. structure-preserving discretization. IFAC-PapersOnLine **52**(7), 57–62 (2019)
16. Vincent, B., Couenne, F., Lefèvre, L., Maschke, B.: Port Hamiltonian systems with moving interface: a phase field approach. IFAC-PapersOnLine **53**(2), 7569–7574 (2020)

Infinite Dimensional Lagrange–Dirac Mechanics with Boundary Conditions

Álvaro Rodríguez Abella[1]([✉]) [iD], François Gay–Balmaz[2] [iD],
and Hiroaki Yoshimura[3] [iD]

[1] Instituto de Ciencias Matemáticas (CSIC–UAM–UC3M–UCM), Calle Nicolás
Cabrera 13–15, Madrid, Spain
alvrod06@ucm.es
[2] Laboratoire de Météorologie Dynamique École Normale Supérieure/CNRS, Paris,
France
francois.gay-balmaz@lmd.ens.fr
[3] School of Science and Engineering Waseda University, 3-4-1, Okubo, Shinjuku,
Tokyo, Japan
yoshimura@waseda.jp

Abstract. The Lagrange–Dirac theory is extended to systems defined on the family of smooth functions on a manifold with boundary, which provides an instance of systems with a Fréchet space as a configuration space. To that end, we introduce the restricted cotangent bundle, a vector subbundle of the topological cotangent bundle which contains the partial derivatives of Lagrangian functions defined through a density. The main achievement of our proposal is that the Lagrange–Dirac equations on the restricted cotangent bundle properly account for the boundary value problem, i.e., the boundary conditions do not need to be imposed *ad hoc*, but they arise naturally from the Lagrange–Dirac formulation. After giving the main theoretical results, and showing how boundary forces can be naturally included in the Lagrange–Dirac formulation, we illustrate our framework with the dynamical equations of a vibrating membrane.

Keywords: Boundary problem · Lagrange–Dirac mechanics · Fréchet space

1 Introduction

Lagrange–Dirac dynamical systems, which were introduced by Yoshimura and Marsden in [15], are based on the use of Dirac structures [2] in conjunction with

ARA is supported by Ministerio de Universidades, Spain, under an FPU grant and partially supported by Ministerio de Ciencia e Innovación, Spain, under grant PID2021-126124NB-I00. FGB is partially supported by CNCS UEFISCDI, project number PN-III-P4-ID-PCE-2020-2888. HY is partially supported by JSPS Grant-in-Aid for Scientific Research (22K03443), JST CREST (JPMJCR1914), Waseda University (SR 2022C-423), and the MEXT "Top Global University Project", SEES.

the Lagrangian formalism. The main advantage of the Lagrange–Dirac approach to dynamical systems is that it provides a unified geometric formulation of systems that are both degenerate and nonholonomic, and that it admits an associated variational formulation. More recently, the Lagrange–Dirac formulation has been extended to a number of situations, including reduction by symmetries [6,16], interconnection [7] and nonequilibrium thermodynamics [4,5]. In addition, different discretizations have been proposed, as in [1] or [9]. Nevertheless, there exists a considerable gap in the literature about infinite dimensional systems from the Lagrange–Dirac point of view.

The aim of this contribution is to propose a Lagrange–Dirac theory for systems defined on the family of smooth functions on a domain with boundary. This family is an instance of an infinite dimensional Fréchet space [11,13]. When the Lagrangian is defined through a density, its partial derivatives lie in a vector subspace of the topological dual space, which will be called the restricted dual. The restricted dual gathers information about the behaviour of the system in both the interior and the boundary of the domain. As a result, the Lagrange–Dirac equations thus obtained incorporate the boundary conditions of the dynamical system, as well as boundary and interior forces, in a natural way. This is illustrated by computing the Lagrange–Dirac equations of a vibrating membrane with free boundary, which leads to the Neumann boundary conditions.

This work constitutes a first step towards a comprehensive Lagrange–Dirac theory for infinite dimensional systems and, more specifically, for systems with a Fréchet manifold as a configuration space. Such a theory would be highly desirable, since it will encompass many physical situations, such as waves, fluid dynamics or continuum mechanics [10].

2   Configuration Space and Induced Dirac Structure

Let $\mathcal{B} \subset \mathbb{R}^m$ be a closed, bounded domain with smooth boundary. In this work, we consider continuum systems with configuration space given by the Fréchet space of smooth functions on $\mathcal{B}$, i.e.,

$$V = C^\infty(\mathcal{B}).$$

2.1   Restricted Cotangent Bundle

As we will see in the next section, the fiber derivatives of Lagrangians defined through a density lie in some subspace of the topological dual V' of V. For this reason, it is convenient to introduce the *restricted dual space*

$$V^\star = C^\infty(\mathcal{B}) \times C^\infty(\partial\mathcal{B}).$$

We regard $V^\star$ as a vector subspace of V' by means of the L^2-dual pairing given by

$$\langle(\alpha, \alpha_\partial), \varphi\rangle = \int_\mathcal{B} \alpha\varphi \, \mathrm{d}x + \int_{\partial\mathcal{B}} \alpha_\partial\varphi \, \mathrm{d}s, \qquad (\alpha, \alpha_\partial) \in V^\star, \ \varphi \in V,$$

where $\mathrm{d}x$ and $\mathrm{d}s$ denote the volume form on $\mathcal{B}$ and the boundary element on $\partial\mathcal{B}$, respectively. Note that this pairing is weakly non-degenerate, i.e., $\langle (\alpha, \alpha_\partial), \varphi \rangle = 0$ for all $\varphi \in V$ implies $(\alpha, \alpha_\partial) = 0$, and $\langle (\alpha, \alpha_\partial), \varphi \rangle = 0$ for all $(\alpha, \alpha_\partial) \in V^\star$ implies $\varphi = 0$. Such a pair $(V, V^\star)$ is referred to as a dual system [11, Chapter 23].

Definition 1. *The **restricted cotangent bundle** of V is given by $T^\star V = V \times V^\star$.*

As a consequence of $V^\star \subset V'$, we observe that the restricted cotangent bundle, $T^\star V$, is naturally included in the topological cotangent bundle, $T'V = V \times V'$. By following the same idea, we define the restricted iterated bundles:

$$
\begin{aligned}
T^\star(TV) &= V \times V \times V^\star \times V^\star \subset T'(TV) &= V \times V \times V' \times V', \\
T(T^\star V) &= V \times V^\star \times V \times V^\star \subset T(T'V) &= V \times V' \times V \times V', \\
T^\star(T^\star V) &= V \times V^\star \times V^\star \times V \subset T'(T^\star V) &= V \times V^\star \times V' \times (V^\star)'.
\end{aligned}
$$

In the last equation, $(V^\star)'$ denotes the topological dual of $V^\star$ and the inclusion $V \subset (V^\star)'$ is understood again by means of the L^2-pairing (cf. [11, Chapter 23]). From the last equation, we also note that we choose to define $(V^\star)^\star = V$. While a more general setting is possible, this choice is enough for our purpose.

Definition 2. *The **restricted Pontryagin bundle** of $T^\star V$ is given by*

$$
T(T^\star V) \oplus T^\star(T^\star V) = V \times V^\star \times (V \times V^\star \times V^\star \times V).
$$

Of course, the restricted Pontryagin bundle is naturally included in the topological Pontryagin bundle $T(T^\star V) \oplus T'(T^\star V)$ of $T^\star V$.

2.2   Restricted Tulczyjew Triple

Following its general definition on cotangent bundles, the *canonical one-form* $\Theta_{T^\star V} \in \Omega^1(T^\star V)$ is given by

$$
\Theta_{T^\star V}(z) \cdot \delta z = \langle z, T_z \pi(\delta z) \rangle, \qquad z \in T^\star V, \ \delta z \in T_z(T^\star V),
$$

where $\pi : T^\star V \to V$ is the (restricted) cotangent bundle projection and $T\pi : T(T^\star V) \to TV$ is the tangent map. In our case, by using $z = (\varphi, \alpha, \alpha_\partial)$ and $\delta z = (\delta\varphi, \delta\alpha, \delta\alpha_\partial)$ one gets the expression

$$
\Theta_{T^\star V}(\varphi, \alpha, \alpha_\partial) \cdot (\delta\varphi, \delta\alpha, \delta\alpha_\partial) = \int_\mathcal{B} \alpha \delta\varphi \, \mathrm{d}x + \int_{\partial\mathcal{B}} \alpha_\partial \delta\varphi \, \mathrm{d}s.
$$

The *canonical symplectic form* on the restricted cotangent bundle is the two-form

$$
\Omega_{T^\star V} = -\mathrm{d}\Theta_{T^\star V} \in \Omega^2(T^\star V),
$$

giving

$$\Omega_{T^\star V}(\varphi, \alpha, \alpha_\partial)((\dot{\varphi}, \dot{\alpha}, \dot{\alpha}_\partial), (\delta\varphi, \delta\alpha, \delta\alpha_\partial))$$
$$= \langle(\delta\alpha, \delta\alpha_\partial), \dot{\varphi}\rangle - \langle(\dot{\alpha}, \dot{\alpha}_\partial), \delta\varphi\rangle$$
$$= \int_B \delta\alpha\dot{\varphi}\,\mathrm{d}x + \int_{\partial B} \delta\alpha_\partial\dot{\varphi}\,\mathrm{d}s - \int_B \dot{\alpha}\delta\varphi\,\mathrm{d}x - \int_{\partial B} \dot{\alpha}_\partial\delta\varphi\,\mathrm{d}s$$

for each $(\varphi, \alpha, \alpha_\partial) \in T^\star V$ and $(\dot{\varphi}, \dot{\alpha}, \dot{\alpha}_\partial), (\delta\varphi, \delta\alpha, \delta\alpha_\partial) \in T_{(\varphi,\alpha,\alpha_\partial)}(T^\star V)$. A straightforward computation leads to the following result.

Proposition 1. *The flat map of the canonical symplectic form,*

$$\Omega^\flat_{T^\star V} : T(T^\star V) \to T'(T^\star V), \quad (\varphi, \alpha, \alpha_\partial; \dot{\varphi}, \dot{\alpha}, \dot{\alpha}_\partial) \mapsto (\varphi, \alpha, \alpha_\partial; -\dot{\alpha}, -\dot{\alpha}_\partial, \dot{\varphi})$$

takes values in the restricted iterated bundle, $T^\star(T^\star V)$. Furthermore, it defines a vector bundle isomorphism over the identity, $\mathrm{id}_{T^\star V}$, between $T(T^\star V)$ and the restricted iterated bundle, $T^\star(T^\star V)$.

Remark 1. Observe that the inclusion $T^\star(T^\star V) \subset T'(T^\star V)$ is strict. Therefore, the canonical symplectic form, $\Omega_{T^\star V}$, is weak, since it does not define an isomorphism between $T(T^\star V)$ and $T'(T^\star V)$. If we confine ourselves to the restricted iterated bundle, then it becomes a strong form. In the following, we focus on the latter situation without further mention.

By mimicking the finite dimensional case, see [15], the following isomorphism over the identity, id_V, is defined

$$\kappa_{T^\star V} : T(T^\star V) \to T^\star(TV), \quad (\varphi, \alpha, \alpha_\partial; \dot{\varphi}, \dot{\alpha}, \dot{\alpha}_\partial) \mapsto (\varphi, \dot{\varphi}; \dot{\alpha}, \dot{\alpha}_\partial, \alpha, \alpha_\partial).$$

In the same vein, we set

$$\gamma_{T^\star V} = \Omega^\flat_{T^\star V} \circ \kappa^{-1}_{T^\star V} : T^\star(TV) \to T^\star(T^\star V).$$

By gathering the previous isomorphisms, we obtain the *restricted Tulczyjew triple* in the family of smooth functions as follows:

$$\gamma_{T^\star V}$$

$$T^\star(TV) \xleftarrow{\quad \kappa_{T^\star V} \quad} T(T^\star V) \xrightarrow{\quad \Omega^\flat_{T^\star V} \quad} T^\star(T^\star V)$$

$$(\varphi, \dot{\varphi}; \dot{\alpha}, \dot{\alpha}_\partial, \alpha, \alpha_\partial) \longleftarrow\!\!\shortmid (\varphi, \alpha, \alpha_\partial; \dot{\varphi}, \dot{\alpha}, \dot{\alpha}_\partial) \longmapsto (\varphi, \alpha, \alpha_\partial; -\dot{\alpha}, -\dot{\alpha}_\partial, \dot{\varphi}).$$

The following result can be shown by using the non-degeneracy of the L^2-dual pairing.

Proposition 2. *The subbundle*

$$D_{T^\star V} = \mathrm{graph}\,\Omega^\flat_{T^\star V} \subset T(T^\star V) \oplus T^\star(T^\star V)$$

is a Dirac structure on the restricted Pontryagin bundle of $T^\star V$.

By using the explicit expression of $\Omega^\flat_{T^*V}$ provided in Proposition 1, we obtain

$$D_{T^*V} = \{(\varphi, \alpha, \alpha_\partial; \dot{\varphi}, \dot{\alpha}, \dot{\alpha}_\partial; \delta\alpha, \delta\alpha_\partial, \delta\varphi) \in T(T^*V) \oplus T^*(T^*V) \mid$$
$$- \dot{\alpha} = \delta\alpha, \ -\dot{\alpha}_\partial = \delta\alpha_\partial, \ \dot{\varphi} = \delta\varphi\}. \tag{1}$$

In the above expression, note that the relations $\dot{\varphi} = \delta\varphi$ and $-\dot{\alpha} = \delta\alpha$ hold in the interior of $\mathcal{B}$, while $-\dot{\alpha}_\partial = \delta\alpha_\partial$ holds on $\partial\mathcal{B}$.

3   Lagrange–Dirac Equations

Let $L : TV \to \mathbb{R}$ be a Lagrangian given by

$$L(\varphi, \dot{\varphi}) = \int_{\mathcal{B}} \mathfrak{L}(\varphi(x), \dot{\varphi}(x), \nabla\varphi(x)) \, dx, \qquad (\varphi, \dot{\varphi}) \in TV, \tag{2}$$

where $\mathfrak{L} : \mathbb{R} \times \mathbb{R} \times \mathbb{R}^m \to \mathbb{R}$ is the Lagrangian density. In the previous expression,

$$\nabla : C^\infty(\mathcal{B}) \to \Omega^1(\mathcal{B}) \simeq C^\infty(\mathcal{B}, \mathbb{R}^m), \quad \varphi \mapsto \left(\frac{\partial\varphi}{\partial x^1}, \dots, \frac{\partial\varphi}{\partial x^m}\right),$$

denotes the gradient, being $(x^1, \dots, x^m)$ the standard (global) coordinates of $\mathbb{R}^m$. In the same vein, the divergence is denoted by

$$\mathrm{div} : \mathfrak{X}(\mathcal{B}) \simeq C^\infty(\mathcal{B}, \mathbb{R}^m) \to C^\infty(\mathcal{B}), \quad \omega = (\omega^1, \dots, \omega^m) \mapsto \sum_{i=1}^m \frac{\partial\omega^i}{\partial x^i}.$$

The partial derivatives of the Lagrangian are the maps

$$\frac{\delta L}{\delta\varphi} : TV \to V', \qquad \frac{\delta L}{\delta\varphi}(\varphi, \dot{\varphi})(\delta\varphi) = \frac{d}{d\epsilon}\Big|_{\epsilon=0} L(\varphi + \epsilon\,\delta\varphi, \dot{\varphi})$$

and

$$\frac{\delta L}{\delta\dot{\varphi}} : TV \to V', \qquad \frac{\delta L}{\delta\dot{\varphi}}(\varphi, \dot{\varphi})(\delta\dot{\varphi}) = \frac{d}{d\epsilon}\Big|_{t=0} L(\varphi, \dot{\varphi} + \epsilon\,\delta\dot{\varphi}),$$

for each $(\varphi, \dot{\varphi}) \in TV$ and $\delta\varphi, \delta\dot{\varphi} \in V$.

Remark 2. The contraction between $\omega = (\omega^1, \dots, \omega^m) \in \mathfrak{X}(\mathcal{B}) \simeq C^\infty(\mathcal{B}, \mathbb{R}^m)$ and $\zeta = (\zeta_1, \dots, \zeta_m) \in \Omega^1(\mathcal{B}) \simeq C^\infty(\mathcal{B}, \mathbb{R}^m)$ is given by the standard inner product on $\mathbb{R}^m$ and is denoted by $\omega \cdot \zeta = \sum_{i=1}^m \omega^i \zeta_i \in C^\infty(\mathcal{B})$.

The following result ensures that the partial derivatives introduced above lie in the restricted dual, $V^* \subset V'$, when the Lagrangian is defined through a density.

Lemma 1. *Let $L : TV \to \mathbb{R}$ be a Lagrangian defined through a density, as in* (2). *Then the partial derivatives of L lie in $V^\star$. Furthermore, they are given by*

$$\frac{\delta L}{\delta \varphi}(\varphi, \dot{\varphi}) = \left(\frac{\partial \mathfrak{L}}{\partial \varphi}(\varphi, \dot{\varphi}, \nabla \varphi) - \operatorname{div} \frac{\partial \mathfrak{L}}{\partial \nabla \varphi}(\varphi, \dot{\varphi}, \nabla \varphi), \left. \frac{\partial \mathfrak{L}}{\partial \nabla \varphi}(\varphi, \dot{\varphi}, \nabla \varphi) \right|_{\partial \mathcal{B}} \cdot n \right),$$

$$\frac{\delta L}{\delta \dot{\varphi}}(\varphi, \dot{\varphi}) = \left(\frac{\partial \mathfrak{L}}{\partial \dot{\varphi}}(\varphi, \dot{\varphi}, \nabla \varphi), 0 \right),$$

for each $(\varphi, \dot{\varphi}) \in TV$, where $n \in C^\infty(\partial \mathcal{B}, \mathbb{R}^m)$ is the outward-pointing, unit, normal vector field on the boundary.

Proof. Let $\delta \varphi \in V$. By definition, we have

$$
\begin{aligned}
\frac{\delta L}{\delta \varphi}(\varphi, \dot{\varphi})(\delta \varphi) &= \left. \frac{d}{d\epsilon} \right|_{\epsilon=0} \int_{\mathcal{B}} \mathfrak{L}(\varphi + \epsilon \, \delta \varphi, \dot{\varphi}, \nabla(\varphi + \epsilon \, \delta \varphi)) \, dx \\
&= \int_{\mathcal{B}} \left(\frac{\partial \mathfrak{L}}{\partial \varphi}(\varphi, \dot{\varphi}, \nabla \varphi) \, \delta \varphi + \frac{\partial \mathfrak{L}}{\partial \nabla \varphi}(\varphi, \dot{\varphi}, \nabla \varphi) \cdot \nabla(\delta \varphi) \right) dx \\
&= \int_{\mathcal{B}} \left(\frac{\partial \mathfrak{L}}{\partial \varphi}(\varphi, \dot{\varphi}, \nabla \varphi) - \operatorname{div} \frac{\partial \mathfrak{L}}{\partial \nabla \varphi}(\varphi, \dot{\varphi}, \nabla \varphi) \right) \delta \varphi \, dx \\
&\quad + \int_{\partial \mathcal{B}} \left(\frac{\partial \mathfrak{L}}{\partial \nabla \varphi}(\varphi, \dot{\varphi}, \nabla \varphi) \cdot n \right) \delta \varphi \, ds,
\end{aligned}
$$

where we have used the standard integration by parts formula. Analogously,

$$\frac{\delta L}{\delta \dot{\varphi}}(\varphi, \dot{\varphi})(\delta \dot{\varphi}) = \left. \frac{d}{d\epsilon} \right|_{\epsilon=0} \int_{\mathcal{B}} \mathfrak{L}(\varphi, \dot{\varphi} + \epsilon \, \delta \dot{\varphi}, \nabla \varphi) \, dx = \int_{\mathcal{B}} \frac{\partial \mathfrak{L}}{\partial \dot{\varphi}}(\varphi, \dot{\varphi}, \nabla \varphi) \, \delta \dot{\varphi} \, dx.$$

Since the previous expressions hold for every $\delta \varphi, \delta \dot{\varphi} \in V$, we conclude.   □

The differential of L is given by

$$\mathrm{d}L : TV \to T'(TV), \quad (\varphi, \dot{\varphi}) \mapsto \left(\varphi, \dot{\varphi}, \frac{\delta L}{\delta \varphi}(\varphi, \dot{\varphi}), \frac{\delta L}{\delta \dot{\varphi}}(\varphi, \dot{\varphi}) \right).$$

The previous lemma ensures that it takes values in $T^\star(TV)$, which enables us to define the *Dirac differential* of L as

$$\mathrm{d}_D L = \gamma_{T^\star V} \circ \mathrm{d}L : TV \to T^\star(T^\star V). \tag{3}$$

By using the explicit expression of $\gamma_{T^\star V}$, it can be checked that

$$\mathrm{d}_D L(\varphi, \dot{\varphi}) = \left(\varphi, \frac{\delta L}{\delta \dot{\varphi}}(\varphi, \dot{\varphi}), -\frac{\delta L}{\delta \varphi}(\varphi, \dot{\varphi}), \dot{\varphi} \right)$$

for each $(\varphi, \dot{\varphi}) \in TV$.

Definition 3. *The **Lagrange–Dirac equations** for a curve*

$$(\varphi, \nu, \alpha, \alpha_\partial) : [t_0, t_1] \to TV \oplus T^\star V = V \times (V \times V^\star)$$

are given by

$$\left((\varphi, \alpha, \alpha_\partial; \dot{\varphi}, \dot{\alpha}, \dot{\alpha}_\partial), \mathrm{d}_D L(\varphi, \nu) \right) \in D_{T^\star V}(\varphi, \alpha, \alpha_\partial). \tag{4}$$

A straightforward computation from (1) and (3), as well as Lemma 1, gives the explicit expression of the equations.

Theorem 1. *Let* $L : TV \to \mathbb{R}$ *be a Lagrangian, as in* (2). *The Lagrange–Dirac Eqs.* (4) *are equivalent to the following system:*

$$
\begin{cases}
\dot{\varphi} = \nu, \\
\alpha = \dfrac{\partial \mathfrak{L}}{\partial \dot{\varphi}}(\varphi, \nu, \nabla\varphi), \qquad \dot{\alpha} = \dfrac{\partial \mathfrak{L}}{\partial \varphi}(\varphi, \nu, \nabla\varphi) - \operatorname{div} \dfrac{\partial \mathfrak{L}}{\partial \nabla\varphi}(\varphi, \nu, \nabla\varphi), \\
\alpha_\partial = 0, \qquad \dot{\alpha}_\partial = \dfrac{\partial \mathfrak{L}}{\partial \nabla\varphi}(\varphi, \nu, \nabla\varphi)\bigg|_{\partial B} \cdot n.
\end{cases}
$$

Observe that the two conditions

$$
\alpha = \frac{\partial \mathfrak{L}}{\partial \dot{\varphi}}(\varphi, \nu, \nabla\varphi), \qquad \alpha_\partial = 0,
$$

above arise from Lagrange–Dirac equations (4) by noting that $(\varphi, \alpha, \alpha_\partial; \dot{\varphi}, \dot{\alpha}, \dot{\alpha}_\partial)$ and $\mathrm{d}_D L(\varphi, \nu)$ must have the same footpoint. These conditions can be written as $(\varphi, \alpha, \alpha_\partial) = \mathbb{F}L(\varphi, \nu)$, where

$$
\mathbb{F}L : TV \to T^\star V, \quad (\varphi, \nu) \mapsto \left(\varphi, \frac{\delta L}{\delta \dot{\varphi}}(\varphi, \nu)\right) = \left(\varphi, \frac{\partial \mathfrak{L}}{\partial \dot{\varphi}}(\varphi, \nu, \nabla\varphi), 0\right)
$$

is the Legendre transform of L. Note also that the boundary condition is part of the Lagrange–Dirac equations.

Example 1 (External forces). The above theory may be easily extended to systems with external forces. More specifically, given a Lagrangian, as in (2), and an external force with values on the restricted cotangent bundle,

$$
F : TV \to T^\star V,
$$

the *forced Lagrange–Dirac equations* for a curve $(\varphi, \nu, \alpha, \alpha_\partial) : [t_0, t_1] \to TV \oplus T^\star V$ are given by

$$
\left(((\varphi, \alpha, \alpha_\partial; \dot{\varphi}, \dot{\alpha}, \dot{\alpha}_\partial), \mathrm{d}_D L(\varphi, \nu) - \widetilde{F}_L(\varphi, \nu)\right) \in D_{T^\star V}(\varphi, \alpha, \alpha_\partial),
$$

where $\widetilde{F}_L : TV \to T^\star(T^\star V)$ is defined as

$$
\tilde{F}_L(\varphi, \nu) = (T\pi)^\star_{\mathbb{F}L(\varphi, \nu)}(F(\varphi, \nu)) \in T^\star_{\mathbb{F}L(\varphi, \nu)}(T^\star V), \qquad (\varphi, \nu) \in TV,
$$

being $\pi : T^\star V = V \times V^\star \to V$ the projection onto the first component, and $(T\pi)_{\mathbb{F}L(\varphi, \nu)} : T_{\mathbb{F}L(\varphi, \nu)}(T^\star V) \to T_\varphi V$ its tangent map at $\mathbb{F}L(\varphi, \nu)$, i.e.,

$$
\left\langle \tilde{F}_L(\varphi, \nu), \delta z \right\rangle = \left\langle F(\varphi, \nu), (T\pi)_{\mathbb{F}L(\varphi, \nu)}(\delta z) \right\rangle, \qquad \delta z \in T_{\mathbb{F}L(\varphi, \nu)}(T^\star V).
$$

Since $V^\star = C^\infty(\mathcal{B}) \times C^\infty(\partial\mathcal{B})$, we may write

$$
F(\varphi, \nu) = (\varphi, \mathcal{F}(\varphi, \nu), \mathcal{F}_\partial(\varphi, \nu)), \qquad (\varphi, \nu) \in TV,
$$

for some $\mathcal{F} : TV \to C^\infty(\mathcal{B})$ and $\mathcal{F}_\partial : TV \to C^\infty(\partial\mathcal{B})$. Subsequently, the forced Lagrange–Dirac equations are equivalent to the following system:

$$
\begin{cases}
\dot{\varphi} = \nu, \\
\alpha = \dfrac{\partial \mathfrak{L}}{\partial \dot{\varphi}}(\varphi, \nu, \nabla\varphi), & \dot{\alpha} = \dfrac{\partial \mathfrak{L}}{\partial \varphi}(\varphi, \nu, \nabla\varphi) - \operatorname{div} \dfrac{\partial \mathfrak{L}}{\partial \nabla\varphi}(\varphi, \nu, \nabla\varphi) + \mathcal{F}(\varphi, \nu), \\
\alpha_\partial = 0, & \dot{\alpha}_\partial = \dfrac{\partial \mathfrak{L}}{\partial \nabla\varphi}(\varphi, \nu, \nabla\varphi)\Big|_{\partial\mathcal{B}} \cdot n + \mathcal{F}_\partial(\varphi, \nu).
\end{cases}
$$

Example 2 (Vibrating membrane with free boundary). The Lagrangian of a vibrating membrane is given by (2) with the following Lagrangian density,

$$
\mathfrak{L}(\varphi, \dot{\varphi}, \nabla\varphi) = \frac{1}{2}\rho_0 \dot{\varphi}^2 - \frac{1}{2}\tau |\nabla\varphi|^2, \qquad (\varphi, \dot{\varphi}, \nabla\varphi) \in \mathbb{R} \times \mathbb{R} \times \mathbb{R}^m,
$$

where $\rho_0 \in \mathbb{R}^+$ is the *density* and $\tau \in \mathbb{R}^+$ is the *tension*. Since the partial derivatives of $\mathfrak{L}$ are given by

$$
\frac{\partial \mathfrak{L}}{\partial \varphi}(\varphi, \dot{\varphi}, \nabla\varphi) = 0, \qquad \frac{\partial \mathfrak{L}}{\partial \dot{\varphi}}(\varphi, \dot{\varphi}, \nabla\varphi) = \rho_0 \dot{\varphi}, \qquad \frac{\partial \mathfrak{L}}{\partial \nabla\varphi}(\varphi, \dot{\varphi}, \nabla\varphi) = -\tau \nabla\varphi,
$$

for each $(\varphi, \dot{\varphi}, \nabla\varphi) \in \mathbb{R} \times \mathbb{R} \times \mathbb{R}^m$, then the dynamical equations given in Theorem 1 for a curve $(\varphi, \nu, \alpha, \alpha_\partial) : [t_0, t_1] \to TV \oplus T^*V$ read

$$
\begin{cases}
\dot{\varphi} = \nu, \\
\alpha = \rho_0 \dot{\varphi}, & \dot{\alpha} = \operatorname{div}(\tau \nabla\varphi), \\
\alpha_\partial = 0, & \dot{\alpha}_\partial = -\tau \nabla\varphi|_{\partial\mathcal{B}} \cdot n.
\end{cases}
$$

Observe that we obtain the *wave equation*,

$$
\ddot{\varphi} = (\tau/\rho_0)\nabla^2 \varphi,
$$

where ∇^2 denotes the Laplacian, together with the *Neumann boundary condition*,

$$
\nabla\varphi|_{\partial\mathcal{B}} \cdot n = 0,
$$

as expected for a vibrating membrane with free boundary. The forced case can be treated similarly, following Example 1.

4   Conclusions and Prospective Work

We have made a proposal for treating Lagrangian systems defined on $C^\infty(\mathcal{B})$ from the Dirac point of view. As we have seen, the main advantage is the incorporation of the boundary conditions in the Dirac structure. We have chosen to carry out the development on a closed, bounded domain, $\mathcal{B} \subset \mathbb{R}^m$, to keep the exposition simple, but our setting can be extended to m-dimensional, (pseudo-)Riemannian manifolds with boundary, (M, g).

On the other hand, we are currently working on the generalization to more general Fréchet spaces, such as the family of k-forms, $V = \Omega^k(\mathcal{B})$, for $0 \leq k \leq m$ (note that we have presented here the case $k = 0$), as well as on the relation between our approach and that using Stokes–Dirac structures [14]. The next step would be to consider the nonlinear case, i.e., dealing with Fréchet manifolds of smooth maps (cf., for example, [8, Chapter IX]) instead of Fréchet spaces. This will allow for treating more general physical systems, such as fluid dynamics in a fixed domain, where the configuration manifold is the family of diffeomorphisms of $\mathcal{B}$, Diff($\mathcal{B}$), or continuum mechanics with moving boundary, where the configuration manifold is the family of embeddings of $\mathcal{B}$ in $\mathbb{R}^m$, Emb($\mathcal{B}, \mathbb{R}^m$). In addition, reduction by symmetries could be performed thanks to the relabeling symmetry and the material frame indifference, yielding the so-called spatial and convective representations.

In addition, it would be desirable to construct a discrete counterpart of this theory in order to obtain numerical integrators that preserve the geometric structures at the discrete level. In [9] and, more recently, in [1], variational discretizations in time for Lagrange–Dirac and Hamilton–Dirac systems have been proposed. We aim to extend them to account for infinite-dimensional systems by following for instance the variational approach to spatial discretization, such as considered in [3,12].

References

1. Caruso, M.I., Fernández, J., Tori, C., Zuccalli, M.: Discrete mechanical systems in a Dirac setting: a proposal (2022). https://arxiv.org/abs/2203.05600
2. Courant, T.J.: Dirac manifolds. Trans. Am. Math. Soc. **319**(2), 631–661 (1990)
3. Gawlik, E.S., Gay-Balmaz, F.: A variational finite element discretization of compressible flow. Found. Comput. Math. **21**, 961–1001 (2012)
4. Gay-Balmaz, F., Yoshimura, H.: Dirac structures in nonequilibrium thermodynamics. J. Math. Phys. **59**, 012701 (2018)
5. Gay-Balmaz, F., Yoshimura, H.: Dirac structures and variational formulation of port-Dirac systems in nonequilibrium thermodynamics. IMA J. Math. Control. Inf. **37** (2020). https://doi.org/10.1093/imamci/dnaa015
6. Gay-Balmaz, F., Yoshimura, H.: Dirac reduction for nonholonomic mechanical systems and semidirect products. Adv. Appl. Math. **63**, 131–213 (2015)
7. Jacobs, H.O., Yoshimura, H.: Tensor products of Dirac structures and interconnection in Lagrangian mechanics. J. Geom. Mech. **6**(1), 67–98 (2014)
8. Kriegl, A., Michor, P.: The Convenient Setting of Global Analysis. American Mathematical Society, Mathematical Surveys (1997)
9. Leok, M., Ohsawa, T.: Variational and geometric structures of discrete Dirac mechanics. Found. Comput. Math. **11**, 529–562 (2011)
10. Marsden, J.E., Hughes, T.J.R.: Mathematical Foundations of Elasticity. Dover Publications, Inc. Mineola (1994)
11. Meise, R., Vogt, D.: Introduction to Functional Analysis. Oxford Graduate Texts in Mathematics, Clarendon Press, Oxford (1997)
12. Pavlov, D., Mullen, P., Tong, Y., Kanso, E., Marsden, J., Desbrun, M.: Structure-preserving discretization of incompressible fluids. Physica D: Nonlinear Phenomena **240**(6), 443–458 (2011)

13. Rudin, W.: Functional Analysis. International Series in Pure and Applied Mathematics, McGraw-Hill, New York (1991)
14. van der Schaft, A.J., Maschke, B.: Hamiltonian formulation of distributed-parameter systems with boundary energy flow. J. Geom. Phys. **42**, 166–194 (2002)
15. Yoshimura, H., Marsden, J.: Dirac structures in Lagrangian mechanics Part I: Implicit Lagrangian systems. J. Geom. Phys. **57**(1), 133–156 (2006)
16. Yoshimura, H., Marsden, J.E.: Dirac cotangent bundle reduction. J. Geom. Mech. **1** (2009). https://doi.org/10.3934/jgm.2009.1.87

Variational Integrators for Stochastic Hamiltonian Systems on Lie Groups

Meng Wu and François Gay-Balmaz[✉]

Laboratoire de Météorologie Dynamique, École Normale Supérieure/CNRS,
Paris, France
meng.wu@lmd.ipsl.fr, francois.gay-balmaz@lmd.ens.fr

Abstract. Motivated by recent advances in stochastic geometric modelling in fluid dynamics, we derive a variational integrator for stochastic Hamiltonian systems on Lie groups by using a discrete version of the stochastic phase space principle. The structure preserving properties of the resulting scheme, such as its symplecticity and preservation of coadjoint orbits are given, as well as a discrete Noether theorem associated to subgroup symmetries. Preliminary numerical illustrations are provided.

Keywords: Hamiltonian systems on Lie groups · Stochastic Hamiltonian systems · Variational integrators

1 Introduction and Preliminaries

We consider stochastic Hamiltonian systems [2,10] which, on the cotangent bundle $T^*V = V \times V^*$ of a vector space V, take the form

$$dq = \frac{\partial H}{\partial p} dt + \sum_i \frac{\partial H_i}{\partial p} \circ dW_i(t), \qquad dp = -\frac{\partial H}{\partial q} dt - \sum_i \frac{\partial H_i}{\partial q} \circ dW_i(t). \quad (1)$$

Here $H, H_i : T^*V \to \mathbb{R}$, $i = 1, ..., N$, are given Hamiltonians, $W_i(t)$, $i = 1, ..., N$ are independent Brownian motions, and the equations are understood in the Stratonovich sense. It is well-known that the flow of (1) almost surely preserves the canonical symplectic form on T^*V, [2,10]. Symplectic integrators for the simulation of stochastic Hamiltonian systems on vector spaces have been derived in [8,11,12] and references therein.

Besides vector spaces, stochastic Hamiltonian systems have also proved useful on Lie groups, for example for stochastic modeling in fluid mechanics [5–7] and for geometric statistical mechanics [1], which form the main motivation for the present work.

Our goal is to derive stochastic variational integrators for stochastic Hamiltonian systems on Lie groups and to study some of their structure preserving properties. In order to achieve this aim, we will first focus on the vector space

© The Author(s), under exclusive license to Springer Nature Switzerland AG 2023
F. Nielsen and F. Barbaresco (Eds.): GSI 2023, LNCS 14072, pp. 212–220, 2023.
https://doi.org/10.1007/978-3-031-38299-4_23

case and the stochastic midpoint method and provide a variational derivation of this symplectic integrator that will be shown to admit a natural analogue for systems on Lie groups in Sect. 2. In particular we will show that the stochastic midpoint method arises from an appropriate discrete version of the stochastic Hamilton phase space principle. A thorough development and extensions will be the subject of a forthcoming paper.

1.1 Stochastic Discrete Phase Space Principle on Vector Spaces

Let us consider the following stochastic phase space principle

$$\delta \int_0^T \langle p, \circ \, \mathrm{d}q \rangle - H(q, p)\mathrm{d}t - \sum_i H_i(q, p) \circ \mathrm{d}W_i(t) = 0 \tag{2}$$

understood in the Stratonovich sense and where q has fixed endpoints. A formal application of this principle, see [4,8,10] for the precise statement of such stochastic variational principles, yields the stochastic Hamiltonian system (1).

In order to obtain the stochastic midpoint method, we propose the following discrete version of the stochastic phase space principle (2):

$$\mathrm{d}\mathcal{S}_d(c_d) \cdot \delta c_d = 0 \tag{3}$$

for the discrete curve of random variables and action functional

$$c_d = \{(\tilde{q}_k, \tilde{p}_k, p_k, q_k) \in (T^*V \oplus T^*V) \times V \mid k = 0, ..., N\} \tag{4}$$

$$\begin{aligned}
\mathcal{S}_d(c_d) = \sum_{k=0}^{N-1} \Delta t \Big[&\Big\langle p_{k+1}, \frac{q_{k+1} - \tilde{q}_{k+1}}{\Delta t} \Big\rangle + \Big\langle \tilde{p}_{k+1}, \frac{\tilde{q}_{k+1} - q_k}{\Delta t} \Big\rangle \\
&- H\Big(\tilde{q}_{k+1}, \frac{\tilde{p}_{k+1} + p_{k+1}}{2}\Big) - \sum_i H_i\Big(\tilde{q}_{k+1}, \frac{\tilde{p}_{k+1} + p_{k+1}}{2}\Big) \frac{\Delta W_i}{\Delta t} \Big],
\end{aligned} \tag{5}$$

where ΔW_i are the increments of the Brownian motion and both q and p have fixed endpoints. We state the following result, whose proof is given later in the more general case of Lie groups.

Remark 1. In the discrete action functional, we can also evaluate both the deterministic and stochastic Hamiltonians at $(\frac{q_k + \tilde{q}_{k+1}}{2}, \tilde{p}_{k+1})$, and the critical point condition will yield the same equations.

Proposition 1. *The critical point condition for (5) yields the stochastic midpoint method*

$$\begin{cases}
\dfrac{q_k - q_{k-1}}{\Delta t} = \dfrac{\partial H}{\partial p}\Big(\dfrac{q_k + q_{k-1}}{2}, \dfrac{p_k + p_{k-1}}{2}\Big) + \sum_i \dfrac{\partial H_i}{\partial p}\Big(\dfrac{q_k + q_{k-1}}{2}, \dfrac{p_k + p_{k-1}}{2}\Big) \dfrac{\Delta W_i}{\Delta t} \\
\dfrac{p_k - p_{k-1}}{\Delta t} = -\dfrac{\partial H}{\partial q}\Big(\dfrac{q_k + q_{k-1}}{2}, \dfrac{p_k + p_{k-1}}{2}\Big) - \sum_i \dfrac{\partial H_i}{\partial q}\Big(\dfrac{q_k + q_{k-1}}{2}, \dfrac{p_k + p_{k-1}}{2}\Big) \dfrac{\Delta W_i}{\Delta t}.
\end{cases}$$

This method is known to be symplectic [11]. Following the usual way to prove the symplectic property of variational integrators [13] and their stochastic version [4], it turns out that the symplecticity of this method can also be obtained by exploiting the discrete stochastic phase space principle stated in (3)–(5). More precisely, one considers the discrete action functional (5) evaluated on the space of solutions of the variational scheme. Computing the critical point allows one to identify from the boundary terms the following two one-forms $\theta^{\pm} \in \Omega^1(M \times M)$ with $M = V \times V^* \times V^* \times V$:

$$\theta^- = \frac{1}{\Delta t}\tilde{p}_1 \cdot \mathrm{d}q_0, \quad \theta^+ = \Big(\frac{\tilde{p}_1 - p_1}{\Delta t} - \frac{\partial H_1}{\partial q}\Big) \cdot \mathrm{d}\tilde{q}_1 + \Big(\frac{\tilde{q}_1 - q_0}{\Delta t} - \frac{1}{2}\frac{\partial H_1}{\partial p}\Big) \cdot \mathrm{d}\tilde{p}_1$$
$$+ \frac{p_1}{\Delta t} \cdot \mathrm{d}q_1 + \Big(\frac{q_1 - \tilde{q}_1}{\Delta t} - \frac{1}{2}\frac{\partial H_1}{\partial p}\Big) \cdot \mathrm{d}p_1,$$

(6)

where we use the notations $(\tilde{q}_0, \tilde{p}_0, p_0, q_0, \tilde{q}_1, \tilde{p}_1, p_1, q_1) \in M \times M$ and

$$H_k := H\Big(\tilde{q}_k, \frac{\tilde{p}_k + p_k}{2}\Big) + \sum_i H_i\Big(\tilde{q}_k, \frac{\tilde{p}_k + p_k}{2}\Big)\frac{\Delta W_i}{\Delta t}.$$

Then, one notes that $\mathrm{d}\theta^- = \mathrm{d}\theta^+$ so that we can define $\omega \in \Omega^2(M \times M)$ as

$$\omega = -\mathrm{d}\theta^- = -\mathrm{d}\theta^+ = \frac{1}{\Delta t}\mathrm{d}q_0 \wedge \mathrm{d}\tilde{p}_1.$$

(7)

Symplecticity then follows by taking the second exterior derivative of the discrete action functional evaluated on the space of solutions. The advantage of this variational approach to symplecticity is that it admits a natural extension to Lie groups that will be used in Sect. 2.

2   Stochastic Variational Integrators on Lie Groups

We now consider stochastic Hamiltonian systems on the cotangent bundle T^*G of a Lie group G, associated to given Hamiltonians $H, H_i : T^*G \to \mathbb{R}$, $i = 1, ..., N$. The corresponding stochastic phase space principle reads

$$\delta \int_0^T \langle p, \mathrm{o}\mathrm{d}g \rangle - H(g,p)\mathrm{d}t - \sum_i H_i(g,p) \circ \mathrm{d}W_i(t) = 0.$$

(8)

To benefit from the vector space structure of the Lie algebra $\mathfrak{g}$ of G, it is convenient to consider the associated trivialized Hamiltonians $h, h_i : \mathfrak{g}^* \times G \to \mathbb{R}$ defined by $H(g,p) = h(g^{-1}p, g)$, $H_i(g,p) = h_i(g^{-1}p, g)$ and to use the trivialized version of the phase space principle (8) given by

$$\delta \int_0^T \langle \mu, \mathrm{o}g^{-1}\mathrm{d}g \rangle - h(\mu, g)\mathrm{d}t - \sum_i h_i(\mu, g) \circ \mathrm{d}W_i(t) = 0.$$

(9)

It yields the stochastic Hamiltonian system on $\mathfrak{g}^* \times G$ as

$$g^{-1}dg = \frac{\delta h}{\delta \mu}dt + \sum_i \frac{\delta h_i}{\delta \mu} \circ dW_i(t), \quad d\mu - ad^*_{g^{-1}dg}\,\mu = -g^{-1}\frac{\delta h}{\delta g}dt - g^{-1}\frac{\delta h_i}{\delta g}\circ dW_i(t).$$

Here $ad^*_\xi : \mathfrak{g}^* \to \mathfrak{g}^*$ denotes the infinitesimal coadjoint operator $\langle ad^*_\xi \mu, \eta \rangle = \langle \mu, [\xi, \eta] \rangle$, for all $\mu \in \mathfrak{g}^*$ and $\xi, \eta \in \mathfrak{g}$. In order to discretize the stochastic phase space principle (9) on Lie groups, we consider a local diffeomorphism $\tau : \mathfrak{g} \to G$ with $\tau(0) = e$, referred to as a retraction map, which is used to express small discrete changes in the group configuration through unique Lie algebra elements, see [3,9]. With such a retraction map at hand, we propose the following Lie group version of the stochastic discrete phase space principle (5):

$$d\mathcal{S}_d(c_d) \cdot \delta c_d = 0 \tag{10}$$

for the discrete curve of random variables

$$c_d = \{(\tilde{g}_k, \tilde{\mu}_k, \mu_k, g_k) \in (G \times \mathfrak{g}^* \oplus \mathfrak{g}^*) \times G \mid k = 0, ..., N\} \tag{11}$$

with action functional

$$\mathcal{S}_d(c_d) = \sum_{k=0}^{N-1} \Delta t \Big[\Big\langle \mu_{k+1}, \frac{\tau^{-1}(\tilde{g}_{k+1}^{-1}g_{k+1})}{\Delta t} \Big\rangle + \Big\langle \tilde{\mu}_{k+1}, \frac{\tau^{-1}(g_k^{-1}\tilde{g}_{k+1})}{\Delta t} \Big\rangle \\ - h\Big(\tilde{g}_{k+1}, \frac{\tilde{\mu}_{k+1} + \mu_{k+1}}{2}\Big) - \sum_i h_i\Big(\tilde{g}_{k+1}, \frac{\tilde{\mu}_{k+1} + \mu_{k+1}}{2}\Big)\frac{\Delta W_i}{\Delta t} \Big], \tag{12}$$

where both g and μ have fixed endpoints.

In this principle, the reduced momenta $\mu_k, \tilde{\mu}_k \in \mathfrak{g}^*$ are related to the momenta $p_k, \tilde{p}_k \in T^*G$ as $\mu_k = \tilde{g}_k^{-1}p_k$ and $\tilde{\mu}_k = \tilde{g}_k^{-1}\tilde{p}_k$. Therefore, for the associated discrete stochastic system on T^*G the discrete curve of random variables is

$$c_d = \{(\tilde{g}_k, \tilde{p}_k, p_k, g_k) \in (T^*G \oplus T^*G) \times G \mid k = 0, ..., N\}$$

where $T^*G \oplus T^*G \to G$ denotes the vector bundle with vector fiber at g given by $T_g^*G \oplus T_g^*G$. This explains the notation $(G \times \mathfrak{g}^* \oplus \mathfrak{g}^*) \times G$ and $(T^*V \oplus T^*V) \times V$ used earlier in (4) and (11). We will use below the trivialized derivative of $\tau^{-1} : G \to \mathfrak{g}$ defined by $d_\xi \tau^{-1}(\eta) = D\tau^{-1}(g) \cdot (\eta g) \in \mathfrak{g}$, for $\xi, \eta \in \mathfrak{g}$, $g = \tau(\xi) \in G$, and its dual map $[d_\xi \tau^{-1}]^* : \mathfrak{g}^* \to \mathfrak{g}^*$.

Proposition 2. *The critical point condition for* (12) *yields the stochastic midpoint Lie group method*

$$
\begin{cases}
\dfrac{1}{\Delta t}\left(\mathrm{Ad}^*_{\tau(\Delta t\tilde{\xi}_k)}\left[\mathrm{d}_{\Delta t\tilde{\xi}_k}\tau^{-1}\right]^*\tilde{\mu}_k - \left[\mathrm{d}_{\Delta t\xi_k}\tau^{-1}\right]^*\mu_k\right) \\
\qquad = \tilde{g}_k^{-1}\dfrac{\delta h}{\delta g}\left(\tilde{g}_k, \dfrac{\tilde{\mu}_k+\mu_k}{2}\right) + \sum_i \tilde{g}_k^{-1}\dfrac{\delta h_i}{\delta g}\left(\tilde{g}_k, \dfrac{\tilde{\mu}_k+\mu_k}{2}\right)\dfrac{\Delta W_i}{\Delta t} \\
\dfrac{1}{\Delta t}\left(\mathrm{Ad}^*_{\tau(\Delta t\xi_k)}\left[\mathrm{d}_{\Delta t\xi_k}\tau^{-1}\right]^*\mu_k - \left[\mathrm{d}_{\Delta t\tilde{\xi}_{k+1}}\tau^{-1}\right]^*\tilde{\mu}_{k+1}\right) = 0 \qquad (13)\\
\tilde{\xi}_k = \dfrac{1}{2}\dfrac{\delta h}{\delta\mu}\left(\tilde{g}_k, \dfrac{\tilde{\mu}_k+\mu_k}{2}\right) + \dfrac{1}{2}\sum_i \dfrac{\delta h_i}{\delta\mu}\left(\tilde{g}_k, \dfrac{\tilde{\mu}_k+\mu_k}{2}\right)\dfrac{\Delta W_i}{\Delta t} \\
\xi_k = \dfrac{1}{2}\dfrac{\delta h}{\delta\mu}\left(\tilde{g}_k, \dfrac{\tilde{\mu}_k+\mu_k}{2}\right) + \dfrac{1}{2}\sum_i \dfrac{\delta h_i}{\delta\mu}\left(\tilde{g}_k, \dfrac{\tilde{\mu}_k+\mu_k}{2}\right)\dfrac{\Delta W_i}{\Delta t},
\end{cases}
$$

where $\Delta t\xi_k = \tau^{-1}(\tilde{g}_k^{-1}g_k)$, $\Delta t\tilde{\xi}_k = \tau^{-1}(g_{k-1}^{-1}\tilde{g}_k)$, *and* $\mathrm{Ad}^*_g : \mathfrak{g}^* \to \mathfrak{g}^*$ *is the coadjoint action.*

Proof. We compute the derivative of the action as follows

$$
\mathrm{d}S_d(c_d)\cdot\delta c_d =
$$

$$
\sum_{k=1}^{N-1}\left[\left\langle \frac{1}{\Delta t}\mathrm{Ad}^*_{g_{k-1}^{-1}\tilde{g}_k}\left[\mathrm{d}_{\tau^{-1}(g_{k-1}^{-1}\tilde{g}_k)}\tau^{-1}\right]^*\tilde{\mu}_k - \frac{1}{\Delta t}\left[\mathrm{d}_{\tau^{-1}(\tilde{g}_k^{-1}g_k)}\tau^{-1}\right]^*\mu_k - \tilde{g}_k^{-1}\frac{\delta h_k}{\delta g}, \tilde{g}_k^{-1}\delta\tilde{g}_k \right\rangle\right.
$$

$$
+ \left\langle \frac{1}{\Delta t}\tau^{-1}(g_{k-1}^{-1}\tilde{g}_k) - \frac{1}{2}\frac{\delta h_k}{\delta\mu}, \delta\tilde{\mu}_k \right\rangle + \left\langle \frac{1}{\Delta t}\tau^{-1}(\tilde{g}_k^{-1}g_k) - \frac{1}{2}\frac{\delta h_k}{\delta\mu}, \delta\mu_k \right\rangle
$$

$$
\left. + \left\langle \frac{1}{\Delta t}\mathrm{Ad}^*_{\tilde{g}_k^{-1}g_k}\left[\mathrm{d}_{\tau^{-1}(\tilde{g}_k^{-1}g_k)}\tau^{-1}\right]^*\mu_k - \frac{1}{\Delta t}\left[\mathrm{d}_{\tau^{-1}(g_k^{-1}\tilde{g}_{k+1})}\tau^{-1}\right]^*\tilde{\mu}_{k+1}, g_k^{-1}\delta g_k \right\rangle\right]
$$

$$
+ \left\langle \frac{1}{\Delta t}\mathrm{Ad}^*_{g_{N-1}^{-1}\tilde{g}_N}\left[\mathrm{d}_{\tau^{-1}(g_{N-1}^{-1}\tilde{g}_N)}\tau^{-1}\right]^*\tilde{\mu}_N - \frac{1}{\Delta t}\left[\mathrm{d}_{\tau^{-1}(\tilde{g}_N^{-1}g_N)}\tau^{-1}\right]^*\mu_N - \tilde{g}_N^{-1}\frac{\delta h_N}{\delta g}, \tilde{g}_N^{-1}\delta\tilde{g}_N \right\rangle
$$

$$
+ \left\langle \frac{1}{\Delta t}\tau^{-1}(g_{N-1}^{-1}\tilde{g}_N) - \frac{1}{2}\frac{\delta h_N}{\delta\mu}, \delta\tilde{\mu}_N \right\rangle + \left\langle \frac{1}{\Delta t}\tau^{-1}(\tilde{g}_N^{-1}g_N) - \frac{1}{2}\frac{\delta h_N}{\delta\mu}, \delta\mu_N \right\rangle
$$

$$
+ \left\langle \frac{1}{\Delta t}\mathrm{Ad}^*_{\tilde{g}_N^{-1}g_N}\left[\mathrm{d}_{\tau^{-1}(\tilde{g}_N^{-1}g_N)}\tau^{-1}\right]^*\mu_N, g_N^{-1}\delta g_N \right\rangle - \left\langle \frac{1}{\Delta t}\left[\mathrm{d}_{\tau^{-1}(g_0^{-1}\tilde{g}_1)}\tau^{-1}\right]^*\tilde{\mu}_1, g_0^{-1}\delta g_0 \right\rangle,
$$

$$
(14)
$$

where we defined

$$
\mathsf{h}_k := h\left(\tilde{g}_k, \frac{\tilde{\mu}_k+\mu_k}{2}\right) + \sum_i h_i\left(\tilde{g}_k, \frac{\tilde{\mu}_k+\mu_k}{2}\right)\frac{\Delta W_i}{\Delta t}.
$$

By collecting the terms proportional to the variations and defining $\Delta t\xi_k = \tau^{-1}(\tilde{g}_k^{-1}g_k)$, $\Delta t\tilde{\xi}_k = \tau^{-1}(g_{k-1}^{-1}\tilde{g}_k)$ we get the stochastic midpoint method. $\square$

Remark 2. In the discrete action functional, we can also evaluate the trivialized Hamiltonians at $(g_k\tau(\frac{1}{2}\tau^{-1}(g_k^{-1}\tilde{g}_{k+1})), \tilde{\mu}_{k+1})$, the Lie group analogue to $(\frac{q_k+\tilde{q}_{k+1}}{2}, \tilde{p}_{k+1})$. The resulting integrator will have a more cumbersome form. The reason is simply that the Lie algebra $\mathfrak{g}$ is a vector space itself, so it is easy to evaluate the 'mean' of two given values. While we need the retraction map to express the mean of two elements in the Lie group G.

3 Symplecticity, Discrete Momentum Maps, and Coadjoint Orbits

The endpoint terms in the derivative of the discrete action functional in (14) naturally yield the definition of the two one-forms $\theta^\pm \in \Omega^1(M \times M)$ with $M = (G \times \mathfrak{g}^* \oplus \mathfrak{g}^*) \times G$:

$$\theta^- = \frac{1}{\Delta t}\tilde{\nu}_1 \cdot g_0^{-1}\mathrm{d}g_0$$

$$\theta^+ = \left(\frac{1}{\Delta t}(\mathrm{Ad}^*_{g_0^{-1}\tilde{g}_1}\tilde{\nu}_1 - \nu_1) - \tilde{g}_1^{-1}\frac{\delta h_1}{\delta g}\right) \cdot \tilde{g}_1^{-1}\mathrm{d}\tilde{g}_1 + \frac{1}{\Delta t}\mathrm{Ad}^*_{\tilde{g}_1^{-1}g_1}\nu_1 \cdot g_1^{-1}\mathrm{d}g_1$$

$$+ \left(\frac{1}{\Delta t}\tau^{-1}(g_0^{-1}\tilde{g}_1) - \frac{1}{2}\frac{\delta h_1}{\delta \mu}\right) \cdot \mathrm{d}\tilde{\mu}_1 + \left(\frac{1}{\Delta t}\tau^{-1}(\tilde{g}_1^{-1}g_1) - \frac{1}{2}\frac{\delta h_1}{\delta \mu}\right) \cdot \mathrm{d}\mu_1,$$
(15)

where $(\tilde{g}_0, \tilde{\mu}_0, \mu_0, g_0, \tilde{g}_1, \tilde{\mu}_1, \mu_1, g_1) \in M \times M$ and we defined

$$\nu_k = [\mathrm{d}_{\tau^{-1}(\tilde{g}_k^{-1}g_k)}\tau^{-1}]^*\mu_k \quad \text{and} \quad \tilde{\nu}_k = [\mathrm{d}_{\tau^{-1}(g_{k-1}^{-1}\tilde{g}_k)}\tau^{-1}]^*\tilde{\mu}_k. \tag{16}$$

It can be checked that $\mathrm{d}\theta^+ = \mathrm{d}\theta^-$ so that the two-form $\omega := -\mathrm{d}\theta^+ = -\mathrm{d}\theta^-$ can be defined, which is noted to be independent of the Hamiltonians.

Proposition 3. *The scheme* (13) *is symplectic.*

Proof. Assuming the Hamiltonian is hyperregular, the scheme induces a discrete flow on a submanifold of $M \times M$, $M = (G \times \mathfrak{g}^* \oplus \mathfrak{g}^*) \times G$, which is in bijection with $G \times \mathfrak{g}^*$. Denoting by $\iota : G \times \mathfrak{g}^* \to M \times M$ the associated injection, by $F_k : G \times \mathfrak{g}^* \to G \times \mathfrak{g}^*$ the discrete stochastic k-step flow, $F_k(g_0, \mu_0) = (g_k, \mu_k)$, and by $\widehat{\mathcal{S}}_d$ the discrete action functional evaluated on the space of solutions, we have from (14) and (15) $\mathrm{d}\widehat{\mathcal{S}}_d = F_N^*(\iota^*\theta^+) - \iota^*\theta^-$, where ι^* denotes the pull-back of forms $\theta^\pm \in \Omega^1(M \times M)$ according to the inclusion ι. Taking the exterior derivative we get $F_N^*(\iota^*\omega) = \iota^*\omega$ which proves the result. □

Proposition 4 (Discrete momentum map and Noether theorem). *Let $K \subset G$ be a subgroup, let $\mathfrak{k}$ be its Lie algebra, and assume that the Hamiltonians h and h_i are left K-invariant, i.e. $h(kg, \mu) = h(g, \mu)$ and $h_i(kg, \mu) = h_i(g, \mu)$ for all $k \in K$, $g \in G$, and $\mu \in \mathfrak{g}^*$. Then the discrete momentum map is given by*

$$J_d(\tilde{g}_0, \tilde{\mu}_0, g_0, \mu_0, \tilde{g}_1, \tilde{\mu}_1, g_1, \mu_1) = \frac{1}{\Delta t}i_{\mathfrak{k}}^*\left(\mathrm{Ad}^*_{g_0^{-1}}\left([\mathrm{d}_{\tau^{-1}(g_0^{-1}\tilde{g}_1)}\tau^{-1}]^*\tilde{\mu}_1\right)\right) \in \mathfrak{k}^*,$$
(17)

where $i_{\mathfrak{k}}^* : \mathfrak{g}^* \to \mathfrak{k}^*$ is the dual map to the Lie algebra inclusion. It is preserved by the discrete stochastic flow of (13).

Proof. We note that if h and h_i are left K-invariant, then the discrete stochastic action functional $\mathcal{S}_d$ is also left K-invariant: $\mathcal{S}_d(kc_d) = \mathcal{S}_d(c_d)$, where the left action of K on the discrete curve of random variables $c_d = \{(\tilde{g}_k, \tilde{\mu}_k, \mu_k, g_k) \mid k = 0, ..., N\}$ is defined by $kc_d = \{(k\tilde{g}_k, \tilde{\mu}_k, \mu_k, kg_k) \mid k = 0, ..., N\}$. Computing the derivative of the map $\mathsf{k} \to \mathcal{S}_d(\mathsf{k}c_d)$ at the identity in some direction $\zeta \in \mathfrak{k}$ and evaluating the result at a solution c_d of (13) we get, from (14) the expression

$$d\mathcal{S}_d(c_d) \cdot \zeta c_d =$$
$$+ \langle \frac{1}{\Delta t} \mathrm{Ad}^*_{\tilde{g}_{N-1}^{-1}\tilde{g}_N} \left[d_{\tau^{-1}(g_{N-1}^{-1}\tilde{g}_N)}\tau^{-1} \right]^* \tilde{\mu}_N - \frac{1}{\Delta t} \left[d_{\tau^{-1}(\tilde{g}_N^{-1}g_N)}\tau^{-1} \right]^* \mu_N - \tilde{g}_N^{-1} \frac{\delta h_N}{\delta g}, \tilde{g}_N^{-1}\zeta\tilde{g}_N \rangle$$
$$+ \langle \frac{1}{\Delta t} \mathrm{Ad}^*_{\tilde{g}_N^{-1}g_N} \left[d_{\tau^{-1}(\tilde{g}_N^{-1}g_N)}\tau^{-1} \right]^* \mu_N, g_N^{-1}\zeta g_N \rangle - \langle \frac{1}{\Delta t} \left[d_{\tau^{-1}(g_0^{-1}\tilde{g}_1)}\tau^{-1} \right]^* \tilde{\mu}_1, g_0^{-1}\zeta g_0 \rangle$$
$$= \frac{1}{\Delta t} \langle \mathrm{Ad}^*_{\tilde{g}_{N-1}^{-1}} \left[d_{\tau^{-1}(g_{N-1}^{-1}\tilde{g}_N)}\tau^{-1} \right]^* \tilde{\mu}_N - \mathrm{Ad}^*_{g_0^{-1}} \left[d_{\tau^{-1}(g_0^{-1}\tilde{g}_1)}\tau^{-1} \right]^* \tilde{\mu}_1, \zeta \rangle,$$
(18)

where we used that $\langle \tilde{g}_N^{-1} \frac{\delta h_N}{\delta g}, \tilde{g}_N^{-1}\zeta\tilde{g}_N \rangle$ vanishes from the K-invariance of the Hamiltonians. From the K-invariance of the action we have $d\mathcal{S}_d(c_d) \cdot \zeta c_d = 0$, for all $\zeta \in \mathfrak{k}$, hence (18) directly shows that the discrete momentum map is given by the expression (17) and is preserved by solution of (13).   □

Proposition 5. *When the Hamiltonians are G-invariant, the scheme preserves the coadjoint orbits in* $\mathfrak{g}^*$

Proof. In the G-invariant case, we have $\delta h/\delta g = \delta h_i/\delta g = 0$ in (13). By combining the first two equations in (13) and using the definitions (16), we get

$$\tilde{\nu}_{k+1} = \mathrm{Ad}^*_{g_{k-1}^{-1}g_k} \tilde{\nu}_k \quad \text{and} \quad \nu_{k+1} = \mathrm{Ad}^*_{\tilde{g}_k^{-1}\tilde{g}_{k+1}} \nu_k.$$

Hence, the scheme preserves the coadjoint orbits $\mathcal{O} = \{\mathrm{Ad}^*_g \mu_0 \mid g \in G\} \subset \mathfrak{g}^*$ in the sense that $\nu_0 \in \mathcal{O} \to \nu_k \in \mathcal{O}$ similarly for $\tilde{\nu}_k$.   □

4   Numerical Examples

As a classical example, we consider a rigid body, with reduced Hamiltonian

$$h(R, \Pi) = \frac{1}{2}\Pi \cdot (\mathbb{I}^{-1}\Pi),$$

where $R \in SO(3)$ is the attitude of the body, $\hat{\Pi} \in \mathfrak{so}^*(3)$ the body angular momentum, and $\mathbb{I}$ the spatial moment of inertia tensor.

In the following we propose that the stochastic Hamiltonians have the form $h_i(R, \Pi) = \Pi \cdot \chi_i$ for $\chi_i \in \mathbb{R}^3$, and we take τ to be the Cayley transform $\tau(A) = \left(I - \frac{A}{2}\right)^{-1}\left(I + \frac{A}{2}\right)$. Its trivialized derivative reads $\left[d_{\Delta t\xi}\tau^{-1} \right]^* \Pi = \Pi + \frac{\Delta t}{2}\xi \times \Pi - \frac{(\Delta t)^2}{4}(\xi \cdot \Pi)\xi$. Then, the stochastic variational integrator derived above reads:

$$\left(\frac{\tilde{\Pi}_k}{\Delta t} - \frac{\tilde{\xi}_k \times \tilde{\Pi}_k}{2} - \frac{\Delta t}{4}(\tilde{\xi}_k \cdot \tilde{\Pi}_k)\tilde{\xi}_k \right) = \left(\frac{\Pi_k}{\Delta t} + \frac{\xi_k \times \Pi_k}{2} - \frac{\Delta t}{4}(\xi_k \cdot \Pi_k)\xi_k \right),$$
(19)

$$\left(\frac{\Pi_k}{\Delta t} - \frac{\xi_k \times \Pi_k}{2} - \frac{\Delta t}{4}(\xi_k \cdot \Pi_k)\xi_k \right) = \left(\frac{\tilde{\Pi}_{k+1}}{\Delta t} + \frac{\tilde{\xi}_{k+1} \times \tilde{\Pi}_{k+1}}{2} - \frac{\Delta t}{4}(\tilde{\xi}_{k+1} \cdot \tilde{\Pi}_{k+1})\tilde{\xi}_{k+1} \right),$$
(20)

$$\tilde{\xi}_k = \xi_k = \frac{\Delta t}{2}\mathbb{I}^{-1}\frac{\tilde{\Pi}_k + \Pi_k}{2} + \sum_i \frac{1}{2}\chi_i \Delta W_k.$$
(21)

Recall that by definition, $\Delta t \xi_k = \tau^{-1}(\tilde{R}_k^{-1} R_k)$ and $\Delta t \tilde{\xi}_k = \tau^{-1}(R_{k-1}^{-1} \tilde{R}_k)$. Thus the reconstruction equations for the rotation matrix R are: $R_k = \tilde{R}_k \tau(\Delta t \xi_k)$ and $\tilde{R}_k = R_{k-1} \tau(\Delta t \tilde{\xi}_k)$.

The momentum map obtained in (17) has the following form for $K = SO(3)$:

$$\frac{1}{\Delta t} \mathrm{Ad}^*_{R_0^{-1}} \left([\mathrm{d}_{\tau^{-1}(R_0^{-1} \tilde{R}_1)} \tau^{-1}]^* \tilde{\Pi}_1 \right) = \frac{1}{\Delta t} \mathrm{Ad}^*_{R_0^{-1}} \tilde{\nu}_1 = R_0 \left(\frac{\tilde{\Pi}_1}{\Delta t} + \frac{\tilde{\xi}_1 \times \tilde{\Pi}_1}{2} - \frac{\Delta t}{4} (\tilde{\xi}_1 \cdot \tilde{\Pi}_1) \tilde{\xi}_1 \right).$$

This quantity can be directly shown to be conserved by adding $\mathrm{Ad}^*_{\Delta t \xi_k}$ (19) to (20) and by noting that $\tau(-\Delta t \xi_k) \tau(-\Delta t \tilde{\xi}_k) = R_k^{-1} R_{k-1}$.

The orbit $\mathcal{O} = \{ \mathrm{Ad}^*_R \tilde{\nu}_0 = R^{-1} \tilde{\nu}_0 \mid R \in SO(3) \}$ is the sphere in $\mathbb{R}^3$ of radius $|\tilde{\nu}_0|$. The scheme preserves the coajoint orbit means that the quantity $|\tilde{\nu}_k|$ is preserved.

In the following experiments, we take $\mathbb{I} = \mathrm{diag}(1, 2, 3)$, $\xi_1 = (0.02, 0, 0)$, $\xi_2 = (0, 0.02, 0)$, $\xi_3 = (0, 0, 0.02)$ and $\Delta t = 0.01$. Starting from the position $\tilde{\Pi}_0 = (0.5, 2, 0)$, the plots show that the evolution of the quantity $\tilde{\nu}$ (Fig. 1). The quantity $\tilde{\nu}$ adheres to the coadjoint orbit in both deterministic and stochastic experiments. The momentum map $\frac{1}{\Delta t} \mathrm{Ad}^*_{R_0^{-1}} \tilde{\nu}_1$ is conserved up to order $E - 10$. The energy is also conserved in the deterministic case up to order $E - 15$.

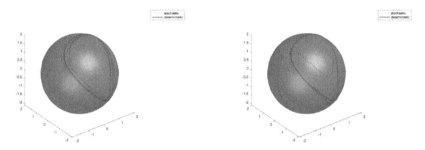

Fig. 1: The first experiment has the stochastic path close to the deterministic one, while the second experiment has the stochastic path surpassing the saddle point.

References

1. Barbaresco, F., Gay-Balmaz, F.: Lie group Cohomology and (Multi)Symplectic integrators: new geometric tools for lie group machine learning based on Souriau geometric statistical mechanics. Entropy **22**, 498 (2020)
2. Bismut, J.: Mécanique aléatoire. In: Bennequin, P. (ed.) Ecole d'Été de Probabilités de Saint-Flour X - 1980. Lecture Notes in Mathematics, vol. 929, pp. 1–100. Springer, Berlin (1982)

3. Bou-Rabee, N., Marsden, J.E.: Hamilton-Pontryagin integrators on Lie groups. Found. Comput. Math. **9**, 197–219 (2009)
4. Bou-Rabee, N., Owhadi, H.: Stochastic variational integrators. IMA J. Numer. Anal. **29**, 421–443 (2009)
5. Holm, D.D.: Variational principles for stochastic fluid dynamics. Proc. R. Soc. Lond. Ser. A Math. Phys. Eng. Sci. **471**, 2176 (2015)
6. Gay-Balmaz, F., Holm, D.D.: Stochastic geometric models with non-stationary spatial correlations in Lagrangian fluid flows. J. Nonlin. Sci. **28**, 873–904 (2018)
7. Gay-Balmaz, F., Holm, D.D.: Predicting uncertainty in geometric fluid mechanics. Disc. Cont. Dyn. Syst. Ser. S **13**, 1229–1242 (2020)
8. Holm, D.D., Tyranowski, T.M.: Stochastic discrete Hamiltonian variational integrators. BIT Numer. Math. **58**(4), 1009–1048 (2018). https://doi.org/10.1007/s10543-018-0720-2
9. Iserles, A., Munthe-Kaas, H.Z., Norsett, S.P., Zanna, A.: Lie-group methods. Acta Numer. **9**, 215–365 (2000)
10. Lazaro-Cami, J.A., Ortega, J.P.: Stochastic Hamiltonian dynamical systems. Rep. Math. Phys. **61**(1), 65–122 (2008)
11. Milstein, G.N., Repinand, Y.M., Tretyakov, M.V.: Numerical methods for stochastic systems preserving symplectic structure. SIAM J. Numer. Anal. **40**, 1583–1604 (2002)
12. Ma, Q., Ding, D., Ding, X.: Symplectic conditions and stochastic generating functions of stochastic Runge-Kutta methods for stochastic Hamiltonian systems with multiplicative noise. Appl. Math. Comput. **219**(2), 635–643 (2012)
13. Marsden, J.E., West, M.: Discrete mechanics and variational integrators. Acta Numer. **10**, 357–514 (2001)

Hamiltonian Variational Formulation for Non-simple Thermodynamic Systems

Hiroaki Yoshimura[1][⊠][iD] and François Gay-Balmaz[2][iD]

[1] School of Science and Engineering, Waseda University, 3–4–1, Okubo, Shinjuku, Tokyo, Japan
yoshimura@waseda.jp
[2] Laboratoire de Météorologie Dynamique École Normale Supérieure/CNRS, Paris, France
francois.gay-balmaz@lmd.ens.fr

Abstract. A Lagrangian variational formulation for nonequilibrium thermodynamics was proposed in [2–4]. In this paper, we develop a Hamiltonian analogue of the Lagrangian variational formulation for non-simple thermodynamic systems [6,8]. We start with the Lagrangian variational formulation for simple systems, where the Lagrangian is degenerate. Under some assumption, we show how to construct the Hamiltonian variational formulation for nonequilibrium thermodynamics for the simple case. Then, we extend it to the case of adiabatically closed non-simple systems, in which there exists several entropy variables in addition to the mechanical variables. Finally, we illustrate our theory of the Hamiltonian variational formulation by an example of the adiabatic piston problem.

Keywords: Hamiltonian variational formulation · Nonequilibrium thermodynamics · Non-simple systems · degenerate Lagrangians

1 The Lagrangian Variational Formulation for Simple Thermodynamic Systems

The Lagrangian variational formulation of nonequilibrium thermodynamics is a generalization of Hamilton's variational principle in mechanics to the case in which there exist nonlinear nonholonomic constraints associated to the entropy production due to irreversible processes. We start below with a short review of this Lagrangian variational formulation by following [2–4].

1.1 Setting for Simple Thermodynamic Systems

Consider a simple thermodynamic system with Lagrangian $L = L(q, \dot{q}, S)$ on $TQ \times \mathbb{R}$, where TQ denotes the tangent bundle of an n-dimensional mechanical configuration space Q and $\mathbb{R}$ is the space of entropy. We write $q = (q^1, ..., q^n) \in Q$ the mechanical configuration variables and $S \in \mathbb{R}$ the entropy. Suppose that the thermodynamic system is subject to external forces $F^{\text{ext}} : TQ \times \mathbb{R} \to T^*Q$ and that there exists an internal friction force $F^{\text{fr}} : TQ \times \mathbb{R} \to T^*Q$ which induces an irreversible process to the system. Here T^*Q is the cotangent bundle of Q.

F. Nielsen and F. Barbaresco (Eds.): GSI 2023, LNCS 14072, pp. 221–230, 2023.
https://doi.org/10.1007/978-3-031-38299-4_24

1.2   The Lagrangian Variational Formulation for Simple Systems

The Lagrangian variational formulation consists in finding a critical curve $(q(t), S(t)) \in Q \times \mathbb{R}$, $t \in [a, b] \subset \mathbb{R}$ for the variational condition

$$\delta \int_a^b L(q, \dot{q}, S)dt + \int_a^b \left\langle F^{\text{ext}}(q, \dot{q}, S), \delta q \right\rangle dt = 0, \tag{1}$$

with respect to all chosen variations $\delta q(t)$ and $\delta S(t)$ that are subject to the *variational constraint*

$$\frac{\partial L}{\partial S}(q, \dot{q}, S)\delta S = \left\langle F^{\text{fr}}(q, \dot{q}, S), \delta q \right\rangle, \tag{2}$$

with $\delta q(t_1) = \delta q(t_2) = 0$, and where the following *nonlinear phenomenological constraint* is imposed on the curve $(q(t), S(t))$ as

$$\frac{\partial L}{\partial S}(q, \dot{q}, S)\dot{S} = \left\langle F^{\text{fr}}(q, \dot{q}, S), \dot{q} \right\rangle. \tag{3}$$

The variational formulation (1)-(3) yields the following evolution equations

$$\frac{d}{dt}\frac{\partial L}{\partial \dot{q}} - \frac{\partial L}{\partial q} = F^{\text{fr}} + F^{\text{ext}}, \quad \frac{\partial L}{\partial S}\dot{S} = \left\langle F^{\text{fr}}, \dot{q} \right\rangle. \tag{4}$$

1.3   The Fundamental Laws in Nonequilibrium Thermodynamics

The First Law. Recall that the Lagrangian energy is defined as

$$E(q, \dot{q}, S) = \left\langle \frac{\partial L}{\partial \dot{q}}(q, \dot{q}, S), \dot{q} \right\rangle - L(q, \dot{q}, S)$$

and a direct computation shows that the 1st law of thermodynamics holds along the solution curve $(q(t), S(t))$ of (4) as

$$\frac{d}{dt}E = \left\langle F^{\text{ext}}, \dot{q} \right\rangle.$$

The Second Law. Here we assume that the friction force is given by $F^{\text{fr}}(q, \dot{q}, S) = -R(q, S)\dot{q}$, where $R(q, S) = [R_{ij}(q, S)], i, j = 1, ..., n$ is the friction coefficient that is experimentally determined and where the symmetric part of the matrix $R = [R_{ij}]$ is positive semi-definite. From the second equation of (4), it follows that the 2nd law is recovered as

$$I = \frac{1}{T}\left\langle R\dot{q}, \dot{q} \right\rangle \geq 0,$$

where $I = \dot{S}$ is the internal entropy production rate and $T := -\frac{\partial L}{\partial S}$ is the temperature of the system [1].

2   Hamiltonian Variational Formulation for Simple Systems

The Lagrangian $L = L(q, \dot{q}, S)$ on $TQ \times \mathbb{R}$ is always degenerate with respect to the thermodynamic variable S, which implies that there exist some difficulties to make the Legendre transform to the Hamiltonian side in general. Nevertheless, by assuming that $L(q, \dot{q}, S)$ is nondegenerate in terms of mechanical state variables $(q, \dot{q}) \in TQ$, we will show that the Hamiltonian $H = H(q, p, S)$ can be constructed by introducing the partial Legendre transform $\mathbb{F}L : TQ \times \mathbb{R} \to T^*Q \times \mathbb{R}$, for each fixed S, between the mechanical variables $(q, \dot{q}) \in TQ$ and $(q, p) \in T^*Q$.

2.1   Hamiltonian Setting for Simple Thermodynamic Systems

The Partial Legendre Transform. Assume here that L is hyperregular with respect to the mechanical part $(q, v) \in TQ$, which means that the partial Legendre transform $\mathbb{F}L : TQ \times \mathbb{R} \to T^*Q \times \mathbb{R}$ defined by, for each fixed $S \in \mathbb{R}$,

$$(q, \dot{q}, S) \mapsto \left(q, \frac{\partial L}{\partial \dot{q}}(q, \dot{q}, S), S \right),$$

is a diffeomorphism. Hence we can define a *Hamiltonian* $H = H(q, p, S)$ on $T^*Q \times \mathbb{R}$ by

$$H(q, p, S) = \langle p, \dot{q} \rangle - L(q, \dot{q}, S),$$

where $\dot{q}$ is uniquely determined from (q, p, S) by the condition $\frac{\partial L}{\partial \dot{q}}(q, \dot{q}, S) = p$. Formally, the Hamiltonian is defined, for each fixed $S \in \mathbb{R}$, by $H = E \circ (\mathbb{F}L)^{-1}$, where $E(q, \dot{q}, S) = \left\langle \frac{\partial L}{\partial \dot{q}}(q, \dot{q}, S), \dot{q} \right\rangle - L(q, \dot{q}, S)$.

Hamiltonian Constraints in Thermodynamics. On the Hamiltonian side, we define the external force $\mathcal{F}^{\text{ext}} : T^*Q \times \mathbb{R} \to T^*Q$ and the friction force $\mathcal{F}^{\text{fr}} : T^*Q \times \mathbb{R} \to T^*Q$ respectively by $\mathcal{F}^{\text{ext}} \circ \mathbb{F}L = F^{\text{ext}}$ and $\mathcal{F}^{\text{fr}} \circ \mathbb{F}L = F^{\text{fr}}$. Let us further introduce the thermodynamic configuration manifold $\mathcal{Q} := Q \times \mathbb{R}$. Then, the variational constraint is described by the set

$$\mathscr{C}_V = \left\{ (q, S, p, \Lambda, \delta q, \delta S) \in T^*\mathcal{Q} \times_{\mathcal{Q}} T\mathcal{Q} \;\middle|\; -\frac{\partial H}{\partial S}(q, p, S)\delta S = \langle \mathcal{F}^{\text{fr}}(q, p, S), \delta q \rangle \right\},$$

where $(q, S) \in \mathcal{Q}$, $(p, \Lambda) \in T^*_{(q,S)}\mathcal{Q}$, and $(\delta q, \delta S) \in T_{(q,S)}\mathcal{Q}$. Note that the temperature of the system is now given by $\mathcal{T}(q, p, S) = \frac{\partial H}{\partial S}(q, p, S)$ since it is defined by $\mathcal{T}(q, p, S) := -\frac{\partial L}{\partial S}(q, \dot{q}, S)$, where $\dot{q}$ is uniquely determined from the condition $p = \frac{\partial L}{\partial \dot{q}}(q, \dot{q}, S)$. By hypothesis $\frac{\partial H}{\partial S}(q, p, S) \neq 0$, hence $\mathscr{C}_V$ is a submanifold of $T^*\mathcal{Q} \oplus T\mathcal{Q}$ of codimension one.

Then, the phenomenological constraint is given by

$$(\dot{q}, \dot{S}) \in \mathscr{C}_V(q, S, p, \Lambda) := \left(\{(q, S, p, \Lambda)\} \times T_{(q,S)}\mathcal{Q} \right) \cap \mathscr{C}_V \subset T_{(q,S)}\mathcal{Q}$$

which is locally given by

$$-\frac{\partial H}{\partial S}(q, p, S)\dot{S} = \langle \mathcal{F}^{\text{fr}}(q, p, S), \dot{q} \rangle.$$

2.2 Hamiltonian Variational Formulation for Simple Systems

The Hamiltonian variational formulation for simple adiabatically closed systems is shown in the following proposition [9].

Proposition 1. *Suppose that a curve* $(q(t), p(t), S(t))$ *on* $T^*Q \times \mathbb{R}$ *is critical for the variational condition*

$$\delta \int_{t_1}^{t_2} \Big[\langle p, \dot{q} \rangle - H(q, p, S) \Big] dt + \int_{t_0}^{t_1} < \mathcal{F}^{\mathrm{ext}}(q, p, S), \delta q > dt = 0$$

for $(\delta q, \delta p, \delta S)$ *satisfying the variational constraint*

$$-\frac{\partial H}{\partial S}(q, p, S)\delta S = \big\langle \mathcal{F}^{\mathrm{fr}}(q, p, S), \delta q \big\rangle$$

with $\delta q(t_1) = \delta q(t_2) = 0$, *and that the curve* $(q(t), p(t), S(t))$ *is subject to the phenomenological constraint*

$$-\frac{\partial H}{\partial S}(q, p, S)\dot{S} = \big\langle \mathcal{F}^{\mathrm{fr}}(q, p, S), \dot{q} \big\rangle .$$

Then, $(q(t), p(t), S(t))$ *satisfies the Hamilton-d'Alembert equations:*

$$\begin{cases} \dot{q} = \dfrac{\partial H}{\partial p}(q, p, S), \quad \dot{p} + \dfrac{\partial H}{\partial q}(q, p, S) = \mathcal{F}^{\mathrm{ext}}(q, p, S) + \mathcal{F}^{\mathrm{fr}}(q, p, S), \\ -\dfrac{\partial H}{\partial S}(q, p, S)\dot{S} = \big\langle \mathcal{F}^{\mathrm{fr}}(q, p, S), \dot{q} \big\rangle . \end{cases} \tag{5}$$

2.3 The Fundamental Laws in the Hamiltonian Description for Simple Systems

The First Law. We can easily check that along the solution curve $(q(t), p(t), S(t))$ of (5), we have the first law

$$\frac{d}{dt} H(q(t), p(t), S(t)) = P_W^{\mathrm{ext}},$$

where $P_W^{\mathrm{ext}} = \langle \mathcal{F}^{\mathrm{ext}}(q, p, S), \dot{q} \rangle$ is the mechanical power associated with the work done on the system.

The Second Law. Recall that the temperature is given by $T = \frac{\partial H}{\partial S}$ so it follows from the third equation in (5) that

$$T\dot{S} = -\big\langle \mathcal{F}^{\mathrm{fr}}(q, p, S), \dot{q} \big\rangle .$$

From the second law we have $\langle \mathcal{F}^{\mathrm{fr}}(q, p, S), \dot{q} \rangle \leq 0$ for all $(q, p, S) \in T^*Q \times \mathbb{R}$, i.e. the friction force $\mathcal{F}^{\mathrm{fr}}$ is dissipative. For the linear phenomenological relation, we have $\mathcal{F}_i^{\mathrm{fr}} = -\lambda_{ij} \frac{\partial H}{\partial p_j}$, where λ_{ij}, $i, j = 1, ..., n$ are functions of (q, S) and where the symmetric part of the matrix $\lambda = [\lambda_{ij}]$ is positive semi-definite.

3 Hamiltonian Variational Formulation for Non-simple Systems

3.1 Fundamental Setting of Non-simple Systems

Now we shall extend the Hamiltonian variational formulation to the case of non-simple systems $\Sigma = \cup_{A=1}^{P} \Sigma_A$, in which each Σ_A, $A = 1, ..., P$ is assumed to be a simple system. In particular, non-simple systems have several entropy variables.

Setting for Non-simple Systems. Consider an adiabatically closed non-simple system Σ. We assume that there exist irreversible processes due to internal friction and heat conduction between subsystems. Let Q be the mechanical configuration manifold of Σ with mechanical variables $q = (q^1, ..., q^n) \in Q$. Let $F^{\text{ext}\to A} : TQ \times \mathbb{R}^P \to T^*Q$ be an external force that acts on each Σ_A and define the total exterior force by $F^{\text{ext}} = \sum_{A=1}^{P} F^{\text{ext}\to A}$. Associated to the irreversible processes, we consider the friction forces $F^{\text{fr}(A)} : TQ \times \mathbb{R}^P \to T^*Q$ for each subsystem Σ_A, which yields an entropy production in Σ_A. Associated to the internal heat exchange between Σ_A and Σ_B, we define the fluxes J_{AB} with $J_{AB} = J_{BA}$. Following the general variational approach to nonequilibrium thermodynamics,[2–4], new variables called *thermal displacements* Γ^A, $A = 1, ..., P$ are defined such that its time rate $\dot{\Gamma}^A$ is the temperature of the subsystem Σ_A. Furthermore, the internal entropy variables Σ_A are introduced, associated to the total entropy production for subsystem A.

Thus, the thermodynamic configuration space is given by $\mathcal{Q} = Q \times V$, with $(q, S_A, \Gamma^A, \Sigma_A) \in \mathcal{Q}$, where $V = \mathbb{R}^P \times \mathbb{R}^P \times \mathbb{R}^P$ is the thermodynamic space with thermodynamic variables $(S_A, \Gamma^A, \Sigma_A) \in V$.

The Partial Legendre Transform. Let $L = L(q, \dot{q}, S_1, ..., S_P)$ be a given Lagrangian on $TQ \times \mathbb{R}^P$, which is assumed to be hyperregular with respect to the mechanical part $(q, \dot{q}) \in TQ$. Then the partial Legendre transform $\mathbb{F}L : TQ \times \mathbb{R}^P \to T^*Q \times \mathbb{R}^P$ given by

$$(q, \dot{q}, S_1, ..., S_P) \mapsto \left(q, \frac{\partial L}{\partial \dot{q}}(q, \dot{q}, S_1, ..., S_P), S_1, ..., S_P \right)$$

is a diffeomorphism. Therefore we can define a *Hamiltonian* $H = H(q, p, S_1, ..., S_P)$ on $T^*Q \times \mathbb{R}^P$ by

$$H(q, p, S_1, ..., S_P) = \langle p, \dot{q} \rangle - L(q, \dot{q}, S_1, ..., S_P),$$

where $\dot{q} \in T_qQ$ is uniquely determined from $(q, p, S_1, ..., S_P) \in T^*Q \times \mathbb{R}^P$ by the condition $\frac{\partial L}{\partial \dot{q}}(q, \dot{q}, S_1, ..., S_P) = p$.

Similarly, we can define the external force $\mathcal{F}^{\text{ext}\to A} : T^*Q \times \mathbb{R}^P \to T^*Q$ and the friction forces $\mathcal{F}^{\text{fr}(A)} : T^*Q \times \mathbb{R}^P \to T^*Q$ such that, for fixed $(S_1, ..., S_P) \in \mathbb{R}^P$,

$$\mathcal{F}^{\text{ext}\to A} \circ \mathbb{F}L = F^{\text{ext}\to A} \quad \text{and} \quad \mathcal{F}^{\text{fr}(A)} \circ \mathbb{F}L = F^{\text{fr}(A)},$$

and the total exterior force is given by $\mathcal{F}^{\text{ext}} = \sum_{A=1}^{P} \mathcal{F}^{\text{ext}\to A}$.

3.2   Hamiltonian Variational Formulation for Non-simple Systems

The Hamiltonian variational formulation for non-simple adiabatically closed systems is shown in the following proposition.

Proposition 2. *Suppose that the curves* $q(t), p(t),\ S_A(t),\ \Gamma^A(t),\ \Sigma_A(t),\ A = 1, ..., P$ *on* $T^*Q \times V$ *are critical for the variational condition*

$$\delta \int_a^b \left[\langle p, \dot{q} \rangle - H\left(q, p, S_1, ..., S_P\right) + \sum_{A=1}^P \dot{\Gamma}^A (S_A - \Sigma_A) \right] dt + \int_a^b \langle \mathcal{F}^{\text{ext}}, \delta q \rangle\, dt = 0,$$

with variations subject to the variational constraint

$$-\frac{\partial H}{\partial S_A} \delta \Sigma_A = \langle \mathcal{F}^{\text{fr}(A)}, \delta q \rangle + \sum_{B=1}^P J_{AB} \delta \Gamma^B, \ for\, A = 1, ..., P, \qquad (6)$$

with $\delta q(t_1) = \delta q(t_2) = 0$ *and* $\delta \Gamma^A(t_1) = \delta \Gamma^A(t_2) = 0,\ A = 1, ..., P,$ *and where the critical curves are subject to the phenomenological constraint*

$$-\frac{\partial H}{\partial S_A} \dot{\Sigma}_A = \langle \mathcal{F}^{\text{fr}(A)}, \dot{q} \rangle + \sum_{B=1}^P J_{AB} \dot{\Gamma}^B, \ for\, A = 1, ..., P. \qquad (7)$$

Then, the curves $q(t), p(t),\ S_A(t),\ \Gamma^A(t),\ \Sigma_A(t),\ A = 1, ..., P$ *satisfy the Hamilton-d'Alembert equations:*

$$\begin{cases} \dfrac{dq}{dt} = \dfrac{\partial H}{\partial p}, \quad \dfrac{dp}{dt} = -\dfrac{\partial H}{\partial q} + \sum_{A=1}^P \mathcal{F}^{\text{fr}(A)} + \mathcal{F}^{\text{ext}}, \\[2ex] -\dfrac{\partial H}{\partial S_A} \dot{S}_A = \langle \mathcal{F}^{\text{fr}(A)}, \dot{q} \rangle + \sum_{B=1}^P J_{AB} \left(\dfrac{\partial H}{\partial S_B} - \dfrac{\partial H}{\partial S_A} \right), \quad A = 1, ..., P. \end{cases} \qquad (8)$$

Proof. By direct computations, we get the following evolution equations:

$$\begin{cases} \dfrac{dq}{dt} = \dfrac{\partial H}{\partial p}, \quad \dfrac{dp}{dt} = -\dfrac{\partial H}{\partial q} + \sum_{A=1}^P \dfrac{\dot{\Gamma}^A}{\frac{\partial H}{\partial S_A}} \mathcal{F}^{\text{fr}(A)} + \mathcal{F}^{\text{ext}}, \\[2ex] -\dfrac{\partial H}{\partial S_A} + \dot{\Gamma}^A = 0, \quad \dot{S}_A - \dot{\Sigma}_A - \sum_{B=1}^P \dfrac{\dot{\Gamma}^A}{\frac{\partial H}{\partial S_A}} J_{BA} = 0, \quad A = 1, ..., P. \end{cases} \qquad (9)$$

Noting the third equation in (9), the temperature of Σ_A, i.e., T^A is given as

$$T^A := \dot{\Gamma}^A = \frac{\partial H}{\partial S_A}, \quad A = 1, ..., P.$$

The fourth equation in (9) becomes

$$\dot{S}_A = \dot{\Sigma}_A + \sum_{B=1}^P J_{BA}, \quad A = 1, ..., P. \qquad (10)$$

Using (7), we can finally get the evolution equations as in (8).

Note that the physical meaning of the phenomenological constraint in (7) is the power associated with the internal entropy production due to friction and heat conduction. Note also that in (10) $\dot{S}_A$ is the rate of entropy change, while $\dot{\Sigma}_A$ is interpreted as the internal entropy production for the subsystem Σ_A, so that we must have $\dot{\Sigma}_A \geq 0$. We refer to [4,5] for further explanation of the difference between the two entropy variables S_A and Σ_A, as well as for the link with Prigogine's equations. In particular, noting that $\dot{S}_A = \dot{\Sigma}_A + \sum_{B=1}^{P} J_{BA}$, we get

$$\dot{S}_A = -\frac{1}{T^A}\langle \mathcal{F}^{\mathrm{fr}(A)}, \dot{q}\rangle + \sum_{B=1}^{P} J_{AB}\frac{T^A - T^B}{T^A},$$

$$\dot{\Sigma}_A = -\frac{1}{T^A}\langle \mathcal{F}^{\mathrm{fr}(A)}, \dot{q}\rangle - \sum_{B=1}^{P} J_{AB}\frac{T^B}{T^A}.$$

(11)

3.3   The 1st Law of Energy Balance

The Hamiltonian $H : T^*Q \times \mathbb{R}^P \to \mathbb{R}$ satisfies the following balance law along the solution curve $(q(t), p(t), S_A(t))$ of (8):

$$\frac{d}{dt}H(q, p, S_1, ..., S_P) = \langle \mathcal{F}^{\mathrm{ext}}, \dot{q}\rangle = P_W^{\mathrm{ext}}.$$

If the Hamiltonian is given by the sum of the Hamiltonian $H_A(q, p, S_A)$ of Σ_A

$$H(q, p, S_1, ..., S_P) = \sum_{A=1}^{P} H_A(q, p, S_A),$$

then, the mechanical equations for Σ_A are given as

$$\frac{dq}{dt} = \frac{\partial H_A}{\partial p}, \qquad \frac{dp}{dt} = -\frac{\partial H_A}{\partial q} + \mathcal{F}^{\mathrm{fr}(A)} + \mathcal{F}^{\mathrm{ext}\to A} + \sum_{B=1}^{P} \mathcal{F}^{B\to A},$$

where $\mathcal{F}^{B\to A}$ denotes the internal force from Σ_B acting on Σ_A. Note that $\mathcal{F}^{B\to A} = -\mathcal{F}^{A\to B}$ from Newton's third law of *action and reaction forces*.

For the Hamiltonian H_A of Σ_A, one finds the balance equation

$$\frac{d}{dt}H_A = P_W^{\mathrm{ext}\to A} + \sum_{B=1}^{P} P_W^{B\to A} + \sum_{B=1}^{P} P_H^{B\to A}.$$

(12)

In the above, we denote by $P_W^{\mathrm{ext}\to A} =< \mathcal{F}^{\mathrm{ext}\to A}, \dot{q} >$ the mechanical power that flows from the exterior into Σ_A, by $P_W^{B\to A} =< \mathcal{F}^{B\to A}, \dot{q} >$ the internal mechanical power that flows from Σ_B into Σ_A, and also by

$$P_H^{B\to A} = J_{AB}\left(\frac{\partial H}{\partial S_A} - \frac{\partial H}{\partial S_B}\right) = J_{AB}(T^A - T^B)$$

the internal heat power from Σ_B to Σ_A.

3.4 The Second Law of Entropy Production

The total entropy of the system is $S = \sum_{A=1}^{P} S_A$. Therefore, from (8) the rate of total entropy production is given, in view of $\dot{\Gamma}^A = \frac{\partial H}{\partial S_A} = T^A$, by

$$\dot{S} = -\sum_{A=1}^{P} \frac{1}{T^A} \left\langle \mathcal{F}^{\mathrm{fr}(A)}, \dot{q} \right\rangle + \sum_{A<B}^{P} J_{AB} \left(\frac{1}{T^B} - \frac{1}{T^A} \right) (T^B - T^A),$$

which is always positive (greater than or equal to zero). In fact, the second law suggests the linear phenomenological relations given by

$$\mathcal{F}_i^{\mathrm{fr}(A)} = -\lambda_{ij}^A p_j \quad \text{and} \quad J_{AB} \frac{T^A - T^B}{T^A T^B} = \mathcal{L}_{AB}(T^B - T^A). \tag{13}$$

In the above, λ_{ij}^A and $\mathcal{L}_{AB}$ are given as functions of q, S_A, S_B, where the symmetric parts of λ_{ij}^A are positive semi-definite and $\mathcal{L}_{AB} \geq 0$, $\forall A, B$. It is apparent that

$$J_{AB} = -\mathcal{L}_{AB} T^A T^B = -\kappa_{AB},$$

with $\kappa_{AB} = \kappa_{AB}(q, S_A, S_B)$ the heat conduction coefficients between subsystem Σ_A and subsystem Σ_B. In particular, $\dot{\Sigma}_A \geq 0$ in (11), while the sign of $\dot{S}_A$ is arbitrary.

4 Examples of Non-simple Systems

The adiabatic piston. We illustrate our theory by the example of the adiabatic piston problem, see [7]. As in Fig. 1, we consider a non-simple thermodynamic system Σ, in which two piston-cylinder systems Σ_1, Σ_2 with mass m_1, m_2 are connected by a rigid movable rod Σ_3 with mass m_3 and an ideal gas is contained in each cylinder. Further q and $r = D - \ell - q$ denote the distances between the top and the bottom of the piston 1 and 2 respectively, and D indicates some constant. Since the state variables for the thermodynamic system Σ are given by (q, v_q, S_1, S_2), the Lagrangian of the system is

$$L(q, v_q, S_1, S_2) = \frac{1}{2} M v_q^2 - U_1(q, S_1) - U_2(q, S_2),$$

where $M := m_1 + m_2 + m_3$, $U_1(q, S_1) := \mathsf{U}_1(S_1, V_1 = a_1 q, N_{10})$, and $U_2(q, S_2) := \mathsf{U}_2(S_2, V_2 = a_2 r, N_{20})$, and where $\mathsf{U}_i(S_i, V_i, N_{i0})$ denotes the internal energy of the fluid, N_{i0} the constant number of moles, and a_i the constant area of the cylinders for $i = 1, 2$.

Now, the Hamiltonian is given by

$$H(q, p, S_1, S_2) = \frac{1}{2} M^{-1} p^2 + U_1(q, S_1) + U_2(q, S_2),$$

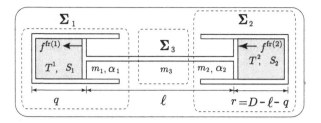

Fig. 1. The adiabatic piston problem

and the friction force is given by $\mathcal{F}^{\text{fr}(A)}(q,p,S_A) = -\lambda^A p$, where $\lambda^A = \lambda^A(q,S^A) \geq 0$. Now we have $\mathcal{J}_{AB} = -\kappa_{AB} =: -\kappa$, where $\kappa = \kappa(S_1,S_2,q) \geq 0$ denotes heat conductivity of the rigid rod. It follows that the Hamiltonian evolution equations for the adiabatic piston are obtained by

$$\begin{cases} \dot{p}_q = \Pi_1(q,S_1)\alpha_1 - \Pi_2(q,S_2)\alpha_2 - (\lambda^1 + \lambda^2)\dot{q}, \\ \dot{q} = M^{-1}p_q, \\ T^1\dot{S}_1 = \lambda^1\dot{q}^2 + \kappa\left(T^2(q,S_2) - T^1(q,S_1)\right), \\ T^2\dot{S}_2 = \lambda^2\dot{q}^2 + \kappa\left(T^1(q,S_1) - T^2(q,S_2)\right), \end{cases}$$

where

$$T^i = \frac{\partial U_i}{\partial S_i}(q,S_i), \quad \Pi_1(q,S_1) = -a_1^{-1}\frac{\partial U_1}{\partial q}$$

and

$$\Pi_2(q,S_2) = a_2^{-1}\frac{\partial U_2}{\partial q}.$$

The Hamiltonian is preserved along the solution curve:

$$\frac{d}{dt}H(q(t),p(t),S_1(t),S_2(t)) = 0.$$

It goes without saying that the conservation of the Hamiltonian agrees with the fact that the system is isolated, namely, the first law. Furthermore, the rate of total entropy production is obtained by

$$\frac{d}{dt}S = \left(\frac{\lambda^1}{T^1} + \frac{\lambda^2}{T^2}\right)\dot{q}^2 + \kappa\frac{(T^2 - T^1)^2}{T^1 T^2} \geq 0,$$

consistently with the second law.

5 Conclusions

We have developed the Hamiltonian variational formulation for non-simple systems. While the Lagrangian $L = L(q,\dot{q},S_1,...,S_P)$ for thermodynamic systems are always degenerate, we can construct the Hamiltonian by using a partial

Legendre transform under the assumption that $L_{(S_1,...,S_P)} = L_{(S_1,...,S_P)}(q,\dot{q})$ is hyperregular in terms of mechanical state variables $(q,\dot{q})$, for each fixed $(S_1,...,S_P)$. We have illustrated our theory with the example of the adiabatic piston.

For the more general cases of non-simple systems in which the Lagrangian is totally degenerate in terms of both thermodynamic variables $(S_1,...,S_P)$ and mechanical variables $(q,\dot{q})$, we need to employ Dirac theory of constraints, which we will explore as future works.

Acknowledgement. HY is partially supported by JSPS Grant-in-Aid for Scientific Research (22K03443), JST CREST (JPMJCR1914), Waseda University (SR 2023C-089), and the MEXT "Top Global University Project", SEES. FGB is partially supported by CNCS UEFISCDI, project number PN-III-P4-ID-PCE-2020-2888.

References

1. de Groot, S. R and Mazur, P.: Nonequilibrium Thermodynamics, North-Holland (1969)
2. Gay-Balmaz, F. and Yoshimura, H.: A Lagrangian variational formulation for nonequilibrium thermodynamics. Part I: discrete systems. J. Geom. Phys. **111**, 169–193 (2017)
3. Gay-Balmaz, F. and Yoshimura, H.: A Lagrangian variational formulation for nonequilibrium thermodynamics. Part II: continuum systems, J. Geom. Phys. **111**, 194–212 (2017)
4. Gay-Balmaz, F., Yoshimura, H.: A variational formulation of nonequilibrium thermodynamics for discrete open systems with mass and heat transfer. Entropy **20**(3), 163 (2018)
5. Gay-Balmaz, F., Yoshimura, H.: Systems, variational principles and interconnections in nonequilibrium thermodynamics (2023)
6. Gruber, C.: Thermodynamique et Mécanique Statistique. Institut de physique théorique, EPFL (1997)
7. Gruber, C.: Thermodynamics of systems with internal adiabatic constraints: time evolution of the adiabatic piston. Eur. J. Phys. **20**, 259–266 (1999)
8. Stueckelberg, E.C.G., Scheurer, P. B.: T:hermocinétique phénoménologique galiléenne, Birkhäuser (1974)
9. Yoshimura, H., Gay-Balmaz, F.: Hamiltonian variational formulation for nonequilibrium thermodynamics of simple closed systems. In: Proceedings of 4th IFAC Workshop on Thermodynamics Foundations of Mathematical Systems Theory TFMST 2022 (Montreal, Canada, 25–27 July 2022), vol. 55, no. 18, pp. 81–86 (2022)

Madelung Transform and Variational Asymptotics in Born-Oppenheimer Molecular Dynamics

Paul Bergold[1]([✉])[ID] and Cesare Tronci[1,2][ID]

[1] Department of Mathematics, University of Surrey, Guildford, UK
p.bergold@surrey.ac.uk, c.tronci@surrey.ac.uk
[2] Department of Physics and Engineering Physics, Tulane University, New Orleans, USA

Abstract. While Born-Oppenheimer molecular dynamics (BOMD) has been widely studied by resorting to powerful methods in mathematical analysis, this paper presents a geometric formulation in terms of Hamilton's variational principle and Euler-Poincaré reduction by symmetry. Upon resorting to the Lagrangian hydrodynamic paths made available by the Madelung transform, we show how BOMD arises by applying asymptotic methods to the variational principles underlying different continuum models and their particle closure schemes. In particular, after focusing on the hydrodynamic form of the fully quantum dynamics, we show how the recently proposed bohmion scheme leads to an on-the-fly implementation of BOMD. In addition, we extend our analysis to models of mixed quantum-classical dynamics.

Keywords: Born–Oppenheimer approximation · Variational principle · Mixed quantum-classical dynamics

1 Introduction

Although many-body quantum simulations have greatly benefited from high-performance computing, large molecular systems continue to pose formidable challenges. In particular, molecular dynamics deals with systems comprising N nuclei with masses $M_n > 0$ and L electrons with masses m_e. The position coordinates of the nuclei and electrons are $r = (r_1, \ldots, r_N) \in \mathbb{R}^{3N}$ and $x = (x_1, \ldots, x_L) \in \mathbb{R}^{3L}$, respectively. The time evolution of the molecular wavefunction $\Psi \in L^2(\mathbb{R}^{3N+3L})$ describing the system is determined by the time-dependent Schrödinger equation (TDSE) $i\hbar\partial_t\Psi = \widehat{H}_{mol}\Psi$. In the absence of external fields, the molecular Hamiltonian reads $\widehat{H}_{mol} = \widehat{T}_e + \widehat{T}_n + \widehat{V}_{ee} + \widehat{V}_{en} + \widehat{V}_{nn}$.

This work was made possible through the support of Grant 62210 from the John Templeton Foundation. The opinions expressed in this publication are those of the authors and do not necessarily reflect the views of the John Templeton Foundation.

The first two terms are the electronic and nuclear kinetic energy operators (i.e., $\widehat{T}_n = -\hbar^2 \sum_{n=1}^{N} M_n^{-1} \Delta_{r_n}/2$, and similarly for $\widehat{T}_e$), and the remaining terms comprise the Coulomb interactions.

Born–Oppenheimer Theory. Given the computational complexity of this problem, one is forced to look for additional assumptions alleviating the curse of dimensions. One of the best-known approaches is the (time-dependent) Born–Oppenheimer (BO) theory from 1927, which stands as the cornerstone of modern quantum chemistry. This theory is based on the observation that the large nuclear-to-electron mass ratio yields a large difference between their kinetic energies, thereby suggesting that nuclei move much slower than the electrons. In the literature, this argument leads to the *BO approximation* $\Psi(r, x, t) \approx \Omega(r, t)\phi(x; r)$, where $\int_{\mathbb{R}^{3L}} |\phi(x; r)|^2 \, dx = 1$ so that ϕ is a conditional wavefunction. Over the decades, the first and second factor have been dubbed *nuclear wavefunction* and *electronic wavefunction*, respectively, even though the nuclear and electronic density matrices are not actually projection operators [2]. Notice that here the electronic factor is time-independent, which is the essence of the BO approximation. In particular, $\phi(r) := \phi(\_; r)$ is chosen as an eigenstate of the electronic Hamiltonian $\widehat{H}_e(r) := \widehat{T}_e + \widehat{V}_{ee} + \widehat{V}_{en}(r) + \widehat{V}_{nn}(r)$, that is

$$\widehat{H}_e(r)\phi(r) = E(r)\phi(r). \tag{1}$$

The resulting eigenvalues for the different nuclear positions, i.e. the energies $E(r)$, are combined to obtain what is known as a *potential energy surface* (PES) $E: \mathbb{R}^{3N} \to \mathbb{R}$. Then, combining the BO approximation and (1), the TDSE can be separated into two smaller, consecutive subproblems. In the first step, one solves a time-independent eigenvalue problem for each fixed nuclear configuration. In the second step, one solves the TDSE $i\hbar \partial_t \Omega = \widehat{T}_n \Omega + E\Omega$ to determine the nuclear motion.

Born–Oppenheimer Molecular Dynamics. Despite its well-known limitations, the Born–Oppenheimer approximation (BOA) provides a valuable tool allowing for a significant reduction in computational costs. Yet, the high-dimensional nature of molecular systems motivates the search for further approximations. Based on the small value of the electron-to-nuclear mass ratio, one can perform a classical limit on the nuclear evolution, whose classical dynamics then takes place on the PES made available by the electronic eigenvalue problem (1). In this picture, one is left with Newton's equations

$$M_n \ddot{R}_n = -\nabla_{R_n} E(R), \qquad n = 1, \dots, N, \tag{2}$$

for the nuclear motion. Known under the name of *Born–Oppenheimer molecular dynamics* (BOMD), the system (1)-(2) is probably the most widely used model in computational quantum chemistry and represents the object of our investigation.

Over the years, the BO theory and its molecular dynamics counterpart have been established by using several powerful methods from mathematical analysis

[5]. Yet, a geometric characterization in terms of reduction by symmetry and variational principles is still lacking. Motivated by the success of Geometric Mechanics in a variety of modeling efforts, here we present a series of possible geometric derivations of BOMD coming from quite different perspectives. Much hinging on the use of diffeomorphic Lagrangian paths in continuum theories, our presentation applies asymptotic analysis to Hamilton's action principle rather than to the TDSE itself. Differing from conventional techniques, not only does our approach enable a deeper geometric understanding, but also provides new tools for the asymptotic analysis of numerical schemes beyond the BO regime.

Exact Factorization and Madelung Transform. Among the tools used in our discussion, the *exact factorization* (XF) plays a prominent role. This technique extends the BO approximation to consider a time-dependent electronic factor, i.e.,

$$\Psi(r, x, t) = \Omega(r, t)\phi(x, t; r), \quad \text{where} \quad \int_{\mathbb{R}^{3L}} |\phi(x, t; r)|^2 \, dx = 1. \tag{3}$$

Going back to von Neumann's celebrated book, the XF has become increasingly popular in quantum chemistry . For example, BOMD was shown to be recovered from the XF equations for Ω and ϕ in [1]. Notice that (3) provides a convenient representation of exact solutions of the TDSE. Based on the variational setting of the TDSE, here we will combine the XF with the *Madelung transform* $\Omega = \sqrt{D}e^{iS/\hbar}$, in such a way that the resulting hydrodynamic formulation allows to take advantage of the associated Lagrangian trajectories.

This paper presents four different variational approaches to the formulation of BOMD. The first two are based on a quantum hydrodynamic description of the molecular evolution. While one of these approaches involves the continuum PDE setting, the other relies on a particle closure ODE system resulting from a suitable regularization of the continuum description. The second part of the paper focuses on a very different approach, which treats the original molecular system as an intrinsically mixed quantum-classical system. Also in this case, we deal with both the continuum PDE setting and its particle ODE closure.

2   Exact Factorization and Variational Asymptotics

Upon exploiting the XF of the molecular wavefunction, this section shows how BOMD can be derived by combining variational asymptotics and Euler–Poincaré reduction by symmetry, within the hydrodynamic formulation of the TDSE. Our point of departure is the Dirac–Frenkel (DF) action principle $\delta \int_{t_1}^{t_2} \int_{\mathbb{R}^{3N}} \langle \Psi, i\hbar\partial_t \Psi - \widehat{H}_{mol}\Psi \rangle \, dr \, dt = 0$ underlying the TDSE. Here, we introduced the real-valued pairing $\langle \Psi_1, \Psi_2 \rangle = \mathrm{Re}\langle \Psi_1|\Psi_2 \rangle$, with $\langle \Psi_1|\Psi_2 \rangle = \int_{\mathbb{R}^{3L}} \Psi_1(x, r)^*\Psi_2(x, r) \, dx$. A direct verification shows that the molecular TDSE is recovered from the DF action principle by using arbitrary variations. As we are dealing with several nuclear masses, it is convenient to introduce the block-diagonal matrix $G := \mathrm{diag}(M_1, \ldots, M_N) \otimes \mathrm{Id}_{3\times 3}$, which induces a metric structure $g: \mathbb{R}^{3N} \times \mathbb{R}^{3N} \to \mathbb{R}$, operating as $g(u, v) := Gu \cdot v$. In turn, this metric induces the norm $\|u\|_g := \sqrt{g(u,u)}$ and its inverse $\|\sigma\|_{g^{-1}} := \sqrt{g^{-1}(\sigma, \sigma)}$ on covectors $\sigma \in \mathbb{R}^{3N}$. For example, the nuclear kinetic energy reads $\hbar^2 \int_{\mathbb{R}^{3N}} g^{-1}(\nabla \Psi^\dagger, \nabla \Psi) \, dr / 2$, where the adjoint $\dagger$ is defined by the electronic inner product $\langle \cdot | \cdot \rangle$ above, so that $g^{-1}(\nabla \Psi^\dagger, \nabla \Psi) = \int_{\mathbb{R}^{3L}} \nabla \Psi^*(x) \cdot G^{-1} \nabla \Psi(x) \, dx$.

Non-dimensionalization of the Action Principle. In order to obtain a hydrodynamic formulation of the molecular system, we proceed by applying the XF in (3) and using the Madelung transform on the nuclear factor. Before implementing these steps, however, it is convenient to perform a non-dimensionalization of the DF Lagrangian in order to reveal the role of the electron-to-nuclear mass ratio. For this purpose, we introduce the unit system in [1] by replacing the electronic mass m_e by the average nuclear mass $M_0 = \sum_n M_n / N$. Each physical unit is then expressed in terms of the mass M_0, the Bohr radius λ_0, the Hartree energy E_h and the elementary charge e. In particular, the units of time and action are given by $t_0 = \sqrt{M_0/E_h}\lambda_0$ and $a_0 = \sqrt{M_0 E_h}\lambda_0$. In addition, the reduced version of Planck constant $\hbar$ reads $\sqrt{\mu}a_0$, where we introduced $\mu := m_e/M_0$. Notice that we refrain from introducing an additional symbol to distinguish between original and dimensionless variables. From now on all quantities are suitably non-dimensionalized. At this point, we are left with the DF Lagrangian $L_{DF} = \int_{\mathbb{R}^{3N}} \langle \Psi, i\sqrt{\mu}\partial_t \Psi - \widehat{H}_{mol}\Psi \rangle \, dr \, dt$, where we have conveniently divided by the Hartree energy E_h. The remainder of this paper will deal only with this non-dimensionalized version of the DF action principle and its variants.

Madelung Transform and Euler-Poincaré Variations. As a next step, we combine the XF (3) of the dimensionless molecular wavefunction with the Madelung transform of the nuclear factor, that is $\Omega = \sqrt{D}e^{iS/\sqrt{\mu}}$. These steps take the DF Lagrangian $\int_{\mathbb{R}^{3N}} \langle \Psi, i\hbar\partial_t \Psi - \widehat{H}_{mol}\Psi \rangle \, dr$ into the form:

$$L_{XF}(S, D, \phi, \partial_t \phi) = \int D\Big(\langle \phi, i\sqrt{\mu}\partial_t \phi - \widehat{H}_e \phi \rangle - \partial_t S$$
$$- \frac{\mu}{2D}\|\nabla\sqrt{D}\|_{g^{-1}}^2 - \frac{1}{2}\|\nabla S + \mathcal{A}_B\|_{g^{-1}}^2 - \epsilon(\phi) \Big) dr, \quad (4)$$

where $\mathcal{A}_B := \langle \phi \,|\, -i\sqrt{\mu}\nabla\phi \rangle$ denotes the Berry connection and

$$\epsilon(\phi) := \frac{\mu}{2}g^{-1}(\nabla\phi^\dagger, \nabla\phi) - \frac{1}{2}\|\mathcal{A}_B\|_{g^{-1}}^2 \quad (5)$$

is usually referred to as the *electronic potential.* We notice that the latter can also be written as $\epsilon(\phi) = \mu \, \mathrm{Tr}(G^{-1} \, \mathrm{Re} \, Q)/2$, where $Q_{jk} := \langle \partial_j \phi | (1 - \phi\phi^\dagger)\partial_k \phi \rangle$ is the *quantum geometric tensor.* Also, in the reminder of this paper all integrals are on $\mathbb{R}^{3N}$, unless otherwise specified.

We observe that, so far, there seems to be nothing in the variational principle ensuring that the electronic factor is normalized at all times. While this condition could be easily enforced by resorting to a Lagrange multiplier, this appears unnecessary at the current level. Indeed, it follows from a direct verification that the normalization of ϕ is preserved in time by the equations resulting from $\delta \int_{t_1}^{t_2} L_{XF}\, dt = 0$. Nevertheless, for later purpose, here we need to encode this normalization in the variational principle. This is due to the fact that the time-conservation of the normalization condition may be lost when attempting to perform suitable approximations on the action, such as the asymptotic expansion that is presented below. Instead of using Lagrange multipliers, here we follow the Euler–Poincaré reduction method in geometric mechanics [4]. In particular, we restrict the electronic factor to evolve on orbits of the infinite-dimensional group $\mathcal{F}(\mathbb{R}^{3N}, \mathcal{U}(L^2(\mathbb{R}^{3L})))$ of mappings $U : \mathbb{R}^{3N} \to \mathcal{U}(L^2(\mathbb{R}^{3L}))$ into the unitary operators on the electronic Hilbert space $L^2(\mathbb{R}^{3L})$. This amounts to setting $\phi(t; r) = U_t(r)\phi_0(r)$ for some curve $U_t \in \mathcal{F}(\mathbb{R}^{3N}, \mathcal{U}(L^2(\mathbb{R}^{3L})))$, so that the normalization of ϕ remains a preserved initial condition. As customary in Euler–Poincaré theory, taking the relevant derivatives of ϕ yields $\partial_t \phi = \xi\phi$ and $\delta\phi = \gamma\phi$, where we have dropped the explicit time-dependence for convenience and both $\xi = \dot{U}U^{-1}$ and $\gamma = \delta U U^{-1}$ are skew-Hermitian operators on $L^2(\mathbb{R}^{3L})$. With this in mind, the Lagrangian (4) becomes

$$
\ell_{XF}(S, D, \xi, \phi) = \int \Big(\langle \phi, i\sqrt{\mu}\xi\phi - \widehat{H}_e\phi \rangle - \partial_t S
$$
$$
- \frac{\mu}{2D}\|\nabla\sqrt{D}\|_{g^{-1}}^2 - \frac{1}{2}\|\nabla S + \mathcal{A}_B\|_{g^{-1}}^2 - \epsilon(\phi) \Big) D dr, \quad (6)
$$

where both δS and δD are arbitrary. Having gone through several preparatory steps, we are ready to show how the BOMD equations can be derived from the action principle associated to (6).

Variational Asymptotics for BOMD. We will now prove that BOMD is recovered from the lowest-order asymptotic expansion of the Lagrangian (6). This expansion is obtained in the limit of a small electron-to-nuclear mass ratio, i.e., $\mu \to 0$.

Proposition 1. *Consider the variational problem* $\delta \int_{t_1}^{t_2} \ell_{XF}\, dt = 0$ *associated to* (6), *with* $\delta\phi = \gamma\phi$ *and arbitrary* γ, δD *and* δS. *In the limit* $\mu \to 0$, *this action principle yields the following continuum PDE system:*

$$
i)\ \frac{\partial D}{\partial t} + \mathrm{div}(DG^{-1}\nabla S) = 0, \quad ii)\ \frac{\partial S}{\partial t} + \frac{1}{2}\|\nabla S\|_{g^{-1}}^2 = -E, \quad iii)\ \widehat{H}_e\phi = E\phi.
$$

Proof. To derive the above equations, we first consider the scaled Lagrangian arising in the limit $\mu \to 0$, which, up to an overall sign, is given by

$$
\tilde{\ell}_{XF}(S, D, \phi) = \int D\Big(\partial_t S + \langle \phi, \widehat{H}_e\phi \rangle + \frac{1}{2}\|\nabla S\|_{g^{-1}}^2 \Big) dr.
$$

The arbitrary variations in δS yield the continuity equation i), while variations in δD yield the nuclear Hamilton–Jacobi equation $\partial_t S + \|\nabla S\|_{G^{-1}}^2/2 =$

$-\langle\phi, \widehat{H}_e\phi\rangle$. Although the Lagrangian $\tilde{\ell}_{XF}$ does not depend on ξ because the term $\sqrt{\mu}D\langle\phi, i\xi\phi\rangle$ has vanished, we notice that the variations of the Euler–Poincaré variational principle $\delta\int_{t_1}^{t_2}\tilde{\ell}_{XF}\,\mathrm{d}t = 0$ still contain the condition $\delta\phi = \gamma\phi$, which yields $[\phi\phi^\dagger, \widehat{H}_e] = 0$. Applying both sides on ϕ results in the eigenvalue problem iii), thereby taking the Hamilton–Jacobi equation into the form ii).

We remark that this conclusion can be equivalently reached by using arbitrary variations $\delta\phi$ and resorting to Lagrange multipliers. The transport equation i) may be problematic due to the presence of second-order gradients of Hamilton's principal function S, which is known to develop caustic singularities in realistic scenarios. At this point, one may introduce particle trajectories by proceeding formally and substituting $D(r, t) = \delta(r - R(t))$ into the transport equation i), thereby obtaining $G\dot{R} = \nabla S(R)$. Then, taking the gradient of equation ii) and evaluating $\mathrm{d}\nabla S(R)/\mathrm{d}t = -\nabla E(R)$ leads to the Newtonian equations (2). Alternatively, one can apply the standard method of characteristics, which provides solutions of the Hamilton–Jacobi equation the initial condition $S(r, t_0) = S_0(r)$. In the present case, the characteristic trajectories of the Hamilton–Jacobi equation are given by the extremals of $\delta\int_{t_1}^{t_2}(\|\dot{q}\|_g^2/2 - E(q))\,\mathrm{d}t$, which in particular satisfy the Newtonian equations (2). Moreover, it is easy to prove that the momentum of the characteristics is given by the gradient ∇S of the solution S.

3   Quantum Hydrodynamics and the Bohmion Method

In this section we show how the variational asymptotic method presented above may be applied in the context of quantum hydrodynamics and one of its particle closure schemes, recently proposed under the name of *bohmion method* [2]. The latter hinges on a sampling process involving the Lagrangian paths underlying the relevant Madelung hydrodynamic equations. These Lagrangian paths identify the well-known *Bohmian trajectories* from quantum theory, thereby explaining the term 'bohmion'. The discussion will proceed in two stages. First, we will present the hydrodynamic formulation of the XF equations as they arise from the action principe associated to (4). In the second step, we will introduce the sampling process leading to the bohmion closure scheme and illustrate how the latter recovers BOMD by variational asymptotics.

Exact Factorization and Hydrodynamics. It is convenient to transform the Lagrangian (4) by observing that the equation $\partial_t D + \mathrm{div}(DG^{-1}(\nabla S + \mathcal{A}_B)) = 0$, obtained from the variations δS, identifies a Lie-transport evolution of the type $\mathrm{d}D/\mathrm{d}t = 0$ along the vector field $u = G^{-1}(\nabla S + \mathcal{A}_B)$. Consequently, if the Lagrangian path $\eta_t \in \mathrm{Diff}(\mathbb{R}^{3N})$ is defined by the instantaneous integral curves of u via the relation $\dot{\eta}_t(r_0) = u \circ \eta_t(r_0)$, then one can write $D(t) = \eta_*D_0$, where η_* denotes the push-forward. If the Lagrangian path is considered as a dynamical variable, then we can make the replacement $-\int_{\mathbb{R}^{3N}} D\partial_t S\,\mathrm{d}r = \int_{\mathbb{R}^{3N}} D\nabla S \cdot u\,\mathrm{d}r =: \int_{\mathbb{R}^{3N}} M \cdot u\,\mathrm{d}r$, in (6). Here, we have dropped an irrelevant total time derivative and we have introduced the hydrodynamic momentum $M = D\nabla S$. Then, the action principle associated to (6) becomes

$$\delta \int_{t_1}^{t_2} \int \left(M \cdot u + D\langle \phi, i\sqrt{\mu}\partial_t \phi - \widehat{H}_e \phi \rangle \right.$$
$$\left. - \frac{\mu}{2}\|\nabla\sqrt{D}\|_{g^{-1}}^2 - \frac{1}{2D}\|M + D\mathcal{A}_B\|_{g^{-1}}^2 - D\epsilon(\phi) \right) dr = 0, \quad (7)$$

where δM and $\delta\phi$ are both arbitrary. Also, the variations

$$\delta D = -\operatorname{div}(Dw), \qquad \delta u = \partial_t w + u \cdot \nabla w - w \cdot \nabla u, \qquad (8)$$

are obtained directly from the relations $D(t) = \eta_* D_0$ and $u = \dot\eta_t \circ \eta_t^{-1}$ above, and by defining the arbitrary vector field $w = \delta\eta_t \circ \eta_t^{-1}$.

While the variational principle (7) succeeds in unfolding the role of Lagrangian paths through the variations (8), we will perform two more steps that allow eliminating the explicit appearance of both the momentum variable M and the Berry connection $\mathcal{A}_B$ in the variational problem. On the one hand, eliminating the momentum allows formulating the problem only in terms of the hydrodynamic velocity u. On the other hand, eliminating $\mathcal{A}_B$ is particularly convenient in obtaining a *gauge-independent* formulation that removes the phase arbitrariness introduced by the XF decomposition (3).

Quantum Motion in the Hydrodynamic Frame. We will proceed by expressing the quantum evolution of ϕ in the frame moving with the path η_t. This process is performed as follows. First, we observe that the quantum evolution reads [2]

$$i\hbar(\partial_t + u \cdot \nabla)\phi = \frac{1}{2D}\frac{\delta F}{\delta\phi} + \widehat{H}_e \phi, \quad \text{where} \quad F(D, \phi) := \int D\epsilon(\phi) \, dr. \quad (9)$$

As a further step, we define $\rho = \phi\phi^\dagger$ and write $\epsilon(\phi) = \mu\|\nabla\rho\|_{g^{-1}}^2/4 =: \tilde\epsilon(\rho)$, where $\epsilon(\phi)$ is given in (5) and $\|\nabla\rho\|_{g^{-1}}^2 = \langle\nabla\rho, G^{-1}\nabla\rho\rangle$. Also, we have denoted by $\langle A, B\rangle = \operatorname{Re}\langle A|B\rangle = \operatorname{Re}\operatorname{tr}(A^\dagger B)$ the pairing between trace-class operators on $L^2(\mathbb{R}^{3L})$. The chain-rule relation $\delta F/\delta\phi = 2(\delta\tilde F/\delta\rho)\phi$ takes (9) into the form

$$i\hbar(\partial_t + u \cdot \nabla)\phi = \left(\frac{1}{D}\frac{\delta\tilde F}{\delta\rho} + \widehat{H}_e\right)\phi, \quad \text{where} \quad \tilde F(D, \rho) := \int D\tilde\epsilon(\rho) \, dr. \quad (10)$$

Since the operator $\widehat{H}_e + D^{-1}\delta\tilde F/\delta\rho$ is Hermitian, the equation above unfolds the nature of the quantum evolution: a quantum state evolves unitarily in the frame moving with the hydrodynamic velocity u. This conclusion motivates us to write the evolution of ϕ as $\phi(t) = (\tilde U_t \phi_0) \circ \eta_t^{-1}$, with $\tilde U_t \in \mathcal{F}(\mathbb{R}^{3N}, \mathcal{U}(L^2(\mathbb{R}^{3L})))$. Notice that the above relation identifies an action of the semidirect-product group $\operatorname{Diff}(\mathbb{R}^{3N}) \circledS \mathcal{F}(\mathbb{R}^{3N}, \mathcal{U}(L^2(\mathbb{R}^{3L})))$ that is constructed by the natural pull-back action of $\operatorname{Diff}(\mathbb{R}^{3N})$ on $\mathcal{F}(\mathbb{R}^{3N}, \mathcal{U}(L^2(\mathbb{R}^{3L})))$. Consequently, the quantum evolution occurs on orbits of this semidirect-product group. Also, the evolution law $\phi = (\tilde{U}_t \phi_0) \circ \eta_t^{-1}$ implies $\partial_t \phi = \tilde{\xi} \phi - u \cdot \nabla \phi$, with $\tilde{\xi} = \partial_t \tilde{U} \tilde{U}^{-1} \circ \eta_t^{-1}$, and we observe that this is precisely the same form as in (10).

To proceed further, we replace $\langle \phi, i \partial_t \phi \rangle = \langle \phi, i \tilde{\xi} \phi \rangle + u \cdot \mathcal{A}_B$ in (7). Then, upon defining $m := M + D \mathcal{A}_B$, and by inverting the Legendre transform $m = D G u$, the action principle (7) becomes $\delta \int_{t_1}^{t_2} \ell_{EP}(u, D, \tilde{\xi}, \rho) \, dt = 0$ with the Euler-Poincaré Lagrangian

$$\ell_{EP} = \int \left(\frac{1}{2} D \|u\|_g^2 + D \langle \rho, i \sqrt{\mu} \tilde{\xi} - \widehat{H}_e \rangle - \frac{\mu}{8D} \|\nabla D\|_{g^{-1}}^2 - \frac{\mu}{4} D \|\nabla \rho\|_{g^{-1}}^2 \right) dr. \quad (11)$$

Here, the variations $\delta \rho = -\operatorname{div}(\rho w) + [\tilde{\gamma}, \rho]$ and $\delta \tilde{\xi} = \partial_t \tilde{\gamma} + [\tilde{\gamma}, \tilde{\xi}] - w \cdot \nabla \tilde{\xi} + u \cdot \nabla \tilde{\gamma}$ follow from the definitions $\rho = \phi \phi^\dagger$ and $\tilde{\xi} = \partial_t \tilde{U} \tilde{U}^{-1} \circ \eta_t^{-1}$, respectively. Eventually, together with (8), the action principle associated to ℓ_{EP} leads to

$$(\partial_t + u \cdot \nabla) u^\flat = -\nabla V_Q - \langle \rho, \nabla \widehat{H}_e \rangle - \frac{\mu}{2D} \operatorname{div} \operatorname{tr}((D \nabla \rho)^\sharp \otimes \nabla \rho),$$

$$i\hbar (\partial_t + u \cdot \nabla) \rho = \left[\widehat{H}_e - \frac{\mu}{2D} \operatorname{div}(D \nabla \rho)^\sharp, \rho \right], \qquad \partial_t D + \operatorname{div}(Du) = 0. \quad (12)$$

Here, we used the musical isomorphisms induced by the metric g while tr is the quantum trace. Also, $V_Q = -\mu \operatorname{div}(\nabla \sqrt{D})^\sharp / (2\sqrt{D})$ is the *quantum potential*.

The system (12) represents the hydrodynamic form of the equations of motion resulting from the variational principle for (4). The second-order gradients in the first two equations make the level of complexity of this system rather intimidating. A finite-dimensional closure scheme becomes necessary in order for these equations to be used in the context of molecular dynamics simulations.

Regularization and the Bohmion Scheme. An immediate consequence of the second-order gradients in (12) is the lack of delta-type solutions of the form $D(r,t) = \delta(r - R(t))$, which are instead allowed after taking the classical limit $\mu \to 0$ as shown in Proposition 1. To overcome this limitation, the bohmion method exploits a variational regularization to restore point-particle trajectories. This regularization is applied after rewriting the Euler-Poincaré Lagrangian (11) in terms of the weighted variable $\tilde{\rho} = D\rho$. Then, one smoothens by replacing $\|\nabla D\|_{g^{-1}}^2 / D \to \|\nabla \bar{D}\|_{g^{-1}}^2 / \bar{D}$ and $\|\nabla \tilde{\rho}\|_{g^{-1}}^2 / D \to \|\nabla \bar{\rho}\|_{g^{-1}}^2 / \bar{D}$, where $\bar{D} = K_\alpha * D$ and $\bar{\rho} = K_\alpha * \tilde{\rho}$, for some normalized convolution kernel K_α depending on a modeling length-scale $\alpha > 0$. In the limit $\alpha \to 0$, we ask for K_α to tend to a delta function thereby recovering the original Lagrangian.

Due to this smoothing process, the resulting regularized equations allow for singular delta-like expressions of D and $\tilde{\rho}$, thereby returning point trajectories called *bohmions*. The trajectory equations may be found by replacing the ansatz $D(r,t) = \sum_{a=1}^{P} w_a \delta(r - q_a(t))$ and $\tilde{\rho}(r,t) = \sum_{a=1}^{P} w_a \varrho_a(t) \delta(r - q_a(t))$ in the regularized Lagrangian, where the positive weights w_a satisfy $\sum_a w_a = 1$. Also, we have $\varrho_a = \varphi_a \varphi_a^\dagger$ and $\varrho_a(t) = U_a(t) \varrho_{0a} U_a(t)^\dagger$, so that $\partial_t \varrho_a = [\xi_a, \varrho_a]$. Upon denoting $\widehat{H}_a = \widehat{H}_e(q_a)$, this process leads to the Lagrangian

$$L(q, \dot{q}, \rho) = \sum_{a=1}^{P} w_a \left(\frac{1}{2} \|\dot{q}_a\|_g^2 + \langle \varrho_a, i\sqrt{\mu}\xi_a - \widehat{H}_a \rangle \right.$$

$$+ \frac{\mu}{8} \sum_{b=1}^{P} w_b (1 - 2\langle \varrho_a, \varrho_b \rangle) \int \frac{g^{-1}(\nabla K_\alpha(r - q_a), \nabla K_\alpha(r - q_b))}{\sum_c w_c K_\alpha(r - q_c)} \, dr \right),$$

where $\delta\xi_a = \partial_t \gamma_a - [\xi_a, \gamma_a]$, $\delta\varrho_a = [\gamma_a, \varrho_a]$, while δq_a and γ_a are arbitrary. We emphasize that these bohmions do not correspond to physical particles, but rather to *computational particles* that sample nuclear hydrodynamic paths.

It is now easy to see how the bohmion scheme reduces to BOMD in the classical limit. Indeed, letting $\mu \to 0$ and taking variations of the resulting Lagrangian $\tilde{L} = \sum_a w_a (\|\dot{q}_a\|_g^2/2 - \langle \varrho_a, \widehat{H}_a \rangle)$ yields

$$G\ddot{q}_a = -\nabla \langle \varphi_a, \widehat{H}_e(q_a)\varphi_a \rangle, \qquad \widehat{H}_e(q_a)\varphi_a = E(q_a)\varphi_a. \tag{13}$$

Here, the second equation follows from the Euler-Poincaré equation $[\widehat{H}_a, \varrho_a] = 0$, which arises from the variations $\delta\xi_a$ in Hamilton's principle. We observe that BOMD is recovered in the case of only one bohmion, that is $P = 1$. Importantly, we have obtained an on-the-fly implementation scheme for BOMD in which the electronic structure problem associated to (1) (usually very challenging) is replaced by a finite-dimensional eigenvalue problem to be solved at each time-step along trajectories. In the general case $P > 1$, the bohmion method recovers an on-the-fly model to BOMD in which the single particle distribution $D(r, t) = \delta(r - R(t))$ is replaced by the statistical sampling $D(r, t) = \sum_{a=1}^{P} w_a \delta(r - q_a(t))$ of the entire density associated to classical nuclear motion.

4   Mixed Quantum-Classical Dynamics

This section extends the variational asymptotic methods discussed earlier to the case of *mixed quantum-classical* (MQC) models. This type of models are motivated by the need to go beyond BOMD when the BO approximation fails to hold. Inspired by the result from BO theory that nuclear dynamics can be approximated as classical one seeks a model in which classical nuclear motion is coupled to fully quantum electronic evolution. Here, we focus on the model presented in [3], to which we refer for a thorough discussion. The main ingredient of this model resides in *Koopman wavefunctions*, that is functions $\chi(r, p) \in L^2(\mathbb{R}^{6N})$ such that $\rho_c = |\chi|^2$ obeys the classical Liouville equation $\partial_t \rho_c = \{H, \rho_c\}$. Then, one can take the tensor-product space to describe MQC dynamics in terms of hybrid wavefunctions $\Upsilon \in L^2(\mathbb{R}^{6N}) \otimes L^2(\mathbb{R}^{3L})$ so that $\rho_c = \Upsilon^\dagger \Upsilon$ and $\hat{\rho}_q = \int \Upsilon \Upsilon^\dagger \, drdp$ are the classical Liouville density and the quantum density matrix, respectively.

Instead of presenting the model equations, here we follow the procedure outlined before and consider their underlying variational principle [3]. For the model under consideration, the latter involves the following Lagrangian $L_{\mathrm{QC}}(\Upsilon, \partial_t \Upsilon, \mathcal{X})$:

$$L_{\mathrm{QC}} = \int \left\langle \Upsilon, i\hbar\partial_t\Upsilon + (\mathcal{A} - \mathcal{A}_B)\cdot\mathcal{X}\Upsilon - (\widehat{H} - \mathcal{A}_B\cdot X_{\widehat{H}})\Upsilon + i\hbar X_{\widehat{H}}\cdot\nabla\Upsilon \right\rangle \mathrm{d}r\mathrm{d}p,$$

where we have introduced $\mathcal{A}_B = \langle \Upsilon, -i\hbar\nabla\Upsilon\rangle/(\Upsilon^\dagger\Upsilon)$ and the MQC Hamiltonian vector field $X_{\widehat{H}} := (\partial_p\widehat{H}, -\partial_r\widehat{H})$ of the operator-valued Hamiltonian function $\widehat{H}(r,p)$. Also, $\mathcal{A} = (p, 0)$ is the coordinate representation of the canonical Liouville one-form $\mathcal{A} = p\cdot\mathrm{d}r$. In the above Lagrangian, the variation $\delta\Upsilon$ is arbitrary, while the variation of the vector field $\mathcal{X}$ arises from its definition in terms of the Lagrangian phase-space paths $\eta_t \in \mathrm{Diff}(\mathbb{R}^{6N})$, that is $\dot{\eta}_t(r_0, p_0) = \mathcal{X}\circ\eta_t(r_0, p_0)$. Then, we have the Euler-Poincaré variation

$$\delta\mathcal{X} = \partial_t\mathcal{Y} + \mathcal{X}\cdot\nabla\mathcal{Y} - \mathcal{Y}\cdot\nabla\mathcal{X}, \tag{14}$$

where $\mathcal{Y}$ is an arbitrary displacement vector field.

Performing the non-dimensionalization of the MQC Lagrangian, applying the XF $\Upsilon(r, p, t) = \Omega(r, p, t)\phi(x, t; r, p)$ as in (3), and using the Madelung transform $\Omega = \sqrt{\rho_c}e^{iS/\sqrt{\mu}}$, we rewrite L_{QC} into the form

$$\ell_{\mathrm{QC}} = \int \left(\rho_c\left(\dot{S} + (\nabla S - \mathcal{A})\cdot\mathcal{X} + \langle\phi, \widehat{H}_e\phi\rangle + \frac{1}{2}\|p\|_{g^{-1}}^2\right) + \mathcal{O}(\sqrt{\mu}) \right)\mathrm{d}r\mathrm{d}p, \tag{15}$$

where we decomposed the MQC Hamiltonian as $\widehat{H} = \|p\|_{g^{-1}}^2/2 + \widehat{H}_e$ into the classical kinetic energy operator for the nuclei and the quantum electronic part. Similarly to the discussion preceding Proposition 1, here we have retained the normalization condition for ϕ by letting $\phi = U\phi_0$. The latter implies $\delta\phi = \gamma\phi$, where γ is an arbitrary skew-Hermitian operator. A direct verification leads to

Proposition 2. *Consider the variational problem $\delta\int_{t_1}^{t_2}\ell_{\mathrm{QC}}\,\mathrm{d}t = 0$ associated to (15), with $\delta\phi = \gamma\phi$, $\delta\mathcal{X}$ as in (14), and arbitrary δD and δS. In the limit $\mu \to 0$, this action principle yields the following continuum PDE system:*

$$i)\ \partial_t\rho_c + \mathrm{div}(\rho_c X_{H_{cl}}) = 0, \quad ii)\ \partial_t S + \{S, H_{cl}\} = p\partial_p H_{cl} - H_{cl}, \quad iii)\ \widehat{H}_e\phi = E\phi,$$

where $H_{cl}(z) = \frac{1}{2}\|p\|_{g^{-1}}^2 + E(q)$.

We observe that the phase S decouples completely and the classical Liouville equation is solved by $\rho_c(r, p, t) = \delta(r - R(t))\delta(p - P(t))$, thereby recovering the phase-space formulation of the BOMD equations (2).

Similarly to the construction of the bohmion code, one can formulate a trajectory-based closure for the MQC model. In particular, we refer to [6] for details on the *koopmon method*. In short, a regularization process similar to the one in Sect. 3 leads to the koopmon Lagrangian

$$L(r, p, \varrho) = \sum_a w_a\left(p_a\dot{q}_a + \left\langle \rho_a, i\sqrt{\mu}\xi_a - \widehat{H}(q_a, p_q) - \frac{i\sqrt{\mu}}{2}\sum_b w_b\,[\rho_b, \mathcal{I}_{ab}] \right\rangle \right),$$

where $\mathcal{I}_{ab} := \int K_a\{K_b, \widehat{H}\}/(\sum_c w_c K_c)\,\mathrm{d}r\mathrm{d}p$ and $K_s(r,p) := K(r - q_s, p - p_s)$. Performing the limit $\mu \to 0$ gives us $\tilde{L} = \sum_a w_a(p_a\dot{q}_a - \langle \rho_a, \widehat{H}(q_a, p_a)\rangle)$, which returns the phase-space form of the bohmion implementation (13) of BOMD.

References

1. Eich, F.G., Agostini, F.: The adiabatic limit of the exact factorization of the electron-nuclear wave function. J. Chem. Phys. **145**(5), 054110 (2016)
2. Foskett, M.S., Holm, D.D., Tronci, C.: Geometry of nonadiabatic quantum hydrodynamics. Acta Appl. Math. **162**(1), 63–103 (2019)
3. Gay-Balmaz, F., Tronci, C.: Evolution of hybrid quantum-classical wavefunctions. Phys. D **440**, 133450 (2022)
4. Holm, D.D., Marsden, J.E., Ratiu, T.S.: The Euler-Poincaré equations and semidirect products with applications to continuum theories. Adv. Math. **137**(1), 1–81 (1998)
5. Lasser, C., Lubich, C.: Computing quantum dynamics in the semiclassical regime. Acta Numer. **29**, 229–401 (2020)
6. Tronci, C., Gay-Balmaz, F. Lagrangian trajectories and closure models in mixed quantum-classical dynamics. Lecture Notes in Comput. Sci. (2023). These proceedings. arXiv:2303.01975

Conservation Laws as Part of Lagrangian Reduction. Application to Image Evolution

Marco Castrillón López[✉]

Depto. Álgebra, Geometría y Topología, Facultad de Ciencias Matemáticas,
Universidad Complutense de Madrid, 28040 Madrid, Spain
mcastri@mat.ucm.es

Abstract. This note collects a series of results into a single formulation. Namely, it is proved that, when reduction is performed to a symmetric Lagrangian, the reduced variational equations can be split into two parts, one of them is exactly the Noether theorem. Conservation laws enter into reduction as part of the new variational equations. We give a short description of this situation in the case of evolution of circles in the plane.

Keywords: Conservation law · reduction · symmetry

1 Introduction

The introduction of Differential Geometry in the formulation of variational problems like those in Mechanics or Field theories has proved to be an essential point of view to understand many properties and results in depth. The so-called Geometric Variational Calculus exploits the notions of calculus in tensors to give a better formulation of essential problems both in pure or applied settings. In this framework the notion of symmetry plays a key role. The action of symmetry groups give, when written in terms of this geometric language, a fruitful description of two fundamental concepts in variational calculus: reduction of the configuration space and the existence of conservation laws. In the first case, with the action of these groups we can eliminate variables of the original problem. In the second case, we have the celebrated result of Noether. The goal of this short note is to give an explicit relationship between these two results.

Reduction theory has a long history in Mechanics, and a more recent formulation in Field Theories (see, for example, [2–7] and many of the references therein). The use of a connection is the way to split the variational equations into two sets: the vertical and the horizontal equations. It has been noticed in many of the references mentioned above that the vertical equations are forms of the Noether conservation law. Here, that equivalence is proven for variational problems of mappings from a finite dimensional manifold to a configuration space (another finite or infinite dimensional manifold). A short example is carried out for variational problems on closed curves, a setting closely related with image processing (see, for example, [1]).

© The Author(s), under exclusive license to Springer Nature Switzerland AG 2023
F. Nielsen and F. Barbaresco (Eds.): GSI 2023, LNCS 14072, pp. 242–250, 2023.
https://doi.org/10.1007/978-3-031-38299-4_26

2 Preliminaries

2.1 Bundles

Let G be a Lie group acting on a manifold Q. The action can be right or left but we choose right action in the following. The choice between right or left actions will induce a change of sign in the results below, but they remain essentially the same. We assume that the action is free and proper, so that the quotient set $\Sigma = Q/G$ can be endowed with a structure of smooth manifold and the natural projection $\pi : Q \to \Sigma$ is a principal G-bundle. In many of the main instances, the manifold Q is equipped with a (pseudo-)Riemannian metric g and the group G is usually taken to be a subgroup of the groups of isometries $\mathcal{I}(Q, g)$.

Recall that a principal connection on a principal bundle $Q \to \Sigma$ is a smooth distribution H such that $T_q Q = H_q \oplus V_q Q$, $q \in Q$, and $(R_g)_* H = H$, where $V_q Q = \ker d\pi$ and $R_g : Q \to Q$ stands for the (right) action by $g \in G$. The map

$$\rho : Q \times \mathfrak{g} \to VQ$$

$$(q, B) \mapsto B_q^Q = \frac{d}{d\varepsilon}\Big|\, R_{\exp(\varepsilon B)}(q)$$

is a vector bundle isomorphism over Q. Given a principal connection, we define its associated $\mathfrak{g}$-valued 1-form as $\omega(X) = \rho^{-1}(X^v)$, for $X \in T_q Q$, $q \in Q$, where $X = X^v + X^h$ with $X^h \in H_q$. This form satisfies that

$$\omega(B_q^Q) = B, \qquad \forall B \in \mathfrak{g},$$

and

$$R_g^* \omega = \mathrm{Ad}_{g^{-1}} \circ \omega, \qquad g \in G,$$

where $\mathrm{Ad}_g : \mathfrak{g} \to \mathfrak{g}$ stands for the adjoint representation. The form ω characterizes the connection and it is common to give a connection by means of the form only. This is particularly convenient when introducing the curvature of a connection ω. It is defined as the $\mathfrak{g}$-valued 2-form on Q given by

$$\Omega = d\omega + [\omega, \omega].$$

The distribution H is integrable in the sense of Frobenius (that is, for every $q \in Q$ there is a local section s of $Q \to \Sigma$ around q such that $\mathrm{im}(ds)_x = H_{s(x)}$ for all x in the domain) if and only if $\Omega = 0$. The adjoint bundle is the associated vector bundle $\tilde{\mathfrak{g}} = (Q \times \mathfrak{g})/G$, where the G action is

$$(q, B) \cdot g = (R_g(q), \mathrm{Ad}_{g^{-1}} B), \qquad q \in Q, B \in \mathfrak{g}, g \in G.$$

The curvature form can be regarded as a 2-form $\tilde{\Omega}$ on Σ with values in the adjoint bundle as

$$\tilde{\Omega}(X, Y) = [q, \Omega(X^h, Y^h)]_G \in \tilde{\mathfrak{g}},$$

for any vectors $X, Y \in T_\sigma \Sigma$, where $X^h, Y^h \in H_q$, for any $g \in \pi^{-1}(\sigma)$, are the horizontal lifts, that is $d\pi_q(X^h) = X$, $d\pi_q(Y^h) = Y$.

When the manifold Q is equipped with a Riemannian metric g and G is a group of isometries, then one can choose a natural horizontal distribution

$$H_q = V_q Q^\perp, \qquad q \in Q.$$

This principal connection plays an important role in different situations and it is usually known as the Mechanical connection.

2.2   Variational Calculus

Let N be a manifold and $J^1(N, Q)$ be the jet space of maps from N to Q. That is, the elements of this jet manifold are linear maps

$$\lambda_{(x,q)} : T_x N \to T_q Q,$$

for any $x \in N$, $q \in Q$. The map

$$p : J^1(N, Q) \to N \times Q \tag{1}$$
$$\lambda_{(x,q)} \mapsto (x, q) \tag{2}$$

is a vector bundle. Actually

$$J^1(N, Q) \simeq T^* N \otimes TQ,$$

where the tangent bundles are pullbacked to $N \times Q$. We could consider a more general class of jet spaces in the case of fiber bundles $P \to N$. In this situation, one deals with the bundle $J^1 P \to P$ of 1 -jets of local sections of $P \to N$. In that case, the bundle is affine. However, for the sake of clarity, we will confine ourselves in the case where $P = N \times Q \to N$ is a trivial bundle. In particular one can think of $N = \mathbb{R}$, so that

$$J^1(\mathbb{R}, Q) \simeq \mathbb{R} \times TQ.$$

This is the geometric setting for one particle Mechanics. If $N = \mathbb{R}^n$, $n \geq 2$, we have

$$J^1(\mathbb{R}^n, Q) \simeq \mathbb{R}^n \times ((\mathbb{R}^n)^* \otimes TQ)$$

that is the phase space of variational field theories in the Euclidean space. If we deal with finite dimensional manifolds and $(x^i)_{i=1}^n$, $(q^\alpha)_{\alpha=1}^m$ are coordinate on domains in N and Q respectively, the, we can build a coordinate system $(x^i, q^\alpha, q_i^\alpha)$ on $J^1(N, Q)$ by setting

$$q_i^\alpha(j_x^1 s) = \frac{\partial q^\alpha \circ s}{\partial x^i}(x), \qquad j_x^1 s \in J^1(N, Q).$$

Let $L : J^1(N, Q) \to \mathbb{R}$ be a smooth function, called Lagrangian. Given a fixed volume form $\mathbf{v} \in \Omega^n(N)$, a map $s : N \to Q$ is said to be critical for the variational problem defined by $L\mathbf{v}$ if for every compactly supported variation $\{s_\varepsilon\}_{\varepsilon \in \mathbb{R}}$ (that is, a smooth map $\hat{s} : N \times \mathbb{R} \to Q$, $\hat{s}(\cdot, \varepsilon) =: s_\varepsilon$, such that $s_0 = s$

and that $s_\varepsilon|_{N-U} = s|_{N-U}$, for an open domain $U \subset N$ with $\bar{U}$ compact) we have that

$$\frac{d}{d\varepsilon}\bigg|_{\varepsilon=0} \int_{\bar{U}} L(j^1 s_\varepsilon)v = 0.$$

Given coordinates $(x^i, q^\alpha, q^\alpha_i)$ as above, with $\mathbf{v} = dx^1 \wedge \cdots \wedge dx^n$, a section is critical if and only if

$$\frac{\partial}{\partial x^i}\left(\frac{\partial L}{\partial q^\alpha_i} \circ j^1 s\right) - \frac{\partial L}{\partial q^\alpha} = 0, \qquad \alpha = 1, ..., m.$$

These are the well known Euler Lagrange equations. Note that each variation $\{s_\varepsilon\}_{\varepsilon\in\mathbb{R}}$ define an infinitesimal variation

$$\delta s = \frac{\partial s_\varepsilon}{\partial \varepsilon}\bigg|_{\varepsilon=0},$$

that is, a vector field along s. Different variations may induce the same infinitesimal variation. It is important to know that the variational principle need not be applied for any variation, but just a family of variations such that their infinitesimal variation δs is arbitrary. This plays an important role when considering reduction.

The variational equations can be cast in a different way as follows. One can check that the n-form on $J^1(N, Q)$ defined as

$$\Theta_L = \left(\frac{\partial L}{\partial q^\alpha_i}dq^\alpha - q^\alpha_j dx^j\right) \wedge \mathbf{v}_i + L\mathbf{v},$$

where $\mathbf{v}_i = dx^1 \wedge ... \wedge dx^{i-1} \wedge dx^{i+1} \wedge ... \wedge dx^n$ is covariant and does not depend on the coordinate system. Then, it is easy to check that a mapping s is critical if and only if

$$(j^1 s)^* i_X d\Theta_L = 0, \tag{3}$$

for any vector field X belonging to the kernel of the differential of the natural map $J^1(N, Q) \to N$. The form Θ_L is called the Poincaré Cartan form of L. It is important to note that the variational equations as well as this Poincaré Cartan form require special attention when we have infinite dimensional manifolds. In that case, the equations may be written in other intrinsic terms.

We now assume that a connected Lie group G acts freely and properly on Q . This action can be naturally lift to the jet spaces as

$$J^1(N, Q) \times G \longrightarrow J^1(N, Q)$$
$$(j^1_x s, g) \mapsto j^1_x(R_g \circ s).$$

If the action on Q is proper and free, so is it on the jet space. Then $J^1(N, Q) \to J^1(N, Q)/G$ is a principal bundle. If a Lagrangian $L : J^1(M, Q) \to \mathbb{R}$ is an invariant function with respect to this action, it will define a so-called *reduced Lagrangian*

$$l : J^1(N, Q)/G \to \mathbb{R}.$$

Furthermore, the Lagrangian L is G-invariant if and only if the Lie derivative $L_{B^Q}\Theta_L = 0$, for any $B \in \mathfrak{g}$. Since $L_{B^Q} = d \circ i_{B^Q} + i_{B^Q} \circ d$, and according to 3, for a G-invariant Lagrangian L and any critical mapping $s : N \to Q$, we have that

$$d[(j^1s)^* i_{B^Q}\Theta_L] = 0,$$

on N. From the expression of the Poincaré-Cartán form, this is equivalent to

$$d[(j^1s)^* \langle \frac{\delta L}{\delta j^1 s}, B^Q \rangle \lrcorner \mathbf{v}] = 0,$$

where $\delta L/\delta j^1 s$ is the so called vertical derivative of L, that is, $\delta L/\delta j^1 s$ is the restriction of the differential of L to vectors tangent to the fibers of the fibration $J^1(N,Q) \to N \times Q$. Since $J^1(N,Q) \simeq T^*N \otimes TQ$ we have that

$$\delta L/\delta j^1 s \in TN \otimes T^*Q.$$

Therefore, when coupling the vertical derivative with B^Q we have that

$$(j^1s)^* \langle \delta L/\delta j^1 s, B^Q \rangle \in \mathfrak{X}(N)$$

is a vector field on N that is contracted with the volume form $\mathbf{v}$ to give a form. This is Noether theorem. And from it one usually defines the momentum map

$$\bar{\mu} : J^1(N,Q) \longrightarrow \wedge^{n-1}T^*N \otimes \mathfrak{g}^*$$
$$\langle \bar{\mu}(j^1_x s), B \rangle = \langle \frac{\delta L}{\delta j^1 s}, B^Q \rangle \lrcorner \mathbf{v}$$

It is useful to keep track of the points in Q. In addition the bundle $\wedge^{n-1}T^*N \to N$ is isomorphic to $TN \to N$ by means of the volume form $\mathbf{v}$. Taking into account these two remarks, we have the following expression for the momentum map

$$\mu : J^1(N,Q) \longrightarrow TN \otimes (Q \times \mathfrak{g}^*). \tag{4}$$

The Noether conservation law is simply $div(\mu \circ j^1 s) = 0$.

3   Reduction and Conservation Laws

We fix a principal connection form ω on the bundle $Q \to \Sigma$. It is not complicated to prove that the morphism of bundles over $N \times \Sigma$

$$J^1(N,Q)/G \longrightarrow J^1(N,\Sigma) \times (T^*N \otimes \tilde{\mathfrak{g}})$$
$$[j^1_x s]_G \mapsto (j^1_x[s]_G, [s(x), \omega \circ ds]_G)$$

is a diffeomorphism. For a mapping $s : N \to Q$, we write $\sigma = [s]_G$ and $\upsilon = [s(x), \omega \circ ds]_G$. Note that υ determines σ. The right hand side of this identification is more advantageous for the reduced Lagrangians l since, in particular, it reveals the jet bundle of of the reduced space Σ. However, the new variational principle will not be standard in the sense that this phase space has something more: the vector bundle of covectors on N with values in $\tilde{\mathfrak{g}}$. To clarify the nature of the variations of υ, we can consider arbitrary variations δs, and project its lift from $J^1(N, Q)$ to $J^1(N, \Sigma) \times (T^*N \otimes \tilde{\mathfrak{g}})$. For that purpose, we make use of the connection form ω again. In particular, we consider horizontal variations (that is, $\omega(\delta s) = 0$) and vertical variations (that is, $\delta s = \eta^Q$ for a map $\eta : N \to \mathfrak{g}$) separately, since any variation will be the sum of two of these. Then we have that:

– For infinitesimal vertical variations $\delta s = \eta^Q$, we obtain that the infinitesimal variations of *upsilon* are of the type

$$\delta \upsilon = \nabla \tilde{\eta} + [\tilde{\eta}, \upsilon], \tag{5}$$

where $\tilde{\eta} : N \to \tilde{\mathfrak{g}}$ is the map $\tilde{\eta}(x) = [s(x), \eta(x)]_G$. The bracket $[\tilde{\eta}, \upsilon]$ is the one of the Lie algebra bundle structure of $\tilde{\mathfrak{g}} \to \Sigma$, and the covariant differential is performed with respect to the connection ω, since $\tilde{\mathfrak{g}}$ is an associated bundle. Note that this infinitesimal variation takes values in $T^*N \otimes \tilde{\mathfrak{g}}$, since this variation is vertical with respect to the vector bundle $T^*N \otimes \tilde{\mathfrak{g}} \to N \times Q$. Furthermore, not any possible infinitesimal variation of υ are of the type (5). The reduced variational problem is constrained.

– For infinitesimal horizontal variations δs ($\omega(\delta s) = 0$), on one hand we have the infinitesimal variation $\delta \sigma$ of $\sigma = [s]_G$. The projection of these δs provides any infinitesimal variation $\delta \sigma$ with no constraint. However, we also have a part of $\delta \upsilon$. More precisely, we get

$$\delta \upsilon = \delta \sigma^h + \tilde{\Omega}(\delta \sigma, \cdot),$$

where $\delta \sigma^h$ is the horizontal lift of $\delta \sigma$ with respect to the connection ω, and $\tilde{\Omega}$ its curvature.

These special type of variations define new variational equations. Again, since the study has been split into vertical and horizontal δs, we have two set of equations for each type. They are the horizontal equations

$$\frac{\delta^\nabla l}{\delta \sigma} - div^\nabla \left(\frac{\delta l}{\delta j^1 \sigma} \right) = \langle \frac{\delta l}{\delta \upsilon}, \tilde{\Omega}(d\delta, \cdot) \rangle,$$

and the vertical equations

$$div^\nabla \frac{\delta l}{\delta \upsilon} = ad_\upsilon^* \frac{\delta l}{\delta \upsilon}.$$

The notation $\delta l / \delta \upsilon$ stands for the vertical differential with respect to the fibration $T^*N \otimes \tilde{\mathfrak{g}}^* \to N \times \Sigma$. On the other hand, the operator div^∇ is the divergence operator defined by the connection ω for $\tilde{\mathfrak{g}}$-valued vector fields on N, that is, for $\mathcal{X} \in C^\infty(N \to TN \otimes \tilde{\mathfrak{g}}^*)$ it is characterized by

$$div\langle \mathcal{X}, \eta \rangle = \langle div^\nabla \mathcal{X}, \eta \rangle + \langle \mathcal{X}, \nabla \eta \rangle, \qquad \forall \eta \in C^\infty(N, \tilde{\mathfrak{g}}). \tag{6}$$

We are interested in the vertical part of these reduced variational equations. We have the main result.

Theorem 1. *Let N be a manifold and let $Q \to \Sigma$ be a G-principal bundle endowed with a principal connection form ω. Given a G-invariant Lagrangian $L : J^1(N, Q) \to \mathbb{R}$, the momentum map $\mu : J^1(N, Q) \to TN \times (Q \times \mathfrak{g})$ introduced in (4) is equivariant and projects to the quotient*

$$\tilde{\mu} : J^1(N, \Sigma) \times (T^*N \otimes \tilde{\mathfrak{g}}) \longrightarrow TN \otimes \tilde{\mathfrak{g}}^*.$$

For a map $s : N \to Q$, we denote by $\upsilon : N \to T^*N \otimes \tilde{\mathfrak{g}}$ the reduced section $\upsilon = [s(x), \omega \circ ds]_G$. Then the condition $div(\mu \circ j^1 s) = 0$ projects to the vertical equation

$$div^\nabla (\tilde{\mu} \circ (j^1\sigma, \upsilon)) = div^\nabla \frac{\delta l}{\delta \upsilon} - \mathrm{ad}^*_\upsilon \frac{\delta l}{\delta \upsilon} = 0.$$

Proof. Given a map s, since $J^1(N, Q) \simeq T^*N \otimes TQ$, and $VQ \simeq Q \times \mathfrak{g}$, we have that $\mu = \delta L / \delta j^1 s \in C^\infty(N, TN \otimes (Q \times \mathfrak{g}^*))$. The G invariance of L provides the equivariance of μ to the corresponding reduced spaces. We also get that $\tilde{\mu} = \delta l / \delta \upsilon$. Furthermore, for any map $\eta : N \to \mathfrak{g}$ we have that

$$div\langle \mu \circ j^1 s, \eta \rangle = div\langle \tilde{\mu} \circ \upsilon, \tilde{\eta} \rangle.$$

Noether theorem consists of the vanishing of the previous expression for maps η that are constant. The corresponding condition for not constant maps η is

$$div\langle \mu \circ j^1 s, \eta \rangle - \langle \mu \circ j^1 s, \nabla^s \eta \rangle = div\langle \tilde{\mu} \circ \upsilon, \tilde{\eta} \rangle - \langle \tilde{\mu} \circ \upsilon, \nabla^{\omega+\upsilon}\tilde{\eta} \rangle = 0.$$

From (6)

$$div\langle \mu \circ j^1 s, \eta \rangle - \langle \tilde{\mu} \circ \upsilon, \nabla^{\omega+\upsilon}\tilde{\eta} \rangle = \langle div^\nabla(\tilde{\mu} \circ \upsilon), \tilde{\eta} \rangle + \langle \tilde{\mu} \circ \upsilon, \nabla\tilde{\eta} \rangle - \langle \tilde{\mu} \circ \upsilon, \nabla^{\omega+\upsilon}\tilde{\eta} \rangle \quad (7)$$

$$= \langle div^\nabla(\tilde{\mu} \circ \upsilon), \tilde{\eta} \rangle - \langle \tilde{\mu} \circ \upsilon, [\upsilon, \tilde{\eta}] \rangle \quad (8)$$

$$= \langle div^\nabla(\tilde{\mu} \circ \upsilon) - \mathrm{ad}^*_\upsilon(\tilde{\mu} \circ \upsilon), \tilde{\eta} \rangle, \quad (9)$$

and the proof is complete.

4   Evolution of Plane Curves

We consider the (infinite) dimensional manifold $Q = \mathrm{Emb}(S^1, V)$ of embeddings of the canonical unitary circle in a vector space V, which is usually taken to be the Euclidean plane $V = \mathbb{R}^2$. We consider $N = \mathbb{R}$ or $\mathbb{R}^2$. Mappings from N to Q are to be thought as time evolution of circles (images) in V for $N = \mathbb{R}$, or the time evolution of a continuously parametrized family of curves for $N = \mathbb{R}^2$. This last model is found, for example, when one has a series of tomographies of a subject taken at different times. The group of symmetries under consideration is going to be $G = \mathrm{Diff}(S^1)$ so that the quotient Q/G is the set Σ of smooth closed curves in V without self intersections. In many applications, the interesting variational problems on closed curves do not depend on the parametrization of the curve but just on its trace in V, that is, we are in the context defined by the quotient $\Sigma = \mathrm{Emb}(S^1, V)/\mathrm{Diff}(S^1)$. There is a canonical connection in the principal bundle $\mathrm{Emb}(S^1, V) \to \Sigma$. Since $T_c\mathrm{Emb}(S^1, V) = C^\infty(S^1, V)$ and the Lie algebra of $\mathrm{Diff}(S^1)$ is $\mathfrak{X}(S^1)$, we can consider

$$H_c = \{X \in C^\infty(S^1, V) : X(\theta) \perp c'(\theta)\}.$$

We write elements of $J^1(\mathbb{R}^2, \mathrm{Emb}(S^1, V))$ as

$$j^1_{(x,t)}c = c_t(\theta)(x,t)dt + c_x(\theta)(x,t)dx, \quad \theta \in S^1, (x,t) \in \mathbb{R}^2.$$

There are many interesting Lagrangians $L : J^1(\mathbb{R}^2, \mathrm{Emb}(S^1, V)) \to \mathbb{R}$ that are $\mathrm{Diff}(S^1)$-invariant. In particular, the Lagrangians $L = L_1 + L_2$ with

$$L_1(j^1 c) = \int_{S^1} (1 + \kappa(\theta)^2)((c_t^h)^2 + (c_x^h)^2)d\theta,$$

$$L_2(j^1 c) = \int_{S^1} (1 + \kappa(\theta)^2)((c_t^v)^2 + (c_x^v)^2)d\theta,$$

or

$$L_1(j^1 c) = \tfrac{1}{2}\int_{S^1} ((c_t^h)^2 + (c_x^h)^2 + (\partial_\theta c_t^h)^2 + (\partial_\theta c_x^h)^2)d\theta,$$

$$L_2(j^1 c) = \tfrac{1}{2}\int_{S^1} ((c_t^v)^2 + (c_x^v)^2 + (\partial_\theta c_t^v)^2 + (\partial_\theta c_x^v)^2)d\theta.$$

For the reduced Lagrangian l, we have that

$$\frac{\delta l}{\delta \upsilon} = \frac{\delta l_2}{\delta \upsilon}$$

where $\upsilon \in T^*\mathbb{R}^2 \times \widetilde{\mathfrak{X}(S^1)}$, where the fibers of adjoint bundle $\widetilde{\mathfrak{X}(S^1)} \to \Sigma$ for $\sigma \in \Sigma$ is simply $\mathfrak{X}(\sigma)$, a vector field along the trace of σ. In particular

$$\frac{\delta l_2}{\delta \upsilon} = \int_\sigma (1 + \kappa(\theta)^2)(\upsilon_t \frac{\partial}{\partial t} + \upsilon_x \frac{\partial}{\partial x})d\sigma, \qquad \text{or}$$

$$\frac{\delta l_2}{\delta \upsilon} = \int_\sigma (\upsilon_t \frac{\partial}{\partial t} + \upsilon_x \frac{\partial}{\partial x} + (\partial_\theta \upsilon_t)\partial_\theta \frac{\partial}{\partial t} + (\partial_\theta \upsilon_x)\partial_\theta \frac{\partial}{\partial x})d\sigma.$$

In the second case, for example, the vertical reduced equation (the Noether conservation law) is

$$\upsilon_{tt} + \upsilon_{xx} + \partial_\theta \upsilon_{tt} + \partial_\theta \upsilon_{xx} = 0,$$

since in this case $\upsilon = \delta l_2 / \delta \upsilon$.

Acknowledgements. Work partially funded by Agencia Estatal de Investigación (Spain), under grantno. PID2021-126124NB-I00.

References

1. Bauer, M., Bruveris, M., Marsland, S., Michor, P.: Constructing reparameterization invariant metrics on spaces of plane curves. Differ. Geom. Appl. **34**, 139–165 (2014)
2. Castrillón López, M., García Pérez, P.L., Ratiu, T.S: Euler-Poincaré reduction on principal bundles. Lett. Math. Phys. **58**, 167–180 (2001). https://doi.org/10.1023/A:1013303320765
3. Castrillón López, M., Marsden, J.E.: Covariant and dynamical reduction for principal bundle field theories. Ann. Global Anal. Geom. **34**(3), 263–285 (2008)
4. Castrillón López, M., Ratiu, T.S.: Reduction in principal bundles: covariant lagrange-poincaré equations. Commun. Math. Phys. **236**, 223250 (2003). https://doi.org/10.1007/s00220-003-0797-5
5. Cendra, H., Marsden, J.E., Ratiu, T.S.: Lagrangian reduction by stages. Mem. Amer. Math. Soc. **152**(722) (2001)
6. Ellis, D., Gay-Balmaz, F., Holm, D., Ratiu, T.S.: Lagrange-Poincaré field equations. J. Geom. Phys. **61**(11), 2120–2146 (2011)
7. Marsden, J.E., Ratiu, T.S.: Introduction to Mechanics and Symmetry, 2nd edn. Texts in Applied Mathematics, vol. 17. Springer, New York (1999). https://doi.org/10.1007/978-0-387-21792-5

Fluid Mechanics and Symmetry

Casimir-Dissipation Stabilized Stochastic Rotating Shallow Water Equations on the Sphere

Werner Bauer[1]([⊠])[ID] and Rüdiger Brecht[2][ID]

[1] University of Surrey, Guildford GU2 7XH, UK
w.bauer@surrey.ac.uk
[2] Universität Hamburg, Hamburg, Germany

Abstract. We introduce a structure preserving discretization of stochastic rotating shallow water equations, stabilized with an energy conserving Casimir (i.e. potential enstrophy) dissipation. A stabilization of a stochastic scheme is usually required as, by modeling subgrid effects via stochastic processes, small scale features are injected which often lead to noise on the grid scale and numerical instability. Such noise is usually dissipated with a standard diffusion via a Laplacian which necessarily also dissipates energy. In this contribution we study the effects of using an energy preserving selective Casimir dissipation method compared to diffusion via a Laplacian. For both, we analyze stability and accuracy of the stochastic scheme. The results for a test case of a barotropically unstable jet show that Casimir dissipation allows for stable simulations that preserve energy and exhibit more dynamics than comparable runs that use a Laplacian.

Keywords: Stochastic flow model · Rotating shallow water · Casimir dissipation · structure preservation

1 Introduction

Stochastic modeling allows one to not only simulate the time evolution of a dynamical system but also to estimate the reliability of the prediction through ensemble runs. When estimating the related uncertainty, it takes into account errors in measurements, in the mathematical model, and via numerical approximations. This is an important advantage over traditional approaches of deterministic models where such uncertainty estimates are not available.

Apart from uncertainty estimates, it is important to preserve quantities such as mass, energy, or circulation to guarantee that the statistics of the solutions is correctly represented and that the numerical schemes remain stable and accurate. Deriving consistent stochastic models that preserve such invariants is challenging, but can be achieved using e.g. the Location Uncertainty (LU) framework [6]. For example, in [3] we derived energy preserving stochastic rotating shallow

© The Author(s), under exclusive license to Springer Nature Switzerland AG 2023
F. Nielsen and F. Barbaresco (Eds.): GSI 2023, LNCS 14072, pp. 253–262, 2023.
https://doi.org/10.1007/978-3-031-38299-4_27

water (RSW) equations. However, there are no satisfying structure preserving discretization methods available for such stochastic flow models.

As a first step towards a consistent, energy preserving discretization of these stochastic RSW equations, we applied in [3] a variational discretization for the deterministic part (introduced in [1]) while treating the stochastic terms using standard finite difference (FD) operators. In the following, we refer to this discrete system as stochastic RSW-LU scheme. As shown in [3], for homogeneous noise where no additional diffusion (in form of a Laplace operator) was required, this system preserved spatially the energy (currently there is no suitable energy preserving stochastic time integrator available). However, when modeling subgrid effects via stochastic processes, small scale features are injected on the grid level leading to numerical instabilities, especially for inhomogeneous noise. Thus, we applied a biharmonic Laplacian for stabilization, which dissipates energy.

As a remedy to this latter point, we suggest here to use Casimir dissipation for stabilization to avoid the dissipative effect of the Laplacian on the total energy while keeping the stochastic scheme stable also in case of inhomogenous noise. The suggested combination is possible because the derivation of the Casimir dissipation for the RSW equations in [2] follows the same variational discretization principle of [1] applied to the RSW equations that form the deterministic core of the stochastic RSW-LU model of [3].

In Sect. 2, we introduce briefly the main idea and some steps of the derivation of the stochastic flow model. In Sect. 3, we explain shortly how the variational principle can be used to derive a scale selective Casimir dissipation, and we combine it with the stochastic model. In Sect. 4, we test the novel scheme on a standard test cases focusing on the energy preserving properties. In Sect. 5, we draw some conclusions.

2   Stochastic RSW Equations Under Location Uncertainty

We briefly introduce the stochastic RSW equations under Location Uncertainty [6] (RSW-LU) used for this study. The full derivation can be found in [3].

Let $\boldsymbol{u} = (u, v)$ be the 2D velocity and h the water depth of the RSW flow. The unresolved random flow is modeled by the random flow term $\boldsymbol{\sigma} d\boldsymbol{B}_t$ (of null ensemble mean) that can be computed with the spectral decomposition: $\boldsymbol{\sigma}(\boldsymbol{x}, t) \, d\boldsymbol{B}_t = \sum_{n \in \mathbb{N}} \boldsymbol{\Phi}_n(\boldsymbol{x}, t) \, d\beta_t^n$, $\boldsymbol{a}(\boldsymbol{x}, t) = \sum_{n \in \mathbb{N}} \boldsymbol{\Phi}_n(\boldsymbol{x}, t) \boldsymbol{\Phi}_n^T(\boldsymbol{x}, t)$, where β^n denotes n independent and identically distributed (i.i.d.) one-dimensional standard Brownian motions. Here, we used the fact that the noise $\boldsymbol{\sigma} d\boldsymbol{B}_t$ and its variance $\boldsymbol{a}$ (measuring the strength of the noise) can be represented by an orthogonal eigenfunction basis $\{\boldsymbol{\Phi}_n(\bullet, t)\}_{n \in \mathbb{N}}$ weighted by the eigenvalues $\Lambda_n \geq 0$ such that $\sum_{n \in \mathbb{N}} \Lambda_n < \infty$. Note that the matrix $\boldsymbol{a}$ has the unit of a diffusion (i.e. $\mathrm{m}^2 \cdot \mathrm{s}^{-1}$). The basis functions can be constructed by numerical or observational data and various forms of noise can be used in the simulations, see e.g. [3]. This allows us to formulate the RSW-LU as:

$$\mathrm{d}_t\boldsymbol{u} = \overbrace{\Big(-\boldsymbol{u}\cdot\boldsymbol{\nabla}\,\boldsymbol{u} - \boldsymbol{f}\times\boldsymbol{u} - g\boldsymbol{\nabla}h \Big)}^{:=\mathbf{det}}\,\mathrm{d}t + \overbrace{\Big(\frac{1}{2}\boldsymbol{\nabla}\cdot\boldsymbol{\nabla}\cdot(\mathrm{a}\mathrm{u})\,\mathrm{d}t - \sigma\mathrm{d}\boldsymbol{B}_t\cdot\boldsymbol{\nabla}\,\mathrm{u}\Big)}^{:=\mathbf{sto}^V}, \quad (1)$$

$$\mathrm{d}_t h = -\,\boldsymbol{\nabla}\cdot(\boldsymbol{u}h)\,\mathrm{d}t + \underbrace{\Big(\frac{1}{2}\,\boldsymbol{\nabla}\cdot\boldsymbol{\nabla}\cdot(ah)\,\mathrm{d}t - \sigma\mathrm{d}\boldsymbol{B}_t\cdot\boldsymbol{\nabla}\,h\Big)}_{:=\mathbf{sto}^h}, \quad (2)$$

where f is the Coriolis parameter and g the gravitational acceleration and where $\frac{1}{2}\boldsymbol{\nabla}\cdot\boldsymbol{\nabla}\cdot(\mathbf{a}\theta) = \frac{1}{2}(\boldsymbol{\nabla}\cdot\mathbf{a})\cdot\nabla\theta + \frac{1}{2}\boldsymbol{\nabla}\cdot(\mathbf{a}\nabla\theta)$ with $\boldsymbol{\nabla}\cdot\mathbf{a} = \sum_i \frac{\partial a_{ik}}{\partial x_i}$ for the coordinate functions $\theta = u, v, h$.

As suggested in [3], our approach is to discretize the deterministic parts (the det and $-\nabla\cdot(\boldsymbol{u}h)$ terms) with a variational discretization [1] and the stochastic terms with standard finite difference operators. Then, we add an energy preserving Casimir dissipation [2] or biharmonic Laplacian dissipation.

3   Discretization of RSW-LU with Casimir Dissipation

In this section we first introduce the variational principle that underpins the derivation of the discrete dissipative variational equations before we combine this deterministic scheme with approximations of the stochastic terms such that together they form a semi-discrete approximation of the RSW-LU equations.

3.1   Discrete Variational Equations

We consider a two-dimensional simplicial mesh $\mathbb{M}$ with n cells on the fluids domain, where triangles (T) are used as the primal grid, and the circumcenter dual (ζ) as the dual grid. The notation is explained in Fig. 1.

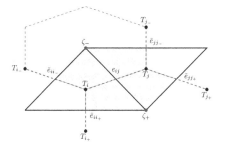

T_i	triangle with area Ω_{ii}
$\zeta_\pm$	dual cell
e_{ij}	primal edge $T_i \cap T_j$
$\tilde{e}_{ij}$	dual edge $\zeta_+ \cap \zeta_-$
$\mathbf{n}_{ij}$	e_{ij} normal vector towards T_j
h_i	waterdepth on T_i
$\bar{h}_{ij}$	$\frac{1}{2}(h_i + h_j)$
V_{ij}	$(\mathbf{u}\cdot\mathbf{n})(e_{ij})$

Fig. 1. Notation for 2D simplicial mesh.

Following [1,3], let matrix A approximate the vector field $\boldsymbol{u}$ and vector $h \in \mathbb{R}^n$ the fluid depths from RSW-LU. Here, piecewise constant functions on the domain

$\mathbb{M}$ are represented by vectors $F \in \mathbb{R}^n$, with value F_i on cell i being the cell average of the continuous function on cell i. The space of discrete functions is denoted by $\Omega_d^0(\mathbb{M})$, and the space of discrete densities $\mathrm{Den}_d(\mathbb{M}) \simeq \mathbb{R}^n$ is defined as the dual space to $\Omega_d^0(\mathbb{M})$ relative to the pairing: $\langle F, G \rangle_0 = F^\top \Omega G$, where Ω is a nondegenerate diagonal $n \times n$ matrix with elements $\Omega_{ii} = \mathrm{Vol}(T_i)$, the area of triangle T_i.

A is an element of the Lie algebra $\mathfrak{d}(\mathbb{M}) = \{A \in \mathfrak{gl}(n) \mid A \cdot \mathbf{1} = 0\}$ with the matrix commutator $[A, B] = AB - BA$ as the Lie bracket, where $\mathfrak{gl}(n)$ is the Lie algebra of $n \times n$ real matrices. To obtain approximations of vector fields we consider a subspace $\mathcal{R} \subset \mathfrak{d}(\mathbb{M})$ in which $\mathcal{R} = R_1 \cap R_2$, $R_1 = \{A \in \mathfrak{d}(\mathbb{M}) \mid A^\top \Omega + \Omega A^\top \text{ is diagonal}\}$ and $R_2 = \{A \in \mathfrak{d}(\mathbb{M}) \mid A_{ij} = 0 \quad \forall j \notin N(i)\}$ with $N(i)$ being the set of cells sharing an edge with the cell T_i. Further, $A_{ij} = -\frac{|e_{ij}|}{2\Omega_{ii}} V_{ij}$, $j \in N(i)$ and $A_{ii} = \frac{1}{2\Omega_{ii}} \sum_{k \in N(i)} |e_{ik}| V_{ik}$, where the former is an off diagonal element associated to the triangular edge with normal pointing from cells i to j, while the latter is a diagonal element associated to cell i. We identify the dual space $\mathcal{R}^*$ with the space $\Omega_d^1(\mathbb{M})$ of discrete one-forms relative to the duality pairing on $\mathfrak{gl}(n)$: $\langle L, A \rangle_1 = \mathrm{Tr}(L^\top \Omega A)$. To obtain an element in $\mathcal{R}^*$, $P \colon \mathfrak{gl}(n) \to \Omega_d^1(\mathbb{M})$ is used, defined by $P(L)_{ij} = \frac{1}{2}(L_{ij} - L_{ji} - L_{ii} + L_{jj})$, which satisfies $\langle L, A \rangle_1 = \langle P(L), A \rangle_1$, for all $A \in \mathcal{R}$, cf. [1] for details.

Then, let $\ell \colon \mathfrak{d}(\mathbb{M}) \times \mathrm{Den}_d(\mathbb{M}) \to \mathbb{R}$ be a semi-discrete Lagrangian and $C \colon \mathfrak{d}(\mathbb{M}) \times \mathrm{Den}_d(\mathbb{M}) \to \mathbb{R}$ be a semi-discretized approximation of a Casimir.

Theorem 1 (Discrete dissipative variational equations). *For a semi-discrete Lagrangian $\ell(A, h)$, the curves $A(t), h(t) \in \mathcal{R}$ are critical for the variational principle of Eq. (28) in [2] if and only if they satisfy*

$$P\left(\frac{d}{dt}\frac{\delta\ell}{\delta A} + \mathcal{L}_A\left(\frac{\delta\ell}{\delta A}\right) - \theta\mathcal{L}_A\left(h\left[\!\left[\frac{\delta C}{\delta M}, A\right]\!\right]^\flat\right) + h\frac{\delta\ell}{\delta h}^\top\right)_{ij} = 0, \qquad (3)$$

where $\mathcal{L}$ is the discrete analog to the Lie derivative $\mathfrak{L}$ and it is defined by the commutator via the following relation $\langle \mathcal{L}_A M, B \rangle_1 = \langle M, [A, B] \rangle_1$. The proof can be found in [2].

Note that for $A, B \in \mathcal{R}$ we have $[A, B]_{ij} = 0$ for all $j \in N(i)$. Since elements of $\mathcal{R}$ are zero for non-neighboring cells, we get $[\mathcal{R}, \mathcal{R}] \cap \mathcal{R} = \{0\}$. In particular $[\mathcal{R}, \mathcal{R}] \neq \mathcal{R}$ hence the subspace $\mathcal{R} \subset \mathfrak{d}(\mathbb{M})$ corresponds to a nonholonomic constraint. Consequently, we need to define a discrete commutator $[\![,]\!]$ such that $[\![A, B]\!] \in \mathcal{R}$, so that we can directly apply the definition of the discrete flat operator to the discrete commutator, cf. [3] for details on this construction.

In the case of the RSW equations, we have the discrete Lagrangian

$$\ell(A, h) = \frac{1}{2}\sum_{i,j=1}^n h_i A_{ij}^\flat A_{ij}\Omega_{ii} + \sum_{i,j=1}^n h_i R_{ij}^\flat A_{ij}\Omega_{ii} - \frac{1}{2}\sum_{i=1}^n g(h_i + (\eta_b)_i)^2\Omega_{ii},$$

$$(4)$$

where R is the vector potential of the angular velocity of the Earth and $\flat$ is the discrete flat operator [1]. The functional derivatives are given by $\frac{\delta \ell}{\delta A}_{ij} = h_i(A_{ij}^\flat + R_{ij}^\flat)$ and $\frac{\delta \ell}{\delta h}_i = \frac{1}{2} \sum_j A_{ij}^\flat A_{ij} + \sum_j R_{ij}^\flat A_{ij} - g(h_i + (\eta_b)_i)$.

We define the discrete enstrophy Casimir (i.e. the discrete potential enstrophy) as

$$\mathcal{C}(M,h) = \frac{1}{2} \sum_\zeta h_\zeta \left(q(M,h)_\zeta \right)^2 |\zeta|, \tag{5}$$

for $M = \frac{\delta \ell}{\delta A}$, with discrete potential vorticity $q(M,h)_\zeta = \frac{(\text{Curl } V)_\zeta + f}{h_\zeta}$ and discrete fluid depth (mass) $h_\zeta = \sum_{T_i \cap \zeta \neq \emptyset} \frac{|T_i \cap \zeta|}{|\zeta|} h_i$, which are discrete realizations of the mass, potential vorticity and potential enstrophy Casimirs of the continuous RSW equations. The functional derivative of $\mathcal{C}$ reads $\left(\frac{\delta \mathcal{C}}{\delta M} \right)_{ij} = -\frac{|e_{ij}|}{2\Omega_{ii}} \frac{2 \text{ Grad}_t q}{\bar{h}_{ij}}$ with Grad_t defined below, see computations in [2].

3.2 Semi-discrete RSW-LU Scheme

The resulting system of semi-discrete equations can be split into a **det**erministic, **diff**usive and **sto**chastic part. In the momentum equation we include the two different diffusion methods (i.e. Casimir or biharmonic Laplacian) which can be applied by setting $\theta > 0$ or $\nu > 0$:

$$\begin{aligned}
\partial_t V_{ij} &= -\mathbf{det}_{ij} - \theta \, \mathbf{diff}_{ij}^{\text{CD}} - \nu \, \mathbf{diff}_{ij}^{\text{BD}} + \mathbf{sto}_{ij}^V, \\
\partial_t h_i &= -\text{Div}(\bar{h}V)_i + \mathbf{sto}_i^h.
\end{aligned} \tag{6}$$

Note that the terms $\mathbf{det}_{ij}$ and $\theta \, \mathbf{diff}_{ij}^{\text{CD}}$ in the momentum equation together with $\text{Div}(\bar{h}V)_i$ in the continuity equation constitute the Casimir-dissipation stabilized RSW equations that result from Eq. (3).

Before specifying these terms, we define the following finite differences operators:

$$(\text{Grad}_n F)_{ij} = \frac{F_{T_j} - F_{T_i}}{|\tilde{e}_{ij}|}, \qquad (\text{Div } V)_i = \frac{1}{|T_i|} \sum_{k \in \{j, i_-, i_+\}} |e_{ik}| V_{ik},$$

$$(\text{Grad}_t F)_{ij} = \frac{F_{\zeta_-} - F_{\zeta_+}}{|e_{ij}|}, \qquad (\text{Curl } V)_\zeta = \frac{1}{|\zeta|} \sum_{nm \in \partial \zeta} nm |V_{nm}|.$$

Predicting only the normal velocity component V_{ij} for all edges, for some terms we need the full velocity field; for this we apply the reconstruction:
$\mathbf{u}_i = \frac{1}{\Omega_{ii}} \sum_{k \in \{j, i_-, i_+\}} |e_{ik}| (\mathbf{x}_{e_{ik}} - \mathbf{x}_{T_i}) V_{ik}$.

The various deterministic terms in Eq. (6) are given by:

$$
\mathbf{det}_{ij} := -\frac{(\mathrm{Curl}\,V + f)_{\zeta_-}}{\overline{h}_{ij}|\tilde{e}_{ij}|}\left(\frac{|\zeta_- \cap T_i|}{2\Omega_{ii}}\overline{h}_{ji_-}|e_{ii_-}|V_{ii_-} + \frac{|\zeta_- \cap T_j|}{2\Omega_{jj}}\overline{h}_{ij_-}|e_{jj_-}|V_{jj_-}\right)
$$
$$
+ \frac{(\mathrm{Curl}\,V + f)_{\zeta_+}}{\overline{h}_{ij}|\tilde{e}_{ij}|}\left(\frac{|\zeta_+ \cap T_i|}{2\Omega_{ii}}\overline{h}_{ji_+}|e_{ii_+}|V_{ii_+} + \frac{|\zeta_+ \cap T_j|}{2\Omega_{jj}}\overline{h}_{ij_+}|e_{jj_+}|V_{jj_+}\right)
$$
$$
- \frac{1}{2}\left(\mathrm{Grad}_n\left(\sum_{k\in\{j,i_-,i_+\}}\frac{|\tilde{e}_{ik}|\,|e_{ik}|(V_{ik})^2}{2\Omega_{kk}}\right)\right)_{ij} - g(\mathrm{Grad}_n\,(h+\eta_b))_{ij}\,,
$$

$$
\mathbf{diff}_{ij}^{\mathrm{CD}} := -\frac{(\mathrm{Curl}\,\widetilde{W})_{\zeta_-}}{\overline{h}_{ij}|\tilde{e}_{ij}|}\left(\frac{|\zeta_- \cap T_i|}{2\Omega_{ii}}\overline{h}_{ji_-}|e_{ii_-}|V_{ii_-} + \frac{|\zeta_- \cap T_j|}{2\Omega_{jj}}\overline{h}_{ij_-}|e_{jj_-}|V_{jj_-}\right)
$$
$$
+ \frac{(\mathrm{Curl}\,\widetilde{W})_{\zeta_+}}{\overline{h}_{ij}|\tilde{e}_{ij}|}\left(\frac{|\zeta_+ \cap T_i|}{2\Omega_{ii}}\overline{h}_{ji_+}|e_{ii_+}|V_{ii_+} + \frac{|\zeta_+ \cap T_j|}{2\Omega_{jj}}\overline{h}_{ij_+}|e_{jj_+}|V_{jj_+}\right)
$$
$$
+ \frac{(2\widetilde{W}_{ij})}{\overline{h}_{ij}}\frac{\mathrm{Div}(V\overline{h})_i + \mathrm{Div}(V\overline{h})_j}{2} - \frac{1}{2}\left(\mathrm{Grad}_n\left(\sum_{k\in\{j,i_-,i_+\}}\frac{|\tilde{e}_{ik}|\,|e_{ik}|(V_{ik}\widetilde{W}_{ik})}{\Omega_{kk}}\right)\right)_{ij}\,,
$$

$$
\mathbf{diff}_{ij}^{\mathrm{BD}} := \mathrm{Lap}(\mathrm{Lap}(V))_{ij} \text{ with } \mathrm{Lap}(V)_{ij} := (\mathrm{Grad}_n(\mathrm{Div}(V))_{ij} - \mathrm{Grad}_t(\mathrm{Curl}V)_{ij})\,,
$$

where $\widetilde{W}_{ij}$ is given by

$$
\widetilde{W}_{ij} = \left(\frac{\delta C}{\delta M}\right)_{ij}\left(\frac{\mathrm{Div}(V)_i + \mathrm{Div}(V)_j}{2}\right) - V_{ij}\left(\frac{\mathrm{Div}\left(\frac{\delta C}{\delta M}\right)_i + \mathrm{Div}\left(\frac{\delta C}{\delta M}\right)_j}{2}\right)
$$
$$
- \mathrm{Grad}_t\left(\left(\frac{\delta C}{\delta \mathbf{m}_\zeta} \times \mathbf{u}_\zeta\right)\cdot \mathbf{k}_\zeta\right)_{ij} \text{ with } \mathbf{u}_\zeta = \sum_{i\in N(\zeta)}\frac{|\zeta \cap T_i|}{|\zeta|}\mathbf{u}_i.
$$

The stochastic term $\mathbf{sto}_{ij}^V := \mathbf{sto}^V(ij)\cdot\mathbf{n}_{ij}$ follows from an evaluation of the stochastic terms in (1) at the edge midpoints ij and a projection onto the edges' normals, and $\mathbf{sto}_i^h := \mathbf{sto}^h(i)$ from an evaluation of the terms in (2) at cell centers i. They further apply a reconstruction of the 3D velocity field in Cartesian coordinates and the evaluation of the partial derivatives using the approximation: $(\partial_{x_m}F)_{ij} = (\mathrm{Grad}_n\,F)n_{ij}^m + (\mathrm{Grad}_t\,F)t_{ij}^m$, $m = 1,2,3$, where n and t are the edge normal and tangential vector components. Full details can be found in [3].

4   Numerical Results

We illustrate on a selected test case (TC), namely a barotropically unstable jet at mid-latitude on the sphere, proposed by [4] and here referred to as Galewsky TC, the superior performance of our Casimir-dissipation (**CD**) stabilized stochastic flow model over a comparable version that applies instead a conventional biharmonic Laplacian diffusion (**BD**). This test case is particularly suited to study the performance of stochastic schemes for low resolution (LR) grids: as small scale features trigger and determine the large scale flow pattern, the former have to be either well resolved (normally with sufficiently high resolution) or accurately represented by the stochastic model. Here, comparable deterministic LR simulations usually fail to show the correct large scale pattern, cf. [3]. Therein, also the initialization of the water depths and velocity fields can be found.

As this manuscript compares the use of two different dissipation methods, one might ask if there is the need for dissipation at all. This is a valid question, especially when considering that we use a spatially energy conserving discretization of a consistent stochastic flow model (see above) which should, in principle, be stable even without diffusion. For the given test case and integrations up to 20 days, our RSW-LU scheme is indeed numerically stable, but we did not explore longer integration times because some kind of dissipation is in general anyway needed. This is mostly due to the accumulation of small scale potential enstrophy (PE) at the grid level – caused in 2D simulations by the PE cascade towards small scales. If not dissipated, either by CD or BD diffusion, this accumulation of PE leads to noisy and nonphysical fields or even to numerical instabilities (shown in [2] for deterministic models, but also valid for stochastic ones).

To support these points, we first show in Fig. 2 the results of a simulation over 12 days of the Galewsky TC on a grid with 20480 triangles and without CD or BD dissipation. Looking at the left panel, we see that the potential vorticity (PV) field exhibits mostly nonphysical small scale noise that results from the accumulation of PE at the grid scale and does not agree with the physics of fluid flows. Looking at the total PE time evolution (right), we observe that with the beginning of day 5, when the large scale flow pattern starts to emerge, PE starts to increase significantly. This increase might eventually lead to a crash of the numerical simulation. Using a spatially energy conserving scheme, the slight drift in total energy (middle) is due to the time integration scheme of our stochastic system. For further discussions and more details about how total energy (E) and PE are computed, see [3].

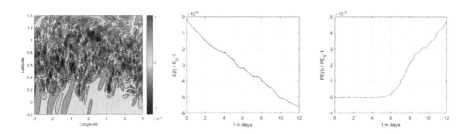

Fig. 2. PV field at day 12 (left), and total energy (middle) and potential enstropy (right) development up to day 12 for the stochastic RSW-LU without CD or BD dissipation. The contour interval in the left panel is $2 \times 10^{-5} \mathrm{s}^{-1}$.

Having illustrated the necessity of dissipating PE, we next compare the BD and CD dissipation methods. To make this comparison fair, we proceed as follows. We first tune the BD diffusion coefficient: we let the low resolution BD simulation run for 12 days with a sufficiently low value for ν and then increase it until we suppress numerical ringing in the PV field. This gives us an optimal value of about $\nu = 3.1 \times 10^{16} \mathrm{m}^4/\mathrm{s}$ for this TC. As can be inferred from Fig. 4, this dissipates beside PE also the total energy E. Considering the frequency spectrum for PE (calculated using the method from [2] but not shown here), we now choose the Casimir dissipation coefficient $\theta = 5 \cdot 10^{21} \mathrm{m}^5 \mathrm{s}$ such that both PE frequency spectra agree, especially at small scales, which indicates that both schemes sufficiently well dissipate the small scale PE noise.

In Fig. 3, we compare for the Galewsky TC the PV fields at day 6 obtained from a high resolution reference large-eddy simulation (Ref) on a grid with 327680 triangles (as used in [3]) (first column) with two low resolution runs where one uses only CD dissipation (second column) and the other only BD dissipation (third column). Note that these fields show the ensemble mean of an ensemble run with 20 members. We observe that the mean flow field of the CD ensemble is closer to Ref than that of the BD ensemble: in the former case, the eddies have been better developed while the distances between vortex (blue) and anti-vortex (red) pairs are larger, hence better agree with those from the reference run. In contrast, vortex anti-vortex pairs are less well developed in the BD runs. As the latter dissipate, besides PE, also total energy, it seems that using BD dissipation removes also dynamics from the system.

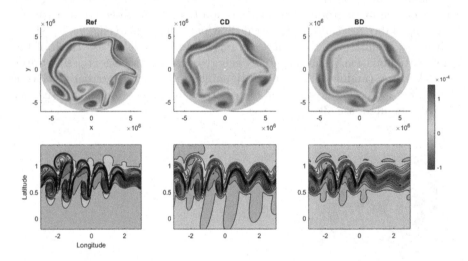

Fig. 3. Ensemble mean of PV fields at day 6 for high resolution (Ref) simulation with 327680 triangles (first column) and for low resolution CD (second column) and BD (third column) simulations. Upper row shows the northern hemisphere from above, lower row shows the projections of these fields onto latitude-longitude grid representations. The contour interval in the lower row pictures is $2 \times 10^{-5} \mathrm{s}^{-1}$.

This conclusion is supported by the time evolution of E and PE of the 20 ensemble members, shown in Fig. 4. For both quantities, we show in red results for CD and in blue for BD. All ensemble members of the CD runs preserve total energy at the order of 10^{-6}. Comparing them with the corresponding curve in

Fig. 2 confirms that our implementation of CD does not diffuse energy (diffusion is only due to time integration). Besides, all members of the CD ensemble diffuse almost the same amount of PE. In contrast, all members of the BD runs diffuse besides PE also a substantial amount of energy (at the order of 10^{-4}), leading to the reduces dynamics as stated above.

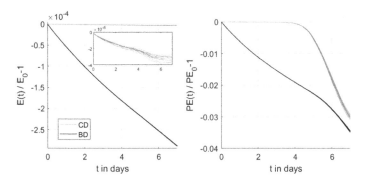

Fig. 4. Time evolution of total energy (left) and potential enstrophy (right) for an integration time of 7 days for CD (red) and BD (blue) for all 20 ensemble members. (Color figure online)

5   Conclusions

We presented a novel structure preserving discretization of a Location Uncertainty version of stochastic rotating shallow water equations, applying an energy conserving Casimir diffusion (CD) to dissipate small scale potential enstrophy. We compared this scheme with a version that applies biharmonic Laplacian diffusion (BD) for stabilization, which is usually used for such models. Our results show clearly that the CD runs achieve a spatial conservation of energy, which, in contrast, is strongly dissipated for BD. Consequently, compared to BD, the dynamics of CD simulations are significantly closer to a high resolutions reference solution, underpinning the use of our CD over standard BD diffusion.

Acknowledgements. RB is funded by the Deutsche Forschungsgemeinschaft (DFG, German Research Foundation) - Project-ID 274762653 - TRR 181.

References

1. Bauer, W., Gay-Balmaz, F.: Towards a geometric variational discretization of compressible fluids: The rotating shallow water equations. J. Comput. Dyn. **6**(1), 1–37 (2019)
2. Brecht, R., Bauer, W., Bihlo, A., Gay-Balmaz, F., MacLachlan, S.: Selective decay for the rotating shallow-water equations with a structure-preserving discretization. Phys. Fluids **33**, 116604 (2021)

3. Brecht, R., Li, L., Bauer, W., Mémin, E.: Rotating shallow water flow under location uncertainty with a structure-preserving discretization. J. Adv. Model. Earth Syst. **13**, e2021MS002492 (2021)
4. Galewsky, J., Scott, R.K., Polvani, L.M.: An initial-value problem for testing numerical models of the global shallow-water equations. Tellus A: Dyn. Meteorol. Oceanogr. **56**(5), 429–440 (2004)
5. Gay-Balmaz, F., Holm, D.: Selective decay by Casimir dissipation in inviscid fluids. Nonlinearity **26**(2), 495 (2013)
6. Mémin, E.: Fluid flow dynamics under location uncertainty. Geophys. Astrophys. Fluid Dy. **108**(2), 119–146 (2014)

High-Order Structure-Preserving Algorithms for Plasma Hybrid Models

Stefan Possanner[1]($\boxtimes$), Florian Holderied[1], Yingzhe Li[1], Byung Kyu Na[1], Dominik Bell[2], Said Hadjout[1], and Yaman Güçlü[1]

[1] Max Planck Institute for Plasma Physics, 85748 Garching, Germany
stefan.possanner@ipp.mpg.de
[2] Technical University of Munich, 85748 Garching, Germany

Abstract. Wave-particle resonance plays a crucial role for the stability of burning plasma in magnetically confined fusion. We present provably stable algorithms for the accurate simulation of such (nonlinear) processes on long time scales. Our approach combines several recent advances in theoretical and numerical research: on the theoretical side, we rely on Hamiltonian fluid-kinetic hybrid models, largely based on the works of Tronci [37]. To achieve high-order discretization, we use finite element exterior calculus (FEEC) introduced by Arnold et al. [5] based on B-splines coupled with particle-in cell for the resonating particles. Last but not least, structure-preservation (in a sense to be defined more clearly in the text) is achieved by discretization of Poisson brackets, rather than PDEs - following the ideas of Kraus et al. [28]. These efforts culminate in the creation of the open-source software package STRU-PHY (STRUcture-Preserving HYbrid codes) [1] which makes available to the scientific community a growing number of plasma hybrid codes, ready for use.

Keywords: geometric finite elements · Hamiltonian systems · plasma models

1 Challenges is Fusion Plasma: Alfvén eigenmodes

The magneto-hydrodynamic (MHD) equations provide an intuitive description of the collective excitations (waves) occurring in magnetized plasma. They describe three types of waves: shear-Alfvén, fast and slow magnetosonic waves. The shear-Alfvén wave with dispersion $\omega = v_\mathrm{A} k_\|$, where $v_\mathrm{A} = B/\sqrt{\rho\mu_0}$ denotes the Alfvén velocity and $k_\|$ is the wave-vector component along the magnetic field, is an incompressible wave that travels parallel to the magnetic field lines. Therefore, it is susceptible to resonant interactions with charged particles, which also travel along the magnetic field and may exchange energy with the wave if certain resonance conditions are met. One necessary conditions is that the phase velocity of the wave and the particle velocity are similar, $v_\mathrm{A} \sim v$. Fusion-born α-particles

Supported by Max Planck Institute for Plasma Physics.

in Tokamaks and Stellarators, for example, satisfy this condition after their initial slowing down phase; this can lead to the excitation of Alfvén Eigenmodes (AEs), which are special kinds of Alfvén waves arising in toroidally confined plasma, usually at extrema of the Alfvén continuum [21,35]. The growing AEs degrade the confinement of the α-particles, which are needed to heat the plasma. This process has been identified as one of the biggest challenges towards sustainable burning plasma conditions in future experiments [16].

The issue of AE excitation is nowadays addressed on the experimental, theoretical as well as on the computational level. The challenges for an accurate description on the computational side are diverse: nonlinear dynamics (AE saturation via phase space particle trapping), complex geometries (toroidal), multi-species (thermal and non-thermal plasma), and multi-scale (Tokamak transport timescales as compared to cyclotron periods). There has been considerable progress in the simulation of AE wave-particle resonance over the past years. Applied models range from fully kinetic descriptions (e.g. the codes ORB5 [9], EUTERPE [25], TRIMEG [29]) over MHD-kinetic hybrid models (JOREK [22], MEGA [36], M3D-K [7], HMGC [10], NIMROD [33]) to pure fluid descriptions (TAEFL [34]) of the wave-particle system. A recent benchmark study of these approaches can be found in Koenies et al. [26]. Meanwhile, there have been new insights into Hamiltonian plasma hybrid models, mainly due to Tronci and collaborators [12,37–39], and also steady progress in the development of "structure-preserving" numerical methods, of which a summary has been gathered by Morrison [30]. These methods can lead to provably stable algorithms due to exact, discrete conservation laws that are built upon the underlying Poisson- or variational structure. They an be viewed as an extension of geometric integration of ODEs [20] to the realm of PDEs. Many groups have worked on the development of structure-preserving methods for plasma equations and the literature is quite extensive; we can possibly mention only a few works here and point the reader to references therein [13–15,31,32,40].

2   STRUPHY - STRUcture-Preserving HYbrid Codes

2.1   Overview

In this work we present the open-source Python package STRUPHY [1,24] which is specifically designed for simulating wave-particle interactions in plasma. STRUPHY provides an API (application programming interface) that allows for the efficient implementation of Hamiltonian plasma hybrid models of an arbitrary number of fluid and/or kinetic species. Moreover, the package provides libraries for MHD equilibria (including interfaces to EQDSK files [2] and GVEC output [3]), kinetic background distributions, mapped domains (see Fig. 1), as well as input/output and diagnostic tools. The package comes with an extensive documentation [4], enabling its use for a wide variety of applications, and intended to be grown by its users (open-source).

STRUPHY is based on the idea of Kraus et al. [28] to discretize Poisson brackets by means of geometric finite elements combined with particle-in-cell

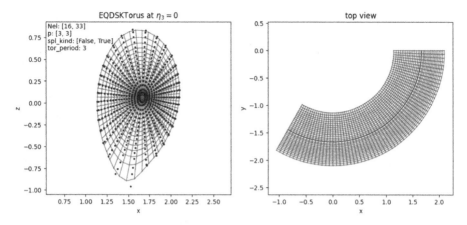

Fig. 1. Cross sections of a STRUPHY mapped simulation domain obtained from an EQDSK file. The red dot represents a polar singularity. Only a third of the torus is simulated (three-fold periodicity assumed). (Color figure online)

(PIC) for the resonating particles. In STRUPHY, this idea has been extended from electromagnetic Vlasov-Maxwell to a variety of fluid-kinetic hybrid models with a Poisson structure. Non-Hamiltonian models, such as linearized plasma hybrid models, have also been successfully implemented. Field and fluid variables are represented in tensor-product, high-order B-spline spaces of finite element exterior calculus (FEEC) [5], on three-dimensional mapped domains [11]. The open source package PSYDAC [19] has been integrated for this purpose. The problem of discontinuous solutions at polar singularities (as in Fig. 1) is mitigated by the use of "polar splines" [23] at the axis. The resonant particles (or "hot" particles) are treated within the PIC paradigm [8], and are thus represented by markers and weights, leading to the Monte-Carlo interpretation of phase space integrals [6].

STRUPHY model equations typically possess a non-canonical Poisson structure (non-Hamiltonian models are not considered in what follows). On the semi-discrete level, this leads to a large system of ODEs of the form

$$\frac{\mathrm{d}}{\mathrm{d}t}\mathbf{X} = Q(\mathbf{X})\nabla_{\mathbf{X}}H(\mathbf{X}).\tag{1}$$

Here, $\mathbf{X} : \mathbb{R} \to \mathbb{R}^N$ are the degrees of freedom of the semi-discrete model, comprising finite element coefficients and markers, such that N is typically of the order of millions, $Q : \mathbb{R}^N \to \mathbb{R}^{N \times N}$ is skew-symmetric, $Q^\top = -Q$, and $H : \mathbb{R}^N \to \mathbb{R}$ is the discrete Hamiltonian. As a consequence,

$$\frac{\mathrm{d}}{\mathrm{d}t}H(\mathbf{X}) = 0,\tag{2}$$

on the semi-discrete level. If H was quadratic, one could use the mid-point rule to transfer the exact conservation of energy to the fully discrete scheme. For non-quadratic H, one could use for instance the discrete gradient method [18] to salvage energy conservation. However, in practice the system (1) is usually too big to apply such implicit methods directly. Rather, one must use splitting schemes [17, 27] to decrease the system size. Via splitting it is usually possible to arrive at exact energy conservation (or at least a uniformly bounded energy error) with reasonable numerical cost of solving (1). On top of energy conservation, which is important for numerical stability, the FEEC paradigm allows for the exact conservation of other physically relevant quantities; for instance, because the finite element spaces form a complex, it is automatic to guarantee a solenoidal magnetic field at all times. In general, it becomes possible to conserve several Casimir invariants of the Poisson structure on the discrete level [30].

An important feature of the FEEC-PIC paradigm employed in STRUPHY is that on the semi-discrete level, all models look like (1), regardless of the mapped domain used or the chosen mesh resolution. On the one hand, this underlines the stability of the algorithm, which is guaranteed even for long-time nonlinear simulations of wave-particle interactions, and should represent an important improvement over conventional simulation methods. On the other hand, it allows for a considerable amount of abstraction and unification in the code itself - a fact that will benefit the implementation of other Hamiltonian hybrid models in the future. In what follows we will discuss some of the models either already or soon to be implemented in STRUPHY. For a current update on available models please consult the documentation [1].

2.2   Hamiltonian Hybrid Models

The generic form of the MHD-kinetic hybrid models in STRUPHY is as follows:

$$
\text{generic MHD hybrid}
\begin{cases}
\dfrac{\partial \rho}{\partial t} + \nabla \cdot (\rho \mathbf{U}) = 0, \\[2mm]
\rho\left(\dfrac{\partial \mathbf{U}}{\partial t} + \mathbf{U} \cdot \nabla \mathbf{U}\right) + \nabla p = \dfrac{(\nabla \times \mathbf{B}) \times \mathbf{B})}{\mu_0} + K(f_{\mathrm{h}}; \mathbf{E}, \mathbf{B}), \\[2mm]
\dfrac{\partial \mathbf{B}}{\partial t} + \nabla \times \mathbf{E} = 0, \\[2mm]
\dfrac{\mathrm{d}}{\mathrm{d}t} f_{\mathrm{h}}(t, \mathbf{z}(t; \mathbf{z}_0)) = 0, \quad \forall \mathbf{z}_0 \in \mathbb{R}^3 \times \mathbb{R}^{d_v}.
\end{cases}
\tag{3}
$$

Here, ρ is the plasma mass density, $\mathbf{U}$ the plasma macroscopic velocity, p denotes the plasma pressure, $\mathbf{B}, \mathbf{E}$ are the electromagnetic fields and μ_0 is the magnetic constant; moreover, f_{h} denotes the phase space distribution function of the "hot" plasma species (resonating with bulk plasma waves), $\mathbf{z}(t; \mathbf{z}_0)$ is a phase space characteristic with foot $\mathbf{z}_0$, and d_v stands for the dimension of velocity space. The term $K(f_{\mathrm{h}}; \mathbf{E}, \mathbf{B})$ on the right-hand side of the momentum conservation law stands for coupling terms due to wave-particle interactions; it usually depends on velocity moments of f_{h} and on the electromagnetic fields. The electric field, the pressure evolution law and the explicit form of the kinetic equation vary from model to model. It is important to keep in mind that the above hybrid model describes two separate species, namely the thermal (bulk) plasma in terms of MHD equations, and the "hot" particles in terms of a kinetic equation. There could be multiple such species, however we shall restrict the discussion two just two for clarity. The generic MHD hybrid model conserves the total energy

$$\mathcal{H} = \underbrace{\int \frac{\rho |\mathbf{U}|^2}{2} d\mathbf{x}}_{=\mathcal{H}_{\text{kinetic}}} + \underbrace{\frac{1}{\gamma - 1} \int p \, d\mathbf{x}}_{\mathcal{H}_{\text{internal}}} + \underbrace{\frac{1}{\mu_0} \int \frac{|\mathbf{B}|^2}{2} d\mathbf{x}}_{\mathcal{H}_{\text{EM}}} + \mathcal{H}_{\text{h}}, \tag{4}$$

where the "hot" energy $\mathcal{H}_{\text{h}}$ is model dependent.

MHD-Vlasov Current/pressure Coupling. These models and their Hamiltonian structure have been derived in [12,37]. The electric field and pressure are given by

$$\mathbf{E} = -\mathbf{U} \times \mathbf{B}, \qquad p = p_{\text{i}}, \qquad \frac{d}{dt}\bigg|_{\mathbf{U}} \left(\frac{p_{\text{i}}}{\rho^\gamma}\right) = 0, \tag{5}$$

where we use the notation

$$\frac{d}{dt}\bigg|_{\mathbf{U}} = \frac{\partial}{\partial t} + \mathbf{U} \cdot \nabla. \tag{6}$$

The coupling terms read

$$\text{current coupling:} \qquad K(f_{\text{h}}; \mathbf{E}, \mathbf{B}) = q_{\text{h}} n_{\text{h}}(\mathbf{U} - \mathbf{u}_{\text{h}}) \times \mathbf{B} \tag{7}$$

$$\text{pressure coupling:} \qquad K(f_{\text{h}}; \mathbf{E}, \mathbf{B}) = -\nabla \cdot \mathbb{P}_{\text{h}, \perp}. \tag{8}$$

where we encounter the density, flux and pressure tensor of the hot species:

$$n_{\text{h}} = \int f_{\text{h}} d\mathbf{v}, \qquad n_{\text{h}} \mathbf{u}_{\text{h}} = \int \mathbf{v} f_{\text{h}} d\mathbf{v}, \qquad \mathbb{P}_{\text{h}, \perp} = m_{\text{h}} \int \mathbf{v}_\perp \mathbf{v}_\perp^\top f_{\text{h}} d\mathbf{v}. \tag{9}$$

The dimension of the velocity space is $d_v = 3$. In current coupling, the kinetic equation reads

$$\frac{\partial f_{\text{h}}}{\partial t} + \mathbf{v} \cdot \nabla f_{\text{h}} + \frac{q_{\text{h}}}{m_{\text{h}}}(\mathbf{v} - \mathbf{U}) \times \mathbf{B} \cdot \frac{\partial f_{\text{h}}}{\partial \mathbf{v}} = 0. \tag{10}$$

In pressure coupling, the hot particles satisfy a "shifted" Vlasov equation:

$$\frac{\partial f_{\text{h}}}{\partial t} + (\mathbf{v} + \mathbf{U}_\perp) \cdot \nabla f_{\text{h}} + \left[\frac{q_{\text{h}}}{m_{\text{h}}}(\mathbf{v} \times \mathbf{B}) - \nabla \mathbf{U}_\perp \cdot \mathbf{v}\right] \cdot \frac{\partial f_{\text{h}}}{\partial \mathbf{v}} = 0. \tag{11}$$

We note that only the perpendicular component of the fluid velocity is retained in Eq. (11). This is different from the original formulation in [37], where the full velocity appears. Our new fomrulation still features a Hamiltonian structure, but has the advantage of less noise in PIC simulations.

In both the current- and the pressure coupling cases, the hot particle energy reads

$$\mathcal{H}_{\text{h}} = \int \frac{m_{\text{h}} |\mathbf{v}|^2}{2} f_{\text{h}} d\mathbf{x} d\mathbf{v}. \tag{12}$$

MHD-drift-kinetic Current Coupling. This model and its variational structure have been derived in [12,38]. The pressure and electric field are as in (5). The coupling term reads

$$K(f_\mathrm{h}; \mathbf{E}, \mathbf{B}) = (q_\mathrm{h} n_\mathrm{h} \mathbf{U} - \mathbf{J}_\mathrm{gc} - \nabla \times \mathbf{M}_\mathrm{gc}) \times \mathbf{B} ,\tag{13}$$

where

$$\mathbf{J}_\mathrm{gc} = q_\mathrm{h} \int \frac{(v_\| \mathbf{B}^* - \mathbf{b}_0 \times \mathbf{E}^*)}{B_\|^*} f_\mathrm{h} \mathrm{d}v_\| \mathrm{d}\mu ,\tag{14}$$

$$\mathbf{M}_\mathrm{gc} = - \int \mu \mathbf{b}_0 f_\mathrm{h} \mathrm{d}v_\| \mathrm{d}\mu ,\tag{15}$$

with the starred fields

$$\mathbf{B}^* = \mathbf{B} + \frac{m_\mathrm{h}}{q_\mathrm{h}} v_\| \nabla \times \mathbf{b}_0 , \qquad B_\|^* = \mathbf{B}^* \cdot \mathbf{b}_0 ,\tag{16}$$

$$\mathbf{E}^* = \mathbf{E} - \frac{\mu}{q_\mathrm{h}} \nabla |B| .\tag{17}$$

The hot particle distribution function $f_\mathrm{h} = f_\mathrm{h}(t, \mathbf{x}, v_\|, \mu)$ $(d_v = 2)$ satisfies the drift-kinetic equation

$$\frac{\partial f_\mathrm{h}}{\partial t} + \frac{(v_\| \mathbf{B}^* - \mathbf{b}_0 \times \mathbf{E}^*)}{B_\|^*} \cdot \nabla f_\mathrm{h} + \frac{q_\mathrm{h}}{m_\mathrm{h}} \frac{\mathbf{B}^* \cdot \mathbf{E}^*}{B_\|^*} \frac{\partial f_\mathrm{h}}{\partial v_\|} = 0 .\tag{18}$$

The hot particle energy reads

$$\mathcal{H}_\mathrm{h} = \int \left(\frac{m_\mathrm{h} v_\|^2}{2} + \mu |B| \right) f_\mathrm{h} B_\|^* \, \mathrm{d}\mathbf{x} \mathrm{d}v_\| \mathrm{d}\mu\tag{19}$$

Hall-MHD-Vlasov Current Coupling. This is an unpublished model that is currently under investigation. The electric field stems from an extended Ohm's law featuring the Hall term (with hot particle contribution) and the electron pressure gradient:

$$\mathbf{E} = -\mathbf{U} \times \mathbf{B} + \frac{m_\mathrm{i}}{q_\mathrm{i} \rho} \left[\frac{(\nabla \times \mathbf{B} - \mu_0 \mathbf{j}_\mathrm{h}) \times \mathbf{B}}{\mu_0} - \nabla p_\mathrm{e} \right] .\tag{20}$$

The plasma pressure is the sum of ion and electron pressure:

$$p = p_\mathrm{i} + p_\mathrm{e} , \qquad \frac{\mathrm{d}}{\mathrm{d}t}\bigg|_\mathbf{U} \left(\frac{p_\mathrm{i}}{\rho^\gamma} \right) = 0 \qquad \frac{\mathrm{d}}{\mathrm{d}t}\bigg|_{\mathbf{U} - \frac{m_\mathrm{i}}{q_\mathrm{i} \rho} \frac{(\nabla \times \mathbf{B} - \mu_0 \mathbf{j}_\mathrm{h}) \times \mathbf{B}}{\mu_0}} \left(\frac{p_\mathrm{e}}{\rho^\gamma} \right) = 0 .\tag{21}$$

The electron pressure is transported with the electron velocity, which couples to the hot particles. The dimension of velocity space is $d_v = 3$ and the hot particles satisfy the standard Vlasov equation

$$\frac{\partial f_\mathrm{h}}{\partial t} + \mathbf{v} \cdot \nabla f_\mathrm{h} + \frac{q_\mathrm{h}}{m_\mathrm{h}} (\mathbf{E} + \mathbf{v} \times \mathbf{B}) \cdot \frac{\partial f_\mathrm{h}}{\partial \mathbf{v}} = 0 .\tag{22}$$

The energy is given in (12). This model is interesting because it features some two-fluid effects (electron dynamics) that are relevant in fusion plasma.

Acknowledgements. We are thankful for the valuable input of E. Sonnendrücker, M. Campos Pinto, X. Wang and F. Hindenlang to the STRUPHY project.

References

1. https://pypi.org/project/struphy/
2. https://w3.pppl.gov/ntcc/TORAY/G_EQDSK.pdf
3. https://gitlab.mpcdf.mpg.de/gvec-group/gvec
4. https://struphy.pages.mpcdf.de/struphy/index.html
5. Arnold, D., Falk, R., Winther, R.: Finite element exterior calculus: from Hodge theory to numerical stability. Bull. Am. Math. Soc. **47**(2), 281–354 (2010)
6. Aydemir, A.Y.: A unified Monte Carlo interpretation of particle simulations and applications to non-neutral plasmas. Phys. Plasmas **1**(4), 822–831 (1994)
7. Belova, E., Denton, R., Chan, A.: Hybrid magnetohydrodynamic-gyrokinetic simulation of toroidal alfvén modes. J. Comput. Phys. (2), 324–336
8. Birdsall, C.K., Langdon, A.B.: Plasma Physics via Computer Simulation. CRC Press, Boca Raton (2018)
9. Bottino, A., et al.: Global nonlinear electromagnetic simulations of tokamak turbulence. IEEE Trans. Plasma Sci. **38**(9), 2129–2135 (2010)
10. Briguglio, S., Vlad, G., Zonca, F., Kar, C.: Hybrid magnetohydrodynamic-gyrokinetic simulation of toroidal alfvén modes. Phys. Plasmas **2**(10), 3711–3723 (1995)
11. Buffa, A., Rivas, J., Sangalli, G., Vázquez, R.: Isogeometric discrete differential forms in three dimensions. SIAM J. Numer. Anal. **49**(2), 818–844 (2011)
12. Burby, J.W., Tronci, C.: Variational approach to low-frequency kinetic-MHD in the current coupling scheme. Plasma Phys. Control. Fusion **59**(4), 045013 (2017)
13. Campos Pinto, M., Kormann, K., Sonnendrücker, E.: Variational framework for structure-preserving electromagnetic particle-in-cell methods. J. Sci. Comput. **91**(2), 46 (2022)
14. Chen, G., Chacón, L.: An energy-and charge-conserving, nonlinearly implicit, electromagnetic 1D–3V Vlasov-Darwin particle-in-cell algorithm. Comput. Phys. Commun. **185**(10), 2391–2402 (2014)
15. Chen, G., Chacón, L., Barnes, D.C.: An energy-and charge-conserving, implicit, electrostatic particle-in-cell algorithm. J. Comput. Phys. **230**(18), 7018–7036 (2011)
16. Chen, L., Zonca, F.: Physics of alfvén waves and energetic particles in burning plasmas. Rev. Mod. Phys. **88**(1), 015008 (2016)
17. Crouseilles, N., Einkemmer, L., Faou, E.: Hamiltonian splitting for the vlasov-maxwell equations. J. Comput. Phys. **283**, 224–240 (2015)
18. Gonzalez, O.: Time integration and discrete hamiltonian systems. J. Nonlinear Sci. **6**, 449–467 (1996)
19. Güçlü, Y., Hadjout, S., Ratnani, A.: Psydac: a high-performance IGA library in python. eccomas2022. https://github.com/pyccel/psydac
20. Hairer, E., Lubich, C., Wanner, G.: Geometric Numerical Integration, vol. 31. Springer, Berlin Series in Computational Mathematics (2006). https://doi.org/10. 1007/3-540-30666-8

21. Heidbrink, W.: Basic physics of alfvén instabilities driven by energetic particles in toroidally confined plasmas. Phys. Plasmas 15(5), 055501 (2008)

22. Hoelzl, M., et al.: The JOREK non-linear extended MHD code and applications to large-scale instabilities and their control in magnetically confined fusion plasmas. Nuclear Fusion 61(6), 065001 (2021)

23. Holderied, F., Possanner, S.: Magneto-hydrodynamic eigenvalue solver for axisymmetric equilibria based on smooth polar splines. J. Comput. Phys. 464, 111329 (2022)

24. Holderied, F., Possanner, S., Wang, X.: Mhd-kinetic hybrid code based on structure-preserving finite elements with particles-in-cell. J. Comput. Phys. 433, 110143 (2021)

25. Jost, G., Tran, T., Cooper, W., Villard, L., Appert, K.: Global linear gyrokinetic simulations in quasi-symmetric configurations. Phys. Plasmas 8(7), 3321–3333 (2001)

26. Könies, A., Briguglio, S., Gorelenkov, N., Fehér, T., Isaev, M., Lauber, P., Mishchenko, A., Spong, D.A., Todo, Y., Cooper, W.A., et al.: Benchmark of gyrokinetic, kinetic MHD and GYROFLUID codes for the linear calculation of fast particle driven TAE dynamics. Nuclear Fusion 58(12), 126027 (2018)

27. Kormann, K., Sonnendrücker, E.: Energy-conserving time propagation for a structure-preserving particle-in-cell vlasov-maxwell solver. J. Comput. Phys. 425, 109890 (2021)

28. Kraus, M., Kormann, K., Morrison, P.J., Sonnendrücker, E.: GEMPIC: geometric electromagnetic particle-in-cell methods. J. Plasma Phys. 83(4), 905830401 (2017)

29. Lu, Z., Meng, G., Hoelzl, M., Lauber, P.: The development of an implicit full f method for electromagnetic particle simulations of alfven waves and energetic particle physics. J. Comput. Phys. 440, 110384 (2021)

30. Morrison, P.J.: Structure and structure-preserving algorithms for plasma physics. Phys. Plasmas 24(5), 055502 (2017)

31. Perse, B., Kormann, K., Sonnendrücker, E.: Geometric particle-in-cell simulations of the vlasov-maxwell system in curvilinear coordinates. SIAM J. Sci. Comput. 43(1), B194–B218 (2021)

32. Qin, H., Liu, J., Xiao, J., Zhang, R., He, Y., Wang, Y., Sun, Y., Burby, J.W., Ellison, L., Zhou, Y.: Canonical symplectic particle-in-cell method for long-term large-scale simulations of the vlasov-Maxwell equations. Nuclear Fusion 56(1), 014001 (2015)

33. Sovinec, C., et al.: Nonlinear magnetohydrodynamics with high-order finite elements. J. Comput. Phys. 195, 355 (2004)

34. Spong, D., Carreras, B., Hedrick, C.: Linearized gyrofluid model of the alpha-destabilized toroidal alfvén eigenmode with continuum damping effects. Phys. Fluids B: Plasma Phys. 4(10), 3316–3328 (1992)

35. Todo, Y.: Introduction to the interaction between energetic particles and alfvén eigenmodes in toroidal plasmas. Rev. Mod. Plasma Phys. 3(1), 1 (2018)

36. Todo, Y., Sato, T.: Linear and nonlinear particle-magnetohydrodynamic simulations of the toroidal alfven eigenmode. Phys. Plasmas 5, 1321–1327 (1998)

37. Tronci, C.: Hamiltonian approach to hybrid plasma models. J. Phys. A: Math. Theor. 43(37), 375501 (2010). http://stacks.iop.org/1751-8121/43/i=37/a=375501

38. Tronci, C.: Variational mean-fluctuation splitting and drift-fluid models. Plasma Phys. Control. Fusion 62(8), 085006 (2020)

39. Tronci, C., Tassi, E., Camporeale, E., Morrison, P.J.: Hybrid vlasov-mhd models: Hamiltonian vs. non-hamiltonian. Plasma Phys. Control. Fusion **56**(9), 095008 (2014)
40. Xiao, J., Qin, H., Liu, J.: Structure-preserving geometric particle-in-cell methods for vlasov-maxwell systems. Plasma Sci. Technol. **20**(11), 110501 (2018). https://doi.org/10.1088/2F2058-6272/2Faac3d1

Hydrodynamics of the Probability Current in Schrödinger Theory

Mauro Spera[(⊠)]

Dipartimento di Matematica e Fisica "Niccolò Tartaglia", Università Cattolica del
Sacro Cuore, Brescia, Italy
mauro.spera@unicatt.it
https://docenti.unicatt.it/ppd2/it/docenti/21065/mauro-spera/profilo

Abstract. The present note explores some hydrodynamical aspects of
the probability current in Schrödinger's theory based on the observation
that the latter shares the same trajectories with the Madelung velocity, whilst exhibiting a regular behaviour. This appears to be useful in
analyzing the motion of the zero set of the wave function.

Keywords: Quantum hydrodynamics · Vortex motion

1 Introduction

Vortex structures arise in various physical contexts as invariant exact solutions
of the Euler equations in fluid dynamics and their study is at the same time
difficult and fascinating. In particular, they emerge as *nodal lines* (zero sets)
of the wave function of a massive spinless particle governed by the Schrödinger
equation, where the probability density vanishes and the phase is undetermined
(see e.g. [11,32]), giving rise to both physical and mathematical subtleties. The
present contribution investigates some hydrodynamical aspects of the *probability current* in Schrödinger's theory based on the observation that, outside the
nodal line, the latter shares the same (Bohm) trajectories with the Madelung
velocity, whilst exhibiting a regular behaviour; moreover, the nodal line motion
is closely related to the time derivative of the probability current and the nodal
line itself becomes a fibre of a Clebsch-type fibration and it is advected via the
hydrodynamical Schrödinger-Madelung equation; this possibly answers a question posed by Kauffman and Lomonaco [21] in their approach to the so-called
"Berry problem", consisting in finding a wave function possessing a prescribed
nodal line (see [2] and Sect. 5.1 for amplification).

2 Basics of Quantum Hydrodynamics

We mostly refer to [33] for quantum hydrodynamics and we act within the
geometric formalism set up in [14–16,19,31]). See [20] (and [1] for topological
aspects) for a comprehensive modern geometric treatment of fluid mechanics.

Supported by UCSC, D1 funds.

F. Nielsen and F. Barbaresco (Eds.): GSI 2023, LNCS 14072, pp. 272–281, 2023.
https://doi.org/10.1007/978-3-031-38299-4_29

2.1 Schrödinger and Madelung Pictures

Let us discuss the simplest case consisting of the motion of a spinless particle of mass $m > 0$ in 3-space. The quantum wave function depends on x and t: $\psi = \psi(x,t)$, with $x \in \mathbf{R}^3$, $t \in \mathbf{R}$ and obeys the *Schrödinger equation* (set $\hbar = m = 1$)

$$\partial_t \psi = -i\hat{H}\psi := -i\left(-\frac{1}{2}\Delta + V\right)\psi \tag{1}$$

with Δ denoting the Laplace operator and $V = V(x)$ a classical potential. Its polar decomposition reads

$$\psi = \sqrt{\rho}\, e^{iS}, \qquad \rho = |\psi|^2 \tag{2}$$

The Schrödinger equation can be cast into the *Madelung-Bohm* [5,6,25] hydrodynamical form (setting $\mathbf{u} = \nabla S$)

$$\begin{cases} \frac{\partial \mathbf{u}}{\partial t} + (\mathbf{u} \cdot \nabla)\mathbf{u} = -\nabla(V + V_q) \\ \frac{\partial \rho}{\partial t} + \mathrm{div}(\rho\,\mathbf{u}) = 0 \end{cases} \tag{3}$$

where $V_q = -\frac{1}{2}\frac{\Delta\sqrt{\rho}}{\sqrt{\rho}}$ is the *quantum potential*, the first equation is an *Euler equation* for a *compressible* irrotational fluid (div $\mathbf{u} = \Delta S \neq 0$ in general) and the second one is the *continuity equation*, involving the *probability current*

$$\mathbf{j} = \rho\,\mathbf{u} = \mathrm{Im}(\psi^\dagger \nabla\psi) = \frac{1}{2i}(\psi^\dagger \nabla\psi - \psi\nabla\psi^\dagger) \tag{4}$$

We shall freely switch vector fields and differential 1-forms via the musical isomorphisms induced by the Euclidean metric in $\mathbf{R}^3$, so we write for instance $j = \rho\, dS$ and so on. After setting

$$\mathcal{H} := \langle\psi|\hat{H}\psi\rangle = \int_{\mathbf{R}^3} \psi^\dagger\left(-\frac{1}{2}\Delta + V(x)\right)\psi\, d^3x = \int_{\mathbf{R}^3} \left\{\frac{1}{2}\,|\nabla\psi|^2 + V(x)|\psi|^2\right\} d^3x, \tag{5}$$

one can rephrase, following Bohm [5,6], the above equations in a Hamiltonian fashion:

$$\frac{\partial\rho}{\partial t} = \frac{\delta\mathcal{H}}{\delta S}, \qquad \frac{\partial S}{\partial t} = -\frac{\delta\mathcal{H}}{\delta\rho} \tag{6}$$

2.2 Symplectic Geometric Portrait

We can geometrically reformulate the above as follows, closely following [31], to which we refer for full details and further bibliography (see in particular [10] for a general theory). Let $\Phi = \{\mathbf{R}^3 \ni x \mapsto (\rho(x), S(x)) \in \mathbf{R}^2\}$ be a smooth map (use polar coordinates: $\sqrt{\rho} \geq 0$, $S \in \mathbf{R}/2\pi\mathbf{Z} = S^1$). The set $\mathcal{M}$ of such maps becomes a symplectic manifold as soon as the target space is equipped with the symplectic structure $d\rho \wedge dS$ (and we tacitly compactify $\mathbf{R}^3$ to S^3). The symplectic form Ω on $\mathcal{M}$ reads

$$\Omega = \int_{\mathbf{R}^3} d^3x\, \delta\rho(x) \wedge \delta S(x) \tag{7}$$

or (*Kähler* structure)

$$\Omega = -i \int_{\mathbf{R}^3} \delta\psi^\dagger(x) \wedge \delta\psi(x) d^3x \tag{8}$$

Let G denote the (connected component of the identity of the) group of *volume preserving* diffeomorphisms of $\mathbf{R}^3$ which rapidly approach the identity at infinity. The space $\mathbf{R}^3$ is equipped with the standard Euclidean metric (allowing the customary identification of vector fields and 1-forms). The symplectic form Ω is *G-invariant* under the natural a *right* action of G on $\Phi \in \mathcal{M}$ via

$$g(\Phi)(x) = \Phi(g^{-1}(x)) \tag{9}$$

(since the Jacobian $J(g) = 1$).

Notice that the probability current can be viewed itself as the velocity field of a *new* fluid (again compressible), with vorticity

$$\mathbf{W} = \operatorname{curl} \mathbf{j} = \nabla\rho \times \nabla S, \quad w = dj = d\rho \wedge dS \tag{10}$$

The variables ρ and S actually provide a Clebsch variable description thereof ([23, 24, 26, 28]), see also below, Subsect. 5.2.

The Hamiltonian algebra (Rasetti-Regge (RR) current algebra, cf. [17, 27–29]) associated to the G-coadjoint orbit pertaining to the divergence-free vector field $\mathbf{W}$ consists of functions $\lambda_{\mathbf{b}}$ - for any $\mathbf{b}$ divergence-free - defined as

$$\lambda_{\mathbf{b}} = \int_{\mathbf{R}^3} \mathbf{j} \cdot \mathbf{b} = \int_{\mathbf{R}^3} \mathbf{W} \cdot \mathbf{B} \tag{11}$$

where $\operatorname{curl} \mathbf{B} = \mathbf{b}$. One checks the Lie algebra structure of the RR-current algebra, namely

$$\{\lambda_{\mathbf{b}}, \lambda_{\mathbf{c}}\} = \lambda_{[\mathbf{b},\mathbf{c}]} \tag{12}$$

the vector field bracket being *minus* the usual one and reading, for divergence-free vector fields

$$[\mathbf{b}, \mathbf{c}] = \operatorname{curl}(\mathbf{b} \times \mathbf{c}) \tag{13}$$

and where the Poisson bracket is the one induced by the symplectic form. Explicitly:

$$\{\lambda_{\mathbf{b}}, \lambda_{\mathbf{c}}\} = \int_{\mathbf{R}^3} \mathbf{j} \cdot [\mathbf{b}, \mathbf{c}] = \int_{\mathbf{R}^3} \mathbf{j} \cdot \operatorname{curl}(\mathbf{b} \times \mathbf{c}) = \int_{\mathbf{R}^3} \mathbf{W} \cdot (\mathbf{b} \times \mathbf{c}) \tag{14}$$

One has the following

Theorem 1. *[31]*

(i) The action of G on $(\mathcal{M}, \Omega)$ gives rise to an equivariant moment map

$$\mu : \mathcal{M} \to \operatorname{Lie}(G)^*$$
$$\Phi = \{x \mapsto (\rho(x), S(x)\} \mapsto \mu(\Phi) := \mathbf{W}_\Phi = \nabla\rho \times \nabla S$$

(ii) In terms of the ensuing Hamiltonian algebra (RR-current algebra), the Schrödinger equation can be written compactly in the Hamiltonian form

$$\dot{\lambda}_{\mathbf{b}} = \{\lambda_{\mathbf{b}}, \mathcal{H}\} \tag{15}$$

This emphasizes the special role played by the probability current. See [14, 16, 19] for a different moment map interpretation of j (together with ρ).

Notice that, while the equations can be written in a geometric form, the ensuing evolution is *not* a coadjoint motion since the Hamiltonian *does not collectivize* via j (in the sense of Guillemin and Sternberg, see e.g. [18]).

3   Vortex Line Motion

The polar decomposition breaks down at the zeros of ψ: if $\psi(\tilde{x}) = 0$, then $\rho(\tilde{x}) = 0$ and S is *undetermined*. If $K := \psi^{-1}(0)$ (at $t = 0$; also assume it is a knot in 3-space and call it *nodal line*), definiteness of the wave function requires *quantization* of circulation [31, 32]:

$$\int_{\mathcal{C}} dS = 2\pi n, \qquad n \in \mathbf{Z} \tag{16}$$

with $\mathcal{C}$ a closed loop encircling K.

A standard computation yields

$$\partial_t \rho = \frac{i}{2} [\Delta \psi^\dagger \psi - \Delta \psi \, \psi^\dagger] \tag{17}$$

therefore, if $\psi(\tilde{x}) = 0$, $\partial_t \rho(\tilde{x}) = 0$ as well. Notice that by virtue of the continuity equation one finds that

$$\lim_{x \to \tilde{x}} \nabla \rho \cdot \nabla S = 0 \tag{18}$$

A detailed analysis of the evolution of the nodal line is then complicated by the fact that the Madelung velocity ∇S is *singular* thereon, whereby a naive manipulation of the hydrodynamical equations can be misleading and calls for recourse to "multivalued" fields (cf. [11, 22]) or, mathematically, to appropriate de Rham currents [9]. Nevertheless, in the paper [3] the following formula for the velocity $\mathbf{u}_{BBS}$ of the nodal line (up to a vector tangent to it) has been worked out:

$$\mathbf{u}_{BBS} = \frac{1}{2i} \frac{\mathbf{w} \times \mathbf{w}^\dagger}{|\mathbf{w} \times \mathbf{w}^\dagger|^2} \times (\mathbf{w} \Delta \psi^\dagger + \mathbf{w}^\dagger \Delta \psi) \tag{19}$$

with $\mathbf{w} = \nabla \psi|_K$ ($\neq 0$) and its authors pointed out that it bore no direct relation to the fluid velocity (our $\mathbf{u}$). However, we shall soon verify that it is indeed related to the time derivative of the probability current.

4   The Schrödinger Evolution of the Probability Current

In order to overcome the above hindrance, we may turn to exploiting the more regular behaviour of the probability current j together with its associated vorticity w. The current j vanishes on K and outside K it is proportional to dS, whence *they share the same integral curves (Bohm trajectories)*, so one can resort to the analysis of the time evolution of the probability current. One computes:

$$\partial_t[\psi^\dagger d\psi] = i[(-\frac{1}{2}\Delta + V)\psi^\dagger)d\psi - i\psi^\dagger d(-\frac{1}{2}\Delta + V)\psi] \tag{20}$$

leading to a formula for $\partial_t j$ which, when restricted to the nodal line K (at $t = 0$)) yields

$$\partial_t j|_K = -\frac{1}{4}[(\Delta\psi^\dagger)d\psi + (\Delta\psi)d\psi^\dagger] \tag{21}$$

Equation (20) then shows that, *up to a $\pi/2$-rotation and scaling*, $\partial_t \mathbf{j}|_K$ coincides with $\mathbf{u}_{BBS}$, which is physically quite plausible. Notice that at any point of K, the vectors $\mathbf{u}_{BBS}$, $\partial_t\mathbf{j}$ and the tangent $\mathbf{t}$ form an orthogonal frame.

Let us also record the ensuing formula for the time derivative of the vorticity $w = dj = -i[d\psi^\dagger \wedge d\psi]$. We find

$$\partial_t w = -[d\psi^\dagger \wedge d(\frac{1}{2}\Delta\psi) - d(\frac{1}{2}\Delta\psi^\dagger) \wedge d\psi + \psi^\dagger dV \wedge d\psi + \psi\, dV \wedge d\psi^\dagger] \tag{22}$$

which, on K, simplifies to

$$\partial_t w|_K = -\frac{1}{2}[d\psi^\dagger \wedge d(\Delta\psi) - d(\Delta\psi^\dagger) \wedge d\psi] \tag{23}$$

and may turn useful in relationship with the Clebsch description (cf. Subsect. 5.2).

5   Miscellanea

This section is partly speculative and hints at possible connections among different research strands.

5.1   On Berry's Problem

The paper [21] outlines a general algebraic-topological solution of the so-called "Berry's problem" ([2]): *To construct a wave function ψ pertaining to the hydrogen atom (or to a generic quantum system) having a prescribed nodal line K.* Then, the question was asked concerning the possible *time evolution* of K. A first answer comes directly from [3], see the above Sect. 3, and it may be complemented by the following remarks.

Let us view the wave function (at $t = 0$) as a map $\psi : S^3 \to D \subset \mathbf{C} \subset S^2$ (via stereographic projection, D a disc): $\psi = \sqrt{\rho_0} \cdot e^{iS}$. We may then adjust its phase so as to produce a map

$$\psi_K(x) = \sqrt{\rho_0} \cdot e^{iS_K} \tag{24}$$

with S_K the *solid angle* function attached to K (defined up to integral multiples of 4π), which is *harmonic* outside K. Then, in terms of *de Rham currents* (distribution-valued forms)

$$dS_K = B_K, \qquad\qquad dB_K = 2\pi\delta_K \tag{25}$$

with B_K the magnetic field (or fluid velocity) generated by the "wire" (or vortex) K (see e.g. [4,7,12,30] for a careful reconstruction of the historical background surrounding the topological outcomes of electromagnetism). Moreover, the level surfaces $S_K = c$ provide *Seifert surfaces* for K, i.e. they are oriented surfaces with boundary K (a smooth extension thereof to K being possible). In particular, for $c = 0$ we get a canonical framing, the so-called *solid angle framing* ([4]). Also recall that, in terms of *any* Seifert surface Σ, we have the explicit (Maxwell-Gauss) formula

$$S_K(\mathbf{x}) = \int_\Sigma \frac{\mathbf{x} - \mathbf{y}}{|\mathbf{x} - \mathbf{y}|^3} \cdot \mathbf{n} \, d\sigma \tag{26}$$

for $\mathbf{y}$ varying on Σ and $d\sigma$ being the area element.

Let us then take ψ_K as the initial wave function. Then its initial density $\rho_0 \, d^3x$ (which vanishes exactly on K) evolves via the Schrödinger-Madelung equation (*advection*) according to [14], and so does the nodal line. Explicitly, closely following [14] up to inessential notational changes:

$$\rho(x, t) = \eta_* \rho_0 := \int \rho_0(y)\delta(x - \eta(y, t))d^3y \tag{27}$$

(Eulerian density) where $\eta \in \text{Diff}(\mathbf{R}^3)$ – with velocity vector field $\mathbf{u} = \dot{\eta} \circ \eta^{-1}$ – and η_* denotes push-forward (advection) via η. The curves $t \mapsto \eta(y, t)$ are the Bohmian trajectories of the quantum fluid.

One concludes that the Schrödinger-Madelung evolution via diffeomorphisms η_t (where $\eta_t(y) = \eta(y, t)$ – a Lagrangian path, see [14]) produces knots K_t together with their Seifert surfaces Σ_t.

As we have seen in Sect. 4, the singular behaviour of the Madelung velocity on the nodal line can be bypassed via the time evolution of the probability current.

5.2   Clebsch Portrait

The variables ρ and S are of *Clebsch* type (cf. [31]): the level surfaces $\rho = c_1$ and $S = c_2$ (c_1 and $c_2 \in \mathbf{R}$) (i.e. the level surfaces of ψ) give rise, generically, to one-dimensional fibres $S^1 \subset S^3$ everywhere tangent to the vorticity field

$$\mathbf{W} = \nabla\rho \times \nabla S \tag{28}$$

In particular, if $\rho = 0$, S is undetermined but we still obtain $\psi^{-1}(0) \approx S^1$. In other words, *the nodal line becomes a fibre of the Clebsch fibration* (see e.g. [23,24,26,28]).

Let us substantiate the preceding observation by a few computations: upon taking for instance (cf. [32]) K as the z-axis and employing cylindrical coordinates, one easily sees that $|\nabla S| \sim \frac{1}{r}$, $\rho \sim r^2 = x^2 + y^2$. In more detail, around the nodal line (represented by the z-axis) we have, at first order in x and y and requiring circular symmetry,

$$\psi = A(z)\,(x + iy) = A(z)\,r\,e^{i\varphi}, \quad A(z) > 0 \tag{29}$$

Thus ($' = d/dz$):

$$\nabla\psi = (A,\, iA,\, (x + iy)A')$$
$$\rho = A^2(x^2 + y^2), \quad d\rho = 2A^2(x\,dx + y\,dy) + 2AA'(x^2 + y^2)\,dz \tag{30}$$
$$S = \varphi, \quad dS = d\arctan \tfrac{y}{x} = \tfrac{x\,dy - y\,dx}{x^2 + y^2}$$

and one immediately gets

$$j = A^2(x\,dy - y\,dx) \tag{31}$$

and, eventually, in vector terms

$$\mathbf{W} = \nabla\rho \times \nabla S = -AA'x\,\hat{\mathbf{i}} - AA'y\,\hat{\mathbf{j}} + 2\,A^2\,\hat{\mathbf{z}} \tag{32}$$

which is everywhere finite. Obviously

$$\mathbf{W}|_K = 2\,A^2\,\hat{\mathbf{z}}. \tag{33}$$

Notice that $\nabla\rho \cdot \nabla S = 0$ identically and that K does indeed bound the level surfaces of S (planes through the z-axis).

Also notice that, from

$$d\psi = A(dx + idy) + A'(x + iy)dz, \quad d\psi^\dagger = A(dx - idy) + A'(x - iy)dz$$
$$\Delta\psi = A''(x + iy), \quad \Delta\psi^\dagger = A''(x - iy) \tag{34}$$

Application of (23) yields, after a simple calculation

$$\partial_t w\,|_K = 0 \tag{35}$$

consistently with (19), which gives $\mathbf{u}_{BBS} = 0$ in the present case.

5.3   Incompressibility

In this subsection we briefly investigate the possibility of describing the vortex line motion via a *volume-preserving* evolution.

5.3.1. First, the probability current $\mathbf{j}$ may be rendered divergence-free by adding a suitable gradient $\nabla\chi$ upon solving an appropriate Poisson equation: indeed, from

$$\tilde{\mathbf{j}} := \mathbf{j} + \nabla\chi, \quad \operatorname{div}\tilde{\mathbf{j}} = 0 \tag{36}$$

we get

$$\Delta\chi = -\mathrm{div}\,\mathbf{j}, \qquad \chi = -\Delta^{-1}\mathrm{div}\,\mathbf{j} \tag{37}$$

This does not affect (globally) the vortex lines since $\tilde{\mathbf{W}} := \mathrm{curl}\,\tilde{\mathbf{j}} = \mathbf{W}$, albeit changing the single Bohmian trajectories. Also, the RR currents do not changeKindly note that section numbers "5.3.1...were given but respective section heading" missing. Please amend if necessary.. leading to a new fluid which may exhibit a non trivial *helicity* (Hopf invariant, Chern-Simons action) as in [23,28].

5.3.2. The vortex velocity $\mathbf{u}_{BBS}$ is defined on a surface (a "tube" spanned by K_t) can be locally extended to a divergence-free vector field $\tilde{\mathbf{u}}_{BBS}$. Indeed, one may reduce to the case wherein one wishes to find a vector field

$$\mathbf{v} = (v_1(x,y), v_2(x,y), v_3(x,y,z)) \tag{38}$$

which is divergence-free and restricts to $(v_1(x,y), v_2(x,y), 0)$ on $z = 0$: one immediately finds

$$v_3(x,y,z) = f(x,y)z + \alpha(x,y) \tag{39}$$

with $f(x,y) = -(\partial_x v_1 + \partial_y v_2)$ and α an arbitrary smooth function.

5.3.3. Alternatively, one may require that, at each instant t, via a $U(1)$-*gauge* (connection), the phase of the wave function equals the solid angle S_{K_t} (K_t being as before the Schrödinger evolved knot): the ensuing velocity $\hat{\mathbf{u}} := \nabla S_{K_t}$ (which is singular on K_t) is then divergence-free as S_{K_t} is harmonic.

5.3.4. The above observation is compatible with Brylinski's theorem ([8], Theorem 3.7.4) stating that, if Y denotes the space of oriented knots in a 3-fold, then its connected components coincide with the orbits of the connected component G^0 of the group G of volume preserving diffeomorphisms of Y and, moreover, any component $\mathcal{C}$ of Y is a homogeneous space G^0/G_C^0 for any $C \in \mathcal{C}$ (G_C^0 being the isotropy group of C). This means that, *so long as the nodal line is concerned, its (compressible) Madelung evolution can be traded for an incompressible one.*

6   Conclusions and Outlook

In this note we sketched a geometric-hydrodynamical portrait describing the evolution of the nodal line of the simplest Schrödinger system focussed on the probability current, which nevertheless needs further amplification and detail to be ultimately understood in terms of modern Euler-Poincaré theory (see e.g. [20]), possibly enhanced by the gauge theoretic approach developed in [13,15,31]. The regular behaviour of the probability current, together with the ensuing geometric framework may provide a clue to overcome the foundational and mathematical intricacies caused by the singularities of the Madelung velocity.

Acknowledgements. The present work has been carried out within the framework of activities of INDAM-GNSAGA. Extensive and enlightening discussions with *D.D. Holm and C. Tronci* are gratefully acknowledged. Thanks are also given to the Referees for their careful reading and suggestions.

References

1. Arnold, V.I., Khesin, B.A.: Topological Methods in Hydrodynamics, 2nd edn. Springer, Cham (2021)
2. Berry, M.: Knotted Zeros in the quantum states of hydrogen. Found. Phys. **31**(4), 659–667 (2001)
3. Bialynicki-Birula, I., Bialynicka-Birula, Z., Śliwa, C.: Motion of vortex lines in quantum mechanics. Phys. Rev. A **61**, 032110 (7 pages) (2000)
4. Binysh, J., Alexander, G.P.: Maxwell's theory of solid angle and the construction of knotted fields. J. Phys. A: Math. Theor. **51**, 385202 (20 pages) (2018)
5. Bohm, D.: A Suggested Interpetation of the Quantum Theory in Terms of "Hidden" Variables I Phys. Rev. **85**(2), 166–179 (1952)
6. Bohm, D.: A suggested interpetation of the quantum theory in terms of "hidden" variables II. Phys. Rev. **85**(2), 180–193 (1952)
7. Borodzik, M., Dangskul, S., Ranicki, A.: Solid angles and Seifert hypersurfaces. Ann. Glob. Anal. Geom. **57**, 415–454 (2020)
8. Brylinski, J.L.: Loop spaces. Characteristic Classes and Geometric Quantization. Birkhäuser, Boston (1993)
9. de Rham, G.: Differentiable Manifolds. Springer, Berlin (1984). https://doi.org/10.1007/978-0-8176-4767-4
10. Donaldson, S.K.: Moment maps and diffeomorphisms. Asian J. Math. **3**(1), 1–16 (1999)
11. dos Santos, F.E.A.: Hydrodynamics of vortices in Bose-Einstein condensates: a defect-gauge field approach. Phys. Rev. A **94**, 063633 (6 pages) (2016)
12. Fenn, R.: Tech. Geom. Topol. Cambridge University Press, Cambridge (1983)
13. Foskett, M.S.: Geometry of quantum hydrodynamics in theoretical chemistry, PhD thesis. University of Surrey 2020; arXiv:2009.13601v1 [math-ph] 28 Sep 2020
14. Foskett, M.S., Holm, D.D., Tronci, C.: Geometry of nonadiabatic quantum hydrodynamics. Acta Appl. Math. **162**, 63–103 (2019)
15. Foskett, M.S., Tronci, C.: Holonomy and vortex structures in quantum hydrodynamics. In: Fathi, A., Morrison, P.J., M-Seara, P.T., Tabachnikov, S. (eds.). Hamiltonian Systems: Dynamics, Analysis, Applications, Mathematical Sciences Research Institute Publication, 72p, Cambridge University Press, Cambridge (2022)
16. Fusca, D.: The madelung transform as a momentum map. J. Geom. Mech. **9**(2), 157–165 (2017)
17. Goldin, G.A., Menikoff, R., Sharp, D.H.: Particle statistics from induced representations of a local current group. J. Math. Phys. **21**(4), 650–664 (1980)
18. Guillemin, V., Sternberg, S.: Symplectic Techniques in Physics. Cambridge University Press, Cambridge (1984)
19. Khesin, B., Misiołek, G., Modin, K.: Geometric hydrodynamics via madelung transform. Proc. Natl. Aacd. Sci. **115**(24), 6165–6170 (2018)
20. Holm, D.D., Schmah, T., Stoica, C.: Geometric Mechanics and Symmetry: From Finite to Infinite Dimensions. Oxford University Press, Oxford (2009)
21. Kauffman, L.H., Lomonaco, S.J. Jr.: Quantum knots and knotted zeros. In: Proceedings of SPIE 10984, Quantum Information Science, Sensing, and Computation XI, 109840A (13 May 2019)
22. Kleinert, H.: Multivalued Fields In Condensed Matter, Electromagnetism, and Gravitation. World Scientific, Singapore (2008)

23. Kuznetsov, E.A., Mikhailov, A.V.: On the topological meaning of canonical Clebsch variables. Phys. Lett. A **77**(1), 37–38 (1980)
24. Lamb, H.: Hydrodynamics. Cambridge University Press, Cambridge (1932)
25. Madelung, E.: Quantentheorie in hydrodynamischer form. Zeit. F. Phys. **40**, 322–326 (1927)
26. Marsden, J., Weinstein, A.: Coadjoint orbits, vortices and Clebsch variables for incompressible fluids. Physica **7D**, 305–323 (1983)
27. Penna, V., Spera, M.: A geometric approach to quantum vortices. J. Math. Phys. **30**(12), 2778–2784 (1989)
28. Penna, V., Spera, M.: On coadjoint orbits of rotational perfect fluids. J. Math. Phys. **33**(3), 901–909 (1992)
29. Rasetti, M., Regge, T.: Vortices in He II, current algebras and quantum knots. Phys. A Statistical Mechanics and its Applications **80**(3), 217–233 (1975)
30. Ricca, R.L., Nipoti, B.: Gauss' linking number revisited. J. Knot Theory Ramif. **20**(10), 1325–1343 (2011)
31. Spera, M.: Moment map and gauge geometric aspects of the Schrödinger and Pauli equations. Int. J. Geom. Methods Mod. Phys. **13**(4), 1630004 (36 pages) (2016)
32. Takabayasi, T.: Hydrodynamical formalism of quantum mechanics and Aharonov-Bohm Effect. Prog. Theor. Phys. **69**(5), 1323–1344 (1983)
33. Wyatt, R.E.: Quantum Dynamics with Trajectories. Springer, New York (2005)

Variational Geometric Description
for Fluids with Permeable Boundaries

François Gay-Balmaz[1(✉)], Meng Wu[1], and Chris Eldred[2]

[1] Laboratoire de Météorologie Dynamique, Ecole Normale Supérieure/CNRS, Paris,
France
francois.gay-balmaz@lmd.ens.fr, meng.wu@lmd.ipsl.fr
[2] Org 01442 (Computational Science), Sandia National Laboratories, Paris,
France
celdred@sandia.gov

Abstract. Motivated by modelling and numerical applications in geophysical fluid dynamics, such as the outflow of free or forced waves, we present a Lagrangian variational formulation for fluids exchanging energy with its surrounding through the boundary of its spatial domain. We give the variational formulation in the material description and deduce the Eulerian variational formulation by applying reduction by symmetry in the Lagrangian framework. In the material description we use the classical Hamilton principle applied to fluid trajectories, appropriately amended to incorporate boundary forces via a Lagrange-d'Alembert approach, and to take into account only the fluid particles present in the fluid domain. In particular, our approach extends to the case of permeable domains the well-known geometric description of fluid motion via diffeomorphism groups.

Keywords: Permeable boundaries · Lagrangian variational
formulation · Reduction by symmetry

1 Introduction

Open boundary conditions for fluid dynamics arise naturally in various contexts and configurations for many real-world applications, [7]. For instance, it is the natural abstract setting for flows in turbines, pipelines, wind tunnels, and jet engines. Open boundary conditions are also a necessity in any non-global ocean or atmosphere circulation model because of the need to truncate the domains of interest and use some artificial boundaries due to limitations of computational resources, [2]. The same need evidently appears in various other fields such as acoustics, solid mechanics, plasma physics, civil engineering, geophysics, and environmental science.

In such configurations, the well-known geometric description of fluid dynamics as a mechanical system on a certain diffeomorphism group of the fluid domain, [1], is lost. Indeed, this description assumes from the start that the

F. Nielsen and F. Barbaresco (Eds.): GSI 2023, LNCS 14072, pp. 282–289, 2023.
https://doi.org/10.1007/978-3-031-38299-4_30

fluid motion preserves the boundary $\partial\Omega$ of the fluid domain Ω, i.e., for any fluid motion $\varphi(t, \cdot) \in \text{Diff}(\Omega)$ one has $\varphi(t, \partial\Omega) = \partial\Omega$ and, consequently, $\dot{\varphi}(t, X) \cdot n(t, x) = 0$ with $n(t, x)$ the outward pointing unit normal to $\partial\Omega$ and $x = \varphi(t, X)$. This last condition is the tangential boundary condition for the fluid velocity $u(t, x) = \frac{d}{dt}\varphi(t, X)$.

In this abstract, we shall show that a geometric and variational description based on diffeomorphism groups is possible in the case of open boundary conditions. In particular, we shall use as configuration manifold the group of diffeomorphisms of the whole ambient Euclidean space while the information of the fluid domain will be taken care in the choice of the Lagrangian. The information concerning the boundary conditions will be inserted in the most natural way from the point of view of Lagrangian mechanics, namely, by using the Lagrange-d'Alembert principle for systems with exterior forces. Recall that it is given as

$$\delta \int_0^T L(q, \dot{q})\mathrm{d}t + \int_0^T \langle F(q, \dot{q}), \delta q \rangle \, \mathrm{d}t = 0,$$

with $L : TQ \to \mathbb{R}$ the Lagrangian function defined on the tangent bundle TQ of the configuration manifold Q and $F : TQ \to T^*Q$ the exterior force, assumed to be a smooth fiber preserving map covering q.

We present in Sect. 2 below the geometric setting and the variational principle for open fluid motion in the material (or Lagrangian) description. This should be understood as the primary description, since it is in the material frame that the variational principle takes its simplest form as well as the form of a first principle from the physical point of view. The equations of motion and boundary conditions, however, take a simpler form in the spatial (or Eulerian) description. This motivates the development of a geometric and variational setting in the spatial description. It is described in Sect. 3, and is obtained by a reduction by symmetry, as it is usually done when passing from the Lagrangian to the spatial frame, [3]. Future directions of research and a conclusion are given in Sect. 4. A thorough development and extensions will be the subject of a forthcoming paper.

2   Lagrange-d'Alembert Principle for Open Fluids

2.1   Geometric Setting and Lagrangian

We consider the motion of a fluid in a domain $\Omega \subset \mathbb{R}^n$, $n = 2, 3$ with permeable boundary. Since the fluid can leave or enter the domain, we cannot use the diffeomorphism group $\text{Diff}(\Omega)$ as the configuration Lie group of the system, as it is done for the case of an impermeable boundary. We shall instead consider the group $\text{Diff}(\mathbb{R}^n)$ of diffeomorphisms of the whole Euclidean space $\mathbb{R}^n$. For a given Lagrangian density $\mathfrak{L}(\varphi, \dot{\varphi}, \nabla\varphi)$, expressed in terms of the position $\varphi(t, X)$, velocity $\dot{\varphi}(t, X)$, and deformation gradient $\nabla\varphi(t, X)$ of the fluid configuration $\varphi(t, X)$, as well as on the mass density $\varrho(t, X)$, the Lagrangian function is defined as

$$L : T\operatorname{Diff}(\mathbb{R}^n) \times \operatorname{Den}(\mathbb{R}^n) \to \mathbb{R}, \qquad L(\varphi, \dot{\varphi}, \varrho) = \int_{\varphi^{-1}(\Omega)} \mathfrak{L}(\varphi, \dot{\varphi}, \nabla\varphi, \varrho)\mathrm{d}X.$$

Here $T\operatorname{Diff}(\mathbb{R}^n)$ denotes the tangent bundle to $\operatorname{Diff}(\mathbb{R}^n)$ and $\operatorname{Den}(\mathbb{R}^n)$ the space of densities, identified with the space of functions here by using the Euclidean volume $\mathrm{d}X$. We note that L is obtained by integrating the Lagrangian density $\mathfrak{L}$ on the domain $\varphi^{-1}(\Omega)$ given by the set of all the labels $X \in \mathbb{R}^n$ whose current position lies in the fluid domain Ω.

2.2   Boundary Forces and the Lagrange-d'Alembert Principle

We shall focus on boundary forces associated to both the fluid momentum and the mass density. By following the setting of the Lagrange-d'Alembert principle, we thus have to augment the Hamilton principle with virtual work terms associated to both these forces. Regarding the boundary force associated to fluid momentum, it can be naturally inserted by pairing it with the variation, or virtual displacement, $\delta\varphi$ of the fluid flow. Regarding the boundary force associated to mass density, an additional scalar field F has to be introduced, in order for this force to be paired with the virtual variations δF. It turns out that such a field F is also needed to impose the continuity equation $\frac{d}{dt}\varrho = 0$ for mass in the Lagrangian description when ϱ is considered as an independent field to be varied in the variational formulation.

Based on these considerations, we consider the action functional $\mathcal{A}$ defined by

$$\mathcal{A}(\varphi, F, \varrho) = \int_0^T \int_{\varphi^{-1}(\Omega)} \left[\mathfrak{L}(\varphi, \dot{\varphi}, \nabla\varphi, \varrho) + \varrho\dot{F} \right] \mathrm{d}X\mathrm{d}t$$

as well as the following boundary forces:

(i) A boundary source of momentum $\mathfrak{J}\mathrm{d}A = \mathfrak{J}(\varphi, \dot{\varphi}, \nabla\varphi, \varrho)\mathrm{d}A$, given as a map

$$\mathfrak{J}\mathrm{d}A : \varphi^{-1}(\partial\Omega) \to T^*\mathbb{R}^n \otimes \Lambda^{n-1}(\varphi^{-1}(\partial\Omega));$$

(ii) A boundary source of mass $\mathrm{j}\mathrm{d}A = \mathfrak{l}(\varphi, \dot{\varphi}, \nabla\varphi, \varrho)\mathrm{d}A$, given as a map

$$\mathrm{j}\mathrm{d}A : \varphi^{-1}(\partial\Omega) \to \Lambda^{n-1}(\varphi^{-1}(\partial\Omega)).$$

In particular, for $X \in \varphi^{-1}(\partial\Omega)$, we have

$$\mathfrak{J}(X)\mathrm{d}A \in T^*_{\varphi(X)}\mathbb{R}^n \otimes \Lambda^{n-1}_X(\varphi^{-1}(\partial\Omega))$$
$$\mathrm{j}(X)\mathrm{d}A \in \Lambda^{n-1}_X(\varphi^{-1}(\partial\Omega)).$$

We have denoted by $\mathrm{d}A$ the area element induced by $\mathrm{d}X$ on $\varphi^{-1}(\partial\Omega)$.

This results in a Lagrange-d'Alembert principle of the form

$$\frac{d}{d\varepsilon}\bigg|_{\varepsilon=0} \mathcal{A}(\varphi_\varepsilon, F_\varepsilon, \varrho_\varepsilon) + \int_0^T \int_{\varphi^{-1}(\partial\Omega)} (\mathfrak{J} \cdot \delta\varphi + \mathrm{j}\delta F)\mathrm{d}A\mathrm{d}t = 0, \qquad (1)$$

for arbitrary variations $\delta\varphi$, δF, $\delta\varrho$ vanishing at the endpoints. Above and below, we use the simplified notation $\mathfrak{J}$ instead of $\mathfrak{J}(\varphi, \dot{\varphi}, \nabla\varphi, \varrho)$, similarly for j.

2.3   Equations for Open Fluids in the Lagrangian Frame

The computation of the critical condition for the variational principle (1) is slightly more involved than the case of closed fluids since it also involves variations of the domain of labels. The resulting system of Euler-Lagrange equations and boundary conditions for open fluids with a general Lagrangian density is stated in the following proposition. An intermediate step in the proof gives the condition $\dot{F} = -\frac{\partial \mathcal{L}}{\partial \varrho}$ which allows to completely eliminate the variable F.

Proposition 1. *The variational principle* (1) *gives the following Euler-Lagrange equations and boundary conditions:*

$$
\begin{cases}
\partial_t \dfrac{\partial \mathcal{L}}{\partial \dot{\varphi}} + \mathrm{DIV}\, \dfrac{\partial \mathcal{L}}{\partial \nabla \varphi} - \dfrac{\partial \mathcal{L}}{\partial \varphi} = 0, \qquad \partial_t \varrho = 0 \quad on \quad \varphi^{-1}(\Omega) \\[2mm]
N \cdot \left(\dfrac{\partial \mathcal{L}}{\partial \nabla \varphi} + ((\nabla \varphi)^{-1} \cdot \dot{\varphi}) \dfrac{\partial \mathcal{L}}{\partial \dot{\varphi}} + \left(\varrho \dfrac{\partial \mathcal{L}}{\partial \varrho} - \mathcal{L} \right) (\nabla \varphi)^{-1} \right) = -\mathfrak{J} \quad on \quad \varphi^{-1}(\partial \Omega) \\[2mm]
N \cdot ((\nabla \varphi)^{-1} \cdot \dot{\varphi}) \varrho = -\mathfrak{j} \quad on \quad \varphi^{-1}(\partial \Omega),
\end{cases}
\tag{2}
$$

where N is the outward pointing unit normal vector field to the boundary of $\varphi^{-1}(\Omega)$.

Remark. We have written DIV, as opposed to div, the divergence operator appearing in (2) to emphasize that it is considered in the material description. Also we have denoted by N the unit normal vector field to the domain $\varphi^{-1}(\Omega)$, to distinguish it from the unit normal vector field to Ω denoted n.

Remark. Writing X^A and x^i the local coordinates in the material and spatial frames, the system (2) takes the following local form:

$$
\begin{cases}
\partial_t \dfrac{\partial \mathcal{L}}{\partial \dot{\varphi}^i} + \partial_A \dfrac{\partial \mathcal{L}}{\partial \varphi^i_{,A}} - \dfrac{\partial \mathcal{L}}{\partial \varphi^i} = 0, \qquad \partial_t \varrho = 0 \\[2mm]
N_A \left(\dfrac{\partial \mathcal{L}}{\partial \varphi^i_{,A}} + (\varphi^{-1})^A_{,j} \dot{\varphi}^j \dfrac{\partial \mathcal{L}}{\partial \dot{\varphi}^i} + \left(\varrho \dfrac{\partial \mathcal{L}}{\partial \varrho} - \mathcal{L} \right) (\varphi^{-1})^A_{,i} \right) = -\mathfrak{J}_i \quad on \quad \varphi^{-1}(\partial \Omega) \\[2mm]
N_A (\varphi^{-1})^A_{,j} \varphi^j \varrho = -\mathfrak{j} \quad on \quad \varphi^{-1}(\partial \Omega),
\end{cases}
\tag{3}
$$

which helps clarifying the form of the equation in the second line.

Remark. Despite their involved general expressions, the equations (2) take a particularly simple form for standard Lagrangians, due to the relabelling symmetry of $\mathcal{L}$. For instance, we note that for the Lagrangian density of a barotropic fluid

$$
\mathcal{L}(\varphi, \dot{\varphi}, \nabla \varphi, \varrho) = \frac{1}{2} \varrho |\dot{\varphi}|^2 - \varepsilon (\varrho / \mathsf{J}\varphi) \, \mathsf{J}\varphi,
$$

with ε the internal energy density and $\mathsf{J}\varphi$ the Jacobian of φ, they simplify as

$$
\begin{cases}
\partial_t (\varrho \dot{\varphi}) = -\nabla p \circ \varphi, \qquad \partial_t \varrho = 0 \quad on \quad \Omega \\[1mm]
N \cdot ((\nabla \varphi)^{-1} \cdot \dot{\varphi}) \, \varrho \dot{\varphi} = -\mathfrak{J} \quad on \quad \partial \Omega \\[1mm]
N \cdot ((\nabla \varphi)^{-1} \cdot \dot{\varphi}) \varrho = -\mathfrak{j} \quad on \quad \partial \Omega,
\end{cases}
\tag{4}
$$

with $p = \frac{\partial \varepsilon}{\partial \rho} \rho - \varepsilon$ the pressure.

3   Eulerian Variational Formulation for Open Fluids

In this section we derive the Eulerian form of the Lagrange-d'Alembert principle with boundary forces (1). While this form of the variational principle is more involved, it produces directly the desired spatial form of the equations, for an arbitrary Lagrangian density.

3.1   Relabelling Symmetry of the Lagrangian and Forces

The Eulerian description is possible only if the Lagrangian density satisfies the following material covariance property:

$$\mathfrak{L}\big(\varphi \circ \psi, \dot{\varphi} \circ \psi, \nabla(\varphi \circ \psi), (\varrho \circ \psi)J\psi\big) = \mathfrak{L}\big(\varphi, \dot{\varphi}, \nabla\varphi, \varrho\big) \circ \psi J\psi,$$

for all diffeomorphisms $\psi \in \mathrm{Diff}(\mathbb{R}^n)$, see [5]. From this assumption there exists a Lagrangian density $\mathfrak{l}(u, \rho)$ in the Eulerian description, such that

$$\mathfrak{L}\big(\varphi, \dot{\varphi}, \nabla\varphi, \varrho\big) \circ \varphi^{-1} J\varphi^{-1} = \mathfrak{l}(u, \rho), \tag{5}$$

for all $\varphi \in \mathrm{Diff}(\mathbb{R}^n)$, where $u = \dot{\varphi} \circ \varphi^{-1}$ and $\rho = (\varrho \circ \varphi^{-1})J\varphi^{-1}$.

Similarly, we assume that the boundary forces satisfy the following material covariance property

$$\mathfrak{J}\big(\varphi \circ \psi, \dot{\varphi} \circ \psi, \nabla(\varphi \circ \psi), (\varrho \circ \psi)J\psi\big) = \mathfrak{J}\big(\varphi, \dot{\varphi}, \nabla\varphi, \varrho\big) \circ (\psi|_{\varphi^{-1}(\partial\Omega)})J(\psi|_{\varphi^{-1}(\partial\Omega)})$$

on $\psi^{-1}(\varphi^{-1}(\partial\Omega))$, for all $\psi \in \mathrm{Diff}(\mathbb{R}^n)$, similarly for j. From this assumption, there exists the boundary fluxes $J = J(u, \rho)$ and $j = j(u, \rho)$, given as maps

$$J da : \partial\Omega \to T^*\mathbb{R}^n \otimes \Lambda^{n-1}(\partial\Omega)$$
$$j da : \partial\Omega \to \Lambda^{n-1}(\partial\Omega),$$

such that

$$\mathfrak{J}\big(\varphi, \dot{\varphi}, \nabla\varphi, \varrho\big) \circ (\varphi^{-1}|_{\partial\Omega})J(\varphi^{-1}|_{\partial\Omega}) = J(u, \rho) \tag{6}$$

similarly for j and j. Here da denotes the area element on $\partial\Omega$ induced by dx.

3.2   Eulerian Variational Formulation for Open Fluids

By using the material covariance assumptions and the relations (5) and (6) it is possible to rewrite the Lagrange-d'Alembert principle (1) exclusively in terms the Eulerian fields defined by $u = \dot{\varphi} \circ \varphi^{-1}$, $\rho = (\varrho \circ \varphi^{-1})J\varphi^{-1}$, and $f = \dot{F} \circ \varphi^{-1}$. This is the purpose of the next proposition.

Proposition 2. *The variational principle (1) yields the following Euler-Poincaré-d'Alembert formulation in the spatial frame:*

$$\delta \int_0^T \int_\Omega \big[\mathfrak{l}(u, \rho) + \rho f\big] dx dt + \int_0^T \int_{\partial\Omega} (J \cdot \zeta + j\nu) da = 0 \tag{7}$$

with respect to the variations

$$\delta u = \partial_t \zeta + \pounds_u \zeta$$
$$\delta f = \partial_t \nu + u \cdot \nabla \nu - \zeta \cdot \nabla f \qquad (8)$$
$$\delta \rho : free$$

with $\delta \rho$, ζ and ν vanishing at the endpoints.

Proof. From (5), for a path φ_ε of fluid configuration we have

$$\frac{d}{d\varepsilon}\Big|_{\varepsilon=0} \int_{\varphi_\varepsilon^{-1}(\Omega)} \mathfrak{L}(\varphi_\varepsilon, \dot{\varphi}_\varepsilon, \nabla\varphi_\varepsilon, \varrho_\varepsilon) dX$$
$$= \frac{d}{d\varepsilon}\Big|_{\varepsilon=0} \int_\Omega \mathfrak{L}(\varphi_\varepsilon, \dot{\varphi}_\varepsilon, \nabla\varphi_\varepsilon, \varrho_\varepsilon) \circ \varphi_\varepsilon^{-1} J\varphi_\varepsilon^{-1} dx = \int_\Omega \frac{d}{d\varepsilon}\Big|_{\varepsilon=0} \ell(u_\varepsilon, \rho_\varepsilon) dx, \qquad (9)$$

where $u_\varepsilon = \dot{\varphi}_\varepsilon \circ \varphi_\varepsilon^{-1}$ and $\rho_\varepsilon = \varrho_\varepsilon \circ \varphi_\varepsilon^{-1} J\varphi_\varepsilon^{-1}$. From these relations, we have the usual Euler-Poincaré variational constraint $\delta u = \partial_t \zeta + \pounds_u \zeta$ with $\zeta = \delta\varphi \circ \varphi^{-1}$, while $\delta\rho$ is free since $\delta\varrho$ is free. Similarly, for paths φ_ε and F_ε, we have

$$\frac{d}{d\varepsilon}\Big|_{\varepsilon=0} \int_{\varphi_\varepsilon^{-1}(\Omega)} \varrho_\varepsilon \dot{F}_\varepsilon dX$$
$$= \frac{d}{d\varepsilon}\Big|_{\varepsilon=0} \int_\Omega \varrho_\varepsilon \circ \varphi_\varepsilon^{-1} J\varphi_\varepsilon^{-1} \dot{F}_\varepsilon \circ \varphi_\varepsilon^{-1} dx = \int_\Omega \frac{d}{d\varepsilon}\Big|_{\varepsilon=0} \rho_\varepsilon f_\varepsilon dx, \qquad (10)$$

where $f_\varepsilon = \dot{F}_\varepsilon \circ \varphi_\varepsilon^{-1}$. From this relation, the variation δf is computed as

$$\delta f = \partial_t \nu + \nabla \nu \cdot u - \nabla f \cdot \zeta,$$

where we have defined $\nu = \delta F \circ \varphi^{-1}$. We have thus derived the expression of the action functional in (7) and the corresponding constrained variations. Finally, by using (6) and the corresponding relation for j, the virtual force terms in (1) can be transformed into their corresponding Eulerian form in (7).

3.3   Equations for Open Fluids in the Spatial Frame

From the variational principle (7)–(8) one directly gets the equations of motion for an open fluid with arbitrary Lagrangian density $\mathfrak{l}(u, \rho)$. The result is stated in the next proposition.

Proposition 3. *The Euler-Poincaré-d'Alembert principle (7) yields the following equations for open fluids:*

$$\begin{cases} \partial_t \dfrac{\partial \mathfrak{l}}{\partial u} + \pounds_u \dfrac{\partial \mathfrak{l}}{\partial u} - \rho\nabla \dfrac{\partial \mathfrak{l}}{\partial \rho} = 0 \quad on \quad \Omega \\ \partial_t \rho + \mathrm{div}(\rho u) = 0 \quad on \quad \Omega \\ (u \cdot n)\dfrac{\partial \mathfrak{l}}{\partial u} = -J \quad on \quad \partial\Omega \\ \rho u \cdot n = -j \quad on \quad \partial\Omega, \end{cases}$$

where n is the outward pointing unit normal vector to $\partial\Omega$, and $\pounds_u m = \nabla m \cdot u + \nabla u^\mathsf{T} m + m\,\mathrm{div}\,u$ is the Lie derivative of the fluid momentum $m = \frac{\partial \mathfrak{l}}{\partial u}$.

Balance of Mass and Energy. From the second and fourth equations, we get the total mass balance equation

$$\frac{d}{dt} \int_\Omega \rho \mathrm{d}x = \int_{\partial\Omega} j \mathrm{d}a.$$

Defining the total energy density associated to $\mathfrak{l}$ as $\mathfrak{e} = \frac{\partial \mathfrak{l}}{\partial u} \cdot u - \mathfrak{l}$, we get its time evolution as

$$\partial_t \mathfrak{e} = \mathrm{div} \left(\left(\rho \frac{\partial \mathfrak{l}}{\partial \rho} - u \cdot \frac{\partial \mathfrak{l}}{\partial u} \right) u \right)$$

from which the general form of the total energy balance is found as

$$\frac{d}{dt} \int_\Omega \mathfrak{e} \mathrm{d}x = \int_{\partial\Omega} \left(J \cdot u - j \frac{\partial \mathfrak{l}}{\partial \rho} \right) \mathrm{d}a.$$

This equation, which is given here for a general Lagrangian density, plays a main role in the choice of appropriate expressions for J and j for geophysical fluid models for modelling, numerical and well-posedness purposes.

It is well-known that great care must be taken when designing the boundary fluxes in order to have a well-posed problem, especially regarding outflow and inflow conditions on different parts of the boundary, and regarding the number of boundary conditions which must be applied in accordance with the subsonic or supersonic character of the flow, [9].

Example: Barotropic Fluid and Boundary Conditions. For the Lagrangian density $\mathfrak{l}(u, \rho) = \frac{1}{2}\rho|u|^2 - \varepsilon(\rho)$ of the barotropic fluid, one gets from Propositon 3 the equations $\rho(\partial_t u + u \cdot \nabla u) = -\nabla p$ and $\partial_t \rho + \mathrm{div}(\rho u) = 0$ on Ω, together with the boundary conditions

$$(\rho u \cdot n)u = -J \quad \text{and} \quad \rho u \cdot n = -j. \tag{11}$$

If we assume some given expressions for J and j, independent of u and ρ, there are two main situations, which can happen on parts of the boundary. First, there is the case $j = 0$, giving $u \cdot n = 0$, which corresponds to an impermeable boundary. From the relation $j^2/\rho = -J \cdot n$ which follows from (11), we must have $J \cdot n = 0$ while there is no boundary condition for $\rho|_{\partial\Omega}$ consistently with the treatment of fluids with impermeable boundaries. From the first relation in (11) we must take $J = 0$ in this case. A second case arises if j never vanishes, then also $J \cdot n$ never vanishes, which gives the boundary condition $\rho|_{\partial\Omega} = -j^2/J \cdot n$ for the mass density. In this case, we also get the Dirichlet boundary condition $u|_{\partial\Omega} = J/j$ for the velocity, imposing in particular that $u \cdot n$ is never zero, meaning that the fluid must cross the boundary. This is the case of an open fluid with inflow boundary condition, for which we choose $j > 0$, and which includes boundary conditions for both u and ρ.

The treatment of an outflow boundary condition needs to consider j given in terms of u and ρ as $j = -|\rho u \cdot n|$, so that the second condition in (11) imposes $u \cdot n > 0$. Then, we set $J = -\rho(u_0 \cdot n)u_0$ for some vector field u_0, which imposes $u|_{\partial\Omega} = u_0$, without specifying $\rho|_{\partial\Omega}$, consistently with outflow boundary conditions.

4 Conclusion and Future Directions

In this abstract, we have presented an extension of the variational geometric description of fluid dynamics on diffeomorphism groups to the case of open fluids. This situation is common in a wide range of real world applications in which there is an exchange of mass and energy of the fluid with its surrounding. We are currently applying this variational approach for several geophysical fluid models in order to develop the associated structure preserving discretization for open fluids in these cases, following [4]. We also project to analyse the Hamiltonian side of the framework in order to extend the Lie-Poisson bracket formulation for fluid dynamics to the case of permeable boundaries. This approach should provide a first principle derivation of the Poisson bracket formulation amended with boundary terms proposed for open systems, see [8].

Acknowledgements. This article has been co-authored by an employee of National Technology & Engineering Solutions of Sandia, LLC under Contract No. DE-NA0003525 with the U.S. Department of Energy (DOE). The employee owns right, title and interest in and to the article and is responsible for its contents. The United States Government retains and the publisher, by accepting the article for publication, acknowledges that the United States Government retains a non-exclusive, paid-up, irrevocable, world-wide license to publish or reproduce the published form of this article or allow others to do so, for United States Government purposes. The DOE will provide public access to these results of federally sponsored research in accordance with the DOE Public Access Plan https://www.energy.gov/downloads/doe-public-access-plan.

References

1. Arnold, V.I.: Sur la géométrie différentielle des groupes de Lie de dimenson infinie et ses applications à l'hydrodynamique des fluides parfaits. Ann. Inst. Fourier **16**, 319–361 (1966)
2. Cushman-Roisin, B., Beckers, J.M.: Introduction to Geophysical Fluid Dynamics: Physical and Numerical Aspects. Academic Press, Cambridge (2011)
3. Holm, D.D., Marsden, J.E., Ratiu, T.S.: The Euler-Poincaré equations and semidirect products with applications to continuum theories. Adv. Math. **137**, 1–81 (1998)
4. Gawlik, E.S., Gay-Balmaz, F.: Variational finite element discretization of compressible flow. Found. Comput. Math. **21**, 961–1001 (2021)
5. Gay-Balmaz, F.: General relativistic Lagrangian continuum theories-Part I: reduced variational principles and junction conditions for hydrodynamics and elasticity (2022). http://arxiv.org/pdf/2202.06560'
6. Gay-Balmaz, F., Marsden, J.E., Ratiu, T.S.: Reduced variational formulations in free boundary continuum mechanics. J. Nonlinear Sci. **22**(4), 463–497 (2012)
7. Johnson, R.: Handbook of Fluid Dynamics (2nd ed.). CRC Press, Aylor & Francis Group (2016)
8. Öttinger, H.C.: Nonequilibrium thermodynamics for open systems. Phys. Rev. E **73**, 036126 (2006)
9. Poinsot, T.J., Lelef, S.K.: Boundary conditions for direct simulations of compressible viscous flows. J. Comput. Phys. **101**(1), 104–129 (1992)

Lagrangian Trajectories and Closure Models in Mixed Quantum-Classical Dynamics

Cesare Tronci[1,2(✉)] and François Gay-Balmaz[3]

[1] Department of Mathematics, University of Surrey, Guildford, UK
c.tronci@surrey.ac.uk
[2] Department of Physics and Engineering Physics, Tulane University, New Orleans, USA
[3] CNRS & Laboratoire de Météorologie Dynamique, École Normale Supérieure, Paris, France
francois.gay-balmaz@lmd.ens.fr

Abstract. Mixed quantum-classical models have been proposed in several contexts to overcome the computational challenges of fully quantum approaches. However, current models typically suffer from long-standing consistency issues, and, in some cases, invalidate Heisenberg's uncertainty principle. Here, we present a fully Hamiltonian theory of quantum-classical dynamics that appears to be the first to ensure a series of consistency properties, beyond positivity of quantum and classical densities. Based on Lagrangian phase-space paths, the model possesses a quantum-classical Poincaré integral invariant as well as infinite classes of Casimir functionals. We also exploit Lagrangian trajectories to formulate a finite-dimensional closure scheme for numerical implementations.

Keywords: Mixed quantum-classical dynamics · Lagrangian trajectory · Koopman wavefunction · Hamilton's variational principle · group action

1 Introduction

The search for a mixed quantum-classical description of many-body quantum systems is motivated by the formidable challenges posed by the curse of dimensionality appearing in fully quantum approaches. For example, it is common practice in molecular dynamics to approximate nuclei as classical particles while retaining a fully quantum electronic description. Similar mixed quantum-classical approximations have also been proposed in quantum plasmas and, more recently, in magnon spintronics.

This work was made possible through the support of Grant 62210 from the John Templeton Foundation. The opinions expressed in this publication are those of the authors and do not necessarily reflect the views of the John Templeton Foundation.

Hybrid Quantum-Classical Models. Despite the computational appeal, the interaction dynamics of quantum and classical degrees of freedom continues to represent a challenging question since the currently available models suffer from several consistency issues. In some cases, the Heisenberg principle is lost due to the fact that the quantum density matrix is allowed to change its sign. In some other cases, the model does not reduce to uncoupled quantum and classical dynamics in the absence of a quantum-classical interaction potential. At a computational level, the most popular approach is probably the Ehrenfest model, which reads

$$\partial_t D + \mathrm{div}\big(D\langle \mathbf{X}_{\widehat{H}}\rangle\big) = 0, \qquad i\hbar\big(\partial_t \psi + \langle \mathbf{X}_{\widehat{H}}\rangle \cdot \nabla\psi\big) = \widehat{H}\psi, \qquad (1)$$

where $\mathbf{X}_{\widehat{H}} = \big(\partial_p \widehat{H}, -\partial_q \widehat{H}\big)$. Here, $D(q,p)$ is the classical density, while $\psi(x;q,p)$ is a wavefunction depending on the quantum $x-$coordinate and parameterized by the classical coordinates (q,p). Also, $\widehat{H}(q,p)$ is a quantum Hamiltonian operator depending on (q,p) and we have resorted to the usual notation $\langle \widehat{A}\rangle = \langle\psi|\widehat{A}(q,p)\psi\rangle$, where $\langle\psi_1|\psi_2\rangle = \int \psi_1^*(x)\psi_2(x)\,\mathrm{d}x$. In this setting, the matrix elements of the quantum density operator are given as

$$\hat{\rho}(x,x') = \int D(q,p)\psi(x;q,p)\psi^*(x';q,p)\,\mathrm{d}q\mathrm{d}p.$$

Despite its wide popularity, the Ehrenfest model (1) fails to reproduce realistic levels of decoherence, which is usually expressed in terms of the norm squared $\|\hat{\rho}\|^2$ of the density operator, a quantity also known as *quantum purity*.

Any quantum-classical description beyond the Ehrenfest model must still ensure its five consistency properties: 1) the classical system is identified by a phase-space probability density at all times; 2) the quantum system is identified by a positive-semidefinite density operator $\hat{\rho}$ at all times; 3) the model is covariant under both quantum unitary transformations and classical canonical transformations; 4) in the absence of an interaction potential, the model reduces to uncoupled quantum and classical dynamics; 5) in the presence of an interaction potential, the *quantum purity* $\|\hat{\rho}\|^2$ is not a constant of motion (decoherence property). A model satisfying properties 1)-4), but not 5) is the *mean-field model*

$$\frac{\partial D}{\partial t} + \big\{D, \langle\widehat{H}\rangle\big\} = 0, \qquad i\hbar\frac{\mathrm{d}\hat{\rho}}{\mathrm{d}t} = \left[\int D\widehat{H}\,\mathrm{d}q\mathrm{d}p, \hat{\rho}\right], \qquad (2)$$

where $\{\cdot,\cdot\}$ denotes the canonical Poisson bracket. Here, we notice that the quantum density matrix $\hat{\rho}$ does not carry any dependence on the phase-space coordinates. Most recent efforts in quantum-classical methods are addressed to the design of new models beyond the Ehrenfest system that can better capture decoherence effects and still retain all the consistency properties above.

Beyond the Ehrenfest Model. Blending Koopman wavefunctions in classical mechanics with the geometry of prequantum theory, we recently formulated a quantum-classical model [3,4] which was developed in two stages. First, we provided an early quantum-classical model [1] that succeeded in satisfying only the properties 2)-5). Then, more recently, we upgraded this model in such a way that property 1) is also secured [3,4]. This upgrade was achieved by combining Lagrangian trajectories on the classical phase-space with a gauge principle which ensures that classical phases are *unobservable*, that is they do not contribute to measurable expectation values. Inspired by Sudarshan's work [6], this combination leads naturally to crucial properties such as the characterization of entropy functionals and the Poincaré integral invariant in the context of quantum-classical dynamics. Nevertheless, the model is nonlinear and its explicit form is rather intricate due to the appearance of the non-Abelian gauge connection $i[P, \nabla P]$, where $P(x, x'; q, p) = \psi(x; q, p)\psi^*(x'; q, p)$. In particular, this gauge connection emerges through the (Hermitian) operator-valued vector field $\widehat{\boldsymbol{\Gamma}} = i[P, \mathbf{X}_P]$ in such a way that the model proposed in [3,4] reads

$$\partial_t D + \mathrm{div}(D\boldsymbol{\mathcal{X}}) = 0, \qquad i\hbar(\partial_t \psi + \boldsymbol{\mathcal{X}} \cdot \nabla \psi) = \widehat{\mathcal{H}}\psi, \qquad (3)$$

with

$$\boldsymbol{\mathcal{X}} = \langle \mathbf{X}_{\widehat{H}} \rangle + \frac{\hbar}{2D} \mathrm{Tr}\big(\mathbf{X}_{\widehat{H}} \cdot \nabla(D\widehat{\boldsymbol{\Gamma}}) - (D\widehat{\boldsymbol{\Gamma}}) \cdot \nabla \mathbf{X}_{\widehat{H}}\big), \qquad (4)$$

and

$$\widehat{\mathcal{H}} = \widehat{H} + i\hbar\Big(\{P, \widehat{H}\} + \{\widehat{H}, P\} - \frac{1}{2D}[\{D, \widehat{H}\}, P]\Big). \qquad (5)$$

Thus, we conclude that the vector field $\boldsymbol{\mathcal{X}}$ and the Hermitian generator $\widehat{\mathcal{H}}$ can be regarded as $\hbar$−modifications of the original Ehrenfest quantities $\langle \mathbf{X}_{\widehat{H}} \rangle$ and $\widehat{H}$, respectively. While Eqs. (3)–(5) appear hardly tractable at first sight, a direct calculation of $\mathrm{div}\boldsymbol{\mathcal{X}}$ reveals that no gradients of order higher than two appear in the Eq. (3). In addition, the Hamiltonian/variational structure of this system unfolds much of the features occurring in quantum-classical coupling. Thus, we consider the equations above as a platform for the formulation of simplified closure models that can be used in physically relevant cases.

Trajectory-Based Numerical Algorithms. The presence of transport terms in Eq. (3) results from the predominant role played by Lagrangian trajectories on the classical phase-space. These terms hint to the possibility of using characteristic curves to design trajectory-based schemes for mixed quantum-classical simulation codes in molecular dynamics [5]. However, the presence of several gradients in the expression of the transport vector field $\boldsymbol{\mathcal{X}}$ prevents the direct application of trajectory-based methods, which instead can be readily used for the Ehrenfest Eq. (1). In the latter case, if we denote $\mathbf{z} = (q, p)$, we observe that the first equation is solved by $D(\mathbf{z}, t) = \sum_{a=1}^{N} w_a \delta(\mathbf{z} - \boldsymbol{\zeta}_a(t))$ with $\dot{\boldsymbol{\zeta}}_a = \langle \mathbf{X}_{\widehat{H}} \rangle|_{\mathbf{z}=\boldsymbol{\zeta}_a}$. Here, the quantity $\langle \mathbf{X}_{\widehat{H}} \rangle|_{\mathbf{z}=\boldsymbol{\zeta}_a}$ requires evaluating $\psi_a(t) := \psi(\boldsymbol{\zeta}_a(t), t)$ at all times and this can indeed be done by multiplying the second in (1) by D and then integrating, so that $i\hbar\dot{\psi}_a(t) = \widehat{H}(\zeta_a(t))\psi_a(t)$. Eventually, direct application of the trajectory method to the Ehrenfest model leads to the equations

$$\dot{q}_a = \partial_{p_a}\langle\psi_a|\widehat{H}_a\psi_a\rangle, \qquad \dot{p}_a = -\partial_{q_a}\langle\psi_a|\widehat{H}_a\psi_a\rangle, \qquad i\hbar\dot{\psi}_a = \widehat{H}_a\psi_a, \qquad (6)$$

where $\widehat{H}_a := \widehat{H}(q_a, p_a)$ and $\zeta_a = (q_a, p_a)$. Then, for a finite-dimensional quantum Hilbert space, $\psi(\mathbf{z}, t) \in \mathbb{C}^n$ and the quantum density matrix is $\hat{\rho} = \int D\psi\psi^\dagger d^2z = \sum_a w_a\psi_a\psi_a^\dagger$.

The same approach may be applied to the mean-field model by writing $\hat{\rho} = \sum_{i=1}^N w_a\psi_a\psi_a^\dagger$ and $D = \sum_{i=1}^N w_a\delta(\mathbf{z} - \zeta_a(t))$ in (2). Then, in the case $N = 1$ (only one trajectory) the closure of the Ehrenfest equations coincides with the closure of the mean-field model for the interaction of a classical particle with a pure quantum state. Notice, however, that the Eq. (6) generally account for decoherence effects when $N > 1$, while the same is not true for the mean-field model. Thus, while the Ehrenfest and the mean-field models are regarded as equivalent in the chemistry literature, this alleged equivalence is actually a mere resemblance that arises from the fact that the equations to be implemented in the closure scheme are the same in the case of only one trajectory. We also mention that quantum-classical algorithms alternative to the Ehrenfest model are widely available. The most popular is the *surface hopping method*, which however does not retain positivity of the quantum density matrix and thus invalidates Heisenberg's uncertainty principle.

In this paper, we propose to exploit the geometric variational structure of the new model (3)–(5) in order to make it amenable to trajectory-based closures associated to the Lagrangian paths in the classical phase-space. Upon regularizing a suitable term in Hamilton's action principle, we will obtain a closure scheme that is formally the same as (6), although the Hamiltonian $\widehat{H}_a$ is replaced by an effective Hamiltonian retaining correlation effects beyond the Ehrenfest theory. The resulting variational closure scheme will be illustrated after reviewing the formulation of the system in (3)–(5) and its geometric properties.

2   Formulation of Mixed Quantum-Classical Models

As mentioned in the Introduction, the model (3)–(5) was formulated in [3] by blending the symplectic geometry of Koopman's wavefunctions in classical mechanics with a gauge-invariance principle that arises from physical arguments.

Koopman Wavefunctions. As shown by Koopman in 1931, classical mechanics may be formulated as a unitary flow on the Hilbert space of square-integrable functions on phase-space. The main observation is that the *Koopman-von Neumann equation* (KvN) $i\hbar\partial_t\chi = \{i\hbar H, \chi\}$ yields the classical Liouville equation $\partial_t D = \{H, D\}$ for $D(\mathbf{z}) = |\chi(\mathbf{z})|^2$. Importantly, the Liouvillian operator $\widehat{L}_H = \{i\hbar H, \}$ is self-adjoint, thereby identifying a unitary evolution for χ.

Since both quantum and classical dynamics are written as unitary dynamics on Hilbert spaces, Sudarshan suggested to consider unitary evolution on the tensor-product space [6]. However, this turns out to be a difficult task and the first difficulty resides in the way phases are treated in KvN theory. Indeed, writing $\chi = \sqrt{D}e^{iS/\hbar}$ gives $dD/dt = 0$ and $dS/dt = 0$ along $\dot{z} = \mathbf{X}_H(\mathbf{z})$, so that the KvN phase evolution fails to reproduce the usual prescription arising from Hamilton-Jacobi theory, that is $dS/dt = \mathcal{L}$, where $\mathcal{L}$ is the Lagrangian. This issue is readily addressed by modifying the Liouvillian $\hat{L}_H$ to include a phase term, so that the resulting *Koopman-van Hove equation* (KvH) reads

$$i\hbar\partial_t\chi = \{i\hbar H, \chi\} - (p\partial_p H - H)\chi, \tag{7}$$

where the terms in parenthesis evidently comprise the phase-space expression of the particle Lagrangian $\mathcal{L}$. In this case, the unitary time propagator is of the particular type $\chi(t) = (e^{-i\phi(t)/\hbar}\chi_0/\sqrt{\det\nabla\boldsymbol{\eta}(t)}) \circ \boldsymbol{\eta}(t)^{-1}$, where $(\boldsymbol{\eta}(t), e^{i\phi(t)/\hbar})$ is an element of the infinite-dimensional group $\{(\boldsymbol{\eta}, e^{i\phi/\hbar}) \in \text{Diff}(T^*Q) \circledS \mathcal{F}(T^*Q, S^1) \mid \boldsymbol{\eta}^*\mathcal{A} + d\phi = \mathcal{A}\}$. Here, $\mathcal{F}(T^*Q, S^1)$ denotes the space of functions on the phase space T^*Q taking values in the unit circle, $\circledS$ denotes the semidirect-product, $\boldsymbol{\eta}^*$ is the pullback, and $\mathcal{A} = pdq$ is the Liouville one-form so that the canonical symplectic two-form reads $\omega = -d\mathcal{A}$. Notice that the relation $\boldsymbol{\eta}^*\mathcal{A} + d\phi = \mathcal{A}$ amounts to preservation of the Liouville one-form under the action of the semidirect-product group $\text{Diff}(T^*Q) \circledS \mathcal{F}(T^*Q, S^1)$. Also, we observe that $\boldsymbol{\eta}$ is a symplectic diffeomorphism, i.e. $\boldsymbol{\eta}^*\omega = \omega$. The main advantage of the KvH equation (7), first arisen in prequantization theory, is that it includes the correct prescription for the phase evolution as well as reproducing the Liouville equation for the density.

At this point, a first quantum-classical theory is obtained by starting with two classical systems and then quantizing one of them. This leads to the *quantum-classical wave equation* (QCWE) for the hybrid wavefunction $\Upsilon(\mathbf{z}, x)$ [1]:

$$i\hbar\partial_t\Upsilon = \{i\hbar\widehat{H}, \Upsilon\} - (p\partial_p\widehat{H} - \widehat{H})\Upsilon. \tag{8}$$

As before, $\mathbf{z} = (q, p)$ are classical coordinates, x is the quantum configuration coordinate, and $\widehat{H}(\mathbf{z})$ is an operator-valued function. Once again, the right-hand side of equation (8) identifies a self-adjoint operator which leads to a unitary evolution of the hybrid wavefunction. The action principle $\delta \int_{t_1}^{t_2} \int \text{Re}\langle\Upsilon|i\hbar\partial_t\Upsilon - \{i\hbar\widehat{H}, \Upsilon\} + (p\partial_p\widehat{H} - \widehat{H})\Upsilon\rangle\, d^2z\, dt = 0$ underlying (8) identifies a Hamiltonian functional $h = \int\langle\widehat{\mathcal{D}}|\widehat{H}\rangle\, d^2z$, where $\langle A|B\rangle = \text{Tr}(A^\dagger B)$ and $\widehat{\mathcal{D}}(\mathbf{z}) := \Upsilon(\mathbf{z})\Upsilon^\dagger(\mathbf{z}) + \partial_p(p\Upsilon(\mathbf{z})\Upsilon^\dagger(\mathbf{z})) + i\hbar\{\Upsilon(\mathbf{z}), \Upsilon^\dagger(\mathbf{z})\}$ is a measure-valued von Neumann operator. Then, $\text{Tr}\,\widehat{\mathcal{D}}$ is the classical density and $\int\widehat{\mathcal{D}}\, d^2z$ is the quantum density matrix.

Phase Symmetry in Classical Dynamics. While the QCWE has been studied extensively, the unitary dynamics of hybrid wavefunctions does not appear sufficient for a consistent theory. For example, the classical density $\text{Tr}\,\widehat{\mathcal{D}}$ associated to (8) is generally sign-indefinite [1]. As a further step, Sudarshan pointed out that classical phases, while crucial to retain quantum-classical correlations, should eventually be made 'unobservable'. We applied this idea by resorting to a 'gauge principle' [3], that is by enforcing a symmetry under the group $\mathcal{F}(T^*Q, S^1)$ of phase transformations, in such a way that the latter are treated as a 'gauge freedom'. For this, one first needs to extract the classical phase from the hybrid wavefunction Υ. This is accomplished by writing $\Upsilon(\mathbf{z}, x) = \sqrt{D(\mathbf{z})} e^{iS(\mathbf{z})/\hbar} \psi(x; \mathbf{z})$, so that the last factor is a conditional quantum wavefunction and $S(\mathbf{z})$ is the classical phase. Replacing this factorization in the action principle underlying the QCWE (8) (see previous paragraph) yields $\delta \int_{t_1}^{t_2} L(D, S, \partial_t S, \psi, \partial_t \psi) \, \mathrm{d}t = 0$, with

$$L = \int D \Big(\partial_t S - \mathrm{Re} \, \langle \psi | i\hbar \partial_t \psi - \{i\hbar \widehat{H}, \psi\} + (p\partial_p \widehat{H} - \widehat{H})\psi + \nabla S \cdot \mathbf{X}_{\widehat{H}} \psi \rangle \Big) \, \mathrm{d}^2 z \quad (9)$$

and arbitrary variations δD, δS, and $\delta \psi$. Upon denoting $\langle\,,\rangle = \mathrm{Re}\langle\,|\,\rangle$, one realizes [4] that replacing $\nabla S \to \boldsymbol{\mathcal{A}} + \langle \psi, i\hbar \nabla \psi \rangle$ makes the Hamiltonian functional $h(D, S, \psi) = \int D \langle \psi, (\widehat{H} - p\partial_p \widehat{H})\psi \rangle + \{i\hbar \widehat{H}, \psi\} - \nabla S \cdot \mathbf{X}_{\widehat{H}} \psi \rangle \, \mathrm{d}^2 z$ gauge-invariant, i.e., invariant with respect to $(D, S, \psi) \mapsto (D, S + \varphi, e^{-i\varphi/\hbar} \psi)$ for all $\varphi(\mathbf{z}, t)$. Here, $\boldsymbol{\mathcal{A}} = (p, 0)$ is the coordinate representation of the one-form $\mathcal{A} = \boldsymbol{\mathcal{A}} \cdot \mathrm{d}\mathbf{z} = p \mathrm{d}q$.

In order to obtain an entire phase-invariant variational principle (not just a Hamiltonian functional), one transforms the term $\int D\partial_t S \, \mathrm{d}^2 z$ in such a way to make ∇S appear explicitly and then replaces $\nabla S \to \boldsymbol{\mathcal{A}} + \langle \psi, i\hbar \nabla \psi \rangle$. This was done in [4] by noting that the equation $\partial_t D + \mathrm{div}(D\langle \mathbf{X}_{\widehat{H}} \rangle) = 0$ resulting from the variations (9) allows to use the dynamical relation $D(t) = \eta(t)_* D_0$, that is the density evolves by the push-forward of the initial condition D_0 by a time-dependent Lagrangian path $\boldsymbol{\eta}(t) \in \mathrm{Diff}(T^*Q)$. Integration by parts with respect to time and phase-space leads to $\delta \int_{t_1}^{t_2} \int D\partial_t S \, \mathrm{d}^2 z \, \mathrm{d}t = -\delta \int_{t_1}^{t_2} \int D \nabla S \cdot \boldsymbol{\mathcal{X}} \, \mathrm{d}^2 z \, \mathrm{d}t$, where we have used $\partial_t D = -\mathrm{div}(D\boldsymbol{\mathcal{X}})$ and the vector field $\boldsymbol{\mathcal{X}}$ is such that $\dot{\boldsymbol{\eta}} =: \boldsymbol{\mathcal{X}} \circ \boldsymbol{\eta}$. Then, a phase-invariant action principle $\delta \int_{t_1}^{t_2} l(\boldsymbol{\mathcal{X}}, D, \psi, \partial_t \psi) \, \mathrm{d}t = 0$ is obtained upon replacing (9) by the Euler-Poincaré Lagrangian [3]

$$l = \int D \Big(\boldsymbol{\mathcal{X}} \cdot \big(\boldsymbol{\mathcal{A}} + \langle \psi, i\hbar \nabla \psi \rangle \big) + \langle \psi, i\hbar \partial_t \psi - \widehat{H} \psi + i\hbar (\mathbf{X}_{\widehat{H}} - \langle \mathbf{X}_{\widehat{H}} \rangle) \cdot \nabla \psi \rangle \Big) \, \mathrm{d}^2 z. \quad (10)$$

Here, the variations δD and $\delta \boldsymbol{\mathcal{X}}$ are found to be constrained so that

$$\delta D = -\mathrm{div}(D\boldsymbol{\mathcal{Y}}), \qquad \delta \boldsymbol{\mathcal{X}} = \partial_t \boldsymbol{\mathcal{Y}} + \boldsymbol{\mathcal{X}} \cdot \nabla \boldsymbol{\mathcal{Y}} - \boldsymbol{\mathcal{Y}} \cdot \nabla \boldsymbol{\mathcal{X}}, \qquad (11)$$

where $\boldsymbol{\mathcal{Y}} = \delta \boldsymbol{\eta} \circ \boldsymbol{\eta}^{-1}$ is arbitrary. Finally, one last convenient step consists in writing $\psi(t) = (U(t)\psi_0) \circ \boldsymbol{\eta}(t)^{-1}$, without loss of generality [3]. Here, $U(t) = U(\mathbf{z}, t)$ is a unitary operator on the quantum Hilbert space that is parameterized by phase-space coordinates. This step amounts to expressing the quantum unitary dynamics in the frame of Lagrangian classical paths. In this way, the Lagrangian (10) is entirely expressed in terms of $P = \psi\psi^\dagger$, i.e. it becomes *gauge-independent*.

3 Geometry of Quantum-Classical Dynamics

The quantum-classical model (3)–(5) follows from the variational principle associated to (10). We will now review the high points of its underlying geometry.

Euler-Poincaré Variational Principle. Expressing the quantum evolution in the classical frame, or, equivalently, setting $\psi(t) = (U(t)\psi_0) \circ \eta(t)^{-1}$ in (10), leads to the action principle $\delta \int_{t_1}^{t_2} \ell\, dt = 0$ for the following Lagrangian:

$$\ell(\boldsymbol{\mathcal{X}}, D, \xi, \mathcal{P}) = \int \left(D\boldsymbol{\mathcal{A}} \cdot \boldsymbol{\mathcal{X}} + \langle \mathcal{P}, i\hbar\xi - \widehat{H} - i\hbar D^{-1}\{\mathcal{P}, \widehat{H}\}\rangle \right) d^2 z. \qquad (12)$$

Here, $\langle\,,\rangle = \mathrm{Re}\langle\,|\,\rangle$, $\mathcal{P} = D\psi\psi^\dagger$, and $\xi = (\dot{U}U^\dagger)\circ\eta^{-1}$ is skew-Hermitian, so that

$$\delta\mathcal{P} = [\Sigma, \mathcal{P}] - \mathrm{div}(\mathcal{P}\boldsymbol{\mathcal{Y}}), \qquad \delta\xi = \partial_t\Sigma + [\Sigma, \xi] + \boldsymbol{\mathcal{X}} \cdot \nabla\Sigma - \boldsymbol{\mathcal{Y}} \cdot \nabla\xi, \quad (13)$$

where $\Sigma = (\delta U U^\dagger)\circ\eta^{-1}$ is skew-Hermitian and arbitrary. These variations arise by standard Euler-Poincaré reduction from Lagrangian to Eulerian variables. Indeed, Lagrangian trajectories play a crucial role in the variational problem associated to (12). In particular, if $\boldsymbol{\eta}(\mathbf{z}_0, t)$ is the diffeomorphic Lagrangian path on phase-space and $U(\mathbf{z}, t)$ is a unitary operator, we define the Eulerian quantities

$$D := \eta_* D_0, \qquad \mathcal{P} := \eta_*(U\mathcal{P}_0 U^\dagger), \qquad \boldsymbol{\mathcal{X}} := \dot{\boldsymbol{\eta}}\circ\eta^{-1}, \qquad \xi := \dot{U}U^\dagger\circ\eta^{-1}. \tag{14}$$

Then, taking the time derivative of the first two in (14) yields $\partial_t D + \mathrm{div}(D\boldsymbol{\mathcal{X}}) = 0$ and $\partial_t\mathcal{P} + \mathrm{div}(\boldsymbol{\mathcal{X}}\mathcal{P}) = [\xi, \mathcal{P}]$, respectively. Furthermore, upon taking variations of (12), the action principle $\delta\int_{t_1}^{t_2} \ell\, dt = 0$ yields

$$\boldsymbol{\mathcal{X}} = \mathbf{X}_{\frac{\delta h}{\delta D}} + \left\langle \mathbf{X}_{\frac{\delta h}{\delta \mathcal{P}}}\right\rangle, \quad \left[i\hbar\xi - \frac{\delta h}{\delta\mathcal{P}}, \mathcal{P}\right] = 0, \quad \text{where} \quad h = \int \langle D\widehat{H} + i\hbar\{\mathcal{P}, \widehat{H}\}\rangle d^2 z \tag{15}$$

and we have used $\langle\widehat{A}\rangle := \langle\psi, \widehat{A}\psi\rangle = D^{-1}\langle\mathcal{P}, \widehat{A}\rangle$. Then, after various manipulations we recover the system (3)–(5). The purely quantum and classical cases are recovered by restricting to the cases $\mathbf{X}_{\widehat{H}} = 0$ and $\widehat{H} = H\mathbf{1}$, respectively [3]. In addition, if one neglects the $\hbar$–terms in the Hamiltonian functional h, then the variational principle (12) recovers the Ehrenfest model.

Notice that the first two in (14) indicate that the evolution of D and $\mathcal{P}$ occurs on orbits of the semidirect-product group $\mathrm{Diff}(T^*Q)\circledS\mathcal{F}(T^*Q, \mathcal{U}(\mathscr{H}))$, where $\mathcal{F}(T^*Q, \mathcal{U}(\mathscr{H}))$ denotes the space of phase-space functions taking values in the group $\mathcal{U}(\mathscr{H})$ of unitary operators on the quantum Hilbert space $\mathscr{H}$. In particular, these orbits are determined by the group action given by the composition of the standard conjugation representation of $\mathcal{F}(T^*Q, \mathcal{U}(\mathscr{H}))$ and the pushforward action of $\mathrm{Diff}(T^*Q)$. The latter diffeomorphism group comprises Lagrangian paths on the classical phase-space.

Hamiltonian Structure. While the Hamiltonian structure of the model (3)–(5) is not necessary towards the development of the trajectory-based closure presented later, we quickly review it here as we are not aware of similar structures occurring elsewhere in continuum mechanics. Notice that the same Hamiltonian structure also applies to the Ehrenfest model (1), which indeed is recovered by neglecting the $\hbar$−terms in the Hamiltonian functional h in (15). Thus, all the considerations in this discussion apply equivalently to the Ehrenfest model (1).

First, we observe that the variable D may be written as $D = \text{Tr}\,\mathcal{P}$ so that the Euler-Poincaré Lagrangian (12) is expressed entirely in terms of the variables $(\mathcal{X}, \xi, P)$. Going through the same steps as above leads to rewriting (15) as $\mathcal{X} = \langle \mathbf{X}_{\delta h/\delta \mathcal{P}} \rangle$ and $[i\hbar\xi - \delta h/\delta\mathcal{P}, \mathcal{P}] = 0$, where $h = \int \langle \widehat{H}\,\text{Tr}\,\mathcal{P} + i\hbar\{\mathcal{P}, \widehat{H}\}\rangle\, \mathrm{d}^2 z$. Then, the Hamiltonian equation

$$i\hbar \frac{\partial \mathcal{P}}{\partial t} + i\hbar\,\text{div}\left(\mathcal{P}\left\langle \mathbf{X}_{\frac{\delta h}{\delta \mathcal{P}}}\right\rangle\right) = \left[\frac{\delta h}{\delta \mathcal{P}}, \mathcal{P}\right] \tag{16}$$

leads directly to the following bracket structure via the usual relation $\dot{f} = \{\!\{f, h\}\!\}$:

$$\{\!\{f, h\}\!\}(\mathcal{P}) = \int \frac{1}{\text{Tr}\,\mathcal{P}}\left(\mathcal{P}:\left\{\frac{\delta f}{\delta \mathcal{P}}, \frac{\delta h}{\delta \mathcal{P}}\right\}:\mathcal{P}\right)\mathrm{d}^2 z - \int \left\langle \mathcal{P}, \frac{i}{\hbar}\left[\frac{\delta f}{\delta \mathcal{P}}, \frac{\delta h}{\delta \mathcal{P}}\right]\right\rangle \mathrm{d}^2 z, \tag{17}$$

where we have introduced the convenient notation $A : B = \text{Tr}(AB)$. The proof that (17) is Poisson involves a combination of results in Lagrangian and Poisson reduction [4]. The first term in (17) is related to the Lagrangian classical paths.

Equation (16) easily leads to characterizing the Casimir invariant $C_1 = \text{Tr} \int D\Phi(\mathcal{P}/D)\,\mathrm{d}^2 z$ for any matrix analytic function Φ. Also, upon writing $\mathcal{P} = D\psi\psi^\dagger$, one finds the quantum-classical Poincaré integral invariant

$$\frac{\mathrm{d}}{\mathrm{d}t} \oint_{c(t)} \left(p\mathrm{d}q + \langle \psi, i\hbar\mathrm{d}\psi\rangle\right) = 0$$

for any loop $c(t) = \boldsymbol{\eta}(c_0, t)$ in phase-space. Here, we notice the important role of the Berry connection $\langle \psi, -i\hbar\mathrm{d}\psi\rangle$. By Stokes theorem, the above relation also allows to identify a Lie-transported quantum-classical two-form on T^*Q, that is

$$\Omega(t) = \eta_* \Omega(0), \qquad \text{with} \qquad \Omega(t) := \omega + \hbar\,\text{Im}\,\langle \mathrm{d}\psi(t)| \wedge \mathrm{d}\psi(t)\rangle,$$

so that $\Omega(t)$ remains symplectic in time if it is so initially. As a result, if $\dim Q = n$, one finds the additional class of Casimirs $C_2 = \int D\Lambda(D^{-1}\Omega^{\wedge n})\mathrm{d}^{2n} z$, where $\Omega^{\wedge n} = \Omega \wedge \cdots \wedge \Omega$ (n times) is a volume form and Λ is any scalar function of one variable. These Casimirs may be used to construct quantum-classical extensions of Gibbs/von Neumann entropies [3]. If $\dim Q = 1$, then $\Omega = (1 + \hbar\,\text{Im}\{\psi^\dagger, \psi\})\omega$.

Quantum-Classical Von Neumann Operator. We observe that the Hamiltonian energy functional h in (15) is not simply given by the usual average of the Hamiltonian operator $\widehat{H}$. Indeed, the $\hbar$−term seems to play a crucial role in taking the model (3)–(5) beyond simple Ehrenfest dynamics. As discussed in [3], this suggests that the quantum-classical correlations trigger extra energy terms that are not usually considered. Alternatively, one may insist that the total energy must be given by an average of $\widehat{H}$. Following this route leads to rewriting the last in (15) as $h = \text{Tr} \int \widehat{\mathcal{D}}\widehat{H}\,\mathrm{d}^2 z$, where

$$\widehat{\mathcal{D}} = DP + \frac{\hbar}{2}\,\text{div}(D\widehat{\boldsymbol{\Gamma}}) = DP + \frac{i\hbar}{2}\,\text{div}(D[P, \mathbf{X}_P])$$

is a measure-valued von Neumann operator and we recall $P = \psi\psi^\dagger$. Then, classical and quantum densities are given by taking the trace and integral of $\widehat{\mathcal{D}}$, respectively. Unlike the quantum density operator, the hybrid operator $\widehat{\mathcal{D}}$ is not sign-definite. Remarkably, however, $\widehat{\mathcal{D}}$ enjoys the equivariance properties

$$\widehat{\mathcal{D}}(\boldsymbol{\eta}_*D, \boldsymbol{\eta}_*P) = \boldsymbol{\eta}_*\widehat{\mathcal{D}}(D, P), \qquad \text{and} \qquad \widehat{\mathcal{D}}(D, \mathscr{U}P\mathscr{U}^\dagger) = \mathscr{U}\widehat{\mathcal{D}}(D, P)\mathscr{U}^\dagger,$$

where $\boldsymbol{\eta}$ is a symplectic diffeomorphism on T^*Q and $\mathscr{U} \in \mathcal{U}(\mathscr{H})$. These two properties ensure the following dynamics in the classical and quantum sector [4]:

$$\frac{\partial D}{\partial t} = \mathrm{Tr}\{\widehat{H}, \widehat{\mathcal{D}}\}, \qquad i\hbar\frac{\mathrm{d}\hat{\rho}}{\mathrm{d}t} = \int [\widehat{H}, \widehat{\mathcal{D}}]\,\mathrm{d}^2z\,.$$

For example, the first can be verified directly upon writing $\boldsymbol{\mathcal{X}} = D^{-1}\langle\widehat{\mathcal{D}}, \mathbf{X}_{\widehat{H}}\rangle + \hbar D^{-1}\mathrm{div}\big(D\,\mathrm{Tr}(\mathbf{X}_{\widehat{H}} \wedge \widehat{\boldsymbol{\Gamma}})\big)$, where $(\mathbf{X}_{\widehat{H}} \wedge \widehat{\boldsymbol{\Gamma}})^{jk} := \big(X_{\widehat{H}}^j\widehat{\Gamma}^k - \widehat{\Gamma}^jX_{\widehat{H}}^k\big)/2$ identifies a bivector.

4   Trajectory-Based Closure

As discussed in the Introduction, the quantum-classical model (3)–(5) does not immediately allow for the application of the trajectory-based closure typically adopted for the Ehrenfest equations. This is due to the appearance of several gradients in the expressions (4) and (5). A similar situation also occurs in quantum hydrodynamics, thereby preventing the existence of particle solutions in Bohmian mechanics [2]. In the latter case, a regularization technique was recently introduced to allow the standard application of trajectory-based closures. Unlike common regularizations, this particular one was introduced at the level of the variational principle and its successful implementation was presented in [5]. The resulting closure arises from a sampling process at the level of the classical Lagrangian paths. This section exploits this approach in such a way to formulate a trajectory-based closure of the quantum-classical model (3)–(5). In particular, we will devise a computational method that inherits basic conservation laws, such as energy and total probability, and retains decoherence effects beyond the standard Ehrenfest model.

Variational Regularization. The present method arises from the observation that the singular solution ansatz $\mathcal{P}(\mathbf{z}, t) = \sum_{a=1}^{N} w_a\rho_a(t)\delta(\mathbf{z} - \boldsymbol{\zeta}_a(t))$ is prevented by the last term in the Hamiltonian h in (15). Thus, if a regularization needs to be introduced at the variational level, it has to be introduced in that term. Here, we will replace the Lagrangian (12) by the *regularized Lagrangian*

$$\bar{\ell} = \int \Big(D\boldsymbol{\mathcal{A}} \cdot \boldsymbol{\mathcal{X}} + \langle\mathcal{P}, i\hbar\xi - \widehat{H}\rangle - \frac{1}{2}\langle\bar{\mathcal{P}}, i\hbar\bar{D}^{-1}[\nabla\bar{\mathcal{P}}, \mathbf{X}_{\widehat{H}}]\rangle\Big)\mathrm{d}^6z\,,$$

where the commutator arises from conveniently projecting $i\{\mathcal{P}, \widehat{H}\}$ on its Hermitian part, and we have introduced the regularized quantities

$$\bar{D} = \int K_\alpha(\mathbf{z} - \mathbf{z}')D(\mathbf{z}')\,\mathrm{d}^2z'\,, \qquad \bar{\mathcal{P}} = \int K_\alpha(\mathbf{z} - \mathbf{z}')\mathcal{P}(\mathbf{z}')\,\mathrm{d}^2z'\,.$$

The *mollifier* K_α is chosen as a smooth convolution kernel that is invariant under phase-space translations and tends to the delta function as $\alpha \to 0$, that is the limit in which one recovers the original model. For example, K_α may be a Gaussian kernel with variance α, although here we will keep it general. With this regularization, one allows to consider the singular solution ansatz

$$D = \sum_{a=1}^{N} w_a \delta(\mathbf{z} - \boldsymbol{\zeta}_a), \qquad \mathcal{P} = \sum_{a=1}^{N} w_a \rho_a \delta(\mathbf{z} - \boldsymbol{\zeta}_a). \tag{18}$$

For example, the first in (3) now leads to $\dot{\boldsymbol{\zeta}}_a = \boldsymbol{\mathcal{X}}_a$ with $\boldsymbol{\mathcal{X}}_a := \boldsymbol{\mathcal{X}}(\boldsymbol{\zeta}_a)$. Similarly, the equation $i\hbar \partial_t \mathcal{P} + i\hbar \operatorname{div}(\mathcal{P}\boldsymbol{\mathcal{X}}) = [\xi, \mathcal{P}]$ (see equation (16) above) leads to $\dot{\rho}_a = [\xi_a, \rho_a]$ with $\xi_a := \xi(\boldsymbol{\zeta}_a)$. Here, we will set $\rho_a = \psi_a \psi_a^\dagger$ so that $\dot{\psi}_a = \xi_a \psi_a$. We remark that, as in the case of the Ehrenfest model, the trajectories $\boldsymbol{\zeta}_a(t)$ in (18) are *not* physical particles, but rather arise from a sampling process of the Lagrangian classical paths underlying the Eulerian action principle associated to (12).

Trajectory Equations. At this stage, we are ready to replace the ansatz (18) in the regularized Lagrangian, thereby obtaining the finite-dimensional Lagrangian

$$L(\{\boldsymbol{\zeta}_a\}, \{\xi_a\}, \{\rho_a\}) = \sum_a w_a \left(p_a \dot{q}_a + \left\langle \rho_a, i\hbar \xi_a - \widehat{H}_a - i\hbar \sum_b w_b [\rho_b, \mathcal{I}_{ab}] \right\rangle \right). \tag{19}$$

Here, $\delta \xi_a = \dot{\Sigma}_a + [\Sigma_a, \xi_a]$, with Σ_a arbitrary, and we have denoted

$$\mathcal{I}_{ab} := \frac{1}{2} \int \frac{K_a \{K_b, \widehat{H}\}}{\sum_c w_c K_c} \, \mathrm{d}^2 z, \qquad \text{and} \qquad K_s(\mathbf{z}, t) := K(\mathbf{z} - \boldsymbol{\zeta}_s(t)).$$

Once again, we observe that if the $\hbar$–terms are neglected in (19), then the associated variational principle recovers the closure Eq. (6) associated to the Ehrenfest model. Instead, in the general case each trajectory is directly coupled to all the others via the $\hbar$–term. The equations of motion read

$$\dot{q}_a = w_a^{-1} \partial_{p_a} h, \qquad \dot{p}_a = -w_a^{-1} \partial_{q_a} h, \qquad i\hbar \dot{\rho}_a = w_a^{-1} [\partial_{\rho_a} h, \rho_a], \tag{20}$$

where

$$h = \sum_a w_a \left\langle \rho_a, \widehat{H}_a + i\hbar \sum_b w_b [\rho_b, \mathcal{I}_{ab}] \right\rangle, \qquad \partial_{\rho_a} h = \widehat{H}_a + i\hbar \sum_b w_b [\rho_b, \mathcal{I}_{ab} - \mathcal{I}_{ba}].$$

In analogy to the discussion in the previous section, we can rearrange the Hamiltonian h above as $h = \operatorname{Tr} \int \widehat{\mathcal{D}} \widehat{H} \, \mathrm{d}^2 z$ with the hybrid von Neumann operator

$$\widehat{\mathcal{D}}(\mathbf{z}, t) = \sum_a w_a \hat{\rho}_a(t) \delta(\mathbf{z} - \boldsymbol{\zeta}_a(t)) + i\hbar \sum_{a,b} w_a w_b \mathcal{J}_{ab}(\mathbf{z}, t) [\hat{\rho}_a(t), \hat{\rho}_b(t)],$$

where

$$\mathcal{J}_{ab} := \frac{1}{4}\left(\left\{K_a, \frac{K_b}{\sum_c w_c K_c}\right\} - \left\{K_b, \frac{K_a}{\sum_c w_c K_c}\right\}\right).$$

The implementation of this closure scheme is currently underway. We observe that the canonical Hamiltonian structure underlying this scheme may pave the way to the application of symplectic integration techniques for the long-time simulation of fully nonlinear processes.

References

1. Bondar, D.I., Gay-Balmaz, F., Tronci, C.: Koopman wavefunctions and classical-quantum correlation dynamics. Proc. R. Soc. A **475**, 20180879 (2019)
2. Foskett, M.S., Holm, D.D., Tronci, C.: Geometry of nonadiabatic quantum hydro-dynamics. Acta Appl. Math. **162**, 1–41 (2019)
3. Gay-Balmaz, F., Tronci, C.: Evolution of hybrid quantum-classical wavefunctions. Phys. D **440**, 133450 (2022)
4. Gay-Balmaz, F., Tronci, C.: Koopman wavefunctions and classical states in hybrid quantum-classical dynamics. J. Geom. Mech. **14**(4), 559–596 (2022)
5. Holm, D.D., Rawlinson, J.I., Tronci, C.: The Bohmion method in nonadiabatic quantum hydrodynamics. J. Phys. A: Math. Theor. **54**, 495201 (2021)
6. Sudarshan, E.C.G.: Interaction between classical and quantum systems and the measurement of quantum observables. Pramana Pramana **6**(3), 117–126 (1976)

A Discrete Version for Vortex Loops in 2D Fluids

Cornelia Vizman[(✉)]

West University of Timişoara, bld. Vasile Pârvan no. 4,
300223 Timişoara, Timiş County, Romania
cornelia.vizman@e-uvt.ro

Abstract. The manifold of weighted vortex loops in the plane that enclose a fixed area and have a constant total vorticity is known to be a coadjoint orbit of the area preserving diffeomorphism group, obtained by symplectic reduction of the space of parametrized loops. Moreover, it admits a polarization that allows the decomposition into "coordinates" - the unparametrized loop - and "momenta" - the vorticity density [5]. We give discrete versions of all these results, starting from the configuration space of k-tuples of points in the plane, endowed with an infinitesimal Hamiltonian $\mathbb{R}$-action having the area function as momentum map.

Keywords: Singular vorticity · Discretization · Symplectic reduction · Polarization

1 Introduction

The symmetry group for ideal fluids in 2D is the group of area preserving diffeomorphisms of $\mathbb{R}^2$, which is the same as the Hamiltonian group. The fluid vorticity lives in the dual of its Lie algebra, the Lie algebra of compactly supported smooth functions $C_c^\infty(\mathbb{R}^2)$ endowed with Poisson bracket, and it is confined to a coadjoint orbit. Several types of singular vorticity in 2D have been studied from the point of view of their coadjoint orbits: point vortices $h \mapsto \sum_{i=1}^k \Gamma_i h(\mathsf{x}_i)$, for k-tuples of distinct points (x_i) endowed with circulations (Γ_i) [8], vortex loops $h \mapsto \int_C h\beta$, for non-degenerate vorticity 1-forms β on closed curves C [2,5], or pointed vortex loops which combines both of them [1].

In this paper we aim at finding a discrete version of coadjoint orbits for vortex loops. Compared to the coadjoint orbits for point vortices, there are similitudes: the orbit symplectic form for point vortices pops up, and there are differences: a symplectic reduction on the configuration space Conf_k of k-tuples of ordered points (viewed as the space of not necessarily simple k-polygons) is needed.

In contrast to the vortex loop case, where the coadjoint orbit is the reduced symplectic manifold for the circle action by reparametrizations on parametrized

Supported by a grant of the Romanian Ministry of Education and Research, CNCS-UEFISCDI, project number PN-III-P4-ID-PCE-2020-2888, within PNCDI III.

loops, in its discrete version only the infinitesimal side survives. The main object here is a vector field ζ on the configuration space Conf_k, pictured in Fig. 1, which plays the role of infinitesimal rotation of polygons. Its components are

$$\zeta_i((\mathsf{x}_i)) = \frac{1}{2\mathsf{w}_i}(\mathsf{x}_{i+1} - \mathsf{x}_{i-1}), \quad i = 1, \ldots, k,$$

where each weight w_i is the vorticity concentrated at the point $\mathsf{x}_i \in C$. In particular, $\sum_{i=1}^k \mathsf{w}_i = \int_C \beta$ is the total vorticity. The infinitesimal rotation ζ turns out to be a Hamiltonian vector field for the polygon area function.

In [5] it is shown that, within the framework of the 2D Euler equations, point vortices cannot be consistently quantized, while vortex loops admit natural polarizations. In the same article the authors attach vortex dipoles to the point vortices, so that the additional degrees of freedom allow for a polarization, hence a decomposition of the phase space into "coordinates" and "momenta". For vortex loops the decomposition is into unparametrized loops and vorticity densities, while for vortex dipoles one gets the decomposition into point vortices and dipole vectors. That a natural polarization for the coadjoint orbit of pointed vortex loops has the same configuration space (namely the space of unparametrized loops) as that for the coadjoint orbit of vortex loops, independently of the number of attached points to the loop, is shown in [1].

The discrete counterpart to vortex loops discussed here also admits a polarization: the "coordinates" correspond to variations tangent to the polygon, which means in the ζ-direction, while the "momenta" correspond to variations normal to the vector field ζ (with respect to the Euclidean metric).

2   Vortex Loops

We consider $\mathbb{R}^2$ endowed with $\omega = dx_1 \wedge dx_2$, a volume form and a symplectic form at the same time, and the 1-form

$$\nu = \tfrac{1}{2}(x_1 dx_2 - x_2 dx_1) \tag{1}$$

satisfying $d\nu = \omega$. We consider also the circle $S^1 = \mathbb{R}/\mathbb{Z}$ endowed with the volume form $\mu = \mathsf{w}dt$ of total volume w.

The space of embeddings $\mathrm{Emb}(S^1, \mathbb{R}^2)$ is a Fréchet manifold whose tangent space at f consists of vector fields $v_f : S^1 \to \mathbb{R}^2$ along f. It carries a symplectic form naturally defined with the symplectic form ω on $\mathbb{R}^2$ and with the volume form μ on S^1:

$$\Omega_f(u_f, v_f) = \mathsf{w}\int_{S^1} \omega(u_f, v_f)(t)dt, \quad u_f, v_f : S^1 \to \mathbb{R}^2. \tag{2}$$

The group $\mathrm{Diff}(S^1, \mu)$ of volume preserving diffeomorphisms of the circle coincides with the group $\mathrm{Rot}(S^1)$ of rigid rotations of the circle. For its 1-dimensional Lie algebra $\mathfrak{X}(S^1, \mu)$ we fix as generator the vector field $Z = \frac{1}{\mathsf{w}}\frac{d}{dt}$, thus identifying it with the Lie algebra $\mathbb{R}$. In particular Z is dual to the 1-form μ, i.e. $i_Z\mu = 1$.

Proposition 1. *The rotation group* $\mathrm{Rot}(S^1)$ *acts on* $\mathrm{Emb}(S^1, \mathbb{R}^2)$ *in a Hamiltonian way, with equivariant momentum map given by the enclosed area.*

The proof is easy, especially when using the hat calculus in [10]. The symplectic form is written as $\Omega = \widehat{\omega \cdot \mu} = d(\widehat{\nu \cdot \mu})$. We get that the infinitesimal generator of Z, which is $\zeta_Z(f) = Tf \circ Z$ for all embeddings f, satisfies

$$i_{\zeta_Z}\Omega = \widehat{\omega \cdot i_Z\mu} = \widehat{\omega} = \widehat{d\nu} = d(\widehat{\nu}).$$

Thus ζ_Z is a Hamiltonian vector field with Hamiltonian function the area enclosed by the embedded curve $f(S^1)$:

$$\widehat{\nu}(f) = \int_{S^1} f^*\nu. \tag{3}$$

For the Hamiltonian action in Proposition 1, the symplectically reduced manifold at non-zero $a \in \mathbb{R}$ is

$$(\mathrm{Emb}_a(S^1, \mathbb{R}^2)/\mathrm{Rot}(S^1), \Omega_a^w), \tag{4}$$

with $\mathrm{Emb}_a(S^1, \mathbb{R}^2)$ denoting the submanifold of embeddings that enclosed the fixed (signed) area a, and Ω_a^w denoting the reduced symplectic form. There is a one-to-one correspondence

$$([f]) \leftrightarrow (f(S^1), f_*\mu) = (C, \beta)$$

between the reduced space and the space $\mathcal{O}_a^w$ of all weighted closed curves (vortex loops) (C, β) with total weight (total vorticity) $\int_C \beta = w$ and enclosed area $\int_C \nu = a$.

The symmetry group for ideal fluids in 2D is the group of area preserving diffeomorphisms of $\mathbb{R}^2$, which is the same as the group $\mathrm{Ham}_c(\mathbb{R}^2)$ of Hamiltonian diffeomorphisms (the compact support ensures a reasonable Lie group structure). Its Lie algebra of compactly supported Hamiltonian vector fields coincides with the Lie algebra of compactly supported smooth functions endowed with Poisson bracket [9]. The pair (C, β) corresponds to an element of the dual Lie algebra

$$\langle (C, \beta), h \rangle = \int_C h\beta, \quad h \in C_c^\infty(\mathbb{R}^2). \tag{5}$$

Proposition 2. [3–5,7,11] *The symplectic manifold* $(\mathcal{O}_a^w, \Omega_a^w)$ *of weighted vortex loops in the plane that enclose a fixed area* a *and have a constant total vorticity* w, *identified with the symplectically reduced manifold* (4), *is a coadjoint orbit of* $\mathrm{Ham}_c(\mathbb{R}^2)$ *via* (5).

3   Discrete Version of Vortex Loops

3.1   The Discrete Objects

By considering consecutive points $t_1, \ldots, t_k \in S^1$, we get a discrete version of the parametrized curve f, namely the simple polygon with k vertices $x_i = f(t_i)$. Here (x_i) lie in the configuration space of k ordered points in the plane

$$\text{Conf}_k = (\mathbb{R}^2)^k \setminus \Delta_k, \tag{6}$$

where $\Delta_k = \{(x_1, ..., x_k) \in (\mathbb{R}^2)^k : x_i = x_j \text{ for some } i \neq j\}$ is the fat diagonal, By allowing also polygons with self-intersections (non-simple polygons), we can work with the entire configuration space Conf_k.

The discrete version of the vorticity density β is the ordered set of partial vorticities (w_i) of the curve segment between the halfway points $f(\frac{t_{i-1}+t_i}{2})$ and $f(\frac{t_i+t_{i+1}}{2})$ on the curve $C = f(S^1)$:

$$w_i := \int_{f(\frac{t_{i-1}+t_i}{2})}^{f(\frac{t_i+t_{i+1}}{2})} \beta = \int_{\frac{t_{i-1}+t_i}{2}}^{\frac{t_i+t_{i+1}}{2}} w\, dt = w\frac{t_{i+1} - t_{i-1}}{2}. \tag{7}$$

They satisfy $w_1 + \cdots + w_k = w\sum_{i=1}^k (t_{i+1} - t_i) = w$.

The discrete version of the symplectic form Ω on the space of embeddings involves these partial vorticities. It is the symplectic form σ on Conf_k given by

$$\sigma = \sum_{i=1}^k w_i \operatorname{pr}_i^* \omega, \quad \operatorname{pr}_i : \text{Conf}_k \to \mathbb{R}^2. \tag{8}$$

Indeed, for the discretized vector fields $(\xi_i) = (u_f(t_i))$ and $(\eta_i) = (v_f(t_i))$, we get that

$$\Omega_f(u_f, v_f) = w \int_{S^1} \omega(u_f, v_f)(t)dt \sim w \sum_{i=1}^k \omega(u_f(t_i), v_f(t_i))\frac{t_{i+1} - t_{i-1}}{2}$$

$$= \sum_{i=1}^k w_i \omega(\xi_i, \eta_i) = \sigma_{(x_i)}((\xi_i), (\eta_i)),$$

using the expression (7) for w_i.

Remark 1. We notice the resemblance of the symplectic form σ in (8) with the orbit symplectic form for point vortices

$$\omega^\Gamma = \sum_{i=1}^k \Gamma_i \operatorname{pr}_i^* \omega, \tag{9}$$

with $\Gamma_i \in \mathbb{R}$ denoting the circulations. We recall that, in the generic case when all Γ_i are distinct, the momentum map for the $\text{Ham}_c(\mathbb{R}^2)$ action

$$J_\Gamma : \text{Conf}_k \to C_c^\infty(\mathbb{R}^2)^*, \quad J_\Gamma(x_1, \ldots, x_k) = \sum_{i=1}^k \Gamma_i \delta_{x_i} \tag{10}$$

is one-to-one onto a coadjoint orbit of $\text{Ham}_c(\mathbb{R}^2)$.

The next object to be discretized is the infinitesimal generator of the action by rigid rotations. Only the infinitesimal action survives in this context: there is no circle action on Conf_k that can copy the $\mathrm{Rot}(S^1)$ action on $\mathrm{Emb}(S^1, \mathbb{R}^2)$. We get the vector field $\zeta = (\zeta_1, \ldots, \zeta_k)$ on Conf_k drawn in Fig. 1 and given by

$$\zeta_i((\mathsf{x}_i)) = \frac{1}{2\mathsf{w}_i}(\mathsf{x}_{i+1} - \mathsf{x}_{i-1}). \qquad (11)$$

Indeed, denoting by ζ_Z the infinitesimal generator of $Z = \frac{1}{\mathsf{w}}\frac{d}{dt}$, we get

$$\zeta_Z(f)(t_i) = (Tf \circ Z)(t_i) = \frac{1}{\mathsf{w}}f'(t_i) \sim \frac{1}{\mathsf{w}}\frac{f(t_{i+1}) - f(t_{i-1})}{t_{i+1} - t_{i-1}} \overset{(7)}{=} \frac{1}{2\mathsf{w}_i}(\mathsf{x}_{i+1} - \mathsf{x}_{i-1}).$$

The components ζ_i of the vector field ζ will play the role of the tangent directions in the generalized polygon.

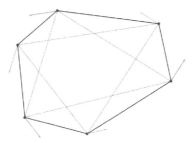

Fig. 1. Infinitesimal generator $\zeta = (\zeta_i)$ on Conf_k.

Remark 2. It would be interesting to further study the vector field ζ, per se. For instance, to better understand its integral curves, e. g. in the special case of triangles, when the vector field ζ at each vertex is parallel to the opposite edge.

The integral curves consist of polygons of constant area. The integral curve that starts at a regular polygon is obtained by rotating the polygon around its center.

It is clear that the enclosed area of the curve has as discrete counterpart the (signed) area of the polygon:

$$A((\mathsf{x}_i)) := \frac{1}{2}\sum_{i=1}^{k} \omega(\mathsf{x}_i, \mathsf{x}_{i+1}).$$

By this we mean the following approximation of the area enclosed by the curve $f(S^1)$:

$$\int_{S^1} f^* \nu \overset{(1)}{=} \frac{1}{2} \int_{S^1} \omega(f(t), f'(t)) dt \sim \frac{1}{2} \sum_{i=1}^{k} \omega(f(t_i), f'(t_i)) \frac{t_{i+1} - t_{i-1}}{2}$$

$$\sim \frac{1}{2} \sum_{i=1}^{k} \omega(x_i, \frac{x_{i+1} - x_{i-1}}{2}) = \frac{1}{4} \sum_{i=1}^{k} (\omega(x_{i-1}, x_i) + \omega(x_i, x_{i+1}))$$

$$= \frac{1}{2} \sum_{i=1}^{k} \omega(x_i, x_{i+1}) = A((x_i)),$$

where $f'(t_i)$ has been approximated with $\frac{f(t_{i+1}) - f(t_{i-1})}{t_{i+1} - t_{i-1}}$.

Now the expected result, similar to Proposition 1, holds:

Proposition 3. *The vector field ζ on the symplectic manifold $(\mathrm{Conf}_k, \sigma)$ is the Hamiltonian vector field with Hamiltonian function A, the area of generalized polygons. In particular, ζ is tangent to the level sets $A^{-1}(a)$ of polygons with constant area a.*

Proof. For (ξ_i) a tangent vector at $(x_i) \in \mathrm{Conf}_k$,

$$d_{(x_i)} A((\xi_i)) = \frac{1}{2} \sum_{i=1}^{k} (\omega(x_i, \xi_{i+1}) + \omega(\xi_i, x_{i+1})) = -\frac{1}{2} \sum_{i=1}^{k} \omega(x_{i+1} - x_{i-1}, \xi_i)$$

$$= -\frac{1}{2} \sum_{i=1}^{k} i_{x_{i+1} - x_{i-1}} \omega(\xi_i) = -\sum_{i=1}^{k} w_i \, \omega(\zeta_i((x_i)), \xi_i)$$

$$= -(i_\zeta \sigma)_{(x_i)}((\xi_i)),$$

hence the conclusion $i_\zeta \sigma = dA$.

3.2   Symplectically Reduced Space

As pointed before, unlike the space of embeddings $\mathrm{Emb}(S^1, \mathbb{R}^2)$, on the symplectic manifold $(\mathrm{Conf}_k, \sigma)$ there is no natural circle action. The vector field ζ in (11) defines an infinitesimal action of the Lie algebra $\mathbb{R}$ on Conf_k by

$$t \in \mathbb{R} \mapsto t\zeta \in \mathfrak{X}(\mathrm{Conf}_k),$$

but also an $\mathbb{R}$-action on every level set of the area function, by Proposition 3. Consequently, the infinitesimal side of the symplectically reduced manifold $\mathrm{Emb}_a(S^1, \mathbb{R}^2)/\mathrm{Rot}(S^1)$ in (4) is the symplectic vector bundle

$$E_a := \ker dA / \mathbb{R}\zeta$$

of rank $2k - 2$ over the submanifold $A^{-1}(a) \subset \mathrm{Conf}_k$ of generalized polygons of fixed area a, where the symplectic form descends from the symplectic form σ on Conf_k, written in (8).

Remark 3. If the flow of ζ on $A^{-1}(a)$ would admit a quotient manifold, i.e. if the topological space whose points are the trajectories of the flow is nice [6], then this quotient manifold would be a discrete counterpart of the reduced manifold $\text{Emb}_a(S^1, \mathbb{R}^2)/\text{Rot}(S^1)$ in (4).

Remark 4. Let $h \in C_c^\infty(\mathbb{R}^2)$. A discrete version of the identity $\int_{S^1} d(h \circ f) = 0$ for $f \in \text{Emb}(S^1, \mathbb{R}^2)$ is

$$\sum_{i=1}^k w_i d_{x_i} h(\zeta_i) \sim 0,$$

since

$$0 = \int_{S^1} d_{f(t)} h(f'(t)) dt \sim \sum_{i=1}^k d_{f(t_i)} h(f'(t_i)) \frac{t_{i+1} - t_{i-1}}{2} \sim \sum_{i=1}^k d_{x_i} h(\frac{x_{i+1} - x_{i-1}}{2})$$

$$= \sum_{i=1}^k w_i d_{x_i} h(\zeta_i).$$

This reflects the fact that ζ almost lies in the kernel of the function $(x_i) \mapsto \sum_{i=1}^k w_i h(x_i)$, similarly to $\langle J_\Gamma, h \rangle$ for the momentum map J_Γ in (10), thus E_a almost plays the role of a coadjoint orbit of the Hamiltonian group.

Similarly to the orthonormal frame $\{T, N\}$ along a closed curve in the plane, we define an orthonormal frame $\{(T_i), (N_i)\}$ along a generalized polygon (x_i) by

$$T_i((x_i)) = \frac{1}{\|x_{i-1} - x_{i-1}\|}(x_{i+1} - x_{i-1}), \quad N_i((x_i)) = \rho_{\pi/2}(T_i((x_i))), \qquad (12)$$

with $\rho_{\pi/2}$ denoting the rotation by $90°$ in the plane. In this frame, each tangent vector $(\xi_i) \in T_{(x_i)} \text{Conf}_k$ has $2k$ coordinates $(a_i), (b_i)$, where

$$\xi_i = a_i T_i((x_i)) + b_i N_i((x_i)).$$

In particular, the coordinates of $\zeta((x_i))$ are $(\frac{\|x_{i+1} - x_{i-1}\|}{2w_i}), (0)$. Moreover,

$$d_{(x_i)} A((\xi_i)) = \frac{1}{2} \sum_{i=1}^k b_i \|x_{i+1} - x_{i-1}\|.$$

Indeed,

$$d_{(x_i)} A((a_i T_i)) = \frac{1}{2} \sum_{i=1}^k a_i \omega(x_{i+1} - x_{i-1}, T_i) = 0$$

$$d_{(x_i)} A((b_i N_i)) = \frac{1}{2} \sum_{i=1}^k b_i \omega(x_{i+1} - x_{i-1}, N_i) = \frac{1}{2} \sum_{i=1}^k b_i \|x_{i+1} - x_{i-1}\|.$$

In particular all the vectors that are "tangent" to the polygon, i.e. with coordinates $((a_i), (0))$, belong to the kernel of dA.

Proposition 4. *The fiber at $(x_i) \in A^{-1}(a)$ of the symplectic vector bundle $E_a =$ $\ker dA/\mathbb{R}\zeta$, over the manifold of generalized polygons of fixed area a, can be identified with the $2k - 2$-dimensional symplectic space*

$$\{([a_i],(b_i)) \in \mathbb{R}^k/\mathbb{R}(\tfrac{\|x_{i+1}-x_{i-1}\|}{2w_i}) \times \mathbb{R}^k : \sum_{i=1}^{k} b_i \|x_{i+1}-x_{i-1}\| = 0\}. \tag{13}$$

With the non-degenerate pairing between the linear spaces $\mathbb{R}^k/\mathbb{R}(\tfrac{\|x_{i+1}-x_{i-1}\|}{2w_i})$ and $\{(b_i) \in \mathbb{R}^k : \sum_{i=1}^{k} b_i\|x_{i+1}-x_{i-1}\| = 0\}$ given by

$$\langle [a_i],(b_i)\rangle = \sum_{i=1}^{k} w_i a_i b_i,$$

the symplectic form that descends from the symplectic form σ in (8) takes Darboux coordinates.

This is the discrete counterpart to the Darboux coordinates on the coadjoint orbit $\mathcal{O}_a^w$ in Proposition 2. It provides a splitting of the tangent space

$$T_{(C,\beta)}\mathcal{O}_a^w = C^\infty(C)/\mathbb{R}\|\beta\|^{-1} \times C_0^\infty(C)$$

where $\|\beta\| \in C^\infty(C)$ denotes the Euclidean norm of β and $C_0^\infty(C)$ denotes the space of zero integral functions with respect to the volume form μ_C on C. The non-degenerate pairing

$$\langle [\lambda],\rho\rangle = \int_C (\lambda\rho)\beta$$

provides Darboux coordinates for the reduced (orbit) symplectic form Ω_a^w. Because $\beta(f') = w$, one gets the approximation $\|\beta\|^{-1} = \frac{1}{w}\|f'\| \sim \frac{1}{2w_i}\|x_{i+1} - x_{i-1}\|$ compatible with (13).

3.3   Polarization

Within the framework of the 2D Euler equations, point vortices cannot be consistently quantized, while vortex loops admit natural polarizations, hence a decomposition of the phase space into "coordinates" and "momenta", as shown in [5]. A natural polarization for the coadjoint orbit of pointed vortex loops, identified with the space of parameterized loops $\mathrm{Emb}(S^1,\mathbb{R}^2)$, has the same configuration space as that for the coadjoint orbit of vortex loops, namely the space of unparametrized loops, independently of the number of attached points to the loop [1]. The discrete counterpart to vortex loops discussed here also admits a polarization: the "coordinates" correspond to variations tangent to the polygon, which means in the ζ-direction, while the "momenta" correspond to normal variations with respect to the Euclidean metric.

For a vortex loop (C,β) in $\mathcal{O}_a^w$, the "coordinate" is the loop C and the "momentum" is the vorticity density β. At the infinitesimal level the "coordinates" are $\rho \in C_0^\infty(C)$ and the "momenta" are $[\lambda] \in C^\infty(C)/\mathbb{R}\|\beta\|^{-1}$. Their discrete counterparts are written in (13): the "coordinates" $(b_i) \in \mathbb{R}^k$ satisfying the linear condition $\sum_{i=1}^{k} b_i \|\mathsf{x}_{i+1} - \mathsf{x}_{i-1}\| = 0$, while the "momenta" are $[a_i] \in \mathbb{R}^k / \mathbb{R}(\frac{\|\mathsf{x}_{i+1} - \mathsf{x}_{i-1}\|}{2\mathsf{w}_i})$.

Remark 5. One could use the same idea as for the discrete counterpart to vortex loops discussed here, to compensate the missing polarization for the point vortices in Remark 1. More precisely, since each $(\mathsf{x}_i) \in \mathrm{Conf}_k$ comes with an ordering, the "coordinates" would correspond to variations in the ζ-direction (tangent) and the "momenta" to variations normal to ζ with respect to the Euclidean metric.

Acknowledgements. I would like to thank Gerald Goldin and Francois Gay-Balmaz for thought-provoking discussions. The author was supported by a grant of the Romanian Ministry of Education and Research, CNCS-UEFISCDI, project number PN-III-P4-ID-PCE-2020-2888, within PNCDI III.

References

1. Ciuclea, I., Vizman, C.: Pointed vortex loops in ideal 2D fluids, to appear in J. Phys. A. arXiv:2212.02612 [math.SG]
2. Gay-Balmaz, F., Vizman, C.: Coadjoint orbits of vortex sheets in ideal fluids, Preprint
3. Gay-Balmaz, F., Vizman, C.: Isotropic submanifolds and coadjoint orbits of the Hamiltonian group. J. Symp. Geom. **17**(3), 663–702 (2019)
4. Gay-Balmaz, F., Vizman, C.: Vortex sheets in ideal 3D fluids, coadjoint orbits, and characters (2020). arXiv:1909.12485
5. Goldin, G.A., Menikoff, R., Sharp, D.H.: Diffeomorphism groups and quantized vortex filaments. Phys. Rev. Lett. **58**, 2162–2164 (1987)
6. Gompf, R.E.: Quotient manifolds of flows. J. Knot Theory Ramifications **26**, 1740005 (2017)
7. Lee, B.: Geometric structures on spaces of weighted submanifolds, SIGMA **5**, 099, 46 (2009)
8. Marsden, J.E., Weinstein, A.: Coadjoint orbits, vortices, and Clebsch variables for incompressible fluids. Phys. D **7**, 305–323 (1983)
9. McDuff, D., Salamon, D.: Introduction to Symplectic Topology, Second Edition, Oxford Graduate Texts in Math, Oxford University Press, Oxford, p. 27 (2005)
10. Vizman, C.: Induced differential forms on manifolds of functions. Archivum Mathematicum **47**, 201–215 (2011)
11. Weinstein, A.: Connections of Berry and Hannay type for moving Lagrangian submanifolds. Adv. Math. **82**, 133–159 (1990)

Learning of Dynamic Processes

Expressiveness and Structure Preservation in Learning Port-Hamiltonian Systems

Juan-Pablo Ortega and Daiying Yin[✉]

Division of Mathematical Sciences, School of Physical and Mathematical Sciences,
Nanyang Technological University, Singapore, Singapore
Juan-Pablo.Ortega@ntu.edu.sg, yind0004@e.ntu.edu.sg

Abstract. A well-specified parametrization for single-input/single-output (SISO) linear port-Hamiltonian systems amenable to structure-preserving supervised learning is provided. The construction is based on controllable and observable normal form Hamiltonian representations for those systems, which reveal fundamental relationships between classical notions in control theory and crucial properties in the machine learning context, like structure-preservation and expressive power. The results in the paper suggest parametrizations of the estimation problem associated with these systems that amount, at least in the canonical case, to unique identification and prove that the parameter complexity necessary for the replication of the dynamics is only $\mathcal{O}(n)$ and not $\mathcal{O}(n^2)$, as suggested by the standard parametrization of these systems.

Keywords: Linear port-Hamiltonian system · machine learning · structure-preserving algorithm · systems theory · physics-informed machine learning

1 Introduction

Machine learning has experienced substantial development in recent years due to significant advances in algorithmics and a fast growth in computational power. In physics and engineering, machine learning is called to play an essential role in predicting and integrating the equations associated with physical dynamical systems. Nevertheless, for physics-related problems, like in mechanics or optics, it is natural to build into the learning algorithm any prior knowledge that we may have about the system based on physics' first principles. This may include specific forms of the laws of motion, conservation laws, symmetry invariance, as well as other underlying geometric and variational structures. This observation regarding the construction of structure-preserving schemes has been profusely exploited with much success before the emergence of machine learning in the field of numerical integration [1,5–7]. Many examples in that context show how the failure to maintain specific conservation laws can lead to physically inconsistent solutions. The translation of this idea to the context of machine learning has led to the emergence of a new domain collectively known as *physics-informed machine learning* (see [4,10,14] and references therein).

© The Author(s), under exclusive license to Springer Nature Switzerland AG 2023
F. Nielsen and F. Barbaresco (Eds.): GSI 2023, LNCS 14072, pp. 313–322, 2023.
https://doi.org/10.1007/978-3-031-38299-4_33

The approach that we propose in this contribution differs from the references mentioned above in two ways. First, these methods are designed to learn the state evolution of Hamiltonian systems, whereas our approach focuses on *learning the input-output dynamics of port-Hamiltonian systems while forgetting about the physical state space*. As will be introduced later, these systems have an underlying Dirac structure that describes the geometry of numerous physical systems with external inputs [11] and includes the dynamics of the observations of Hamiltonian systems as a particular case. This is a significant difference with respect to the results available in the literature, which mostly deal with autonomous Hamiltonian systems on which one assumes access to the entire phase space and not only to its observations. Second, instead of a general nonlinear system for which only approximation error can be possibly estimated, we consider, as a first approach *exclusively linear systems*, for which we can obtain explicit representations in normal form, which allows us to propose a structure-preserving learning paradigm with a provable minimal parameter space (in the canonical case).

The contributions in this paper are mostly based on classical techniques in control theory, the Cayley-Hamilton theorem, and symplectic linear algebra that allow us to define the notion of *normal-form controllable and observable Hamiltonian representations*. Our main result (Theorem 1) shows the existence of system morphisms that allow us to represent any linear port-Hamiltonian system in normal form as the image of a normal-form controllable Hamiltonian representation of the same dimension. Note that since the original port-Hamiltonian system and the new linear system are linked by a system morphism, the image of the input/output relations of the latter is the input/output relations of the former. In particular, the new system can be used to learn to reproduce the input/output dynamics of the original port-Hamiltonian system (for a subspace of initial conditions) and *this learning paradigm is structure-preserving by construction*. Analogously, we introduce another type of system morphisms that link any linear port-Hamiltonian system to some normal-form observable Hamiltonian representation of the same dimension. Consequently, the input-output relations of the original port-Hamiltonian system with respect to any initial condition can be reproduced using the observable Hamiltonian representation.

The controllable and observable representations are closely related to each other, and both system morphisms become isomorphisms for canonical port-Hamiltonian systems. For the purpose of learning a general port-Hamiltonian system that may not be canonical, we reveal that there is a trade-off between the *structure-preserving property* and the *expressive power*. These results establish a strong link between classical notions in control theory, e.g., controllability and observability, and those in machine learning, e.g., structure-preservation and expressive power.

The results in the paper point at parametrizations of the estimation problem associated with these systems that amount, at least in the canonical case, to unique identification and prove that the parameter complexity necessary for the replication of the dynamics is only $\mathcal{O}(n)$ and not $\mathcal{O}(n^2)$, as suggested by the standard parametrization of these systems. We shall sketch some results that point to a characterization of the unique identification problem in terms of groupoid/group orbit spaces that will be studied in detail in a forthcoming publication.

2 Preliminaries

State-Space Systems and Morphisms. A continuous time state-space system is given by the following two equations

$$\begin{cases} \dot{\mathbf{z}} = F(\mathbf{z}, u), \\ y = h(\mathbf{z}), \end{cases} \tag{1}$$

where $u \in \mathcal{U}$ is the *input*, $\mathbf{z} \in \mathcal{Z}$ is the *internal state*, $y \in \mathcal{Y}$ is the *output*, $F : \mathcal{Z} \times \mathcal{U} \rightarrow \mathcal{Z}$ is called the *state map*, and $h : \mathcal{Z} \rightarrow \mathcal{Y}$ is the *readout* or *output map*. The first equation is called the *state equation* while the second is usually called the *observation equation*. State-space systems will sometimes be denoted using the triplet $(\mathcal{Z}, F, h)$. The solutions of (1) (when available and unique) yield an *input/output map* or *filter* $U_{F,h} : C^1(I, \mathcal{U}) \times \mathcal{Z} \longrightarrow C^1(I, \mathcal{Y})$ that is by construction causal and time-invariant. The filter $U_{F,h}$ associates to each pair $(\mathbf{u}, z_0) \in C^1(I, \mathcal{U}) \times \mathcal{Z}$ the solution $U_{F,h}(\mathbf{u}, z_0) \in C^1(I, \mathcal{Y})$ of the non-autonomous differential equation (1). Different state-space systems may induce the same input/output system. This is what we call the *identification problem*, which is related to the next definition (see [2]).

Definition 1. *A map* $f : \mathcal{Z}_1 \rightarrow \mathcal{Z}_2$ *is called a system morphism between the continuous-time state-space systems* $(\mathcal{Z}_1, F_1, h_1)$ *and* $(\mathcal{Z}_2, F_2, h_2)$ *if it satisfies the following two properties:*

(i) *System equivariance:* $f(F_1(\mathbf{z}_1, u)) = F_2(f(\mathbf{z}_1), u)$, *for all* $\mathbf{z}_1 \in \mathcal{Z}_1$ *and* $u \in \mathcal{U}$.
(ii) *Readout invariance:* $h_1(\mathbf{z}_1) = h_2(f(\mathbf{z}_1))$ *for all* $\mathbf{z}_1 \in \mathcal{Z}_1$.

An elementary but very important fact is that if $f : \mathcal{Z}_1 \rightarrow \mathcal{Z}_2$ is a linear system-equivariant map between $(\mathcal{Z}_1, F_1, h_1)$ and $(\mathcal{Z}_2, F_2, h_2)$ ($\mathcal{Z}_1$ and $\mathcal{Z}_2$ are in this case vector spaces) then, for any solution $\mathbf{z}_1 \in C^1(I, \mathcal{Z}_1)$ of the state equation associated to F_1 with initial condition $\mathbf{z}_1^0 \in \mathcal{Z}$ and to the input $\mathbf{u} \in C^1(I, \mathcal{U})$, its image $f \circ \mathbf{z}_1 \in C^1(I, \mathcal{Z}_2)$ is a solution for the state space system associated to F_2 with the same input and initial condition $f(\mathbf{z}_1^0) \in \mathcal{Z}_2$. This implies that

$$U_{F_2,h_2} \circ \left(\mathbb{I}_{C^1(I,\mathcal{U})} \times f \right) = U_{F_1,h_1}. \tag{2}$$

This fact has as an important consequence that, in general, input/output systems *are not uniquely identified* since all the system-isomorphic state-space systems via a map f yield the same input/output map, with initial conditions related by the map f according to (2).

Hamiltonian and Port-Hamiltonian Systems. The *Hamiltonian system* determined by the *Hamiltonian function* $H \in C^1(\mathbb{R}^{2n})$ is given by the differential equation

$$\dot{\mathbf{z}} = \mathbb{J}\frac{\partial H}{\partial \mathbf{z}}, \tag{3}$$

where $\mathbb{J} = \begin{bmatrix} 0 & \mathbb{I}_n \\ -\mathbb{I}_n & 0 \end{bmatrix}$ is the so-called the *canonical symplectic matrix*. A *linear* Hamiltonian system is determined by a quadratic Hamiltonian function $H(\mathbf{z}) = \frac{1}{2}\mathbf{z}^T Q \mathbf{z}$, where $\mathbf{z} \in \mathbb{R}^{2n}$ and $Q \in M_{2n}$ is a square matrix that without loss of generality can be assumed to be symmetric. In this case, Hamilton's equations (3) reduce to $\dot{\mathbf{z}} = \mathbb{J}Q\mathbf{z}$.

Port-Hamiltonian systems (see [11]) are state-space systems that generalize autonomous Hamiltonian systems to the case in which external signals or inputs control in a time-varying way the dynamical behavior of the Hamiltonian system. The family of input-state-output port-Hamiltonian systems have the explicit form (see [11]):

$$\begin{cases} \dot{\mathbf{x}} = [J(\mathbf{x}) - R(\mathbf{x})]\dfrac{\partial H}{\partial \mathbf{x}}(\mathbf{x}) + g(\mathbf{x})u, \\ y = g^T(\mathbf{x})\dfrac{\partial H}{\partial \mathbf{x}}(\mathbf{x}), \end{cases} \tag{4}$$

where (u, y) is the input-output pair (corresponding to the control and output conjugated ports), $J(\mathbf{x})$ is a skew-symmetric interconnection structure and $R(\mathbf{x})$ is a symmetric positive-definite dissipation matrix. Our work concerns *linear* port-Hamiltonian systems in *normal form*, that is, the skew-symmetric matrix J is constant and equal to the canonical symplectic matrix $\mathbb{J}$, the Hamiltonian matrix Q is symmetric positive-definite, and the energy dissipation matrix $R = 0$, in which case (4) takes the form:

$$\begin{cases} \dot{\mathbf{z}} = \mathbb{J}Q\mathbf{z} + \mathbf{B}u, \\ y = \mathbf{B}^T Q\mathbf{z}, \end{cases} \tag{5}$$

with $\mathbf{z} \in \mathbb{R}^{2n}$, $u, y \in \mathbb{R}$, and where $\mathbf{B} \in \mathbb{R}^{2n}$ specifies the interconnection structure simultaneously at the input and output levels. In all that follows, we denote the family of state-space systems that can be written as (5) with the symbol PH_n, and we refer to the elements of this set as *normal form port-Hamiltonian systems*. All these systems have the existence and uniqueness of solutions property and hence determine a family of input/output systems that will be denoted by $\mathcal{PH}_n$. By definition, the systems in PH_n are fully determined by the pairs $(Q, \mathbf{B})$, and hence we define the parameter space

$$\Theta_{PH_n} := \left\{ (Q, \mathbf{B}) \mid 0 < Q \in M_{2n}, Q = Q^T, \mathbf{B} \in \mathbb{R}^{2n} \right\}.$$

Let $\theta_{PH_n} : \Theta_{PH_n} \to PH_n$ the map that associates to the parameter $(Q, \mathbf{B}) \in \Theta_{PH_n}$ the corresponding port-Hamiltonian state space system (5). We emphasize that two different elements in Θ_{PH_n} can determine the same input/output system in $\mathcal{PH}_n$; said differently, the parameter set Θ_{PH_n} does not uniquely specify the elements in $\mathcal{PH}_n$.

3   Controllable and Observable Hamiltonian Systems

Definition 2. *Given* $\mathbf{d} = (d_1, \ldots, d_n)^T \in \mathbb{R}^n$, *with* $d_i > 0$, *and* $\mathbf{v} \in \mathbb{R}^{2n}$, *we say that a $2n$-dimensional linear state space system is a controllable Hamiltonian (respectively, observable Hamiltonian) representation if it takes the form*

$$
\begin{cases} \dot{\mathbf{s}} = g_1^{ctr}(\mathbf{d}) \cdot \mathbf{s} + (0,0,\cdots,0,1)^T \cdot u, \\ y = g_2^{ctr}(\mathbf{d},\mathbf{v}) \cdot \mathbf{s}, \end{cases} \left(resp., \begin{cases} \dot{\mathbf{s}} = g_1^{obs}(\mathbf{d}) \cdot \mathbf{s} + g_2^{obs}(\mathbf{d},\mathbf{v}) \cdot u, \\ y = (0,0,\cdots,0,1) \cdot \mathbf{s}, \end{cases} \right) \quad (6)
$$

where $g_1^{ctr}(\mathbf{d}) \in \mathbb{M}_{2n}$ *and* $g_2^{ctr}(\mathbf{d},\mathbf{v}) \in \mathbb{M}_{1,2n}$ *(respectively,* $g_1^{obs}(\mathbf{d}) \in \mathbb{M}_{2n}$ *and* $g_2^{obs}(\mathbf{d},\mathbf{v}) \in \mathbb{R}^{2n}$ *) are constructed as follows:*

(i) *Given* $\mathbf{d} \in \mathbb{R}^n$, *let* $\{a_0, a_1, \ldots, a_{2n-1}\}$ *be the real coefficients that make* $\lambda^{2n} + \sum_{i=0}^{2n-1} a_i \cdot \lambda^i = (\lambda^2 + d_1^2)(\lambda^2 + d_2^2) \ldots (\lambda^2 + d_n^2)$ *an equality between the two polynomials in* λ. *Let* $a_{2n} = 1$ *by convention. Note that the entries* a_i *with an odd index i are zero. Define:*

$$
g_1^{ctr}(\mathbf{d}) := \begin{bmatrix} 0 & 1 & 0 & \ldots & 0 \\ 0 & 0 & 1 & \ldots & 0 \\ \vdots & \vdots & \ddots & \vdots & \vdots \\ 0 & 0 & 0 & \ldots & 1 \\ -a_0 & -a_1 & -a_2 & \ldots & -a_{2n-1} \end{bmatrix}_{2n \times 2n},
$$

(respectively, $g_1^{obs}(\mathbf{d}) = g_1^{ctr}(\mathbf{d})^\top$ *).*

(ii) *Given* $\mathbf{d}$ *and* $\mathbf{v}$, *then*

$$
g_2^{ctr}(\mathbf{d},\mathbf{v}) := \begin{bmatrix} 0 & c_{2n-1} & 0 & c_{2n-3} & \ldots & 0 & c_1 \end{bmatrix}, \quad (resp., g_2^{obs}(\mathbf{d},\mathbf{v}) = g_2^{ctr}(\mathbf{d},\mathbf{v})^\top)
$$

where

$$
c_{2k+1} = \mathbf{v}^T \begin{bmatrix} F_k & 0 \\ 0 & F_k \end{bmatrix} \mathbf{v},
$$

for $k = 0, \ldots, n-1$, *and*

$$
F_k = \begin{bmatrix} f_1 & & & & \\ & f_2 & & \text{\Large 0} & \\ & & \ddots & & \\ & \text{\Large 0} & & f_{n-1} & \\ & & & & f_n \end{bmatrix}
$$

with $f_l = d_l \cdot \sum_{\substack{j_1,\ldots,j_k \neq l \\ 1 \le j_1 < \cdots < j_k \le n}} (d_{j_1} d_{j_2} \cdots d_{j_k})^2$, $l = 1, \ldots, n$.

We denote CH_n *(respectively,* OH_n*) the set of all systems of the form (6), and we call them controllable Hamiltonian (respectively, observable Hamiltonian) representations. The symbol* $\mathcal{CH}_n$ *(respectively,* $\mathcal{OH}_n$*) denotes the set of input/output systems induced by the state space systems in* CH_n *(respectively,* OH_n*). We emphasize that the elements of both* CH_n *and* OH_n *can be parameterized with the set* $\Theta_{CH_n} = \Theta_{OH_n} := \{(\mathbf{d},\mathbf{v}) | d_i > 0, \mathbf{v} \in \mathbb{R}^{2n}\}$.

The maps $\theta_{CH_n} : \Theta_{CH_n} \to CH_n$ and $\theta_{OH_n} : \Theta_{OH_n} \to OH_n$ associate to each parameter the corresponding state space system. Note that the elements in CH_n (respectively, in OH_n) of the form (6) are in canonical controllable (respectively, observable) form in the sense of [12] and they are hence controllable (respectively, observable). Our main result establishes a relationship between port-Hamiltonian systems and controllable (respectively, observable) Hamiltonian representations as defined above, which will be used later on for considerations on the structure preservation and expressiveness modeling of PH_n.

Theorem 1. (i) *There exists, for each $S \in Sp(2n, \mathbb{R})$, a map*

$$\varphi_S : \quad CH_n \quad \longrightarrow \quad PH_n$$
$$\theta_{CH_n}(\mathbf{d}, \mathbf{v}) \longmapsto \theta_{PH_n}\left(S^T \begin{bmatrix} D & 0 \\ 0 & D \end{bmatrix} S, S^{-1}\mathbf{v}\right),$$

with $D = \mathrm{diag}(\mathbf{d})$, such that the controllable Hamiltonian system $\theta_{CH_n}(\mathbf{d}, \mathbf{v}) \in CH_n$ and the port-Hamiltonian image $\varphi_S(\theta_{CH_n}(\mathbf{d}, \mathbf{v})) \in PH_n$ are linked by a linear system morphism $f_S^{(\mathbf{d},\mathbf{v})} : \mathbb{R}^{2n} \to \mathbb{R}^{2n}$.

(ii) *Given a port-Hamiltonian system $\theta_{PH_n}(Q, \mathbf{B}) \in PH_n$, there exists an explicit linear system morphism $f^{(Q, \mathbf{B})} : \mathbb{R}^{2n} \to \mathbb{R}^{2n}$ between the state space of $\theta_{PH_n}(Q, \mathbf{B}) \in PH_n$ and that of an observable Hamiltonian system $\theta_{OH_n}(\mathbf{d}, \mathbf{v}) \in OH_n$, where $(\mathbf{d}, \mathbf{v}) \in \Theta_{OH_n}$ is determined by the Williamson's normal form decomposition [13] of Q determined by $S \in Sp(2n, \mathbb{R})$, that is,*

$$Q = S^T \begin{bmatrix} D & 0 \\ 0 & D \end{bmatrix} S, \ D = \mathrm{diag}(\mathbf{d}) \ and \ \mathbf{v} = S \cdot \mathbf{B}.$$

Controllability, Observability, and Invertibility. The linear system morphism $f_S^{(\mathbf{d},\mathbf{v})} : \mathbb{R}^{2n} \to \mathbb{R}^{2n}$ in part **(i)** is implemented by a matrix L, that is, $\mathbf{z} = f_S^{(\mathbf{d},\mathbf{v})}(\mathbf{s}) = L\mathbf{s}$ and L can be explicitly computed once $(\mathbf{d}, \mathbf{v})$ has been fixed. Moreover, L can be transformed by elementary column operations into the controllability matrix of $\varphi_S(\theta_{CH_n}(\mathbf{d}, \mathbf{v})) \in PH_n$. Consequently, the condition of L being invertible, i.e., that the two systems in CH_n and PH_n linked by φ_S being isomorphic, is equivalent to the controllability matrix of $\varphi_S(\theta_{CH_n}(\mathbf{d}, \mathbf{v}))$ having full rank (regardless of the choice of $S \in Sp(2n, \mathbb{R})$), which is again equivalent to $\varphi_S(\theta_{CH_n}(\mathbf{d}, \mathbf{v}))$ being canonical since controllability and observability are intertwined concepts in the linear port-Hamiltonian category. Indeed, it can be proved (see [8]) that if a linear port-Hamiltonian system without dissipation is controllable and $\det(Q) \neq 0$, then it is also observable. Conversely, if it is observable, then this implies that $\det(Q) \neq 0$ and it is also controllable. As it is customary in systems theory, we say a linear port-Hamiltonian system in normal form is *canonical* if it is both controllable and observable. In view of the results that we just recalled, the condition $\det(Q) \neq 0$, which is part of the definition of PH_n implies that in this context, either controllability or observability is equivalent to the system being canonical. Thus, in the remaining, we only refer to controllability.

The systems in CH_n are by construction in controllable canonical form and are therefore always controllable. If the image system $\varphi_S\left(\theta_{CH_n}(\mathbf{d},\mathbf{v})\right)$ that we want to learn is controllable (or equivalently, observable), then by the previous argument, L is necessarily an invertible matrix which means that $\theta_{CH_n}(\mathbf{d},\mathbf{v})$ and $\varphi_S\left(\theta_{CH_n}(\mathbf{d},\mathbf{v})\right)$ are necessarily isomorphic systems by construction. As a consequence, $\theta_{CH_n}(\mathbf{d},\mathbf{v})$ is, in such case, not only controllable but also observable.

Application to Structure-Preserving System Learning. As a corollary of the previous result, we can use controllable Hamiltonian systems to learn port-Hamiltonian systems in an efficient and structure-preserving fashion. Indeed, given a realization of a port-Hamiltonian system, a system of the type $\theta_{CH_n}(\mathbf{d},\mathbf{v}) \in CH_n$ can be estimated using an appropriate loss. A controllable Hamiltonian representation is more advantageous than the original port-Hamiltonian one for two reasons:

(i) The *model complexity* of the controllable Hamiltonian representation is only of order $\mathcal{O}(n)$, as opposed to $\mathcal{O}(n^2)$ for the original port-Hamiltonian one.
(ii) This learning scheme is automatically *structure-preserving*. Indeed, once a system $\theta_{CH_n}(\mathbf{d},\mathbf{v}) \in CH_n$ has been estimated for a given realization, we have shown that there exists a family of linear morphisms, each of which is between the state space of $\theta_{CH_n}(\mathbf{d},\mathbf{v})$ and some $\theta_{PH_n}(Q,\mathbf{B}) \in PH_n$, such that any solution of (6) is automatically a solution of some system in PH_n. Hence, *even in the presence of estimation errors* for the parameters $(\mathbf{d},\mathbf{v}) \in \Theta_{CH_n}$, its solutions correspond to a port-Hamiltonian system. Hence, this structure is *preserved* by the learning scheme.

System Learning and Expressive Power. In our context, we refer to the expressive power of a model as the extent to which port-Hamiltonian dynamics can be replicated. There is an important relation between the controllability of a system in PH_n and the expressive power of the corresponding representation in CH_n. Indeed, if a system of the form (5) in PH_n is controllable (i.e. $f_S^{(\mathbf{d},\mathbf{v})}$ invertible), then for an arbitrary initial state of PH_n, we can find the corresponding initial state for CH_n simply by taking the preimage under $f_S^{(\mathbf{d},\mathbf{v})}$. Therefore, the preimage of (5) by the map φ_S in CH_n can reproduce all possible solutions of (5), which amounts to the learning scheme based on the systems in CH_n having full expressive power. However, if $\theta_{PH_n}(Q,\mathbf{B}) \in PH_n$ fails to be controllable (i.e. $f_S^{(\mathbf{d},\mathbf{v})}$ not invertible), then the full expressive power is not guaranteed. As a rule of thumb, the more controllable a system of the type $\theta_{PH_n}(Q,\mathbf{B}) \in PH_n$ is, the higher the rank of $f_S^{(\mathbf{d},\mathbf{v})}$ is, and then the more expressive the corresponding controllable Hamiltonian representation is.

On the other hand, we have shown that for each $\theta_{PH_n}(Q,\mathbf{B}) \in PH_n$, there exists a linear morphism between the state space of $\theta_{PH_n}(Q,\mathbf{B}) \in PH_n$ and some $\theta_{OH_n}(\mathbf{d},\mathbf{v})$, such that any solution of (5) is automatically a solution of some system in OH_n. Hence, *no matter the original port-Hamiltonian system is controllable or not*, its solution can always be reproduced by an element in OH_n by learning a correct parameter $(\mathbf{d}, \mathbf{v}) \in \Theta_{OH_n}$. Hence, this learning scheme has *full expressive power.*

4   Unique Identifiability of Port-Hamiltonian Systems

The main theorem in the previous section provides a structure-preserving learning scheme of optimal complexity. It does not, however, solve the unique identifiability problem for the family of filters $\mathcal{PH}_n$ induced by the linear port-Hamiltonian systems in normal form (5) that we have set to learn in this paper. Even though this problem will be studied in detail in a forthcoming publication [9], we put forward some preliminary results that can be obtained as a consequence of Theorem 1.

We recall that the unique identification problem in our case has its source in the fact that system isomorphic elements in PH_n induce exactly the same input/output system in $\mathcal{PH}_n$. It is easy to show that system isomorphisms induce an equivalence relation on PH_n that we shall denote by $\sim_{sys}$.

One may hope that $\mathcal{PH}_n \simeq PH_n/\sim_{sys}$, so that the quotient space $PH_n/\sim_{sys}$ is the space in which unique identifiability is achieved. Unfortunately, this holds only when we are restricting to canonical systems, which form an open and dense subset. More precisely, if we denote by PH_n^{can} the subset of canonical port-Hamiltonian systems in PH_n and denote by $\mathcal{PH}_n^{can}$ the corresponding filters, then it holds that

$$\mathcal{PH}_n^{can} \simeq PH_n^{can}/\sim_{sys} \simeq \Theta_{CH_n}^{can}/(S_n \rtimes \mathbb{T}^n).$$

Furthermore, this unique identification space can be characterized as the orbit space of certain action of the semi-direct product group $S_n \rtimes \mathbb{T}^n$ (S_n is the permutation group and $\mathbb{T}^n$ is the n-tori) on some dense subset $\Theta_{CH_n}^{can}$ of Θ_{CH_n}. The orbit space has a natural smooth structure and global charts that can be used in the numerical implementation of learning problems.

Generally speaking, in the presence of non-canonical port-Hamiltonian systems, the relationship between $\mathcal{PH}_n$ and $PH_n/\sim_{sys}$ is complicated. In such a scenario, two distinct elements in PH_n that are $\sim_{sys}$-equivalent always induce the same filter in $\mathcal{PH}_n$, whereas a filter in $\mathcal{PH}_n$ could be realized by two elements in PH_n that are not $\sim_{sys}$-equivalent, since a filter identifies the canonical part (i.e. minimal realization) and that part only, see [3]. Said differently, by going to the quotient space $PH_n/\sim_{sys}$, we remove some redundancies of the set PH_n that yield the same input-output dynamics, but not all. In what follows, we give a characterization of $PH_n/\sim_{sys}$ while we emphasize again that $PH_n/\sim_{sys}$ is not the same as the unique identifiability space $\mathcal{PH}_n$ in general.

In order to make the quotient space $PH_n/\sim_{sys}$ manageable, it is desirable to characterize it in terms of the parameter set Θ_{PH_n} or, even better, by the less complex Θ_{CH_n}. This is indeed possible. More explicitly, an equivalence relation $\sim_\star$ on Θ_{CH_n} can be defined such that

$$PH_n/\sim_{sys} \simeq \Theta_{CH_n}/\sim_\star. \tag{7}$$

The equivalence relation $\sim_\star$ is defined as: the pairs $(\mathbf{d}_1, \mathbf{v}_1)$ and $(\mathbf{d}_2, \mathbf{v}_2)$ in Θ_{CH_n} are $\sim_\star$-equivalent, i.e. $(\mathbf{d}_1, \mathbf{v}_1) \sim_\star (\mathbf{d}_2, \mathbf{v}_2)$, if there exists a permutation matrix $P_\sigma \in \mathbb{M}_n$ and an invertible matrix A such that, for $D_i = \mathrm{diag}(\mathbf{d}_i)$, $i \in \{1, 2\}$ and $P = \begin{bmatrix} P_\sigma & 0 \\ 0 & P_\sigma \end{bmatrix}$, the following conditions hold true:

$$P \begin{bmatrix} D_1 & 0 \\ 0 & D_1 \end{bmatrix} P^T = \begin{bmatrix} D_2 & 0 \\ 0 & D_2 \end{bmatrix}, \quad A^T \begin{bmatrix} D_1 & 0 \\ 0 & D_1 \end{bmatrix} A\mathbf{v}_1 = \begin{bmatrix} D_1 & 0 \\ 0 & D_1 \end{bmatrix} \mathbf{v}_1,$$

$$A\mathbb{J} \begin{bmatrix} D_1 & 0 \\ 0 & D_1 \end{bmatrix} = \mathbb{J} \begin{bmatrix} D_1 & 0 \\ 0 & D_1 \end{bmatrix} A, \quad \text{and} \quad \mathbf{v}_2 = PA\mathbf{v}_1$$

Characterization of $PH_n/\sim_{sys}$ in terms of Lie groupoid orbit spaces.
As it is customary, groupoids will be denoted with the symbol $s, t : \mathcal{G} \rightrightarrows M$ (or simply $\mathcal{G} \rightrightarrows M$), where s and t are the *source* and the *target* maps, respectively. Given $m \in M$, the *groupoid orbit* that contains this point is given by $\mathcal{O}_m = t\left(s^{-1}(m)\right) \subset M$. The *orbit space* associated to $\mathcal{G} \rightrightarrows M$ is denoted by $M/\mathcal{G}$.

In this paragraph, we show how the quotient spaces in (7) can be characterized in terms of two Lie groupoid orbits. More precisely, the set of equivalence classes $PH_n/\sim_{sys}$ (resp. $\Theta_{CH_n}/\sim_\star$) is the orbit space $PH_n/\mathcal{G}_n$ (resp. $\Theta_{CH_n}/\mathcal{H}_n$) of a groupoid $\mathcal{G}_n \rightrightarrows PH_n$ (resp. $\mathcal{H}_n \rightrightarrows \Theta_{CH_n}$) which we now construct. The statement (7) is hence equivalent to stating that the orbit spaces $PH_n/\sim_{sys}$ and $\Theta_{CH_n}/\mathcal{H}_n$ of the two groupoids coincide.

The total space $\mathcal{G}_n$ of the first groupoid is given by

$$\mathcal{G}_n := \{(L, (Q, \mathbf{B})) \,|\, L \in GL(2n, \mathbb{R}), (Q, \mathbf{B}) \in PH_n \text{ such that}$$
$$\mathbb{J}^T L \mathbb{J} Q L^{-1} \text{ is symmetric and positive-definite and } B = \mathbb{J}^T L^T \mathbb{J} LB\}.$$

The target and source maps $\alpha, \beta : \mathcal{G}_n \to PH_n$ are defined by $\alpha(L, (Q, \mathbf{B})) := (\mathbb{J}^T L \mathbb{J} Q L^{-1}, L\mathbf{B})$ and $\beta(L, (Q, \mathbf{B})) := (Q, \mathbf{B})$. It can be proved that the orbit space of this groupoid $PH_n/\mathcal{G}_n$ coincides with $PH_n/\sim_{sys}$.

The total space $\mathcal{H}_n$ of the second groupoid is given by

$$\mathcal{H}_n := \Big\{ ((P_\sigma, A), (\mathbf{d}, \mathbf{v})) \,|\, P_\sigma \in \mathbb{M}_n \text{ is a permutation matrix}, A \in GL(2n, \mathbb{R}),$$
$$(\mathbf{d}, \mathbf{v}) \in \Theta_{CH_n}, \text{ such that } A^T \begin{bmatrix} D & 0 \\ 0 & D \end{bmatrix} A\mathbf{v} = \begin{bmatrix} D & 0 \\ 0 & D \end{bmatrix} \mathbf{v}$$
$$\text{and } A\mathbb{J} \begin{bmatrix} D & 0 \\ 0 & D \end{bmatrix} = \mathbb{J} \begin{bmatrix} D & 0 \\ 0 & D \end{bmatrix} A, \text{ with } D = \mathrm{diag}(\mathbf{d}) \Big\}.$$

The target and source maps $\alpha, \beta : \mathcal{H}_n \to \Theta_{CH_n}$ are defined as $\alpha((P_\sigma, A), (\mathbf{d}, \mathbf{v})) := (\mathbf{d}, \mathbf{v})$ and $\beta((P_\sigma, A), (\mathbf{d}, \mathbf{v})) := (P_\sigma \mathbf{d}, PA\mathbf{v})$, where $P = \begin{bmatrix} P_\sigma & 0 \\ 0 & P_\sigma \end{bmatrix}$.

The orbit space of this groupoid $\Theta_{CH_n}/\mathcal{H}_n$ coincides with $\Theta_{CH_n}/\sim_\star$ and, moreover, the orbit spaces of the Lie groupoids $\mathcal{G}_n \rightrightarrows PH_n$ and $\mathcal{H}_n \rightrightarrows \Theta_{CH_n}$ are isomorphic, as desired.

Acknowledgements. The authors thank Lyudmila Grigoryeva for helpful discussions and remarks and acknowledge partial financial support from the Swiss National Science Foundation (grant number 175801/1) and the School of Physical and Mathematical Sciences of the Nanyang Technological University. DY is funded by the Nanyang President's Graduate Scholarship of Nanyang Technological University.

References

1. Gonzalez, O.: Time integration and discrete Hamiltonian systems. In: Mechanics: From Theory to Computation. Springer, New York, NY (2000). https://doi.org/10.1007/978-1-4612-1246-1_10

2. Grigoryeva, L., Ortega, J.P.: Dimension reduction in recurrent networks by canonicalization. J. Geom. Mech. **13**(4), 647–677 (2021). https://doi.org/10.3934/jgm.2021028

3. Kalman, R.E.: Mathematical description of linear dynamical systems. J. Soc. Indus. Appl. Math. Ser. A Control. **1**(2), 152–192 (1963). https://doi.org/10.1137/0301010

4. Karniadakis, G.E., Kevrekidis, I.G., Lu, L., Perdikaris, P., Wang, S., Yang, L.: Physics-informed machine learning. Nat. Rev. Phys. **3**(6), 422–440 (2021)

5. Leimkuhler, B., Reich, S.: Simulating Hamiltonian Dynamics. Cambridge University Press, Cambridge (2004)

6. Marsden, J.E., West, M.: Discrete mechanics and variational integrators. Acta Numer. **10**, 357–514 (2001). https://doi.org/10.1017/S096249290100006X

7. McLachlan, R.I., Quispel, G.R.W.: Geometric integrators for ODEs. J. Phys. A Math. Gener. **39**(19), 5251 (2006)

8. Medianu, S., Lefevre, L., Stefanoiu, D.: Identifiability of linear lossless Port-controlled Hamiltonian systems. In: 2nd International Conference on Systems and Computer Science, pp. 56–61 (2013). https://doi.org/10.1109/IcConSCS.2013.6632023

9. Ortega, J.P., Yin, D.: Learnability of linear port-Hamiltonian systems. arXiv preprint arXiv:2303.15779 (2023)

10. Raissi, M., Perdikaris, P., Karniadakis, G.E.: Physics informed deep learning (part i): Data-driven solutions of nonlinear partial differential equations. arXiv preprint arXiv:1711.10561 (2017)

11. van der Schaft, A., Jeltsema, D.: Port-Hamiltonian systems theory: an introductory overview, vol. 1 (2014).https://doi.org/10.1561/2600000002

12. Sontag, E.: Mathematical Control Theory: Deterministic Finite Dimensional Systems. Springer-Verlag, Heidelberg (1998). https://doi.org/10.1007/978-3-540-69532-5_16

13. Williamson, J.: On the algebraic problem concerning the normal forms of linear dynamical systems. Am. J. Math. **58**(1), 141–163 (1936)

14. Wu, J.L., Xiao, H., Paterson, E.: Physics-informed machine learning approach for augmenting turbulence models: a comprehensive framework. Phys. Rev. Fluids **3**(7), 74602 (2018)

Signature Estimation and Signal Recovery Using Median of Means

Stéphane Chrétien[1(✉)] and Rémi Vaucher[1,2]

[1] Laboratoire ERIC, 69676 Bron, France
stephane.chretien@univ-lyon2.fr, remi.vaucher@halias.fr
[2] Halias Technologies, 38240 Meylan, France
https://sites.google.com/site/stephanegchretien

Abstract. The theory of Signatures [1,7] is a fast growing field which has demonstrated wide applicability to a large range of fields, from finance to health monitoring [2,6,10]. Computing signatures often relies on the assumptions that the signal under study is not corrupted by noise, which is rarely the case in practice. In the present paper, we study the influence of noise on the computation of Signatures via the theory of anti-concentration. We then propose a median of means (MoM) approach to the estimation problem and give a bound on the estimation error using Rademacher complexity.

Keywords: Signature Theory · Median of Means · Anticoncentration

1 Introduction: Using the Signature of a Function Dictionary to Get the Signature of a Path

Signatures, a transform that applies to time dependent signals, have become a tool of choice for the analysis of multidimensional dynamical phenomena which are pervasive in many applications of machine learning. Computing Signatures allows to extract meaningful features about the various time dependencies of the components of the signal in a natural way, even when sampling may possibly be irregular and at different time stamps for different components.

1.1 Background on the Signature Transform

The signature of order k (denoted by $S_{[0,t]}^{(k)}(X) \in \mathbb{R}^{\overbrace{d \times d \times \ldots \times d}^{k \text{ times}}}$) of a signal $X(t) = (X(t)^1, \ldots, X(t)^d)$, $t \in [0,T]$ is defined for every word $i_1 i_2 \ldots i_k$ from $\{1, \ldots, d\}$ by

$$S(X)_{0,t}^{i_1, \ldots, i_k} = \int_{0 < t_k < t} \cdots \int_{0 < t_1 < t_2} dX_{t_1}^{i_1} \ldots dX_{t_k}^{i_k}. \tag{1}$$

F. Nielsen and F. Barbaresco (Eds.): GSI 2023, LNCS 14072, pp. 323–331, 2023.
https://doi.org/10.1007/978-3-031-38299-4_34

Signatures have very useful properties that make them ideal for feature extraction. First, they are invariant with respect to reparametrisation. Indeed, for any two indices i_1 and i_2 in $\{1, \ldots, d\}$, consider $\tilde{X}^{i_1}$ and $\tilde{X}^{i_2}$ defined by $\tilde{X}_s^{i_j} = X_{\phi(s)}^{i_j}$ for $j = 1, 2$ and ϕ a surjective, increasing differentiable function $\phi : [0, T] \mapsto [0, T]$. Then

$$\int_0^t \tilde{X}_s^{i_1} \, d\tilde{X}_s^{i_2} = \int_0^t X_s^{i_1} \, dX_s^{i_2} \tag{2}$$

for all $t \in [0, T]$.

Another very important property is the Shuffle product identity, which states that for a path $X : [0, T] \mapsto \mathbb{R}^d$ and two multi-indexes $I = (i_1, \ldots, i_k)$ and $J = (j_1, \ldots, j_m)$ with $i_1, \ldots, i_k, j_1, \ldots, j_m \in \{1, \ldots, d\}$, it holds that

$$S(X)_{0,T}^I S(X)_{0,T}^J = \sum_{K \in I \sqcup\!\sqcup J} S(X)_{0,T}^K \tag{3}$$

where $\sqcup\!\sqcup$ denotes the shuffle product. Finally, Chen's property completes the presentation of the most elementary properties of Signatures. Let $X : [0, T] \mapsto \mathbb{R}^d$ and $Y : [T, T'] \mapsto \mathbb{R}^d$ be two paths. Then, for the concatenation of X and Y, we have

$$S(X * Y)_{0,T'} = S(X)_{0,T} \otimes S(Y)_{T,T'}. \tag{4}$$

Given a path $X : [0, T] \mapsto \mathbb{R}^d$ and a set of timestamps $T_\nu = \{t_1, \ldots, t_\nu\}$, one defines X_{T_ν} as the piecewise linear path obtained by linearly interpolating the observed values of X at times $t_1, \ldots, t_\nu$.

Assume that there exists a basis (i.e. a dictionary of piecewise differentiable functions) $\psi = (\psi_1, \ldots, \psi_m) : [0; 1] \to \mathbb{R}^m$ and a linear map $A = (a_{ij}) \in \mathbb{R}^{n \times m}$ such that

$$X = A\psi : t \mapsto \left(\sum_{i=1}^m a_{1,i} \psi_i(t), \ldots, \sum_{i=1}^m a_{n,i} \psi_i(t) \right).$$

Starting from (1), it was proved in [11] that

$$S^{(3)}(X) = S^{(3)}(A\psi) = [\![C_\psi; A, A, A]\!] \in \mathbb{R}^{n \times n \times n} \tag{5}$$

where $C_\psi = S^{(3)}(\psi) = (c_{ijk})_{i,j,k \in [\![1;n]\!]}$ and

$$[\![\bullet_1; \bullet_2, \bullet_2, \bullet_2]\!] : (C, A) \in \mathbb{R}^{n \times n \times n} \times \mathbb{R}^{n \times m} \mapsto ([\![C; A, A, A]\!])_{\alpha, \beta, \gamma} \tag{6}$$

is defined for all $\alpha, \beta, \gamma \in \{1, ..., n\}$ by

$$[\![C_\psi; A, A, A]\!]_{\alpha, \beta, \gamma} = \sum_{i=1}^m \sum_{j=1}^m \sum_{k=1}^m c_{ijk} a_{\alpha,i} a_{\beta,j} a_{\gamma,k}. \tag{7}$$

1.2   Goal of the Paper

In practice, signals are often corrupted by additive observation noise, and we can write

$$X = X^* + \epsilon. \tag{8}$$

X is a discrete path, as observations occurs at a finite number of time $\{t_1, .., t_\nu\}$, but can be viewed as continuous by linearly interpolating between two observations.

In the present paper, we will assume for the sake of simplicity that ϵ is a Gaussian white noise function.

The main problems studied in the present paper are:

– estimating the Signature tensor of X^*
– recovering the original signal X^* in an appropriate basis, based on the Signature tensor of a subsampled noisy version of X^*.

Under observation noise, the computed signature may not be an accurate estimation of the signature of the true signal and one needs to resort to an appropriate regularisation procedure in order to recover the seeked signature tensor.

In the first part of this paper, we study anticoncentration of the components of the noisy signature tensor, a result which raises the questions of how reliable estimation problems for Signature-based machine learning can be. In the second part, we show how to use a new technique from the field of Robust Statistics, named Median of Means, in order to estimate a robust version of the signature tensor.

2   Anti-concentration for the 3-Signature Coefficients

One standard approach to estimating $S^{(3)}(\mathbb{E}[X])$, i.e. the signature of the original signal is to solve the following least-square regression problem:

$$\min_{A \in \mathbb{R}^{n \times m}} \left\| S^{(3)}(X) - [\![C; A, A, A]\!] \right\|_F^2. \tag{9}$$

Using that, after [4], for an orthogonal basis ψ,

$$\epsilon = E\psi \tag{10}$$

where E is a i.i.d. Gaussian matrix, and using the expansion

$$X^* = \mathbb{E}[X] = A^*\psi \tag{11}$$

of X^* in the basis ψ, we obtain

$$S^{(3)}(X^* + \epsilon) = S^{(3)}(A^*\psi + E\psi)$$
$$= S^{(3)}((A^* + E)\psi)$$
$$= [\![C; A^* + E, A^* + E, A^* + E]\!].$$

For the sake of making the problem finite dimensional, we will further approximate $S^{(3)}(X^* + \epsilon)$ with $S^{(3)}(XT_\nu{}^* + \epsilon T_\nu)$.

2.1 Expanding the Expression of the Signature Tensors

To simplify, we will note

– $[\![C]\!]_{A+E} = [\![C; A + E, A + E, A + E]\!]$
– $[\![C]\!]_{A+E,\alpha,\beta,\gamma} = [\![C; A + E, A + E, A + E]\!]_{\alpha,\beta,\gamma}$

First, it is easy to see that

$$[\![C]\!]_{A+E} = [\![C; A, A, A]\!] + [\![C; E, E, E]\!]$$
$$+ \sum_{i=0}^{2} \left([\![C; \sigma^i(A, A, E)]\!] + [\![C; \sigma^i(A, E, E)]\!] \right) \tag{12}$$

with $\sigma = (1, 2, 3) \in S_3$. So each coefficients from $[\![C]\!]_{A+E}$ takes the form of a polynomial of coefficients $e_{\alpha,i}$ with $\alpha \in [\![1; n]\!], i \in [\![1, m]\!]$.

We deduce from (7) and (12):

$$[\![C]\!]_{A+E,\alpha,\beta,\gamma} = P_1(E) + P_2(E) + P_3(E) + R = P(E) \tag{13}$$

where:

- $P_1(E) = \sum_{i=1}^{m} \sum_{j=1}^{m} \sum_{k=1}^{m} c_{ijk} e_{\alpha,i} e_{\beta,j} e_{\gamma,k}$
- $P_2(E) = \sum_{i=1}^{m} \sum_{j=1}^{m} \sum_{k=1}^{m} c_{ijk} (a_{\alpha,i} a_{\beta,j} e_{\gamma,k} + a_{\alpha,i} e_{\beta,j} a_{\gamma,k} + e_{\alpha,i} a_{\beta,j} a_{\gamma,k})$
- $P_3(E) = \sum_{i=1}^{m} \sum_{j=1}^{m} \sum_{k=1}^{m} c_{ijk} (a_{\alpha,i} e_{\beta,j} e_{\gamma,k} + e_{\alpha,i} e_{\beta,j} a_{\gamma,k} + e_{\alpha,i} a_{\beta,j} e_{\gamma,k})$
- $R = \sum_{i=1}^{m} \sum_{j=1}^{m} \sum_{k=1}^{m} c_{ijk} a_{\alpha,i} a_{\beta,j} a_{\gamma,k}$

2.2 Anti-concentration of Coefficients

We use here Theorem 1.8 from [8] or Theorem 1.2 from [5] depending on multi-linearity (or not) of P (i.e. depending on $\alpha \neq \beta \neq \gamma$ or not):

The case $\alpha \neq \beta \neq \gamma$ (P is multilinear):

Theorem 1. *There is an absolute constant B such that the following holds. Let $\xi_1, \ldots, \xi_n$ be independent (but not necessarily iid) random variables. Let P be a polynomial of degree d with the form*

$$P(\xi_1, \ldots, \xi_n) = \sum_{S \subset 1,\ldots,n, |S| \leq d} a_S \prod_{i \in S} \xi_i$$

whose rank $r \geq 2$. Assume that there are positive numbers p and ε such that for each $1 \leq i \leq n$, there is a number y_i such that $\min \{\mathbf{P}(\xi_i \leq y_i), \mathbf{P}(\xi_i > y_i)\} = p$ and $\mathbf{P}(|\xi_i - y_i| \geq 1) \geq \varepsilon$. Assume furthermore that $\tilde{r} := (p\varepsilon)^d r \geq 3$. Then for any interval I of length 1

$$\mathbf{P}(P(\xi_1, \ldots, \xi_n) \in I) \leq \min \left(\frac{Bd^{4/3}(\log \tilde{r})^{1/2}}{(\tilde{r})^{1/(4d+1)}}, \frac{\exp\left(Bd^2 \left(\log \log(\tilde{r})^2\right)\right)}{\sqrt{\tilde{r}}} \right) \tag{14}$$

The application here is simple. P is here a multilinear polynomial of degree $d = 3$, and all $e_{\alpha,i}$ are independants. The existence of p and ε such that, for all $e_{\alpha,i}$ (with $(\alpha, i) \in [\![1, n]\!] \times [\![1, m]\!]$) there exist $y_{\alpha,i}$ verifying:

$$\min \{\mathbb{P}(e_{\alpha,i} \leq y_{\alpha,i}), \mathbb{P}(e_{\alpha,i} \geq y_{\alpha,i})\} = p \tag{15}$$

and

$$\mathbb{P}(|e_{\alpha,i} - y_{\alpha,i}| \geq 1) = \varepsilon \tag{16}$$

depends on the $e_{\alpha,i}$ distributions.

Finally, the rank r depends on the dictionary ψ: $r = \#\{c_{ijk}, \ |c_{ijk}| > 1\}$. For the last hypothesis ($\tilde{r} \geq 3$), we need $r \geq \frac{3}{(p\varepsilon)^3}$. Under these assumptions on ψ and $a = \{a_{\alpha,i}\}$, the equation (14) applies.

The Two Other Cases: For the two other cases, as the polynomials are no more multilinear, we need an anti-concentration inequality for multivariate polynomials. With this in mind, we use Theorem 2. First, we need to define PSD anti-concentration:

A distribution $\mathcal{D}$ has *PSD* anti-concentration if there exist $C, c > 0$ such that the following holds. Let A be an $n \times n$ positive semi-definite matrix with $\mathrm{Tr}(A) = 1$. Then for any $\varepsilon > 0$,

$$\mathbb{P}_{\mathbf{x} \in \mathcal{D}^n} \left[\mathbf{x}^t A \mathbf{x} \leq \varepsilon\right] \leq C \cdot \varepsilon^c.$$

We now fix a notation for sum and subtraction of independent variables for the same distribution:

Define $d\mathcal{D} := \mathcal{D} + \ldots + \mathcal{D}$ to be the distribution of the sum of d independent elements sampled from D, and $\mathcal{D} - \mathcal{D}$ to be the distribution of the difference of two independent elements sampled from D.

And now the theorem:

Theorem 2. *Let $\mathcal{D}$ be a distribution over $\mathbb{R}$ such that $\mathcal{D} - \mathcal{D}$ has PSD anti-concentration. Then there exist $C_d, c_d > 0$ such that the following holds. Let $f(\mathbf{x}) = f(x_1, \ldots, x_n)$ be a degree d polynomial, normalized to have $\mathrm{Var}_{(d\mathcal{D})^n}[f] = 1$. Then for any $t \in \mathbb{R}$ and $\varepsilon > 0$,*

$$\mathbb{P}_{\mathbf{x} \sim (d\mathcal{D})^n}[|f(x) - t| \leq \varepsilon] \leq C_d \cdot \varepsilon^{1/c_d}, \tag{17}$$

where $c_d = O\left(d \cdot 2^{O(d)}\right)$.

Under this assumption, which is clearly satisfied for i.i.d. Gaussian matrices, on

$$E = \{e_{\alpha,i}, e_{\beta,j}, e_{\gamma,k}, \quad \alpha, \beta, \gamma \in [\![1, n]\!], \quad i, j, k[\![1, m]\!]\}, \tag{18}$$

we can apply (17).

3   Estimation of the Signal Decomposition Using Median of Means (MoM)

We now turn to the problem of estimating the Signature coefficients.

For this purpose, we will first consider the general problem of estimating the expectation $\mu_P = P[X]$ of a distribution P from the observation of an i.i.d. sample $\mathcal{D}_N = (X_1, \ldots, X_N)$ of real valued random variables with common distribution P.

In this part, we will note ϵ for the noise of an observation (so $X = X^* + \epsilon$) to avoid confusion with Rademacher variables.

3.1   The MoM Principle

Let K and b such that $N = Kb$ and let $B_1, \ldots, B_K$ denote a partition of $\{1, \ldots, N\}$ into subsets of cardinality b. For any $k \in \{1, \ldots, K\}$, let $P_{B_k} X = b^{-1} \sum_{i \in B_k} X_i$. The MOM estimators of μ_P are defined by

$$\mathrm{MOM}_K[X] \in \mathrm{median}\left\{P_{B_k} X, k \in \{1, \ldots, K\}\right\}.$$

Recall that the Rademacher complexity of a class $\mathcal{F}$ of functions $f : \mathcal{X} \to \mathbb{R}$ is defined by

$$D(\mathcal{F}) = \left(\mathbb{E}\left[\sup_{f \in \mathcal{F}} \left\{ \frac{1}{\sqrt{N}} \sum_{i=1}^{N} \xi_i f(X_i) \right\} \right] \right)^2. \tag{19}$$

where ξ_i, $i = 1, \ldots, N$ are independant ± 1 Rademacher random variables.

Theorem 1. *(Concentration for suprema of MOM processes). Let $\mathcal{F}$ denote a separable set of functions $f : \mathcal{X} \to \mathbb{R}$ such that $\sup_{f \in \mathcal{F}} \sigma^2(f) = \sigma^2 < \infty$, where $\sigma^2(f) = \mathrm{Var}(f(X))$. Then, for any $K \in \{1, \ldots, N/2\}$,*

$$\mathbb{P}\left(\sup_{f \in F} |\mathrm{MOM}_K[f] - Pf| \geq 128\sqrt{\frac{D(\mathcal{F})}{N}} \vee 4\sigma\sqrt{\frac{2K}{N}} \right) \leq e^{-K/32}. \tag{20}$$

3.2   Application to Signal Decomposition

Let $X = X^* + \epsilon$ be a continuous noisy observation of X^*, i.e. corrupted by a continuous noise ϵ on the interval $[0, T]$. and C_Ψ be the third degree signature of an orthogonal basis ψ on $[0, T]$. We will assume that C_Ψ is computable without subsampling; otherwise, we can approximate C_Ψ by subsampling the basis functions in Ψ as well. Given an positive integer ν, and a random set of timestamps T_ν, our goal is to estimate the quantity

$$\mathbb{E}_{T_n, \epsilon}\left[\|S^{(3)}(X^*_{T_\nu} + \epsilon_{T_\nu}) - [\![C_\psi; A, A, A]\!]\|_F^2 \right] \tag{21}$$

In order to put the MoM principle to work, we need n samples of the variable

$$Y = \left\| S^{(3)}(X^*_{T_\nu} + \epsilon_{T_\nu}) - [\![C_\Psi; A, A, A]\!] \right\|^2_F. \tag{22}$$

Let $\{T^{(1)}_\nu, T^{(2)}_\nu, \ldots, T^{(N)}_\nu\}$ be a set of N sets of timestamps with same distribution as T_ν. For all $i = 1, \ldots, N$, define

$$Y^{(i)} = \left\| S^{(3)}(X^*_{T^{(i)}_\nu} + \epsilon_{T^{(i)}_\nu}) - [\![C_\Psi; A, A, A]\!] \right\|^2_F. \tag{23}$$

Notice that the vectors $\epsilon^{(i)}$, $i = 1, \ldots, N$ with

$$\epsilon^{(i)} = \begin{bmatrix} \epsilon^{(i)}_{t^{(i)}_1} \\ \vdots \\ \epsilon^{(i)}_{t^{(i)}_\nu} \end{bmatrix} \tag{24}$$

are i.i.d. and therefore, the signals

$$X^*_{T^{(i)}_\nu} + \epsilon_{T^{(i)}_\nu}, \tag{25}$$

$i = 1, \ldots, N$ are i.i.d. as well. Based on these assumptions, we can use the MoM approach to estimate the expectation (22) Instead of estimating (22) using the mean of $Y^{(i)}$, $i = 1, \ldots, N$, we will turn to the Median of Means technique. Let $N = Kb$, Writing $P_{B_k}Y = \frac{1}{b} \sum_{i \in B_k} Y^{(i)}$:

$$MOM(Y) = \text{median}\{P_{B_k}Y, k \in \{1, \ldots, K\}\} \tag{26}$$

$$= \text{median}\left\{ \frac{1}{b} \sum_{i \in B_k} \left\| S^{(3)}(X^*_{T^{(i)}_\nu} + \epsilon_{T^{(i)}_\nu}) - [\![C_\Psi; A, A, A]\!] \right\|^2_F, k \in \{1, \ldots, K\} \right\}$$

In order to apply Theorem 1, we need to compute $D(\mathcal{F})$, where

$$\mathcal{F} = \left\{ (\epsilon, T) \mapsto \|S^{(3)}(X^*_T + \epsilon)\|^2_F - 2\left\langle S^{(3)}(X^*_T + \epsilon), [\![C_\Psi; A, A, A]\!]\right\rangle \right.$$

$$\left. + \|[\![C_\Psi; A, A, A]\!]\|^2_F \right\}. \tag{27}$$

Thus, we have

$$\mathbb{E}\left[\sup_{f \in \mathcal{F}} \left\{ \frac{1}{\sqrt{N}} \sum_{i=1}^N \xi_i f(X_i) \right\} \right] \leq \mathbb{E}\left[\sup_{A \in \mathbb{R}^{n \times m}, b \in \mathbb{R}^n} \left\{ \frac{1}{\sqrt{N}} \sum_{i=1}^N \xi_i \left(\|S^{(3)}(b + \epsilon)\|^2_F \right.\right.\right.$$

$$\left.\left.\left. - 2\left\langle S^{(3)}(b + \epsilon), [\![C_\Psi; A, A, A]\!]\right\rangle + \|[\![C_\Psi; A, A, A]\!]\|^2_F \right) \right\} \right] \tag{28}$$

conditioning on ϵ gives

$$
\mathbb{E}\left[\sup_{f \in \mathcal{F}}\left\{\frac{1}{\sqrt{N}}\sum_{i=1}^{N}\xi_i f(X_i)\right\}\right] \leq \mathbb{E}\left[\mathbb{E}\left[\sup_{A \in \mathbb{R}^{n \times m}, b \in \mathbb{R}^n}\left\{\frac{1}{\sqrt{N}}\sum_{i=1}^{N}\xi_i\left(\|S^{(3)}(b+\epsilon)\|_F^2\right.\right.\right.\right.
$$
$$
\left.\left.\left.\left. - 2\left\langle S^{(3)}(b+\epsilon), [\![C_\Psi; A, A, A]\!]\right\rangle + \|[\![C_\Psi; A, A, A]\!]\|_F^2\right)\right\}\right]\mid\overline{\epsilon}\right]
$$
$$
= \mathbb{E}[\mathcal{R}_\epsilon(\mathcal{F}) \mid \epsilon] \tag{29}
$$

In (29), $\mathcal{R}_\epsilon(\mathcal{F})$ is the empirical Rademacher complexity. In order to bound this quantity, it is important to note that (22) is the squared Frobenius norm of a matrix where every coefficient is a degree 3 polynomial of ϵ. Hence, the final quantity is a degree 6 polynomial in the variable ϵ. Rademacher complexity of polynomials has been addressed in earlier sources such as, e.g. [9].

At this point, we are referring to [3] to "convert" a polynomial function to a polynomial network and [12] for the Rademacher complexity.

Consider $\mathcal{E} = (\epsilon_1, \ldots, \epsilon_n)$ a set of n i.i.d. samples from the same distribution as $\epsilon \in \mathbb{R}^d$. Let $\mathcal{C}$ denote the event

$$
\mathcal{C} = \{\|\epsilon_k\|_\infty \leq 1, k \in \{1, \ldots, n\}\} \tag{30}
$$

Then, it is proved in [12] that there exist two constants μ and λ such that

$$
\mathcal{R}_\mathcal{E}(\mathcal{F}) \leq 2\mu\lambda\sqrt{\frac{12\log(d)}{n}} \tag{31}
$$

on $\mathcal{C}$. It follows from (29) that

$$
D(\mathcal{F}) \leq \mathbb{E}\left[\mathbb{E}[\mathcal{R}_\mathcal{E}(\mathcal{F}) \mid \mathcal{E}]\right]
$$
$$
\leq \mathbb{E}\left[\mathbb{E}[\mathcal{R}_\mathcal{E}(\mathcal{F}) \mid \mathcal{E}] \mid \mathcal{C}\right]\mathbb{P}(\mathcal{C}) + \mathbb{E}\left[\mathbb{E}[\mathcal{R}_\mathcal{E}(\mathcal{F}) \mid \mathcal{E}] \mid \overline{\mathcal{C}}\right]\mathbb{P}(\overline{\mathcal{C}})
$$
$$
\leq 2\mu\lambda\sqrt{\frac{12\log(d)}{n}}\mathbb{P}(\mathcal{C}) + \frac{c}{\sqrt{n}} \tag{32}
$$

where we used that $\sqrt{n}\mathbb{E}\left[\mathbb{E}[\mathcal{R}_\mathcal{E}(\mathcal{F}) \mid \mathcal{E}] \mid \overline{\mathcal{C}}\right]\mathbb{P}(\overline{\mathcal{C}})$ can be shown to be bounded by a constant using a peeling argument. Combining (1), (32), we obtain the following result

Theorem 2. *Let Y be defined by (22), $Y^{(I)}$, $i = 1, \ldots, N$ be defined by (23), and the MoM estimator be defined by (26). Then, we have*

$$
\mathbb{P}\left(\sup_{Y \in \mathcal{F}}|\mathrm{MOM}_K[Y] - PY| \geq 128\sqrt{\frac{2\mu\lambda\sqrt{\frac{12\log(d)}{n}}\mathbb{P}(\mathcal{C}) + \frac{c}{\sqrt{n}}}{N}} \vee 4\sigma\sqrt{\frac{2K}{N}}\right)
$$
$$
\leq e^{-K/32}. \tag{33}
$$

Complete proof details will be provided in an extended version of the paper.

References

1. Chen, K.T.: Iterated integrals and exponential homomorphisms. Proc. Lond. Math. Soc. **3**(1), 502–512 (1954)
2. Chevyrev, I., Kormilitzin, A.: A primer on the signature method in machine learning. arXiv preprint arXiv:1603.03788 (2016)
3. Chrysos, G.G., Moschoglou, S., Bouritsas, G., Panagakis, Y., Deng, J., Zafeiriou, S.: P-nets: deep polynomial neural networks. In: Proceedings of the IEEE/CVF Conference on Computer Vision and Pattern Recognition, pp. 7325–7335 (2020)
4. Johnstone, I.M.: Function estimation and gaussian sequence models, vol. 2, no. 5.3, p. 2 (2002, unpublished)
5. Lovett, S.: An elementary proof of anti-concentration of polynomials in gaussian variables. In: Electronic Colloquium on Computational Complexity, vol. 17, p. 182 (2010)
6. Lyons, T., McLeod, A.D.: Signature methods in machine learning. arXiv preprint arXiv:2206.14674 (2022)
7. Lyons, T.J.: Differential equations driven by rough signals. Revista Matemática Iberoamericana **14**(2), 215–310 (1998)
8. Meka, R., Nguyen, O., Vu, V.: Anti-concentration for polynomials of independent random variables. arXiv preprint arXiv:1507.00829 (2015)
9. Mohri, M., Rostamizadeh, A., Talwalkar, A.: Foundations of Machine Learning. MIT Press, Cambridge (2018)
10. Schell, A., Oberhauser, H.: Nonlinear independent component analysis for discrete-time and continuous-time signals. Ann. Stat. **51**(2), 487–518 (2023)
11. Seigal, A.L.: Structured tensors and the geometry of data. University of California, Berkeley (2019)
12. Zhu, Z., Latorre, F., Chrysos, G.G., Cevher, V.: Controlling the complexity and lipschitz constant improves polynomial nets. arXiv preprint arXiv:2202.05068 (2022)

Learning Lagrangian Fluid Mechanics with E(3)-Equivariant Graph Neural Networks

Artur P. Toshev[1]([✉]), Gianluca Galletti[1], Johannes Brandstetter[2], Stefan Adami[1], and Nikolaus A. Adams[1]

[1] School of Engineering and Design, Chair of Aerodynamics and Fluid Mechanics, Technical University of Munich, Garching, Germany
`artur.toshev@tum.de, g.galletti@tum.de`
[2] Microsoft Research AI4Science, Amsterdam, The Netherlands

Abstract. We contribute to the vastly growing field of machine learning for engineering systems by demonstrating that equivariant graph neural networks have the potential to learn more accurate dynamic-interaction models than their non-equivariant counterparts. We benchmark two well-studied fluid-flow systems, namely 3D decaying Taylor-Green vortex and 3D reverse Poiseuille flow, and evaluate the models based on different performance measures, such as kinetic energy or Sinkhorn distance. In addition, we investigate different embedding methods of physical-information histories for equivariant models. We find that while currently being rather slow to train and evaluate, equivariant models with our proposed history embeddings learn more accurate physical interactions.

Keywords: Graph Neural Networks · Equivariance · Fluid mechanics · Lagrangian Methods · Smoothed Particle Hydrodynamics

1 Particle-Based Fluid Mechanics

The Navier-Stokes equations (NSE) are omnipresent in fluid mechanics. However, for the majority of problems, solutions are analytically intractable, and obtaining accurate solutions necessitates numerical approximations. Those can be split into two categories: grid/mesh-based (Eulerian description) and particle-based (Lagrangian description).

Smoothed Particle Hydrodynamics. In this work, we investigate Lagrangian methods, more precisely the Smoothed Particle Hydrodynamics (SPH) approach, which was independently developed by [16] and [23] to simulate astrophysical systems. Since then, SPH has established itself as the preferred approach in various applications ranging from free-surface flows such as ocean waves [36]

Our code will be released under https://github.com/tumaer/sph-hae.

© The Author(s), under exclusive license to Springer Nature Switzerland AG 2023
F. Nielsen and F. Barbaresco (Eds.): GSI 2023, LNCS 14072, pp. 332–341, 2023.
https://doi.org/10.1007/978-3-031-38299-4_35

to selective laser melting in additive manufacturing [40]. The main idea behind SPH is to represent fluid properties at discrete points in space and to use truncated radial interpolation kernel functions to approximate them at any arbitrary location. The kernel functions can be interpreted as state-statistics estimators which define continuum-scale interactions between particles. The justification for truncating the kernel support is the assumption of the locality of interactions between particles. The resulting discretized equations are integrated in time using numerical integration techniques such as the symplectic Euler scheme, by which the particle positions are updated.

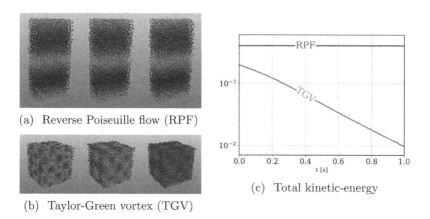

(a)   Reverse Poiseuille flow (RPF)

(b)   Taylor-Green vortex (TGV)

(c)   Total kinetic-energy

Fig. 1. Time snapshots of x-velocity of reverse Poiseuille flow (a), velocity magnitude of Taylor-Green vortex flow (b), and kinetic-energy evolution (c).

To generate training data for our machine learning tasks, we implemented our own fully-differentiable SPH solver in JAX [6] based on the transport velocity formulation of SPH by [1], which achieves homogeneous particle distributions over the domain. We then selected two flow cases, which have been extensively studied in fluid mechanics: 3D Taylor-Green vortex and 3D reverse Poiseuille flow. We expect to open-source the datasets in the near future.

Taylor-Green Vortex. The Taylor-Green vortex system (TGV, see Fig. 1(a)) was introduced by Taylor & Green in 1937 to study turbulence [34]. We investigate the TGV with Reynolds number of Re = 100, which is neither laminar nor turbulent, i.e. there is no layering of the flow (typical for laminar flows), but also the small scales caused by vortex stretching do not lead to a fully developed energy cascade (typical for turbulent flows) [5]. We compute the Reynolds number $Re = UL/\eta$ as in [1] with domain size $L = 1$, reference velocity $U = 1$, and dynamic viscosity $\eta = 0.01$. We note that this setup differs from the one in [5], where the domain is $L = 2\pi$. We use the initial velocity field from [5]:

$$u = \sin(kx)\cos(ky)\cos(kz) \,, \tag{1a}$$
$$v = -\cos(kx)\sin(ky)\cos(kz) \,, \tag{1b}$$
$$w = 0 \,, \tag{1c}$$

where $k = 2\pi/L$. The TGV dataset used in this work consists of 80/10/10 trajectories for training/validation/testing, where each trajectory comprises 8000 particles. Each trajectory spans 1s physical time and was simulated with $dt = 0.001$s starting from a random initial particle distribution. We choose to train the learned solver on 10x larger time steps, i.e. temporal coarsening, which we implement by subsampling every 10th frame resulting in 100 samples per trajectory.

Reverse Poiseuille Flow. The Poiseuille flow, i.e. laminar channel flow, is another well-studied fluid mechanics problem. However, the channel flow requires the treatment of wall-boundary conditions, which is beyond the focus of the current work. Therefore, in this work, we consider data obtained by reverse Poiseuille flow (RPF, see Fig. 1(b)) [11], which essentially consists of two opposing streams in a fully periodic domain. In terms of the SPH implementation, the flow is exposed to opposite force fields, i.e. the upper and lower half are accelerated in negative x direction and positive x direction, respectively. Here we also choose to work with Re = 100, in which case the flow is not purely laminar and there is no analytical solution for the velocity profile. The domain has size 1/2/0.5 in x/y/z directions (width, height, depth), and for the computation of the Reynolds number $Re = UL/\eta$ we use $U = 1$, $L = 1$, $\eta = 0.01$.

Due to the statistically stationary [27] solution of the flow, the RPF dataset consists of one long trajectory spanning 100s. The flow field is again discretized by 8000 particles and simulated with $dt = 0.001$, followed by subsampling at every 10th step. Thus, we again aim to train models to perform temporal coarsening. The resulting number of training/validation/testing instances is the same as for TGV, namely 8000/1000/1000.

2   (Equivariant) Graph Network-Based Simulators

We first formalize the task of autoregressive prediction of the next state of a Lagrangian flow field based on the notation from [31]. If X^t denotes the state of a particle system at time t, one full trajectory of $K + 1$ steps can be written as $\mathbf{X}^{t_0:K} = (\mathbf{X}^{t_0}, \dots, \mathbf{X}^{t_K})$. Each state $\mathbf{X}^t$ is made up of N particles, namely $\mathbf{X}^t = (\mathbf{x}_1^t, \mathbf{x}_2^t, \dots \mathbf{x}_N^t)$, where each $\mathbf{x}_i$ is the state vector of the i-th particle. However, the inputs to the learned simulator can span multiple time instances. Each node $\mathbf{x}_i^t$ can contain node-level information like the current position $\mathbf{p}_i^t$ and a time sequence of H previous velocity vectors $\dot{\mathbf{p}}^{t_k - H:k}$, as well as global features like the external force vector $\mathbf{F}_i$ in the reverse Poiseuille flow. To build the connectivity graph, we use an interaction radius of ~ 1.5 times the average interparticle distance, which results in around 10–20 one-hop neighbors.

Graph Network-Based Simulator. The GNS framework [31] is one of the most popular learned surrogates for engineering particle-based simulations. The main idea of the GNS model is to use the established encoder-processor-decoder architecture [3] with a processor that stacks several message passing layers [15]. One major strength of the GNS model lies in its simplicity given that all its building blocks are regular MLPs. However, the performance of GNS when predicting long trajectories strongly depends on the choice of Gaussian noise to perturb the input data. Additionally, GNS and other non-equivariant models are less data-efficient [4]. For these reasons, we implement and tune GNS as a comparison baseline, and employ it as an inspiration for which setup, features, and hyperparameters to use for equivariant models.

Steerable E(3)-equivariant Graph Neural Network. SEGNNs [8] are an instance of E(3)-equivariant GNNs, i.e. GNNs that are equivariant with respect to isometries of the Euclidean space (rotations, translations, and reflections). Most E(3)-equivariant GNNs tailored for prediction of molecular properties [2, 4, 35] parametrize Clebsch-Gordan tensor products using a learned embedding of pairwise distances. In contrast, the SEGNN model uses general steerable node and edge attributes ($\hat{\mathbf{a}}_i$ and $\hat{\mathbf{a}}_{ij}$ respectively) to condition the layers directly. In particular, SEGNNs introduce the concept of steerable MLPs, which are linear Clebsch-Gordan tensor products $\otimes_{CG}^{\mathcal{W}_{\hat{\mathbf{a}}}}$ parametrized by learnable parameters $\mathcal{W}_{\hat{\mathbf{a}}}$ interleaved with gated non-linearities σ [38]. Namely, the hidden state $\hat{\mathbf{f}}$ at layer $l+1$ is updated as

$$\hat{\mathbf{f}}^{l+1} := \sigma(\mathcal{W}_{\hat{\mathbf{a}}}\hat{\mathbf{f}}^l) \quad \text{with} \quad \mathcal{W}_{\hat{\mathbf{a}}}\hat{\mathbf{f}} := \hat{\mathbf{f}} \otimes_{CG}^{\mathcal{W}_{\hat{\mathbf{a}}}} \hat{\mathbf{a}} \ . \tag{2}$$

Due to these design choices, SEGNNs are well suited for a wide range of engineering problems, where various vector-valued features need to be modeled in an E(3) equivariant way. In practice, SEGNNs extend upon the message passing paradigm [15] using steerable MLPs of Eq. (2) for both message $M_{\hat{\mathbf{a}}_{ij}}$ and node update functions $U_{\hat{\mathbf{a}}_i}$. The i-th node steerable features $\hat{\mathbf{f}}_i$ are updated as

$$\hat{\mathbf{m}}_{ij} = M_{\hat{\mathbf{a}}_{ij}} \left(\hat{\mathbf{f}}_i, \hat{\mathbf{f}}_j, \|x_i - x_j\|^2 \right) , \tag{3}$$

$$\hat{\mathbf{f}}_i' = U_{\hat{\mathbf{a}}_i} \left(\hat{\mathbf{f}}_i, \sum_{j \in \mathcal{N}(i)} \hat{\mathbf{m}}_{ij} \right) , \tag{4}$$

where $\mathcal{N}(i)$ is the neighborhood of node i. In Eq. (3), $M_{\hat{\mathbf{a}}_{ij}}$ has the subscript $\hat{\mathbf{a}}_{ij}$ because it is conditioned on the edge attributes, whereas $U_{\hat{\mathbf{a}}_i}$ is conditioned on the node attributes $\hat{\mathbf{a}}_i$.

Historical Attribute Embedding (HAE). Finding physically meaningful edge and node attributes is crucial for good performance since every Clebsch-Gordan tensor product is conditioned on them. For the problems at hand, we empirically found that a strong choice for steerable attributes is

$$\hat{\mathbf{a}}_{ij} = Y\left(\mathbf{p}_i - \mathbf{p}_j\right), \tag{5}$$

$$\hat{\mathbf{a}}_i = A_{\mathcal{W}}\left(Y\left(\dot{\mathbf{p}}_i^{(1:H)}\right) + \sum_{j \in \mathcal{N}(i)} \hat{\mathbf{a}}_{ij}\right) = A_{\mathcal{W}}\left(\hat{\mathbf{a}}_i^{(1:H)}\right), \tag{6}$$

where $\hat{\mathbf{a}}_i$ and $\hat{\mathbf{a}}_{ij}$ are the node and edge attributes respectively, $Y_m^{(l)} : S^2 \to \mathbb{R}$ is the spherical harmonics embedding, and $A_{\mathcal{W}}$ is a function parameterized by $\mathcal{W}$ that embeds the historical node attributes $\hat{\mathbf{a}}_i^{(h)}$. In particular, we investigate the averaging $A_{\mathcal{W},avg}$, weighted averaging $A_{\mathcal{W},lin}$, and steerable MLP embedding conditioned on the most recent velocity $A_{\mathcal{W},\otimes}$.

$$A_{\mathcal{W},avg} := \frac{1}{H}\sum_h \hat{\mathbf{a}}_i^{(h)}, \tag{7}$$

$$A_{\mathcal{W},lin} := \sum_h w_h \hat{\mathbf{a}}_i^{(h)}, \tag{8}$$

$$A_{\mathcal{W},\otimes} := \sigma\left(\hat{\mathbf{a}}_i^{(1:H)} \otimes_{CG}^{\mathcal{W}} \hat{\mathbf{a}}_i^H\right). \tag{9}$$

Figure 2 sketches the HAE-SEGNN architecture. Subfigure 2a connects past particle positions $\mathbf{p}^{(1:H)}$ and their embeddings $A_{\mathcal{W}}^n$ within the updated architecture, whereas Subfig. 2b shows the effect of $A_{\mathcal{W},avg}$ and $A_{\mathcal{W},lin}$.

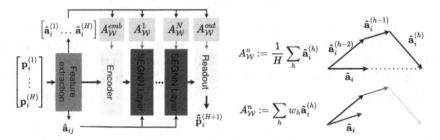

(a) Layer-wise attribute embeddings   (b) Examples of embedding effects

Fig. 2. SEGNN architecture with Historical Attribute Embedding. (a): $A_{\mathcal{W}}^n$ is a learnable embedding of previous velocities for node attributes $\hat{\mathbf{a}}_i^{(h)}$. (b): effects of $A_{\mathcal{W}}^n$ as arithmetic mean (top) and weighted mean of past attributes (bottom).

We found that initializing the embedding weights for $A_{\mathcal{W},lin}$ and $A_{\mathcal{W},\otimes}$ with $N(\mu = \frac{1}{\#\mathcal{W}}, \sigma = \frac{1}{\sqrt{fan_{in}}})$ (shifted initialization to resemble the average $A_{\mathcal{W},avg}$) makes training quicker and also slightly improves final performance. This behavior reiterates the significance of attributes in conditioning the architecture, and the inclusion of potentially less relevant attributes can lead to a substantial decrease in performance.

We currently don't have a systematic way of finding the attributes, but we found that in engineering systems having physical features (such as velocity and force) in the attributes has a positive effect, as one could see it as conditioning the network on system dynamics. Exploring an algorithmic framework for finding attributes in the broader context was not investigated and is left to future work.

Related Work. Related steerable E(3)-equivariant GNNs, such as [2,4,25,35] are mostly tailored towards molecular property prediction tasks, and thus restrict the parametrization of tensor products to an MLP-embedding of pairwise distances. This is a reasonable design choice since distances are crucial information for molecules, but not straightforward to adapt to fluid dynamics problems where prevalent quantities are e.g. force and momentum. Another family of E(3)-equivariant GNNs are models that use invariant quantities, such as distances and angles [12,13,32]. Although these models have an advantage concerning runtimes since no Clebsch-Gordan tensor product is needed, they cannot a priori model vector-valued information in an E(3) equivariant way. On a slightly more distant note, there has been a rapid raise in physics-informed neural networks (PINNs) [28] and equivariant counterparts thereof [19], as well as operator learning [17,20–22], where functions or surrogates are learned in an Eulerian (grid-based) way. Furthermore, equivariant models have been applied to grid-based data [19,37] utilizing group-equivariant CNNs [10,39]. Recently, Clifford algebra-based layers [7] have been proposed on grids as well as graph-structured data [29,30], but exploring their performance on SPH data is left to future work. Non-equivariant deep learning surrogates for Lagrangian dynamics were introduced for particles [31], meshes [26], and within complex geometries [24].

3   Results

The task we train on is the autoregressive prediction of accelerations $\ddot{\mathbf{p}}$ given the current position $\mathbf{p}_i$ and $H = 5$ past velocities of the particles $\dot{\mathbf{p}}_i^{(1:H)}$. The influence of the choice of H is discussed in detail in the supplementary materials to [31] and we use the same value as suggested in this paper. For training SEGNNs, we verified that adding Gaussian noise to the inputs [31] indeed significantly improves performance. In addition, we train both models by employing the pushforward trick [9] with up to five pushforward steps and an exponentially decaying probability with regard to the number of steps. We measured the performance of the GNS and the SEGNN models in four aspects when evaluating on the test datasets:

1. *Mean-squared error* (MSE) of particle positions MSE_p when rolling out a trajectory over 100 time steps (1 physical second for both flow cases). This is also the validation loss during training.
2. *Sinkhorn distance* as an optimal transport distance measure between particle distributions. Lower values indicate that the particle distribution is closer to the reference one.
3. *Kinetic energy* E_{kin} $(= 0.5mv^2)$ as a global measure of physical behavior.

Table 1. Performance measures on the Taylor-Green vortex and reverse Poiseuille flow. The Sinkhorn distance is averaged over test rollouts.

	Taylor-Green vortex			reverse Poiseuille flow		
	MSE_p	$MSE_{E_{kin}}$	Sinkhorn	MSE_p	$MSE_{E_{kin}}$	Sinkhorn
GNS	6.7e-6	7.1e-3	1.2e-7	1.4e-6	2.2e-2	4.1e-7
$SEGNN_{avg}$	1.6e-6	8.4e-3	2.9e-8	1.4e-6	8.2e-3	1.4e-7
$SEGNN_{lin}$	**1.4e-6**	**3.1e-4**	2.0e-8	**1.3e-6**	2.0e-2	1.2e-7
$SEGNN_\otimes$	**1.4e-6**	1.9e-3	**1.6e-8**	**1.3e-6**	**9.4e-4**	**9.1e-8**

Performance comparisons are summarized in Table 1. GNS and SEGNN have 1.2 M and 360 k parameters respectively for both Taylor-Green and reverse Poiseuille (both have 10 layers, but 128 vs 64-dim features). For all SEGNN models, we used maximum spherical harmonics order $l_{max} = 1$ attributes as well as features; we found that in our particular case, higher orders become computationally unfeasible to train and evaluate. With regards to runtime for 8000 particles dynamics, the GNS model takes around 35 ms per step, all SEGNN models take roughly 150 ms per step, and the original SPH solver takes around 100 ms per 10 steps (note: we learn to predict every 10th step). Both our SPH solver as well as GNS and SEGNN models are implemented in the Python library JAX [6], and both use the same neighbors-search implementation from the JAX-MD library [33], making for a fair runtime comparison. It is known that steerable equivariant models are slower than non-equivariant ones, which is related to how the Clebsch-Gordan tensor product is implemented on accelerators like GPUs. However, we observed that equivariant models reach their peak performance with fewer parameters, and they often significantly outperform GNS, especially when measuring physics quantities like kinetic energy or Sinkhorn distances.

Taylor-Green Vortex. One of the major challenges of the Taylor-Green dataset is the varying input and output scales throughout a trajectory, in our case by up to one order of magnitude. This results in the larger importance of initial frames in the loss even after data normalization. Figure 3 (top) summarizes the performance properties of the Taylor-Green vortex experiment. Both models are able to match the ground truth kinetic energy. However, all SEGNN models achieve 20 times lower $MSE_{E_{kin}}$ errors, and regarding MSE_p, GNS predictions drift away from the reference SPH trajectory much earlier.

Reverse Poiseuille Flow. The challenge of the reverse Poiseuille case lies in the different velocity scales between the main flow direction (x-axis) and the y and z components of the velocity. In contrast to GNS, whose inputs we can normalize with the direction-dependent dataset statistics, this breaks equivariance and we are forced to normalize the SEGNN inputs only in magnitude. Although such unbalanced velocities are used as inputs, target accelerations in x-, y-, and z-direction all have similar distributions. This, combined with temporal coarsening makes the problem sensitive to input deviations. Additionally,

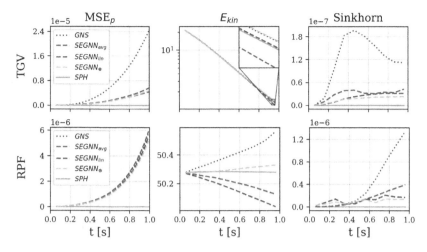

Fig. 3. Evolution of performance measures over time on the Taylor-Green vortex (top) and reverse Poiseuille flow (bottom).

including the external force vector $\mathbf{F}_i$ to either the node features or SEGNN attributes has a positive impact on the results. Figure 3 (bottom) shows that SEGNNs reproduce the particle distributions quite well, whereas GNS show signs of particle-clustering artifacts, leading to a much larger Sinkhorn distance.

4   Future Work

In this work, we demonstrate that equivariant models are well suited to capture the underlying physical properties of particle-based fluid mechanics systems. We found that conditioning on physical quantities through our tensor product historical embedding increases expressive power at almost no additional cost. Moreover, employing more recent training strategies, such as the pushforward trick, has proven to be helpful in stabilizing training and improving performance. Finally, selecting suitable (physical) performance measures different than plain MSE errors is crucial for assessing and improving deep learning models.

Interesting directions for future work include accelerating the inference time of equivariant GNNs as well as developing more specialized and expressive equivariant building blocks. We conjecture that together with such extensions, equivariant models offer a promising direction to tackle some of the long-standing problems in fluid mechanics, such as the learning of coarse-grained representations of turbulent flow problems, e.g. Taylor-Green [5], or learning the multi-resolution dynamics of NSE problems [18].

Acknowledgements. We are thankful to the developers of the `e3nn-jax` library [14], which offers efficient implementations of E(3)-equivariant building blocks.

References

1. Adami, S., Hu, X., Adams, N.A.: A transport-velocity formulation for smoothed particle hydrodynamics. J. Comput. Phys. **241**, 292–307 (2013)
2. Batatia, I., Kovács, D.P., Simm, G.N.C., Ortner, C., Csányi, G.: Mace: higher order equivariant message passing neural networks for fast and accurate force fields (2022)
3. Battaglia, P.W., et al.: Relational inductive biases, deep learning, and graph networks. arXiv (2018)
4. Batzner, S., et al.: E(3)-equivariant graph neural networks for data-efficient and accurate interatomic potentials. Nat. Commun. **13**(1), 2453 (2022)
5. Brachet, M.E., Meiron, D., Orszag, S., Nickel, B., Morf, R., Frisch, U.: The taylor-green vortex and fully developed turbulence. J. Stat. Phys. **34**(5–6), 1049–1063 (1984)
6. Bradbury, J., et al.: JAX: composable transformations of Python+NumPy programs (2018)
7. Brandstetter, J., Berg, R.V.D., Welling, M., Gupta, J.K.: Clifford neural layers for PDE modeling. arXiv preprint arXiv:2209.04934 (2022)
8. Brandstetter, J., Hesselink, R., van der Pol, E., Bekkers, E.J., Welling, M.: Geometric and physical quantities improve E(3) equivariant message passing. In: ICLR (2022)
9. Brandstetter, J., Worrall, D.E., Welling, M.: Message passing neural PDE solvers. In: ICLR (2022)
10. Cohen, T.S., Welling, M.: Group equivariant convolutional networks. In: Proceedings of the 33rd ICML, ICML 2016, vol. 48, pp. 2990–2999. JMLR.org (2016)
11. Fedosov, D.A., Caswell, B., Em Karniadakis, G.: Reverse poiseuille flow: the numerical viscometer. In: AIP Conference Proceedings, vol. 1027, pp. 1432–1434. American Institute of Physics (2008)
12. Gasteiger, J., Becker, F., Günnemann, S.: Gemnet: universal directional graph neural networks for molecules. NeurIPS **34**, 6790–6802 (2021)
13. Gasteiger, J., Groß, J., Günnemann, S.: Directional message passing for molecular graphs. In: ICLR (2020)
14. Geiger, M., Smidt, T.: e3nn: Euclidean neural networks (2022)
15. Gilmer, J., Schoenholz, S.S., Riley, P.F., Vinyals, O., Dahl, G.E.: Neural message passing for quantum chemistry. In: ICML, pp. 1263–1272. PMLR (2017)
16. Gingold, R.A., Monaghan, J.J.: Smoothed particle hydrodynamics: theory and application to non-spherical stars. Mon. Not. R. Astron. Soc. **181**(3), 375–389 (1977)
17. Gupta, J.K., Brandstetter, J.: Towards multi-spatiotemporal-scale generalized PDE modeling. arXiv preprint arXiv:2209.15616 (2022)
18. Hu, W., Pan, W., Rakhsha, M., Tian, Q., Hu, H., Negrut, D.: A consistent multi-resolution smoothed particle hydrodynamics method. Comput. Methods Appl. Mech. Eng. **324**, 278–299 (2017)
19. Lagrave, P.Y., Tron, E.: Equivariant neural networks and differential invariants theory for solving partial differential equations. In: Physical Sciences Forum, vol. 5, p. 13. MDPI (2022)
20. Li, Z., et al.: Neural operator: graph kernel network for partial differential equations. arXiv preprint arXiv:2003.03485 (2020)
21. Li, Z., et al.: Fourier neural operator for parametric partial differential equations. In: ICLR (2021)

22. Lu, L., Jin, P., Karniadakis, G.E.: DeepONet: learning nonlinear operators for identifying differential equations based on the universal approximation theorem of operators. arXiv preprint arXiv:1910.03193 (2019)
23. Lucy, L.B.: A numerical approach to the testing of the fission hypothesis. Astron. J. **82**, 1013–1024 (1977)
24. Mayr, A., Lehner, S., Mayrhofer, A., Kloss, C., Hochreiter, S., Brandstetter, J.: Boundary graph neural networks for 3D simulations. arXiv preprint arXiv:2106.11299 (2021)
25. Pezzicoli, F.S., Charpiat, G., Landes, F.P.: Se (3)-equivariant graph neural networks for learning glassy liquids representations. arXiv:2211.03226 (2022)
26. Pfaff, T., Fortunato, M., Sanchez-Gonzalez, A., Battaglia, P.W.: Learning mesh-based simulation with graph networks. arXiv preprint arXiv:2010.03409 (2020)
27. Pope, S.B.: Turbulent Flows. Cambridge University Press, Cambridge (2000)
28. Raissi, M., Perdikaris, P., Karniadakis, G.E.: Physics-informed neural networks: a deep learning framework for solving forward and inverse problems involving nonlinear partial differential equations. J. Comput. Phys. **378**, 686–707 (2019)
29. Ruhe, D., Brandstetter, J., Forré, P.: Clifford group equivariant neural networks. arXiv preprint arXiv:2305.11141 (2023)
30. Ruhe, D., Gupta, J.K., de Keninck, S., Welling, M., Brandstetter, J.: Geometric clifford algebra networks. arXiv preprint arXiv:2302.06594 (2023)
31. Sanchez-Gonzalez, A., Godwin, J., Pfaff, T., Ying, R., Leskovec, J., Battaglia, P.: Learning to simulate complex physics with graph networks. In: ICML, pp. 8459–8468. PMLR (2020)
32. Satorras, V.G., Hoogeboom, E., Welling, M.: E(n) equivariant graph neural networks. In: ICML, pp. 9323–9332. PMLR (2021)
33. Schoenholz, S.S., Cubuk, E.D.: JAX M.D. a framework for differentiable physics. In: NeurIPS, vol. 33. Curran Associates, Inc. (2020)
34. Taylor, G.I., Green, A.E.: Mechanism of the production of small eddies from large ones. Proc. Roy. Soc. London Ser. A Math. Phys. Sci. **158**(895), 499–521 (1937)
35. Thomas, N., et al.: Tensor field networks: rotation- and translation-equivariant neural networks for 3D point clouds. CoRR abs/1802.08219 (2018)
36. Violeau, D., Rogers, B.D.: Smoothed particle hydrodynamics (SPH) for free-surface flows: past, present and future. J. Hydraul. Res. **54**(1), 1–26 (2016)
37. Wang, R., Walters, R., Yu, R.: Incorporating symmetry into deep dynamics models for improved generalization. In: ICLR (2021)
38. Weiler, M., Geiger, M., Welling, M., Boomsma, W., Cohen, T.S.: 3D steerable CNNs: learning rotationally equivariant features in volumetric data. In: Bengio, S., Wallach, H., Larochelle, H., Grauman, K., Cesa-Bianchi, N., Garnett, R. (eds.) NeurIPS, vol. 31. Curran Associates, Inc. (2018)
39. Weiler, M., Geiger, M., Welling, M., Boomsma, W., Cohen, T.S.: 3D steerable CNNs: learning rotationally equivariant features in volumetric data. In: Advances in Neural Information Processing Systems, vol. 31 (2018)
40. Weirather, J., et al.: A smoothed particle hydrodynamics model for laser beam melting of NI-based alloy 718. CMA **78**(7), 2377–2394 (2019)

Forward and Inverse Approximation Theory for Linear Temporal Convolutional Networks

Haotian Jiang[1] and Qianxiao Li[1,2]

[1] Department of Mathematics, National University of Singapore, Singapore,
Singapore
qianxiao@nus.edu.sg
[2] Institute for Functional Intelligent Materials, National University of Singapore,
Singapore, Singapore

Abstract. We present a theoretical analysis of the approximation properties of convolutional architectures when applied to the modeling of temporal sequences. Specifically, we prove an approximation rate estimate (Jackson-type result) and an inverse approximation theorem (Bernstein-type result), which together provide a comprehensive characterization of the types of sequential relationships that can be efficiently captured by a temporal convolutional architecture. The rate estimate improves upon a previous result via the introduction of a refined complexity measure, whereas the inverse approximation theorem is new.

Keywords: Approximation · Temporal convolutional neural networks · Sequence Modeling

1 Introduction

Although convolutional neural networks (CNNs) are commonly used for image inputs, they are shown to be effective in tackling a range of temporal sequence modeling problems compared to the traditional recurrent neural networks [2,13]. For example, WaveNet [11] is a convolution-based model for generating audio from text. It utilizes a multilayer dilated convolutional structure, resulting in a filter with a large receptive field. The approximation properties of the dilated convolutional architectures applied to sequence modeling have been studied in [7]. A complexity measure is defined to characterize the types of targets that can be efficiently approximated.

This work aims to enhance prior research and contribute new findings. We formulate the approximation of sequence modeling in a manner that parallels classic function approximation, where we consider three distinct types of approximation results, including universal approximation, approximation rates, and inverse approximation. Our main contributions are summarized as follows: **1.** We refine the complexity measure to make it naturally adapted to the approximation of convolutional architectures. The resulting approximation rate in the forward approximation theorem is tighter compared to results in [7]. **2.** We prove a Bernstein-type inverse approximation result. It states that a target can be effectively approximated only if its complexity measure is small, which presents a converse of the forward approximation theorem.

The readers may refer to [6] for a detailed discussion of related research on the approximation theory for sequence modeling.

2   Mathematical Formulation of Sequence Modeling Problems

This section provides a theoretical formulation of sequence modeling problems and defines the architecture of the dilated convolutional model. Specifically, a temporal sequence is defined as a function $\boldsymbol{x} : \mathcal{I} \to \mathbb{R}^d$ that maps an index set $\mathcal{I}$ to real vectors. The value of the sequence at time index t is denoted by $x(t) \in \mathbb{R}^d$. Additionally, a finitely-supported discrete sequence can also be considered as a matrix, where each column denotes a time step and the total number of columns is referred to as the length of the filter.

In this section, we theoretically formulate the sequence modeling problems and define the dilated convolutional model architecture. A temporal sequence $\boldsymbol{x} : \mathcal{I} \to \mathbb{R}^d$ is regarded as a function from the index set $\mathcal{I}$ to real vectors. The index set can be either $\mathbb{Z}$ or $\mathbb{N}$, where $\mathbb{Z}$ represent the set of integers and $\mathbb{N}$ represent the set of non-negative integers. The value at time index t is denoted by $x(t) \in \mathbb{R}^d$. We may also consider a finitely-supported discrete sequence as a vector. We use $\boldsymbol{x}_{[t_1,t_2]}$ to denote a truncation of $\boldsymbol{x}$ in the interval $[t_1, t_2]$. For a discrete sequence $\boldsymbol{\rho} : \mathbb{N} \to \mathbb{R}^d$, define the radius of $\boldsymbol{\rho}$ by $r(\boldsymbol{\rho}) = \sup\{s : \rho(s) \neq 0\}$. The interval $[0, r(\boldsymbol{\rho})]$ contains non-zero parts of $\boldsymbol{\rho}$, which we refer as its support. We use $|\cdot|$ to denote the Euclidean norm of a vector.

Sequence to Sequence Modeling. In this study, we consider the supervised learning problem of sequence to sequence (seq2seq) modeling. That is, given an input $\boldsymbol{x}$, we want to predict a corresponding output sequence $\boldsymbol{y}$ at each time step t. For an output at time $t \in \mathbb{Z}$, the relationship between the output $y(t) \in \mathbb{R}$ and $\boldsymbol{x}$ can be considered as a functional H_t such that $y(t) = H_t(\boldsymbol{x})$. Thus, the mapping between two temporal sequences $\boldsymbol{x}$ and $\boldsymbol{y}$ can be regarded as a sequence of functionals $\boldsymbol{H} = \{H_t : t \in \mathbb{Z}\}$. As an example, let's consider the shift of sequences. Given $\boldsymbol{x}$, the corresponding output $\boldsymbol{y}$ is a shift of $\boldsymbol{x}$ in time such that $y(t) = x(t - k)$. Then this sequence to sequence mapping can be written as $y(t) = H_t(\boldsymbol{x}) = \sum_{s=-\infty}^{\infty} \rho(s)x(t - s)$, where $\rho(s) = 1$ when $s = k$ and $\rho(s) = 0$ otherwise.

We now specify the input and output spaces. We use $\boldsymbol{c}(\mathbb{Z}, \mathbb{R}^d)$ to denote the set of temporal sequences from the integer index set $\mathbb{Z}$ to d dimensional real vectors, and use $\boldsymbol{c}_0$ to denote temporal sequences with compact support. The input space is defined as $\mathcal{X} = \left\{ \boldsymbol{x} \in \boldsymbol{c}(\mathbb{Z}, \mathbb{R}^d) : \sum_{s \in \mathbb{Z}} |x(s)|^2 < \infty \right\}$. This is indeed the ℓ^2 sequence space. It forms a Banach space under the norm $\|x\|_\mathcal{X}^2 := \sum_{s \in \mathbb{Z}} |x(s)|^2$. For the output space, we consider scalar temporal sequences, $\mathcal{Y} = c(\mathbb{Z}, \mathbb{R})$. To deal with vector-valued outputs, we can consider each dimension separately.

Dilated Convolution Architecture. We now define the dilated convolutional architectures.

Definition 1. *Let* $f : \mathbb{Z} \to \mathbb{R}^d$, $g : \mathbb{Z} \to \mathbb{R}^d$ *be two discrete sequences. The discrete causal dilated convolution with dilation rate r is defined as* $(f *_r g)(t) = \sum_{s \geq 0} f(s)^\top g(t - rs)$. *When $r = 1$, this is the usual convolution. , and we denote it by $*_1$.*

A general dilated temporal CNN model with K layers and M channels at each layer is

$$h_{1,i} = \sum_{j=1}^d w_{0ji} *_1 x_j, \quad h_{k+1,i} = \sigma \left(\sum_{j=1}^M w_{kji} *_{d_k} h_{k,j} \right),$$

$$\hat{y} = \sum_{i=1}^M h_{K,i}, \tag{1}$$

where $i = 1, \ldots, M$, $k = 1, \ldots, K - 1$. Here, x_j denotes the scalar sequence corresponding to the jth dimension of x. Furthermore, $w_{kji} : \mathbb{N} \to \mathbb{R}$ is the convolutional filter at layer k, mapping from channel j at layer k to channel i at layer $k + 1$, which has compact support $[0, l - 1]$. We assume all the filters w_{kji} have the same length $l \geq 2$. We set the dilation rate to be $d_k = l^K$, which is the standard practice for dilated convolution architecture [11,14]. The element-wise activation function is denoted as σ. The dilated CNN hypothesis space is defined as follows

$$\mathcal{H}_{\text{CNN}} = \bigcup_{K,M} \mathcal{H}_{\text{CNN}}^{(M,K)} = \bigcup_{K,M} \left\{ x \mapsto \hat{y} \text{ in Eq. (1)} \right\}. \tag{2}$$

3 Approximation Results

In this section, we present the approximation results. We consider a linear setting where the activation σ is linear. Despite the linearity of the activation function, the dependence on time remains nonlinear. As shown in [10] and [7], this particular setting is meaningful in that it captures the intrinsic structure of the architectures. The CNN hypothesis space with linear activation is defined as

$$\mathcal{H}_{\text{L-CNN}} = \bigcup_{K,M} \mathcal{H}_{\text{L-CNN}}^{(M,K)} = \left\{ \hat{H} : \hat{H}_t(x) = \sum_{s=0}^\infty \rho^{(\hat{H})}(s)^\top x(t - s) \right\}, \tag{3}$$

where $\rho^{(\hat{H})} : [0, l^K - 1] \to \mathbb{R}$ is defined by

$$\rho_i^{(\hat{H})} = \sum_{i_1,\ldots,i_K=1}^{M} w_{K-1,i_{K-1},i_K} *_{l^{K-1}} \ldots *_{l^1} w_{0,i,i_1}. \tag{4}$$

It is important to note that due to the non-associativity of dilated convolution, the order of the convolutions mentioned above should be evaluated from left to right. In this work, we consider target seq2seq mappings satisfying the following conditions, which are standard in functional analysis.

Definition 2. *Let $H = \{H_t : t \in \mathbb{Z}\}$ be a sequence of functionals.*

1. **H** *is causal if it does not depend on the future inputs: for any $x_1, x_2 \in \mathcal{X}$ and any $t \in \mathbb{Z}$ such that $x_1(s) = x_2(s)$ for all $s \leq t$, we have $H_t(x_1) = H_t(x_2)$.*
2. *For all t, $H_t \in H$ is a continuous linear functional if H_t is continuous and for any $x_1, x_2 \in \mathcal{X}$ and $\lambda_1, \lambda_2 \in \mathbb{R}$, $H_t(\lambda_1 x_1 + \lambda_2 x_2) = \lambda_1 H_t(x_1) + \lambda_2 H_t(x_2)$.*
3. **H** *is time-homogeneous if for any $t, \tau \in \mathbb{Z}$, $H_t(x) = H_{t+\tau}(x^{(\tau)})$ where $x^{(\tau)}(s) := x(s - \tau)$ for all $s \in \mathbb{Z}$.*

Following Definition 2, we define the target space as

$$\mathcal{C} = \{H : H \text{ satisfies conditions in Definition 2}\}. \tag{5}$$

The following lemma is a result of Riesz representation theorem for Hilbert space [9], which shows that the target space admits a convolutional representation.

Lemma 1. *For any $H \in \mathcal{C}$, there exists a unique ℓ^2 sequence $\rho^{(H)} : \mathbb{N} \to \mathbb{R}^d$ such that*

$$H_t(x) = \sum_{s=0}^{\infty} \rho^{(H)}(s)^\top x(t - s), \quad t \in \mathbb{Z}. \tag{6}$$

For a target H, the corresponding representation is denoted as $\rho^{(H)}$. Compared to the hypothesis space in equation Eq. (3), we note that approximating H with CNNs is equivalent to approximating $\rho^{(H)}$ with $\rho^{(\hat{H})}$. The hypothesis space in this study is linked to the concept of linear time-invariant systems (LTI systems). Specifically, a causal LTI system will produce a linear functional that adheres to the conditions outlined in Definition 2. It is important to note, however, that not all linear functionals that satisfy Definition 2 necessarily correspond to an LTI system. The objective of our investigation is to examine general functionals without assuming that the data is generated by a hidden linear system.

Tensor Structure of Dilated Convolution Architectures. To begin with, we discuss the tensor structures related to dilated convolution architectures. Dilated convolutional architectures are designed with the goal of achieving long filters with few parameters. A model with K layers and one channel produces a filter with length l^K using only lK parameters. We are interested in how the number of channels affects the expressiveness of the model, and in this regard, we observe that the dilated convolution architectures indeed possess a tensor structure. By treating the filters w_{kji} in Eq. (4) as length l vectors, we replace the convolution operation with tensor products, resulting in a tensor

$$T(\rho_i^{(\hat{H})}) = \sum_{i_1,\ldots,i_K=1}^{M} w_{K-1,i_{K-1},i_K} \otimes w_{K-2,i_{K-2},i_{K-1}} \otimes \cdots \otimes w_{0,i,i_1}, \quad (7)$$

where $T(\rho_i^{(\hat{H})}) \in \mathbb{R}^{l \times \cdots \times l}$. From this point of view, the number of channels M determines the rank of the tensor $T(\rho_i^{(\hat{H})})$. This inspires us to reshape the target representation $\rho^{(H)}$ into a tensor and consider the approximation between tensors, where this approach was first introduced in our previous work [7] We make use of higher-order singular value decomposition (HOSVD) [4] to achieve this. According to HOSVD, a tensor $A \in \mathbb{R}^{I_1 \times \cdots \times I_K}$ can be decomposed into $A = \sum_{i_1}^{I_1} \cdots \sum_{i_K}^{I_K} s_{i_1 \ldots i_K} u_{i_1} \otimes \cdots \otimes u_{i_K}$, where $s_{i_1 \ldots i_K} \in \mathbb{R}$ are scalars and $u_{i_k} \in \mathbb{R}^{I_k}$ forms a set of orthonormal bases for $\mathbb{R}^{I_k}$. We are now ready to discuss the approximation results.

3.1 Jackson-Type and Bernstein-Type Results

The universal approximation property (UAP) of $\mathcal{H}_{\text{L-CNN}}$ is proved in [7], which ensures that the hypothesis space $\mathcal{H}_{\text{L-CNN}}$ is dense in the target space $\mathcal{C}$. However, it does not guarantee the quality of the approximation. Specifically, given a target $H \in \mathcal{H}_{\text{L-CNN}}$, we are concerned with how the approximation error behaves as we increase K and M. This is precisely what the approximation rate considers. As a demonstration, we use m to denote the complexity of a candidate in $\mathcal{H}_{\text{L-CNN}}$. This complexity is typically measured by the number of parameters needed, also called the approximation budget.

In general, a hypothesis $\mathcal{H} = \bigcup_m \mathcal{H}^{(m)}$ is usually built up by candidates with different approximation budgets, where a larger m typically results in better approximation quality. The UAP ensures that the hypothesis space is dense in the target space. However, it does not guarantee the quality of the approximation. For instance, given a target $H \in \mathcal{H}$, we are concerned with how the approximation error behaves as we increase K, M. This is exactly what approximation rate is considering. As a demonstration, we use m to denote the complexity of a candidate in $\mathcal{H}$. It usually measure the number of parameters needed, which is also called approximation budget. A hypothesis $\mathcal{H} = \bigcup_m \mathcal{H}^{(m)}$ is usually built up by candidates with different approximation budget, larger m results in better approximation quality. We may write an approximation rate in the following form

$$\inf_{\hat{H} \in \mathcal{H}^{(m)}} \left\| H - \hat{H} \right\| \leq C_{\mathcal{H}}(H, m). \quad (8)$$

The error bound $C_{\mathcal{H}}(\cdot, m)$ decreases as $m \to \infty$, and the speed of decay is the approximation rate, which quantitatively shows how the error behaves. $C_{\mathcal{H}}(H, \cdot)$

is a complexity measure for the target H. If $C_{\mathcal{H}}(H, \cdot)$ decays rapidly, the target H can be easily approximated. As an example, let's consider the approximation of $f \in C^{\alpha}[0, 1]$ using polynomials. The Jackson theorem [1] states that $C_{\mathcal{H}}(f, m) = \frac{c_{\alpha}}{m^{\alpha}} \max_{r=1\ldots\alpha} \|f^{(r)}\|_{\infty}$. This implies that a function can be efficiently approximated by a polynomial if it has a small Sobolev norm. We call these kinds of approximation rates Jackson-type results.

Jackson-type results are considered forward approximation results, as they allow us to determine the approximation rate given the complexity measure. However, the approximation rate is usually an upper bound on the error. We may ask whether the target can be approximated faster than the given rate. This leads us to the inverse approximation problem, where we are given the approximation rate of a target and want to determine its complexity. In the context of polynomial approximation, the Bernstein theorem [1] states that only functions at least α times differentiable can be approximated faster than $\frac{A}{m^{\alpha}}$. Bernstein-type results are useful for determining the suitability of an architecture for a particular problem theoretically. It is worth mentioning that the choice of complexity measure usually depends on the hypothesis space. In order to investigate the approximation with dilated convolutional architectures, we need to define suitable complexity measures.

Complexity Measure. Based on the previous discussion, we define the complexity measure of targets suitable for the hypothesis space $\mathcal{H}_{\text{L-CNN}}$. Considering the spectrum of $T(\rho^{(H)})$, we define the following complexity measure

$$C_1^{(g)}(H) = \inf \left\{ c : \sum_{j=1}^{d} \sum_{i=s}^{l^K} |s_i^{(j,K)}|^2 \le cg(s-1), s \ge 0, K \ge 1 \right\}, \qquad (9)$$

where $|s_i^{(j,K)}|$ comes from the HOSVD of $T((\rho_j^{(H)})_{[0,l^K-1]})$ in decreasing order, and $g \in c(\mathbb{N}, \mathbb{R}_+)$ is a non-increasing function converges to zero at infinity. This complexity measure takes into account the decay of the spectrum of the target. Likewise, define a complexity measure considering the tail sum of $\rho^{(H)}$

$$C_2^{(f)}(H) = \inf \left\{ c : \sum_{i=s}^{\infty} |\rho^{(H)}(i)|^2 \le cf(s), s \ge 0 \right\}. \qquad (10)$$

We restrict ourselves to considering targets that satisfy the above two complexity measures.

$$\mathcal{C}^{(g,f)} = \{ H \in \mathcal{C} : C_1^{(g)}(H) + C_2^{(f)}(H) < \infty \}. \qquad (11)$$

Having established the complexity measures, we next present the approximation results.

Theorem 1 (Jackson-type). *Fix $l \ge 2$ and $g, f \in c_0(\mathbb{N}, \mathbb{R}_+)$. For any $H \in \mathcal{C}^{(g,f)}$ and any M, K we have*

$$\inf_{\hat{H} \in \mathcal{H}_{L\text{-}CNN}^{(M,K)}} \left\| H - \hat{H} \right\|^2 \le C_1^{(g)}(H)g(M) + C_2^{(f)}(H)f(l^K). \qquad (12)$$

Sketch of proof. For $\boldsymbol{x} \in \mathcal{X}$ with $\|\boldsymbol{x}\| \leq 1$, we have

$$\left|H_t(\boldsymbol{x}) - \hat{H}_t(\boldsymbol{x})\right|^2 \leq \sum_{s=0}^{l^K-1} \left|\boldsymbol{\rho}^{(H)}(s) - \boldsymbol{\rho}^{(\hat{H})}(s)\right|^2 + \sum_{s=l^K}^{\infty} \left|\boldsymbol{\rho}^{(H)}(s)\right|^2. \tag{13}$$

For the second term of the error, we have $\sum_{s=l^K}^{\infty} |\boldsymbol{\rho}^{(H)}(s)|^2 \leq C_2^{(f)}(\boldsymbol{H})f(l^K)$ directly from the definition of C_2. While for the first term, we have

$$\sum_{s=0}^{l^K-1} \left|\boldsymbol{\rho}^{(H)}(s) - \boldsymbol{\rho}^{(\hat{H})}(s)\right|^2 = \sum_{j=1}^{d} \left\|T(\boldsymbol{\rho}_j^{(H)}) - T(\boldsymbol{\rho}_j^{(\hat{H})})\right\|_F^2 \tag{14}$$

$$= \sum_{j=1}^{d} \sum_{i=M+1}^{l^K} |s_i^{(j,K)}|^2 \leq C_1^{(g)}(\boldsymbol{H})g(M). \tag{15}$$

We construct $\boldsymbol{\rho}^{(\hat{H})}$ such that it matches the bases of the HOSVD decomposition of $\boldsymbol{\rho}^{(H)}$, then the second equality results from the orthogonality of the bases in the HOSVD decomposition.

This theorem shows that a target can be efficiently approximated by a dilated CNN if it has a fast decaying spectrum (small C_1) and fast decaying memory (small C_2). It is notable that the two terms in the rate are independent of each other, where K controls the tail error and M controls the error resulting from the spectrum. This is considered an improvement over the previous Jackson-type result presented in [7], where the first term has the form $C(\boldsymbol{H})g(KM^{\frac{1}{K}} - K)$. In that case, the first term increases with K, which may result in the overall error bound not decreasing as K increases. The current result solves this problem, ensuring that the overall error bound always decreases as the number of parameters increases. Moreover, under the same g, the bound in our current work is tighter than the one in [7]. Next, we present the Bernstein-type result.

Theorem 2 (Bernstein-type). *Let $\mathcal{D}$ is a distribution such that at every time step t we have $x(t) \sim \mathcal{N}(0,1)$. Fix $l \geq 2$ and $g, f \in c_0(\mathbb{N}, \mathbb{R}_+)$. Given $\boldsymbol{H} \in \mathcal{C}^{(g,f)}$ and $A, B > 0$, suppose for any M, K there exists $\hat{\boldsymbol{H}}^{(M,K)} \in \mathcal{H}_{L\text{-}CNN}^{(M,K)}$ such that*

$$\sup_t \mathop{\mathbb{E}}_{\boldsymbol{x} \sim \mathcal{D}} [|H_t(\boldsymbol{x}) - \hat{H}_t(\boldsymbol{x})|^2] \leq Ag(M) + Bf(l^K), \tag{16}$$

then we have $\boldsymbol{H} \in C^{(g,f)}$ and $C_1^{(g)}(\boldsymbol{H}) \leq A$, $C_2^{(f)}(\boldsymbol{H}) \leq B$.

Sketch of proof. Since the above inequality holds for any (K, M), we can again consider the two terms separately. For any K, consider a model $\hat{\boldsymbol{H}}$ with a large M such that $Ag(M)$ goes to zero. We conclude that for any K, there exists $\hat{\boldsymbol{H}}$ such that

$$\sum_{s=l^K}^{\infty} \left|\boldsymbol{\rho}^{(H)}(s)\right|^2 = \|\boldsymbol{H} - \hat{\boldsymbol{H}}\|^2 \leq Bf(l^K), \tag{17}$$

which implies $C_2^{(f)}(\boldsymbol{H}) \leq B$. The intuition of the proof is that in $[l^K, \infty]$, the error is completely determined by the tail error of $\boldsymbol{\rho}^{(H)}$. Thus, the error rate directly implies the decay rate of the tail. Similarly, for any M, we pick a large enough K such that $Bf(l^K)$ goes to zero. Then, there exists $\hat{\boldsymbol{H}}$ such that

$$\sum_{j=1}^{d} \sum_{s=M+1}^{\infty} |s_i^{(j,K)}|^2 = \|\boldsymbol{H} - \hat{\boldsymbol{H}}\|^2 \leq Ag(M), \tag{18}$$

which implies $C_1^{(g)}(\boldsymbol{H}) \leq A$.

The Bernstein-type result serves as a converse to the Jackson-type Theorem 1. It suggests that a target can be well approximated by the model only if it has small complexity measure, which in turn implies that the target exhibits good spectrum regularity and rapid decay of memory. Combining the two results, we obtain a complete characterization of $\mathcal{H}_{\text{L-CNN}}$: a target in $\mathcal{H}$ can be efficiently approximated by: linear temporal CNNs if and only if it has fast decaying spectrum and memory.

4   Related Works

Tensor decomposition of convolutional filters have also been discussed in [3,5, 8,12]. In these works, the problem considered is convolution applied to image inputs with shape (H, W, C). At a specific layer, the convolution kernel applies to it have shape (I, J, C, C'), where I, J is the size of the convolution kernel and C, C' is input channels and output channels, respectively. They mainly focus on decomposition of the four dimensional convolutional kernel, which is different from what we do in this current work and the previous work [7]. Instead of considering decomposition of a kernel at a specific layer, we consider the tensor structure of the entire model resulting from the multilayer dilated convolutions. In our case, the convolution kernel for sequence modeling is a one dimensional array with length l^K. We tensorize it into a K dimensional tensor and consider its decompositions. This enables us to analyze the precise relationship between the target and the convolutional model.

5   Conclusion

In practical machine learning applications, approximation results provide useful insights into model selection. Temporal CNNs are adapted to approximate sequential relationships which have regular spectrum (small C_1) and decaying memory (small C_2), that is, we have a rate estimate as in Theorem 1 if and only if the target has small complexity measure C_1 and C_2. This highlights the importance of both Jackson-type and Bernstein-type results in machine learning, as they enable us to assess the suitability of an architecture prior to investing computational resources and time in a trial-and-error process. By doing so, we can increase the efficiency and effectiveness of the model selection process.

References

1. Achieser, N.I.: Theory of Approximation. Courier Corporation (2013)
2. Bai, S., Kolter, J.Z., Koltun, V.: An Empirical Evaluation of Generic Convolutional and Recurrent Networks for Sequence Modeling (2018). https://doi.org/10.48550/arXiv.1803.01271
3. Chu, B.S., Lee, C.R.: Low-rank Tensor Decomposition for Compression of Convolutional Neural Networks Using Funnel Regularization (2021). https://doi.org/10.48550/arXiv.2112.03690
4. De Lathauwer, L., De Moor, B., Vandewalle, J.: A multilinear singular value decomposition. SIAM J. Matrix Anal. Appl. **21**(4), 1253–1278 (2000). https://doi.org/10.1137/S0895479896305696
5. Hayashi, K., Yamaguchi, T., Sugawara, Y., Maeda, S.I.: Exploring unexplored tensor network decompositions for convolutional neural networks. In: Advances in Neural Information Processing Systems, vol. 32. Curran Associates, Inc. (2019)
6. Jiang, H., Li, Q., Li, Z., Wang, S.: A Brief Survey on the Approximation Theory for Sequence Modelling (2023). https://doi.org/10.48550/arXiv.2302.13752
7. Jiang, H., Li, Z., Li, Q.: Approximation theory of convolutional architectures for time series modelling. In: Proceedings of the 38th International Conference on Machine Learning, pp. 4961–4970. PMLR (2021)
8. Kim, Y.D., Park, E., Yoo, S., Choi, T., Yang, L., Shin, D.: Compression of Deep Convolutional Neural Networks for Fast and Low Power Mobile Applications (2016). https://doi.org/10.48550/arXiv.1511.06530
9. Kreyszig, E.: Introductory Functional Analysis with Applications. Wiley Classics Library, Wiley, New York, wiley classics library ed edn. (1989)
10. Li, Z., Han, J., E, W., Li, Q.: Approximation and optimization theory for linear continuous-time recurrent neural networks. J. Mach. Learn. Res. **23**(42), 1–85 (2022)
11. van den Oord, A., et al.: WaveNet: A Generative Model for Raw Audio. arXiv:1609.03499 (2016)
12. Phan, A.H., et al.: Stable Low-rank Tensor Decomposition for Compression of Convolutional Neural Network (2020). https://doi.org/10.48550/arXiv.2008.05441
13. Yin, W., Kann, K., Yu, M., Schütze, H.: Comparative Study of CNN and RNN for Natural Language Processing (2017). https://doi.org/10.48550/arXiv.1702.01923
14. Yu, F., Koltun, V.: Multi-Scale Context Aggregation by Dilated Convolutions (2016). https://doi.org/10.48550/arXiv.1511.07122

The Geometry of Quantum States

Monotonicity of the Scalar Curvature of the Quantum Exponential Family for Transverse-Field Ising Chains

Takemi Nakamura[✉][iD]

Department of Complex Systems Science, Graduate School of Informatics,
Nagoya University, Nagoya 464-8601, Japan
nakamura.takemi.d7@s.mail.nagoya-u.ac.jp

Abstract. The monotonicity of the scalar curvature of the state space equipped with the Bogoliubov-Kubo-Mori metric for the mixing process of the state was conjectured by Petz. From the standpoint of quantum statistical mechanics, the quantum exponential family, a special parametric family of the state space, plays a central role. In this contribution, we investigate the monotonicity of the scalar curvature of the family with respect to temperature for transverse-field Ising chains in various sizes and consequently, find that the monotonicity breaks down for the chains in finite sizes, whereas the monotonicity seems to hold if the chain is non-interacting or infinite-size. Our results suggest that finite-size effects can appear in the curvature through monotonicity with respect to majorization.

Keywords: Scalar curvature · Quantum exponential family · Ising model

1 Introduction

Investigating the state space of a physical system is often useful for considering its physical properties. A physically important example of this is the Riemannian-geometric or information-geometric formulation of thermodynamics and statistical physics [3,11]. In information geometry [15], the Gibbs distribution in classical statistical physics belongs to the exponential family, and a set of macroscopic variables or parameters in the distribution can be regarded as a coordinate system of the statistical manifold. The scalar curvature induced from the Fisher metric of the manifold is considered to be a physically significant quantity because it is related to thermal phase transitions and correlation [12,13]. Also in the quantum setting, a few works discuss the scalar curvature from the viewpoint of physics and suggest its physical significance [16,17].

The Bogoliubov-Kubo-Mori (BKM) metric or the canonical correlation is one of the distinctive monotone metrics on the finite-dimensional quantum state space [9,10]. The scalar curvature induced from the metric attracts interest in the

F. Nielsen and F. Barbaresco (Eds.): GSI 2023, LNCS 14072, pp. 353–362, 2023.
https://doi.org/10.1007/978-3-031-38299-4_37

mathematical context due to issues around the so-called Petz conjecture. The conjecture states the monotonicity of the scalar curvature of the state space under majorization [1,4,5,9]. In other words, it is conjectured that the scalar curvature of the state space shows entropy-like behavior. However, the monotonicity of the scalar curvature of the quantum exponential family, a parametric family of the state space, is little discussed though in quantum statistical mechanics the Gibbs state is relevant. The curvature tensor of a manifold is usually different from that of a submanifold of the manifold (e.g., a sphere in Euclidean space), hence one may wonder whether the scalar curvature of the quantum exponential family has such monotonicity and how it can be related to thermodynamic entropy.

The purpose of this study is to investigate the monotonicity property of the scalar curvature of the quantum exponential family with simple physical systems, transverse-field (TF) Ising chains, in various sizes. This is motivated by Ref. [2], which discusses finite-size effects on the scalar curvature.

It should be remarked here that this is a physically oriented work just inspired by the Petz conjecture and is not intended to aim at proving or disproving the conjecture itself.

This paper is organized into three sections: preliminaries, behaviors of scalar curvatures, and discussion and conclusions.

2   Preliminaries

This section is devoted to introducing the concepts used here and fixing their notations.

2.1   Scalar Curvature of the Quantum Exponential Family

Let the Hilbert space of a finite-dimensional quantum system be $\mathcal{H}$. The quantum state of the system is generally described by a density operator on $\mathcal{H}$. The set of all invertible density operators is denoted by $\mathcal{S}(\mathcal{H}) := \{\hat{\rho} \mid \text{Tr}\hat{\rho} = 1,\ \hat{\rho} > 0\}$.

Our concern is a quantum statistical manifold, a parametric family of density operators

$$\{\hat{\rho}(x) \subset \mathcal{S}(\mathcal{H}) \mid x \in \mathcal{X} \subset \mathbb{R}^n\}, \tag{1}$$

where $x = (x^1, x^2, ..., x^n)$ denotes a set of the parameters of a model forming a coordinate system whose domain denoted by $\mathcal{X}$ is a subspace of $\mathbb{R}^n$. In this context, the Gibbs state at an inverse temperature[1] $\beta = 1/T$ for a Hamiltonian $\hat{H}$, $e^{-\beta\hat{H}}/Z$, where $Z := \text{Tr}[e^{-\beta\hat{H}}]$ is the partition function, belongs to the following quantum exponential family

$$\hat{\rho}(\theta) = \exp\left[\theta^i \hat{\mathcal{O}}_i - \psi(\theta)\right] \tag{2}$$

with the potential function

$$\psi(\theta) := \ln \text{Tr}\left[\exp\left[\theta^i \hat{\mathcal{O}}_i\right]\right] = \ln Z(\theta). \tag{3}$$

---

[1] Here we set the Boltzmann constant to unity.

Here $\theta = (\theta^1, \theta^2, ..., \theta^n)$ and $\{\hat{O}_i\}_{i=1}^n$ denote a set of the natural parameters forming an e-affine coordinate system and a set of self-adjoint operators representing some physical observables, respectively. We here assume that the self-adjoint operators are linearly independent of each other so that the family is n-dimensional.

If density operators of a quantum system of interest are modeled in this way, the metric on the family may be the pullback of the BKM metric defined on the whole space of invertible states $\mathcal{S}(\mathcal{H})$. The components of this induced metric can be given by the Hessian of the potential:

$$g_{ij}(\theta) = \partial_i \partial_j \psi(\theta), \tag{4}$$

where ∂_i denotes $\partial/\partial\theta^i$ [6,7]. This means that once we obtain the potential of a quantum system, we can effectively have a Riemannian metric on the quantum exponential family of the system.

The Christoffel symbols Γ_{ijk} and the Riemannian curvature tensor R_{ijkl} induced from this Hessian metric can be calculated as [7]

$$\Gamma_{ijk}(\theta) = \frac{1}{2}\psi_{ijk}(\theta), \tag{5}$$

$$R_{ijkl}(\theta) = \frac{1}{4}g^{ab}(\psi_{aik}\psi_{bjl} - \psi_{ail}\psi_{bjk}), \tag{6}$$

where $\partial_i \partial_j \partial_k \psi(\theta)$ is abbreviated as $\psi_{ijk}(\theta)$. Here we define the Riemannian curvature tensor as the curvature of a sphere becomes negative, in accordance with Ruppeiner [11–13]. For the two-dimensional quantum exponential family, the scalar curvature can be calculated by the formula [7]

$$R(\theta^1, \theta^2) = \frac{2R_{1212}}{\det g} = \frac{\begin{vmatrix} \psi_{11} & \psi_{12} & \psi_{22} \\ \psi_{111} & \psi_{112} & \psi_{122} \\ \psi_{112} & \psi_{122} & \psi_{222} \end{vmatrix}}{2\begin{vmatrix} g_{11} & g_{12} \\ g_{21} & g_{22} \end{vmatrix}^2}. \tag{7}$$

2.2   TF Ising Chains

The Hamiltonian of the TF Ising chain [14] with nearest-neighbor interactions composed of N qubits is given by

$$\hat{H}_N = -J \sum_{i=1}^{N-1} \hat{\sigma}_i^z \hat{\sigma}_{i+1}^z - \Gamma \sum_{i=1}^{N} \hat{\sigma}_i^x, \tag{8}$$

or when we impose the periodic boundary condition $\hat{\sigma}_{N+1}^z = \hat{\sigma}_1^z$ for $N \geq 3$, it can be also given by

$$\hat{H}_N = -J \sum_{i=1}^{N} \hat{\sigma}_i^z \hat{\sigma}_{i+1}^z - \Gamma \sum_{i=1}^{N} \hat{\sigma}_i^x, \tag{9}$$

where Pauli matrices $\hat{\sigma}_i^z$ and $\hat{\sigma}_i^x$ are represented as

$$\hat{\sigma}_i^z := \begin{pmatrix} 1 & 0 \\ 0 & -1 \end{pmatrix}, \quad \hat{\sigma}_i^x := \begin{pmatrix} 0 & 1 \\ 1 & 0 \end{pmatrix}. \tag{10}$$

J and Γ represent a ferromagnetic interaction and a transverse field, respectively. The natural parameters are $(\theta, x) := (\beta J, \beta \Gamma)$.

For emphasizing the novelty of this work, it would be better to state here physically in which sense this TF model differs from the one-dimensional classical (i.e., longitudinal-field) Ising model,

$$\hat{H}_N^{\text{cl.}} = -J \sum_{i=1}^{N} \hat{\sigma}_i^z \hat{\sigma}_{i+1}^z - h \sum_{i=1}^{N} \hat{\sigma}_i^z, \tag{11}$$

which has been investigated in Ref. [7]. At first glance, both models consist of qubits alike, and hence no major difference seems to appear. However, at least in a physical sense, there is a big difference between the two models, that is, with or without quantum fluctuation. The classical model is a sum of only one directional component of spin $1/2$, thus it can not have quantum fluctuation coming from the non-commutativity among directional components of spin $1/2$. On the other hand, the TF model is a sum of two directional components of spin $1/2$, z- and x-directional components as shown in Eq. (8) or Eq. (9), therefore quantum fluctuation arises for the TF model. In other words, the Gibbs state for the TF model is a quantum state whereas that for the classical model is essentially a classical probability. This difference in the existence of quantum fluctuation affects physical properties such as critical exponents [8]. Our work suggests that quantumness can affect the monotonicity of the scalar curvature.

3   Behaviors of Scalar Curvatures

In this section, using Eq. (7), we plot scalar curvatures for TF Ising chains in sizes of $N = 1, 2, 3, \infty$ as a function of temperature $T = 1/\beta$ and check if they are monotone with respect to temperature. It should be stressed that the monotonicity discussed here has nothing to do with the original Petz conjecture, because we here focus not on the whole space but on a subfamily of it.

3.1 $N = 1$ TF Ising Chain: Non-interacting Case

This system consists of only one qubit, thus it can not have interactions. Instead, we add a contribution from a longitudinal field h into the Hamiltonian:

$$\hat{H}_1 = -h\hat{\sigma}^z - \Gamma \hat{\sigma}^x. \tag{12}$$

This is often called the zero-dimensional TF Ising model and has physical correspondence with the one-dimensional classical Ising model [8]. We can regard this system as a non-interacting chain just by arranging independent qubits in a line because the two systems are essentially the same. The potential for the model is

$$\psi_1(z, x) = \ln(2 \cosh r), \tag{13}$$

where $(z, x) := (\beta h, \beta \Gamma)$ are the natural parameters, and $r := \sqrt{z^2 + x^2}$. Therefore, the scalar curvature can be analytically obtained as [6,16]

$$R_1 = \frac{2r - \tanh r}{2r^2 \tanh r} \cosh^2 r - \frac{1 + \tanh^2 r}{2 \tanh^2 r}. \tag{14}$$

Figure 1 is a semi-log plot of this scalar curvature as a function of temperature. As we can see, $R_1(T)$ decreases monotonically with respect to temperature. In fact, we can check directly from Eq. (14) that $R_1(T)$ is a monotonically decreasing function of temperature and hence $R_1(T) \geq 0$ since $R_1(T \to \infty) \to 0$.

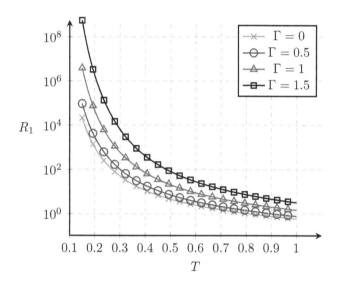

Fig. 1. $R_1(T)$ with $J = 1$ and different values of Γ

3.2 $N = 2$ and $N = 3$ TF Ising Chains: Finite Size Case

The potential is calculated respectively as

$$\psi_2(\theta, x) = \ln\left[2\cosh(\theta) + 2\cosh\left(\sqrt{\theta^2 + 4x^2}\right)\right], \tag{15}$$

$$\psi_3(\theta, x) = \ln\left[2e^{-\theta}\cosh(x) + 2e^{\theta - x}\cosh\left(2\sqrt{\theta^2 + \theta x + x^2}\right)\right.$$
$$\left. + 2e^{\theta + x}\cosh\left(2\sqrt{\theta^2 - \theta x + x^2}\right)\right], \tag{16}$$

where the periodic boundary condition $\hat{\sigma}_4^z = \hat{\sigma}_1^z$ is imposed for $N = 3$. The analytic expressions of the scalar curvatures R_2, R_3 are brutally complicated to compute, and we just give plots: Fig. 2 for $N = 2$ and Fig. 3 for $N = 3$.

Surprisingly, these scalar curvatures can move up and down.

At higher temperatures, they seem to decrease monotonically and converge to zero. In the high-temperature limit where thermal fluctuation dominates over quantum fluctuation, the behavior of a quantum system approaches its classical nature. This might be reflected in this high-temperature behavior of the scalar curvature. Note that the behavior seems to be common in all three cases.

In the middle, however, the monotonicity apparently breaks down: they moderately decrease and then increase. They can have even negative values during decreasing. As seen in the figures, this behavior seems to vary depending largely on the strength of a transverse field and to appear only with particular values of Γ/J. Otherwise, it disappears, and the monotonicity recovers.

Around zero temperature, they seem to spike up to infinity as $T \to 0$. In any case, the scalar curvature may diverge as $T \to 0$.

3.3 $N \to \infty$ TF Ising Chain: Thermodynamic Limit

In the thermodynamic limit $N \to \infty$, only the potential per site or the potential density is meaningful because the potential itself has an extensive property. In this case, we refer to the potential as the potential density.

The potential for the model can be obtained as

$$\psi_\infty(\theta, x) := \frac{1}{N} \ln Z = \int_0^\pi \frac{dk}{\pi} \ln(2 \cosh f(k; \theta, x)), \tag{17}$$

where $f(k; \theta, x) := \sqrt{\theta^2 + x^2 + 2\theta x \cos k}$. Once we can obtain a physically relevant potential like Eq. (17), we can compute naively the metric through Eq. (4) as susceptibilities or correlations and hence can also obtain the scalar curvature by Eq. (7), even though the corresponding Hibert space $\mathcal{H}$ is not finite-dimensional. What is meaningful here is that geometrical quantities calculated in this way are found to be physically significant since they exhibit physical information such as quantum or thermal phase transitions, basically in the context of the thermodynamic limit [11–13,16,17]. Thus, despite little suitable mathematical discussion, we will follow this approach here.

The plot of the scalar curvature $R_\infty(T)$ for different Γ are shown in Fig. 4. Unlike $R_2(T)$ and $R_3(T)$, $R_\infty(T)$ decreases monotonically, which confirms the monotonicity property. In addition, these exponential behaviors are similar to $R_1(T)$ whereas Γ dependence seems a little different. Note that $\Gamma = J$ is the quantum phase transition point, hence $R_\infty(T)$ may expect to follow a power law, which is also confirmed in Fig. 4.

For a comparison with the scalar curvature for the classical Ising chain Eq. (11) in the thermodynamic limit $N \to \infty$, which has been analytically obtained in Ref. [7], we plot this scalar curvature $R_\infty^{cl.}(T)$ with different values of the longitudinal field h in Fig. 5. Numerical evaluation for the case $h = 0$ has been

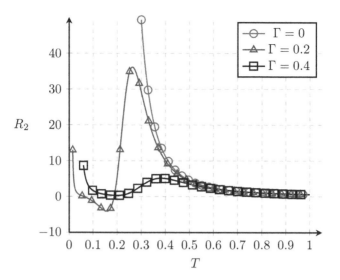

Fig. 2. $R_2(T)$ with $J = 1$ and different values of Γ

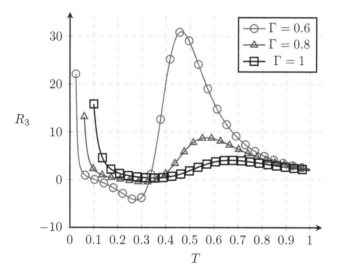

Fig. 3. $R_3(T)$ with $J = 1$ and different values of Γ

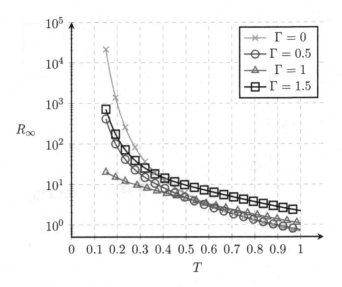

Fig. 4. $R_\infty(T)$ with $J = 1$ and different values of Γ

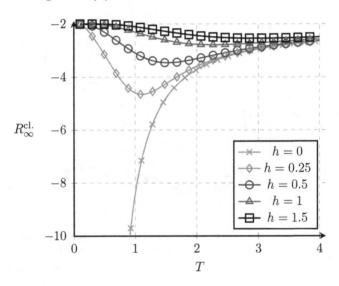

Fig. 5. $R_\infty^{\mathrm{cl.}}(T)$ with $J = 1$ and different values of h

addressed explicitly in Ref. [7], and $R_\infty^{\mathrm{cl.}}(T; h = 0)$ seems monotone with respect to temperature. From this, one might think that it is monotone in general. However, as seen in Fig. 5, this case of $h = 0$ is special rather than general. For example, for $h/J = 0.25$, the scalar curvature is not monotone apparently. In addition to this difference in monotonicity, the signs of R_∞ and $R_\infty^{\mathrm{cl.}}$ are also different: positive for the TF model but negative for the classical model.

4   Discussion and Conclusions

We investigated the scalar curvature of the quantum exponential family for TF Ising chains in various sizes of $N = 1$ (this case can be regarded as a non-interacting chain), 2, 3, and ∞ as a function of temperature, and checked whether it is monotone with respect to temperature. Consequently, we can expect the monotonicity for $N = 1$ and ∞. However, in finite-size cases $N = 2$ and 3 with an appropriate strength of a transverse field, the scalar curvature no longer possesses the monotonicity property and also can have negative values[2].

Here we should note again that the result that the scalar curvature for finite-size chains is not monotone does not give any piece of contributing information to the Petz conjecture because we here discuss different spaces.

We can provide the scalar curvature with another physical interpretation from these results. Comparing the case of a non-interacting chain with the two cases of the finite-size interacting chains, we may deduce that in addition to finite-size effects, ferromagnetic interactions can cause the monotonicity to break down as well. In other words, quantum fluctuations urge the scalar curvature to be positive whereas ferromagnetic interactions urge it to be negative. We may expect that this balance determines the behavior of the scalar curvature.

Our results imply that

(i) $R(T \to \infty) = 0$ and $R(T \to 0) = \infty$ in any size,
(ii) $R(T)$ of the quantum exponential family for an infinite-size interacting quantum system is monotone with respect to temperature T.
(iii) finite-size effects can appear in the non-monotonicity of the scalar curvature with respect to majorization,
(iv) we might be able to utilize the scalar curvature to judge whether an interacting quantum equilibrium system is in a finite size effectively.

Note that combining (i) and (ii) yields the non-negativity of the scalar curvature: $R(T) \geq 0$. This non-negativity and the monotonicity (ii) may suggest that the scalar curvature of the quantum exponential family has some similarity to the thermodynamic entropy. We hope this contribution serves for establishing the solid physical interpretation of the scalar curvature of the family.

Acknowledgments. The present author appreciates Prof. Shogo Tanimura for reviewing the draft of this paper.

References

1. Andai, A.: On the monotonicity conjecture for the curvature of the kubo-mori metric. arXiv preprint math-ph/0310064 (2003). https://doi.org/10.48550/arXiv.math-ph/0310064
2. Brody, D.C., Ritz, A.: Information geometry of finite ising models. J. Geom. Phys. **47**(2–3), 207–220 (2003). https://doi.org/10.1016/S0393-0440(02)00190-0

---

[2] Again the curvature is defined here as that of a 2-sphere becomes negative.

3. Crooks, G.E.: Measuring thermodynamic length. Phys. Rev. Lett. **99**, 100602 (2007). https://doi.org/10.1103/PhysRevLett.99.100602

4. Dittmann, J.: On the curvature of monotone metrics and a conjecture concerning the kubo-mori metric. Linear Algebra Appl. **315**(1–3), 83–112 (2000). https://doi.org/10.1016/S0024-3795(00)00130-0

5. Gibilisco, P., Isola, T.: On the monotonicity of scalar curvature in classical and quantum information geometry. J. Math. Phys. **46**(2), 023501 (2005). https://doi.org/10.1063/1.1834693

6. Ingarden, R.S., Janyszek, H., Kossakowski, A., Kawaguchi, T.: Information geometry of quantum statistical systems. Tensor **37**, 105–111 (1982)

7. Janyszek, H., Mrugała, R.: Riemannian geometry and the thermodynamics of model magnetic systems. Phys. Rev. A **39**(12), 6515–6523 (1989). https://doi.org/10.1103/PhysRevA.39.6515

8. Masuo, S.: Relationship between d-dimensional quantal spin systems and (d+1)-dimensional ising systems: equivalence, critical exponents and systematic approximants of the partition function and spin correlations. Progress Theoret. Phys. **56**(5), 1454–1469 (1976). https://doi.org/10.1143/PTP.56.1454

9. Petz, D.: Geometry of canonical correlation on the state space of a quantum system. J. Math. Phys. **35**(2), 780–795 (1994). https://doi.org/10.1063/1.530611

10. Petz, D., Toth, G.: The bogoliubov inner product in quantum statistics: dedicated to j. merza on his 60th birthday. Lett. Math. Phys. **27**(3), 205–216 (1993). https://doi.org/10.1007/BF00739578

11. Ruppeiner, G.: Riemannian geometry in thermodynamic fluctuation theory. Rev. Mod. Phys. **67**(3), 605 (1995). https://doi.org/10.1103/RevModPhys.67.605

12. Ruppeiner, G.: Thermodynamic curvature measures interactions. Am. J. Phys. **78**(11), 1170–1180 (2010). https://doi.org/10.1119/1.3459936

13. Ruppeiner, G.: Thermodynamic curvature and black holes. In: Bellucci, S. (ed.) Breaking of Supersymmetry and Ultraviolet Divergences in Extended Supergravity. SPP, vol. 153, pp. 179–203. Springer, Cham (2014). https://doi.org/10.1007/978-3-319-03774-5_10

14. Sei, S., Jun-ichi, I., Chakrabarti, B.K.: Quantum Ising Phases and Transitions in Transverse Ising Models, vol. 862. Springer, Heidelberg (2012). https://doi.org/10.1007/978-3-642-33039-1

15. Shun-ichi, A., Hiroshi, N.: Methods of Information Geometry, vol. 191. American Mathematical Soc. (2000). https://doi.org/10.1090/mmono/191

16. Takemi, N.: Scalar curvature of the quantum exponential family for the transverse-field ising model and the quantum phase transition. arXiv preprint arXiv:2212.12919 (2022). https://doi.org/10.48550/arXiv.2212.12919

17. Zanardi, P., Giorda, P., Cozzini, M.: Information-theoretic differential geometry of quantum phase transitions. Phys. Rev. Lett. **99**, 100603 (2007). https://doi.org/10.1103/PhysRevLett.99.100603

Can Čencov Meet Petz

F. M. Ciaglia[1]([⊠]) [iD], F. Di Cosmo[1,2] [iD], and L. González-Bravo[1] [iD]

[1] Departamento de Matemáticas, Universidad Carlos III de Madrid,
Avenida de la Universidad 30, 28911 Leganés, Madrid, Spain
{fciaglia,fcosmo,lauragon}@math.uc3m.es
[2] ICMAT, Instituto de Ciencias Matemáticas (CSIC-UAM-UC3M-UCM),
Madrid, Spain

Abstract. We discuss how to exploit the recent formulation of classical and quantum information geometry in terms of normal states on W^*-algebras to formulate a problem that unifies Čencov's theorem and Petz's theorem.

Keywords: Čencov's theorem · Petz's theorem · Information Geometry · W^*-algebras

1 Introduction

Čencov's[1] and Petz's theorems are two pillars of classical and quantum information geometry, respectively, but their mathematical formulation is quite different as it relies on the mathematics of probability distributions in the classical case, and on the mathematics of quantum states in the quantum case. Recently, W^*-algebras and normal states on them are being employed as a common mathematical framework unifying classical and quantum information geometry [8–12,15,16], and in this short contribution we would like to exploit this mathematical framework in order to formulate a problem that unifies those answered by Čencov and Petz, respectively. At this purpose, let us briefly recall Čencov's and Petz's results.

Čencov's theorem proves that, among all possible families of Riemannian metric tensors on (the interior of) finite-dimensional simplexes, the Fisher-Rao metric tensor is the only one (up to an overall positive constant) which is invariant with respect to the class of Markov maps known as congruent embeddings

---

[1] Sometimes, the name Čencov is also spelled Chentsov.

FMC and LG-B acknowledge that this work has been supported by the Madrid Government (Comunidad de Madrid-Spain) under the Multiannual Agreement with UC3M in the line of "Research Funds for Beatriz Galindo Fellowships" (C&QIG-BG-CM-UC3M), and in the context of the V PRICIT (Regional Programme of Research and Technological Innovation). FDC thanks the UC3M, the European Commission through the Marie Sklodowska-Curie COFUND Action (H2020-MSCA-COFUND-2017- GA 801538)for their financial support through the CONEX-Plus Programme.

F. Nielsen and F. Barbaresco (Eds.): GSI 2023, LNCS 14072, pp. 363–371, 2023.
https://doi.org/10.1007/978-3-031-38299-4_38

[25]. Specifically, let $\mathcal{X}_{(n+1)}$ denote a discrete set with $(n+1)$-elements. It is well-known that the space of probability measures on $\mathcal{X}_{(n+1)}$ can be identified with the n-simplex

$$\overline{\Delta}_n := \left\{ \mathbf{p} = (p^1, \cdots, p^{n+1}) \in \mathbb{R}^{n+1} \mid p^j \geq 0, \sum_{j=1}^{n+1} p^j = 1 \right\}. \qquad (1)$$

The n-simplex is a manifold with corners and its highest-dimensional stratum Δ_n is composed by all those $\mathbf{p} \in \overline{\Delta}_n$ such that $p^j > 0$ for $j \in [1, \cdots, n+1]$. A linear map $M \colon \mathbb{R}^{n+1} \to \mathbb{R}^{m+1}$ is called a **Markov map** if $M(\overline{\Delta}_n) \subseteq \overline{\Delta}_m$, and a Markov map M is called faithful if $M(\Delta_n) \subseteq \Delta_m$. A faithful Markov map is called a **congruent embedding** if $M(\Delta_n) \subseteq \Delta_m$ and it admits a left-inverse N which is still a faithful Markov map. A characterization of congruent embedding can be found in [4,25].

Problem 1. Determine all families $\{G_n\}$, where $n > 0$ and G_n is a Riemannian metric tensor on Δ_n, satisfying the invariance property

$$M^* G_m = G_n \qquad (2)$$

for all congruent embeddings M.

Čencov investigated Problem 1 and showed that, up to an overall positive multiplicative constant, there is only one such family determined by

$$G_n = \sum_{j=1}^{n+1} \frac{\mathrm{d}p^j \otimes \mathrm{d}p^j}{p^j}, \qquad (3)$$

where the p^j's are the standard Cartesian coordinates on $\mathbb{R}^n$ associated with the standard basis and with respect to which the n-simplex is defined (see Eq. (1)) [13,25]. The Riemannian metric G_n coincides with the Fisher-Rao metric tensor on Δ_n, ubiquitous in classical information geometry, statistics, and estimation theory [1].

After tackling the classical case, Čencov and Morozova started an investigation of the quantum case [17,18]. They provided a preliminary partial solution of the problem that was later fully developed by Petz in the finite-dimensional case. Petz's theorem is a sort of quantum counterpart of Čencov's result for it classifies, among all possible families of Riemannian metric tensors on the manifolds of faithful quantum states in finite dimensions, the ones that are monotone with respect to the completely-positive and trace-preserving (CPTP) maps [22]. Specifically, let us consider a quantum system with a finite-dimensional Hilbert space $\mathcal{H} \cong \mathbb{C}^n$. A quantum state ρ is then a density operator on $\mathcal{H}$, namely, a self-adjoint operator in the algebra $\mathcal{B}(\mathcal{H})$ of bounded linear operators which has unit trace and is positive semi-definite, meaning that $\langle \psi \mid \rho \mid \psi \rangle \geq 0$ for every $\psi \in \mathcal{H}$. A quantum state is called **faithful** if $\langle \psi \mid \rho \mid \psi \rangle = 0$ implies that ψ is the zero vector in $\mathcal{H}$. Note that, in finite dimensions, a faithful quantum state is invertible as a linear operator (the same is no-longer true in infinite dimensions because density operators are required to be trace-class so that they never admit a bounded inverse). The space of faithful quantum states is denoted by $\mathscr{S}_f(\mathcal{H})$, and it is a smooth manifold which is also an homogeneous space for the Lie group $GL(\mathcal{H})$ [6,7,14]. A linear map $\Phi\colon \mathcal{B}(\mathcal{H}) \to \mathcal{B}(\mathcal{K})$ is called positive if it sends positive operators to positive operators, and it is called completely-positive (CP) if the map $\Phi \otimes 1_n\colon \mathcal{B}(\mathcal{H}) \otimes M_n(\mathbb{C}) \to \mathcal{B}(\mathcal{K}) \otimes M_n(\mathbb{C})$, where 1_n is the identity map on the space of square complex matrices $M_n(\mathbb{C}) = \mathcal{B}(\mathbb{C}^n)$, is positive for all n [5]. A CP maps Φ is called trace-preserving (TP) if $\mathrm{Tr}_{\mathcal{K}}(\Phi(\mathbf{x})) = \mathrm{Tr}_{\mathcal{H}}(\mathbf{x})$ for all $\mathbf{x} \in \mathcal{B}(\mathcal{H})$. Of course, a completely-positive and trace-preserving (CPTP) map Φ sends quantum states into quantum states, and if Φ sends faithful states into faithful states, meaning that $\Phi(\mathscr{S}_f(\mathcal{H})) \subseteq \mathscr{S}_f(\mathcal{K})$, then it is called **faithful** and referred to as a fCPTP map.

It is worth noting that faithful positive maps are also called *strictly positive* [2], and have been recently shown to possess a sort of "universal kernel" [26].

Problem 2. Determine all families $\{G_{\mathcal{H}}\}$, where $\mathcal{H}$ is a finite-dimensional Hilbert space and $G_{\mathcal{H}}$ is a Riemannian metric tensor on $\mathscr{S}_f(\mathcal{H})$, satisfying the monotonicity property

$$\Phi^* G_{\mathcal{K}} \leq G_{\mathcal{H}} \tag{4}$$

for all fCPTP maps.

Remark 1. A quantum congruent embedding would be an fCPTP map Φ possessing a left-inverse Ψ which is itself an fCPTP map. The most general form of such a map is $\Phi(\rho) = \mathbf{U}(\rho \otimes \sigma)\mathbf{U}^\dagger$, where σ is a fixed faithful quantum state on $\mathcal{K}$ and $\mathbf{U}$ a unitary operator on $\mathcal{H} \otimes \mathcal{K}$ [19]. A double application of Eq. (4) forces every monotone family to be invariant with respect to quantum congruent embeddings.

Čencov and Morozova gave a preliminary definition and partial solution of Problem 2 [17,18]. Petz investigated Problem 2 and showed that every admissible monotone family is of the type $\{G_{\mathcal{H}}^f\}$ where f is an operator monotone function $f\colon [0,\infty) \to [0,\infty)$ satisfying $f(t) = tf(t^{-1})$ [22]. We refer to this family as the Morozova-Čencov-Petz family of monotone metric tensors. In particular, the explicit expression of the Riemannian metric tensor $G_{\mathcal{H}}^f$ reads

$$(G_{\mathcal{H}}^f)_\rho(\mathbf{a}, \mathbf{b}) = \mathrm{Tr}_{\mathcal{H}}\left(\mathbf{a}(K_\rho^f)^{-1}(\mathbf{b})\right), \tag{5}$$

where $\mathbf{a}, \mathbf{b} \in T_\rho \mathscr{S}_f(\mathcal{H}) \subset \mathcal{B}(\mathcal{H})$ are trace-less, and the superoperator K_ρ^f is given by

$$K_\rho^f = f\left(L_\rho R_\rho^{-1}\right) R_\rho, \tag{6}$$

with $L_\rho(\mathbf{b}) = \rho\mathbf{b}$ and $R_\rho(\mathbf{b}) = \mathbf{b}\rho$.

In the rest of this work we exploit the formalism of W^*-algebras in order to formulate a problem that encompasses and unifies Problem 1 and Problem 2, and discuss a conjecture concerning a possible solution of said problem.

2   Information Geometry and W^*-algebras

Even if the mathematical frameworks underlying Čencov's and Petz's theorems may seem quite different at first sight, they can be unified in the context of W^*-algebras, namely, C^*-algebras admitting a pre-dual [24]. Roughly speaking, a C^*-algebra is a functional-analytic abstraction of the algebra $\mathrm{M}_n(\mathbb{C})$ of square complex matrices. Specifically, a C^*-algebra is a quintuple $\mathscr{A} \equiv (A, +, \cdot, \|\cdot\|, \dagger)$ where $(A, +, \cdot, \|\cdot\|)$ is a complex, associative Banach algebra endowed with a map $\dagger\colon A \to A$, often written as $\dagger(\mathbf{x}) \equiv \mathbf{x}^\dagger$, which is norm-continuous and satisfies the following properties:

$$(\alpha \mathbf{x} + \beta \mathbf{y})^\dagger = \overline{\alpha}\mathbf{x}^\dagger + \overline{\beta}\mathbf{y}^\dagger \quad \forall \mathbf{x}, \mathbf{y} \in A, \ \ \forall \alpha, \beta \in \mathbb{C} \tag{7}$$

$$(\mathbf{xy})^\dagger = \mathbf{y}^\dagger \mathbf{x}^\dagger \quad \forall \mathbf{x}, \mathbf{y} \in A \tag{8}$$

$$(\mathbf{x}^\dagger)^\dagger = \mathbf{x} \quad \forall \mathbf{x} \in A \tag{9}$$

$$\|\mathbf{x}\mathbf{x}^\dagger\| = \|\mathbf{x}\|^2. \tag{10}$$

A prototypical examples of unital, Abelian C^*-algebra is the algebra $C(\mathcal{X})$ of complex-valued, continuous function on the topological space $\mathcal{X}$, where the norm is the sup-norm and $\dagger$ is just complex conjugation, while the prototypical example of unital, non-Abelian C^*-algebra is the algebra $\mathcal{B}(\mathcal{H})$ of bounded linear operators on the complex Hilbert space $\mathcal{H}$, where the norm is the operator norm and $\dagger$ is the operator adjoint (in particular, when $\mathcal{H} \cong \mathbb{C}^n$ then $\mathcal{B}(\mathcal{H}) \cong \mathrm{M}_n(\mathbb{C})$).

An element $\mathbf{x} \in \mathscr{A}$ is called self-adjoint if $\mathbf{x}^\dagger = \mathbf{x}$, while it is called positive if $\mathbf{x} = \mathbf{yy}^\dagger$ for some $\mathbf{y} \in \mathscr{A}$. Note that positive elements are necessarily self-adjoint. Positive elements are positive functions when $\mathscr{A} = C(\mathcal{X})$, and are positive semidefinite operators when $\mathscr{A} = \mathcal{B}(\mathcal{H})$.

A W^*-algebra $\mathscr{A}$ is a C^*-algebra which is itself the dual space of a Banach space called the **pre-dual** of $\mathscr{A}$ and denoted by $\mathscr{A}_*$. The notion of W^*-algebra is an algebraic abstraction of that of a von Neumann algebra, namely, an involutive subalgebra of some $\mathcal{B}(\mathcal{H})$ which is equal to its double commutant (or closed in the weak operator topology), in the sense that every von Neumann algebra is a W^*-algebra and every W^*-algebra can be concretely represented as a von Neumann algebra [23]. Consequently, every W^*-algebras $\mathscr{A}$ is unital because, when represented as a von Neumann algebra, it must be equal to its double commutant which always contains the identity operator. Of course, every finite-dimensional C^*-algebra is a W^*-algebra, while the same is not true in infinite dimensions. For instance, the C^*-algebra $\mathcal{K}(\mathcal{H})$ of compact linear operators on the complex Hilbert space $\mathcal{H}$ is a C^*-algebra which is not a W^*-algebra (for instance, it does not contain the identity operator). The prototypical example of an Abelian W^*-algebra is $\mathcal{L}^\infty(\mathcal{X}, \mu)$, that is, the algebra of (equivalence classes of) complex-valued measurable functions on the measure space $(\mathcal{X}, \mu)$ which are μ-essentially bounded. From standard results in functional analysis, the pre-dual of $\mathcal{L}^\infty(\mathcal{X}, \mu)$ can be identified with $\mathcal{L}^1(\mathcal{X}, \mu)$ according to

$$\langle f \mid \xi \rangle := \int_{\mathcal{X}} f(x)\, \xi(x)\, \mathrm{d}\mu(x), \tag{11}$$

where $f \in \mathcal{L}^\infty(\mathcal{X}, \mu)$ and $\xi \in \mathcal{L}^1(\mathcal{X}, \mu)$. The prototypical example of non-Abelian W^*-algebra is again $\mathcal{B}(\mathcal{H})$, which is the dual space of the space $\mathcal{TC}(\mathcal{H})$ of trace-class linear operators on $\mathcal{H}$ (therefore, $\mathcal{B}(\mathcal{H})$ is also the double dual of the C^*-algebra $\mathcal{K}(\mathcal{H})$ of compact linear operators on $\mathcal{H}$). In this case, the dual pairing is given by the trace on $\mathcal{H}$ according to

$$\langle \mathbf{x} \mid \rho \rangle := \mathrm{Tr}_\mathcal{H}(\mathbf{x}\,\rho), \tag{12}$$

where $\mathbf{x} \in \mathcal{B}(\mathcal{H})$ and $\rho \in \mathcal{TC}(\mathcal{H})$.

The way in which W^*-algebras provide a unifying framework for classical and quantum information geometry is through the notion of normal states. Given a C^*-algebra $\mathscr{A}$, a positive linear functional ρ is a linear functional in the dual space $\mathscr{A}^*$ such that

$$\rho(\mathbf{x}^\dagger \mathbf{x}) \geq 0 \quad \forall \mathbf{x} \in \mathscr{A}, \tag{13}$$

which means that ρ takes non-negative values on positive elements. A positive linear functional is called **faithful** if $\rho(\mathbf{x}^\dagger \mathbf{x}) = 0$ implies $\mathbf{x} = \mathbf{0}$. The space of positive linear functionals on $\mathscr{A}$ is denoted by $\mathscr{P}(\mathscr{A})$, while the space of faithful positive linear functionals is denoted by $\mathscr{P}_f(\mathscr{A})$. A positive linear functional ρ is a **state** if it is normalized in the sense that $\|\rho\| = 1$. In the particular case in which $\mathscr{A}$ admits a unit $\mathbb{I}$, the norm of a positive linear functional can be proved to satisfy the equality $\|\rho\| = \rho(\mathbb{I})$ [24]. A state is called **faithful** if it is faithful as a positive linear functional. The space of states on $\mathscr{A}$ is denoted by $\mathscr{S}(\mathscr{A})$, while the space of faithful states is denoted by $\mathscr{S}_f(\mathscr{A})$.

Given a state $\rho \in \mathscr{S}(\mathscr{A})$, we can always build a Hilbert space $\mathcal{H}_\rho$ which is naturally associated with ρ by means of the so-called Gelfand-Naimark-Segal (GNS) construction [3]. Specifically, we introduce a pre-Hilbert space structure on $\mathscr{A}$ setting

$$\langle \mathbf{x} \mid \mathbf{y} \rangle_\rho := \rho(\mathbf{x}^\dagger \mathbf{y}), \tag{14}$$

and then introduce the subset $N_\rho = \{\mathbf{x} \in \mathscr{A} \mid \rho(\mathbf{x}^\dagger \mathbf{x}) = 0\}$, which is a left ideal in $\mathscr{A}$ called the Gel'fand ideal. The scalar product $\langle \cdot \mid \cdot \rangle_\rho$ descends to the quotient $\mathscr{A}/N_\rho$ on which it is non-degenerate by construction. Finally, we obtain the (GNS) Hilbert space $\mathcal{H}_\rho$ as the Hilbert-space-completion of $\mathscr{A}/N_\rho$.

If $\mathscr{A}$ is a W^*-algebra, a linear functional $\hat{\xi} \in \mathscr{A}^*$ is called **normal** if it is the image $\hat{\xi} = i(\xi)$ of an element $\xi \in \mathscr{A}_*$ through the canonical immersion i of the pre-dual $\mathscr{A}_*$ into its double dual $\mathscr{A}^*$. In the following, we will often identify $\mathscr{A}_*$ with its image in $\mathscr{A}^*$ through i for the sake of notational simplicity. The space of normal (faithful) positive linear functionals on $\mathscr{A}$ is denoted by $\mathscr{P}_n(\mathscr{A})$ ($\mathscr{P}_{nf}(\mathscr{A})$), while the space of normal (faithful) states on $\mathscr{A}$ is denoted by $\mathscr{S}_n(\mathscr{A})$ ($\mathscr{S}_{nf}(\mathscr{A})$). Of course, in the finite-dimensional case, all linear functionals are normal.

A linear map $\Psi\colon \mathscr{A} \to \mathscr{B}$ between C^*-algebras is called **normal** if its dual map Ψ^* is such that $\Psi^*(\mathscr{B}_*) \subseteq \mathscr{A}_*$; it is called positive if it sends positive elements into positive elements, and it is called completely-positive (CP) if the map $\Psi \otimes 1_n\colon \mathscr{A} \otimes \mathrm{M}_n(\mathbb{C}) \to \mathscr{B} \otimes \mathrm{M}_n(\mathbb{C})$, where 1_n is the identity map on the W^*-algebra of square complex matrices, is positive for all n [5]. When either $\mathscr{A}$ or $\mathscr{B}$

is Abelian, all positive maps are automatically completely-positive. The positive map Ψ is called unital if $\Psi(\mathbb{I}_{\mathscr{A}}) = \mathbb{I}_{\mathscr{B}}$, and a completely-positive map which is unital will be denoted called a CPU map. The dual map Ψ^* of the normal CPU map Ψ preserves normal states in the sense that $\Psi^*(\mathscr{S}_n(\mathscr{B})) \subseteq \mathscr{S}_n(\mathscr{A})$, and Ψ is in addition called faithful if it preserves faithful states. Maps that are normal, CPU and faithful will be called fnCPU maps.

3   Can Čencov Meet Petz?

We are now in a position to present a unified mathematical framework for classical and quantum information geometry in which Čencov's and Petz's theorems would appear as particular cases of a yet to be proved general theorem. The idea is to consider finite-dimensional W^*-algebras and the manifolds of normal faithful states as the unifying objects for probability distributions and quantum states. Indeed, the connection with classical information geometry follows from the fact that the space of normal states of $\mathscr{A} = \mathcal{L}^\infty(\mathcal{X}, \mu)$ is identified with the space of probability measures on $(\mathcal{X}, \mu)$ which are absolutely continuous with respect to μ. Consequently, when considering the finite discrete space $\mathcal{X}_n$ endowed with the counting measure μ, the space $\mathscr{S}_{nf}(\mathscr{A})$ of faithful normal states on $\mathscr{A} = \mathcal{L}^\infty(\mathcal{X}_n, \mu)$ coincides with the interior Δ_n of the n-simplex appearing as the geometrical background of Čencov's theorem discussed in Sect. 1. Analogously, the connection with quantum information geometry follows from the fact that the space of normal states of $\mathscr{A} = \mathcal{B}(\mathcal{H})$ is identified with the space of density operators on $\mathcal{H}$, that is, with the space of quantum states of quantum mechanics. Therefore, when $\mathcal{H} \cong \mathbb{C}^n$, the space $\mathscr{S}_{nf}(\mathscr{A})$ of faithful normal states on $\mathscr{A} = \mathcal{B}(\mathcal{H})$ coincides with the space of faithful (invertible) quantum states appearing as the geometrical background of Petz's theorem discussed in Sect. 1. Then, instead of faithful Markov maps and fCPTP maps, we consider the class of maps which are dual to nfCPU maps. When $\mathscr{A} \cong \mathcal{L}^\infty(\mathcal{X}_n, \mu)$ then these maps reduce to the faithful Markov maps appearing in Čencov's theorem, while when $\mathscr{A} \cong \mathcal{B}(\mathcal{H})$, with $\mathcal{H} \cong \mathbb{C}^n$, these maps reduce to the fCPTP maps appearing in Petz's theorem. Finally, we may formulate the following problem that unifies the ones discussed by Čencov and Petz in a single framework:

Problem 3. Determine all families $\{G_{\mathscr{A}}\}$, where $\mathscr{A}$ is a finite-dimensional W^*-algebra and $G_{\mathscr{A}}$ is a Riemannian metric tensor on $\mathscr{S}_{nf}(\mathscr{A})$, satisfying the monotonicity property

$$\Phi^* G_{\mathscr{A}} \leq G_{\mathscr{B}} \tag{15}$$

for every map Φ which is the dual of a nfCPU map $\Psi \colon \mathscr{A} \to \mathscr{B}$.

Remark 2. If Φ admits a left-inverse which is still the dual of a nfCPU, then Eq. (15) becames an equality.

At the moment, we do not have a complete solution to Problem 3, but what we can show is that Petz's classification, once suitably reformulated in the W^*-algebraic language, provides us with families of Riemannian metric tensors satisfying the monotonicity property in Eq. (15). At this purpose, we note that, being ρ invertible as a density operator, both the left and right multiplication operators L_ρ and R_ρ are invertible as superoperators acting on $\mathcal{B}(\mathcal{H})$. Therefore, every $\mathbf{a} \in T_\rho \mathscr{S}_f(\mathcal{H})$ can be written as

$$\mathbf{a} = \{\rho, \mathbf{v}_\rho\} \equiv \frac{1}{2}\left(\rho\,\mathbf{v}_\rho + \mathbf{v}_\rho\,\rho\right) = \frac{R_\rho}{2}(L_\rho R_\rho^{-1} + \mathbf{1}_{\mathcal{B}(\mathcal{H})})(\mathbf{v}_\rho), \tag{16}$$

and $\mathrm{Tr}_{\mathcal{H}}(\rho\,\mathbf{v}_\rho) = 0$. Then, we recall that the superoperator $L_\rho R_\rho^{-1}$ appears in the framework of W^*-algebras as the generator Δ_ρ of the modular automorphism associated with the state ρ on the GNS Hilbert space associated with ρ [21]. Therefore, we can rewrite Eq. (5) as

$$(G_{\mathcal{H}}^f)_\rho(\mathbf{a}, \mathbf{b}) = \mathrm{Tr}_{\mathcal{H}}\left(\mathbf{a}(K_\rho^f)^{-1}(\mathbf{b})\right) = \frac{1}{2}\mathrm{Tr}_{\mathcal{H}}\left(\rho\{\mathbf{v}_\rho, F(\Delta_\rho)(\mathbf{w}_\rho)\}\right) \tag{17}$$

where

$$F(\Delta_\rho) := (f(\Delta_\rho))^{-1}(L_\rho R_\rho^{-1} + \mathbf{1}_{\mathcal{B}(\mathcal{H})}), \tag{18}$$

so that, recalling the GNS scalar product in Eq. (14), we conclude that

$$(G_{\mathcal{H}}^f)_\rho(\mathbf{v}_\rho, \mathbf{w}_\rho) = \Re\left(\langle \mathbf{v}_\rho \mid F(\Delta_\rho)(\mathbf{w}_\rho)\rangle_\rho\right). \tag{19}$$

The right-hand-side of Eq. (19) is well-defined for every W^*-algebra so that we can define the family $\{G_{\mathscr{A}}^F\}$ of Riemannian metric tensors setting

$$(G_{\mathscr{A}}^F)_\rho(\mathbf{v}_\rho, \mathbf{w}_\rho) = \Re\left(\langle \mathbf{v}_\rho \mid F(\Delta_\rho)(\mathbf{w}_\rho)\rangle_\rho\right) \tag{20}$$

where $\mathbf{v}_\rho, \mathbf{w}_\rho \in T_\rho \mathscr{S}_{nf}(\mathscr{A})$ and $\langle \cdot \mid \cdot \rangle_\rho$ is the GNS Hilbert space product associated with ρ. By suitably adapting the proof of theorem 4 in [20], it is possible to prove that the elements in the family $\{G_{\mathscr{A}}^F\}$ satisfy the monotonicity property in Eq. (15), and thus are admissible monotone families (we omit the proof here because of space constraints and we refer to a forthcoming work for a detailed proof).

Remark 3. After completing this work, it has been pointed out to us that Eq. (20) essentially coincides with the real part of equation [146] in [16], where, however, the author introduces a family of Morozova-Čencov-Petz-like inner products on a possibly infinite-dimensional W^*-algebra. A deeper investigation of the connections between our work an [16] will be the object of future works.

4   Conclusions

The formalism of W^*-algebras and normal states on them provides a natural framework for discussing Čencov's and Petz's problems simultaneously. In particular, solving Problem 3 would allow for a simultaneous generalization of both Problem 1 and Problem 2 in such a way that the solutions of both these problems, as provided by Čencov and Petz respectively, appear as particular cases. We conjecture that the expression appearing in Eq. (20) provides a solution for problem 3. One of the reasons behind our conjecture is that the appearance of a function of Δ_ρ as the unique departure from the GNS Hilbert product is compatible with Čencov's uniqueness result in the classical case because Δ_ρ reduces to the identity for every faithful normal state on a commutative W^*-algebra. We are actively investigating the validity of this conjecture in a setting that extends the one presented here because it deals with not-necessarily finite-dimensional W^*-algebras, thus bringing to the table a whole lot of functional-analytic technicalities.

References

1. Ay, N., Jost, J., Lê, H.V., Schwachhöfer, L.: Information Geometry. EMG-FASMSM, vol. 64. Springer, Cham (2017). https://doi.org/10.1007/978-3-319-56478-4
2. Bhatia, R.: Positive Definite Matrices. Princeton University Press, Princeton (2007)
3. Bratteli, O., Robinson, D.W.: Operator algebras and quantum statistical mechanics I. Springer, Berlin, 2nd edn (1987). https://doi.org/10.1007/978-3-662-03444-6
4. Campbell, L.L.: An extended Čencov characterization of the information metric. Proc. Am. Math. Soc. **98**(1), 135–141 (1986)
5. Choi, M.: A schwarz inequality for positive linear maps on C^*-algebras. Illinois J. Math. **4**(3), 565–574 (1974). https://doi.org/10.1215/ijm/1256051007
6. Chruściński, D., Ciaglia, F.M., Ibort, A., Marmo, G., Ventriglia, F.: Stratified manifold of quantum states, actions of the complex special linear group. Ann. Phys. **400**, 221–245 (2019). https://doi.org/10.1016/j.aop.2018.11.015, https://arxiv.org/abs/1811.07406arXiv:1811.07406 [quant-ph]
7. Ciaglia, F.M., Di Cosmo, F., Ibort, A., Laudato, M., Marmo. G.: Dynamical vector fields on the manifold of quantum states. Open Sys. Inf. Dyn. **24**(3), 1740003–38 (2017). https://doi.org/10.1142/S1230161217400030, https://arxiv.org/abs/1707.00293arXiv:1707.00293 [quant-ph]
8. Ciaglia, F.M., Di Nocera, F., Jost, J., Schwachhöfer, L.: Parametric models and information geometry on W^*-algebras. Inf. Geometry **5**(1) (2023). https://doi.org/10.1007/s41884-022-00094-6, https://arxiv.org/abs/2207.09396arXiv:2207.09396 [math-ph]
9. Ciaglia, F.M., Ibort, A., Jost, J., Marmo, G.: Manifolds of classical probability distributions and quantum density operators in infinite dimensions. Inf. Geom. **2**(2), 231–271 (2019). https://doi.org/10.1007/s41884-019-00022-1
10. Ciaglia, F.M., Jost, J., Schwachhöfer, L.: Differential geometric aspects of parametric estimation theory for states on finite-dimensional C*-algebras. Entropy **22**(11), 1332 (2020). https://doi.org/10.3390/e22111332, https://arxiv.org/abs/2010.14394arXiv:2010.14394 [math-ph]
11. Ciaglia, F.M., Jost, J., Schwachhöfer, L.: From the Jordan product to Riemannian geometries on classical and quantum states. Entropy **22**(06), 637–27 (2020). https://doi.org/10.3390/e22060637, https://arxiv.org/abs/2005.02023arXiv:2005.02023 [math-ph]
12. Ciaglia, F.M., Jost, J., Schwachhöfer, L.: Information geometry, Jordan algebras, and a coadjoint orbit-like construction. arXiv (2021). https://arxiv.org/abs/2112.09781arXiv:2112.09781 [math.DG]

13. Fujiwara, A.: Hommage to Chentsov's theorem. Inf. Geom. (2023). https://doi.org/10.1007/s41884-022-00077-7
14. Grabowski, J., Kuś, M., Marmo, G.: Geometry of quantum systems: density states and entanglement. J. Phys. A: Math. Gener. **38**(47), 10217–10244 (2005). https://doi.org/10.1088/0305-4470/38/47/011.3
15. Kostecki, R.P.: Quantum theory as inductive inference. In: AIP Conference Proceedings, vol. 1305, no. 33 (2011). https://aip.scitation.org/doi/10.1063/1.3573636, https://arxiv.org/abs/1009.2423arXiv:1009.2423 [math-ph]
16. Kostecki, R.P.: Local quantum information dynamics (2016). https://arxiv.org/abs/1605.02063arXiv:1605.02063 [quant-ph]
17. Morozova, E.A., and N. N. Čencov. Markov maps in noncommutative probability theory and mathematical statistics. In Prokhorov, Y.V., Statulevičius, V.A., Sazonov, V.V., Grigelionis, B. (eds.), Probability theory and mathematical statistics: proceedings of the Fourth Vilnius Conference, vol. 2, pp. 287–310. VNU Science Press, Utrecht (1987). https://doi.org/10.1515/9783112313985-026
18. Morozova, E.A., Čencov, N.N.: Markov invariant geometry on manifolds of states. J. Soviet Math. **56**(5), 2648–2669 (1991). https://doi.org/10.1007/BF01095975
19. Nayak, A., Sen, P.: Invertible quantum operations and perfect encryption of quantum states. Quantum Inf. Comput. **07**(01), 103–110 (2007). https://doi.org/10.5555/2011706.2011712
20. Petz, D.: Quasi-entropies for states of a von Neumann Algebra. Publications of the RIMS, Kyoto Univ. **21**, 787–800 (1985). https://doi.org/10.2977/prims/1195178929
21. Petz, D.: Quasi-entropies for finite quantum systems. Rep. Math. Phys. **23**(1), 57–65 (1986). https://doi.org/10.1016/0034-4877(86)90067-4
22. Petz, D.: Monotone metrics on matrix spaces. Linear Algebra Appl. **244**, 81–96 (1996). https://doi.org/10.1016/0024-3795(94)00211-8
23. Sakai, S.: A characterization of W^*-algebras. Pac. J. Math. **6**(4), 763–773 (1956)
24. Takesaki, M.: Theory of Operator Algebra I. Springer-Verlag, Berlin (2002)
25. Čencov, N.N.: Statistical decision rules and optimal inference. Am. Math. Soc. Providence, RI (1982). https://doi.org/10.1090/mmono/053
26. vom Ende, F.: Strict positivity and D-majorization. Linear Multilinear Algebra **70**(19), 4023–4048 (2020). https://arxiv.org/abs/2004.05613arXiv:2004.05613 [quant-ph]. https://doi.org/10.1080/03081087.2020.1860887

Souriau's Geometric Principles for Quantum Mechanics

Frederic Barbaresco[✉]

THALES Land & Air Systems, Velizy-Villacoublay, France
frederic.barbaresco@thalesgroup.com

Abstract. As Landau and Lifchitz have pointed out, the relations between quantum mechanics and classical mechanics are of a very particular type: they coexist instead of succeeding each other. Any analysis of quantum structure is necessarily twofold; especially geometric analysis. We will avoid any ambiguity by putting in the first part, "classic", everything that can be put there. In particular the study of symmetries; and also, as we shall see, the role of Planck's constant. Then two "classically" defined geometric structures (prequantization, generator group) make it possible to determine what the "quantum states" are: the solutions of a certain system of inequalities. In this sense therefore, the program of geometric quantization is carried out. A few examples show that the usual quantum structures are indeed found in all their details. But the choice and the own existence of a generator group possessing quantum states pose to the physicist a new problem for each concrete system; problem which is solved here only in a few cases.

Keywords: Quantum state · Geometric prequantization · Generator group

Souriau work on "Structure of Dynamical Systems" and his symplectic model of mechanics and statistical mechanics were elaborated as preamble of his geometric model of quantum mechanics as explained in his interview: *"In 1958, I returned to France, to Marseille. And there I found myself confronted with theoretical physicists and with the problems of quantum mechanics which had disturbed me during my studies like all students, I think. I realized that symplectic geometry was an indispensable tool for quantum mechanics. And that in fact it was even more appropriate to quantum mechanics than it was to classical mechanics. When I wrote my book on the subject I wanted to write a book on quantum mechanics and I realized that I had to present all classical mechanics in detail, as well as statistical mechanics. They were not foreign theories since they were connected by the symplectic structure and by the symmetries. You take two particles that revolve around each other according to Newton's laws, and then you take a hydrogen atom of which you only see the spectrum. These are two objects that have a priori nothing to do with each other; but they have symplectic symmetries in common. A door is ajar."*
This article is a translation of part 2 of French paper [1] by Jean-Marie Souriau.

© The Author(s), under exclusive license to Springer Nature Switzerland AG 2023
F. Nielsen and F. Barbaresco (Eds.): GSI 2023, LNCS 14072, pp. 372–381, 2023.
https://doi.org/10.1007/978-3-031-38299-4_39

1 Quantum Mechanics

We are going to propose a "quantization" algorithm for a dynamical system whose motion space X has a prequantization Ξ and a generator group S. Classical mechanics suffices to define the symplectic structure of X; the choice of Ξ and of S will therefore be the "additional structure" necessary and sufficient to quantize. But not all concrete dynamical systems possess an S obvious generating group. How to do? Perhaps by giving up demanding that S there be a Lie group; the construction [of generator group] still applies to "diffeological groups" and produces generator groups for all connected symplectic manifolds. But we content ourselves here with the case of Lie groups. A reinterpretation of the states of statistical mechanics suggests the axiomatics of quantum states, functions defined on the quantum extension of S and solutions of a simple set of inequalities. This axiomatic is enough to bring out the common objects of quantum mechanics (probabilistic interpretation, uncertainty relations; Hilbert spaces, operators, unitary representations; mixed states; etc.) in a rigorous mathematical framework, sometimes broader than the usual framework. A few examples illustrate this theory.

1.1 Statistical Mechanics and Probabilities

Liouville Equation: Traditionally, the "states" of classical statistical mechanics are characterized by a distribution function, defined on the phase space, positive solution of the Liouville evolution equation, and whose integral (for the Liouville measure) is equal to 1. Experience shows that such "statistical states (especially Gibbs states) are sometimes a better description of physical reality than the only "movements" of classical mechanics.

Statistical Mechanics on the Space of Motions: Liouville's equation is here a simple expedient to compensate for the non-covariant character of the phase space; there is a strictly equivalent covariant formulation: a "statistical state" is a completely continuous probability law μ on the space of motions X. Moreover, we do without the "completely continuous" restriction, which in fact only serves to write Liouville's equation without qualms; in this case, we can identify each classical movement x with the law of probability $\delta(x)$ concentrated at the point x.

Probability Laws and Groups: But what is the definition of a probability law μ on a set X? One of the standard definitions is to make μ a functional, associating certain functions f defined on X a number noted: $\int_X f(x) d\mu(x)$ or simply $\mu(f)$, and which is called the mean value of f.

In the case of a X locally compact topological space, a probability law μ(in the standard sense of the term) makes it possible in particular to calculate the average values of the continuous functions on X values in the torus T, the set of complex numbers of modulus 1. We note that these functions constitute a multiplicative group Γ, and we can show that the values of the functional μ restricted to Γ suffice to characterize the law of probability. On the other hand, it is easy to establish the following properties of this restricted functional:

$$\mu(e) = 1 (e: \text{ neutral element of } \Gamma)$$

$$0 \le \sum_{jk} \overline{C}_j C_k \mu \left(\gamma_j^{-1} \gamma_k \right); \left| \sum_j C_j \mu(\gamma_j) \right| \le \sup_{x \in X} \left| \sum_j C_j \gamma_j(x) \right|$$

valid whatever the C_j in C, the γ_j in Γ.

New Writing of the Calculus of Probabilities: We are going to take these properties as axioms, not only of classical statistical mechanics, but more generally of the calculus of probabilities: a set X will be made probabilistic by the arbitrary choice of a group Γ of applications $X \rightarrow T$; a probability law will be a complex function μ on Γ verifying the above axioms.

The essential properties of the probabilities are saved:

- μ is extendable, in a single way, by a positive linear functional of mass 1 on the set A of "trial functions", uniform limits of the complex linear combinations of elements of Γ.
- At each point x, the concentrated probability law in x is its functional:

$$\gamma \mapsto \gamma(x)$$

- The set of probability laws is convex.
- The (immediate) definition of the tensor product of the groups Γ, Γ' of two probabilistic spaces X, X'; which defines by duality the tensorial product of two probability laws, that is to say the "compound probabilities".

Etc (A is the C^* algebra generated by Γ; it is obviously commutative. Its Γ Gelfand spectrum $\mathfrak{S}$ is a compact topological spectrum; there exists a canonical map of X on a dense subset of $\mathfrak{S}$. Each Γ probability law can be interpreted as a classical probability law on the compact $\mathfrak{S}$).

Separation: Probability Γ of a set X will be said to be separate if the elements of Γ separate the points of X; in other words if the function: $x \mapsto [\gamma \mapsto \gamma(x)]$ is an injection of X in the group of characters of Γ.

Otherwise, the set of these characters is a quotient of X, where the group Γ and the Γ probability laws descend; this probabilisation of the quotient is separate.

Example: Let us probabilize the numerical space $\mathbb{R}^n$ with the group Γ of "harmonic" functions: $x \mapsto e^{i\langle \omega, x \rangle}$ ($\omega \in \mathbb{R}^n$) which is isomorphic to the additive group of ω. It is clear that Γ separates the points.

Classical probability laws (derived from the topology of $\mathbb{R}^n$) do indeed define Γ probability laws, since the harmonic functions are continuous; they are moreover characterized by these laws: this is Γ Bochner's theorem. We recognize in these functions on Γ the "characteristic functions" in the sense of Poincaré-Levy: $\omega \mapsto \int e^{i\langle \omega, x \rangle} d\mu(x)$.

But there are other Γ laws, for example the characteristic function of a closed subgroup: they are interpreted in the space of x as a law of equal probability on the dual lattice;

the simplest case being the characteristic function of the subgroup {0}, which describes an evenly matched probability in $\mathbb{R}^n$. Objects of this type are obviously prohibited by the classical calculus of probabilities; but they will be useful to us.

1.2 Quantum States and Representations

Statistical and Hamiltonian States: Consider X the space of motions of a system; as mentioned in the introduction, we assume that X is a pre-quantized symplectic manifold, possessing a generating group S, which allows to use all the constructions [and notations of Hamiltonian actions on a pre-quantizable manifold]. In particular, quantum extension G will play an essential role. It can be shown that any classical probability distribution of X is determined by the mean values of the functions $\exp i_{\odot} H$: $\gamma : x \mapsto e^{iH(x)}$. H traversing the space of Hamiltonian functions $\mathfrak{H}$; they form a group Γ. Therefore any classical statistical state of the system is characterized by a functional μ on Γ, satisfying the previous inequalities.

A Pseudo Morphism: We know that there is an isomorphism of the Lie algebra $\mathfrak{H}$ of the group G with $\mathfrak{g}$: $z \mapsto H_z$. Reeb vector R belongs to the center of $\mathfrak{g}$, and satisfies $\exp(2\pi R) = e$, it follows that $\exp(Z)$ depends on the function H_Z only through its additive class modulo 2π; that is to say that there is a map $I : \Gamma \to G$, characterized by the formula: $I(\exp i_{\odot} H_Z) = \exp(Z)$. In the vicinity of the neutral element, I is a diffeomorphism of Γ on G(these two Lie groups have the same dimension); I is not a group morphism (Γ is commutative, but not G); but I induces a morphism on any subgroup of Γ obtained by choosing the Z in a subalgebra of $\mathfrak{g}$, a subgroup that we will call "isotropic". The image by I of an isotropic subgroup is a connected commutative subgroup of G.

Axioms of Quantum States: We are going to define a new type of mathematical object: quantum states. Statistical states μ are complex functions defined on the commutative group Γ, satisfying the previous inequalities; define a quantum state m as a complex function on the noncommutative group satisfying the inequality system:

$$\mu(e)=1 \quad (e: \text{neutral element})$$

$$0 \le \sum_{jk} \bar{C}_j C_k m\left(g_j^{-1} g_k\right), \forall C_i \in C, \forall g_i \in G \text{ in arbitrary finite number}$$

$$\left|\sum_j C_j m\left(\exp(Z_j)\right)\right| \le \sup_{x \in X} \left|\sum_j C_j e^{iH_j(x)}\right|, \forall C_i \in C, \forall Z_j \in \mathfrak{g}, \text{ commuting 2 by 2}$$

The are the H_j Hamiltonian functions corresponding to the Z_j.

Probability Laws Associated with a Quantum State: The last axiom can also be written,

using the application I: $\left|\sum_j C_j [m_{\odot} I](\gamma_j)\right| \le \sup_{x \in X} \left|\sum_j C_j \gamma_j\right|$ as soon as the γ_j belong to the

same isotropic subgroup of Γ. It follows that $m_{\odot} I$ satisfies the axioms of probabilities [previous] not on the group Γ, but on all its isotropic subgroups. Whence the laws of probability in the generalized sense which will be used to "interpret" m. Take for example H any Hamiltonian corresponding to an element Z of the Lie algebra $\mathfrak{g}$, the generating group, of dimension 1, is obviously isotropic; the associated probability law is carried by the variety X, but we know that it descends on the associated separated quotient, which is simply the set $K = H(X)$ of values of H; this portion of the line K is probabilised by the previous "harmonic" functions; the associated probability law will be called the spectrum of the Hamiltonian H in the state m. If K is bounded, the trial functions are all continuous functions on the closure $\overline{K}$ of K(Stone's theorem); the law of probability is therefore a classical probability measure μ supported by $\overline{K}$ and computable by: $\int\limits_{-\infty}^{+\infty} e^{i\omega t} d\mu(\omega) = m(\exp(tZ))$, $\forall t \in \mathbb{R}$. This regardless of the quantum state m. We still obtain a classical spectrum μ whenever the state m is continuous (as a function defined on G), and this whatever the Hamiltonian H. We will see later an example of a generalized spectrum (discontinuous state, unbounded Hamiltonian).

Some Mathematical Properties of Quantum States: We can establish the following propositions:

a) The set M of quantum states of G is a convex and weakly closed part of the space $B(G)$ of bounded functions. $B(G)$ is the dual of a Banach space, namely the space $L^1(G)$ if we endow G with the discrete topology; which defines its weak topology.
b) If M is not empty, M has extremal points, and is equal to the smallest weakly closed convex containing these points.
c) There exists a convex action of the generating group S on M by: $s(m)(g') = m(g^{-1}g'g)$, g denoting a lift of s in G.

   (a) Shows that any uniform, or weak, limit of quantum states is still in quantum state. (c) shows that quantum states are classical objects, in the sense of Felix Klein: they belong to the same S-geometry than classical movements.
   Let m be a quantum state. Then:

a) There is a Hilbert space $\mathbf{H}$, a unitary representation ρ of G on $\mathbf{H}$, a unit vector Ω in $\mathbf{H}$, cyclic for representation, such that, $m(g) = \langle \Omega, \rho(g)\Omega \rangle$ $\forall g \in G$
b) This formula defines $\mathbf{H}$, ρ and Ω up to unit equivalence.
c) Whatever the unit vector $\Psi \in \mathbf{H}$("state vector"), the function m': $m'(g) = \langle \Psi, \rho(g)\Psi \rangle$ is a quantum state. The unitary representations of G thus constructed will be called quantum representations.
d) For the quantum representation associated with a state m to be irreducible, it is necessary and sufficient that m to be extremal in the convex M of the quantum states.
e) Consider ρ a quantum representation, D a "density operator", that is to say a trace operator, positive, of trace 1. Then the formula $m(g) = Trace(D\rho(g))$ defines a quantum state m("mixed state"). If E is a "Gibbs operator", ie a self-adjoint operator such that it exists $\beta_0 \in \mathbb{R}$ satisfying $Trace(e^{-\beta_0 E}) < \infty$ then for all $\beta \geq \beta_0$, the operator $D_\beta = \frac{e^{-\beta E}}{Trace(e^{-\beta E})}$ is a density operator, and therefore defines a quantum state m_β; the limit of m_β(at least for the weak topology) when β tends to $+\infty$ is the quantum state m_∞ associated with the density operator $\frac{\Pi}{r}$, Π denoting the eigenprojector

associated with the smallest eigenvalue of E, and r its rank (the existence of these objects is mandatory for Gibbs operators).

f) For the quantum representation associated with a quantum state m to be continuous (in the sense of the strong topology of the unitary group of H), it is necessary and sufficient that the function m be continuous at the point e; so to any Hamiltonian H is associated a self-adjoint operator $\hat{H}$ of the Hilbert space characterized by: $\rho(\exp(tZ)) = e^{it\hat{H}} \; \forall t \in \mathbb{R}$. If the group G is compact, these operators are bounded, and satisfy the Dirac commutation relations.

Combined with [the previous result (b)], the result (d) shows that the existence of a single quantum state of G is sufficient for there to exist an irreducible quantum representation on a Hilbert space **H**. Then the correspondence "*State vector* $\mapsto$ *State* " produces an injection of the Hilbertian projective space $PC(\mathbf{H})$ into the set of extremal bridges of the convex M (this follows from (b), and from the fact that the unit vectors are all cyclic in the case of an irreducible representation), there exists"many" quantum states. The mixed states defined in (e) obviously have the properties encountered in quantum statistics and quantum chemistry, in particular the Gibbs and Hartree- Fock states. The result (f) obviously makes it possible to establish the link with one of the usual formulas of quantum mechanics; notably by using the spectral decomposition of self-adjoint operators. It will be noted that the application I makes it possible to give to the definition of the "quantified" $\hat{H}$ of a Hamiltonian H suggestive riting: $\left[\rho_\odot I\right]\left(e^{itH}\right) = e^{it\hat{H}}$. But the examples of discontinuous quantum states, below, indicate that the use of " Hamiltonian operators" is only valid in certain cases.

1.3 Examples

We will quickly examine some typical examples of dynamical systems quantizied by the previous method; with, in each case, at least one example of a quantum state, which suffices to establish the existence of an irreducible quantum representation.

Spin: We saw [in first non-translated part] that the space X of "proper motions" of a particle with Galilean spin is a sphere S^2, endowed with the symplectic form $\sigma = s.surf$ that prequantization is only possible if $p = 2s$ is an integer, that one can choose positive; it is unique (up to an equivalence: the quantum manifold is a sphere S^3 if $s = \frac{1}{2}$, a lenticular space if $s \geq 1$). The Galileo group and the reduction rule define a generating group S of X: the group $SO(3)$ of rotations, the associated Hamiltonians are the traces on the sphere of the affine functions of $\mathbb{R}^3$.

Theorem: In any quantum state of this system, the spectrum of the Hamiltonian sx_k (" spin component n°k") is discrete, and contained in the set of $p + 1$ elements:

$$[-s, -s + 1, ..., s - 1, s]$$

The quantum extension G can be constructed as the quotient of the group $U(2)$ by the discrete subgroup $\mathbf{Z}_p$ of the p- th roots of unity. A quantum state m does indeed exist, namely: $m(u\mathbf{Z}_p) = [u_{11}]^p$ with $u = \begin{pmatrix} u_{11} & u_{12} \\ u_{21} & u_{22} \end{pmatrix}$ a unitary matrix; $\mathbf{Z}_p$ is the subgroup of matrices $\begin{pmatrix} z & 0 \\ 0 & z \end{pmatrix}$, with $z^p = 1$. In these states, the spectrum of $s.x_3$ is focused on the value $+s$("spin up"). The spectrum of each of the two other components $s.x_1$ and $s.x_2$ is distributed over the $p + 1$ values allowed according to the symmetric Bernoulli law. The probabilities constitute the expansion of the binomial of $\left[\frac{1}{2} + \frac{1}{2}\right]^p$. It is therefore only for large values of the spin that this state will "look like" a "vertically polarized" statistical state, owing to the "sharp" nature of Bernoulli 's law ("law of large numbers"). The action of the compact group $S = SO(3)$ on the convex M of the quantum states (c), the convexity and the compactness of M, show that the formula: $m'(g) = \underset{s \in S}{mean}\, s(m)(g)$ defines a new quantum state m', manifestly invariant under rotation. The calculation of $m'(uZ_p)$ gives.

- for $s = \frac{1}{2} : m'(uZ_p) = \frac{1}{2} Trace(u)$
- for $s = 1 : m'(uZ_p) = \frac{1}{6}\left[Trace(u^2) + Trace(u)^2\right]$

In this state, the spin projection in any direction has an even spectrum over the $p + 1$ allowed values. This is the "natural" state of spin observed in the Stern and Gerlach experiment from an unpolarized source.

Relativistic Free Particle: A classical motion of the particle is defined by its worldline, which is a timelike curve. These curves constitute a manifold X of dimension 6, on which the Poincaré group S acts transitively. Up to a factor m(the mass of the particle), there exists a single symplectic structure on X which is invariant by the action of S, which becomes generator; the prequantization is unique. In a given inertia reference frame, an element s of the Poincaré group S can be written $s = (L, e, f, g, h)$. L denoting a Lorentz matrix and e, f, g, h four real numbers; acts on s spacetime coordinates t, x, y, z according to the formula: $\begin{pmatrix} t \\ x \\ y \\ y \end{pmatrix} \mapsto L\begin{pmatrix} t \\ x \\ y \\ y \end{pmatrix} + \begin{pmatrix} e \\ f \\ g \\ h \end{pmatrix}$ which defines the group law of S.

Aristotle's subgroup corresponds to the cs where the matrix L"does not mix space and time"; it is then written: $L = \begin{pmatrix} 1 & 0 \\ 0 & R \end{pmatrix}$ (R: rotation matrix). We can take as a basis of the space $\mathcal{H}$of Hamiltonians the following 11 functions:

- energy E and the three components of momentum $\vec{p}$; these quantities are linked by the relation $E^2 - \vec{p}^2 c^2 = m^2 c^4$, $E > 0$;
- The three components of the vector $E\vec{r}$, $\vec{r}$ designating the position of the particle at date 0;
- The three components of angular momentum with respect to the point of origin: $\vec{l} = \vec{r} \times \vec{p}$;

- Finally the constant function 1.

The quantum extension G is trivial: it is the direct product $S \times \mathbf{T}$. An element of G will therefore be written: $g = (s, u)$ $(s \in S, u \in \mathbf{T})$. There is a quantum state m_0, defined by: $m_0(g) = \begin{cases} u \exp(ime) & \text{if } s \text{ is in the Aristotle's group} \\ 0 & \text{otherwise} \end{cases}$.

m is the mass, e the time translation which appears in the preceding equations.

It is visibly invariant by the action of Aristotle's group, that is to say by temporal translations ("stationary states") and by Euclidean displacements ("homogeneous and isotropic state"). Moreover, it is not continuous. One easily calculates the spectra of the Hamiltonians in this state, with the following results:

- The spectra of the energy E and the three components of $\vec{p}$ are simultaneously focused on the values mc^2,0,0,0. These four Hamiltonians generate an isotropic subgroup of Γ, associated with the abelian group of space-time translations. Their image describes a 3-dimensional manifold, the mass hyperboloid K with equation: $E^2 - \vec{p}^2 c^2 = m^2 c^4, E > 0$. Therefore any quantum state defines a probability law on K. Here, this law is concentrated in one point.
- The spectra of E_x, E_y, E_z are evenly distributed on the real line. On the other hand, these three Hamiltonians do not commute, do not belong to the same isotropic subgroup (relativistic effect!); there is no law of probability for the position of the particle in space at a given moment. Similarly, and this is better known, for the angular momentum l.
- The three components of angular momentum each have an $\vec{p}$ equidistributed spectrum on the lattice of integer multiples of $\hbar$.

These equipartitions on $\mathbb{R}$ or $\mathbf{Z}h$ are of course described by the probabilistic moment above. The description of a state endowed with these symmetries (stationary, homogeneous and isotropic) by means of a solution of the Klein-Gordon equation: $\frac{1}{c^2} \frac{\partial^2 \Psi}{\partial t^2} - \Delta \Psi + \frac{m^2 c^2}{\hbar^2} \Psi = 0$ would consist in choosing: $\Psi(x, y, z, t) = e^{\frac{imc^2}{\hbar} t}$ but, even by normalizing it, this function refuses to enter the official Hilbert space. The axiomatization of quantum states, on the other hand, gives a mathematical legitimacy to the state m_0: there is indeed a Hilbert space and an associated quantum representation.

Linear System: Let us consider a linear dynamical system, that is to say whose variety of motions X is a vector space; the symplectic form is invariant under translation. Under these conditions, the prequantization is unique; the group of translations S is generative; the Hamiltonians are the affine functions on X, which can be written using the symplectic form σ as: $H(x) = \sigma(a, x) + b$ $(a \in X, b \in \mathbb{R})$.

Theorem: Given m a quantum state:

- If m has an expansion of order 2 in the neighborhood of the neutral element of G, the spectrum of any Hamiltonian is a classical probability distribution, possessing a finite standard deviation Δ.

- So whatever Hamiltonians H and $H\prime$: $H(x) = \sigma(a, x) + b$, $H\prime(x) = \sigma(a', x) + b'$, the standard deviations Δ and $\Delta\prime$ their spectra satisfy the inequality: $\Delta\Delta\prime \geq \hbar |\sigma(a, a')|$

We recognize, in precise form, the Heisenberg uncertainty relations. The quantum extension G is trivial, which allows us to set: $g = (x, z) x \in X$, $z \in \mathbf{T}$. The calculation gives the group law: $(x, z)(x', z') = \left(x + x', zz' e^{-i\sigma(x,x')/2}\right)$. G is the Heisenberg-Weyl group. We effectively obtain a quantum state by the following procedure: the vector space X (whose dimension is even) can be provided with a complex Hermitian structure $\langle, \rangle$ whose symplectic form is the imaginary part: $\sigma(x, y) = \text{Im}(\langle x, y \rangle)$ then the function m: $m(x, z) = ze^{-\langle x,x \rangle/2}$ is a quantum state of G. In this state m, the spectrum of a Hamiltonian $H(x) = \sigma(a, x) + b$ is the Gaussian measure of center b and standard deviation $\Delta = \|a\|$ ($\|a\|$ denotes Hermitian norm $\|a\|^2 = \langle a, a \rangle$). A simple calculation shows that the Heissenberg uncertainty inequality can be realized strictly, by a suitable choice of a and a'; m is therefore what is called a coherent state. The representation associated with m does not depend on the choice of the Hermitian structure; it is irreducible, it is the "Schrödinger representation of commutation relations". But there are also discontinuous states, which therefore do not belong to the Schrödinger representation.

References

1. Souriau, J.M. : Des principes géométriques pour la mécanique quantique. Act. Acad. Sci. Taurin **124** (Suppl.):296–306. Exposé au colloque du Collège de France : "La Mécanique Analytique de Lagrange et son héritage" (1990)
2. Souriau, J.M.: Des particules aux ondes: quantification géométrique. In: Huygens'principle 1690–1990: theory and applications. Studies in Mathematical Physics, vol. 3, pp. 299–341. North-Holland, Amsterdam (1992)
3. Souriau, JM.: Interpretation geometrique des etats quantiques. In: Bleuler, K., Reetz, A. (eds.) Differential Geometrical Methods in Mathematical Physics. LNM, vol. 570. Springer, Berlin (1977). https://doi.org/10.1007/BFb0087784
4. Souriau, J.M.: Structure des systèmes dynamiques. Dunod, Paris (1970)
5. Souriau, J.M., Groupes différentiels de physique mathématique. In : Feuilletages et quantification géométrique 29 (Journées lyonnaises de la S.M.F., 14–17 juin 1983). Travaux en Cours, vol. 6, pp. 73– 119. Hermann, Paris (1984)
6. Souriau, J.M. : Quantification géométrique. In Physique quantique et géométrie (Colloque Géométrie et Physique, Paris, 16–20 juin 1986), vol. 32, pp. 141–193. Hermann (1988)
7. Souriau, J.M. : Quantique? Alors c'est Géométrique... Exposé au colloque "Feuilletages – Quantification géométrique" (2003). https://hal.campus-aar.fr/medihal-01471022
8. Souriau, J.M. : Indice de Maslov des variétés lagrangiennes orientables, C.R. Acad. Sci. Acad. Sc. Paris, t.276, Série A Physique Mathématique, pp. 1025–1026, 2 avril (1973)
9. Souriau, J.M. : Géométrie globale du problème à deux corps, in IUTAM-ISIMM Symposium in Modern Developments in Analytical Mechanics, Suppl. Atti Acad. Sc. Torino **117**, 369–418 (1983)
10. Souriau, J.M. : Construction explicite de l'indice de Maslov. Applications. Lecture Notes in Phys. vol. 50, pp. 117–148. Springer, Berlin (1976)
11. Souriau, J.M.: Un algorithme générateur de structures quantiques. Astérisque, tome **S131**, 341–399 (1983)
12. Souriau, J.M.: On geometric mechanics. Discrete Cont. Dyn. Syst. **19**(3), 595–607 (2007)

13. Souriau, J.M.: Mécanique statistique, groupes de Lie et cosmologie. In: Colloque International du CNRS "Géométrie symplectique et physique Mathématique", 1974
14. Les formes extérieures en mécanique: Gallisot, F. Annales de I 'Institut Fourier, Grenoble **4**, 145–297 (1952)
15. Barbaresco, F.: Symplectic theory of heat and information geometry. In: Chapter 4, Handbook of Statistics, vol. 46, pp. 107–143, Elsevier (2022)
16. Barbaresco, F.: Jean-Marie Souriau's symplectic model of statistical physics: seminal papers on lie groups thermodynamics - Quod Erat Demonstrandum. In: Barbaresco, F., Nielsen, F. (eds.) Geometric Structures of Statistical Physics, Information Geometry, and Learning. SPIGL 2020. Springer Proceedings in Mathematics & Statistics, vol. 361. Springer, Cham (2020). https://doi.org/10.1007/978-3-030-77957-3_2
17. Barbaresco, F.: Symplectic foliation structures of non-equilibrium thermodynamics as dissipation model: application to metriplectic nonlinear lindblad quantum master equation. Entropy **24**, 1626 (2022)

Unitarity Excess in Schwartzschild Metric

Philippe Jacquet[1]([⊠])[iD] and Véronique Joly[2]

[1] Inria Saclay Ile-de-France, Palaiseau, France
philippe.jacquet@inria.fr
[2] Onera Alumni, Palaiseau, France

Abstract. We refer to the black hole information paradox. We look after the existence of eigenvalues with non zero imaginary part in the Gordon Klein equation with Schwarzschild metric. Such eigenvalues exist because the Schwarzschild metric is singular on the event horizon. The eigenvalues should be proportional to the inverse of black hole radius. The existence has many impacts, among other that black holes should be again eternal. However the effects of the unitary violation should not be detectable within known black holes with existing technologies.

Keywords: Unitarity · Schwartzschild metric · Gordon Klein equation

1 Motivation

It is known that the black hole thermal radiation, described in 1976 [1], leads to the *Black Hole Information Loss Paradox*. This paradox suggests a violation of quantum unitarity. Since then, many solutions have been proposed to dissipate the paradox. In fact collecting all the solutions to the paradox proposed so far is already a challenge. To make it short, the solutions proposed since almost 50 years span over a great variety of ideas:

- (a) information could be encoded in thermal radiation, in correlation between the future and the past of the black hole [2,3];
- (b) information stay hidden in a Planck size Black Hole remain [4];
- (c) information escape in a baby universe [5];
- (d) an hypothetical firewall prevents the information to enter the black hole or simply prevents the formation of the black hole [2,6];
- (e) the black hole has soft hair which make the event horizon to have fluctuations which allow information to escape [7,8];
- (f) a quantum theory of gravity where "graviton" are not massless [9].

Most of the proposed solutions struggle for keeping intact the principle of quantum unitarity. On the other hand the "anti-solutions", *i.e.* the solutions which accept the possibility of an actual violation of unitarity, are by far less numerous. Our aim is to try to recover the simplest non unitary solution. By "simple" we mean that we limit our investigations in a semi-classic Schwarzschild metric. Surprisingly we got imaginary eigenvalues which indeed hold the signature

F. Nielsen and F. Barbaresco (Eds.): GSI 2023, LNCS 14072, pp. 382–391, 2023.
https://doi.org/10.1007/978-3-031-38299-4_40

of non unitary evolution. The reason why we get a non unitary evolution from an equation which is basically unitary, comes from the fact the Schwartzschild metric around a black-hole shows singularities at the event horizon.

However there is a real challenge in the fact that the occurrence of such imaginary eigenvalues should not bring any significant departures from what has been measured via today labs experiments or astronomical observations. To make it simple, if the consequence of non unitary quantum physics would make stones to rush out of our atmosphere toward the closest black hole, then the result of this paper would be highly questionable.

An unexpected outcome of our results (under some hypotheses), is that the average apparent lifetime of a black hole before complete evaporation is made again infinite. This is a kind of paradoxical since it is precisely the evaporation of the black hole which led to the unitary violation hypothesis. Our paper is divided in three main contributions:

1. The handling of the semi classical Klein-Gordon equation of scalar field in Schwarzschild metric for a single particle;
2. the asymptotic estimating of its main eigenvalues when the event horizon radius is large. In particular the eigenvalues with non zero imaginary part have their imaginary part asymptotically smaller than $\frac{1}{4R}$, R being the radius of the black hole, all expressed in Planck unit, the value is rather large since it leads to an unitary excess of one unit per traversal time of the event horizon at light speed;
3. The global behaviour of non unitary black-hole when the number of particles is proportional to the area of the event horizon. Under this hypothesis the non unitary black hole is eternal. But a *doomsday* analysis show that the actual *posterior* effect on quantum measurement is in fact not very important (although real) of the order of $\frac{1}{R^3}$ and might be difficult to detect through measurement.

2   Semi-classic Scalar Field Equation

The scalar field function Ψ under the relativistic Gordon-Klein equation

$$\hbar^2 \Box \Psi + m^2 c^2 \Psi = 0. \tag{1}$$

The d'Alembertian operator $\Box \Psi$ defined as follow

$$\Box \Psi = \frac{1}{\sqrt{\det(\mathbf{g})}} \sum_{ij} \partial_i (g^{ij} \sqrt{\det(\mathbf{g})} \partial_j \Psi)$$

where $\mathbf{g}$ is the metric matrix. The relativistic version of the operator $\partial_t^2 \Psi - \Delta \Psi$, where Δ is the spatial Laplacian.

2.1 Schwarzschild Metric

Assuming an isolated black hole of event horizon radius R and a point with polar coordinates (r, θ, ϕ) taken from black hole center, r is the distance to the center, θ is the polar angle, and ϕ the azimuthal angle, the metric matrix is:

$$\mathbf{g} = [g_{ij}] = \text{diag}(-(1 - \frac{R}{r})^{-1}, -r^2, -r^2 \sin^2 \phi, (1 - \frac{R}{r}))$$

$$\mathbf{g}^{-1} = [g^{ij}] = \text{diag}((-(1 - \frac{R}{r}), r^{-2}, -r^{-2} \sin^{-2} \phi, (1 - \frac{R}{r})^{-1}).$$

where diag(.) is the diagonal matrix. The term in $(1 - \frac{R}{r})^{-1}$ confirms that the metric is singular on the event horizon (the points at distance R from the center, and on the center itself. Under a spherical symmetry hypothesis which makes $\Psi = \Psi(r, t)$ to depend only on parameter r and time t, the Gordon-Klein equation becomes:

$$\frac{1}{1 - R/r} \frac{1}{c^2} \partial_t^2 \Psi - \frac{1}{r^2} \partial_r (r^2 (1 - R/r) \partial_r \Psi) + m^2 \beta^2 \Psi = 0 \qquad (2)$$

with $\beta^2 = c^2 / \hbar^2$.

3 Eigenvectors of Gordon-Klein Equation

In this section we will assume R large compared to Compton wavelength $\lambda_C = \frac{\hbar}{mc}$. The case where R is of the same order than the Compton wavelength will be investigated in a further section.

We investigate the case where $\Psi(r, t) = e^{-i\omega t} \Psi(r)$ thus ω is an eigenvalue and function $\Psi(r)$ is the spatial part at time $t = 0$. Quantity ω is a complex number; in an unitary setting have the imaginary part $\Im(\omega)$ should always be null. In any case the function $\Psi(r)$ satisfies the equation:

$$(1 - R/r) \left(\frac{1}{r^2} \partial_r (r^2 (1 - R/r) \partial_r \Psi) - m^2 \beta^2 \Psi \right) = -\frac{\omega^2}{c^2} \Psi.$$

thus it comes that $\Psi(r, t) = e^{\pm i\omega t} \Psi(r)$ since $\partial_t^2 \Psi(r, t) = -\omega^2 \Psi(r, t)$.

We consider as area of interest the horizon area where $r \approx R$. The central area with $r = 0$ can be investigated by analytical continuation but does not give interesting insight.

3.1 Analysis in the Horizon Area

In the horizon area with $r = R + x$, we have the first order approximation

$$\frac{1}{R^2} x \partial_x (x \partial_x \Psi(r)) - \frac{m^2 \beta^2}{R} x \Psi(r) = -\frac{\omega^2}{c^2} \Psi(r).$$

We introduce the auxiliary function f_a which satisfies for some $a \in \mathbb{C}$,

$$x\partial_x(x\partial_x f_a(x)) - xf_a(x) = a^2 f_a(x).$$

In other words we are looking for the eigenvectors of the operator $x\partial_x(x\partial_x f(x)) - xf(x)$, for the eigenvalue a^2. When $R \to \infty$ we have $\Psi(r) \approx f_a(Ax)$ with $A = Rm^2\beta^2$ and $a = \pm i\frac{R}{c}\omega$. Indeed the above equation can be rewritten in

$$x\partial_x(x\partial_x\Psi(r)) - Ax\Psi(r) = -\frac{R^2}{c^2}\omega^2\Psi(r).$$

where $f_{iR\omega/c}(Ax)$ is clearly the solution. Let $f_a^*(s)$ denotes the Mellin transform of $f_a(x)$, it must satisfy:

$$s^2 f_a^*(s) - f_a^*(s+1) = a^2 f_a^*(s)$$

or $(s-a)(s+a)f_a^*(s) = f_a^*(s+1)$ which has solution $f_a^*(s) = \Gamma(s-a)\Gamma(s+a)$, where $\Gamma(.)$ denotes the Euler function.

Lemma 1. *When $a \neq 0$ we have $f_a(x)$ which behave like $x^{-|\Re(a)|}$ when $x \to 0$, where $\Re(a)$ denotes the real part of a. The quantity $f_0(x)$ behaves like $\log x$ when $x \to 0$. In both case we have $f_a(x)$ which exponentially decays when $x \to +\infty$.*

Proof. For convenience, we assume that $\Re(a) \geq 0$. The Mellin transform f_a^* is defined for all s such that $\Re(s) > \Re(a)$. The inverse Mellin transform tells that $f_a = \frac{1}{2i\pi}\int_{c-i\infty}^{c+i\infty} f_a^*(s)x^{-s}dx$ for c in the definition domain of $f_a^*(s)$. When moving the integration line toward the left one meets a first pole on $s = a$ with residue $x^{-\lambda}\Gamma(2a)$ which gives the leading term. When $a = 0$, the pole becomes a double pole at $s = 0$ which has residue $\log(x)$. See Fig. 1 for a plot of function $f_0(x)$.

Theorem 1. *The eigenvalue ω of the wave function is $\frac{ic}{R}a$ with $\Re(a) \in]-\frac{1}{4}, \frac{1}{4}[$ and the eigenvector is close to $f_a(Ax)$ with $A = Rm^2\beta^2$, m being the mass of the particle.*

Proof. We notice that the local integrability of $|\Psi(r,t)|^2$ on a neighbourhood of the event horizon implies that $|\Re(a)| < \frac{1}{4}$. Indeed in Schwartzschild metric the integral $\int |\Psi(r,t)|^2\sqrt{\det(\mathbf{g}_t)}drd\theta d\phi \approx \pi^2 \int |f_a(Ax)|R^{5/2}x^{-1/2}dx$ ($\mathbf{g}_t$ being $\mathbf{g}$ restricted on its spatial components $\det(\mathbf{g}_t) \approx \frac{R^5}{x}\sin\phi$.

We have $\frac{R}{c}\Im(\omega) \in]-\frac{1}{4}, \frac{1}{4}[$ which allows for the possibility of non zero imaginary part. A more involved analysis shows that when $R \to \infty$ the admissible values of $i\frac{R}{c}\omega$ forms a countable set which ends to be dense in the strip $\{s, \Re(s) \in]-\frac{1}{4}, \frac{1}{4}[\}$. What is remarkable in the above analysis is that the asymptotic eigenvalues do not depend on the mass m.

3.2   Preponderent Solution, Unitarity Excess Rate

From the above we get that $\Psi(r,t) = \Psi(r)e^{\pm i\omega t}$. The quantity $|\Psi(r,t)|^2$ integrated on a whole spatial slice at time t would be in $e^{\pm 2\Im(\omega)t}$. The solutions with positive or negative $\Im(\omega)$ have same spatial expression. However the preponderant solutions are those with the negative imaginary part of factor ω. Indeed if $\Im(\omega) < 0$, then a combination of $e^{-i\Im(\omega)t}$ and $e^{i\Im(\omega)t}$ will have its components other than $e^{-2\Im(\omega)t}$ vanishing exponentially before $e^{-2\Im(\omega)t}$. Therefore the solutions with negative $\Im(\omega)$ are preponderant when $t \to +\infty$. Let's define the *unitarity excess* at time T as the logarithm of the integral of $|\Psi(r,t)|^2$ on a space-time slice defined by $t = T$, and the *unitarity excess rate* the unitarity excess derived by T. When $\Psi(r,t) = e^{-i\omega t}\Psi(r)$ the unitarity excess should be $2\Im(\omega)T + \log \int |\Psi(r)|^2$ and the unitarity excess rate should be $2\Im(\omega)$.

One should notice that since the largest value of $2\Im(\omega)$ is $\frac{c}{2R}$, the unitarity excess rate correspond to one unit per time taken by light to travel over a distance equivalent to the horizon diameter, which is rather considerable. The traversal time of the horizon of a black hole of one solar mass would be 10^{-5} seconds, for the massive black hole in the center of our galaxy it would be 3 min.

The results we have so far, *i.e.* that the imaginary part of the main eigenvalue is asymptotically equal to $\frac{c}{4R}$, are valid as long as the Compton wavelength $\frac{\hbar}{mc}$ of the particle is negligible, compared to horizon radius R. When $m\frac{c}{\hbar}$ is of order $1/R$ we are no longer in this situation. But a careful look leads to the fact that in the case where the Compton wavelength is proportional to $1/R$, the eigenvalues of the Klein-Gordon equation are also proportional to $1/R$ for all radius R. However this non asymptotic case makes the analysis difficult, since the Klein-Gordon equation looks much more complicated to solve. Anyhow by continuity we assume that the main eigenvalue is $\frac{c}{4R}$ as with the small Compton wavelength, although this is not a proof.

4   Impact of Unitarity Excess in Quantum Measurement

From now we switch to *Planck units* to simplify the presentation (for which $c = 1$, $h = 1$, $G = 1$, *etc.*). In Planck unit, black hole mass M and black hole radius R are just linked by the relation $R = 2M$. In this section we assume that the average number N of particles, or independent constituent at the event horizon, depends on the radius R. The classic hypothesis is that this number is proportional to the area of the event horizon: $N = N(R) \approx \gamma R^2$ for some constant γ. There are several possibilities for γ but they are mostly of the same order. The largest estimate would be $\gamma = \frac{\pi}{\log 2}$, as if an event horizon particle would represents the minimal mass for carrying one bit of the entropy of the black hole. This would be in accordance with the conjectured holographic principle [10] based on the remark that the black hole entropy is proportional to the event horizon area. The smallest value would be $\gamma = \frac{1}{\alpha}$ where $\frac{\alpha}{R}$ is the average mass of one evaporated photon when the black hole is of radius R. (see next section for the estimate of α). Whatever the estimate of γ, the consequence of the hypothesis $N = O(R^2)$

makes any black hole to be again eternal with an infinite apparent average delay between each evaporated particle. This will be detailed in a next subsection.

4.1 Posterior Impact of Unitarity Excess on Binary Events

The unitary excess rate of the whole black hole system of radius R is $U(R) = \frac{N(R)}{2R}$, since it is made of $N(R)$ particles, each having the same eigenvalue imaginary part $\frac{1}{4R}$. We define the *final unitarity excess* as the integral of the average unitary excess rate computed at the end of the black hole lifetime, and let's denote it $\Omega(R)$ where R is the black-hole initial radius R. Notice that the integral may not be trivial since the radius R and N may randomly vary with the time.

Let $\Psi_R(r,t)$ be the wave function at time t of a particle of the black hole of radius R, we have $\partial_t \Psi_R(r,t) = -i\omega_R \Psi(r,t)$ with $\Im(\omega(R)) = \frac{1}{4R}$, thus

$$\Psi_R(r,T) = \exp\left(-i\int_0^T \omega(R(t))dt\right)\Psi_R(r,0),$$

where $R(t)$ is the black hole radius at time t. The integral of the squared wave function $\int |\Psi_R(r,t)|^2$ on a space slice $t = T$ is equal to

$$\int |\Psi_R(r,0)|^2 \exp\left(\int_0^T 2\Im(\omega R(t))\right)dt).$$

The density operator $\rho(R)$ of the black hole, made of $N(R)$ particles (*a priori* in mixed state for a non zero entropy, but it does not matter here) satisfies

$$\mathrm{trace}(\rho(R(T))) = \exp\left(\int_0^T N(R)2\Im(\omega(R))dt\right)\left(\int |\Psi_R(r,0)|^2\right)^{N(R)}.$$

This quantity is interpreted as a probability, and after evaporation this quantity is equal to $e^{\Omega(R)}\left(\int |\Psi(r,0)|^2\right)^{N(R)}$ when $T \to \infty$, since $\Im(\omega_R) = \frac{1}{4R}$. By normalization (no unitarity excess at time 0) we assume that $\int |\Psi_R(r,0)|^2 = 1$

Let assume two black holes of respective initial radii R_1 and R_2. We consider the following binary event (*e.g.* a spin measurement over a particle made on Earth with outcome $(-1,+1)$ with fair probability $(\frac{1}{2}, \frac{1}{2})$:

– if the measurement is $+1$ then a mass m is thrown in black hole 1;
– if the measurement is -1, then the mass is thrown in black hole 2.

One should notice that the natural fate of any mass in the galaxy is to be eventually absorbed either by the central black hole of our galaxy or of another galaxy, if the mass is expelled with the liberation speed. The surprising effect of unitary excess is that it may have a posterior impact on the outcome probability of the measurement made on Earth. The time line of the outcome $+1$ is affected with an unitarity excess of $\exp(\Omega(R_1 + 2m))\exp(\Omega(R_2))$, since the first black hole radius rises to $R_1 + 2m$ and the second black hole radius stays at R_2. In Planck unit the radius of a black hole of radius R which absorbs a mass m is exactly $R + 2m$. The timeline of the outcome -1 is affected by a final unitarity excess of $\exp(\Omega(R_1))\exp(\Omega(R_2 + 2m))$. Considering the unitarity excess as a posterior impact on the initial probabilities at time $t = 0$ (the simplest way to interpret unitarity excess), the final probabilities of event -1 and $+1$ are:

$$P(-1) = \frac{\exp(\Omega(R_2 + 2m))\exp(\Omega(R_1))}{\exp(\Omega(R_2 + 2m))\exp(\Omega(R_1)) + \exp(\Omega(R_2))\exp(\Omega(R_1 + 2m))}$$

$$= \frac{1}{1 + \exp\left((\Omega(R_2) - \Omega(R_2 + 2m) + \Omega(R_1 + 2m) - \Omega(R_1)\right)} \tag{3}$$

If this quantity differs of $\frac{1}{2}$, it characterizes a *posterior* effect. However we will show that all unitarity excesses are infinite, and this complicates the analysis.

4.2   The Evaporation Effect Impacted by Unitarity Excess

Let $\beta = \frac{\zeta(3)}{1024\pi^2}$ and $\alpha = \frac{\pi^4}{16\zeta(3)}$ where $\zeta(.)$ is the Riemann *zeta* function. The evaporation is considered like a classical black body thermal radiation. Without taking into account the loss of unitarity we have a rate $\frac{\beta}{R}$ of particle evaporation, mostly a photon with a wavelength $\frac{R}{R}$ in Planck unit (thus equivalent to the subsequent black hole loss of mass). The temperature of a black hole of radius R is $\frac{1}{8\pi R}$ in Planck unit.

Theorem 2. *Let a black hole with large radius, then $\Omega(R) = \infty$ and the black hole has an infinite average lifetime.*

Proof. Let $R(t)$ be the radius of the black hole at time t. In the transition from time t to time $t + dt$ the estimate of the quantity $\Omega(R(t))$ increases of $U(R(t))dt$. Meanwhile, with probability $\frac{\beta}{R}dt$, the black hole loses a mass $\frac{\alpha}{R}$. This translates into the following functional equation where $R = R(t)$

$$\exp(\Omega(R)) = e^{U(R)dt}\left((1 - \frac{\beta}{R}dt)\exp(\Omega(R)) + \frac{\beta}{R}dt\exp(\Omega(R - \alpha/R))\right) \tag{4}$$

which when $dt \to 0$ resolves into

$$0 = U(R) - \frac{\beta}{R}\left(1 - \exp\left(\Omega\left(R - \frac{\alpha}{R}\right) - \Omega(R)\right)\right)$$

Taking $N(R) \approx \gamma R^2$, thus $U(R) \approx \gamma R/2$, the final identity

$$\exp\left(\Omega\left(R - \frac{\alpha}{R}\right) - \Omega(R)\right) = 1 - \frac{\gamma}{2\beta}R^2$$

is clearly not solvable for large enough values of $R(t)$, in theory larger than $\sqrt{2\beta/\gamma}$, sufficient to make the right hand side non positive, but this corresponds to black holes with tiny masses for which the evaporation theory is not yet established.

Notice that in this analysis we assume that the black hole is isolated and does not absorb any other mass (this would not change the nature of the result as long as the quantity of absorbed mass remain the same on the two timelines). The proof of the infinite lifetime is given next with the "Doomsday" analysis.

4.3 Posterior Effect Analysis in "Doomsday" Hypothesis

In this section, we use the "Doomsday" argument to get around the problem of infinite excess unitarity. The trick is similar to the calculation of the Casimir effect where two infinite terms subtracted produce a finite result. We assume that a fictitious term T applies to any object subject to a non-zero unitary excess and we let T grow to infinity. Let $\Omega(T, R)$ be the excess unitarity accumulated by a black hole of radius R during a time interval of length T. This leads to a new expression of the Eq. (4):

$$\exp(\Omega(T, R)) = e^{U(R)dt} \tag{5}$$
$$\times \left(\left(1 - \frac{\beta}{R}dt\right) \exp(\Omega(T - dt, R)) + \frac{\beta}{R}dt \exp(\Omega(T - dt, R - \frac{\alpha}{R})) \right).$$

where $U(R)$ is the unitary excess rate of the black hole system. In fact to avoid trivially not acceptable solutions we should consider a black hole system made by a single mass 2β plus γR^2 particles so that $U(R) = \gamma \frac{R}{2} + \frac{\beta}{R}$. When $dt \to 0$ we get the equation:

$$0 = U(R) - \partial_T \Omega(T, R) - \frac{\beta}{R} \left(1 - \exp(\Omega(T, R - \frac{\alpha}{R}) - \Omega(T, R)) \right) \tag{6}$$

Theorem 3. *Under the doomsday analysis we have the posterior impact when R_1 and R_2 are much larger than unity and m is much smaller than R_1 and R_2:*

$$P(-1) = \cfrac{1}{1 + \exp\left(-\frac{4\beta^2}{\alpha\gamma^2} \left(\frac{1}{(R_1+2m)^2} - \frac{1}{R_1^2} + \frac{1}{R_2^2} - \frac{1}{(R_2+2m)^2}\right) + O(\frac{1}{R_1^4} + \frac{1}{R_2^4})\right)} \tag{7}$$

$$\approx \frac{1}{2} + \frac{m\beta^2}{\alpha\gamma^2} \left(R_2^{-3} - R_1^{-3}\right) \tag{8}$$

Thus the posterior effect is small but not null, however paving the way to a theoretical ability to transmit information backward in time, but not in a practical way with existing technologies.

Proof. To simplify the discussion we approximate $\Omega(T, R - \frac{\alpha}{R}) - \Omega(T, R)$ by $-\frac{\alpha}{R}\partial_R\Omega(T, R)$. Notice that both expressions are wrong for small values of R because the evaporation theory does not apply well to small radii. Thus our analysis will only be valid for large values of radius R. Taking the second expression, Eq. (6) becomes

$$0 = \frac{\gamma R}{2} - \partial_T \Omega(T, R) + \frac{\beta}{R} \exp(-\frac{\alpha}{R}\partial_R\Omega(T, R)). \tag{9}$$

Thus if we integrate, keeping the fact that $\Omega(0, R) = 0$:

$$\Omega(T, R) = \frac{\gamma R}{2}T + \frac{\beta}{R}\int_0^T \exp(-\frac{\alpha}{R}\partial_R\Omega(t, R))dt.$$

Let $C(R) = \frac{\beta}{R}\int_0^\infty \exp(-\frac{\alpha}{R}\partial_R\Omega(t, R))dt$, we have

$$\Omega(T, R) = \frac{\gamma R}{2}T + C(R) - \int_T^\infty \exp(-\frac{\alpha}{R}\partial_R\Omega(t, R))dt.$$

The last term is an exponentially decreasing function of T. It turns out that indeed $\Omega(T, R) \to \infty$ when $T \to \infty$. The consequence is that black holes are now back to be eternal as it was assumed before Hawking.

Under the assumption of an absolute black hole lifetime limit set at T, the quantity $\frac{e^{\Omega(t,R)}}{e^{\Omega(T,R)}}$, for any $t < T$, is interpreted as the probability that the apparent lifetime of the black hole is less than t. The reason is that if the black hole evaporates completely before t, then the system returns to simple unitary mode, which means that between t and T the imaginary parts of its eigenvalues are all zero. But since $\lim_T \Omega(T, R) = \infty$, it follows that the probability that the lifetime of the black hole is less than t is in fact 0 whatever t since for any fixed t $\frac{e^{\Omega(t,R)}}{e^{\Omega(T,R)}} \to 0$ when $T \to \infty$. Moreover, the black hole simply "abstains" from radiating. This surprising effect is not detectable because the thermal radiations of known black holes are already far too weak (in R^{-2}) to be observed, even compared to the cosmic background radiation.

Coming back to the posterior impact analysis, the terms in $\gamma RT/2$ cancel in (3):

$$P(-1) = \lim_{T\to\infty} \frac{1}{1 + \exp\left((\Omega(T, R_2) - \Omega(T, R_2 + 2m) + \Omega(T, R_1 + 2m) - \Omega(T, R_1)\right)}$$

$$= \frac{1}{1 + \exp\left(C(R_1 + 2m) - C(R_1) + C(R_2) - C(R_2 + 2m)\right)}$$

We notice that the expression of $P(-1)$ does not trivially tend to 0 or 1 when $T \to \infty$ because we choose $U(R) = \gamma\frac{R}{2} + \frac{\beta}{R}$. We could have chosen $U(R) = CR + D + \frac{\beta}{R}$, with arbitrary constants C and D. It is particularly important to have these cancellations because a trivial convergence of $P(-1)$ would systematically make all masses in the universe to rush toward a single black hole.

If we make the approximation $\partial_R\Omega(t, R) = \gamma t/2 + \partial_R C(R)$ (ignoring the exponentially decreasing part), and plug it in a Taylor expansion in $\frac{1}{R}$:

$$C(R) = \frac{\beta}{R}\int_0^\infty \exp\left(-\frac{\alpha}{2R}\gamma t - \partial_R C(R)\right) dt = \frac{2\beta}{\gamma\alpha} - \frac{4\beta^2}{\gamma^2\alpha R^2} + O(\frac{1}{R^4}).$$

The first term cancels in the expression of $P(-1)$ and ignoring the $1/R^4$ terms:

$$P(-1) \approx \frac{1}{1 + \exp\left(-\frac{4\beta^2}{\alpha\gamma^2}\left(\frac{1}{(R_1+2m)^2} - \frac{1}{R_1^2} + \frac{1}{R_2^2} - \frac{1}{(R_2+2m)^2}\right)\right)}.$$

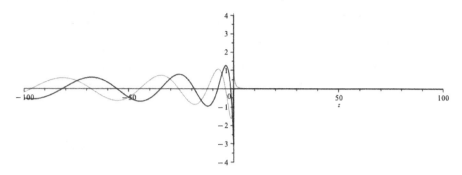

Fig. 1. function $f_0(x)$, real part (brown), imaginary part (blue) (Color figure online)

5 Conclusion

We have investigated the possibility of imaginary eigenvalues for the Gordon Klein equation with the Schwarzschild metric of a black hole. We have found that indeed imaginary eigenvalues exist and have order $\frac{ic}{4R}$, leading to a considerable unitary excess rate, being the signature of an unitary violation which comes along with the black hole information paradox.

However, under appropriate hypotheses, this unitary violation would lead to physical effects ranging from radiation suspension to posterior impact, all effects of order R^{-2} which should not be detectable on known black holes with existing technologies.

References

1. Hawking, S.W.: Breakdown of predictability in gravitational collapse. Phys. Rev. D **14**(10), 2460–2473 (1976)
2. Raju, S.: Lessons from the information paradox. Phys. Rep. **943**, 1–80 (2022)
3. Hartle, J.B.: Generalized Quantum Theory in Evaporating Black Hole Spacetimes. Black Holes and Relativistic Stars, p. 195 (1998)
4. Giddings, S.: Black holes and massive remnants. Phys. Rev. D **46**(4), 1347–1352 (1992)
5. Preskill, J.: Do black holes destroy information? In: International Symposium on Black Holes, Membranes, Wormholes, and Superstrings (1992)
6. Mathur, S.D.: The information paradox: a pedagogical introduction. Class. Quantum Gravity **26**(22), 224001 (2009)
7. Mathur, S.D.: The fuzzball proposal for black holes: an elementary review. Fortschritte der Physik. **53**(7–8), 793–827 (2005)
8. Hawking, S.W., Perry, M.J., Strominger, A.: Soft hair on black holes. Phys. Rev. Lett. **116**(23), 231301 (2016)
9. Geng, H., Karch, A.: Massive islands. J. High Energy Phys. **2020**(9), 121 (2020)
10. Susskind, L.: The World as a Hologram. J. Math. Phys. **36**, 6377–6396 (1995). https://doi.org/10.1063/1.531249. arXiv hep-th/9409089

Modelling and Structure-Preserving Discretization of the Schrödinger as a Port-Hamiltonian System, and Simulation of a Controlled Quantum Box

Gabriel Verrier, Ghislain Haine$^{(\boxtimes)}$ ⓘ, and Denis Matignon ⓘ

ISAE-SUPAERO, Université de Toulouse, Toulouse, France
Gabriel.verrier@student.isae-supaero.fr,
{ghislain.haine,denis.matignon}@isae.fr

Abstract. The modelling of the Schrödinger Equation as a port-Hamiltonian system is addressed. We suggest two Hamiltonians for the model, one based on the probability of presence and the other on the energy of the quantum system in a time-independent potential. In order to simulate the evolution of the quantum system, we adapt the model to a bounded domain. The model is discretized thanks to the structure-preserving Partitioned Finite Element Method (PFEM). Simulations of Rabi oscillations to control the state of a system inside a quantum box are performed. Our numerical experiments include the transition between two levels of energy and the generation of Schrödinger cat states.

Keywords: port-Hamiltonian systems · open quantum systems

1 Introduction

Over the past two decades, the port-Hamiltonian (pH) theory has continued to develop as a preferential paradigm to describe distributed parameter systems [13]. This formalism is judicious to model and control complex dynamical systems: subsystems communicate *via* ports and the power balance is incorporated within the pH structure. It has been successfully applied to many fields such as structural mechanics, electromagnetism and fluid mechanics (see *e.g.* [3,8,13] and references therein).

Controlling and stabilizing a quantum system in a desired state, such as two-state quantum systems called qubits, is the key of quantum information science [12]. A first attempt in applying the pH theory to quantum systems is proposed in the present work, together with promising numerical experiments.

The modelling of the Schrödinger equation including a control is presented in Sect. 2. In Sect. 3, a discrete (in space) port-Hamiltonian system is obtained by applying PFEM [3,5]. Finally, Sect. 4 ends this work with Rabi oscillations simulation results, *i.e.* controlling the state of a quantum box.

More details, developments and discussions on this work can be found in [14].

Supported by the AID from the French Ministry of the Armed Forces.

2   Port-Hamiltonian Modelling

To take advantage of the distributed port-Hamiltonian paradigm, we adopt the wave function representation $\Psi(\mathbf{r}, t)$ of a quantum system [6]:

$$H\Psi(\mathbf{r}, t) = i\hbar \frac{\partial \Psi(\mathbf{r}, t)}{\partial t}. \tag{1}$$

This is the Partial Differential Equation (PDE) form of the Schrödinger Equation. The Hamiltonian operator H is given *via*:

$$H\Psi(\mathbf{r}, t) := -\frac{\hbar^2}{2m}\Delta\Psi(\mathbf{r}, t) + V(\mathbf{r}, t)\Psi(\mathbf{r}, t).$$

The control of the wave function $\Psi(\mathbf{r}, t)$ appears in the potential $V(\mathbf{r}, t)$ of the Hamiltonian operator. To build a port-Hamiltonian system, we have to define ports, and especially those composed of controls and observations. Moreover, a Hamiltonian *functional*, a form depending on the *states*, has to be chosen [13].

2.1   Ports: Control and Observation

By splitting the potential $V(\mathbf{r}, t)$ into two terms, one positive and stationary $V_s(\mathbf{r})$ and the other time-dependent $V_c(\mathbf{r}, t)$, we suggest an extension of the Schrödinger Equation. We call this new equation the Controlled Schrödinger Equation:

$$H_s\Psi(\mathbf{r}, t) + i\hbar u(\mathbf{r}, t) = i\hbar \frac{\partial \Psi(\mathbf{r}, t)}{\partial t}, \tag{2}$$

where $u(\mathbf{r}, t)$ denotes the control. The time-independent part of the Hamiltonian operator is then:

$$H_s\Psi(\mathbf{r}, t) := -\frac{\hbar^2}{2m}\Delta\Psi(\mathbf{r}, t) + V_s(\mathbf{r})\Psi(\mathbf{r}, t).$$

Thanks to this point of view, Ψ follows the Schrödinger Eq. (1) if and only if $u(\mathbf{r}, t) = -\frac{i}{\hbar}V_c(\mathbf{r}, t)\Psi(\mathbf{r}, t)$. Thus, the control u is given by a linear state feedback (see Fig. 2). This imposes the wave function $\Psi(\mathbf{r}, t)$ to be the observation.

2.2   The Hamiltonian of the pH System

The choice of the Hamiltonian functional is led by the Ehrenfest Theorem, which provides the evolution of the expectation value of quantum operators (hereafter the identity and H_s), similarly to the evolution equation of observables in classical Hamiltonian mechanics [6]. We define two possible Hamiltonians:

$$\mathcal{H}_\mathcal{P} := \int_{\mathbf{r}\in\mathbb{R}^3} |\Psi(\mathbf{r}, t)|^2 d\mathbf{r},$$

$$\mathcal{H}_\mathcal{N} := \int_{\mathbf{r}\in\mathbb{R}^3} \frac{\hbar^2}{2m}\|\mathbf{grad}\Psi(\mathbf{r}, t))\|_{\mathbb{C}^3}^2 + V_s(\mathbf{r})|\Psi(\mathbf{r}, t)|^2 d\mathbf{r}.$$

The former $\mathcal{H}_\mathcal{P}$ is based on the *probability of presence*: according to Born rule [6], $\mathcal{H}_\mathcal{P} = 1$. The latter $\mathcal{H}_\mathcal{N}$ is based on the *energy*: it is preserved in absence of a time-dependent potential of control. From an analytic point of view, these two Hamiltonians are norms squared of the wave function w.r.t Hermitian inner products. Indeed, $\mathcal{H}_\mathcal{P} = \langle \Psi | \Psi \rangle$ and $\mathcal{H}_\mathcal{N} = \langle \Psi, \Psi \rangle_\mathcal{N}$ with:

- $\langle u | v \rangle := \int_{\mathbf{r} \in \mathbb{R}^3} u(\mathbf{r}, .)^* v(\mathbf{r}, .) d\mathbf{r}$. We identify the L^2 inner product.
- $\langle u, v \rangle_\mathcal{N} := \int_{\mathbf{r} \in \mathbb{R}^3} \frac{\hbar^2}{2m} \mathbf{grad} u(\mathbf{r}, .))^* . \mathbf{grad} v(\mathbf{r}, .)) + V_s(\mathbf{r}) u(\mathbf{r}, .)^* v(\mathbf{r}, .) d\mathbf{r}$, which is equivalent to the H^1 inner product if $V_s(\mathbf{r})$ is lower bounded by a positive constant and upper bounded. In keeping with a change in the energy origin, any constant can be added to $V_s(\mathbf{r})$ without changing the quantum system.

The balance equations of these two Hamiltonians can be derived by adapting the Ehrenfest Theorem to the Controlled Schrödinger Eq. (2). For $\langle ., . \rangle \in \{\langle . | . \rangle, \langle ., . \rangle_\mathcal{N}\}$, depending on the choice of the Hamiltonian, we can write canonically the evolution of the Hamiltonian $\mathcal{H} = \langle \Psi, \Psi \rangle$ along the trajectories:

$$\frac{d}{dt}\mathcal{H} = \frac{d}{dt}\langle \Psi, \Psi \rangle = 2\Re(\langle \Psi, \frac{\partial \Psi}{\partial t} \rangle) = 2\Re(\langle \Psi, -i\frac{H_s}{\hbar}\Psi + u \rangle). \tag{3}$$

2.3   Port-Hamiltonian Formulation

The Hamiltonian functional is defined on a complex-valued state, which is the wave function. We define the energy variable of the port-Hamiltonian system $\alpha := \Psi$. Thus, the distributed port-Hamiltonian paradigm has to be extended to complex-valued states. The Hamiltonian functionals of distributed port-Hamiltonian systems are written in the form of an integral on a domain Ω of a real-valued function h:

$$\mathcal{H}[\alpha] = \int_{\mathbf{r} \in \Omega} h(\alpha) d\mathbf{r}.$$

The evolution of the Hamiltonian along the trajectories requires the differentiation of the functional w.r.t the energy variable. Using the $\mathbb{CR}$-calculus, we can consider h as a holormorphic function w.r.t the conjugate coordinates (α, α^*) [7]. Thus the differential rule leads to:

$$\mathcal{H}[\alpha + v] - \mathcal{H}[\alpha] = \int_{\mathbf{r} \in \Omega} 2\Re\left(\frac{\partial h(\alpha, \alpha^*)}{\partial \alpha} v\right) d\mathbf{r} + o(v). \tag{4}$$

Using the inner product $\langle ., . \rangle$ on Ω, we can identify the functional derivative (or variational derivative of the functional), denoted $\delta_\alpha \mathcal{H}$, such that:

$$\int_{\mathbf{r} \in \Omega} \frac{\partial h(\alpha, \alpha^*)}{\partial \alpha} v d\mathbf{r} = \langle \delta_\alpha \mathcal{H}, v \rangle.$$

Using the chain rule, the evolution of the Hamiltonian is the symmetrization of the usual balance (see Fig. 1):

$$\frac{d\mathcal{H}}{dt} = 2\Re(\langle \delta_\alpha \mathcal{H}, \frac{\partial \alpha}{\partial t} \rangle). \tag{5}$$

Focusing on the pH modelling of the Controlled Schrödinger Eq. (2) and using the inner products defined in Sect. 2.2, we can define the co-energy variable $e := \delta_\alpha \mathcal{H} = \Psi$. This leads to the complex-valued distributed pH systems:

$$\begin{cases} \partial_t \alpha = \mathcal{J} e + Gu \\ y = G^H e \end{cases} \quad \text{with} \quad \mathcal{J} = -\frac{iH_s}{\hbar} \quad \text{and} \quad G = 1. \tag{6}$$

The structure operator $\mathcal{J}$ is skew-symmetric since H_s is symmetric for both inner products. This property implies that the Hamiltonian balance is achieved *via* the dual controls and observations. Indeed, using (5) (compare with (3)):

$$\frac{d\mathcal{H}}{dt} = 2\Re(\langle \delta_\alpha \mathcal{H}, \mathcal{J} \delta_\alpha \mathcal{H} \rangle + \langle \delta_\alpha \mathcal{H}, Gu \rangle) = 2\Re(\langle G^H \delta_\alpha \mathcal{H}, u \rangle) = 2\Re(\langle y, u \rangle).$$

These properties have been identified in [4] to recover the (complex) Stokes-Dirac structure of distributed port-Hamiltonian systems.

Hamiltonian system		**Distributed pH system**	
Canonical equations		Stokes-Dirac structure	
Hamiltonian functional (typically a quadratic form)	$\mathcal{H}[\Psi] = \frac{1}{2}\|\Psi\|_{\mathcal{A}}^2$	Hamiltonian functional (typically a quadratic form)	$\mathcal{H}[\Psi] = \frac{1}{2}\|\Psi\|_{\mathcal{A}}^2$
Observable $\mathcal{A}$.	$\frac{d\mathcal{A}}{dt} = \{\mathcal{A}, \mathcal{H}\} + \frac{\partial \mathcal{A}}{\partial t}$	Conservation of $\mathcal{H}$ via ports	$\frac{d\mathcal{H}}{dt} = \langle \frac{\partial \Psi}{\partial t}, \delta_\Psi \mathcal{H} \rangle$
Quantum mechanics		**Distributed pH system with complex state**	
Schrödinger equation / Ehrenfest theorem		(Complex) Stokes-Dirac structure	
Hamiltonian operator $\widehat{H} = \frac{\widehat{p}^2}{2m} + \widehat{V}$		Hamiltonian functional $\mathcal{H}[\Psi] = \|\Psi\|_{\widehat{A}}^2 = \langle \Psi | \widehat{A} | \Psi \rangle$	
Observable (operator) $\widehat{A}$. $\frac{d\langle\widehat{A}\rangle_\Psi}{dt} = \frac{1}{i\hbar}\langle[\widehat{A}, \widehat{H}]\rangle_\Psi + \langle\frac{\partial\widehat{A}}{\partial t}\rangle_\Psi$		Conservation of $\mathcal{H}$ via ports $\frac{d\mathcal{H}}{dt} = 2\Re(\langle\frac{\partial\Psi}{\partial t}, \delta_\Psi\mathcal{H}\rangle)$	

Fig. 1. The new framework for distributed port-Hamiltonian with complex-valued state as a trade off between quantum mechanics (canonical quantization of the classical Hamiltonian mechanics) and the existing framework for distributed port-Hamiltonian systems with real-valued state.

2.4   Adaptation of the Model to Simulation

The wave function that represents the quantum system extends to infinity. However, the dynamics of the quantum system is simulated in a *bounded* domain $\Omega \subset \mathbb{R}^3$ only. A *virtual* system has to be defined, such that its solution is the restriction of the wave function to Ω. Let us denote $\partial\Omega$ the boundary, and **n** its outward normal. We consider the *restricted* inner products $\langle u|v \rangle$ and $\langle u, v \rangle_{\mathcal{N}}$:

– $\langle u, v \rangle_\Omega := \int_{\mathbf{r} \in \Omega} u(\mathbf{r}, .)^* v(\mathbf{r}, .) d\mathbf{r},$

– $\langle u, v \rangle_{N\Omega} := \int_{\mathbf{r} \in \Omega} \frac{\hbar^2}{2m} \mathbf{grad} u(\mathbf{r}, .)^* . \mathbf{grad} v(\mathbf{r}, .) + V_s(\mathbf{r}) u(\mathbf{r}, .)^* v(\mathbf{r}, .) d\mathbf{r}.$

The corresponding Hamiltonians are $\mathcal{H}_\mathcal{P}^\Omega := \langle \Psi, \Psi \rangle_\Omega$ and $\mathcal{H}_\mathcal{N}^\Omega := \langle \Psi, \Psi \rangle_{N\Omega}$. The formulation of the port-Hamiltonian systems is the same. H_s becomes formally symmetric for these inner products. Thus, a boundary contribution now takes place in the Hamiltonian balance. Defining the speed operator by $\mathbf{v}\Psi(\mathbf{r}, t) := -i \frac{\hbar}{m} \mathbf{grad} \Psi(\mathbf{r}, t)$, we can derive the equation of probability conservation by integrating by parts (3):

$$\frac{d}{dt} \mathcal{H}_\mathcal{P}^\Omega = 2\Re(\langle \Psi, u \rangle_\Omega) - \Re \left(\int_{\partial \Omega} \Psi^*(\mathbf{r}, t)(\mathbf{v}\Psi(\mathbf{r}, t).\mathbf{n}) d\gamma \right). \tag{7}$$

The first term is the expected flow from the model $2\Re(\langle y, u \rangle_\Omega)$. It is zero when Ψ follows the Schrödinger Equation with potential $V(\mathbf{r}, t)$. The second term is the probability flows across the boundary. Indeed, denoting the real-valued probability current $\mathbf{J} := \frac{i\hbar}{2m}(\Psi \mathbf{grad} \Psi^* - \Psi^* \mathbf{grad} \Psi)$ then $\mathbf{J} = \Re(\Psi^* \mathbf{v}\Psi)$.

The energy balance for $\mathcal{H}_\mathcal{N}^\Omega$ is more difficult to interpret. This is the reason why the pH system based on $\mathcal{H}_\mathcal{P}^\Omega$ is chosen for the sequel of this work.

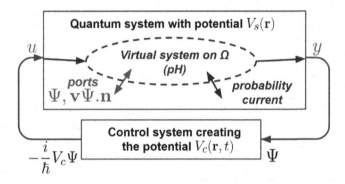

Fig. 2. The Schrödinger Equation as a pH model controlled by a linear state feedback from a control system. The pH model of the virtual system is the restriction of the pH model describing the Controlled Schrödinger Equation to a bounded domain.

3   Structure-preserving Discretization

In this part we apply the PFEM to build a discrete pH representation of the model [3,5]. Let ϕ be a sufficiently smooth complex-valued test function. Writing the weak formulation of (6) with the inner product $\langle ., . \rangle_\Omega$ and integrating by parts, one gets:

$$\langle \phi, \partial_t \Psi \rangle_\Omega = -\frac{i}{\hbar} \langle \phi, \Psi \rangle_{N\Omega} + \langle \phi, u \rangle_\Omega - \frac{1}{2} \int_{\partial\Omega} \phi^* \underbrace{(-i\frac{\hbar}{m}\mathbf{grad}\Psi.\mathbf{n})}_{\mathbf{v}\,\Psi} d\gamma. \qquad (8)$$

The normal speed boundary condition is identified.

As will be seen in Sect. 4.1, Dirichlet boundary condition is essential to model a quantum box. Following [3], such a control can be imposed by using Lagrange multipliers. Splitting the boundary into two distinct and complementary parts $\partial\Omega = \Gamma_N \cup \Gamma_D$, one can distinguish between Dirichlet controls u^D imposed via the Lagrange multipliers $\lambda^D = -\frac{1}{2}(\mathbf{v}\,\Psi).\mathbf{n}|_{\Gamma_D}$ on Γ_D, and Neumann control $u^N = -\frac{1}{2}(\mathbf{v}\,\Psi).\mathbf{n}|_{\Gamma_N}$ on Γ_N.

Let us define the finite real approximation spaces $\overline{W} = \text{span}((\phi_n)_n)$ and $\overline{B} = \text{span}((b_n)_n)$ to discretize the variables defined on Ω and on $\partial\Omega$ respectively. Let us denote $\Psi(\mathbf{r},t) \approx \psi^d(\mathbf{r},t) = \sum_{n=1}^{\dim\overline{W}} \psi_n(t)\phi_n(\mathbf{r}) = \underline{\phi}(\mathbf{r})^T \underline{\psi}(t)$ and similarly for other quantities.

Defining the skew-symmetric complex matrix $(J)_{m,n} = -\frac{i}{\hbar}\langle\phi_m,\phi_n\rangle_{N\Omega}$ of size $\dim\overline{W} \times \dim\overline{W}$, the symmetric real matrices $(M)_{m,n} = \langle\phi_m,\phi_n\rangle_\Omega$ of size $\dim\overline{W} \times \dim\overline{W}$, $(M^D)_{m,n} = \langle b_m,b_n\rangle_{\Gamma_D}$ and $(M^N)_{m,n} = \langle b_m,b_n\rangle_{\Gamma_N}$ of size $\dim\overline{B} \times \dim\overline{B}$, and the real matrices $(G^D)_{m,n} = \langle\phi_m,b_n\rangle_{\Gamma_D}$ and $(G^N)_{m,n} = \langle\phi_m,b_n\rangle_{\Gamma_N}$ of size $\dim\overline{W} \times \dim\overline{B}$, the discrete counterpart of (8) reads:

$$M\frac{d}{dt}\underline{\psi} = J\underline{\psi} + G\underline{u} + G^N\underline{u}^N + G^D\underline{\lambda}^D.$$

The discretized Dirichlet algebraic constraint is:

$$G^D\underline{u}^D = (B^D)^H\underline{\psi}.$$

The discretized Neumann boundary observation (of the wave function) is:

$$M^N\underline{y}^N = (G^N)^H\underline{\psi}.$$

The discretized Dirichlet boundary observation (of the normal speed), through the Lagrange multipliers, is:

$$M^D\underline{y}^D = (B^D)^H\underline{\lambda}^D.$$

And finally, the distributed observation is:

$$M^\Omega\underline{y} = G^H\underline{\psi}.$$

Note that $M = G = M^\Omega$ and $M^D = B^D$ in this work, although this is not mandatory (using $e.g.$ different spaces of approximation for controls and observations).

Altogether, denoting $\mathbb{E} := \text{diag}(M, 0, M^\Omega, M^N, M^D)$ leads to:

$$\mathbb{E}\begin{pmatrix} \frac{d}{dt}\underline{\psi} \\ \frac{d}{dt}\underline{\lambda}^D \\ -\underline{y} \\ -\underline{y}^N \\ -\underline{y}^D \end{pmatrix} = \begin{pmatrix} J & G^D & G & G^N & 0 \\ -(G^D)^H & 0 & 0 & 0 & B^D \\ -(G)^H & 0 & 0 & 0 & 0 \\ -(G^N)^H & 0 & 0 & 0 & 0 \\ 0 & -(B^D)^H & 0 & 0 & 0 \end{pmatrix}\begin{pmatrix} \underline{\psi} \\ \underline{\lambda}^D \\ \underline{u} \\ \underline{u}^N \\ \underline{u}^D \end{pmatrix}. \qquad (9)$$

This Differential Algebraic Equation (DAE) respects the Dirac structure, known as the algebraic structure describing finite-dimensional port-Hamiltonian systems [4, 10, 13].

The Schrödinger Equation is embedded into the system by considering the time-dependant linear feedback from the control system that generates $V_c(\mathbf{r}, t)$. We introduce the symmetric real matrix $(V_c^d)_{m,n} = \langle \phi_m, V_c \phi_n \rangle_\Omega$ of size $\dim \overline{W} \times \dim \overline{W}$. Then the discretization of the feedback associed to the Schrödinger Equation leads to:

$$M^\Omega \underline{u} = -\frac{i}{\hbar} V_c^d(t) \underline{y}. \tag{10}$$

Let us denote $\mathcal{H}_\mathcal{P}^d(t) := \mathcal{H}_\mathcal{P}^\Omega[\psi^d] = \langle \psi^d, \psi^d \rangle_\Omega = \underline{\psi}^H M \underline{\psi}$ the Hamiltonian of the discretized system. Thanks to the Dirac structure enlightened in (9), the corresponding balance follows:

$$\frac{d}{dt} \mathcal{H}_\mathcal{P}^d = 2\Re \left(\underline{y}^H M^\Omega \underline{u} + (\underline{y}^D)^H M^D \underline{u}^D + (\underline{y}^N)^H M^N \underline{u}^N \right).$$

This equality is the discrete counterpart of (7). Similarly to the continuous equation, the first term is equal to zero in the case of the Schrödinger Eq. (10), because the feedback matrix $-\frac{i}{\hbar} V_c^d(t)$ is always skew-symmetric.

To conclude, thanks to the application of the PFEM to the complex-valued distributed pH system, both Dirac structure and probability conservation (Hamiltonian balance) have been preserved at the discrete level.

4   Numerical Experiments

4.1   One-Dimensional Quantum Box

The quantum system is entirely in the domain of the box Ω. We impose a homogeneous Dirichlet condition on the boundary $\partial \Omega$. For $x \in \Omega = [0, L]$, in absence of potential, the stationary states $\Psi_n(x, t)$ are the states of energies $E_n = \frac{\hbar^2 k_n^2}{2m} = \hbar \omega_n = \frac{\hbar^2}{2m} \frac{n^2 \pi^2}{L^2}$ $(n \in \mathbb{N}^\star)$. These states $\Psi_n(x, t) = \sqrt{\frac{2}{L}} \sin(k_n x) \exp(-i\omega_n t)$, analogous to oscillating modes, form an ortho-normal basis of conservative states [6].

For our simulations, the spatial structure-preserving discretization of (6) is performed as in (9), using Lagrange finite elements of order 1. Then, the resulting DAE is solved in time by the *Matlab* DAE solver *ode23t*. Let us point out that defining time schemes for finite-dimensional (real-valued) port-Hamiltonian DAE is a current topic of research, see *e.g.* [9, 10] and references therein.

In Table 1, we reproduce the propagation of the first modes for a sufficiently fine discretization (sampling of oscillations). The Hamiltonian balance is well respected.

Table 1. Results of the simulation of conservative systems over a period T_f. N_x is the number of points of discretization. N_{osc} is the number of temporal oscillations of the analytical solution. N_t is the number of instants at which the numerical solution is computed. The L^2-error corresponds to the maximum in space of the L^2-error between the numerical solution of the wave function and the analytical solution over the time span. The relative error on $\mathcal{H}_\mathcal{P}^d$ compares the value at T_f to the initial value.

Initial state	N_x	T_f	N_{osc}	N_t	L^2 error	Relative error on $\mathcal{H}_\mathcal{P}^d$
State 1 : Ψ_1	51	$2\pi/\omega_1$	1	43	9.7×10^{-3}	4.9×10^{-4}
State 2 : Ψ_2	51	$2\pi/\omega_1$	4	160	1.6×10^{-2}	4.9×10^{-4}
State 3 : Ψ_3	51	$2\pi/\omega_1$	9	357	5.7×10^{-2}	4.9×10^{-4}
States 1 + 2 : $(\Psi_1 + \Psi_2)/\sqrt{2}$	51	$2\pi/\omega_1$	1	203	4.0×10^{-3}	6.8×10^{-7}

4.2 Rabi Oscillations: Control of the Energy Levels and Generation of Schrödinger Cat States

Applying an electric field of amplitude $\mathcal{E}_0$ and angular frequency ω_E on a dipole of charge q results in a potential $V_c(x,t) = -qx\mathcal{E}_0 \sin(\omega_E t)$ [1]. If the initial state of the system is the superposition of the states 1 and 2 and if $\omega_E \approx \omega_2 - \omega_1$, then the system oscillates between state 1 and state 2 at the Rabi period T_{Rabi} that can be computed analytically for the quantum box. The probability of measuring the state n is $P_n = |\langle \Psi_n, \Psi \rangle_\Omega|^2$.

In the following numerical experiment, we apply the control of pulsation $\omega_E = \omega_2 - \omega_1$ on a time interval $[t_{\mathrm{start}}, t_{\mathrm{stop}}]$ of period $T_{\mathrm{Rabi}}/2$ (Fig. 3). We correctly reproduce the Rabi oscillations and we identify the expected high frequency oscillations of period π/ω_E (Fig. 3 on the right). Thanks to the control, we switch from state 1 to state 2. Balanced Schrödinger cat states are reached by applying the control on a period $T_{\mathrm{Rabi}}/4$ (blue dot on Fig. 3). Note that the Hamiltonian balance accuracy is clearly driven by the precision of the time scheme as soon as the control is switched on (Fig. 3 on the left).

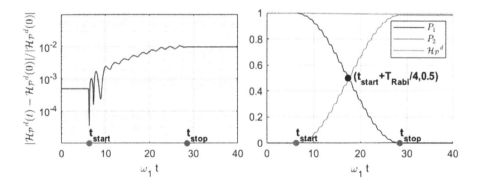

Fig. 3. Simulation of Rabi oscillations and transition from the state 1 to the state 2. $N_x = 51$, $N_t = 1346$. The values of q and $\mathcal{E}_0$ are taken arbitrarily.

5 Conclusion

In this work, we modelled the PDE form of the Schrödinger Equation as a distributed pH system. We proposed a (space-)discretization preserving the probability of presence balance equation at the discrete level. Our numerical experiments show that it is possible to apply simple controls on quantum systems to manipulate their states.

Compared to our real-valued modelling approach [14], the framework we suggested here for modelling distributed port-Hamiltonian systems with complex-valued states describes the system in a more compact way and better emphasizes the formalism used in quantum mechanics.

The interaction of the quantum system with a system of control has been interpreted as the feedback of a potential of control. Thus the system of control is represented by a black box. However, the port specifications of the model remain robust to further pH modelling, *e.g.* the Schrödinger-Newton Equation could be tackled by coupling the model with the Poisson Equation [2].

Modelling interacting quantum systems in the port-Hamiltonian framework is still challenging. It cannot only be the interconnection of individual quantum systems. Indeed, without considering the system dynamics, the space of states of interacting systems is the tensor product of the individual space of states, which deeply restrains the system decomposition [6,11].

Acknowledgements. The authors would like to thank Pr. S. Massenot for his useful advice on the choice of quantum mechanics numerical experiments.

References

1. Auffeves, A.: Oscillation de Rabi à la frontière classique-quantique et génération de chats de Schrödinger. Ph.D. thesis, Université Pierre et Marie Curie-Paris VI (2004)
2. Bahrami, M., Großardt, A., Donadi, S., Bassi, A.: The Schrödinger-Newton equation and its foundations. New J. Phys. **16**(11), 115007 (2014)
3. Brugnoli, A., Haine, G., Matignon, D.: Explicit structure-preserving discretization of port-Hamiltonian systems with mixed boundary control. IFAC-PapersOnLine. **55**(30), 418–423 (2022), Proceedings of 25th International Symposium on Mathematical Theory of Networks and Systems (MTNS)
4. Brugnoli, A., Haine, G., Matignon, D.: Stokes-Dirac structures for distributed parameter port-Hamiltonian systems: an analytical viewpoint. arXiv preprint arXiv:2302.08816 (2023)
5. Cardoso-Ribeiro, F.L., Matignon, D., Lefèvre, L.: A partitioned finite-element method for power-preserving discretization of open systems of conservation laws. IMA J. Math. Control Inf. **38**(2), 493–533 (2021)
6. Hall, B.C.: Quantum Theory for Mathematicians. GTM, vol. 267. Springer, New York (2013). https://doi.org/10.1007/978-1-4614-7116-5
7. Kreutz-Delgado, K.: The complex gradient operator and the CR-calculus. arXiv preprint arXiv:0906.4835 (2009)

8. Macchelli, A., Melchiorri, C.: Modeling and control of the Timoshenko beam: the distributed port-Hamiltonian approach. SIAM J. Control. Optim. **43**(2), 743–767 (2004)
9. Mehrmann, V., Morandin, R.: Structure-preserving discretization for port-Hamiltonian descriptor systems. In: 58th IEEE Conference on Decision and Control (CDC), pp. 6863–6868. IEEE (2019), invited session
10. Mehrmann, V., Unger, B.: Control of port-Hamiltonian differential-algebraic systems and applications. Acta Numer. **32**, 395–515 (2023)
11. Meijer, A.S.: Kinematic decomposition of quantum systems. B.S. thesis, University of Twente (2022)
12. Mirrahimi, M., Rouchon, P.: Dynamics and control of open quantum systems. Lecture notes (2015)
13. Rashad, R., Califano, F., van der Schaft, A., Stramigioli, S.: Twenty years of distributed port-Hamiltonian systems: a literature review. IMA J. Math. Control. Inf. **37**(4), 1400–1422 (2020)
14. Verrier, G., Matignon, D., Haine, G.: Modelling and structure-preserving discretization of the Schrödinger equation as a port-Hamiltonian system, and simulation of a controlled quantum box. Technical report, Toulouse, April 2023. https://oatao.univ-toulouse.fr/29605/

Some Remarks on the Notion of Transition

Florio M. Ciaglia[1] and Fabio Di Cosmo[1,2(✉)]

[1] Departamento de Matemáticas, Universidad Carlos III de Madrid,
Avenida de la Universidad 30, 28911 Leganés, Madrid, Spain
{fciaglia,fcosmo}@math.uc3m.es
[2] ICMAT, Instituto de Ciencias Matemáticas (CSIC-UAM-UC3M-UCM),
Madrid, Spain

Abstract. In this paper some reflections on the concept of transition are presented: groupoids are introduced as models for the construction of a "generalized logic" whose basic statements involve pairs of propositions which can be conditioned. We could distinguish between classical probability theory where propositions can be conditioned if they have a non-zero intersection, from cases where "non-local" conditioning are allowed. The algebraic and geometrical properties of groupoids can be exploited to construct models of such non-local description.

Keywords: Transition · Groupoid · Boolean lattices · grade 2-measures

1 Introduction

One of the founding fathers of information theory is Claude E. Shannon with is work on the mathematical theory of communication [1]. His aim was to introduce a mathematical description of the process of communication: a message elaborated by an information source that reaches the destination via a channel. In Shannon's description both the concepts of states and transitions between states are primitive objects which have a mathematical counterpart in the concept of Markov process (see [2]). This assumption has, then, become the consolidated paradigm when describing the process of communication. From a more philosophical point of view, we could alternatively say that a mathematical description of information sources, and more generally of communication processes would require a way to implement both the concepts of "being" and "becoming". Both these aspects will play a crucial role when passing to the quantum description but with some differences, since there are plenty of experiments showing how distant quantum and classical mechanics could be.

In this paper we are going to present some reflections on the concept of transition and how groupoids could actually encode their properties both from a classical and a quantum point of view. Even if some of these ideas have already been discussed in our previous works on the Schwinger's picture of Quantum Mechanics [9,14], in this paper we are going to revisit those considerations from

© The Author(s), under exclusive license to Springer Nature Switzerland AG 2023
F. Nielsen and F. Barbaresco (Eds.): GSI 2023, LNCS 14072, pp. 402–411, 2023.
https://doi.org/10.1007/978-3-031-38299-4_42

the point of view of a "generalised logic". Indeed, in a previous work [10] the authors showed that the functions on a groupoid can be endowed with a von Neumann algebra [3] structure and the associated lattice of projections is an orthocomplemented lattice. This lattice, therefore, represents an example of a propositional calculus satisfying the axioms introduced by Birkhoff and von Neumann [5]. But what can one say about transitions? With respect to this question, the lattice approach seems to look at the logical structure of the experimental propositions providing information about the state of the system. May it provide information about transitions?

As a first attempt towards a statistical definition of transition, one could say that a transition between two events (which correspond to propositions of the logical calculus) could happen if the occurrence of one could be conditioned by the other. Therefore, one could introduce a "generalized logic" whose basic propositions are pairs of "experimental propositions" that can be conditioned. In classical probability theory events are described by measurable sets of a Borel algebra and one implements the conditioning in a "local" way: the two events involved in the transition must have a non-empty intersection. Therefore, if we assume that the lattice of experimental proposition is atomic[1], the basic atoms cannot be conditioned, since they have empty intersections. However, one could go beyond this "local" approach suitably using the structure of measurable groupoid.

From an algebraic point of view, a groupoid is a set G of morphisms with a partial composition law, on which two maps are defined which associate to each morphism two objects, its source and target. These objects belong to another set, say Ω. If we introduce a Boolean lattice of measurable subsets of the groupoid, say $\mathcal{P}(G)$, we have a measurable groupoid. Using the composition law one can define a product among elements of $\mathcal{P}(G)$. On the other hand, the source and the target maps send every element of $\mathcal{P}(G)$ to some subset of Ω. Both these structures can be used to define a notion of "non-local" conditioning among subsets of Ω. In this paper we are going to show that using a measurable groupoid and introducing a generalized notion of conditioning between sets of objects, a measure on the groupoid determines a grade-2 measure on the set of objects [4]. Grade-2 measures where introduced by Sorkin in his formulation of Quantum Mechanics as a generalization of probability theory which could allow for the description of interference phenomena. Interestingly, the above mentioned grade-2 measure can be also associated to a state on the algebra of the functions on the groupoid and from this perspective the sets of zero 2-measure can be seen as elements in the Gelfand ideal of this state.

---

[1] Anticipating some further definitions, a bounded lattice L is a set with a partial order relation, say $\subset$, two operations, $\wedge$ and $\vee$, and a minimal and a maximal element, say $\emptyset$ and $\mathbb{I}$, respectively. An atom of a lattice is an element a such that there is no element $b \in L$ such that $\emptyset \subset b \subset a$. A bounded lattice L is said to be atomic if every element $c \supset \emptyset$ has an atom below, in the sense of the partial order relation.

2 The Lattice of Propositions

In this section we are going to shortly recall the basic notions introduced by Birkhoff and von Neumann to describe the logic of Quantum Mechanics, which is the set $\mathcal{P}$ of all "experimental proposition" that can be inferred about a system. These propositions are basic ingredients for the formulation of every physical model, and consequently they will share some common properties. In order to convey more clearly the idea we will restrict to a situation where only a finite set of propositions are considered. The basic relation existing on $\mathcal{P}$ is an order relation, say $\subset$, which is the mathematical implementation of the implication relation between propositions. Once we have this order we assume that for every pair of elements $a, b \in \mathcal{P}$ there is a greatest lower bound, $a \wedge b$, and a least upper bound, $a \vee b$, which are, respectively, the greatest (with respect to the previous order) of all the propositions which are implied by a and b, and the least of all the propositions which imply both a and b. Additionally, we can assume $\mathcal{P}$ to contain two special propositions: the one which is always true, denoted by $\mathbb{I}$, and the one which is always false, say $\emptyset$. The final property of the set $\mathcal{P}$ is that for every proposition a there is a complementary proposition $a^{\perp}$ satisfying the conditions $a \vee a^{\perp} = \mathbb{I}$ and $a \wedge a^{\perp} = \emptyset$. A set $\mathcal{P}$ endowed with the previous structures $(\subset, \wedge, \vee, \perp)$ is called an orthocomplemented lattice, and this is commonly considered to be a suitable mathematical model for the set of "experimental propositions" about a physical system. The property that distinguishes the classical setting from the quantum one is the distributive character of the two operations, $\wedge$ and $\vee$, i.e., $a \wedge (b \vee c) = (a \wedge b) \vee (a \wedge c)$. When we have a finite distributive lattice $\mathcal{P}_c$ we can identify $\mathbb{I}$ with the set Γ of all its atoms, and $\mathcal{P}_c$ will be isomorphic to the power set $\mathcal{P}(\Gamma)$, i.e., the set of all subsets of Γ. We would call an orthocomplemented lattice which is distributive a Boolean lattice or Boolean algebra. On the other hand, when we relax the distributive requirement, another condition allows to select lattices which can be realized as the set of vector subspaces of a certain Hilbert space, which is the departure point in the von Neumann description of Quantum Mechanics [6]. If one assumes a weaker form of the distributive identity, called modular identity

$$a \subset c \;\Rightarrow\; a \vee (b \wedge c) = (a \vee b) \wedge (a \vee c) = (a \vee b) \wedge c, \tag{1}$$

it is possible to prove that any complemented modular lattice of finite dimensions is isomorphic to the cartesian product of a finite Boolean algebra and a finite number of projective geometries [5] (the definition of projective geometry will be given below). Moreover, if one additionally assumes irreducibility of the lattice, which means that the only elements $x \in \mathcal{P}$ satisfying

$$a = (a \wedge x) \vee (a \wedge x^{\perp}) \quad \forall a \tag{2}$$

are $x = \emptyset, \mathbb{I}$, then the modular lattice is a finite projective geometry [5]. A projective geometry is a system of k-dimensional elements, where 0-dimensional objects are elementary points and 1-dimensional objects are called lines, satisfying the following axioms [7]:

- Two distinct points are contained in one and only one line
- If A, B, C are three points not on the same line and E ≠ D are points such that B, C, D are on a line and A, C, E are on a line, then there is a point F such that both A, B, F are on a line and D, E, F are on a line
- Every line contains at least three points
- The points on lines through any k-dimensional element and a fixed point not in the element are a (k + 1)-element, and every (k + 1)-element is defined in this way.

The set $\mathcal{C}(\mathcal{H})$ of Hilbert subspaces of the Hilbert space $\mathcal{H}$, is an example of a projective geometry. In this case k-dimensional elements correspond to (k + 1)-dimensional closed subspaces of $\mathcal{H}$. Note that one can also describe such projective geometry via the orthocomplemented lattice of orthogonal projections onto each closed subspace of $\mathcal{H}$, where the operation of meet and join, however, do not coincide with the sum and product between operators on $\mathcal{H}$, generically. Indeed, given the projection P_1 and P_2, the meet $P_1 \wedge P_2$ is the projection onto the intersection of the supports of P_1 and P_2, whereas the join $P_1 \vee P_2$ is the projection onto the vector subspace generated by the supports of P_1 and P_2 (for an algebraic expression of the two operations see [10]).

Once the structure of the experimental propositions is given, it is possible to associate to every proposition its probability: for Boolean lattices, monotone additive functions will play the role. For the modular case, in general, one can always define a dimension function $d : \mathcal{P} \rightarrow [0, 1]$ obeying the following properties:

- $d(a) > d(b)$ iff $b \subset a$
- $d(a) + d(b) = d(a \wedge b) + d(a \vee b)$.

In the case of the lattice of projections mentioned above, the function d is the normalized trace of each projection and represents the dimension of each closed subspace of the Hilbert space. More generally, one can consider real valued functions m on the lattice which are additive only with respect to the meet of orthogonal projections. In this case Gleason theorem [8] ensures the existence of a density matrix ρ such that $m(a) = \text{Tr}(\rho P_a)$ with P_a the projection corresponding to the closed subspace a in the Hilbert space $\mathcal{H}$. This point of view, therefore, allows to give a statistical interpretation to each event, but only to events: we are interpreting only the "static" information. What about transitions? The standard approach introduced by von Neumann, define the probability transitions between two states of the system in terms of the square modulus of the scalar product of the vectors corresponding to the two states. In the rest of this short paper we will try to follow a different approach.

3 Groupoid and Grade 2-Measures

In this section we are going to introduce a generalized model of conditioning via the concept of groupoids. In order to avoid inessential technical difficulties and for the sake of communicability we are going to limit the discussion to finite groupoids and consider the extension to the non-finite case in forthcoming works. The departure point is the definition of groupoids. From the algebraic point of view, a groupoid $G \rightrightarrows \Omega$, is a set G of morphisms, together with a pair of maps $s, t \colon G \to \Omega$ from the set of morphisms to the set of objects Ω, these maps being called source and target, respectively.

Given the groupoid G, we will denote G^j the set of morphisms whose target is $j \in \Omega$ (analogously we denote G_j the set of morphisms whose source is j). The set of morphisms having the same source and target object $j \in \Omega$ is a group, the isotropy group at j, and will be denoted G_j^j. Two morphisms α and β will be said to be composable if $s(\alpha) = t(\beta)$ and their composition gives another morphism denoted by $\alpha \circ \beta$, this operation being associative. Units in the groupoid G will be denoted as $1_j \colon j \to j$ and they satisfy the conditions

$$\alpha \circ 1_i = \alpha, \quad 1_j \circ \alpha = \alpha \tag{3}$$

provided that $\alpha \colon i \to j$. Finally there is an inverse operation $\tau \colon \alpha \mapsto \alpha^{-1}$ such that

$$\alpha^{-1} \circ \alpha = 1_i, \quad \alpha \circ \alpha^{-1} = 1_j. \tag{4}$$

Some basic examples of groupoids, which will be used in the rest of the paper, are provided by the following ones. Firstly, the groupoid of pairs $G(\Omega) = \Omega \times \Omega \rightrightarrows \Omega$ of an arbitrary set Ω: it has source and target maps $s(j, i) = i$, $t(j, i) = j$, respectively, composition law $(k, j) \circ (j, i) = (k, i)$, units $1_i = (i, i)$ and inverse $(j, i)^{-1} = (i, j)$. Secondly, standard sets are groupoids, i.e., if Ω is a set we consider the groupoid $\Omega \rightrightarrows \Omega$ with only units (corresponding to each point of the set), so that a morphism is composable only with itself.

Since a groupoid G is a set of morphisms, we can consider the lattice $\mathcal{P}(G)$ of all its subsets and, as we have already discussed in the previous section, $\mathcal{P}(G)$ is a Boolean algebra. However, apart from this family of subsets, one can use the algebraic structure of the groupoid to generalize the "local" concept of conditioning expressed in terms of non-vanishing intersections. Indeed, if one considers two subsets $A, B \subset G$ it is possible to construct a new subset $C = B \circ A$ as follows:

$$C = \{G \ni \gamma = \beta \circ \alpha \mid \beta \in B, \, \alpha \in A\}. \tag{5}$$

In other words, C is the set of all morphisms obtained by composing one morphism from the set A and one from the set B. One can prove that this product is distributive with respect to the union of disjoint sets, but it is not commutative. In particular, one can take $A = s^{-1}(a)$ and $B = \tau^{-1}(s^{-1}(b)) = t^{-1}(b)$ with a, b being two subsets of the set Ω of all objects of the groupoid. Then, we can define two events $a, b \in \mathcal{P}(\Omega)$ to be conditioned iff $\tau^{-1}(s^{-1}(b)) \circ s^{-1}(a) \neq \emptyset$. Let us write $a\mathcal{R}b$ if a and b are subsets of Ω which are conditioned according to the previous definition. It can be straightforwardly proven from the definition that :

- $a\mathcal{R}a$, i.e., $\tau^{-1}(s^{-1}(a)) \circ s^{-1}(a) \neq \emptyset$;
- $a\mathcal{R}b \Rightarrow b\mathcal{R}a$, since $\tau^{-1}(s^{-1}(a)) \circ s^{-1}(b) = \tau^{-1}(\tau^{-1}(s^{-1}(b)) \circ s^{-1}(a))$;

– in general it fails to be transitive (it fails for the groupoid made only of units in the example below, where if $a \wedge b \neq \emptyset$ and $b \wedge c \neq \emptyset$, then one cannot conclude that $a \wedge c \neq \emptyset$).

Therefore, conditioned pairs form the graphs of a reflexive and symmetric relation, which in general fails to be transitive. In the two main examples of groupoids above introduced, this definition can be specified as follows:

– for the groupoid $G(\Omega) \rightrightarrows \Omega$ of pairs of points of the set Ω, one can see that every pair of atoms $i, j \in \mathcal{P}(\Omega)$ is conditioned because

$$\tau^{-1}(s^{-1}(j)) \circ s^{-1}(i) = (j, i).$$

Since the other sets are obtained via the distributive property of the product with respect to the union of disjoint events, the products between units permit to verify if any pair of subsets $(a, b) \in \mathcal{P}(\Omega) \times \mathcal{P}(\Omega)$ is conditioned.
– for the groupoid $\Omega \rightrightarrows \Omega$ made up of units only, the product

$$c = \tau^{-1}(s^{-1}(b)) \circ s^{-1}(a) = b \wedge a.$$

Using the lattice $\mathcal{P}(G)$ as the algebra of measurable subsets of G, we can introduce a measure on G, say μ[2] How can we use it to give a statistical interpretation to subsets of the groupoid? First of all, due to the algebraic structure of the groupoid, we will consider measures which are compatible with the composition and the inversion, in a way that generalizes the usual Haar measure for a group. Therefore, we will endow the groupoid G with a measure μ which admits the following decomposition[3]

$$\mu(A) = \sum_{j \in \Omega} \lambda(j) \nu^j(A), \tag{6}$$

with respect to a measure λ on the space of objects Ω. The family $\{\nu^j\}$ is a family of measures each one concentrated on the set G^j. It is called a left Haar system of measures if they satisfy the following properties:

– for $\alpha \colon i \to j$,

$$((L_\alpha)_* \nu^{(i)}) = \nu^{(j)},$$

where $(L_\alpha)_* \nu^{(i)}$ is the pushforward of the measure $\nu^{(j)}$ under the map

$$L_\alpha \colon G^i \to G^j, \quad L_\alpha(\beta) = \alpha \circ \beta,$$

representing the left multiplication by the morphism α. This property generalizes the left-invariance of the Haar measure to a framework where the composition among morphisms is only partially defined.

---

[2] This is not the unique measurable structure which G can be endowed with, but in the finite case it is natural to assume that each point of G is a measurable set. If were not the case, we could simply eliminate the atoms which will not be measurable.

[3] in the finite case, but not only, every measure on G can be decomposed in this way so that giving a measure λ on the base Ω and a family of measures on the fibers G^j is equivalent to introducing a measure on G.

– Concerning the inversion map τ,

$$\tau_* \mu = \Delta^{-1} \mu \,,$$

where $\Delta : G \to \mathbb{R}_+$ is called the modular function and it is a homomorphism of the groupoid G to the group of positive real numbers $\mathbb{R}_+$. This property means that by inversion we cannot transform a set of measure zero into a set of non-zero measure or viceversa. Once more, the modular function for groupoids generalizes the modular function for groups.

The triple $(G, \mathcal{P}(G), [\mu])$ with $[\mu]$ denoting the class of all measures equivalent to the left system of Haar measure μ, will be called a measure groupoid. Let us remark that in this class, there is always one measure Λ which is invariant with respect to the inversion, i.e., $\tau_*(\Lambda) = \Lambda$ [11]. Let us consider, now, the set of pairs of conditioned subsets (b, a) in $\mathcal{P}(\Omega) \times \mathcal{P}(\Omega)$. Then, we can define the following two-set function $D : \mathcal{P}(\Omega) \times \mathcal{P}(\Omega) \to \mathbb{R}$:

$$D(b, a) = \Lambda(\tau^{-1}(s^{-1}(b)) \circ s^{-1}(a)) \,. \tag{7}$$

It can be easily seen, that this function satisfies the following properties:

– positivity $D(a, a) \geq 0$
– bi-additivity $D(a, b \cup c) = D(a, b) + D(a, c)$ whenever $b \cap c = \emptyset$
– symmetry $D(b, a) = D(a, b)$.

Let us remark that everything would work in the same manner if the measure Λ would be weighted with a phase factor e^{iS} with the function $S : G \to \mathbb{R}$ satisfying the "logarithmic" properties

$$S(\beta \circ \alpha) = S(\beta) + S(\alpha) \,, \quad S(\alpha^{-1}) = -S(\alpha) \,.$$

The only difference would consists in the replacement of the symmetry condition with the hermiticity, since the function will be complex valued.

The properties of the function D characterize what is called a decoherence functional and the associated quadratic form $\mu_2(a) = D(a, a)$ is what Sorkin [4] defined a grade-2 measure. The concept of grade-2 measure was introduced as a generalization of probability theory in order to account for the description of quantum phenomena. Differently from a traditional measure, a grade 2-measure μ_2 on a Boolean lattice $\mathcal{P}(\Omega)$ satisfies

$$\mu_2(a \vee b) - \mu_2(a) - \mu_2(b) = I(a, b) \neq 0 \tag{8}$$

for every pair (a, b) of disjoint subsets of Ω, where the two-set function $I(a, b)$ is called the interference functional. Even if the interference functional does not vanish, a grade-2 measure satisfies the condition

$$0 = I^{(3)}(a, b, c) = \mu_2(a \vee b \vee c) - \mu_2(a \vee b) - \mu_2(a \vee c) - \mu_2(b \vee c) + \mu_2(a) + \mu_2(b) + \mu_2(c) \,.$$

Summarizing the above discussion, the algebraic structure of groupoids permits to introduce a generalized notion of conditioning between pairs of sets of the

Boolean algebra $\mathcal{P}(\Omega)$. Once a suitable measure is introduced on the groupoid, this measure determines a decoherence functional on the set of conditioned pairs in $\mathcal{P}(\Omega) \times \mathcal{P}(\Omega)$, and, consequently, a grade-2 measure on the space of objects Ω.

The above picture can be analyzed, also, from an algebraic point of view by considering the space of functions on the groupoid. Let $C(G)$ be the space of complex valued continuous functions on the finite groupoid G. On a finite groupoid this set is isomorphic to some vector space, say $\mathbb{C}^N$ where N is the number of morphisms in G. Generalizing the construction for group-algebras, we can introduce a algebra structure on $C(G)$ as follows:

$$
\begin{aligned}
f \star h &= \left(\sum_{\alpha \in G} f_\alpha \delta_\alpha \right) \star \left(\sum_{\beta \in G} h_\beta \delta_\beta \right) \\
&= \sum_{\gamma \in G} \sum_{(\alpha,\beta) \in G^{(2)} | \alpha \circ \beta = \gamma} f_\alpha h_\beta \mu(\gamma) \delta_\gamma ,
\end{aligned}
\tag{9}
$$

where δ_α is the function which takes value 1 at α and 0 everywhere else, μ denotes the left invariant Haar system in the class $[\mu]$, and $G^{(2)}$ denotes the set of composable pairs. This product, called convolution product, is associative. Moreover we can define the following involutive operator on $C(G)$:

$$
f^\dagger = \sum_{\alpha \in G} \Delta^{-1}(\alpha) \overline{f}_\alpha \delta_{\alpha^{-1}} ,
\tag{10}
$$

so that $C(G)$ is endowed with a $*$-algebra structure. Here Δ is the modular function associated with the measure groupoid structure. Moreover, it is possible to introduce a suitable topology [13] on this space so that we eventually get a von Neumann algebra, which we will denote $\nu(G)$. Once we have obtained this algebraic structure it is possible to express the previous decoherence functional in an alternative way. Indeed, it is easy to see that the product $A \circ B = C$ between subsets of G can be expressed in terms of characteristic functions as follows:

$$
\chi_C = \chi_A \star \chi_B ,
\tag{11}
$$

where χ_C is the characteristic function supported on the set $C = A \circ B$, and analogously one defines χ_A and χ_B. It can be proven [12] that the integration of a function $f \in \nu(G)$ with respect to the measure μ determines a positive linear functional on the von Neumann algebra which is normalized whenever the measure λ on the set of objects Ω is a probability measure, i.e., a state. Let us denote it by ω_μ. Then, by direct computation one can show that

$$
D(b,a) = \omega_\mu \left((\chi_{s^{-1}(b)})^\dagger \star \chi_{s^{-1}(a)} \right) .
\tag{12}
$$

Once this algebraic point of view is introduced, we can recognize that the sets whose corresponding characteristic functions are in the Gelfand ideal [3] of the state ω_μ correspond to sets whose grade 2-measure μ_2 is zero. Therefore, we can interpret the GNS construction as the construction of a suitable Hilbert space associated with the set of events of the groupoid.

4 Conclusions

In this paper we have presented some reflections on the notion of transitions, associated with the concept of pairs of conditioned events. After a brief review of the approach to quantum logic introduced by Birkhoff and von Neumann, we have discussed a different model where groupoids are the departure point. Indeed, using the rich algebraic structure defining a groupoid, we can introduce a way of conditioning sets which generalize the "local" approach of classical probability in a straightforward way. Of course, groupoid are not the only way to obtain this non-local conditioning. However, we have shown that using a groupoid this notion can be related to the C^*-algebraic approach to Quantum Mechanics and to the corresponding notion of state.

Acknowledgements. FMC acknowledges that this work has been supported by the Madrid Government (Comunidad de Madrid-Spain) under the Multiannual Agreement with UC3M in the line of "Research Funds for Beatriz Galindo Fellowships" (C&QIG-BG-CM-UC3M), and in the context of the V PRICIT (Regional Programme of Research and Technological Innovation). FDC thanks the UC3M, the European Commission through the Marie Sklodowska-Curie COFUND Action (H2020-MSCA-COFUND-2017- GA 801538) for their financial support through the CONEX-Plus Programme.

References

1. Shannon, C.E.: A mathematical theory of communication. Bell Syst. Tech. J. **27**(3), 379–423 (1948)
2. Gikhman, I.I., Skorokhod, A.V.: The Theory of Stochastic Processes I. Springer, Heidelberg (2004). https://doi.org/10.1007/978-3-642-61943-4
3. Takesaki, M.: Theory of Operator Algebras I. Springer, Heidelberg (2002). https://doi.org/10.1007/978-1-4612-6188-9
4. Sorkin, R.D.: Quantum mechanics as quantum measure theory. Mod. Phys. Lett. A **9**(33), 3119–3127 (1994)
5. Birkhoff, G., von Neumann, J.: The logic of quantum mechanics. Ann. Math. **37**, 823–843 (1936)
6. von Neumann, J.: Mathematical Foundations of Quantum Mechanics. Princeton University Press, Princeton (2018)
7. Birkhoff, G.: Combinatorial relations in projective geometries. Ann. Math. **36**, 743–748 (1935)
8. Gleason, A.M.: Measures on the closed subspaces of a Hilbert space. J. Math. Mech. **6**, 885–893 (1957)
9. Ciaglia, F.M., Di Cosmo, F., Ibort, A., Marmo, G.: Schwinger's picture of quantum mechanics. Int. J. Geom. Methods Mod. Phys. **17**(14), 2050054 (2020)
10. Ciaglia, F.M., Di Cosmo, F., Ibort, A., Marmo, G.: Evolution of classical and quantum states in the groupoid picture of quantum mechanics. Entropy **22**(11), 2050054 (2020)
11. Hahn, P.: Haar measure for measure groupoids. Trans. Am. Math. Soc. **242**, 1–33 (1978)

12. Ciaglia, F.M., Ibort, A., Marmo, G.: Schwinger's picture of quantum mechanics III: the statistical interpretation. Int. J. Geom. Methods Mod. Phys. **16**(11), 1950165 (2019)
13. Ibort, A., Rodriguez, M.A.: An Introduction to the Theory of Groups, Groupoids and Their Representations. CRC, Boca Raton (2019)
14. Ciaglia, F.M., Ibort, A., Marmo, G.: A gentle introduction to Schwinger's formulation of quantum mechanics: the groupoid picture. Mod. Phys. Lett. A **33**(8), 1850122 (2018)

Geometric Quantum States and Lagrangian Polar Duality: Quantum Mechanics Without Wavefunctions

Maurice A. de Gosson[✉]

Institute of Mathematics, University of Vienna, Oskar-Morgenstern-Platz 1,
1090 Vienna, Austria
maurice.de.gosson@univie.ac.at

Abstract. We propose a geometric formulation of Gaussian (and more general) pure quantum states based on an extended version of polar duality based on the notion of Lagrangian frame (a Lagrangian frame in symplectic space is a pair of transverse Lagrangian planes in that space). Our approach leads to the replacement of the usual interpretation of the uncertainty principle without using the *ad hoc* notion of spreading, which might lead to a notion of quantum mechanical phase space.

Keywords: Polar duality · Lagrangian plane · uncertainty principle · Wigner transform · John ellipsoid

1 Introduction

In a recent paper [15] we extended the usual notion of polar duality between convex sets by giving a symplectically invariant fornication in terms of Lagrangian planes in the phase space $T^*\mathbb{R}^n \equiv \mathbb{R}_x^n \times (\mathbb{R}_x^n)^*$ equipped with its standard symplectic structure $\omega = \sum_{j=1}^n dp_j \wedge dx_j$. Doing this, we had in mind a geometric formulation of the uncertainty principle of quantum mechanics. In the present work we go several steps further and show that all Gaussian states can be reconstructed using our notion of Lagrangian polar duality in [15]; also see our preprint [16]. More precisely, we show that to every ellipsoid X_ℓ carried by a Lagrangian plane $\ell \subset (T^*\mathbb{R}^n, \omega)$ we can associate infinitely many Gaussian functions of the type

$$\Psi_{A,B}(x) = \left(\frac{1}{\pi\hbar}\right)^{n/4} (\det A)^{1/4} \exp\left[-\frac{1}{2\hbar}(A + iB)x \cdot x\right] \tag{1}$$

where A and B are real symmetric $n \times n$ matrices, and A is positive definite. Notice that for $A = I_{n\times n}$ and $B = 0_{n\times n}$ this is the standard Gaussian state $\phi_0(x) = (\pi\hbar)^{-n/4}e^{-|x|^2/2\hbar}$. The function $\Psi_{A,B}$ is the ground state of the second order partial differential equation

$$\widehat{H}_{AB}\Psi_{A,B} = \frac{1}{2}\hbar(\text{Tr } A)\Psi_{A,B} \tag{2}$$

© The Author(s), under exclusive license to Springer Nature Switzerland AG 2023
F. Nielsen and F. Barbaresco (Eds.): GSI 2023, LNCS 14072, pp. 412–419, 2023.
https://doi.org/10.1007/978-3-031-38299-4_43

where $\widehat{H}_{AB}$ is the elliptic operator

$$\widehat{H}_{AB} = \frac{1}{2}\left[(-i\hbar\partial_x + Bx)^2 + |Ax|^2\right]\Psi_{A,B}. \tag{3}$$

This operator generalizes the usual n-dimensional harmonic oscillator Schrödinger operator $\frac{1}{2}(-\hbar^2\partial_x^2 + |x|^2)$, to which it reduces when $A = I_{n\times n}$ and $B = 0_{n\times n}$. We will prove the following result, which we glorify as a Theorem because of its importance:

Theorem 1. *Let X_ℓ be a centered ellipsoid carried by Lagrangian plane ℓ and $X_{\ell'}^\hbar \subset \ell'$ its polar dual. with respect to a Lagrangian plane ℓ' such that $\ell \cap \ell' = 0$. The John ellipsoid $\Omega_{AB} = (X_\ell \times X_{\ell'}^\hbar)_{\text{John}}$ of $X_\ell \times X_{\ell'}^\hbar$ is the covariance matrix of the Gaussian $\Psi_{A,B}$ defined by $\Psi_{A,B} = \widehat{S}_{AB}^{-1}\Psi_{I,0}$ where $\widehat{S}_{AB}$ is one of the two metaplectic operators covering the symplectic matrix $S_{AB} \in \text{Sp}(n)$ such that*

$$X_\ell \times X_{\ell'}^\hbar = S_{AB}(B_X^n(\sqrt{\hbar}) \times B_P^n(\sqrt{\hbar}))$$

and we have in this case $\Omega_{AB} = S_{AB}(B^{2n}(\sqrt{\hbar}))$.

We recall [8] that the metaplectic group $Mp(n)$ is the unitary representation in $L^2(\mathbb{R}_x^n)$ of the double covering of the symplectic group $\text{Sp}(n)$ of $(T^*\mathbb{R}^n, \omega)$. The John ellipsoid [2] of a convex body Ω in an Euclidean space is the maximum volume ellipsoid contaied in Ω.

This theorem is a consequence of the following simple geometric result proven in [15]:

Lemma 1. *Let $B_X^n(\sqrt{\hbar})$ and $B_P^n(\sqrt{\hbar})$ be the balls with radius $\sqrt{\hbar}$ centered at the origin in $\mathbb{R}_x^n$ and $\mathbb{R}^n$. The John ellipsoid of the Cartesian product of $B_X^n(\sqrt{\hbar})$ and $B_P^n(\sqrt{\hbar})$ is*

$$\left(B_X^n(\sqrt{\hbar}) \times B_P^n(\sqrt{\hbar})\right)_{\text{John}} = B^{2n}(\sqrt{\hbar}). \tag{4}$$

Lemma 1 is a particular case of Theorem 1 corresponding to the choice $X = B_X^n(\sqrt{\hbar})$: the ball $B^{2n}(\sqrt{\hbar})$ is the covariance ellipsoid of the standard Gaussian $\Psi_{I,0}(x) = (\pi\hbar)^{-n/4}e^{-|x|^2/2\hbar}$, the latter being the ground state of the Schrödinger operator $\widehat{H}_{I,0} = \frac{1}{2}(-\hbar^2\partial_x^2 + |x|^2)$.

2 Lagrangian Dual Polarity

We briefly recall the usual notion of polar duality from convex geometry (we are following our presentation in [15], also see [11]). Let X be a convex body in configuration space $\mathbb{R}_x^n$. We assume in addition that X contains 0 in its interior. The *polar dual* of X is the subset

$$X^\hbar = \{p \in \mathbb{R}_x^n : \sup_{x\in X}(p \cdot x) \le \hbar\} \tag{5}$$

of the dual space $\mathbb{R}_p^n \equiv (\mathbb{R}_x^n)^*$. Notice that it trivially follows from the definition that $X^\hbar$ is convex and contains 0 in its interior. (In the mathematical literature one usually chooses $\hbar = 1$). The following properties of polar duality straightforward:

- *Reflexivity (bipolarity):* $(X^\hbar)^\hbar = X$
- *Antimonotonicity:* $X \subset Y \Longrightarrow Y^\hbar \subset X^\hbar$
- *Scaling property:* $A \in GL(n, \mathbb{R}) \Longrightarrow (AX)^\hbar = (A^T)^{-1}X^\hbar.$

Let A be a real invertible and symmetric $n \times n$ matrix. We have

$$\{x : Ax \cdot x \le \hbar\}^\hbar = \{p : A^{-1}p \cdot p \le \hbar\} . \tag{6}$$

and hence in particular $B_X^n(\sqrt{\hbar})^\hbar = B_P^n(\sqrt{\hbar})$.

Let $\mathrm{Lag}(n)$ be the Lagrangian Grassmannian of the symplectic space $(T^*\mathbb{R}^n, \omega)$: $\ell \in \mathrm{Lag}(n)$ if and only if ℓ is a n-dimensional subspace of $T^*\mathbb{R}^n$ on which ω vanishes identically. Let (ℓ, ℓ') be a Lagrangian frame in $(T^*\mathbb{R}^n, \omega)$, i.e. $(\ell, \ell') \in \mathrm{Lag}(n) \times \mathrm{Lag}(n)$ and $\ell \cap \ell' = 0$. For X_ℓ a centrally symmetric convex body in ℓ the Lagrangian polar dual $X_{\ell'}^\hbar$ of X_ℓ in ℓ' is the convex subset of ℓ' consisting of all $z' \in \ell'$ such that

$$\omega(z, z') \le \hbar \quad \text{for all} \quad z' \in X_\ell . \tag{7}$$

In the particular case $\ell = \ell_X = \mathbb{R}_x^n \times 0$ and $\ell' = \ell_P = 0 \times \mathbb{R}_p^n$ this reduces to ordinary polarity as defined above (we will call (ℓ_X, ℓ_P) the canonical Lagrangian frame). Using the transitivity of the action $\mathrm{Sp}(n) \times \mathrm{Lag}(n) \longrightarrow \mathrm{Lag}(n)$ one has the following symplectic covariance property of of Lagrangian polar duality [15]. It reduces the study of Lagrangian polar duality to that of the canonical frame:

Proposition 1. *Let $S \in \mathrm{Sp}(n)$ be such that $(\ell, \ell') = S(\ell_X, \ell_P)$. The Lagrangian polar dual $X_{\ell'}^\hbar \subset \ell'$ of $X_\ell \subset \ell$ is $X_{\ell'}^\hbar = S(X^\hbar)$. Thus $(X_\ell, X_{\ell'}^\hbar) = S(X, X^\hbar)$ where $X^\hbar$ is the usual polar dual of $X = S^{-1}(X_\ell) \subset \ell_X$.*

This result readily follows from the transitivity of the action of $\mathrm{Sp}(n)$ on the set of all Lagrangian frames (see [8, 15]).

3   Geometric Quantum States

We now consider a quantum system with configuration space $\ell_X = \mathbb{R}_x^n \times 0$. The corresponding momentum space is identified with $\ell_P = 0 \times \mathbb{R}_p^n$. Assume that after a large number of independent position measurements the quantum system has been found to be contained in an ellipsoid $X(x_0) = x_0 + X$ with $X : Ax \cdot x \le \hbar$ where A is a symmetric and positive definite real $n \times n$ matrix. The polar dual of X is $X^\hbar : A^{-1}p \cdot p \le \hbar$; and we consider the ellipsoid $P(p_0) = p_0 + X^\hbar$ centered at some value p_0. We will *identify* the quantum system with the Cartesian product $X(x_0) \times P(p_0)$. This is a convex body in phase space $\ell_X \times \ell_P = \mathbb{R}_x^n \times \mathbb{R}_p^n$, centered at $z_0 = (x_0, p_0)$, so we may consider its John ellipsoid [2] $(X(x_0) \times (P(p_0))_{\mathrm{John}}$. Since a trivial phase space translation reduces these geometric quantum states to those being centered at the origin $(x_0, p_0) = (0, 0)$ we have

$$(X(x_0) \times (P(p_0))_{\mathrm{John}} = (X \times X^\hbar)_{\mathrm{John}} + (x_0, p_0).$$

From now on we limit our study to the centered case $X(0) = X$, $P(0) = X^\hbar$. More generally, we consider the set of all pairs $(X_\ell, X_{\ell'}^\hbar) = S(X, X_{\ell'})$ where $(\ell, \ell') = S(\ell_X, \ell_P)$ for some $S \in \mathrm{Sp}(n)$. This set will be denoted by $\mathrm{Quant}_0(n)$; by construction, and taking into account the transitivity of the action of $\mathrm{Sp}(n)$ on the set of all Lagrangian frames, we have a continuous transitive action

$$\mathrm{Sp}(n) \times \mathrm{Quant}_0(n) \longrightarrow \mathrm{Quant}_0(n). \tag{8}$$

It follows from Theorem 1 that we have a commutative diagram

$$\begin{array}{ccc}
\mathrm{Sp}(n) \times \mathrm{Quant}_0(n) & \longrightarrow & \mathrm{Quant}_0(n) \\
\updownarrow & & \updownarrow \\
\mathrm{Mp}(n) \times \mathrm{Gauss}_0(n) & \longrightarrow & \mathrm{Gauss}_0(n)
\end{array}$$

where $\mathrm{Gauss}_0(n)$ is the set of all Gaussians Ψ_{AB} of the type (1).

Geometric quantum states have interesting interpretations in symplectic topology; for instance using the results in [1] one shows that

$$c_{\mathrm{HZ}}(X_\ell \times X_{\ell'}^\hbar) = c_{\max}(X_\ell \times X_{\ell'}^\hbar) = 4\hbar \tag{9}$$

where c_{HZ} is the Hofer–Zehnder symplectic capacity and $c_{\max}$ is the largest (normalized) symplectic capacity. We have shown in [15] that, in fact,

$$c_{\min}^{\mathrm{lin}}(X_\ell \times X_{\ell'}^\hbar) = \pi\hbar \tag{10}$$

where $c_{\min}^{\mathrm{lin}}$ is the smallest linear symplectic capacity. This relation can be viewed as a and topological version of the uncertainty principle related to our results in [7,17].

4   Proof of Theorem 1

Let Ω be a convex body in $\mathbb{R}^m$ (i.e. a convex and compact set with non-empty interior).. The John ellipsoid Ω_{John} is the unique ellipsoid in $\mathbb{R}^m$ with maximum volume contained Ω. When Ω is symmetric and centered at the origin we have [2] $\Omega_{\mathrm{John}} \subset \Omega \subset \sqrt{m}\,\Omega_{\mathrm{John}}$. The John ellipsoid is linearly covariant: if $L \in GL(m, \mathbb{R})$ then

$$(L(\Omega))_{\mathrm{John}} = L(\Omega_{\mathrm{John}}). \tag{11}$$

In view of Proposition 1 there exists $S' \in \mathrm{Sp}(n)$ such that $(X_\ell, X_{\ell'}^\hbar) = S'(X, X^\hbar)$ where $X \subset \ell_X$ and $X^\hbar \subset \ell_P$. Using formula (11) in the $2n$-dimensional case with $L = S'$ we have

$$(X_\ell \times X_{\ell'}^\hbar)_{\mathrm{John}} = (S'(X \times X^\hbar))_{\mathrm{John}} = S'((X \times X^\hbar)_{\mathrm{John}})$$

so we may assume that there exists $A > 0$ such that $X_\ell = X$ and $X_{\ell'}^\hbar = X^\hbar$ where

$$X = \{x \in \mathbb{R}_x^n : Ax \cdot x \le \hbar\} = A^{-1/2}(B_X^n(\sqrt{\hbar})) \tag{12}$$

$$X^\hbar = \{p \in \mathbb{R}_p^n : A^{-1}p \cdot p \le \hbar\} = A^{1/2}(B_P^n(\sqrt{\hbar})). \tag{13}$$

This can be rewritten as

$$X \times X^\hbar = S''(B_X^n(\sqrt{\hbar}) \times B_P^n(\sqrt{\hbar}))$$

where $S'' = \begin{pmatrix} A^{-1/2} & 0 \\ 0 & A^{1/2} \end{pmatrix} \in \mathrm{Sp}(n)$ and hence

$$(X_\ell \times X_{\ell'}^\hbar)_{\mathrm{John}} = S'S''((B_X^n(\sqrt{\hbar}) \times B_P^n(\sqrt{\hbar})))_{\mathrm{John}}.$$

Setting $S = S'S''$ we thus have

$$(X_\ell \times X_{\ell'}^\hbar)_{\mathrm{John}} = S(B^{2n}(\sqrt{\hbar})) \tag{14}$$

showing that $(X_\ell \times X_{\ell'}^\hbar)_{\mathrm{John}}$ is a quantum blob [8,10,17], and hence the covariance ellipsoid of a Gaussian (1). The latter is determined as follows: using a pre-Iwasawa decomposition [8,12] of symplectic matrices and the invariance of $B^{2n}(\sqrt{\hbar})$ under symplectic rotations one shows that S in (14) can be chosen of the type [12]

$$S_{AB} = \begin{pmatrix} A^{1/2} & 0 \\ A^{-1/2}B & A^{-1/2} \end{pmatrix}, \quad A = A^T > 0, \quad B = B^T \tag{15}$$

so that we have

$$S_{AB}^T S_{AB} = \begin{pmatrix} A + BA^{-1}B & BA^{-1} \\ A^{-1}B & A^{-1/2} \end{pmatrix}.$$

It follows [8,17] that the quantum blob $\Omega_{AB} = S_{AB}(B^{2n}(\sqrt{\hbar}))$ is the covariance ellipsoid of the Gaussian Ψ_{AB} as follows from the fact that the Wigner transform of that function is

$$W\Psi_{A,B}(z) = \left(\frac{1}{\pi\hbar}\right)^n \exp\left(-\frac{1}{\hbar}(S_{AB}^T S_{AB})z \cdot z\right) = W(\widehat{S}_{AB}^{-1}\Psi_{I,0}(z) \tag{16}$$

where $\widehat{S}_{AB} \in \mathrm{Mp}(n)$ covers S_{AB}.

5   Discussion and Perspectives

The Wigner transform and its variants (Bargmann transform, Gabor transform), have in common that they associate to a *function* on configuration space another function defined on phase space. For instance, in quantum mechanics, the Wigner transform takes an arbitrary probability amplitude ψ (the wavefunction) to a quasi-probability density on $W\psi$ phase space whose marginals give the true probabilities of finding the quantum system under consideration in both some localization of configuration space and a localization in momentum space. A similar result is obtained in time-frequency analysis where one uses the Gabor transform to analyze the localization of a signal in the time and frequency space.

In the present work we have suggested a novel purely geometric

"Configuration space <-> phase space"

correspondence of a more general nature: instead of starting with functions to which we associate quasi-probability densities, we started with convex subsets of x-space (ellipsoids or convex bodies) to which we made correspond subsets in phase space obtained using the notion of polar dually familiar from convex geometry. This procedure is somehow related to the so-called "Pauli reconstruction problem" (see our discussion in [14]), which can be formulated in the Wigner formalism as follows: knowing the probability densities $|\psi|^2$ and $|\widehat{\psi}|^2$ ($\widehat{\psi}$ the Fourier transform of ψ) can we determine uniquely ψ? (It is known that the answer to this question is generally negative). What is remarkable and new in our treatment is that we do not assume that the "geometric quantum states" we define are associated to any notion of wavefunction. This is where our approach departs from the usual one. One has first to understand that as opposed to a widespread belief, Quantum Mechanics does not say that particles are waves. While the wave interpretation was de Broglie's original point of view, it is today is untenable. Of course particle dynamics may be described using waves as solutions of Schrödinger's equation, but this is not the same as saying particles *are* waves! Quantum mechanics is a probabilistic theory from which we can extract consequences on the statistical behavior of many measurements performed on a great number of equally prepared systems. The real issue is the calculation of probabilities, and the objects used to study them (variances, spreading, covariances, mean values). Physically speaking, the Robertson–Schrödinger uncertainty inequalities are statements about the variances and covariances of the random variables corresponding to independently measured position and momentum in an ensemble of equally prepared systems. We emphasize that they are neither a statement about the measure of both momentum and position of a single particle nor an effect of the interaction with a measurement device as early interpretations of the Heisenberg inequalities tend to suggest. Uffink and Hilgevoord have analyzed this problem with great care, and they have shown in a series of works [3–6] that the use variances and covariances to describe the statistical properties of a quantum system is valid only for Gaussian or almost Gaussian states; see in this context our discussion in [13]. Our approach avoids this pitfall, because we do not make any assumption of statistical nature other than position measurements are localized in a convex set.

We would like to generalize our constructions to arbitrary convex sets, not necessarily ellipsoids (which correspond, as we have seen, to Gaussian states). This can be done following the lines above, but at the expense of some mathematical difficulties. The first difficulty is the choice of the point with respect to which the polar dual should be defined since there is no privileged "center"; different choices may lead to polar duals with very different volumes. To define the polar dual of X in the general convex case one has to use the *Santaló point* of the convex set under consideration. Santaló proved in [18] the following remarkable result: there exists a *unique* interior point x_S of X (the "Santaló point of X") such that the polar dual $X^\hbar(x_S) = (X - x_S)^\hbar$ has centroid $\overline{p} = 0$ and its volume

$\mathrm{Vol}_n(X^\hbar(x_S))$ is *minimal* for all possible interior points x_0:

$$\mathrm{Vol}_n(X)\,\mathrm{Vol}_n(X^\hbar(x_S)) \le (\mathrm{Vol}_n\, B^n(\sqrt{\hbar})^2 \tag{17}$$

with equality if and only if X is an ellipsoid. We note that the practical determination of the Santaló point is in general difficult and one has to use ad hoc methods in each particular case. It would moreover be interesting to reformulate (17) as a version of the uncertainty principle; this is moreover of a great mathematical interest since it is closely related to the Mahler conjecture (see our discussion in [13]).

Having in mind that the polar dual of X is calculated with respect to its Santaló point—not its centroid—we could define the associated canonical quantum state as follows let (ℓ_X, ℓ_P) be the canonical Lagrangian frame and $X(x_S)$ a convex body carried by ℓ_X and with Santaló point x_S. The associated geometric state would then be

$$X(x_S) \times (X(x_S) - x_S)^\hbar + p_0) = X(x_S) \times X^\hbar(p_0).$$

We also notice that we can associated to every state an ellipsoid using the John ellipsoid method, but the role played by the latter is unclear (it is not quite obvious that it should be a quantum blob; if it were the case it could correspond to the covariance matrix of the state). At this point it would be fruitful to use techniques using the Minkowski functional of convex sets. All these questions are open, and we hope to come with answers and advances in a near future.

– This work has been financed by the Grant P 33447 N of the Austrian Science Fund FWF.

References

1. Artstein-Avidan, S., Karasev, R., Ostrover, Y.: From Symplectic measurements to the Mahler conjecture. Duke Math. J. **163**(11), 2003–2022 (2014)
2. Ball, K.M.: Ellipsoids of maximal volume in convex bodies. Geom. Dedicata. **41**(2), 241–250 (1992)
3. Uffink, J.B.M., Hilgevoord, J.: Uncertainty principle and uncertainty relations. Found. Phys. **15**(9), 925 (1985)
4. Uffink, J.B.M., Hilgevoord, J.: More certainty about the uncertainty principle. Eu. J. Phys. **6**, 165 (1985)
5. Hilgevoord, J.: The standard deviation is not an adequate measure of quantum uncertainty. Am. J. Phys. **70**(10), 983 (2002)
6. Hilgevoord, J., Uffink, J.: The uncertainty principle, the Stanford encyclopedia of philosophy (Winter 2016 Edition). Edward N. Zalta (ed.). URL = <https://plato.stanford.edu/archives/win2016/entries/qt-uncertainty/>
7. de Gosson, M.: Phase space quantization and the uncertainty principle. Phys. Lett. A **317**(5–6), 365–369 (2003)
8. de Gosson, M.: Symplectic Geometry and Quantum Mechanics, vol. 166. Springer Science & Business Media, Cham (2006)

9. de Gosson, M.: The symplectic camel and the uncertainty principle: the tip of an iceberg? Found. Phys. **99**, 194 (2009)
10. de Gosson, M.: Quantum blobs. Found. Phys. **43**(4), 440–457 (2013)
11. Gosson, M.A.: Quantum indeterminacy and polar duality. Math. Phys. Anal. Geom. **18**(1), 1–10 (2015). https://doi.org/10.1007/s11040-015-9175-8
12. de Gosson, M.: Symplectic coarse-grained dynamics: Chalkboard motion in classical and quantum mechanics. Adv. Theor. Math. Phys. **24**(4), 925–977 (2020)
13. de Gosson, M.: Quantum polar duality and the symplectic camel: a new geometric approach to quantization. Found. Phys. **51**, Article number: 60 (2021)
14. de Gosson, M.: The Pauli problem for gaussian quantum states: geometric interpretation. Mathematics **9**(20), 2578 (2021)
15. de Gosson, M.: Polar duality between pairs of transversal Lagrangian planes; applications to uncertainty principles. Bull. Sci. Math. **179**, 103171 (2022)
16. de Gosson, M., de Gosson, C.: Pointillisme à la signac and construction of a quantum fiber bundle over convex bodies. Found. Phys. (2023). arXiv preprint: arXiv:2208.00470 (2022)
17. de Gosson, M., Luef, F.: Symplectic capacities and the geometry of uncertainty: the irruption of Symplectic topology in classical and quantum mechanics. Phys. Reps. **484**, 131–179 (2009)
18. Santaló, L.A.: Un invariante a n para los cuerpos convexos del espacio de n dimensiones. Portugaliae. Math. **8**, 155–161 (1949)

Integrable Systems and Information Geometry (From Classical to Quantum)

Complete Integrability of Gradient Systems on a Manifold Admitting a Potential in Odd Dimension

Prosper Rosaire Mama Assandje[1](✉)[iD], Joseph Dongho[1][iD],
and Thomas Bouetou Bouetou[2][iD]

[1] University of Maroua, 814, Maroua, Cameroon
mamarosaire@yahoo.fr
[2] Higher National School of Polytechnic, 8390 Yaounde, Cameroon
http://www.fs.univ.maroua.cm

Abstract. The aim of this paper is to propose a method to study the complete integrability of gradient systems on a odd dimensional statistical manifold with a potential function. We show that these gradient systems are Hamiltonian and completely integrable.

Keywords: Hamiltonian system · Completely integrable systems · Gradient system

1 Introduction

Systems of non-linear differential equations generally do not find analytical solution. This is because there are no functions with precise names to write the explicit solutions of these systems, and to the fact that, these systems exhibit Chaotic behaviour, a property that makes it possible to know a particular explicit solution. In this case, we use many means analytical, geometric, and topological technic to understand and solve these systems, especially in the field of dynamical systems. A dynamical system is a way of describing the passage in time of all points of a space S. In elementary case, S is a submanifold of $\mathbb{R}^n$; and there is a particular type of system, name gradient system with particularly interesting Liapunov functions. But, in much case, the state space of physical system are not Euclidean submanifold. Such a type of system are related to differential operator call gradient field, which is well define from smooth function in Riemannian manifold. In general, it is well known that the integral curves of a smooth vector field manifold are the solution of a system of differential equations related to the field in question. In Poisson manifold one of important system which modelise a dynamic of particles is a Hamiltonian system. What makes them important is

Supported by UFD-SF-UMa.

the fact that the Hamiltonian function is a first integral or constant of motion. In other words, $\mathcal{H}$ is constant along each solution of the system;

$$dH = 0.$$

The importance of knowing that a given system is Hamiltonian is the fact that we can essentially draw the phase portrait without solving the system. The study of completely integrable Hamiltonian system goes back to Liouville's work who find local solutions by quadratures. In 1993, Nakamura [5], pointed out that certain gradient flows on Gaussian and multinomial distributions can be characterized as completely integrable Hamiltonian systems. This is the first suggestion of the connection between two seemingly unrelated research fields, i.e., information geometry Poisson manifold and its applications. In the same year, Fujiwara [3] stated a theorem giving the method of studying the complete integrability of gradient systems in even dimension on statistic manifolds admitting a potential function. To date however, there is no general method for determining whether or not a given gradient system is integrable in odd dimensions. This work focuses on the method for proving the complete integrability of gradient systems defined on a odd dimensional statistical manifold

$$S = \left\{ p_\theta(x), \begin{array}{c} \theta \in \Theta \\ x \in \mathcal{X} \end{array} \right\}$$

where p_θ is the density function parametrized by

$$\theta = (\theta_1, \ldots, \theta_n).$$

We show that if

$$\dim S = n = 2\ell + 1$$

with $\ell \in \mathbb{N}^*$ and if S admits a pair of dual coordinates (θ, η), with

$$\eta = (\eta_1, \ldots, \eta_n)$$

then it gradient system is a completely integrable Hamiltonian system with position

$$Q_k = \eta_{2k}, \ P_k = 1/\eta_{2k-1}, \ Q'_k = \eta_{2k+1}, \ P'_k = 1/\eta_{2k}$$

and Hamiltonian

$$\mathcal{H} = Q_k P_k + Q'_k P'_k, \ (k = 1, \cdots, \ell).$$

The ℓ quantities

$$\mathcal{H}_k = \eta_{2k}/\eta_{2k-1} + \eta_{2k+1}/\eta_{2k}$$

are mutually independent constants of motion.

2   Preliminary

2.1   Gradient System on Riemannian Manifold

Let

$$S = \left\{ p_\theta(x),\ \begin{matrix} \theta \in \Theta \\ x \in \mathcal{X} \end{matrix} \right\}$$

be a statistical model, with probabilities densities p_θ, parametrized by Θ, open subset of $\mathbb{R}^n$; on the sample space $\mathcal{X} \subseteq \mathbb{R}$. Let $\mathcal{F}(\mathcal{X}, \mathbb{R})$ be the space of real-valued smooth functions on $\mathcal{X}$. According to Ovidiu [6], the log-likelihood function is a mapping defined by:

$$l : S \longrightarrow \mathcal{F}(\mathcal{X}, \mathbb{R})$$
$$p_\theta \longmapsto l\,(p_\theta)\,(x) = \log\,(p_\theta(x))$$

Sometimes, for convenience reasons, this will be denoted by

$$l(x, \theta) = l\,(p_\theta)\,(x).$$

Let

$$\theta = (\theta_1, \ldots, \theta_n)$$

be a coordinate system on S. Let g be a riemannian structure on S. Let $\mathfrak{X}(S)$ denote the module of vector fields on S. Any $X \in \mathfrak{X}(S)$ induce a $C^\infty(S)$-linear form g_X on $\mathfrak{X}(S)$ defined by

$$g_X(Y) = g(X, Y).$$

In other words for all $X \in \mathfrak{X}(S)$, $g_X \in \Omega(S)$. Therefore each riemannian structure g on S induces an isomorphism

$$b_g : \mathfrak{X}(S) \xrightarrow{\ \sim\ } \Omega(S)$$
$$X \longmapsto g_X.$$

Let $\Phi \in C^\infty(S)$, we have $d\Phi \in \Omega(S)$ and then there exist $X_\Phi \in \mathfrak{X}(S)$, such that

$$b_g(X_\Phi) = d\Phi.$$

As a linear mapping, b_g has a matrix $[b_g]$ with respect to the basis pair $(\partial_{\theta_i},\, d\theta_i)$ and it is well known that

$$(g_{ij})_{1 \leq i,j \leq n} = [b_g].$$

Since

$$[b_g]X_\Phi = d\Phi,$$

we have

$$X_\Phi = [b_g]^{-1} d\Phi.$$

We denote

$$X_\Phi = grad_g(\Phi)$$

and Φ is called gradient potential function associated to the gradient field X_Φ with respect to g. By definition, the gradient system on the riemannian manifold (S, g) is the negative flow of the vector field $grad_g(\Phi)$. In other words

$$\overrightarrow{\dot\theta} = -[b_g]^{-1}\partial_\theta\Phi(\theta),$$

with $\Phi(\theta)$ the potential function,

$$\partial_\theta\Phi(\theta) = (\partial_{\theta_1}\Phi(\theta), \ldots, \partial_{\theta_n}\Phi(\theta))^T$$

and

$$[b_g] = (g_{ij})_{1\leq i;j\leq n}.$$

The Fisher information is defined by

$$(g_{ij})_{1\leq i;j\leq n} = \left(-\mathbb{E}[\partial_{\theta_i}\partial_{\theta_j}l(x,\theta)]\right). \tag{1}$$

More explicitly, when $S = \left\{ p_\theta(x), \begin{array}{l} \theta \in \Theta \\ x \in \mathcal{X} \end{array} \right\}$ is a statistical manifold and

$$G = (g_{ij})_{1\leq i;j\leq n}$$

the Fisher information matrix, the gradient systems of S is:

$$\overrightarrow{\dot\theta} = -G^{-1}\partial_\theta\Phi(\theta). \tag{2}$$

In physics mathematics [4], the system (2) is said to be Hamiltonian if it exists a bivector field $\bar\wedge$ and, $\mathcal{H}$, such that:

$$\dot\theta(t) = \bar\wedge\frac{\partial\mathcal{H}}{\partial\theta} \tag{3}$$

where $\mathcal{H}$ is smooth function called hamiltonian function and $\bar\wedge$ is a bivector fields such that

$$[\bar\wedge; \bar\wedge] = 0$$

where [;] denotes the Schouten Nijenhus bracket. In Poisson geometry the equation

$$[\bar\wedge, \bar\wedge] = 0$$

is equivalent to the fact that , operator { ; } defined by:

$$\{\mathcal{H}; F\} = \langle\frac{\partial\mathcal{H}}{\partial x}, \bar\wedge\frac{\partial F}{\partial x}\rangle = \sum_{i,j}\bar\wedge_{ij}\frac{\partial\mathcal{H}}{\partial x_i}\frac{\partial F}{\partial x_j}$$

is Poisson bracket on $C^\infty(S)$. The system (3) is said to be completely integrable in the sense of Liouville-Arnol'd [2] if it has n-prime integrals

$$\mathcal{H} = \mathcal{H}_1, \mathcal{H}_2, \ldots, \mathcal{H}_n$$

i.e.,
$$\{\mathcal{H}_i, \mathcal{H}_j\} = 0;$$
$1 \leqq i, j \leqq n$ and they are functionally independent, i.e;
$$d\mathcal{H}_1 \wedge d\mathcal{H}_2 \wedge \cdots \wedge d\mathcal{H}_n \neq 0.$$

The system (3) is said to be completely integrable in dimensional $n = 2m + c$, $m \in \mathbb{N}^*$ in the sense of Lesfari [4] if it has c-prime integrals $\mathcal{H}_{m+1}, \mathcal{H}_{m+2}, \ldots, \mathcal{H}_{m+c}$ called casimir function such that:
$$\bar{\wedge} \frac{\partial \mathcal{H}_{m+i}}{\partial \theta} = 0, \ 1 \leqq i \leqq c.$$

2.2 Dualistic Geometry

Let (S, g) be an n-dimensional riemannian manifold with metric g. Let $\theta = (\theta_i)$ and $\eta = (\eta_i)$ be two coordinate systems of S.

Let $\{\partial_{\theta_i} = \frac{\partial}{\partial \theta_i}\}_{i=1}^n$ and $\{\partial_{\eta_j} = \frac{\partial}{\partial \eta_j}\}_{j=1}^n$ be the basis of the tangent space $T_p S$ at a point $p \in S$ with respect to the coordinate system θ and η respectively. We define,
$$g_{ij} = \langle \partial_{\theta_i}, \partial_{\theta_j} \rangle_g := g\left(\partial_{\theta_i}, \partial_{\theta_j}\right).$$

Since
$$\partial_{\eta_j} = \sum_k \frac{\partial \theta_k}{\partial \eta_j} \partial_{\theta_k},$$

we have
$$g\left(\partial_{\theta_i}, \partial_{\eta_j}\right) = \frac{\partial \theta_k}{\partial \eta_j} g_{ik}. \tag{4}$$

When
$$\forall \ i, \ j, \ g\left(\partial_{\theta_i}, \partial_{\eta_j}\right) = \delta_i^j,$$

we say that the basis $\{\partial_{\theta_i}\}$ and $\{\partial_{\eta_j}\}$ are biorthogonal with respect to the metric g. Two coordinates systems θ and η are said to be mutually dual when their corresponding basis of $T_p S$ are biorthogonal with respect to g. If
$$S = \left\{ p_\theta(x), \begin{array}{c} \theta \in \Theta \\ x \in \mathcal{X} \end{array} \right\}$$

is a statical manifold. Then we can use the Amari [1], l-representation associated respectively to coordinate θ and η; and define the following scalar production on T_θ where T_θ is the l-representation of the tangent space with respect to the coordinate θ,
$$g_{ij} = \mathbb{E}\left[\partial_{\theta_i} l(\theta, x) \partial_{\theta_j} l(\theta, x)\right].$$

Since
$$\frac{\partial l(\theta, x)}{\partial \eta_j} = \sum_{k=1}^m \frac{\partial \theta_k}{\partial \eta_j} \frac{\partial l(\theta, x)}{\partial \theta_k}.$$

$$g\left(\frac{\partial l(\theta, x)}{\partial \theta_i}, \frac{\partial l(\theta, x)}{\partial \eta_j}\right) = \sum_{k=1}^{m} \frac{\partial \theta_k}{\partial \eta_j} g_{ik}.$$

When

$$S = \left\{p_\theta(x), \begin{array}{c} \theta \in \Theta \\ x \in \mathcal{X} \end{array}\right\}$$

is a statistical manifold the theorem 3.4 of [1] stipulate that there exist one potential functions $\Phi(\theta)$ and $\Psi(\eta)$ such that

$$g_{ij}(\theta) = \partial_{\theta_i} \partial_{\theta_j} \Phi(\theta), \quad g^{ij}(\eta) = \partial_{\eta_i} \partial_{\eta_j} \Psi(\eta). \tag{5}$$

here g^{ij} is the inverse of g_{ij}. Conversely, when either potential function Φ or Ψ exists from which the metric is derived by differentiating it twice, there exist a pair of dual coordinate systems. The dual coordinate systems are related by the following Legendre transformations

$$\theta_i = \partial_{\eta_i} \Psi(\eta), \quad \eta_i = \partial_{\theta_i} \Phi(\theta) \tag{6}$$

where the two potential functions satisfy the identity

$$\Psi(\theta) = \theta_i.\eta_i - \Phi(\eta) \tag{7}$$

and $\theta \cdot \eta = \theta_i \eta_i$. Let us show that the gradient system define by the relation (2) is linearizable. It follow from (6)+(5) that

$$g_{ij} = \frac{\partial \eta_j}{\partial \theta_i}. \tag{8}$$

From these relations, we seduce that

$$g_{ij}\dot{\theta}_i = \dot{\eta}, \tag{9}$$

and the fact that:

$$\dot{\theta}_i = -g^{ij}\partial_{\theta_j}\Phi(\theta),$$

imply

$$\dot{\eta}_i = -\eta_i. \tag{10}$$

Equation (10) is the linear system associated to the above gradient system.

3   Main Theorem

Let

$$S = \left\{p_\theta(x), \begin{array}{c} \theta \in \Theta \\ x \in \mathcal{X} \end{array}\right\}$$

be a statistical model where p_θ is a density function parameterized by

$$\theta = (\theta_1, \ldots, \theta_n)$$

and admitting a potential Φ. Thus, let us consider the pair of dual coordinates (θ, η) with

$$\eta = (\eta_1, \ldots, \eta_n).$$

We have the following theorem

Theorem 1

If $\dim S$ *is odd, say* $2\ell+1$, $\ell \in \mathbb{N}^*$, *then the dynamical system (10) is a completely integrable Hamiltonian system with position*

$$Q_k = \eta_{2k}, \ P_k = 1/\eta_{2k-1}, \ Q'_k = \eta_{2k+1}, \ P'_k = 1/\eta_{2k}$$

and Hamiltonian

$$\mathcal{H} = Q_k P_k + Q'_k P'_k, \ (k = 1, \cdots, \ell).$$

The ℓ *quantities*

$$\mathcal{H}_k = \eta_{2k}/\eta_{2k-1} + \eta_{2k+1}/\eta_{2k}$$

are mutually independent constants of motion.

Proof

It follows from, definition Q_k, P_k, Q'_k and P'_k of the that

$$\mathcal{H} = \frac{\eta_2}{\eta_1} + \frac{\eta_3}{\eta_2} + \cdots + \frac{\eta_{2\ell}}{\eta_{2\ell-1}} + \frac{\eta_{2\ell+1}}{\eta_{2\ell}}. \tag{11}$$

Therefore, from (11), we deduce the following relation

$$\dot{\mathcal{H}} = \frac{\dot{\eta}_2 \cdot \eta_1 - \dot{\eta}_1 \cdot \eta_2}{\eta_1^2} + \cdots + \frac{\dot{\eta}_{2\ell+1} \cdot \eta_{2\ell} - \dot{\eta}_{2\ell} \cdot \eta_{2\ell+1}}{\eta_{2\ell-1}^2}. \tag{12}$$

Using the equation (10), we have:

$$\left. \begin{array}{c} \dot{\eta}_{2k} = -\eta_{2k} \\ \dot{\eta}_{2k} \cdot \eta_{2k-1} = -\eta_{2k} \cdot \eta_{2k-1}, \\ \text{with } (k = 1, \cdots, \ell). \end{array} \right\} \tag{13}$$

Similarly, we also have the same relations when the indices are odd:

$$\left. \begin{array}{c} \dot{\eta}_{2k+1} = -\eta_{2k+1} \\ \dot{\eta}_{2k+1} \cdot \eta_{2k} = -\eta_{2k+1} \cdot \eta_{2k}, \\ \text{with } (k = 1, \cdots, \ell). \end{array} \right\} \tag{14}$$

Using the relation (13) and (14), (12) become

$$\dot{\mathcal{H}} = \frac{-\eta_2 \cdot \eta_1 + \eta_1 \cdot \eta_2}{\eta_1^2} + \cdots + \frac{-\eta_{2\ell+1} \cdot \eta_{2\ell} + \eta_{2\ell} \cdot \eta_{2\ell+1}}{\eta_{2\ell}^2}.$$

So,

$$\frac{d\mathcal{H}}{dt} = 0. \tag{15}$$

We conclude that $\mathcal{H}$ is constant of motion. By putting

$$Q_k = \eta_{2k}, \ P_k = 1/\eta_{2k-1}, \ Q'_k = \eta_{2k+1}, \ P'_k = 1/\eta_{2k}$$

the Hamiltonian can be written as following:

$$\mathcal{H} = Q_1 P_1 + Q'_1 P'_1 + \cdots + Q_\ell P_\ell + Q'_\ell P'_\ell.$$

And then,

$$\frac{\partial \mathcal{H}}{\partial P_k} = Q_k.$$

More explicitly we obtain the following system

$$\frac{\partial \mathcal{H}}{\partial P_k} = \eta_{2k}, \text{ with } (k = 1, \cdots, \ell). \tag{16}$$

Using (13), the relation (16) becomes:

$$-\frac{\partial \mathcal{H}}{\partial P_k} = \frac{d\eta_{2k}}{dt}, \text{ with } (k = 1, \cdots, \ell).$$

Then we have:

$$-\frac{\partial \mathcal{H}}{\partial P_k} = \frac{dQ_k}{dt}.$$

In the same way, we have:

$$\frac{\partial \mathcal{H}}{\partial Q_k} = P_k = \frac{1}{\eta_{2k-1}}. \tag{17}$$

This relation (17) is equivalent to

$$-\frac{\partial \mathcal{H}}{\partial Q_k} = -\frac{\eta_{2k-1}}{\eta_{2k-1}^2}.$$

When we apply (14) in (18) we obtain,

$$-\frac{\partial \mathcal{H}}{\partial Q_k} = \frac{\dot{\eta}_{2k-1}}{\eta_{2k-1}^2} = -\frac{dP_k}{dt}. \tag{18}$$

In the same way,

$$\frac{\partial \mathcal{H}}{\partial P_k'} = Q_k' = \eta_{2k+1}. \tag{19}$$

When we apply (14) in (19), we obtain:

$$-\frac{\partial \mathcal{H}}{\partial P_k'} = \frac{dQ_k'}{dt}.$$

In the same way we have:

$$\frac{\partial \mathcal{H}}{\partial Q_k'} = P_k' = \frac{1}{\eta_{2k}} = \frac{\eta_{2k}}{\eta_{2k}^2}. \tag{20}$$

Using (13) in (20) we have:

$$-\frac{\partial \mathcal{H}}{\partial Q_k'} = \frac{\dot{\eta}_{2k}}{\eta_{2k}^2} = \frac{dP_k'}{dt}.$$

$$\begin{cases} \dfrac{dP_k}{dt} = \dfrac{\partial \mathcal{H}}{\partial Q_k} \\[2mm] \dfrac{dQ_k}{dt} = -\dfrac{\partial \mathcal{H}}{\partial P_k} \\[2mm] \dfrac{dP'_k}{dt} = \dfrac{\partial \mathcal{H}}{\partial Q'_k}. \\[2mm] \dfrac{dQ'_k}{dt} = -\dfrac{\partial \mathcal{H}}{\partial P'_k}. \end{cases} \tag{21}$$

With $(k = 1, \cdots, \ell)$. This system (21) have the following matrix representation:

$$\begin{pmatrix} \dot{P}_k \\ \dot{Q}_k \\ \dot{P'}_k \\ \dot{Q'}_k \end{pmatrix} = \begin{pmatrix} O_{\ell \times \ell} & Id_{\ell \times \ell} & O_{\ell \times \ell} & O_{\ell \times \ell} \\ -Id_{\ell \times \ell} & O_{\ell \times \ell} & O_{\ell \times \ell} & O_{\ell \times \ell} \\ O_{\ell \times \ell} & O_{\ell \times \ell} & O_{\ell \times \ell} & Id_{\ell \times \ell} \\ O_{\ell \times \ell} & O_{\ell \times \ell} & -Id_{\ell \times \ell} & O_{\ell \times \ell} \end{pmatrix} \begin{pmatrix} \frac{\partial \mathcal{H}}{\partial P_k} \\ \frac{\partial \mathcal{H}}{\partial Q_k} \\ \frac{\partial \mathcal{H}}{\partial P'_k} \\ \frac{\partial \mathcal{H}}{\partial Q'_k} \end{pmatrix} \tag{22}$$

where

$$\dot{P}_k = \begin{pmatrix} \dot{P}_1 \\ \vdots \\ \dot{P}_\ell \end{pmatrix}, \quad \dot{Q}_k = \begin{pmatrix} \dot{Q}_1 \\ \vdots \\ \dot{Q}_\ell \end{pmatrix}, \quad \dot{P'}_k = \begin{pmatrix} \dot{P'}_1 \\ \vdots \\ \dot{P'}_\ell \end{pmatrix}, \quad \dot{Q'}_k = \begin{pmatrix} \dot{Q'}_1 \\ \vdots \\ \dot{Q'}_\ell \end{pmatrix}$$

$$\frac{\partial \mathcal{H}}{\partial P_k} = \begin{pmatrix} \frac{\partial \mathcal{H}}{\partial P_1} \\ \vdots \\ \frac{\partial \mathcal{H}}{\partial P_\ell} \end{pmatrix}, \quad \frac{\partial \mathcal{H}}{\partial Q_k} = \begin{pmatrix} \frac{\partial \mathcal{H}}{\partial Q_1} \\ \vdots \\ \frac{\partial \mathcal{H}}{\partial Q_\ell} \end{pmatrix}, \quad \frac{\partial \mathcal{H}}{\partial P'_k} = \begin{pmatrix} \frac{\partial \mathcal{H}}{\partial P'_1} \\ \vdots \\ \frac{\partial \mathcal{H}}{\partial P'_\ell} \end{pmatrix}, \quad \frac{\partial \mathcal{H}}{\partial Q'_k} = \begin{pmatrix} \frac{\partial \mathcal{H}}{\partial Q'_1} \\ \vdots \\ \frac{\partial \mathcal{H}}{\partial Q'_\ell} \end{pmatrix}$$

$$O_{\ell \times \ell} = \begin{pmatrix} 0_{1,1} & \cdots & 0_{1,\ell} \\ \vdots & \ddots & \vdots \\ 0_{\ell,1} & \cdots & 0_{\ell,\ell} \end{pmatrix}, \quad Id_{\ell \times \ell} = \begin{pmatrix} 1_{1,1} & \cdots & 0_{1,\ell} \\ \vdots & \ddots & \vdots \\ 0_{\ell,1} & \cdots & 1_{\ell,\ell} \end{pmatrix}$$

Denote

$$\overline{\wedge} = \begin{pmatrix} O_{\ell \times \ell} & Id_{\ell \times \ell} & O_{\ell \times \ell} & O_{\ell \times \ell} \\ -Id_{\ell \times \ell} & O_{\ell \times \ell} & O_{\ell \times \ell} & O_{\ell \times \ell} \\ O_{\ell \times \ell} & O_{\ell \times \ell} & O_{\ell \times \ell} & Id_{\ell \times \ell} \\ O_{\ell \times \ell} & O_{\ell \times \ell} & -Id_{\ell \times \ell} & O_{\ell \times \ell} \end{pmatrix}$$

This is a matrix of a bivector field. We have just shown that the gradient system defined on S is Hamiltonian of Hamiltonian

$$\mathcal{H} = Q_k P_k + Q'_k P'_k, \ (k = 1, \cdots, \ell)$$

where ℓ quantities are given by:

$$\mathcal{H}_k = \eta_{2k} / \eta_{2k-1} + \eta_{2k+1} / \eta_{2k}.$$

This system (22) equivalent to:

$$\dot{\theta} = \bar{\wedge}\frac{\partial\mathcal{H}}{\partial\theta}.$$

where

$$\bar{\wedge} = \begin{pmatrix} O_{\ell\times\ell} & Id_{\ell\times\ell} & O_{\ell\times\ell} & O_{\ell\times\ell} \\ -Id_{\ell\times\ell} & O_{\ell\times\ell} & O_{\ell\times\ell} & O_{\ell\times\ell} \\ O_{\ell\times\ell} & O_{\ell\times\ell} & O_{\ell\times\ell} & Id_{\ell\times\ell} \\ O_{\ell\times\ell} & O_{\ell\times\ell} & -Id_{\ell\times\ell} & O_{\ell\times\ell} \end{pmatrix}.$$

We conclude on the complete integrability in the sense of Lesfari [4].   □

4   Conclusion

The gradient system defined on an odd-dimensional statistical manifold S is Hamiltonian and completely integrable. In this case, the dynamical system (10) is considered as a sub-dynamics of a completely integrable Hamiltonian system of higher dimensions by combining it with an odd-dimensional independent gradient system.

Acknowledgements. I gratefully acknowledge all my discussions with members of ERAG of the University of Maroua. I also would like to thank A. Souleymanou of the Higher National School of Polytechnic Yaounde I for encouragement. Thanks are due to Dr. Kemajou Theophile for fruitful discussions.

References

1. Amari, S.: Differential-Geometrical Methods in Statistics, 2nd Printing, vol. 28. Springer, Tokyo (2012). https://doi.org/10.1007/978-1-4612-5056-2
2. Arnol'd, V.I., Givental, A.B., Novikov, S.P.: Symplectic Geometry (Dynamical Systems IV), vol. 4. Springer, Heidelberg (2001). https://doi.org/10.1007/978-3-662-06791-8
3. Fujiwara, A.: Dynamical systems on statistical models (state of art and perspectives of studies on nonliear integrable systems). RIMS Kkyuroku **822**, 32–42 (1993). http://hdl.handle.net/2433/83219
4. Lesfari, A.: Théorie spectrale et probléme non-linéaires. Surv. Math. Appl. **5**, 141–180 (2010). http://www.utgjiu.ro/math/sma
5. Nakamura, Y.: Completely integrable gradient systems on the manifolds of gaussian and multinomial distributions. Jpn. J. Ind. Appl. Math. **10**(2), 179–189 (1993). https://doi.org/10.1007/BF03167571
6. Ovidiu, C., Constantin, U.: Geometric Modeling in Probability and Statistic, vol. 121. Springer, Cham (2014). https://doi.org/10.1007/978-3-319-07779-6

Geometry-Preserving Lie Group Integrators for Differential Equations on the Manifold of Symmetric Positive Definite Matrices

Lucas Drumetz[1(✉)], Alexandre Reiffers-Masson[1], Naoufal El Bekri[1,2], and Franck Vermet[2]

[1] IMT Atlantique, Lab-STICC, UMR CNRS 6285, Brest, France
`lucas.drumetz@imt-atlantique.fr`
[2] Univ Brest, UMR CNRS 6205, Laboratoire de Mathématiques de Bretagne Atlantique, Brest, France

Abstract. In many applications, one encounters time series that lie on manifolds rather than a Euclidean space. In particular, covariance matrices are ubiquitous mathematical objects that have a non Euclidean structure. The application of Euclidean methods to integrate differential equations lying on such objects does not respect the geometry of the manifold, which can cause many numerical issues. In this paper, we propose to use Lie group methods to define geometry-preserving numerical integration schemes on the manifold of symmetric positive definite matrices. These can be applied to a number of differential equations on covariance matrices of practical interest. We show that they are more stable and robust than other classical or naive integration schemes on an example.

Keywords: Lie groups · Differential equations · Symmetric positive definite matrices · Stochastic differential equations

1 Introduction

Ordinary Differential Equations (ODEs) arise everywhere in science, and are a fundamental tool to describe continuous-time dynamical systems [14]. However, numerical integration is almost always required to obtain approximate solutions. Most of the time, the variable to integrate lives in a Euclidean space, typically $\mathbb{R}^n$. One can then choose from many methods, ranging from simple explicit/implicit Euler or Runge-Kutta methods to adaptive time step schemes [5].

In a number of situations, one may require that the variable to integrate lies on a manifold [12]. Examples include flows on spheres, rotation or covariance matrices (or other matrix manifolds) [13]... For embedded submanifolds of $\mathbb{R}^n$,

This work was supported by Agence Nationale de la Recherche under grant ANR-21-CE48-0005 LEMONADE.

F. Nielsen and F. Barbaresco (Eds.): GSI 2023, LNCS 14072, pp. 433–443, 2023.
https://doi.org/10.1007/978-3-031-38299-4_45

though the underlying vector space makes it possible to apply classical ODE integration methods, they cannot guarantee that the numerical solution stays on the manifold at each time step. Providing these guarantees is crucial for subsequent uses of the solution, e.g. computing geodesic distances for Riemannian manifolds [1], or simply keeping the structural or physical interpretation of a variable.

Formally, the flow of a smooth vector field on a smooth manifold $\mathcal{M}$ generated by an ODE with the initial condition $x(0) = x_0$, writes:

$$\frac{dx}{dt} = F|_{x(t)}(t), \tag{1}$$

where $x \in \mathcal{M}$ and $F|_{x(t)}(t)$ is a (time dependent) tangent vector to $\mathcal{M}$ at x, and $F : [0, +\infty[\to \mathfrak{X}(\mathcal{M})$, with $\mathfrak{X}(\mathcal{M})$ the set of smooth vector fields on $\mathcal{M}$ [18]. To integrate such ODEs when $\mathcal{M}$ has the additional structure of a Lie group, several frameworks were developed under the umbrella term "Lie group integrators" [6]. Interestingly, these methods can be extended to any smooth manifold acted upon transitively by a Lie group [13], that is to any homogeneous space.

In this paper, we focus more specifically on the manifold of $n \times n$ symmetric positive definite (SPD) matrices, denoted as Sym_n^+ [4]. It is the manifold of (non-degenerate) covariance matrices, fundamental for multivariate statistics. Flows of covariance matrices arise in many applications, such as Brain Computer Interfaces (BCI) [21], Diffusion Tensor Image processing [10], finance [7], control [3], or data assimilation [11], to represent the evolution of second order moments of random variables. For example, the second order moments of the solution of Stochastic Differential Equations (SDEs) provide simplified and interpretable representations of stochastic processes, though partial in general. In data assimilation, quantifying and propagating the uncertainty of the state variable is crucial and is done in practice using covariance matrices [19]. Solutions of covariance matrix ODEs which are not SPD are meaningless in terms of statistical interpretation. Thus, we focus on equations similar to (1), where the manifold $\mathcal{M}$ is Sym_n^+, and the RHS of (1) is a symmetric matrix (an element of the tangent space of Sym_n^+). In spite of the ubiquiteness of covariance matrices, to the best of our knowledge, Lie group integrators have not been considered yet for Sym_n^+.

Our contributions are multiple. i) We show that in the case of Sym_n^+, for small enough time steps, classical methods actually remain in the manifold, but, with moderately big time steps, the iterates may cross the boundary of Sym_n^+, (consisting in positive semidefinite matrices), leading to meaningless solutions or even diverging algorithms. ii) We propose to use a Lie group action of invertible matrices on Sym_n^+ to turn the latter into a homogeneous space, that can be used to integrate many equations of interest. iii) From there, we design Lie group versions of the Euler and Runge-Kutta 4 (RK4) methods (applicable to many other schemes) on Sym_n^+. iv) We conduct experiments an example ODE on Sym_n^+ related to a multivariate SDE. They indicate that our integrators perform better than classical or naive schemes, in particular when the integration step is large.

2   Theoretical Results

We begin with a theoretical result in which we provide sufficient conditions on the integration time step ρ of the form $\mathbf{P}_{i+1} = \mathbf{P}_i + \rho\mathbf{T}$ to either stay in or leave Sym_n^+. The results are stated in the following theorem:

Theorem 1. *Let* $\mathbf{P} \in \mathrm{Sym}_n^+$, *and* $\mathbf{T} \in \mathrm{Sym}_n$. *We denote as* $\lambda_1 \leq ... \leq \lambda_n$ *the eigenvalues of* $\mathbf{P}$ *and as* $\nu_1 \leq ... \leq \nu_n$ *the (real) eigenvalues of* $\mathbf{T}$. *We define a set* $S = \{(i,j) \in \mathbb{N}_*^2, 1 \leq i, j \leq n, i+j = n+1, \nu_i \leq 0\}$. *Then the following holds:*

1. *If* $\mathbf{T}$ *is positive semidefinite, then* $\forall \rho \in \mathbb{R}^+$, $\mathbf{P} + \rho\mathbf{T} \in \mathrm{Sym}_n^+$.
2. *When* $\mathbf{T}$ *has at least one negative eigenvalue,* S *in nonempty and if*

$$\rho \geq \rho_{min} = \min_{(i,j) \in S} -\frac{\lambda_j}{\nu_i},$$

   then $\mathbf{P} + \rho\mathbf{T} \notin \mathrm{Sym}_n^+$.
3. *When* $\mathbf{T}$ *has at least one negative eigenvalue, if*

$$\rho < \rho_{max} = -\frac{\lambda_1}{\nu_1},$$

   then $\mathbf{P} + \rho\mathbf{T} \in \mathrm{Sym}_n^+$.

The proof is relegated to an extended online version of this work [9] (with more supplementary material). This theorem means that when the vector used for the update happens to be positive semidefinite, then a classical Euler or RK step will remain on Sym_n^+. However, in the general case, that vector may have negative eigenvalues, and we have shown that the next iterate will leave (resp. remain on) Sym_n^+ for time steps that are too large (resp. small enough). In practice, the lower bound is obtained by searching the best value in S, whose cardinal is the number of negative eigenvalues of $\mathbf{T}$. Time steps whose values lie in between both bounds may or may not stay on the manifold. Thus, algorithms at least guaranteeing that the trajectory remains on Sym_n^+ are necessary. We stress that even when iterates of classical methods remain on Sym_n^+, the geometry of the manifold is not accounted for, which may lead to low quality solutions nonetheless.

3   Background on Lie Group Integrators

The general idea behind basic Lie group methods relies on the fact that the flow of a simple class of vector fields on the Lie group is easy to compute via the Lie exponential map. The corresponding equations are analogous to the linear ODE $dx/dt = \mathbf{A}x$ in Euclidean spaces. In the case of general ODEs, discretizing by temporarily fixing the vector field of a general equation with a nonconstant "$\mathbf{A}$" (depending on x and t), an approximate Euler-like scheme can be computed step by step. Higher order schemes such as RK4 require a vector space structure to be able to manipulate vector fields at different locations and times. Then, we have to further translate the ODE from the Lie group to the Lie algebra.

Interestingly, all these methods can be effortlessly extended to smooth manifolds on which we can find a *transitive Lie group action* [6,18] (i.e. homogeneous spaces). Throughout this section, we follow [13] (Chap. 2). Here, we limit ourselves to matrix Lie groups for simplicity. Let $\mathcal{M}$ be the smooth manifold, G the matrix Lie group. A smooth map $\Lambda : G \times \mathcal{M} \to \mathcal{M}$ is a Lie group action if and only if

$$\forall x \in \mathcal{M}, \Lambda(\mathbf{I}, x) = x, \tag{2}$$

$$\forall x \in \mathcal{M}, \forall\ \mathbf{A}, \mathbf{B} \in G, \Lambda(\mathbf{A}, \Lambda(\mathbf{B}, x)) = \Lambda(\mathbf{AB}, x). \tag{3}$$

Λ is further said to be transitive if

$$\forall x, y \in \mathcal{M}, \exists \mathbf{A} \in G, \Lambda(\mathbf{A}, x) = y. \tag{4}$$

This means that any point of $\mathcal{M}$ can be reached from any other using the group action with an element of G. Equivalently, the group action has only one orbit.

To every Lie group G is associated a Lie algebra $\mathfrak{g}$, which is a vector space (the tangent space to the Lie group at the identity). For matrix Lie groups, the Lie algebra is also a set of matrices. Associated to a transitive Lie group action on a smooth manifold is a *algebra action* that translates the group action into an infinitesimal action giving an element of tangent space to the manifold at every point. It determines the type of equations that can be dealt with. It is defined [13] (Lemma 2.6) as a map $\lambda_* : \mathfrak{g} \times \mathcal{M} \to \mathfrak{X}(\mathcal{M})$ such that, for a given point $x \in \mathcal{M}$:

$$\lambda_*(\mathbf{A})(x) = \frac{d}{ds}\Lambda(\boldsymbol{\rho}(s), x)|_{s=0}, \tag{5}$$

where $\boldsymbol{\rho}(s)$ is a smooth curve on G, parameterized by a scalar s, with initial value $\boldsymbol{\rho}(0) = \mathbf{I}$ and initial speed $\mathbf{A} \in \mathfrak{g}$ ($\boldsymbol{\rho}'(0) = \mathbf{A}$). On matrix groups, $\boldsymbol{\rho}$ can be written as a Taylor expansion:

$$\boldsymbol{\rho}(s) = \mathbf{I} + s\mathbf{A} + o(s). \tag{6}$$

Intuitively, this curve represents a direction $\mathbf{A}$ on the Lie algebra towards which we can move infinitesimally from any point in G.

The first step is to write the differential equation in terms of the algebra action associated with an adequately chosen group action (Λ, with the associated λ_*):

$$\frac{dx}{dt} = F|_{x(t)}(t) = \lambda_*(\boldsymbol{\xi}(x(t), t))(x(t)), \tag{7}$$

with the initial condition $x(t_i) = x_i$. $\boldsymbol{\xi} : \mathcal{M} \times \mathbb{R}^+ \to \mathfrak{g}$ is a smooth function.

Thanks to [13] (Lemma 2.7), we know there exists $\mathbf{Y}(t) \in G$ following the differential equation (with $\mathbf{Y}(0) = \mathbf{I}$) :

$$\frac{d\mathbf{Y}}{dt} = \boldsymbol{\xi}(x, t)\mathbf{Y}(t). \tag{8}$$

such that $\Lambda(\mathbf{Y}(t), x_0)$ is the solution of Eq. (7). At this step, we can already design a simple Lie-Euler method by temporarily fixing $\boldsymbol{\xi}$ to its current value $\boldsymbol{\xi}(x_i, t_i)$. Then, the solution of the "frozen" ODE (8) on G is [13] (Theorem 2.8):

$$\tilde{\mathbf{Y}}(t) = \exp(t\boldsymbol{\xi}(x_i, t_i)), \tag{9}$$

with exp the matrix (Lie group) exponential. By setting $t = t_{i+1} = t_i + h$, we obtain the next iterate on G. We come back to $\mathcal{M}$ using the group action:

$$x_{i+1} = x(t_i + h) = \Lambda(\tilde{\mathbf{Y}}(t + h), x_i). \tag{10}$$

Algorithm 1. Lie-Euler algorithm

Require: Transitive group action Λ of G on $\mathcal{M}$, associated algebra action λ_*. Smooth function $\boldsymbol{\xi}$ from $\mathcal{M}$ to $\mathfrak{g}$ such that the ODE on $\mathcal{M}$ writes as in 7. Initial condition $x_0 \in \mathcal{M}$. Step size h. Number of time steps N.
$i = 0$
$x_i \leftarrow x_0$
while $i \le N$ **do**
    $\mathbf{A} \leftarrow h\boldsymbol{\xi}(x_i, t_i)$
    $x_{i+1} \leftarrow \Lambda(\exp(\mathbf{A}), x_i))$
    $i \leftarrow i + 1$
end while

The Lie-Euler method is summarized in Algorithm 1. For other schemes requiring to combine the current iterate on the manifold and intermediary values of the flow, e.g. RK4, we cannot form linear combinations of evaluations of the vector field for different times t and locations x, since those live on different tangent spaces, nor can we add them to points on $\mathcal{M}$. Thus, we need an ODE defined on a Euclidean space: the Lie algebra $\mathfrak{g}$. Fortunately, [13] (Lemma 3.1) tells us that the solution of Eq. (8) writes as:

$$\mathbf{Y}(t) = \exp(\boldsymbol{\theta}(t)) \tag{11}$$

where $\boldsymbol{\theta}(t) \in \mathfrak{g}$ solves an ODE in the Lie algebra:

$$\frac{d\boldsymbol{\theta}}{dt} = \operatorname{dexp}_{\boldsymbol{\theta}(t)}^{-1}(\boldsymbol{\xi}(x, t)) \tag{12}$$

with $\boldsymbol{\theta}(0) = \mathbf{0}$. $\operatorname{dexp}_{\boldsymbol{\theta}(t)}^{-1}$ is the inverse of the derivative of the matrix exponential at $\boldsymbol{\theta}(t)$. Its approximation at order 4 (see [13], (Def. 2.18)), for any $\mathbf{A}, \boldsymbol{\theta} \in \mathfrak{g}$ is

$$d\exp_{\boldsymbol{\theta}}^{-1}(\mathbf{A}) = \mathbf{A} + \frac{1}{2}[\boldsymbol{\theta}, \mathbf{A}] + \frac{1}{12}[\boldsymbol{\theta}, [\boldsymbol{\theta}, \mathbf{A}]] - \frac{1}{720}[\boldsymbol{\theta}, [\boldsymbol{\theta}, [\boldsymbol{\theta}, [\boldsymbol{\theta}, \mathbf{A}]]]] + o(\|\boldsymbol{\theta}\|^4) \tag{13}$$

using the matrix commutator $[\mathbf{A}, \mathbf{B}] = \mathbf{A}\mathbf{B} - \mathbf{B}\mathbf{A}$. On Eq. (12), we can apply a classical RK4 scheme. Each necessary evaluation $\mathbf{K}_i \in \mathfrak{g}$ of the vector field

---

Algorithm 2. Runge-Kutta-Munthe-Kaas 4 algorithm

Require: Transitive group action Λ of G on $\mathcal{M}$, associated algebra action λ_*. Smooth
function ξ from $\mathcal{M}$ to $\mathfrak{g}$ such that the ODE on $\mathcal{M}$ writes as in 5. Initial condition
$x_0 \in \mathcal{M}$. Step size h. Number of time steps N.
$i = 0$
$x_i \leftarrow x_0$
while $i \leq N$ **do**
    $\mathbf{A}_1 \leftarrow h\xi(x_i, t_i)$
    $\mathbf{K}_1 \leftarrow \mathbf{A}_1$
    $\mathbf{A}_2 \leftarrow h\xi(\Lambda(\exp(\mathbf{K}_1/2), x_i), t_i + h/2)$
    $\mathbf{K}_2 \leftarrow \mathrm{dexp}^{-1}_{\mathbf{K}_1/2}(\mathbf{A}_2)$
    $\mathbf{A}_3 \leftarrow h\xi(\Lambda(\exp(\mathbf{K}_2/2), x_i), t_i + h/2)$
    $\mathbf{K}_3 \leftarrow \mathrm{dexp}^{-1}_{\mathbf{K}_2/2}(\mathbf{A}_3)$
    $\mathbf{A}_4 \leftarrow h\xi(\Lambda(\exp(\mathbf{K}_3), x_i), t_i)$
    $\mathbf{K}_4 \leftarrow \mathrm{dexp}^{-1}_{\mathbf{K}_3}(\mathbf{A}_4)$
    $\boldsymbol{\Theta} = \mathbf{K}_1/6 + \mathbf{K}_2/3 + \mathbf{K}_3/3 + \mathbf{K}_4/6$ ▷ This operation is possible and stable
because all summands are in $\mathfrak{g}$
    $x_{i+1} \leftarrow \Lambda(\exp(\boldsymbol{\Theta}), x_i))$
    $i \leftarrow i + 1$
end while

---

can be done using the algebra action λ_* at the right location and time. In the
Lie algebra, computing weighted combinations of these evaluations is possible,
giving a final vector field $\boldsymbol{\Theta} \in \mathfrak{g}$. Finally, we obtain x_{i+1} on the manifold via:

$$x_{i+1} = x(t_i + h) = \Lambda(\exp(\boldsymbol{\Theta}), x_i). \tag{14}$$

Error order is guaranteed to be the same as the Euclidean scheme as long as there
are enough terms (we gave enough for RK4) in the approximation (13). The so-
called Runge-Kutta-Munthe-Kaas (RKMS) 4 method is an adaptation of the
general RKMS algorithm [13], and can be extended to any choice of coefficients
defining a classical RK method. This method is summarized in Algorithm 2.

4   Application to the SPD Manifold

Here, we examine a suitable Lie group action to build Lie group integration
schemes on Sym_n^+. First, Sym_n^+ is indeed a smooth manifold, whose tangent
space at each point can be identified with the set of symmetric matrices Sym_n.
Thus, any differential equation on Sym_n^+ has a symmetric matrix as a RHS.

We choose the Lie group to be the general linear group $GL_n(\mathbb{R})$. Its Lie
algebra is simply $\mathbb{R}^{n \times n}$. The group action we consider is:

$$\begin{aligned} \Lambda: \; GL_n(\mathbb{R}) \times \mathrm{Sym}_n^+ &\to \; \mathrm{Sym}_n^+ \\ (\mathbf{M}, \mathbf{P}) &\mapsto \mathbf{MPM}^T. \end{aligned} \tag{15}$$

It is a well known group action, and we can easily check it is indeed transitive [9]. For covariance matrices, it corresponds to the effect of an invertible linear transformation of a random vector on its covariance matrix.

Using (5), we can derive the algebra action $\boldsymbol{\lambda}_* : \mathbb{R}^{n \times n} \times \mathrm{Sym}_n^+ \to \mathfrak{X}(\mathrm{Sym}_n^+)$ [9]:

$$\boldsymbol{\lambda}_*(\mathbf{M})(\mathbf{P}) = \frac{d}{ds}\Lambda(\mathbf{Y}(s), \mathbf{P})|_{s=0} = \mathbf{M}\mathbf{P} + \mathbf{P}\mathbf{M}^T, \tag{16}$$

where $\mathbf{Y}(s) = \mathbf{I} + s\mathbf{M} + o(s)$ is a smooth curve on the Lie group with $\mathbf{Y}(0) = \mathbf{I}$ and initial speed $\mathbf{M} \in \mathbb{R}^{n \times n}$. Following Eq. (7), we can tackle equations of the form

$$\frac{d\mathbf{P}}{dt} = \boldsymbol{\xi}(\mathbf{P}, t)\mathbf{P} + \mathbf{P}\boldsymbol{\xi}(\mathbf{P}, t)^T, \tag{17}$$

with $\boldsymbol{\xi} : \mathrm{Sym}_n^+ \times \mathbb{R}^+ \to \mathbb{R}^{n \times n}$ *any* smooth function. Equation (17) is not very restrictive and many equations of interest can be written this way. For instance, with a constant $\boldsymbol{\xi}$, Eq. (17) governs the dynamics of the covariance of a random variable that propagates via a deterministic linear dynamical system [19]. More complex functions $\boldsymbol{\xi}$ can model more complex situations (see [9] for more examples).

5   Case Study

5.1   Multivariate Geometric Brownian Motion

We are interested here in a multivariate generalization of a Geometric Brownian Motion (GBM), given by the (Itô) SDE [2]:

$$d\boldsymbol{X} = \left(\mathbf{A} + \frac{1}{2}\mathbf{B}^2\right)\boldsymbol{X}dt + \mathbf{B}\boldsymbol{X}dW_t, \tag{18}$$

with $\boldsymbol{X}$ a random vector of size n, $\mathbf{A}, \mathbf{B} \in \mathbb{R}^{n \times n}$ two commuting matrices, such that the eigenvalues of $\mathbf{A} + \frac{1}{2}\mathbf{B}^2$ have a strictly negative real part. $W_t \in \mathbb{R}$ is a Brownian motion. A closed form expression of the trajectories for a deterministic initial condition $\mathbf{x}_0$ is given by $\boldsymbol{X}(t) = \exp(t\mathbf{A} + \mathbf{B}W_t)\mathbf{x}_0$. We can derive ODEs followed by the mean $\mathbf{m}$ (taking expectations in (18)) and covariance matrix $\mathbf{P}$ (using Itô's Lemma [17] on $\boldsymbol{X}\boldsymbol{X}^T$, taking expectations, and a few algebraic manipulations, see [9]). They provide a broad summary of the statistics of the process up to order two (much simpler than e.g. the Fokker-Planck equation):

$$\frac{d\mathbf{m}}{dt} = \left(\mathbf{A} + \frac{1}{2}\mathbf{B}^2\right)\mathbf{m} \tag{19}$$

$$\frac{d\mathbf{P}}{dt} = \left(\mathbf{A} + \frac{1}{2}\mathbf{B}^2\right)\mathbf{P} + \mathbf{P}\left(\mathbf{A} + \frac{1}{2}\mathbf{B}^2\right)^T + \mathbf{B}(\mathbf{P} + \mathbf{m}\mathbf{m}^T)\mathbf{B}^T. \tag{20}$$

Equation (20) can indeed be put in the form of Eq. (17), using

$$\boldsymbol{\xi}(\mathbf{P}) = \left(\mathbf{A} + \frac{1}{2}\mathbf{B}^2\right) + \frac{1}{2}\mathbf{B}(\mathbf{P} + \mathbf{m}\mathbf{m}^T)\mathbf{B}^T\mathbf{P}^{-1}. \tag{21}$$

We can only detail this example (with $n = 2$) here, but our method also applies e.g. to the covariance of a multivariate Ornstein-Uhlenbeck process [15], or to various Riccati equations encountered in control [3], with suitable choices of ξ, see [9]. Still, we will consider two cases, only differing in the way the matrix $\mathbf{A}$ is defined, and the chosen time step, all other parameters remaining the same.

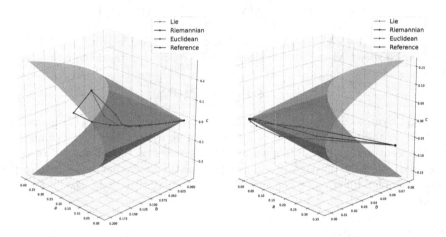

Fig. 1. Trajectories of the covariance matrices in Sym_2^+ obtained by the different RK4 methods for Eq. (20) using the embedding $\left(\begin{smallmatrix} a & c \\ c & b \end{smallmatrix} \right) \mapsto (a, b, c) \in \mathbb{R}^3$. The boundary of Sym_2^+ is given by the surface (in gray) defined by $ab - c^2 = 0$. Left: first case, where the iterates always stay in the manifold but only the Lie-RK4 method is accurate. Right: second case, where the Euclidean method produces iterates outside Sym_2^+.

5.2   Choice of Numerical Values for the SDE (18)

For the case study of Sect. 5, we chose $n = 2$ to be able to visualize the trajectories in the natural embedding of Sym_2^+ in $\mathbb{R}^3$, see Fig. 1. $\mathbf{P}_0$ was generated by using $\texttt{scikit - learn}$'s function $\texttt{make\_spd\_matrix}$ [20], with the same random state in both cases, and multiplying the result by a factor of 0.15. In practice, we have $\mathbf{P}_0 \approx \begin{pmatrix} 0.3383 & -0.0716 \\ -0.0716 & 0.0743 \end{pmatrix}$. We chose commuting $\mathbf{A}$ and $\mathbf{B}$ matrices so as to be able to use the closed form solution given in [2]. We further chose them such that $\mathbf{A} + \frac{1}{2}\mathbf{B}^2$ is negative definite. More in detail, we chose a symmetric $\mathbf{B} = \begin{pmatrix} -0.4 & 0.1 \\ 0.1 & -0.2 \end{pmatrix}$ To make sure $\mathbf{A}$ and $\mathbf{B}$ commute, we chose $\mathbf{A}$ by diagonalizing $\mathbf{B} = \mathbf{ODO}^T$, changing the eigenvalues in $\mathbf{D}$, and then computing $\mathbf{A} = \mathbf{OD'O}^T$. In the first case, we chose $\mathbf{D'} = \begin{pmatrix} -10 & 0 \\ 0 & -2 \end{pmatrix}$, and in the second case we chose $\mathbf{D'} = \begin{pmatrix} -4 & 0 \\ 0 & -8 \end{pmatrix}$.

We set $t \in [0, 2]$ (resp. $t \in [0, 1.5]$), and show results with 30 (resp. 11) evenly spaced time steps in the first (resp. second) case. In the second case, we note that this yields a time step $h = 0.15$. Using the eigenvalues of $\mathbf{P}_0$ and of the RHS of Eq. (20), we obtain $\rho_{max} = 0.02121$, and $\rho_{min} = 0.1345$. For the RK4 algorithm, similarly, we obtain $\rho_{max} = 0.0223$ and $\rho_{min} = 0.1414$. With our choice of time step, the classical Euler and RK4 methods will then fail in the second case. In practice, the upper bound ρ_{max} seems quite loose on that example and it seems experimentally that the iterate stays on the manifold as long as $\rho \leq \rho_{min}$.

5.3   Results

We compare three explicit RK4 schemes for Eq. (20): a Euclidean scheme, a variant where each step is brought back to the manifold using a Riemannian exponential map, with the affine invariant metric of [4] (Riemmanian RK4), and our method (Lie RK4). We compare them to a reference obtained from a classical RK4 method with a small enough time step so we know the trajectory remains on Sym_n^+ and the integration error is small. We start from a Gaussian initial condition $\mathbf{X}_0 \sim \mathcal{N}(\mathbf{m}_0, \mathbf{P}_0)$. Then, the process converges to a distribution given by a Dirac centered at $\mathbf{0}$. We show the trajectories obtained for the three competing methods embedded in $\mathbb{R}^3$ in Fig. 1 (left: first case, right: second case), and plot two distances between the trajectories and the reference in Fig. 2: the Euclidean distance in $\mathbb{R}^{n \times n}$, as well as the affine invariant distance [4] on Sym_2^+.

We see from both Figs. 1 and 2 that in the first case, only the Lie-RK4 method is able to accurately follow the reference dynamics for this value of the integration step. The Euclidean RK4 method is unable to produce a good solution while still producing iterates remaining on the manifold, which may be quite deceptive in applications. The Riemannian RK4 method here actually performs worse than its Euclidean counterpart. Indeed, thus method is not truly intrinsic to the manifold, so a bad step in the Euclidean domain often cannot be completely made up for. In the second case, the configuration is different and the integration step larger, so the result is that Euclidean RK4 produces iterates that immediately leave the manifold, producing a statistically meaningless trajectory. This is why the Riemannian distance cannot be computed for this method in the right panel of Fig. 2. In this case, Riemannian RK4 is able to guarantee that the iterates remain in the manifold, but the solution does not follow accurately the true dynamics. Finally, Lie-RK4 produces a very accurate solution guaranteed to stay in the manifold, even in such a challenging situation.

For smaller time steps, Lie RK4 remains the most precise up to a point, then classical RK4 may becomes marginally better than our approach in some cases, probably due to accumulating errors during the additional computations. However, in many applications, the step size is imposed by the problem, e.g. when observation data have a low sampling rate.

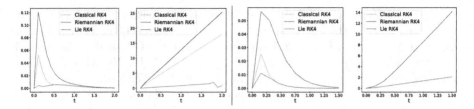

Fig. 2. Euclidean and Riemannian distances between each integrated trajectory and the reference, on the left and right of each panel, respectively. The case corresponding to the left of Fig. 1 is on the left panel, and the other case is in the right panel. In the latter, the Riemannian distance cannot be computed for the Euclidean method since the iterates are outside the manifold.

6 Conclusion

We have presented a Lie group framework to define structure-preserving integration schemes for flows of SPD matrices. Our fully intrinsic integrators keep iterates on the manifold, and provides smaller integration error than classical or naive methods, especially for large time steps. This will be useful in our future work to learn and represent uncertainty in data assimilation [8,16] or controls from observation data when governing equations are unknown (by learning a function ξ matching the data). In such cases, the time step is imposed by data and the training process may lead to ill-conditioned equations.

References

1. Absil, P.A., Mahony, R., Sepulchre, R.: Optimization algorithms on matrix manifolds. In: Optimization Algorithms on Matrix Manifolds. Princeton University Press, Princeton (2009)
2. Barrera, G., Högele, M., Pardo, J.: Cutoff stability of multivariate geometric Brownian motion. arXiv preprint arXiv:2207.01666 (2022)
3. Başar, T., Olsder, G.J.: Dynamic Noncooperative Game Theory. SIAM. Academic Press, New York (1998)
4. Bhatia, R.: Positive Definite Matrices. Princeton University Press, Princeton (2009)
5. Butcher, J.C.: Numerical Methods for ODEs. Wiley, Hoboken (2016)
6. Celledoni, E., Marthinsen, H., Owren, B.: An introduction to lie group integrators-basics, new developments and applications. J. Comput. Phys. **257**, 1040–1061 (2014)
7. Dellaportas, P., Pourahmadi, M.: Large time-varying covariance matrices with applications to finance. Technical report, Athens University of Economics, Department of Statistics (2004)
8. Dridi, N., Drumetz, L., Fablet, R.: Learning stochastic dynamical systems with neural networks mimicking the Euler-Maruyama scheme. In: 2021 29th European Signal Processing Conference (EUSIPCO), pp. 1990–1994. IEEE (2021)

9. Drumetz, L., Reiffers-Masson, A., Bekri, N.E., Vermet, F.: Geometry-preserving lie group integrators for differential equations on the manifold of symmetric positive definite matrices (2022). https://arxiv.org/abs/2210.08842

10. Dryden, I.L., Koloydenko, A., Zhou, D.: Non-euclidean statistics for covariance matrices, with applications to diffusion tensor imaging. Ann. Appl. Stat. **3**(3), 1102–1123 (2009)

11. Evensen, G., Vossepoel, F.C., van Leeuwen, P.J.: Data Assimilation Fundamentals: A Unified Formulation of the State and Parameter Estimation Problem. Springer, Cham (2022). https://doi.org/10.1007/978-3-030-96709-3

12. Hairer, E.: Solving differential equations on manifolds. Université de Geneve, Lecture Notes (2011)

13. Iserles, A., Munthe-Kaas, H.Z., Nørsett, S.P., Zanna, A.: Lie-group methods. Acta numerica **9**, 215–365 (2000)

14. Jordan, D., Smith, P.: Nonlinear Ordinary Differential Equations: An Introduction for Scientists and Engineers. OUP, Oxford (2007)

15. Meucci, A.: Review of statistical arbitrage, cointegration, and multivariate Ornstein-Uhlenbeck. Cointegration, and Multivariate Ornstein-Uhlenbeck (2009)

16. Nguyen, D., Ouala, S., Drumetz, L., Fablet, R.: Variational deep learning for the identification and reconstruction of chaotic and stochastic dynamical systems from noisy and partial observations. arXiv preprint arXiv:2009.02296 (2020)

17. Oksendal, B.: Stochastic Differential Equations: An Introduction with Applications. Springer, Heidelberg (2013)

18. Owren, B.: Lie group integrators. In: Ebrahimi-Fard, K., Liñàn M.B. (eds.) Discrete Mechanics, Geometric Integration and Lie-Butcher Series, pp. 29–69. Springer, Cham (2018). https://doi.org/10.1007/978-3-030-01397-4

19. Pannekoucke, O., Ricci, S., Barthelemy, S., Ménard, R., Thual, O.: Parametric kalman filter for chemical transport models. Tellus A Dyn. Meteorol. Oceanogr. **68**(1), 31547 (2016)

20. Pedregosa, F., et al.: Scikit-learn: machine learning in python. J. Mach. Learn Res. **12**, 2825–2830 (2011)

21. Zanini, P., Congedo, M., Jutten, C., Said, S., Berthoumieu, Y.: Transfer learning: a Riemannian geometry framework with applications to brain-computer interfaces. IEEE Trans. Biomed. Eng. **65**(5), 1107–1116 (2017)

The Phase Space Description
of the Geodesics on the Statistical
Model on a Finite Set
- Trajectory-Confinement and Integrability -

Yoshio Uwano[✉]

Kyoto Pharmaceutical University, Kyoto 607-8414, Japan
uwano@mb.kyoto-phu.ac.jp

Abstract. The geodesics on the statistical model, $\mathcal{S}_{n-1}$, on a finite set with n elements are studied as dynamical systems on the phase space, $T^*\mathcal{S}_{n-1}$, namely the cotangent bundle of $\mathcal{S}_{n-1}$. The Riemannian geodesics are described by a Hamiltonian equation on $T^*\mathcal{S}_{n-1}$ like in the case [10] of the exponential geodesics. In contrast, the mixture geodesics are described by a non-Hamiltonian first-order differential equation on $T^*\mathcal{S}_{n-1}$. Both of the equations are integrable in the sense that they are solvable by quadrature. Through this study, the symplectic reduction to have the phase space, $T^*\mathcal{S}_{n-1}$, works effectively. In particular, a novel clear account is given for confining the mixture geodesics in $\mathcal{S}_{n-1}$ by the use of the reduction.

Keywords: integrability · information geometry · geodesics · phase space

1 Introduction

It is well known that a family of parametric probability distributions is often accompanied with a smooth manifold each of whose point specifies a distribution in the family. Such a manifold is called a statistical model labeled with the name of the distribution family or the range of random variables of the distributions. The statistical models admit a pair of significant geometric objects: One is the Fisher metric and another is the affine connections named the α-connections which are known to allow the duality [1]. Among the α-connections, the exponential, the Riemannian and the mixture ones have taken much interest together with the geodesics associated with them. The quantum statistical models accompanied with the quantum density matrices have been studied also in information geometry [6].

In the paper [9] by the author, an averaged Hebbian learning equation (abbr. AHLE) extended on a quantum statistical model is studied to show that all of whose trajectories are the exponential (e-) geodesics on that model. Further, in a discussion paper [10] after [9], the e-geodesics on the statistical model $\mathcal{S}_{n-1}$ on a finite set with n elements are studied both from a dynamical system viewpoint

F. Nielsen and F. Barbaresco (Eds.): GSI 2023, LNCS 14072, pp. 444–453, 2023.
https://doi.org/10.1007/978-3-031-38299-4_46

and a systems science viewpoint: The Hamiltonian system is found successfully on the phase space, $T^*\mathcal{S}_{n-1}$, namely the cotangent bundle of $\mathcal{S}_{n-1}$, whose flow realizes all the e-geodesics. Furthermore, every e-geodesic is shown to be a trajectory of a parametric AHLE, and vice versa. The Hamiltonian system is shown to be integrable in the sense that it is solvable by quadrature. The commutative conserved quantities giving rise to the integrability are understood to be the learning parameters of the AHLE's. Throughout this paper, the term 'phase space' means the cotangent bundle of a manifold under consideration.

As a continuation of the papers [9,10], this paper aims to study the Riemannian (R-) geodesics and the mixture (m-) geodesics on $\mathcal{S}_{n-1}$ from a dynamical system viewpoint on the phase space, $T^*\mathcal{S}_{n-1}$, namely the cotangent bundle of $\mathcal{S}_{n-1}$. The R-geodesics are described by a Hamiltonian equation on $T^*\mathcal{S}_{n-1}$. In contrast, a non-Hamiltonian first-order differential equation on $T^*\mathcal{S}_{n-1}$ is found, which describes the m-geodesics. Like in the case of the e-geodesics [10], both of the equations on $T^*\mathcal{S}_{n-1}$ for the R-geodesics and for the m-geodesics are solved by quadrature, so that those equations are integrable. To obtain the solutions explicitly by quadrature, the symplectic reduction [7] of the higher dimensional phase space, $T^*\mathbb{R}^n_+$, to $T^*\mathcal{S}_{n-1}$ works very effectively, where $T^*\mathbb{R}^n_+$ denotes the cotangent bundle of the open positive orthant, $\mathbb{R}^n_+$, of $\mathbb{R}^n$. The solution curves describing the geodesics on $\mathcal{S}_{n-1}$ are drawn through the reduction from those of certain first-order differential equations on $T^*\mathbb{R}^n_+$. In particular, a novel clear account for confining the m-geodesics in $\mathcal{S}_{n-1}$ is given through the reduction.

The organization of this paper is outlined as follows. In Sect. 2, geometric setting-up is made for the phase space, $T^*\mathcal{S}_{n-1}$, together with the symplectic reduction of $T^*\mathbb{R}^n_+$ to $T^*\mathcal{S}_{n-1}$ which is used effectively in Sects. 3 and 4. In Sect. 3, the Hamiltonian form for the R-geodesics is obtained on $T^*\mathcal{S}_{n-1}$, which is shown to be integrable by quadrature. In Sect. 4, a non-Hamiltonian first order differential equation on $T^*\mathcal{S}_{n-1}$ is drawn for the m-geodesics from a kinematical understanding of them. The solution to the non-Hamiltonian equation is obtained explicitly, which is naturally equipped with the novel account for the m-geodesics confinement. Section 5 is made for concluding remarks.

2   Geometric Setting-Up

It is widely known that, in the language of differential geometry, the Hamiltonian mechanics is developed on symplectic manifolds [2]. It is also known that there are two ways of describing the Hamiltonian mechanics: One is the coordinate-free description and the other is the local-coordinates one. As written below, the description applied in this paper is *intermediate* between those, which works well for the following analysis, e.g. explict solution, confinement and integrability.

2.1   The Statistical Model $\mathcal{S}_{n-1}$ and Its Cotangent Bundle $T^*\mathcal{S}_{n-1}$

Let us consider the finite-set distributions associated with of n fundamental events ($n \geq 2$, $n \in \mathbb{N}$) with the probability $p(j) = \xi_j$ ($j = 1, 2, \cdots, n$). The family of them is identified with the $n-1$ dimensional standard simplex

$$\mathcal{S}_{n-1} = \left\{ \xi = (\xi_j) \in \mathbb{R}^n \mid \sum_{j=1}^{n} \xi_j = 1, \ \xi_j > 0 \ (j = 1, 2, \cdots, n) \right\}. \tag{1}$$

The coordinate system $\xi = (\xi_j)$ is well-known to be the barycentric coordinate system of the simplex, while the other coordinate system $\tilde{\xi} = (\tilde{\xi}_k)_{k=1,\cdots n-1}$ subject to $\xi_j = \tilde{\xi}_j$ $(j = 1, 2, \cdots, n-1)$ is preferred to be applied to $\mathcal{S}_{n-1}$ in much literature on information geometry, e.g. [1]. Throughout this paper, the barycentric one, $\xi = (\xi_j)$, is chosen for convenience of global analysis on both $\mathcal{S}_{n-1}$ and its cotangent bundle $T^*\mathcal{S}_{n-1}$. Note also that $\xi = (\xi_j)$ works as the mixture coordinate system of $\mathcal{S}_{n-1}$.

The Fisher metric is endowed with $\mathcal{S}_{n-1}$ as a significant information geometric structure, which is expressed to be

$$ds^2 = \sum_{j=1}^{n} (1/\xi_j) d\xi_j \otimes d\xi_j \tag{2}$$

in terms of ξ (cf. [1]). The Riemannian manifold $\mathcal{S}_{n-1}$ endowed with ds^2 is then referred to as the statistical model on the finite set with n elements.

We move to the cotangent bundle, $T^*\mathcal{S}_{n-1}$, of $\mathcal{S}_{n-1}$ endowed with the standard symplectic form. In view of (1), the tangent bundle of $\mathcal{S}_{n-1}$ takes the form

$$T\mathcal{S}_{n-1} = \{(\xi, v) \in \mathcal{S}_{n-1} \times \mathbb{R}^n \mid \sum_{j=1}^{n} v_j = 0\}. \tag{3}$$

Then, using the Fisher metric ds^2, we define the diffeomorphism $\alpha : T\mathcal{S}_{n-1} \to T^*\mathcal{S}_{n-1}$ subject to the relation

$$ds^2((\xi, v), (\xi, v')) = \alpha((\xi, v)) \cdot (\xi, v') \quad (^\forall (\xi, v') \in T\mathcal{S}_{n-1}), \tag{4}$$

where $\cdot$ stands for the natural pairing. By calculation, we have

$$\alpha(\xi, v) = (\xi, \eta) \quad ((\xi, v) \in T\mathcal{S}_{n-1}) \quad \text{with} \quad \eta_j = v_j/\xi_j \quad (j = 1, 2, \cdots n), \tag{5}$$

so that $T^*\mathcal{S}_{n-1}$ takes the form

$$T^*\mathcal{S}_{n-1} = \left\{ (\xi, \eta) \in \mathcal{S}_{n-1} \times \mathbb{R}^n \mid \xi^T \eta = 0 \right\}, \tag{6}$$

where ^{T} denotes the transpose. Note that $T^*\mathcal{S}_{n-1}$ can be identified, as a manifold, with the statistical bundle over the densities of $\mathcal{S}_{n-1}$ [4]: The inverse, α^{-1}, of α has the same role as the trivialization mapping of the statistical bundle to the tangent bundle [4]. By endowing the symplectic form [2]

$$\omega = \sum_{j=1}^{n} d\eta_j \wedge d\xi_j \tag{7}$$

with the phase space $T^*\mathcal{S}_{n-1}$, we make $T^*\mathcal{S}_{n-1}$ into the symplectic manifold $(T^*\mathcal{S}_{n-1}, \omega)$. We will seek for the first order differential equations for the Riemannian and the mixture geodesics on $\mathcal{S}_{n-1}$ in the succeeding pair of sections.

2.2 Symplectic Reduction of $(T^*\mathbb{R}^n_+, \Omega)$ to $(T^*\mathcal{S}_{n-1}, \omega)$

As seen in the previous subsection, we have to handle the constraints

$$\sum_{j=1}^{n} \xi_j = 1, \quad \xi^T \eta = 0, \quad \sum_{j=1}^{n} \dot{\xi}_j = 0, \quad \dot{\xi}^T \eta + \xi^T \dot{\eta} = 0 \tag{8}$$

arising from Eqs. (1) and (6) in calculations on $T^*\mathcal{S}_{n-1}$. To avoid handling those constraints in calculations, the symplectic reduction works effectively, which is outlined below.

Let $\mathbb{R}^n_+$ be the open positive orthant in $\mathbb{R}^n$ and let the Cartesian coordinates $x = (x_j)$ of $\mathbb{R}^n$ be applied to $\mathbb{R}^n_+$ as an open subset of $\mathbb{R}^n$. Both tangent and the cotangent bundles of $\mathbb{R}^n_+$ are diffeomorphic to $\mathbb{R}^n_+ \times \mathbb{R}^n$. Indeed, using the Riemannian metric

$$d\sigma^2 = \frac{1}{\|x\|^2} \sum_{j=1}^{n} dx_j \otimes dx_j \tag{9}$$

of $\mathbb{R}^n_+$ with $\|x\|^2 = x^T x$, we have the diffeomorphism $\beta : (x, u) \in T\mathbb{R}^n_+ \mapsto (x, y) \in T^*\mathbb{R}^n_+$ subject to

$$d\sigma^2((x, u), (x, u')) = \beta(x, u) \cdot (x, u') \quad ({}^{\forall}(x, u') \in T\mathbb{R}^n_+), \tag{10}$$

where $\cdot$ denotes the natural pairing (cf. (4)). A calculation shows

$$\beta(x, u) = (x, y) \quad ((x, u) \in T\mathbb{R}^n_+) \quad \text{with} \quad y_j = \frac{u_j}{\|x\|^2} \quad (j = 1, 2, \cdots, n). \tag{11}$$

The introduction of $d\sigma^2$ helps the phase space description of the R-geodesics in the next section. The phase space $T^*\mathbb{R}^n_+$ is made into the symplectic manifold, $(T^*\mathbb{R}^n_+, \Omega)$, with the symplectic form [2]

$$\Omega = \sum_{j=1}^{n} dy_j \wedge dx_j. \tag{12}$$

The symplectic reduction of $(T^*\mathbb{R}^n_+, \Omega)$ starts with the R-action ϕ_s on $\mathbb{R}^n_+$ and its symplectic lift $\phi_s^{\#}$ on $T^*\mathbb{R}^n_+$ which are defined to be

$$\phi_s(x) = e^s x, \quad \phi_s^{\#}(x, y) = (e^s x, e^{-s} y) \quad (x \in \mathbb{R}^n_+, (x, y) \in T^*\mathbb{R}^n_+, s \in \mathbb{R}). \tag{13}$$

Note that the symplecticity of $\phi_s^{\sharp}$ stands for $\phi_s^{\#*} \Omega = \Omega$, where $*$ denotes the pull-back. The function $J(x, y) = x^T y$ on $T^*\mathbb{R}^n_+$ is taken as the moment map associated with $\phi_s^{\#}$. Since the R-action $\phi_s^{\#}$ leaves $J^{-1}(0)$ invariant, the quotient space $J^{-1}(0)/\mathbb{R}$ is made into the tangent bundle $T^*\mathcal{S}_{n-1}$ along with the diagram

$$T^*\mathbb{R}^n_+ \xleftarrow{\iota} J^{-1}(0) \xrightarrow{\nu} T^*\mathcal{S}_{n-1} \cong J^{-1}(0)/\mathbb{R}, \tag{14}$$

according to the reduction theorem by Marsden and Weinstein [7]. In Eq. (14), $\nu : (x,y) \in J^{-1}(0) \mapsto (\xi,\eta) \in T^*\mathcal{S}_{n-1}$ is the projection in the form

$$\xi_j = \frac{x_j^2}{\|x\|^2}, \quad \eta_j = \frac{\|x\|^2 y_j}{2x_j} \quad (j = 1, 2, \cdots, n), \tag{15}$$

and ι is the inclusion. Along with (14), the reduced symplectic form of $T^*\mathcal{S}_{n-1}$ is determined to satisfy $\iota^*\Omega = \nu^*\omega$, which turns out to be identical with ω defined already by (7). In a summary, we have the following theorem.

Theorem 1. *The symplectic manifold $(T^*\mathbb{R}_+^n, \Omega)$ is reduced by the symplectic $\mathbb{R}$-action $\phi_s^\#$ of (13) to the symplectic manifold $(T^*\mathcal{S}_{n-1}, \omega)$.*

3　The Riemannian Geodesics on $\mathcal{S}_{n-1}$

In much literature on mechanics, we have learned a recipe to have the Hamilton form for the Riemannian (R-) geodesics: In our case, on taking the kinetic energy,

$$H_R(\xi,\eta) = \frac{1}{2}\sum_{j=1}^n \xi_j \eta_j^2 \quad ((\xi,\eta) \in T^*\mathcal{S}_{n-1}), \tag{16}$$

associated with the Fisher metric ds^2 as the Hamiltonian, the Hamiltonian equation on $T^*\mathcal{S}_{n-1}$ is equivalent to the R-geodesic equation on $\mathcal{S}_{n-1}$. However, the constraints (8) arising from Eqs. (1) and (6) may cause a difficulty in solving the Hamiltonian equation on $(T^*\mathcal{S}_{n-1}, \omega, H_R)$. To avoid such a difficulty, we consider the Hamiltonian system $(T^*\mathbb{R}_+^n, \Omega, H_R^\#)$ associated with the Hamiltonian

$$H_R^\#(x,y) = \frac{1}{8}\|x\|^2\|y\|^2, \tag{17}$$

which is reduced by the $\mathbb{R}$-action $\phi_s^\#$ to the Hamiltonian system $(T^*\mathcal{S}_{n-1}, \omega, H_R)$ for the R-geodesics. Note that $H_R^\#(x,y)$ is the kinetic energy associated with the metric $d\sigma^2$. This is the account for endowing $d\sigma^2$ with $\mathbb{R}_+^n$ in Sect. 2. The Hamiltonian equation of $(T^*\mathbb{R}_+^n, \Omega, H_R^\#)$ takes the form

$$\dot{x}_j = \frac{\partial H_R^\#}{\partial y_j} = \frac{\|x\|^2 y_j}{4}, \quad \dot{y}_j = -\frac{\partial H_R^\#}{\partial x_j} = -\frac{\|y\|^2 x_j}{4} \quad (j = 1, 2, \cdots, n). \tag{18}$$

By applying the tangent map $\nu_* : T(J^{-1}(0)) \to T(T^*\mathcal{S}_{n-1})$, of ν to the restriction of (18) to $J^{-1}(0)$, we obtain the Hamiltonian equation of $(T^*\mathcal{S}_{n-1}, \omega, H_R)$ below. In fact, since the tangent map ν_* takes the form

$$\nu_* : ((x,y),(\dot{x},\dot{y})) \in T(J^{-1}(0)) \mapsto (\nu(x,y),(\dot{\xi},\dot{\eta})) \in T\mathcal{S}_{n-1},$$

$$\dot{\xi}_j = \frac{2x_j}{\|x\|^2}\dot{x}_j - \frac{2}{\|x\|^4}(x^T\dot{x}),$$

$$\tag{19}$$

$$\dot{\eta}_j = \frac{\|x\|^2}{2x_j^2}(x_j\dot{y}_j - y_j\dot{x}_j) + \frac{y_j}{x_j}(x^T\dot{x}) \quad (j = 1, 2, \cdots, n)$$

we have

$$\dot{\xi}_j = \xi_j \eta_j, \quad \dot{\eta}_j = -\frac{1}{2}\sum_{k=1}^{n}\xi_k\eta_k^2 - \frac{1}{2}\eta_j^2 \quad (j = 1, 2, \cdots, n) \tag{20}$$

as the Hamiltonian equation of $(T^*\mathcal{S}_{n-1}, \omega, H_R)$ for the R-geodesics on $\mathcal{S}_{n-1}$.

The solution of Eq. (20) is drawn from that of Eq. (18) with $J(x(0), y(0)) = 0$ in the following way. Since the values of $J(x, y) = x^T y$, $\|x\|^2$ and $\|y\|^2$ are kept invariant along every solution curve on $J^{-1}(0)$, the solution curves on $J^{-1}(0)$ are in one-to-one correspondence with the solution curves of the harmonic oscillator whose Hamiltonian is

$$K(x, y) = \frac{a^2}{2}\|y\|^2 + \frac{b^2}{2}\|x\|^2 \quad \left(a = \frac{\|x(0)\|}{2}, b = \frac{\|y(0)\|}{2}\right). \tag{21}$$

Hence, under $J(x(0), y(0)) = x(0)^T y(0) = 0$, we have

$$x_j(t) = x_j(0)\cos(abt) + \frac{a}{b}y_j(0)\sin(abt),$$
$$y_j(t) = -\frac{b}{a}x_j(0)\sin(abt) + y_j(0)\cos(abt) \quad (j = 1, 2, \cdots, n) \tag{22}$$

for $b \neq 0$ and the fixed-point solution for $b = 0$. The solution $(\xi(t), \eta(t))$ to the Hamiltonian Equation (20) is given through $(\xi(t), \eta(t)) = \nu(x(t), y(t))$ with (22), which is explicitly written in the form

$$\xi_j(t) = \xi_j(0)\left(\cos(abt) + \frac{\eta_j(0)}{2ab}\sin(abt)\right)^2,$$
$$\eta_j(t) = \frac{\xi_j(0)}{\xi_j(t)}\left\{\eta_j(0)\cos(2abt) + \left(\frac{\eta_j(0)^2}{ab} - ab\right)\sin(2abt)\right\} \tag{23}$$

$$(0 \leq t < t_R, j = 1, 2, \cdots, n).$$

We note that $H_R(\xi(0), \eta(0)) = H_R^\#(x(0), y(0)) = 2(ab)^2$. The t_R is defined by

$$t_R = \frac{1}{ab}\tan^{-1}\left(\min_{l \in \mathcal{L}}\left|\frac{2ab}{\eta_l(0)}\right|\right), \tag{24}$$

where $\mathcal{L}$ denotes the non-empty set of indices subject to $\eta_l(0) < 0$.

As indicated in (23), $(\xi(t), \eta(t))$ is understood to be valid only in the finite time-interval $0 \leq t < t_R$, since $\xi(t)$ reaches to the boundary $\partial \mathcal{S}_{n-1}$ at $t = t_R$ with divergent momenta. As a similar phenomenon to this, the collison orbits of the two-body problem is worth thought of. In a summary, we have the following theorem.

Theorem 2. *The Riemannian geodesics on $\mathcal{S}_{n-1}$ are described by the solution (23) of the Hamiltonian system $(T^*\mathcal{S}_{n-1}, \omega, H_R)$, which is integrable in the sence that Eq. (20) is solvable by quadrature.*

In view of Eq. (22), $x_j y_k - y_j x_k$ $(j < k)$ are shown to be invariant along any solution curves, (22), on $J^{-1}(0)$, which are in involution. Their $\phi_s^{\#}$-invariance and the reduction procedure are put together to show the following theorem.

Theorem 3. *The Hamiltonian system* $(T^* \mathcal{S}_{n-1}, \omega, H_R)$ *admits the involutive conserved quantities,* $2\sqrt{\xi_j \xi_k}(\eta_k - \eta_j)$ $(j < k)$.

4   The Mixture Geodesics on $\mathcal{S}_{n-1}$

We have learned that the mixture (m-) geodesics are the 'straight lines' in the mixture coordinate system of the statistical model under consideration, e.g. [1]. In the present case, since ξ are the mixture coordinate system, any m-geodesic is naively expected to be written in the form,

$$\xi_j^M(t) = \dot{\xi}_j^M(0)t + \xi_j^M(0) \quad (j = 1, 2, \cdots, n), \tag{25}$$

which might allow any m-geodesic to extend out of $\mathcal{S}_{n-1}$. As far as the author is concerned, it seems that no clear account has been given yet for confining the m-geodesics in $\mathcal{S}_{n-1}$. In what follows, by building up a first order differential equation on the phase space, $T^* \mathcal{S}_{n-1}$, describing the m-geodesics, we succeed to give a novel clear account for the confinement of the m-geodesics in $\mathcal{S}_{n-1}$. To support our account, the reduction of $T^* \mathbb{R}_+^n$ to $T^* \mathcal{S}_{n-1}$ works effectively.

We start with a naive expression (25) of the straight line segments on $\mathcal{S}_{n-1}$. The derivative $\dot{\xi}^M(\tau)$ at $t = \tau$ is of course understood to be a tangent vector at $\xi^M(\tau)$, which is mapped to the cotangent vector $\eta^M(\tau)$ at $\xi^M(\tau)$ by applying the map α given by (5) to $(\xi^M(\tau), \dot{\xi}^M(\tau))$. Hence, we have

$$\eta_j^M(\tau) = \dot{\xi}_j^M(\tau)/\xi_j^M(\tau) \quad (j = 1, 2, \cdots, n) \tag{26}$$

at any $t = \tau$, which yields the equation

$$\dot{\xi}_j = \xi_j \eta_j \quad (j = 1, 2, \cdots, n) \tag{27}$$

on $\dot{\xi}$. Note that if Eq. (27) holds true then so do the latter half of the constraints (8) and vice versa. We move to seek for the equation on $\dot{\eta}$, which has to be compatible with the equations (27) and

$$\frac{d}{dt}(\xi_j \eta_j) = 0 \quad (j = 1, 2, \cdots, n) \tag{28}$$

claiming that $\dot{\xi}$ is kept constant. A calculation shows that only the equation

$$\dot{\eta}_j = -\eta_j^2 \quad (j = 1, 2, \cdots n) \tag{29}$$

on $\dot{\eta}$ is compatible with (27) and (28). Since Eqs. (27) and (29) are put together to draw

$$\ddot{\xi}_j = 0 \quad (j = 1, 2, \cdots, n), \tag{30}$$

on $\mathcal{S}_{n-1}$ for the m-geodesics, we conclude that the first-order differential equation

$$\dot{\xi}_j = \xi_j \eta_j, \quad \dot{\eta}_j = -\eta_j^2 \quad (j = 1, 2, \cdots, n) \tag{31}$$

on $T^*\mathcal{S}_{n-1}$ is what we have sought for as the first order differential equation on the phase space, $T^*\mathcal{S}_{n-1}$, for the m-geodesics. Since all the $\xi_j \eta_j$'s are kept invariant along the solution, Eq. (31) is solved to be

$$\xi_j(t) = \xi_j(0)\eta_j(0)t + \xi_j(0), \quad \eta_j(t) = \frac{\eta_j(0)}{\eta_j(0)t + 1} \tag{32}$$

$$(0 \leq t < t_M, \ j = 1, 2, \cdots, n)$$

with

$$t_M = \min_{l \in \mathcal{L}} \left| \frac{1}{\eta_l(0)} \right|, \tag{33}$$

where $\mathcal{L}$ stands for the non-empty set of indices subject to $\eta_j(0) < 0$ again (cf. Eq. (24)). As seen from (32), $\xi(t)$ reaches to the boundary $\partial\mathcal{S}_{n-1}$ at $t = t_M$ with divergent momenta. Thus, we succeed to give a novel account for confining the m-geodesics in $\mathcal{S}_{n-1}$ by presenting their time evolution in the explicit form (32).

We provide the other strong support for our account for the confinement of the m-geodesics by using the reduction of $T^*\mathbb{R}_+^n$ to $T^*\mathcal{S}_{n-1}$: Let us consider the $\phi_s^\sharp$-invariant equation

$$\dot{x}_j = \frac{\|x\|^2 y_j}{4}, \quad \dot{y}_j = -\frac{\|x\|^2 y_j^2}{4x_j} \quad (j = 1, 2, \cdots, n) \tag{34}$$

on $T^*\mathbb{R}_+^n$, which is reduced to Eq. (31). On $J^{-1}(0)$, Eq. (34) is solved to be

$$x_j(t) = \sqrt{\frac{\|x(0)\|^2 x_j(0)y_j(0)}{2}t + x_j(0)^2}, \quad y_j(t) = \frac{x_j(0)y_j(0)}{x_j(t)} \tag{35}$$

$(j = 1, 2, \cdots, n)$, where $(x(0), y(0)) \in J^{-1}(0)$ with $\nu((x(0), y(0)) = (\xi(0), \eta(0))$. Due to a nature of the square roots in (35), we can say that any solution curve by (35) *terminates* at $(x(t_M), y(t_M))$ (see (33) for t_M). Since any of the m-geodesics is given as the reduced solution curve, $(\xi(t), \eta(t)) = \nu(x(t), y(t))$, of a solution curve $(x(t), y(t))$ given by (35), any m-geodesic *terminates* at $(\xi(t_M), \eta(t_M))$, too. Those terminations support definitely our account for the confinement of the m-geodesics in $\mathcal{S}_{n-1}$.

We show that Eq. (31) is not in Hamiltonian form by partly using coordinate-free description: Let X (resp. $X^\sharp$) denote the vector fields generated by (31) (resp. (34)). Then, as a sufficient condition for (31) to be non-Hamiltonian, we show that X is not a symplectic flow. Namely, the Lie derivative, $\mathcal{L}_X \omega$, to ω along X does not vanish. Due to Cartan's homotopy formula [2], the closedness of ω and Ω, and a reduction formula $\iota^* \Omega = \nu^* \omega$ (see Subsect. 2.2), we have

$$\iota^*(\mathcal{L}_{X^\sharp}\Omega) = \nu^*(\mathcal{L}_X \omega), \tag{36}$$

$$\mathcal{L}_{X^\sharp}\Omega = d(i_{X^\sharp}\Omega) + i_{X^\sharp}(d\Omega) = d(i_{X^\sharp}\Omega) = d\left(\sum_{j=1}^n (\dot{y}_j dx_j - \dot{x}_j dy_j) \right), \tag{37}$$

where $i_{X^\sharp}$ denotes the interior product by $X^\sharp$. By calculation with (34), we see $\iota^*(\mathcal{L}_{X^\sharp}\Omega)$ does not vanish and neither does $\nu^*(\mathcal{L}_X\omega)$. Therefore, Eq. (31) is non-Hamiltonian. In a summary, we have the following theorem.

Theorem 4. *The non-Hamiltonian equation (31) on $T^*\mathcal{S}_{n-1}$ describes the mixture geodesics, which ensures the confinement of the m-geodesics in $\mathcal{S}_{n-1}$. Equation (31) is integrable in the sense that it is solvable by quadrature, which admits $\xi_j\eta_j$ $(j = 1, 2, \cdots, n)$ as commutative conserved quantities.*

5   Concluding Remarks

We have found the first order differential equations on the phase space $T^*\mathcal{S}_{n-1}$ describing the Riemannian (R-) geodesics and the mixture (m-) geodesics on $\mathcal{S}_{n-1}$. Equation (20) for the R-geodesics is in Hamiltonian form, while Eq. (31) for the m-geodesics is not. Both equations are integrable in the sense of their solvability by quadrature. Owing the explicit expression (32) of the m-geodesics obtained as the reduction of the curves by (35), we succeed to give a novel clear account for confining the m-geodesics in $\mathcal{S}_{n-1}$: As far as the author is concerned, no such a clear account for the confinement seems to have been given before.

In reference to those result, we review the Hamiltonian form for the exponential (e-) geodesics on $\mathcal{S}_{n-1}$ which is also given in a condensed form in [10]. The Hamiltonian equation on $T^*\mathcal{S}_{n-1}$ for the e-geodesics takes the form

$$
\begin{aligned}
\dot{\xi}_j &= 2\xi_j \left(\xi_j\eta_j - \sum_{k=1}^{n} \xi_k^2\eta_k \right), \\
\dot{\eta}_j &= -2\eta_j \left(\xi_j\eta_j - \sum_{k=1}^{n} \xi_k^2\eta_k \right) \quad (j = 1, 2, \cdots, n)
\end{aligned}
\tag{38}
$$

associated with the Hamiltonian

$$
H_E(\xi, \eta) = \sum_{j=1}^{n} (\xi_j\eta_j)^2.
\tag{39}
$$

The Hamiltonian equation (38) is solved to be

$$
\begin{aligned}
\xi_j(t) &= \left(\sum_{k=1}^{n} e^{2t\xi_k(0)\eta_k(0)} \xi_k(0) \right)^{-1} e^{2t\xi_j(0)\eta_j(0)} \xi_j(0) \\
\eta_j(t) &= \left(\sum_{k=1}^{n} e^{2t\xi_k(0)\eta_k(0)} \xi_k(0) \right) e^{-2t\xi_j(0)\eta_j(0)} \eta_j(0) \quad (j = 1, 2, \cdots, n).
\end{aligned}
\tag{40}
$$

In contrast with the cases of the R-geodesics and the m-geodesics, the solution by (40) can be prolonged to $t \to \infty$. The functions, $\xi_j\eta_j$'s, are commutative conserved quantities.

Another remark is on an application of Sect. 3 to a new geometric understanding of the Grover-type sequence for an ordered tuple of l states of N-qubit organized by the author [8]: Since the search sequence is understood to be on an R-geodesic on $J^{-1}(0)$ of $(T^*\mathbb{R}_+^{2^{N+1}l}, \Omega, H^\#)$, the reduced search sequence obtained through ν is on an R-geodesic on $\mathcal{S}_{2^{N+1}l-1}$, while the search sequence is shown to be projected on an m-geodesic on the statistical model on the quantum density matrices discussed in [8].

It is known that there are two kinds of phase spaces different from our phase spaces in cotangent bundle form in information geometry: One is any even-dimensional statistical models [5] and another is the square of any statistical models [3]. Either kind of phase spaces, however, does not seem to be applicable to achieve the objective of the present paper.

What will a quantum analogue of this paper be? In fact, this paper can be placed as an etude for composing a quantum statistical version, since $\mathcal{S}_{n-1}$ appears as a submanifold of the quantum statistical model of degree n.

Acknowledgement. The author would like to thank Dr. Daisuke Tarama for supporting this work partly by JSPS KAKENHI Grant number 19K14540.

References

1. Amari, S., Nagaoka, H.: Methods of Information Geometry. AMS, Providence (2000). https://doi.org/10.1090/mmono/191
2. Arnold, V.I.: Mathematical Method of Classical Mechanics, 2nd edn. Springe, New York (1989). https://doi.org/10.1007/978-1-4757-2063-1
3. Boumuki, N., Noda, T.: On gradient and Hamiltonian flows on even dimensional dually flat spaces. Fundam. J. Math. Math. Sci. **6**, 51–66 (2016)
4. Chirco, G., Malagò, L., Pistone, G.: Lagrangian and Hamiltonian dynamics for probabilities on the statistical bundles. Int. J. Geom. Methods Mod. Phys. **19**, 2250214 (2022). https://doi.org/10.1142/S0219887822502140
5. Fujiwara, A., Amari, S.: Gradient systems in view of information geometry. Physica D **80**, 317–327 (1995). https://doi.org/10.1016/0167-2789(94)00175-P
6. Hayashi, M.: Quantum Information. Springer, Heidelberg (2006). https://doi.org/10.1007/3-540-30266-2
7. Marsden, J., Weinstein, A.: Reduction of symplectic manifolds with symmetry. Rep. Math. Phys. **5**, 121–130 (1974). https://doi.org/10.1016/0034-4877(74)90021-4
8. Uwano, Y.: Geometry and dynamics of a quantum search algorithm for an ordered tuple of multi-qubits, chap. 11. InTech, Rijeka, Croatia (2013). https://doi.org/10.5772/53187
9. Uwano, Y.: All the trajectories of an extended averaged Hebbian learning equation on the quantum state space are the e-geodesics. Math. Model. Geom. **4**, 19–33 (2016). https://doi.org/10.26456/mmg/2016-412
10. Uwano, Y.: The averaged Hebbian learning equation, the exponential-type geodesics of the finite discrete distributions, and their quantum statistical analogues. RIMS Kôkyûroku **2137**, 168–182 (2019)

Geodesic Flows of α-connections for Statistical Transformation Models on a Compact Lie Group

Daisuke Tarama[1]([✉])[iD] and Jean-Pierre Françoise[2][iD]

[1] Department of Mathematical Sciences, Ritsumeikan University, 1-1-1 Nojihigashi, Kusatsu, Shiga 525-8577, Japan
dtarama@fc.ritsumei.ac.jp
[2] Laboratoire Jacques-Louis Lions, UMR 7598 CNRS, Sorbonne Université, 4 Place Jussieu, 75252 Paris, France
Jean-Pierre.Francoise@upmc.fr

Abstract. The geodesic flows of the α-connections are studied for a class of statistical transformation models on a compact Lie group. The Fisher-Rao (semi-definite) metric and the Amari-Chentsov cubic tensor, as well as the associated α-connections, are considered for general statistical models. Then, the general framework of statistical transformation models is explained following Barndorff-Nielsen and his coauthors. In particular, a couple of formulae are given for the Fisher-Rao (semi-definite) metric and the Amari-Chentsov cubic tensor for statistical transformation models, which are used in the latter part of the present paper. The α-connections and the associated geodesic flows are considered for the class of statistical transformation models introduced previously by the authors of the present paper. The ordinary differential equations on the corresponding Lie algebras are explicitly obtained to describe the geodesic flows on the basis of Euler-Poincaré and the Lie-Poisson equations on the Lie algebra.

Keywords: Statistical transformation model · information geometry · α-connection · geodesic flow · compact Lie group

1 Introduction

This paper deals with the geodesic flows of the α-connections for a class of statistical transformation models on a compact Lie group.

In information geometry, the Fisher-Rao (semi-definite) metric and the Amari-Chentsov cubic tensor play a very important role. The α-geodesics are particularly studied in view of the dually flat structures which connect the information geometry with the Hessian geometry. The details can be found in the standard textbooks on information geometry such as [1–3], but the ideas also appeared in pioneering works such as [6,7].

Partially supported by JSPS KAKENHI 19K14540, 22H01138.

The relations between information geometry and dynamical systems have been studied by Nakamura, Fujiwara, and Amari in 1990's [8, 10–12]. However, the geodesic flows in the information geometry for general statistical models are still of much interest for further studies both from the viewpoints of geometry and dynamical systems theory.

Among general statistical models, Barndorff-Nielsen and his coauthors [4] have introduced the *statistical transformation model* which consists of a smooth manifold of samples equipped with a Lie group action and a relatively invariant family of probability density functions with respect to the group action. The geodesic flows of the Fisher-Rao (semi-definite) metric have been studied by the authors of the present paper in [15, 16] from the viewpoint of geometric mechanics.

In the present paper, the Amari-Chentsov cubic tensor is studied for a general statistical transformation model and the geodesic flows with respect to the α-connections are considered for the statistical transformation model on a compact Lie group introduced in [15, 16]. As main results, the differential equations for the α-geodesic flows are explicitly deduced.

The structure of the paper is as follows:
In Sect. 2, an overview on the Fisher-Rao (semi-definite) metric and the Amari-Chentsov cubic tensor is given for a general statistical models and then a few concrete formulae of these tensors are given for statistical transformation models. In Sect. 3, the class of the statistical transformation models on a compact Lie group is introduced along the line of [15, 16]. The differential equations for the geodesic flows of α-connections are deduced in view of the Euler-Poincaré and the Lie-Poisson equations which are obtained in [15, 16].

2   General Statistical Transformation Models

In this section, we give an overview on the Fisher-Rao (semi-definite) metric and the Amari-Chentsov cubic tensor for general statistical transformation models.

2.1   Information Geometry of Statistical Transformation Models

We first recall the general description of the Fisher-Rao (semi-definite) metric and the Amari-Chentsov cubic tensor for a given statistical model.

Let M be a smooth manifold of statistical samples equipped with a volume form dvol_M. A statistical model consists of a family of smooth probability density functions $\rho(u, x)$ on M parameterized by the point $u \in U$ in an n-dimensional smooth manifold U of the parameters. For simplicity, we assume $\rho(u, x) > 0$ for all $(u, x) \in U \times M$. Given a random variable $f : M \to \mathbb{R}$, we have the expectation value of f with respect to the probability density function ρ as

$$E\left[f\right] := \int_M f(x)\rho(u, x)\mathrm{dvol}_M(x).$$

Note that $E[f]$ is a function in $u \in U$. It is known and in fact easy to check that the expectations of the logarithmic derivatives of ρ vanish:

$$E\left[\frac{\partial}{\partial u^i}\left(\log \rho\right)\right] = 0, \qquad i = 1, \ldots, n,$$

where $(u^1, \cdots, u^n)$ is a local coordinate system of U. To measure the distance of two probability density functions in the above family, it is convenient to consider the expectation value of the product of two logarithmic derivatives:

$$g_{ij} := E\left[\left(\frac{\partial}{\partial u^i}\log \rho\right) \cdot \left(\frac{\partial}{\partial u^j}\log \rho\right)\right], \qquad i, j = 1, \ldots, n.$$

Then, the symmetric $(0, 2)$-tensor $\sum\limits_{i,j=1}^{n} g_{ij} du^i \otimes du^j$ on U is positive-semi-definite and we call it as *Fisher-Rao semi-definite metric*.

Similarly, the expectation value of the product of three logarithmic derivatives

$$c_{ijk} := E\left[\left(\frac{\partial}{\partial u^i}\log \rho\right) \cdot \left(\frac{\partial}{\partial u^j}\log \rho\right) \cdot \left(\frac{\partial}{\partial u^k}\log \rho\right)\right], \qquad i, j, k = 1, \ldots, n,$$

gives rise to the symmetric $(0, 3)$-tensor $\sum\limits_{i,j,k=1}^{n} c_{ijk} du^i \otimes du^j \otimes du^k$ on U.

If the Fisher-Rao semi-definite metric is a positive-definite (and hence Riemannian) metric, then we have the associate Levi-Civita connection ∇ on U. Further, the Amari-Chentsov cubic tensor induces the α-connections $\nabla^{(\alpha)}$ on U, $\alpha \in \mathbb{R}$, defined through

$$\left\langle \nabla_X^{(\alpha)} Y, Z \right\rangle = \langle \nabla_X Y, Z \rangle + \frac{\alpha}{2} C(X, Y, Z), \tag{1}$$

for all vector fields X, Y, Z on U, which form a pencil of affine connections on U. Here, $\langle \cdot, \cdot \rangle$ denotes the Fisher-Rao metric and C stands for the Amari-Chentsov cubic tensor.

More detailed discussions around the Fisher-Rao (semi-definite) metric, the Amari-Chentsov cubic tensor, and the α-connections, as well as the dually flat structures which play important roles in information geometry, can be found e.g. in [1–3].

2.2   Statistical Transformation Models

In this subsection, we focus on the statistical transformation models initiated by Barndorff-Nielsen and his collaborators [4]. See also [2, §8.3].

We consider the smooth manifold M of statistical samples on which a Lie group G acts smoothly. Suppose that there is given a family of smooth probability density functions $\rho(g, x)$, $x \in M$, on M parameterized by the point $g \in G$ in the

Lie group. For simplicity, we assume that the volume form dvol_M on M satisfies the relative invariance with respect to the Lie group action:

$$\mathrm{dvol}_M(g \cdot x) = \chi(g)\mathrm{dvol}_M(x), \qquad \text{for all } g \in G, x \in M.$$

Here, $\mathrm{dvol}_M(g \cdot x)$ denotes the evaluation of the volume form at the point $g \cdot x$ and the function $\chi : G \to \mathbb{R}_{>0}$ is a one-dimensional representation of the group G. Further, we assume that the relative invariance of the family of probability density functions $\rho(g, x)$ in the sense that

$$\rho(hg \cdot x) = \rho\left(g, h^{-1}x\right) \chi(h)^{-1},$$

for any $g, h \in G$ and any $x \in M$.

Following the general arguments in the previous subsection, we consider the Fisher-Rao (semi-definite) metric and the Amari-Chentsov cubic tensor for the statistical transformation model. They are defined in the framework of left-invariant tensors on the parameter Lie group G as follows:

For an element $X \in \mathfrak{g}$ in the Lie algebra $\mathfrak{g}$ of the Lie group G, we denote the associated left-invariant vector field on G by $X^{(L)}$. We consider the expectation value of the logarithmic derivative of the probability density function ρ with respect to the left-invariant vector fields on G. As in the general case, we have

$$E\left[X^{(L)}(\log \rho)\right] = 0$$

for any $X \in \mathfrak{g}$.

Definition 1. *The positive-semi-definite bilinear form*

$$\langle X, Y \rangle := E\left[\left(X^{(L)} \log \rho\right) \cdot \left(Y^{(L)} \log \rho\right)\right], \qquad X, Y \in \mathfrak{g},$$

on $\mathfrak{g}$ is called the Fisher-Rao (semi-definite) bilinear form and the associated left-invariant $(0, 2)$ tensor on G is called the Fisher-Rao semi-definite metric.

When the Fisher-Rao semi-definite metric is positive-definite, which we cannot assume in general, we call it as Fisher-Rao metric.

We can check that the expectation value in the right hand side of the formula in Definition 1 is left-invariant as follows:
For $g, h \in G$, $x \in M$, we have

$$\begin{aligned}
X_{hg}^{(L)}\left(\log \rho(hg, x)\right) &= X_g^{(L)}\left(\log\left(\rho(g, h^{-1} \cdot x)\chi(h)^{-1}\right)\right) \\
&= X_g^{(L)}\left(\log\left(\rho(g, h^{-1} \cdot x)\right)\right) + X_g^{(L)}\left(\log \chi(h)^{-1}\right) \\
&= X_g^{(L)}\left(\log\left(\rho(g, h^{-1} \cdot x)\right)\right)
\end{aligned}$$

and hence

$$E\left[\left(X_{hg}^{(L)}\log\rho(hg,\cdot)\right)\cdot\left(Y_{hg}^{(L)}\log\rho(hg,\cdot)\right)\right]$$

$$=\int_M\left(X_{hg}^{(L)}\log\rho(hg,x)\right)\cdot\left(Y_{hg}^{(L)}\log\rho(hg,x)\right)\rho(hg,x)\mathrm{dvol}_M(x)$$

$$=\int_M\left(X_g^{(L)}\log\rho(g,h^{-1}\cdot x)\right)\cdot\left(Y_g^{(L)}\log\rho(g,h^{-1}\cdot x)\right)\rho(g,h^{-1}\cdot x)\chi(h)^{-1}\mathrm{dvol}_M(x)$$

$$=\int_M\left(X_g^{(L)}\log\rho(g,h^{-1}\cdot x)\right)\cdot\left(Y_g^{(L)}\log\rho(g,h^{-1}\cdot x)\right)\rho(g,h^{-1}\cdot x)\mathrm{dvol}_M(h^{-1}\cdot x)$$

$$=\int_M\left(X_g^{(L)}\log\rho(g,x)\right)\cdot\left(Y_g^{(L)}\log\rho(g,x)\right)\rho(g,x)\mathrm{dvol}_M(x)$$

$$=E\left[\left(X_g^{(L)}\log\rho(g,\cdot)\right)\cdot\left(Y_g^{(L)}\log\rho(g,\cdot)\right)\right].$$

Similarly, we consider the Amari-Chentsov cubic tensor as follows:

Definition 2. *The left-invariant* $(0,3)$-*tensor*

$$C(X,Y,Z):=E\left[\left(X^{(L)}\log\rho\right)\cdot\left(Y^{(L)}\log\rho\right)\cdot\left(Z^{(L)}\log\rho\right)\right],\qquad X,Y,Z\in\mathfrak{g},$$

on G *is called the Amari-Chentsov cubic tensor.*

The proof of the left-invariance for the right hand side of the expectation value in Definition 2 is similarly carried out as in the case of Fisher-Rao semi-definite metric.

Here, we mention some useful formulae for the Fisher-Rao semi-definite metric and the Amari-Chentsov cubic tensor.

Proposition 1. *For* $X,Y,Z\in\mathfrak{g}$, *the following relations hold:*

$$\langle X,Y\rangle=-E\left[X^{(L)}\left(Y^{(L)}\log\rho\right)\right],\qquad(2)$$

$$C\left(X,Y,Z\right)=E\left[X^{(L)}\left(Z^{(L)}\left(Y^{(L)}\log\rho\right)\right)\right]$$
$$+E\left[Y^{(L)}\left(Z^{(L)}\left(X^{(L)}\log\rho\right)\right)\right].\qquad(3)$$

Proof. (2) Since $0=E\left[Y^{(L)}\log\rho\right]$, we have

$$0=X^{(L)}\left(E\left[Y^{(L)}\log\rho\right]\right)=E\left[X^{(L)}\left(Y^{(L)}\log\rho\right)\right]+E\left[\left(X^{(L)}\log\rho\right)\cdot\left(Y^{(L)}\log\rho\right)\right].$$

Then, the formula (2) follows.

(3) Since the Fisher-Rao metric $\langle X,Y\rangle$ is left-invariant, using (2), we have

$$0=Z^{(L)}\left(\langle X,Y\rangle\right)=-Z^{(L)}\left(E\left[X^{(L)}\left(Y^{(L)}\log\rho\right)\right]\right)$$
$$=-E\left[Z^{(L)}\left(X^{(L)}\left(Y^{(L)}\log\rho\right)\right)\right]-E\left[X^{(L)}\left(Y^{(L)}\log\rho\right)\cdot\left(Z^{(L)}\log\rho\right)\right]$$

and, by the symmetry of the metric,

$$0 = -E\left[Z^{(L)}\left(Y^{(L)}\left(X^{(L)}\log\rho\right)\right)\right] - E\left[Y^{(L)}\left(X^{(L)}\log\rho\right)\cdot\left(Z^{(L)}\log\rho\right)\right].$$

Now, by Definition 1, we have

$$0 = Z^{(L)}\left(\langle X, Y\rangle\right)$$
$$= E\left[Z^{(L)}\left(\left(X^{(L)}\log\rho\right)\right)\cdot\left(Y^{(L)}\log\rho\right)\right] + E\left[\left(X^{(L)}\log\rho\right)\cdot Z^{(L)}\left(\left(Y^{(L)}\log\rho\right)\right)\right]$$
$$+ E\left[\left(X^{(L)}\log\rho\right)\cdot\left(Y^{(L)}\log\rho\right)\cdot Z^{(L)}\left(\log\rho\right)\right].$$

Thus,

$$C(X, Y, Z) = E\left[\left(X^{(L)}\log\rho\right)\cdot\left(Y^{(L)}\log\rho\right)\cdot\left(Z^{(L)}\log\rho\right)\right]$$
$$= -E\left[Z^{(L)}\left(X^{(L)}\log\rho\right)\cdot\left(Y^{(L)}\log\rho\right)\right] - E\left[\left(X^{(L)}\log\rho\right)\cdot Z^{(L)}\left(Y^{(L)}\log\rho\right)\right]$$
$$= E\left[X^{(L)}\left(Z^{(L)}\left(Y^{(L)}\log\rho\right)\right)\right] + E\left[Y^{(L)}\left(Z^{(L)}\left(X^{(L)}\log\rho\right)\right)\right].$$

If the Fisher-Rao semi-definite metric is positive-definite and defines a left-invariant Riemannian metric on G, then the Levi-Civita connection ∇ and the α-connections $\nabla^{(\alpha)}$ are defined as in the end of the previous subsection. Note that ∇ and $\nabla^{(\alpha)}$ are left-invariant.

Even if the Fisher-Rao semi-definite metric is not positive-definite as a tensor on G, there remains the possibility to have an associated Riemannian metric on a quotient manifold of G by a certain subgroup as in the following manner. Assume that the Lie group G is compact and hence the Lie algebra $\mathfrak{g}$ is equipped with a bi-invariant inner product B. Suppose further that there exists a Lie subgroup $H \subset G$ whose Lie algebra $\mathfrak{h} \subset \mathfrak{g}$ has the orthogonal complement $\mathfrak{m}$ with respect to the inner product B such that the restriction $\langle\cdot,\cdot\rangle\,|_{\mathfrak{m}\times\mathfrak{m}}$ is positive-definite and Ad_H-invariant. Under these conditions, the Fisher-Rao semi-definite metric $\langle\cdot,\cdot\rangle$ induces a Riemannian metric on the quotient manifold G/H of the group G relative to the subgroup H with respect to the action from right. We call this Riemannian metric as Fisher-Rao Riemannian metric. Note that the metric on G/H is invariant with respect to the action of the group G from the left. By virtue of the general results by Nomizu in [13, Theorem 13.1], the Levi-Civita connection ∇ on G/H with respect to the Fisher-Rao Riemannian metric can be written as

$$\nabla_X Y = \frac{1}{2}[X, Y]_{\mathfrak{m}} + U(X, Y), \qquad X, Y \in \mathfrak{g}, \tag{4}$$

where $U(X, Y)$ is defined through

$$\langle U(X, Y), Z\rangle = \frac{1}{2}\left(\langle[X, Z]_{\mathfrak{m}}, Y\rangle + \langle X, [Y, Z]_{\mathfrak{m}}\rangle\right), \qquad Z \in \mathfrak{g}.$$

Here, for $W \in \mathfrak{g}$, we denote its $\mathfrak{m}$-component by $W_{\mathfrak{m}}$.

3   Geodesic Flows of α-connections in the Statistical Transformation Model on a Compact Lie Group

Among statistical transformation models, we focus on a specific family of probability density functions on a compact Lie group which play both of the roles of the parameter manifold and the sample manifold.

3.1   A Statistical Transformation Model on a Compact Lie Group

For a statistical transformation model, we further assume that the manifold M of samples coincides with the Lie group G, which is also supposed to be compact. As in the end of previous section, the Lie algebra $\mathfrak{g}$ is equipped with the bi-invariant inner product B. On $M = G$, we take the Haar measure dvol_G as its volume form. As the family of probability density functions on $M = G$, we consider the smooth functions

$$\rho(g, h) := c \cdot \exp\left(F\left(g^{-1}h\right)\right), \qquad (g, h) \in G \times G,$$

where $F(\theta) = B\left(Q, \mathrm{Ad}_\theta N\right)$ with $Q, N \in \mathfrak{g}$ being fixed elements. The constant c is the normalization multiple so that ρ is a probability density function. The function F appears in the description of the generalized Toda lattice equation on the compact real forms of complex semi-simple Lie algebra $\mathfrak{g}$ as double bracket equations in [5]. Clearly, we see that the family of the probability density functions $\rho(g, h)$ are invariant, namely relatively invariant with $\chi \equiv 1$, in view of the Lie group action:

$$\rho(k \cdot g, h) = \rho(g, k^{-1} \cdot h)$$

for all $g, h, k \in G$.

Under the above conditions, we compute the Fisher-Rao semi-definite metric and the Amari-Chentsov cubic tensor. By means of the formula (2), we can give an expression of the Fisher-Rao semi-definite metric as

$$\langle X, Y \rangle = B\left([X, Q], [Y, N']\right), \qquad X, Y \in \mathfrak{g},$$

where $N' = \displaystyle\int_G \mathrm{Ad}_{g^{-1}h} N \rho(g, h) \mathrm{dvol}_G(h) = \int_G \mathrm{Ad}_h N \rho(e, h) \mathrm{dvol}_G(h)$. Suppose that N', Q are sitting in the same Cartan subalgebra $\mathfrak{h} \subset \mathfrak{g}$ and that they are regular elements so that $\mathrm{ad}_{N'}|_\mathfrak{m}$ and $\mathrm{ad}_Q|_\mathfrak{m}$ are invertible, where $\mathfrak{m}$ is the orthogonal complement of $\mathfrak{h}$ with respect to B. Taking the Cartan subgroup $H \subset G$ whose Lie algebra is $\mathfrak{h}$, we see that the restriction $\langle \cdot, \cdot \rangle|_{\mathfrak{m} \times \mathfrak{m}}$ of the Fisher-Rao semi-definite metric to the complement $\mathfrak{m}$ to $\mathfrak{h}$ is positive-definite. Then, by the arguments of Nomizu [13, Theorem 13.1], we obtain a Riemannian metric on $\mathcal{O} := G/H$ which is invariant with respect to the left-action by G.

Similarly, we compute the Amari-Chentsov cubic tensor C for this statistical transformation model through (3) and we have

$$C(X, Y, Z) = B\left(Z, [[X, N'], [Y, Q]]\right) + B\left(Z, [[Y, N'], [X, Q]]\right).$$

As is already obtained in [15,16], we can describe the geodesic flows induced by the Levi-Civita connection for the Fisher-Rao Riemannian metric on $\mathcal{O} = G/H$ as Lagrangian and Hamiltonian systems.

In the Lagrangian formalism, the geodesic flow can be characterized by the Lagrangian function

$$\mathcal{L}(X) := \frac{1}{2} \langle X, X \rangle = \frac{1}{2} B \left([X, Q], [X, N'] \right), \quad X \in \mathfrak{g},$$

which can be regarded as a left-invariant function on TG with respect to the action by G, and hence it can be described by the Euler-Poincaré equation

$$\frac{\mathrm{d}}{\mathrm{d}t}[[X, Q], N'] = [X, [[X, Q], N']], \quad X \in \mathfrak{g}.$$

Recall that the right hand side of this ODE coincides with $\nabla_X X$ given by (4).

Now, by the definition of the α-connection $\nabla^{(\alpha)}$ in (1), we have the following description of the geodesic flow with respect to $\nabla^{(\alpha)}$.

Theorem 1. *In the Lagrangian formalism, the geodesic flow for the α-connection $\nabla^{(\alpha)}$ on $\mathcal{O} = G/H$ with respect to the Fisher-Rao Riemannian metric $\langle \cdot, \cdot \rangle$ and the Amari-Chentsov cubic tensor C is described by the differential equation*

$$\frac{\mathrm{d}}{\mathrm{d}t}[[X, Q], N'] = [X, [[X, Q], N']] + \alpha\, [[X, N'], [Q, X]], \quad X \in \mathfrak{m}. \quad (5)$$

The dynamical properties of (5) are still largely unknown. However, we observe that, when $\alpha = 1$, the equation (5) can be reduced to a simpler form

$$\frac{\mathrm{d}}{\mathrm{d}t}[[X, Q], N'] = [[X, [X, Q]], N'].$$

Similarly, in the Hamiltonian formalism the geodesic flow can be given by the Hamiltonian function

$$\mathcal{H}(X) := \frac{1}{2} B \left(X_{\mathfrak{m}}, \mathrm{ad}_Q^{-1} \circ \mathrm{ad}_{N'}^{-1} X_{\mathfrak{m}} \right), \quad X \in \mathfrak{g},$$

which can be seen as a left-invariant function on T^*G with respect to the action by G. Thus, the geodesic flow can be expressed in terms of the Lie-Poisson equation

$$\frac{\mathrm{d}}{\mathrm{d}t}X = \left[X, \mathrm{ad}_Q^{-1} \circ \mathrm{ad}_{N'}^{-1} X_{\mathfrak{m}} \right], \quad X \in \mathfrak{g}.$$

See [9,14] for the generalities about the Euler-Poincaré and the Lie-Poisson equations.

Finally, through the Legendre transform, the geodesic flow with respect to the $\nabla^{(\alpha)}$ can be described as follows:

Theorem 2. *In the Hamiltonian formalism, the geodesic flow for the α-connection $\nabla^{(\alpha)}$ on $\mathcal{O} = G/H$ with respect to the Fisher-Rao Riemannian metric $\langle \cdot, \cdot \rangle$ and the Amari-Chentsov cubic tensor C is given by the differential equation*

$$\frac{\mathrm{d}}{\mathrm{d}t}X = \left[X, \mathrm{ad}_Q^{-1} \circ \mathrm{ad}_{N'}^{-1} X_{\mathfrak{m}} \right] + \alpha \left[\mathrm{ad}_{N'}^{-1} X_{\mathfrak{m}}, \mathrm{ad}_Q^{-1} X_{\mathfrak{m}} \right], \quad X \in \mathfrak{g}. \quad (6)$$

References

1. Amari, S.-I.: Information Geometry and its Applications. Applied Mathematical Sciences, vol. 194. Springer, Tokyo (2016). https://doi.org/10.1007/978-4-431-55978-8
2. Amari, S.-I., Nagaoka, H.: Methods of information geometry, translated from the 1993 Japanese original by D. Harada, Translations of Mathematical Monographs, vol. 191. American Mathematical Society, Providence; Oxford University Press, Oxford (2000)
3. Ay, N., Jost, J., Vân Lê, H., Schwachhöfer, L.: Information geometry, Ergebnisse der Mathematik und ihrer Grenzgebiete. 3. Folge. A Series of Modern Surveys in Mathematics, vol. 64. Springer, Cham (2017)
4. Barndorff-Nielsen, O.E., Blæsild, P., Eriksen, P.S.: Decomposition and Invariance of Measures, and Statistical Transformation Models. Lecture Notes in Statistics, vol. 58. Springer, New York (1989)
5. Bloch, A., Brockett, R.W., Ratiu, T.S.: Completely integrable gradient flows. Comm. Math. Phys. **147**(1), 57–74 (1992)
6. Chentsov, N.N.: Geometry of the "manifold" of a probability distribution. Dokl. Akad. Nauk SSSR **158**, 543–546 (1963)
7. Fréchet, M.: Sur l'existence de certaines évaluations statistiques au cas de petits échantillons. Rev. Int. Statist. **11**, 182–205 (1943)
8. Fujiwara, A., Amari, S.-I.: Gradient systems in view of information geometry. Physica D **80**, 317–327 (1995)
9. Marsden, J.E., Ratiu, T.S.: Introduction to Mechanics and Symmetry. Texts in Applied Mathematics, 2nd edn., vol. 17, Springer, New York (2003)
10. Nakamura, Y.: Completely integrable gradient systems on the manifolds of Gaussian and Multinomial Distributions. Japan J. Indust. Appl. Math. **10**(2), 179–189 (1993)
11. Nakamura, Y.: Neurodynamics and nonlinear integrable systems of Lax type. Japan J. Indust. Appl. Math. **11**(1), 11–20 (1994)
12. Nakamura, Y.: Gradient systems associated with probability distributions. Japan J. Indust. Appl. Math. **11**(1), 21–30 (1994)
13. Nomizu, K.: Affine connections on homogeneous spaces. Amer. J. Math. **76**(1), 33–65 (1954)
14. Ratiu, T.S., et al.: A crash course in geometric mechanics. In: Montaldi, J., Ratiu, T. (eds.) Geometric Mechanics and Symmetry: The Peyresq Lectures. Cambridge University Press, Cambridge (2005)
15. Tarama, D., Françoise, J.-P.: Information geometry and Hamiltonian systems on lie groups. In: Nielsen, F., Barbaresco, F. (eds.) GSI 2021. LNCS, vol. 12829, pp. 273–280. Springer, Cham (2021). https://doi.org/10.1007/978-3-030-80209-7_31
16. Tarama, D., Françoise, J.-P.: Dynamical systems over lie groups associated with statistical transformation models. Phys. Sci. Forum **5**, 21 (2022)

Neurogeometry

Differential Operators Heterogenous in Orientation and Scale in the V_1 Cortex

Mattia Galeotti[1]($\boxtimes$), Giovanna Citti[1], and Alessandro Sarti[2]

[1] Università di Bologna, Bologna, Italy
galeotti.mattia.work@gmail.com
[2] EHESS, Paris, France

Abstract. In the primary visual cortex V_1 of the human brain, cortical receptive profiles (RPs) changing from point to point act as a differential operator on the visual stimulus, while cortical connectivity inverts the problem and recovers an image called perceived image. We analyze the transform of the visual stimulus performed by the RPs on V_1 in order to reconstruct the image perceived by the subject.

We consider a differential operator determined by a heterogenous distribution of orientation and scale, whose local representation can be a regular Laplacian, a degenerate Laplacian or a degenerate Laplacian with different scale at different points. We model the associated cortical dynamic via discrete differential operators whose coefficients have a stochastic distribution. We prove a homogenization result, showing that a large class of heterogenous operators H-converges to the classic Laplacian.

Keywords: Partial differential equations · visual cortex · discrete differentials

1 Introduction

The visual brain extracts features from the visual stimulus by acting as a differential operator on it and solving an inverse problem on the output function. In this work, that follows [10] by the same authors, is shown a reconstruction process of the perceived image based on the distribution of cells in the primary visual cortex V_1.

A receptive profile (RP) is a function modeling the impulse response of a cortical cell to some impulse $I \colon \mathbb{R}^2 \to [0,1]^3$ defined over the retinal plane (we are considering a colored image, therefore a 3-chain valued function). Statistical studies on the RPs of V_1 cells in macaques show a great heterogeneity of behaviors [9]. Copies of center-surrounds RPs (Mexican hats) are present, as well as a great variety of simple cells with anysotropic RPs that detect the boundary orientation.

– In the case of *Mexican hats*, the RP is axial symmetric and writes

$$RP_h(x,y) = \Delta G(x,y),$$

where G is a 2-dimensional Gaussian and Δ an Euclidean Laplacian.

F. Nielsen and F. Barbaresco (Eds.): GSI 2023, LNCS 14072, pp. 465–473, 2023.
https://doi.org/10.1007/978-3-031-38299-4_48

- For *simple cells* we consider the directional derivative $X_\theta = \cos\theta\partial_x + \sin\theta\partial_y$ and the receptive profile of with preferred orientation $\theta + \pi/2$ and order β writes

$$RP_s(x,y) = X_\theta^\beta G(x,y).$$

- In another version that take scale into accounts, the receptive profile writes

$$RP_{ss}(x,y) = c(x,y)X_\theta^\beta G(x,y),$$

where the function $c\colon \mathbb{R}^2 \to \mathbb{R}^+$ is a scale factor.

The preferred orientation of simple cells changes in the cortex from point to point [3] giving rise to the so called orientation map $\theta\colon \mathbb{R}^2 \to [0,\pi]$. In primates and cats, neurons with similar orientation selectivity are clustered together to constitute the so-called pinwheel organisation [2]. A simple model of pinwheel shaped orientation map is proposed in [7] where the map is obtained through the superposition of randomly weighted complex sinusoids

$$\theta_1(x,y) = \arg\sum_{k=1}^{N} c_k(x,y)e^{i2\pi(x\cos(2\pi k/N)+y\sin(2\pi k/N))}, \tag{1.1}$$

with N denoting the number of orientation samples and where the coefficients $c_k\colon \mathbb{R}^2 \to [0,1]$ are white noise functions. In rodents the orientation information encoded in V_1 is distributed in a less organized fashion known as "salt-and-pepper map" [4], and is modeled by $\theta_2(x,y) = d(x,y)$ where $d\colon \mathbb{R}^2 \to [0,\pi]$ is a white noise function.

2   The Cortical Transform

We introduce the model for the coupled activity of classical receptive profiles and the short range connectivity that modulates the feed forward input giving rise to contextual modulation effects.

2.1   Action of the Receptive Profiles

A receptive profile act on the input stimulus by convolution, inducing an output function

$$O(x,y) := RP_s(x,y) \star I(x,y).$$

In the case of the simple cells RPs, this becomes

$$O_s(x,y,\theta) = X_\theta^\beta G \star I = X_\theta^\beta I_s,$$

where I_s is the smoothed image obtained by convoluting I with the Gaussian, $I_s = G \star I$.

We identify the retinal manifold with the plane $M = \mathbb{R}^2$. Once we fix a $\mathcal{C}^1$ orientation map $\theta\colon M \to [0,2\pi]$, the vector field $X_{\theta(x,y)}$ generates a 1-dimensional distribution $H \subset TM$. By fixing a metric that gives $X_{\theta(x,y)}$ a unitary norm, we give (M, H) a sub-Riemannian structure. For the sake of simplicity, in what follows we will omit the dependence on (x, y) and use the notation X_θ. The sub-Laplacian associated to X_θ is

$$L_{X_\theta} I := X_\theta^* X_\theta I, \tag{2.1}$$

where X_θ^* is the formal adjoint operator to X_θ defined by the equation

$$\int_M (X_\theta f) g = \int f(X_\theta^* g) \quad \forall f, g \in C_0^\infty(M).$$

We will apply the operator in the weak sense of the Sobolev space W_X^1 associated to the operator itself. Therefore the solution of the partial differential equation $L_{X_\theta} u = f$ can be written down as follows.

Definition 1. *A function $u \in W_{X,0}^1$ is a weak solution of the equation $L_{X_\theta} u = f$ in an open set Ω if for every smooth function ϕ compactly supported in Ω*

$$\int (L_{X_\theta} u)\phi = \int X_\theta u X_\theta \phi = -\int f\phi.$$

2.2   The Heterogeneous Poisson's Equation

We consider a modified sub-Laplacian $\widetilde{L}_{X_\theta} = X_\theta^* c(x, y) X_\theta$ where c is a scale random function that take values in a closed interval. As already said, the output of a distribution of V_1 cells in response to the visual input I is the differentiation of the visual input itself. We propose here a procedure to reconstruct the perceived image by considering the integration of the operator, using the short range connectivity of V_1 and heterogenous operators. In particular we model this reconstruction as the heterogenous Poisson's problem

$$\widetilde{L}_{X_\theta} u(x, y) = \widetilde{L}_{X_\theta} I(x, y) \tag{2.2}$$

where the right side term is known and represents the action of receptive profiles on the visual input, while the solution u represents the reconstructed-perceived image. The classic Poisson's problem identifies the stimulus $I(x, y)$ up to a harmonic function (Retinex effect). In the case of the heterogeneous Poisson's problem (2.2), the stimulus is identified up to a sub-harmonic function, meaning that is annulled by $\widetilde{L}_{X_\theta}$.

The existence of a fundamental solution is known for large class of operators with measurable coefficients. In [6] this is proved for second order uniformly elliptic operators, and the same technique ensures the existence of a fundamental solution for uniformly sub-elliptic operators, while the problem is open in full generality. In order to find a numerical solution, in the last section we use the steepest descent method. In particular we consider a function $u_t(x, y)$ that describes the flow associated to Eq. (2.2), meaning the following,

$$\partial_t u(x, y) = \widetilde{L}_{X_\theta} I(x, y) - \widetilde{L}_{X_\theta} u(x, y). \tag{2.3}$$

In order to see the asymptotic behavior of a function verifying the equation above, we consider the scalar product

$$\langle u, v \rangle_{c,X} := \int_{\Omega} c(x,y)(X_\theta u)(X_\theta v) dx dy,$$

the associated norm $\|-\|_{c,X}$ and the associated Sobolev space $H_0^2(\Omega)$ where Ω is a bounded domain in $\mathbb{R}^2$. We consider the convex functional $J(u) := \|u\|_{c,X}^2 - \int_{\Omega} u \cdot \widetilde{L}I$ and a sequence u_{t_n} with $t_n \to +\infty$ for $n \to +\infty$. If $J(u_{t_n}) \to m \in \mathbb{R}$ and $dJ(u_{t_n}) \to 0$, then $\nabla_X J(u_{t_n}) = 2c \cdot X_\theta u_{t_n} - \widetilde{L}I \to 0$. This implies that $\int_{\Omega} c|X_\theta u_{t_n}|^2$ is borned and therefore $c \cdot X_\theta u_{t_n}$ weakly converges. Therefore it must converge to $\widetilde{L}I$. This proves the Palais-Smale condition PS_m for any $m \in \mathbb{R}$ and therefore implies that any sequence u_{t_n} induced by Eq. (2.3) converges to a solution of (2.2), see for example [1, Theorem 7.12].

3   Homogenization Results

3.1   Discrete Differential Operators

The action of the receptive profiles is obviously discrete. Hence we recall here how to discretize second order differential operators, referring to [5] and [8] for a wider introduction.

We consider a smooth bounded domain $Q \subset \mathbb{R}^d$ and set $Q_\varepsilon = Q \cap \varepsilon \mathbb{Z}^d$, with $\varepsilon > 0$. Moreover, we introduce the appropriate L^2-norm

$$\|u^\varepsilon\|_{L^2(Q_\varepsilon)}^2 := \varepsilon^d \cdot \sum_{x \in Q_\varepsilon} |v^\varepsilon(x)|^2.$$

If Λ a finite subset of $\mathbb{Z}^d$ symmetric with respect to 0, we define the boundary $\partial Q_\varepsilon^\Lambda$ of Q_ε by

$$\partial Q_\varepsilon^\Lambda := \{x + \varepsilon z | x \in Q_\varepsilon, z \in \Lambda\} \backslash Q_\varepsilon.$$

We also introduce $\overline{Q}_\varepsilon := Q_\varepsilon \cup \partial Q_\varepsilon^\Lambda$.

If we denote by ∂_z^ε the usual discrete derivative along $z \in \Lambda$, we can introduce a discrete Dirichlet problem with second member $f^\varepsilon : Q_\varepsilon \to \mathbb{R}$. We consider a symmetric $|\Lambda| \times |\Lambda|$ real matrix function $A^\varepsilon = (a_{zz'}^\varepsilon)_{z,z' \in \Lambda}$ and denote the Dirichlet problem with the same symbol A^ε.

$$A^\varepsilon u^\varepsilon(x) = \sum_{z,z' \in \Lambda} \partial_{-z}(a_{zz'}^\varepsilon(x) \partial_{z'} u^\varepsilon(x)) = f^\varepsilon(x) \quad \forall x \in Q_\varepsilon, \qquad (3.1)$$

with $u^\varepsilon(x) = 0$ if $x \in \partial Q_\varepsilon^\Lambda$.

Example 1. In order to discretize Eq. (2.2), we consider $\Lambda = \{e_x, -e_x, e_y, -e_y\} \subset \mathbb{R}^2$ where $e_x = (1,0)$ and $e_y = (0,1)$. Then in the notation of Eq. (3.1), we have

$$a_{e_x,e_x}^\varepsilon = \cos^2 \theta(x,y)$$
$$a_{e_x,e_y}^\varepsilon = a_{e_y,e_x}^\varepsilon = \sin \theta(x,y) \cos \theta(x,y)$$
$$a_{e_y,e_y}^\varepsilon = \sin^2 \theta(x,y),$$

and 0 for the others.

In what follows, we consider a slightly different discretized Dirichlet problem where the only non-null coefficients are the $a_{zz'}^\varepsilon$ with $z = z'$ and we maintain the property that $\sum_{z \in \Lambda} a_{zz} = 1$.

We translate the notion of Sobolev space to the discrete setting. In fact, if $v^\varepsilon \colon \varepsilon \mathbb{Z}^d \to \mathbb{R}$, then v^ε is in $W_0^{1,2}(Q_\varepsilon)$ if $v^\varepsilon(x) = 0$ for any $x \notin Q_\varepsilon$.

Definition 2. *For any $\Lambda \subset \mathbb{Z}^d$ finite and symmetric with respect to 0, if $\eta, \eta' \in \mathbb{R}^{|\Lambda|}$ and A is a $|\Lambda| \times |\Lambda|$ symmetric matrix,*

$$\langle \eta, \eta' \rangle_{A,\Lambda} := \sum_{z,z' \in \Lambda} a_{zz'}^\varepsilon \eta_z \eta_{z'}.$$

This allows to define the associated norm $\|\eta\|_{A,\Lambda}^2 = \langle \eta, \eta \rangle_{A,\Lambda}$.

If $\{e_i | \; i = 1, \ldots, d\}$ is the standard basis of $\mathbb{R}^d$, we consider the standard symmetric set $\Lambda_d := \{\pm e_i | \; i = 1, \ldots, d\}$ and denote by $I_d := (\delta_{zz'})_{z,z' \in \Lambda_d}$ the standard identity matrix associated to Λ_d.

Definition 3. *The problem (3.1) is uniformly elliptic if there exist $c_1, c_2, \varepsilon_0 > 0$ such that for any $\eta \in \mathbb{R}^{|\Lambda|}$ and any $\varepsilon < \varepsilon_0$,*

$$|a_{zz'}^\varepsilon(x)| \leq c_1 \quad \forall x \in \overline{Q}_\varepsilon \; \forall z, z' \in \Lambda \tag{3.2}$$

$$c_2 \cdot \|\eta\|_{I_d, \Lambda_d} \leq \|\eta\|_{A,\Lambda} \tag{3.3}$$

The following result is introduced in [8, Proposition 1.3] and gives a sufficient condition for the uniform ellipticity of a discrete differential problem.

Proposition 1. *Consider a function $p_z^\varepsilon \colon Q_\varepsilon \to \mathbb{R}$ defined for any $z \in \Lambda$, if it satisfies the following three properties*

1. *$p_z^\varepsilon(x) \geq 0$ and $\sum_{z \in \Lambda} p_z^\varepsilon(x) = 1 \; \forall x \in Q_\varepsilon$;*
2. *$\exists \delta > 0$ such that $p_{\pm e_i}^\varepsilon \geq \delta$ for $i = 1, \ldots, d$;*
3. *$p_z^\varepsilon(x) = p_{-z}^\varepsilon(x + \varepsilon z)$;*

then the following problem is uniformly elliptic

$$u^\varepsilon(x) - \sum_{z \in \Lambda} p_z^\varepsilon(x) u^\varepsilon(x + \varepsilon z) = \varepsilon^2 \cdot f^\varepsilon(x) \quad in \; Q_\varepsilon, \quad u^\varepsilon(x) = 0 \; \forall x \in \partial Q_\varepsilon^\Lambda, \tag{3.4}$$

for any $f^\varepsilon \colon Q_\varepsilon \to \mathbb{R}$.

Observe that the operator associated to the previous mean value formula can be rewritten in the form (3.1) with $a_{zz'}^\varepsilon(x) = p_z^\varepsilon(x)$ if $z = z' \neq 0$ and 0 otherwise.

In order to compare the functions defined over Q_ε with those having a continuous argument, we follow Kozlov [5] in defining a mesh completion. If $f^\varepsilon \colon Q_\varepsilon \to \mathbb{R}$ then we denote by $\tilde{f}^\varepsilon \colon Q \to \mathbb{R}$ the function such that

$$\tilde{f}^\varepsilon(x) = f(y_x) \quad \forall x \in Q,$$

where $y_x = (y_1, \cdots, y_d)$ is the point in Q_ε such that $y_i - \frac{\varepsilon}{2} \leq x_i < y_i + \frac{\varepsilon}{2}$. In what follows, if it is clear from the context we will use a little abuse of notation denoting $\tilde{f}^\varepsilon \in W_0^{1,2}(Q)$ by f^ε. Furthermore, we will say that f^ε converges (strongly or weakly) to f in $L^2(Q_\varepsilon)$, $W^{1,2}(Q_\varepsilon)$ or $W^{-1,2}(Q_\varepsilon)$ when the mesh completion $\tilde{f}^\varepsilon$ converges to f in $L^2(Q)$, $W^{1,2}(Q)$, $W^{-1,2}(Q)$. If needed we will replace this mesh completion with a piecewise linear one $\tilde{f}^\varepsilon$ in $W^{1,2}(Q)$.

3.2   Convergence Results

Proposition 2. *If the problem (3.1) is uniformly elliptic and $f^\varepsilon \in L^2(Q_\varepsilon)$, then there exists unique a solution of the problem $u^\varepsilon \in W_0^{1,2}(Q_\varepsilon)$ and there exists $c > 0$ such that*

$$\|u^\varepsilon\|_{W_0^{1,2}} \leq c \cdot \|f^\varepsilon\|_{L^2},$$

uniformly in ε.

In what follows, consider a family of uniformly elliptic discrete Dirichlet problems (3.1). We denote by $A^\varepsilon(x)$ the coefficient matrix functions of the family and by $A(x) = (a_{zz'}(x))$ another $|A| \times |A|$ real matrix function defined for $x \in Q$.

Definition 4. *Denote by u^ε the solution of the Dirichlet problem associated to A^ε and by u^0 the solution of the continuous Dirichlet problem*

$$Au^0 = \sum_{z,z' \in A} -\frac{\partial}{\partial z}\left(a_{zz'}(x)\frac{\partial u^0(x)}{\partial z'}\right) = f(x), \quad u^0 \in W_0^{1,2}(Q).$$

We say that the matrix A^ε H-converges to A, $A^\varepsilon \xrightarrow[\varepsilon \to 0]{H} A$, if for any sequence $f^\varepsilon \in W^{-1,2}(Q_\varepsilon)$ such that $f^\varepsilon \to f \in W^{-1,2}(Q_\varepsilon)$, we have

$$u^\varepsilon \rightharpoonup u^0 \text{ in } W_0^{1,2}(Q_\varepsilon)$$

$$D_{a^\varepsilon}u^\varepsilon = \sum_{z' \in A} a_{zz'}^\varepsilon \partial_{z'} u^\varepsilon \rightharpoonup D_a u^0 = \sum_{z' \in A} a_{zz'}\frac{\partial u^0}{\partial z'} \text{ in } L^2(Q_\varepsilon)$$

Definition 5. *If the discrete Dirichlet problem (3.1) is uniformly elliptic for any $\varepsilon > 0$ and there is a constant matrix A^0 such that $A^\varepsilon \xrightarrow{H} A^0$, then we call the matrix A^0 the homogenized matrix for A^ε.*

In order to describe the action of the cortical receptive profiles, we introduce operators with random coefficients. Let $(\Omega, \mathcal{F}, \mu)$ be a probability space and $\{T_x \colon \Omega \to \Omega \mid x \in \mathbb{Z}^d\}$ a group of $\mathcal{F}$-measurable transformations respecting the following properties:

1. $T_x \colon \Omega \to \Omega$ is $\mathcal{F}$-measurable $\forall x \in \mathbb{Z}^d$;
2. $\mu(T_x \mathcal{B}) = \mu(\mathcal{B})$, $\forall \mathcal{B} \in \mathcal{F}$ and $\forall x \in \mathbb{Z}^d$;
3. $T_0 = \text{id}$ and $T_x \circ T_y = T_{x+y} \; \forall x, y \in \mathbb{Z}^d$.

We recall that the group T_x is called ergodic if for any $f \in L^1(\Omega)$ such that $f(T_x\omega) = f(\omega)$ μ-a.s. for any $x \in \mathbb{Z}^d$, there exists $K \in \mathbb{R}$ such that μ-a.s. $f = K$.

Consider an $\mathcal{F}$-measurable $|\Lambda| \times |\Lambda|$ symmetric matrix function $\mathcal{A}(\omega) := (a_{zz'}(\omega))$, it induces the family of operators

$$A^\varepsilon(x)(\omega) := \mathcal{A}(T_{x/\varepsilon}\omega), \quad \forall x \in \varepsilon\mathbb{Z}^d.$$

Suppose that $\Lambda_d \subset \Lambda$ and there are $c_1, c_2 > 0$ such that the following are true a.s.

$$|a_{zz'}(\omega)| \leq c_1$$
$$\|\eta\|_{I_d, \Lambda_d} \leq c_2 \cdot \|\eta\|_{\mathcal{A}, \Lambda}, \quad \forall \eta \in \mathbb{R}^{|\Lambda|}.$$

Then the Dirichlet problems induced by the A^ε are uniformly elliptic in any regular domain Q.

In order to build a model of the cortical connectivity action, we apply these results about discretized second order operators to random distributions over $\mathbb{R}^2$. In particular we define a function

$$p(x, z) = p_z(x) \colon \mathbb{R}^2 \times \Lambda \to \mathbb{R},$$

where $\Lambda \subset \mathbb{Z}^2$ is a finite set symmetric with respect to 0. In order to describe a uniformly elliptic problem as in the Eq. (3.4), from the $p_z(x)$ we define the transition functions $p_z^\varepsilon(x)$,

$$p_z^\varepsilon(x) := p_z(x/\varepsilon) \quad \forall x \in Q_\varepsilon \subset \varepsilon\mathbb{Z}^2.$$

The associated operators $A^\varepsilon = (a_{zz'}^\varepsilon)$ are defined by $a_{zz}^\varepsilon(x) = p_z^\varepsilon(x)$ if $z \neq 0$ and $a_{zz'}^\varepsilon(x) = 0$ otherwise.

Proposition 3 (See [5, §2]). *If the functions p_z^ε satisfy a.s. the three hypothesis of Proposition 1, then the A^ε H-converge a.s. to an elliptic operator A^0 with constant coefficients and moreover A^0 is isotropic,*

$$A^0 = (a_{zz'}^0) = a^\delta(r) \cdot \mathrm{id},$$

where $a^\delta(r) \in \mathbb{R}$.

In what follows we suppose that $f^\varepsilon \to f$ in $W^{-1,2}(Q_\varepsilon)$ and denote by u^0 the solution to $A^0 u = f$. Therefore by Proposition 3, $u^\varepsilon \rightharpoonup u^0$ in $W_0^{1,2}(Q_\varepsilon)$.

Example 2. One commonly used definition of p comes from a probability distribution over $\mathbb{R}^2$. We split $\mathbb{R}^2$ into squares $\left\{ \left[-\frac{1}{2}, \frac{1}{2} \right] + j \mid j \in \mathbb{Z}^2 \right\}$, and consider a random variable κ defined over $\mathbb{R}^2$, which takes at each square the value $\delta > 0$ or 1 with probability r or $1 - r$ respectively, where $0 < r < 1$. For any $x \in \mathbb{R}^2$ and $z \in \Lambda$, the $p_z(x)$ are functions uniquely determined by the values $\{\kappa(x+z) \mid z \in \Lambda\}$. The independence of $\kappa(j)$ for different $j \in \mathbb{Z}^2$ implies that the transformation group allowing the construction of the families $\{p_z(x) \mid x \in \mathbb{R}^2\}$ is ergodic.

Example 3. The work [8] focus on the case where the functions p_z are defined by

$$p_z(x) := \begin{cases} \frac{2\kappa(x)\cdot\kappa(x+z)}{4(\kappa(x)+\kappa(x+z))} & \text{if } z \in \Lambda\backslash\{0\} \\ 1 - \sum_{z\neq 0} p_z(x) & \text{if } z = 0. \end{cases}$$

Finally, we introduce the possibility of operators of different scale. We recall that $A^\varepsilon u = \sum \partial_{-z}(p_z^\varepsilon \partial_z u)$ and define a modified problem $\tilde{A}^\varepsilon$. Consider a random function $c(x)$ defined over $\mathbb{R}^2$ taking values in $[1, 2] \subset \mathbb{R}$, and define

$$c^\varepsilon(x) := c(x/\varepsilon).$$

The modified problem is defined as

$$\tilde{A}^\varepsilon u(x) = \sum_{z\in\Lambda} \partial_{-z}(c^\varepsilon(x) \cdot p_z^\varepsilon(x)\partial_z u(x)),$$

thus allowing a stochastically defined scale at every point.

Theorem 1. *The Dirichlet problem $\tilde{A}^\varepsilon u = f^\varepsilon$ has a solution $\tilde{u}^\varepsilon$ for every $\varepsilon > 0$. Furthermore*

$$\tilde{u}^\varepsilon \rightharpoonup \ln(2) \cdot u^0 \quad \text{in } L^2(Q_\varepsilon)$$

for $\varepsilon \to 0$.

Proof. Observe that $\sum_{z\in\Lambda} c^\varepsilon p_z^\varepsilon = c^\varepsilon$, therefore we can re-write the problem as

$$c^\varepsilon(x) \cdot \left(\tilde{u}^\varepsilon(x) - \sum_{z\in\Lambda} p_z^\varepsilon(x)\tilde{u}^\varepsilon(x + \varepsilon z)\right) = \varepsilon^2 \cdot f^\varepsilon(x) \quad \forall x \in Q_\varepsilon,$$

with $\tilde{u}^\varepsilon(x) = 0$ for any $x \in \partial Q_\varepsilon^\Lambda$. Therefore the solution $\tilde{u}^\varepsilon$ exists unique for any $\varepsilon > 0$ as a consequence of Proposition 2, and in particular $\tilde{u}^\varepsilon = u^\varepsilon/c^\varepsilon$.

Moreover, by Proposition 3, u^0 weakly converges to u^0 in $W_0^{1,2}(Q_\varepsilon)$ and by definition $\frac{1}{c^\varepsilon}$ is bounded and weakly converges to the constant function $\ln(2)$ in $L^2(Q_\varepsilon)$. The thesis follows as $\tilde{u}^\varepsilon$ is obtained by taking their product. □

4   Numerical Results

We numerically solve Eq. (2.3) after approximation of the operators with finite differences (centered in space, forward in time), with images defined over a bounded domain $\Omega \subset \mathbb{R}^2$ and with Neumann boundary condition $\partial u/\partial n = 0$ on the boundary $\partial\Omega$ (Fig. 1). This is coherent with the theoretic considerations at the end of Sect. 2.2.

Typical time step of the evolution is $dt = 0.1$. As explained above, we consider stochastically defined orientation and scale, with scale taken in the interval $[1, 2] \subset \mathbb{R}$. The three bands of the RGB image stimulus shown in the figure have been processed separately. The forcing term of Eq. (2.3) is given by the differentiation of the RGB image with the mixture of operators, meaning that the differential operator considered may have different degree at every point.

Fig. 1. (Left) Original painting of Piero Della Francesca. (Center) Its second order differentiation with random horizontal or vertical orientation, meaning $\theta = 0$ or $\pi/2$. (Right) Reconstructed image using the steepest descent method. The slight change in color is due to the modulation effect of the cortical action; in fact the input is reconstructed up to a sub-harmonic function, $i.e.$ a function annulled by the operator $\widetilde{L}_{X_\theta}$.

Acknowledgements. Work supported by #NEXTGENERATIONEU (NGEU) and funded by the Ministry of University and Research (MUR), National Recovery and Resilience Plan (NRRP), project MNESYS (PE0000006) – A Multiscale integrated approach to the study of the nervous system in health and disease (DN. 1553 11.10.2022).

References

1. Ambrosetti, A., Malchiodi, A.: Nonlinear Analysis and Semilinear Elliptic Problems, vol. 104. Cambridge University Press, Cambridge (2007)
2. Bonhoeffer, T., Grinvald, A.: ISO-orientation domains in cat visual cortex are arranged in pinwheel-like patterns. Nature **353**(6343), 429–431 (1991)
3. Hubel, D.H., Wiesel, T.N.: Receptive fields of single neurones in the cat's striate cortex. J. Physiol. **148**(3), 574 (1959)
4. Kaschube, M.: Neural maps versus salt-and-pepper organization in visual cortex. Curr. Opin. Neurobiol. **24**, 95–102 (2014)
5. Kozlov, S.: Averaging of difference schemes. Math. USSR-Sbornik **57**(2), 351 (1987)
6. Littman, W., Stampacchia, G., Weinberger, H.F.: Regular points for elliptic equations with discontinuous coefficients. Annali della Scuola Normale Superiore di Pisa-Classe di Scienze **17**(1–2), 43–77 (1963)
7. Petitot, J.: The neurogeometry of pinwheels as a sub-riemannian contact structure. J. Physiol. Paris **97**, 265–309 (2003). https://doi.org/10.1016/j.jphysparis.2003.10.010
8. Piatnitski, A., Remy, E.: Homogenization of elliptic difference operators. SIAM J. Math. Anal. **33**(1), 53–83 (2001)
9. Ringach, D.L.: Spatial structure and symmetry of simple-cell receptive fields in macaque primary visual cortex. J. Neurophysiol. **88**(1), 455–463 (2002)
10. Sarti, A., Galeotti, M., Citti, G.: The cortical V1 transform as a heterogeneous poisson problem. arXiv preprint arXiv:2206.06895 (2022)

Gabor Frames and Contact Structures: Signal Encoding and Decoding in the Primary Visual Cortex

Vasiliki Liontou(✉) (iD)

University of Toronto, Toronto, ON M5T 3J1, Canada
vasiliki.liontou@mail.utoronto.ca

Abstract. This is an overview of an ongoing research project regarding the existence of Gabor frames on manifolds with contact structure, motivated by the need for a mathematical model of V_1 which allows the encoding and decoding of a signal by a discrete family of orientation and position dependent receptive profiles. Contact structures and Gabor functions have been used, independently, to model the activity of the mammalian primary visual cortex. Gabor functions are also used in signal analysis and in particular in signal encoding and decoding. A one-dimensional signal, an L^2 function of one variable, can be represented in two dimensions, with time and frequency as coordinates. The signal is expanded into a series of Gabor functions (an analog of a Fourier basis), which are constructed from a single seed function by applying time and frequency translations. This article summarizes the construction of a framework of signal analysis on unit cotangent bundles, determined by their natural contact structure, joint work with prof. M. Marcolli.

Keywords: Gabor Analysis · Contact Structures · Primary Visual Cortex

1 Motivation

A recurring question in mathematical modeling in neuroscience is whether it is possible to reconstruct visual stimuli by the activity of the visual cortex (V_1). Alternatively stated, could one use firing neurons of the visual cortex as elementary signals to represent and reconstruct the visual stimulus? Although there is a variety of methods in signal analysis that answer this question in $\mathbb{R}^n$, see [1] for example, they cannot be applied to describe the activity of the visual cortex which has a more complicated geometry. More specifically, the neurons of visual cortex are arranged in columns over the points of the retina they are connected to, forming a fiber bundle over $\mathbb{R}^2$. For instance, orientation sensitive neurons are arranged into a circle bundle over the retina, parameterized by the angle $\theta \in \mathbb{S}^1$ which they fire to. However, there is no good construction of filters for signal analysis on manifolds rather than on vector spaces, although partial results exist for splines discretization, diffusive wavelets, or special geometries

© The Author(s), under exclusive license to Springer Nature Switzerland AG 2023
F. Nielsen and F. Barbaresco (Eds.): GSI 2023, LNCS 14072, pp. 474–482, 2023.
https://doi.org/10.1007/978-3-031-38299-4_49

such as spheres and conformally flat manifolds, see for instance [3,7,15]. Thus, the need for the construction of a signal analysis framework on manifolds, and in particular on fiber bundles arises.

The answer to this problem is given by the functional architecture of V_1. Since the '80s, two seemingly unrelated mathematical models have been used to describe the functional architecture of V_1. The first model describes the connectivity between neurons of the visual cortex, organized in hypercolumns over the retina. This connectivity is dictated by the sensitivity of neurons in specific orientations and is mathematically modeled by a co-oriented contact structure, namely a distribution of tangent hyperplanes $\xi \subset TM$, defined as the kernel of a differential 1-form a which is maximally non-degenerate, meaning

$$a \wedge (da)^n \neq 0$$

or equivalently that the 2-form da is non-degenerate on ξ. Considering V_1 as a fiber bundle equipped with a contact structure describes both its hypercolumnar structure and the connections between neurons of different columns [9,14,17].

On the other hand, in order to reconstruct a visual signal out of the activity of the visual cortex, it is necessary to represent it using a discrete set of simple cells. As argued by Daugman in 1985 [6], simple cells of V_1 try localize at the same time the position (x, y) and the frequency of (η_1, η_2) of a signal detected on the retina. However, it is impossible to detect both position and frequency of a signal with arbitrary precision due to space-frequency uncertainty principle. To solve this problem of signal analysis, Dennis Gabor (1946) suggested a method to represent signals in space as a signal with spatial and frequency variables. In particular, Gabor proposed to expand a signal function $\mathcal{I} : \mathbb{R}^2 \to \mathbb{R}$ using space and frequency shifts of the same mother function, the Gaussian $g(x, y) = e^{-\pi |(x,y)|^2}$. The composition of countable space shifts

$$T_{(w_1,w_2)} g(x, y) = e^{-\pi |(x,y)-(w_1,w_2)|^2}$$

and frequency shifts

$$M_{(\eta_1,\eta_2)} g(x, y) = e^{-\pi |(x,y)|^2} e^{2\pi i (\eta_1 \cdot \eta_2) \cdot (x,y)}$$

creates a family of countable elementary functions

$$\{M_\eta T_w g : (w, \eta) \in \Lambda \text{ where } \Lambda = A\mathbb{Z}^4 \text{ and } A \in GL(4, \mathbb{R})\}$$

which can be used to represent the signal $\mathcal{I}$ as a series

$$\mathcal{I}(x, y) = \sum_{(w,\eta) \in \Lambda} C_{(w,\eta)} M_\eta T_w g(x, y) \tag{1}$$

where its coefficient $C_{(w,\eta)}$ represents a quantum of information, associated to an area of the space-frequency plane covered by $M_\eta T_w g$. The mother function g can be replaced by another function $f \in \mathcal{L}^2(\mathbb{R}^2)$, however the choice of $g(x, y) = e^{-\pi |(x,y)|^2}$ is not arbitrary. The function g and its space-frequency shifts minimize the uncertainty principle. Hence, choosing Gabor functions to model receptive profiles of simple cortical cells describes the optimal efficiency with which simple cells process spatiotemporal information, since Gabor functions minimize the uncertainty principle [5,6,12]. Finally, the series (1) does not necessarily converge. It converges when the family of elementary Gabor functions $\{M_\eta T_w g : (w, \eta) \in \Lambda\}$ satisfies a "weak Parseval's identity"

$$A||\mathcal{I}||^2 \leq \sum_{(w,\eta)\in\Lambda} | < \mathcal{I}, M_\eta T_w g > |^2 \leq B||\mathcal{I}||^2 \tag{2}$$

for some $A, B > 0$, which is called **the frame condition**. This condition provides a lot more freedom than the convergence requirement for Fourier series does, making Gabor representations a more suitable tool for biological models.

The same neurons in the visual cortex appear as Gabor elementary functions $M_w T_\eta g$ and as points of a circle bundle equipped with a co-orientable contact structure. Recent work of Petitot and Tondut [14], Citti and Sarti [16] and Sarti, Citti, Petitot [17] has shown that the simple cell's profile shape as Gabor elementary functions and their hypercolumnar arrangement together with their connectivity with respect to orientation sensitivity are governed by the same action or the roto-translation group $SE(2)$, as well as by a principle of selectivity of maximal response. Thus, the co-existence of both structures is not coincidental. Therefore, the signal analysis properties of V_1 are determined by its geometric structure as a contact fiber bundle. This observation leads to the question: can we use the visual cortex as motivation to examine which classes of contact manifolds carry a Gabor analysis framework determined by their geometry? In the following paragraphs a signal analysis framework for unit-cotangent bundles of Riemannian manifolds (B, g) is presented. In paragraph Sect. 2 such a framework for the 3-dim fiber bundle underlying the model of V_1 [11] is summarized. In paragraph Sect. 3 the main points of the generalization of this framework on unit cotangent bundles of arbitrary dimension are listed and the boundary detection property of the filters in use is highlighted.

2   Gabor Frames from Contact Geometry in Models of the Primary Visual Cortex

The contact 3-manifold underlying the model of V_1 as presented in [14,17] by A. Sarti, G. Citti and J. Petitot is of the form $M = \mathbb{S}(T^*\mathbb{R}^2)$, namely the unit cotangent bundle of $\mathbb{R}^2$ with respect to the Euclidean Riemannian metric or equivalently the manifold of co-oriented contact elements of $\mathbb{R}^2$. The 3-dimensional manifold M carries a contact structure, given by the "skating condition": *a tangent vector $X \in T_m M$ at a contact element $m = (b, p)$ belongs in the contact hyperplane $\xi \subset TM$ if the tangent vector $\pi_* X$ to $b \in \mathbb{R}^2$ belongs to the contact element p*, where π is the projection $\pi : \mathbb{S}(T^*\mathbb{R}^2) \to \mathbb{R}^2$. The space of 1-jets of curves on $\mathbb{R}^2$, $J^1(\mathbb{R}, \mathbb{R})$, parametrizes all the non-vertical contact elements of $\mathbb{R}^2$, hence it serves as a chart for M, with coordinates (x, y, θ), namely the contact element at the point (x, y) with slope $tan(\theta) \neq \infty$ is represented by (x, y, θ). The contact structure $\xi \subset TM$ is (non-canonically) described as the kernel of a 1-form a written in local coordinates coordinates as $a = cos(\theta)dx + sin(\theta)dy$, which is exactly the contact structure described in [17]. Using the direction of the maximal response of the simple cells $R_a := cos(\theta)\partial_x + sin(\theta)\partial_y$, which is the direction of the level sets of the signal $I : \mathbb{R}^2 \to \mathbb{R}$, and the direction of the gradient ∇I which is $R_{a_J} := -sin(\theta)\partial_x + cos(\theta)\partial_y$ we can construct a framework of signal encoding and decoding of I. In [11], we construct this framework in a more general setting where the underlying manifold M, is the manifold of oriented contact elements $M = \mathbb{S}(T^*B)$ of a Riemann surface B, equipped with a Riemannian metric g.

Signals on M: When taking into account that the base manifold B, where the signal is detected, is non-flat, one needs to distinguish between the local variables (x, y) on a chart $(U, z = x + iy)$ on the surface B (or the local variables (x, y, θ) on the 3-manifold M) and the linear variables in its tangent space $T_{(x,y)}B$. Thus, we think of the retinal signal as a collection of compatible signals in the planes $T_{(x,y)}B$, as (x, y) varies in B. The retinal signal depends on the position alone, represented by the points (x, y) in local charts of the retina. However the signal analysis is performed on the cortical signal which is obtained from the 2-dimensional retinal signal via the action of receptive profiles. We consider a real 2-plane bundle on the 3-manifold M that describes this geometric space of cortical signals.

Definition 1. Let $\mathcal{E}$ be the real 2-plane bundle on the contact 3-manifold $M = \mathbb{S}(T^*B)$ obtained by pulling back the tangent bundle TB of the surface B to M along the projection $\pi : \mathbb{S}(T^*B) \to B$ of the unit sphere bundle of T^*B,

$$\mathcal{E} = \pi^*TB. \tag{3}$$

At each point $(x, y, \theta) \in M$, with $z = x + iy$ the coordinate in a local chart (U, z) of B, the fiber $\mathcal{E}_{(x,y,\theta)}$ is the same as the fiber of the tangent bundle $T_{(x,y)}B$. Also let $\mathcal{E}^\vee$ be the dual bundle of $\mathcal{E}$, namely the bundle of linear functionals on $\mathcal{E}$,

$$\mathcal{E}^\vee = Hom(\mathcal{E}, \mathbb{R}).$$

Locally the exponential map $exp_g : TB \to B$, induced by the metric g, from TB to B allows for a comparison between the description of signals in terms of the linear variables of TB and the nonlinear variables of B. The linear variables of TB are the ones to which the Gabor filter analysis applies. Thus, in terms of the contact 3-manifold M, we think of cortical signal as a consistent family of L^2 functions on the fibers $\mathcal{E}_{(x,y,\theta)}$, or equivalently a signal on the total space of the 2-plane bundle $\mathcal{E}$

$$\left(\int_M \int_{\mathcal{E}_{(x,y,\theta)}} \mathcal{I}^2_{(x,y,\theta)}(V) \, dvol_{\mathcal{E}_{(x,y,\theta)}}(V) \, dvol_M(x, y, \theta) \right)^{1/2} < \infty, \tag{4}$$

The volume form $dvol_{\mathcal{E}_{(x,y,\theta)}}$ is the pullback of the volume form on $T_{(x,y)}B$, while the volume form $dvol_M$ is the restriction of the volume form of T^*B on M. When

B is compact, for every function $f \in L^\infty(B)$ with respect to the volume form of the Riemannian metric g, there exists a signal $\mathcal{I}(f) \in \mathcal{E} \to \mathbb{R})$ such that f is fully determined by $\mathcal{I}(f)$. The lift $\mathcal{I}(f)$ of the retinal signal f is obtained by precomposition with the exponential map $exp : TB \to B$ and the bundle map $h : \mathcal{E} \to TB$

$$\mathcal{I}(f) = f \circ exp \circ h.$$

Note that the lift depends only on the spatial variables, since, on the one hand the exponential map depends only on the spatial variables and on the other hand the pull-back of TS with respect to the projection $\pi : \mathbb{S}(T^*B) \to B$ is constant along the fibers.

The Bundle of Filters. The construction of Gabor filters we consider here follows closely the model of [17], reformulated in a way that more explicitly reflects the underlying contact geometry. We first show how to obtain the mother function (window function) of the Gabor system and then we will construct the lattice that generates the system of Gabor filters. These functions will have to minimize the uncertainty principle on the time-frequency domain. When the signal analysis takes place on a manifold, the time-frequency domain is represented by the Whitney sum $TB \oplus T^*B$. Let V and η denote, respectively, the linear variables in the fibers $V \in T_{(x,y)}B \simeq \mathbb{R}^2$, $\eta \in T^*_{(x,y)}B \simeq \mathbb{R}^2$, with $\langle \eta, V \rangle_{(x,y)}$ the duality pairing of $T^*_{(x,y)}B$ and $T_{(x,y)}B$. Each tangent space $T_{(x,y)}B$ represents the spatial variables and each cotangent space $T^*_{(x,y)}B$ the spacial frequency variables.

Definition 2. *A window function on the bundle $TB \oplus T^*B$ over B is a smooth real-valued function Φ_0 defined on the total space of $TB \oplus T^*B$, of the form*

$$\Phi_{(x,y)}(V, \eta) := \exp\left(-V^t A_{(x,y)} V - i\langle \eta, V \rangle_{(x,y)}\right), \tag{5}$$

where A is a smooth section of $T^*B \otimes T^*B$ that is symmetric and positive definite as a quadratic form on the fibers of TB, with the property that at all points (x, y) in each local chart U in S the matrix $A_{(x,y)}$ has eigenvalues uniformly bounded away from zero, $Spec(A_{(x,y)}) \subset [\lambda, \infty)$ for some $\lambda > 0$.

By restricting Φ to the fiber bundle $TB \times \mathbb{S}(T^*B) \subset TB \oplus T^*B$ and expressing $\eta \in T^*B$ in polar coordinates over a local chart of B, $\eta_\theta = (cos(\theta), sin(\theta)), \theta \in \mathbb{S}^1$ we obtain a mother function on $TB \times \mathbb{S}(T^*B)$ of the form

$$\Psi_{(x,y)}(V, \eta_\theta) := \exp\left(-V^t A_{(x,y)} V - i\langle \eta_\theta, V \rangle_{(x,y)}\right). \tag{6}$$

By identifying the total spaces of $TB \times \mathbb{S}(T^*B)$ and $\mathcal{E} = \pi^*(TB)$, (7) is a Gabor Function on $\mathcal{E}$

$$\Psi_{(x,y,\theta)}(V) := \exp\left(-V^t A_{(x,y)} V - i\langle \eta_\theta, V \rangle_{(x,y)}\right), \tag{7}$$

namely a family of functions parametrized by $\mathbb{S}(T^*B)$, which are rapidly decaying along the fibers $\mathcal{E}_{(x,y,\theta)} \simeq T_{(x,y)}B$ of $\mathcal{E}$.

The Bundle of Lattices. The contact 1-form a of $M = \mathbb{S}(T^*B)$ together with the complex structure $J \in \Gamma(B, TB \otimes T^*B), J^2 = -I$, of the Riemann surface B determine a canonical choice of a basis $\{R_\alpha, R_{\alpha_J}\}$ for the bundle $\mathcal{E}$ and its dual basis $\{\alpha, \alpha_J\}$ for $\mathcal{E}^\vee$. The bundle is formed as follows. The complex structure, written in local coordinates $(U, x + iy)$,

$$J = -1dy \otimes \partial_x + 1dx \otimes \partial_y,$$

induces a "twisted" tautological 1-form on T^*B, $\lambda_{J((x,y),p)} = \pi^*(Jp)$ where $\pi : T^*B \to B$ and $p \in T^*_{(x,y)}B$. The tautological 1-form $\lambda_{((x,y),p)} = \pi^*(p)$ restricted on $\mathbb{S}(T^*B)$ gives the contact 1-form

$$a = cos(\theta)dx + sin(\theta)dy$$

expressed in local coordinates over a local chart $(V, (x, y, \theta))$ of $\mathbb{S}(T^*B)$. On the other hand, the twisted tautological 1-form restricted to $\mathbb{S}(T^*B)$ gives a different contact 1-form,

$$a_J = -sin(\theta)dx + cos(\theta)dy,$$

expressed in local coordinates over the same local chart. The fibers

$$\{\pi^{-1}_{|_{\mathbb{S}(T^*B)}}(b), b \in B\} \simeq \mathbb{S}^1$$

are Legendrian, regular submanifolds of $\mathbb{S}(T^*B)$ with respect to both a and a_J and R_a is Legendrian and their Reeb vector fields

$$R_a = cos(\theta)\partial_x + sin(\theta)\partial_y \text{ and } R_{a_J} = sin(\theta)\partial_x + cos(\theta)\partial_y$$

are Legendrian with reprect to a_J and a respectively, therefore R_a and R_{a_J} determine a basis on the fibers of $\mathcal{E}$.

We denote by $\{R_\alpha^\vee, R_{\alpha_J}^\vee\}$ the dual basis of $\mathcal{E}^\vee$ (over the same chart U of B) characterized by $\langle R_\alpha^\vee, R_\alpha \rangle = 1, \langle R_\alpha^\vee, R_{\alpha_J} \rangle = 0, \langle R_{\alpha_J}^\vee, R_\alpha \rangle = 0, \langle R_{\alpha_J}^\vee, R_{\alpha_J} \rangle = 1$. By the properties of Reeb and Legendrian vector fields, we can identify the dual basis with the contact forms, $\{R_\alpha^\vee, R_{\alpha_J}^\vee\} = \{\alpha, \alpha_J\}$. This determines bundles of framed lattices (lattices with an assigned basis) over a local chart in M of the form

$$\Lambda_{\alpha,J} := R_\alpha + R_{\alpha_J} \text{ and } \Lambda_{\alpha,J}^\vee := \alpha + \alpha_J. \tag{8}$$

where $\Lambda_{\alpha,J}$ and $\Lambda_{\alpha,J}^\vee$ here can be regarded as a consistent choice of a lattice $\Lambda_{\alpha,J,(x,y,\theta)}$ (respectively, $\Lambda_{\alpha,J,(x,y,\theta)}^\vee$) in each fiber of $\mathcal{E}$ (respectively, of $\mathcal{E}^\vee$). The bundle of framed lattices

$$\Lambda := \Lambda_{\alpha,J} \oplus \Lambda_{\alpha,J}^\vee \tag{9}$$

correspondingly consists of a lattice in each fiber of the bundle $\mathcal{E} \oplus \mathcal{E}^\vee$ over M.

The choice of the window function Ψ_0, together with the lattice (9), determine a Gabor system $\mathcal{G}(\Psi_0, \Lambda_{\alpha,J} \oplus \Lambda_{\alpha,J}^\vee)$ which consists, at each point $(x, y, \theta) \in M$ of the Gabor system

$$\mathcal{G}(\Psi_{0,(x,y,\theta)}, \Lambda_{(x,y,\theta)}) = \{M_\xi T_W \Psi_{(x,y,\theta)} : (\xi, W) \in \Lambda_{\alpha,J,(x,y,\theta)} \oplus \Lambda_{\alpha,J,(x,y,\theta)}^\vee\} \tag{10}$$

in the space $L^2(\mathcal{E}_{(x,y,\theta)})$.

Frame Condition and Lattice Truncation: For a scaled lattice of the form

$$\Lambda^{b,c} := b\Lambda_{a,J} + c\Lambda_{a,J}^{\vee} \tag{11}$$

where b, c are $\mathbb{R}^+$-valued functions on M, we say that the Gabor system $\mathcal{G}(\Psi_0, \Lambda)$ satisfies the smooth Gabor frame condition on M if there are smooth functions $\mathbb{R}_+^*$-valued functions C, C' on the local charts of M, such that the frame condition holds pointwise in (x, y, θ),

$$C_{(x,y,\theta)} \|f\|^2_{L^2(\mathcal{E}_{(x,y,\theta)})} \leq \sum_{(W,\xi) \in \Lambda^{b,c}_{(x,y,\theta)}} |\langle f, M_\xi T_W \Psi_0 \rangle|^2 \leq C'_{(x,y,\theta)} \|f\|^2_{L^2(\mathcal{E}_{(x,y,\theta)})} . \tag{12}$$

The scaling of the lattice by b and c affects its density and determines whether the frame condition is satisfied. For Gabor Systems with Gaussian window where the lattice is a subset of $\mathbb{R}^2$ the condition $bc < 1$ is both necessary and sufficient for the frame condition to be satisfied [10,18]. For Gabor Systems in higher dimensions, such as Gabor Systems (10), much less is known. However, the geometry of (B, g) being a Riemannian manifold, as well as the fact that in reality only a finite amount of receptive profiles contribute in the analysis of retinal signals, suggest a lattice scaling for $\mathcal{G}(\Psi_0, \Lambda_{a,J} \oplus \Lambda_{a,J}^{\vee})$. At a given point (x, y) in a compact manifold B, we consider the supremum R_{ing} of all radii $R > 0$ such that $exp^g_{(x,y)} : B(0, R) \subset T_{(x,y)}B \to B$ is a diffeomorphism of $B(0, R)$. The signals which are detected inside $exp_{(x,y)}(B(0, R_{inj}))$ can be lifted through the diffeomorphism $exp^{-1}_{|B(0,R_{inj})}$ on the fibers of TB and therefore on the fibers of $\mathcal{E}$ where the signal analysis takes place. If R_{inj} is divided by the maximal size R_{max} of the neurons that participate in the analysis, we obtain a function

$$b_M(x, y, \theta) = \frac{R_{inj}(x, y)}{R_{max}} : M \to (0, 1) \tag{13}$$

Theorem 1. *With the scaling by the function $b = b_M(x, y, \theta)$ of (13), the Gabor system $\mathcal{G}(\Psi_0, (b\Lambda_{\alpha,J}) \oplus \Lambda_{\alpha,J}^{\vee})$ satisfies the frame condition.*

Scaling the part of the lattice $\Lambda_{\alpha,J}$ that lies on the fibers of $\mathcal{E}$, ensures that the receptive profiles which participate in the analysis lie inside the domain where the exponential map $exp_{(x,y)} : T_{(x,y)}B \to B$ is a diffeomorphism.

3   Gabor Frames of Higher Dimensions and Boundary Detection

We present here the main points of a generalization of the construction in [11] to the manifold of contact elements $\mathbb{S}(T^*B)$ of a Riemannian manifold (B, g) of arbitrary dimension n, which is a work in progress. The key observation of [11] is that it is important to maintain the distinction between the coordinates of the curved manifolds B and $M = \mathbb{S}(T^*B)$ and the linear coordinates in the fibers of the bundles TB and T^*B. Making this distinction precise geometrically requires introducing a suitable *bundle of signal spaces*, namely the pullback bundle $\mathcal{E} = \pi^*(TB)$ of the tangent bundle TB through the projection $\pi : \mathbb{S}(T^*B) \to B$. This generalizes the bundle of signal planes introduced in [11]. When B is compact, the function $f \in L^\infty(B)$ with respect to the volume form of the Riemannian metric g determines a signal $\mathcal{I}(f) \in L^2(\mathcal{E})$ such that f can be recovered by $\mathcal{I}(f)$. Passing from $f : B \to \mathbb{R}$ to $\mathcal{I}(f) : \mathcal{E} \to \mathbb{R}$ corresponds to replacing a signal detected on B with a consistent collection of signals on its local linearizations where the analysis is possible.

To construct a filter in this context one can follow the same process as in dimension 3, (7), without any problems. Hence the filter is a smooth function $\Psi : \mathcal{E} \to \mathbb{R}$ from the total space of the bundle $\mathcal{E}$ of signal spaces,

$$\Psi_{(b,p)}(V) = exp\big(-V^t A_b V - i\langle \eta_p, V \rangle_b\big), \tag{14}$$

where η_p is just the point $p \in \mathbb{S}^{n-1} \simeq \mathbb{S}(T_b^*B)$ seen as a cotangent vector.

These filters are especially suitable to detect $(n-1)$-dimensional boundaries in a signal $f : B \to \mathbb{R}$ (lifted to a signal $\mathcal{I}(f) : \mathcal{E} \to \mathbb{R}$). The restriction of the window function Ψ to $\mathcal{E}_m$ is a rapid decay function, for $m = (b,p)$, where the unit cotangent vector η_p parameterizes a choice of an oriented hyperplane in $\mathcal{E}_m \simeq \mathbb{R}^n$. Consider a signal $f : B \to \mathbb{R}$ that is a characteristic function $f = \chi_U$ of a bounded open set $U \subset B$ with smooth boundary $\Sigma = \partial U$ given by an $(n-1)$-dimensional smooth hypersurface Σ in B. Let $\mathcal{I}(f)_{\mathcal{E}_m} : \mathcal{E}_m \to \mathbb{R}$ denote the lifted signals on the fibers of the bundle $\mathcal{E}$ of signal spaces. As in [17], the output function is

$$\mathcal{O}_b(f, \eta_p) := \int_{\mathcal{E}_{(b,p)}} \mathcal{I}(f)|_{\mathcal{E}_{(b,p)}}(V) \cdot \Psi_{(b,p)}(V) \; dvol_{\mathcal{E}_{(b,p)}}(V). \tag{15}$$

The output function $\mathcal{O}_b(\eta_p)$ has a local maximum for $p \in \mathbb{S}^{n-1}$ the normal vector $\nu_b(\Sigma)$ at b to the boundary hypersurface $\Sigma = \partial U$. Finally, to construct a lattice bundle $\Lambda \subset \mathcal{E} \oplus \mathcal{E}^\vee$, we consider the action of $SO(n)$ on $\mathbb{S}^{n-1} \simeq \pi^{-1}(b)$ to generate an n-tuple of contact forms, on a local trivialization chart of $\mathbb{S}(T^*B)$, obtained by the standard contact form a of the unit co-tangent bundle.

The bundle of signal planes $\mathcal{E}$, the window function (14) and the lattice constructed by the standard contact form on $\mathbb{S}(T^*B)$ provide a construction of Gabor frames that encodes local linearizations $\mathcal{I}(f)$ of signals f with Gabor filters that are adapted to the detection of smooth hypersurfaces $\Sigma \subset B$.

References

1. Barbieri, D.: Reconstructing group wavelet transform from feature maps with a reproducing kernel iteration. Front. Comput. Neurosci. **12** (2022)
2. Baspinar, E., Sarti, A., Citti, G.: A sub-Riemannian model of the visual cortex with frequency and phase. J. Math. Neurosci. **10**(11) (2020)

3. Bernstein, S., Keydel, P.: Orthogonal wavelet frames on manifolds based on conformal mappings, frames and other bases in abstract and function spaces. Appl. Numer. Harmon. Anal. **1**, 303–332 (2017)

4. Boscain, U., Prandi, D., Sacchelli, L., et al.: A bio-inspired geometric model for sound reconstruction. J. Math. Neurosc. **11**(2) (2021)

5. Daugman, J.G.: Two-dimensional spectral analysis of cortical receptive field profiles. Vis. Res. **20**(10), 847–856 (1980)

6. Daugman, J.G.: Uncertainty relation for resolution in space, spatial frequency, and orientation optimized by two-dimensional visual cortical filters. J. Opt. Soc. Am. A **2**, 1160–1169 (1985)

7. Ebert, S., Wirth, J.: Diffusive wavelets on groups and homogeneous spaces. Proc. R. Soc. Edinb. A Math. **141**(3), 497–520 (2011)

8. Feichtinger, H., Gröchenig, K.: Gabor wavelets and the heisenberg group: gabor expansions and short time fourier transform from the group theoretical point of view. Wavelet Anal. Appl. **2**, 359–398 (1992)

9. Hoffman, W.C.: The visual cortex is a contact bundle. Appl. Math. Comput. **32**(2–3), 137–167 (1989)

10. Lyubarskiĭ, Y.I.: Frames in the Bargmann space of entire functions. Entire and subharmonic functions. Adv. Soviet Math. **11**, 167–180 (1992)

11. Liontou, V., Marcolli, M., Gabor frames from contact geometry in models of the primary visual cortex, arXiv:2111.02307 (2022)

12. Marcelja, S.: Mathematical descriptions of the responses of simple cortical cells. J. Optical Soc. Am. **70**, 297–300 (1980)

13. Petitot, J.: Neurogéométrie de la vision, Les Éditions de l'École Polytechnique (2008)

14. Petitot, J., Tondut, Y.: Vers une neurogéométrie. fibrations corticales, structures de contact et contours subjectifs modaux. Mathématiques informatique et sciences humaines **145**, 5–102 (1999)

15. Pesenson, I.: Variational splines on Riemannian manifolds with applications to integral geometry. Adv. Appl. Math. **33**(3), 548–572 (2004)

16. Sarti, A., Citti, G.: A cortical based model of perceptual completion in the roto-translation space. J. Math. Imaging Vis. **24**(3), 307–326 (2006)

17. Sarti, A., Citti, G., Petitot, J.: Functional geometry of the horizontal connectivity in the primary visual cortex. J. Physiol.-Paris **103**(1–2), 37–45 (2009)

18. Seip, K.: Density theorems for sampling and interpolation in the Bargmann-Fock space I. J. Reine Angew. Math. **429**, 91–106 (1992)

A Sub-Riemannian Model
of the Functional Architecture of M1
for Arm Movement Direction

Caterina Mazzetti[1,2]([✉]), Alessandro Sarti[2], and Giovanna Citti[1,2]

[1] Department of Mathematics, University of Bologna, Bologna, Italy
{caterina.mazzetti2,giovanna.citti}@unibo.it
[2] Centre d'Analyse et de Mathématique Sociales, Sorbonne Université, Paris, France
alessandro.sarti@ehess.fr

Abstract. In this paper we propose a neurogeometrical model of the behaviour of cells of the arm area of the primary motor cortex (M1). We mathematically express the hypercolumnar organization of M1 discovered by Georgopoulos, as a fiber bundle, as in classical sub-riemannian models of the visual cortex (Hoffmann, Petitot, Citti-Sarti). On this structure, we consider the selective tuning of M1 neurons of kinematic variables of positions and directions of movement. We then extend this model to encode the notion of fragments of movements introduced by Hatsopoulos. In our approach fragments are modelled as integral curves of vector fields in a suitable sub-Riemannian space. These fragments are in good agreements with movement decomposition from neural activity data. Here, we recover these patterns through a spectral clustering algorithm in the subriemannian structure we introduced, and compare our results with the neurophysiological ones of Kadmon-Harpaz et al.

Keywords: Primary motor cortex · Movement decomposition · Neurogeometry · Sub-Riemannian geometry

1 Introduction

A fundamental problem regarding the study of motor cortex deals with the information conveyed by the discharge pattern of motor cortical cells. This is a quite difficult topic as the input of primary motor area comes from higher brain cortical regions, whereas the output is movement.

Starting from 1978, a pioneering work for the study of primary motor cortex (M1) was developed by A. Georgopoulos, whose experiments allow to recognize many important features of the arm area functional architecture. In particular, he discovered that cells of this area are sensible to the position and direction of the hand movement [4,6], and are organized in a columnar structure, according to movement directions [5]. After the work of Georgopoulos, other experiments proved that activity of neurons in M1 correlates with a broader variety of movement-related variables, including endpoint position, velocity, acceleration

© The Author(s), under exclusive license to Springer Nature Switzerland AG 2023
F. Nielsen and F. Barbaresco (Eds.): GSI 2023, LNCS 14072, pp. 483–492, 2023.
https://doi.org/10.1007/978-3-031-38299-4_50

(see [17] as a review). This phenomenon, known as "cortical tuning", describes the selective responsiveness of M1 neurons to specific movement features. In other words, M1 neurons become active or "tuned" based on specific movement characteristics. Furthermore, the tuning for movement parameters is not static, but varies with time ([1,15]) and for this reason Hatsopoulos et al. [8] proved that individual motor cortical cells rather encode short movement trajectories, called "movement fragments" (see Fig. 2(a)). Comparable findings were obtained by Kadmon-Harpaz et al. [10] in 2019 who examined the temporal dynamics of neural populations in the primary motor cortex of macaque monkeys performing forelimb reaching movements. Using a hidden Markov model, they found a structure of hidden states in the population activity of neurons in M1, which organizes the behavioural output, in acceleration and deceleration trajectory segments with fixed directional selectivity (see Fig. 1). The data analysis was performed at the neural level, and the authors posed the problem to recover the same decomposition by using only kinematic variables.

Aim of this paper is to answer to this problem, extending a result obtained in [12]. We present a neurogeometrical model inspired by the functional architecture of the arm area of motor cortex referred to a set of cortical tuning parameters in response to point-to-point reaching movements. We modelled the time dependent selectivity of each neuron, through integral curves of a suitable sub-Riemannian vector fields in the space of kinematic variables. The same organization in elementary trajectories can be obtained with a kernel component analysis associated to the sub-Riemannian distance.

The main novelty with respect to [12] is that we apply a second clustering in the space of elementary trajectories and we obtain, using only phenomenological variables, the same neural PCAs provided by Karpaz-Harpaz et al. [10]. In this second step the grouping is carried out in the space of curves (movement fragments), each one identified by its mean orientation and acceleration.

The whole process provides detailed information on which kinematic variables are responsible for the neural process, but also gradually leads to a shift from a space mainly described by kinematic points to a space of movement trajectories.

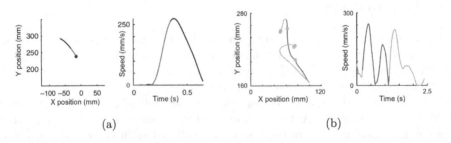

(a)   (b)

Fig. 1. Examples of a center-out task and of a random target pursuit task, with position and speed profile colored according to the identified neural states. Black dot represents the starting position. From [10].

2   A Sub-riemannian Model of M1 Cells Encoding Movement Direction

We aim at realizing a unified neurogeometrical framework that contains both the geometrical findings of Georgopoulos regarding direction of movement and the time dependent model of [8]. The space variables will be the cortical features of time, position, direction of movement, speed and acceleration, the constraint between them will be described via a sub-Riemannian metric and time varying selective behaviour will be represented through integral curves of suitable vector fields.

2.1   A 2D Kinematic Tuning Model of Movement Directions

We first consider that the basic functional properties of cellular activity in the arm area of M1 involve directional and positional tuning, as described in [4,6]. Hence we introduce a variable (x, y) which accounts for hand's position in a two dimensional space, and a variable $\theta \in S^1$, which encodes hand's movement direction on the plane (x, y). The relation between these variables is expressed by the relation $\frac{dy}{dx} = \tan(\theta)$, or equivalently by the vanishing of the 1-form

$$\omega_1 = -\sin\theta \, dx + \cos\theta \, dy = 0. \tag{1}$$

It is worthwhile to note that this first constraint is inspired by the models of visual cortex (see [2,16]).

We also consider that the cortex codes other movement-related variables, including velocity, acceleration (see [17]). Hence we introduce the time variable t, and the variables v and a which represent hand's speed and acceleration along the direction θ.

Recall that Georgopoulos [5] also provided a physiological model of hypercolumnar organization for the cellular arrangement in M1. On the other hand, the hypercolumnar structure is also present in V1 where it has been modeled as a fiber bundle by Hoffmann [9] and Petitot and Tondut [16]. This suggests to use a fiber bundle representation also as a model of M1. In our model, the triple $(t, x, y) \in \mathbb{R}^3$ describes the position of the hand at time t, and it is assumed to belong to the base space of a fiber bundle structure, whereas the variables $(\theta, v, a) \in S^1 \times \mathbb{R}^2$ form the selected features on the fiber over the point (t, x, y) (see [11] for the definition of fiber bundle). We therefore consider the 6D features set

$$\mathcal{M} = \mathbb{R}^3_{(t,x,y)} \times S^1_\theta \times \mathbb{R}^2_{(v,a)}, \tag{2}$$

Let us specify that in V1, the fiber bundle is compatible with the hypercolumnar organization and its correspondence with the retinal plane. Similarly, in M1, the fiber bundle captures the topographic organization resulting from competing mappings related to somatotopy, hand location, and movement organization [7]. Although there are differences in receptive profiles (simple cells in V1) versus "actuator profiles" (cells in M1), both regions evaluate the alignment between preferred features and external input variables.

The choice of the space variables (2) with their differential constraints induce the vanishing of the following 1-forms

$$\omega_2 = \cos\theta\,\mathrm{d}x + \sin\theta\,\mathrm{d}y - v\,\mathrm{d}t = 0, \quad \omega_3 = \mathrm{d}v - a\,\mathrm{d}t = 0. \tag{3}$$

The one-form ω_2 encodes the direction of velocity over time: the unitary vector $(\cos\theta, \sin\theta)$ is the vector in the direction of velocity, and its product with $(\dot{x}, \dot{y})$ yields the speed. We call horizontal distribution $D^{\mathcal{M}}$ the kernel of all three forms, which is the set of vector fields orthogonal to ω_i, $i = 1, ...3$. It turns out to be spanned by the vector fields

$$X_1 = v\cos\theta\frac{\partial}{\partial x} + v\sin\theta\frac{\partial}{\partial y} + a\frac{\partial}{\partial v} + \frac{\partial}{\partial t}, \quad X_2 = \frac{\partial}{\partial\theta}, \quad X_3 = \frac{\partial}{\partial a}. \tag{4}$$

We call horizontal curve an integral curve of these vector fields:

$$\begin{cases} \dot{\gamma}(t) = X_1(\gamma(t)) + \dot{\theta}(t)X_2(\gamma(t)) + \dot{a}(t)X_3(\gamma(t)) \\ \gamma(0) = \eta_0 \in \mathcal{M}. \end{cases} \tag{5}$$

The functions $t \mapsto \dot{\theta}(t)$ and $t \mapsto \dot{a}(t)$ represent, respectively, the rate of change of the selective tuning to movement direction and acceleration variables. We identified each movement fragment detected in [8] as local integral curves of the space $\mathcal{M}$ (Fig. 2(a) and (b)). The whole space of cortical neurons is no more modelled as a set of points, but a set of trajectories, solution of (5) (Fig. 2(c)).

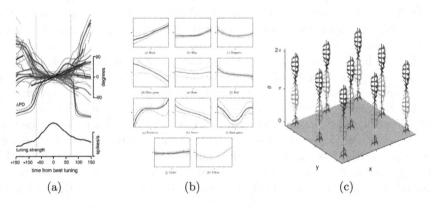

$$(a) \qquad\qquad (b) \qquad\qquad (c)$$

Fig. 2. (a) Temporal evolution of the selective responsiveness to movement direction (direction tuning) in twelve neurons of M1. Below is shown the mean strength of direction tuning, where time 0 is assumed to be the time of strongest tuning. From [8]. (b) Best fit with an integral curve of the space (in blue) of each measured time-dependent direction tuning curve, represented with the same color as in [8]. (c) The cortical space is described as a space of integral curves of system (5). (Color figure online)

3   Spatio-Temporal Grouping Model for M1

We will now recover the coherent behaviors of neural activity obtained in [10] in terms of kinematic parameters. The authors noted that the desired neural decomposition can not be obtained by none of the distances previously proposed in literature. Here we show that a distance that takes into account the differential relations between the variables can provide the correct decomposition. The algorithm we will use to provide the classification is a variant of k-means that considers this distance, which proves that the set of kinematic variables considered is sufficient to recover the cortical decomposition.

3.1   Homogeneous Distance on $\mathcal{M}$

Let us now introduce a natural distance associated to the vector fields (4), with the metric which makes them orthonormal. In this way we endow $\mathcal{M}$ with a sub-Riemannian structure. Note that we only have chosen 3 vector fields at every point in a 6D space, and we will be able to obtain a basis of the space considering their commutators. We will also assign a degree, which is the number of commutators we need to obtain a vector field. Precisely

$$X_1,\ X_2,\ X_3 \in D^{\mathcal{M}} \text{ so that } \deg(X_1) = \deg(X_2) = \deg(X_3) = 1.$$

$$X_4 = [X_1, X_2],\ X_5 = [X_3, X_1],\ \text{so that } \deg(X_4) = \deg(X_5) = 2$$

$$X_6 = [X_5, X_1] = [[X_3, X_1], X_1],\ \text{so that } \deg(X_4) = 3.$$

In particular the vector fields satisfy the Hörmander condition, and a distance $d^{\mathcal{M}}$ (called Carnot-Carathéodory distance) can be defined as follows in the whole cortical feature space $\mathcal{M}$. For all $\eta_0, \eta_1 \in \mathcal{M}$

$$d^{\mathcal{M}}(\eta_0, \eta_1) = \inf \{l(\gamma) : \gamma \text{ is a horizontal curve connecting } \eta_0 \text{ and } \eta_1\}, \quad (6)$$

where the notion of horizontal curve has been introduced in (5). The path that realizes (6) is called geodesic. Geodesics in this space are related to a model of arm-reaching movements proposed by Flash and Hogan [3] (see [13] for further details).

It is possible to provide a local estimate of the distance $d^{\mathcal{M}}$ using an approximation result due to Nagel Stein Wainger [14].

Remark 1. We fix a point $\eta_0 = (x_0, y_0, \theta_0, v_0, a_0, t_0)$ and call canonical coordinates of any other point $\eta_1 = (x_1, y_1, \theta_1, v_1, a_1, t_1)$ the constants e_i which solve the system

$$\begin{cases} \dot{\gamma}(s) = e_1 X_1 + e_2 X_2 + e_3 X_3 + e_4 X_4 + e_5 X_5 + e_6 X_6 \\ \gamma(0) = \eta_0 \quad \gamma(1) = \eta_1. \end{cases} \quad (7)$$

Given a family of constant positive coefficients $\{c_i\}_{i=1}^{6}$, according to the work of Nagel et al. [14], the homogeneous distance between two points $\eta_0, \eta_1 \in \mathcal{M}$ can be estimated as follows

$$d^{\mathcal{M}}(\eta_0, \eta_1) \approx \left(\sum_{i=1}^{6} c_i |e_i|^{6/\deg(X_i)} \right)^{\frac{1}{6}}. \tag{8}$$

In our experiments, we will use this estimate of the distance.

3.2   Model of Movement Decomposition

3.2.1   Clustering for Identification of Elementary Trajectories

We define in the cortical feature space $\mathcal{M}$, a connectivity kernel $\omega_{\mathcal{M}}$ expressed in term of distance (8):

$$\omega_{\mathcal{M}}(\eta_0, \eta) = e^{-d^{\mathcal{M}}(\eta_0, \eta)^2}, \quad \eta_0, \eta \in \mathcal{M}, \tag{9}$$

Then, we discretize (9) on a set of reaching paths, and obtain a real symmetric affinity matrix A:

$$A = \omega_{\mathcal{M}}\left((x_i, y_i, \theta_i, v_i, a_i, t_i), (x_j, y_j, \theta_j, v_j, a_j, t_j)\right), \tag{10}$$

which contains the connectivity information between all the kinematic variables of the reaching trajectory.

We propose to apply a spectral clustering analysis of this affinity matrix to a set of movement trajectories in order to obtain a decomposition in elementary trajectories to be compared with the one obtained in [10] (see Fig. 1 as a reference).

3.2.2   Clustering for Classification of Elementary Trajectories

The next step involves grouping the elementary trajectories based on the properties described in Kadmon Harpaz et al. [10]. Recall that elementary trajectories are regular curves with values in the space $\mathcal{M}$. Up to a change of e parameterization in the variable s, we can assume that the elementary trajectory space $\mathcal{F}(\mathcal{M})$ is a subset of $C^1([0, 1], \mathcal{M})$.

In order to perform a grouping of these elementary trajectories we note that movement direction θ and acceleration a are almost constant on the elementary trajectories recovered in Sects. 3.2.1, and that neurons are invariant with respect to time and position (e.g. [8,10]).

Hence we associate to each elementary trajectory its mean orientation and acceleration. If $\gamma : [0, 1] \rightarrow \mathcal{M}$ is an elementary trajectory, we denote

$$\bar{\theta}(\gamma) = \int_0^1 \theta(s)ds \quad \text{and} \quad \bar{a}(\gamma) = \int_0^1 a(s)ds.$$

Subsequently, we perform a new clustering by adapting the sub-Riemannian distance previously defined. Precisely, if γ_0, γ_1 are elementary trajectories, the new distance will be defined

$$d^{\mathcal{F}}(\gamma_0, \gamma_1) = d^{\mathcal{M}}\Big((0, 0, \bar{\theta}(\gamma_0), 0, \bar{a}(\gamma_0), 0), (0, 0, \bar{\theta}(\gamma_1), 0, \bar{a}(\gamma_1), 0)\Big).$$

The kernel over the elementary trajectories space is given by

$$\omega_{\mathcal{F}}(\gamma_0, \gamma_1) = e^{-d^{\mathcal{F}}(\gamma_0, \gamma_1)^2}, \quad \gamma_0, \gamma_1 \in \mathcal{F}(\mathcal{M}). \tag{11}$$

Affinity matrix turns out to be defined by consequence, and we can apply the previous clustering analysis using the distance between these pairs. Note that even if the elementary trajectories are curves, the clustering algorithm is performed on the averages, so that it takes place in a finite dimensional space. In this way we obtain a classification of elementary trajectories into classes, called fragments, to be compared with the ones in Kadmon-Harpaz et al. [10].

3.3 Results

In the following we will show two test cases. In both cases, as in the paper [10], the motion trajectory is visualized by two graphs, one on the (x, y) plane, the reaching path, and one on the (t, v) plane corresponding to the velocity profile. In Test 1, the clustering method described in [12] is sufficient. However, in Test 2, in order to achieve the classification results obtained by Kadmon-Harpaz et al. [10], it is necessary to utilize the newly adopted algorithm outlined in Sect. 3.2.2.

Test 1: Simulation of a Center-Out Task
As a first example, we will analyze a trajectory of movement performing a center-out task.

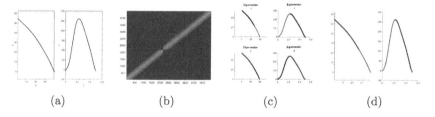

(a)   (b)   (c)   (d)

Fig. 3. (a) Reaching path and speed profile of a center-out task over the (x, y) plane and (t, v) plane. The red dot represents the movement starting position. (b) The Affinity matrix. (c–d) Projections of the eigenvectors. (Color figure online)

In this very simple case, movement direction is almost constant with only one target point to be reached and just one maximum point is present on the speed profile. The affinity matrix which is clearly divided in two blocks (Fig. 3(b)) identifies the eigenspaces associated to the two major eigenvalues. The projection of the eigenvectors over the reaching trajectory (Fig. 3(c) and (d))), corresponds precisely to the acceleration and deceleration phases of the movement task coherently with the neural states found in [10] (see also Fig. 1(a)). In this simple case we do not need to apply the second clustering of our algorithm.

Test 2: Simulation of a Random Target Pursuit Task

In this test, we apply our spectral algorithm on an approximate trajectory of
Fig. 1(b). The analyzed motion is represented in Fig. 4(a). In [10], the experiment
performed by the monkey consists of reaching several targets one after the other.
Here, a trajectory is extrapolated that starts from a fixed point (red point in
Fig. 4(a)), arrives at a second target (blue point in Fig. 4(a)) and comes to an
end. The affinity matrix is divided into four blocks (see Fig. 4(b)). As before we
project the eigenvectors associated with the largest eigenvalues onto the motion
trajectory (see Fig. 4(d)).

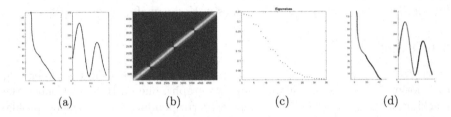

(a)   (b)   (c)   (d)

Fig. 4. Reaching path and speed profile of a random target pursuit task: approximation
of Fig. 1 (b). (b–c) The Affinity matrix and the eigenvalues plot. (d) Projections of the
eigenvectors over the reaching trajectory. (Color figure online)

After that we apply the second clustering in the space of sub-trajectories,
with respect to the θ and a variables. The resulting clusters appropriately group
acceleration and deceleration phases, as well as phases with constant direction.
The eigenvectors colored in green denote the acceleration phase, those colored
in orange the deceleration phase. The resulting decomposition pattern displayed
in Fig. 5 is in agreement with the experimental result of [10] shown in Fig. 1(b).

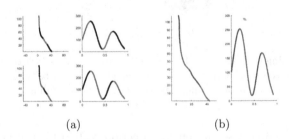

(a)   (b)

Fig. 5. Resulting reaching trajectory segmentation according to spatio-temporal invariant clusters. Acceleration and deceleration phases are respectively identified. See
Fig. 1(b) for a direct comparison.

4   Conclusions

We presented a sub-Riemannian model of the arm area of motor cortex expressed in terms of the kinematic variables experimentally measured in [4] and [17]. The metric of the space was directly deduced from the constraint between these variables, and was expressed in terms of suitable vector fields. We showed that their integral curves provide a good model of the time-dependent directional tuning of neurons in this area, experimentally found in [8]. We finally introduced a distance that allows to perform a kernel component analysis which is the phenomenological counterpart of the neural PCAs provided by Kadmon-Harpaz et al. [10]. In particular, we emphasize that by working only on kinematic variables we recovered the same neural classification acquired by electrode array. This proves that the distance $d^{\mathcal{M}}$ is adequate, not only because of the properties of the kinematic space, but also because of the classification in sub-trajectory fragments given by the clustering algorithm, which has a neural foundation. In particular the kinematic parameters we identified are sufficient to completely explain the process observed in [10].

Acknowledgments. GHAIA project, H2020 MSCA n. 777622; NGEU-MUR-NRRP, project MNESYS n. PE0000006.

References

1. Churchland, M., Shenoy, K.: Temporal complexity and heterogeneity of single-neuron activity in premotor and motor cortex. J. Neuroph. **97**(6), 4235–57 (2007)
2. Citti, G., Sarti, A.: A cortical based model of perceptual completion in the roto-translation space. J. Math. Imag. Vis. **24**(3), 307–326 (2006)
3. Flash, T., Hogan, N.: The coordination of arm movements: an experimentally confirmed mathematical model. J. Neurosci. **5**(7), 1688–1703 (1985)
4. Georgopoulos, A.P., Caminiti, R., Kalaska, J.F.: Static spatial effects in motor cortex and area 5: quantitative relations in a two-dimensional space. Exp. Brain Res. **54**(3), 446–454 (1984)
5. Georgopoulos, A.P.: Columnar organization of the motor cortex: direction of movement. In: Casanova, M.F., Opris, I. (eds.) Recent Advances on the Modular Organization of the Cortex, pp. 123–141. Springer, Dordrecht (2015). https://doi.org/10.1007/978-94-017-9900-3_8
6. Georgopoulos, A.P., Kalaska, J.F., Caminiti, R., Massey, J.T.: On the relations between the direction of two-dimensional arm movements and cell discharge in primate motor cortex. J. Neurosci. **2**(11), 1527–1537 (1982)
7. Graziano, M.S.A., Aflalo, T.N.: Mapping behavioral repertoire onto the cortex. Neuron **56**(2), 239–251 (2007)
8. Hatsopoulos, N.G., Xu, Q., Amit, Y.: Encoding of movement fragments in the motor cortex. J. Neurosci. **27**(19), 5105–5114 (2007)
9. Hoffmann, W.C.: The visual cortex is a contact bundle. Appl. Math. Comput. **32**, 132–167 (1989)
10. Kadmon Harpaz, N., Ungarish, D., Hatsopoulos, N.G., Flash, T.: Movement decomposition in the primary motor cortex. Cerebral Cortex **29**(4), 1619–1633 (2019)

11. Le Donne, E.: Lecture notes on sub-riemannian geometry (2010, preprint)
12. Mazzetti, C., Sarti, A., Citti, G.: Functional architecture of M1 cells encoding movement direction. J. Comput. Neurosci., 1–29 (2023). https://doi.org/10.1007/s10827-023-00850-2
13. Mazzetti, C., Sarti, A., Citti, G.: A model of reaching via subriemannian geodesics in Engel-type group. arXiv preprint arXiv:2301.05765 (2023)
14. Nagel, A., Stein, E.M., Wainger, S.: Balls and metrics defined by vector fields I: basic properties. Acta Math. **155**, 103–147 (1985)
15. Paninski, L., Fellows, M.R., Hatsopoulos, N.G., Donoghue, J.P.: Spatiotemporal tuning of motor cortical neurons for hand position and velocity. J. Neuroph. **91**(1), 515–532 (2004)
16. Petitot, J., Tondut, Y.: Vers une neurogéométrie. fibrations corticales, structures de contact et contours subjectifs modaux. Mathématiques et sciences humaines **145**, 5–101 (1999)
17. Schwartz, A.B.: Useful signals from motor cortex. J. Phys. **579**(3), 581–601 (2007)

Geometry of Saccades and Saccadic Cycles

D. V. Alekseevsky[1,2] and I. M. Shirokov[3]

[1] Institute for Information Transmission Problems, B. Karetnuj per., 19,
Moscow 127051, Russia
dalekseevsky@iitp.ru
[2] Faculty of Science, University of Hradec Králové, Rokitanského 62,
500 03 Hradec Králové, Czech Republic
[3] St. Petersburg Department of Steklov Mathematical Institute, Fontanka, 27,
191023 St. Petersburg, Russia
shirokov.im@phystech.edu

Abstract. The paper is devoted to the development of the differential geometry of saccades and saccadic cycles. We recall an interpretation of Donder's and Listing's law in terms of the Hopf fibration of the 3-sphere over the 2-sphere. In particular, the configuration space of the eye ball (when the head is fixed) is the 2-dimensional hemisphere S_L^+, which is called Listing's hemisphere. We give three characterizations of saccades: as geodesic segment ab in the Listing's hemisphere, as the gaze curve and as a piecewise geodesic curve of the orthogonal group. We study the geometry of saccadic cycle, which is represented by a geodesic polygon in the Listing hemisphere, and give necessary and sufficient conditions, when a system of lines through the center of eye ball is the system of axes of rotation for saccades of the saccadic cycle, described in terms of world coordinates and retinotopic coordinates. This gives an approach to the study the visual stability problem.

Keywords: Donder's and Listing's law · quaternions · Hopf bundle · fixation eyes movements · drift · micrsasccades · remapping · shift of receptive fields · neurogeometry

1 Introduction

Our eyes continually move. Even while we fix our gaze on an object, they participate in fixational eye movements – tremor, drift and microsaccades. The classical experiments by A. Yarbus show that compensation of eye movements leads to loss of vision in 2–3 s. The information about saccades is coded into control commands of the oculomotor system which governs muscles contraction. A copy

D. V. Alekseevsky was supported by the Grant "Basis-foundation (Leader)" 22-7-1-34-1. I.M. Shirokov was supported by the Grant "Basis-foundation (Leader)" 22-7-1-34-1 and Ministry of Science and Higher Education of the Russian Federation, agreement no. 075-15-2022-289.

F. Nielsen and F. Barbaresco (Eds.): GSI 2023, LNCS 14072, pp. 493–500, 2023.
https://doi.org/10.1007/978-3-031-38299-4_51

of these command (efference copy or corollary discharge) is send to the frontal eye field of the frontal cortex [2], where it meets the statistical information about retina images, described in coordinates associated to the end gaze direction of the previous saccade.

The vision is the result of interaction of these two types of information. The time synchronization of oculomotor information about eye rotations and retinal information is very important for the correct decoding.

In [1] a new interpretation of the Donder's and Listing's laws in terms of the Hopf bundle had been given and used for geometrical descriptions of saccades and drift. This paper, which is a continuation of [1], is devoted to development of the differential geometry of saccades and saccadic cycles. We give three characterizations of saccades – as geodesic segment ab in the Listing's hemisphere (which is the configuration space of the eye), as the gaze curve $Sac(AB) = \chi(a)\chi(b) \subset S_E^2 \cap span\,(A, B, -i)$, where $\chi : S^3 \to S^2$ is the Hopf projection, and as a piecewise geodesic curve of the orthogonal group. Then we study the geometry of saccadic cycle which is represented by a geodesic polygon in the Listing hemisphere. We give necessary and sufficient condition, when a system of lines $([\Omega_1], \cdots, [\Omega_n])$, where $\Omega_\ell \in S_E^2$, is the system of axes for saccades of the saccadic cycle, described in terms of world coordinates and retinotopic coordinates.

This clarifies the relation between oculomotor information about saccadic rotation the gaze direction and hence the eye position and shows that all visual information may be transformed into world-centered coordinates, which gives an approach to the study of the visual stability problem.

2   The Hopf Bundle and the Listing's Section

Let
$$\mathbb{H} = \mathbb{R}^4 = span\,(1, i, j, k) = \mathbb{R}1 + E^3, \; E^3 = Im\mathbb{H} = span\,(i, j, k)$$

be the algebra of quaternions with the standard Euclidean metric $|q|^2 =< q, q >= q\bar{q}$, where $\bar{q} = q_R - q_E$ is the conjugation.

The sphere $\mathbb{H}_1 = S^3$ of unit quaternions is the Lie group and the adjoint action
$$\mathrm{Ad} : \mathbb{H}_1 \to SO(E^3), \; a \mapsto \mathrm{Ad}_a, \; \mathrm{Ad}_a x = axa^{-1} = ax\bar{a}$$

is the universal $\mathbb{Z}_2$-covering.

The orbit $\mathrm{Ad}_{\mathbb{H}_1} i = S_E^2 \simeq \mathbb{H}_1/SO(2)$ is the unit sphere and $\chi : S^3 \to S_E^2$ is the principal bundle.

Definition 1. *The Hopf bundle is the natural $\mathbb{H}_1$-equivariant map*

$$\chi : S^3 = \mathbb{H}_1 \to S_E^2, \; a \mapsto A = \chi(a) := \mathrm{Ad}_a i = aia^{-1}.$$

Denote by $S_L^2 = S^3 \cap span\,(1, j, k) = S^3 \cap i^\perp$ the equatorial 2-sphere of S^3 w.r.t. the poles $\pm i$ and by $S_L^+ = \{a = a_0 1 + a_2 j + a_3 k \in S_L^2, a_0 > 0\}$ the north hemisphere. It is called the Listing hemisphere [1] and its boundary S_L^1 is called the Listing's circle.

Proposition 1. *The (restricted to S_L^+) Hopf map*

$$\chi : S_L^+ \to S_E^2, \ a \mapsto A = \chi(a) = \mathrm{Ad}_a i = a i a^{-1}$$

is a diffeomorphism of S_L^+ onto the punctured sphere $\tilde{S}_E^2 = S_E^2 \setminus \{-i\}$ and $\chi(S_L^1) = \{-i\}$.

Definition 2. *i) The map $\sigma = \chi^{-1} : \tilde{S}_E^2 \to S_L^+$ is called the Listing's section.*
ii) The frame $f^1 = (i, j, k)$ is called the standard and the frame $f^a := \mathrm{Ad}_a f^1 = (aia^{-1}, aja^{-1}, aka^{-1})$ is called admissible.

Proposition 2. *[1] i)Any two points $a, b \in S_L^+$ determine a unique (oriented) geodesic of the Listing's sphere with canonical parametrization*

$$\gamma_{a,b} = \gamma_{p,m}(t) = \cos t\, p + \sin t\, m = e^{tv} p$$

where $p = \cos\theta\, j + \sin\theta\, k$ is the intersection of $\gamma_{a,b}$ with S_L^1, $m = e^{\psi q} = \cos\psi + \sin\psi q$, $q = ip$ is the top point of the geodesic and $v = m\bar{p} = \sin\psi\, i - \cos\psi p \in S_E^2$.
ii) If $m = 1$, then $\gamma_{a,b} = \gamma_{p,1} = e^{tp}$, $p \in S_L^1$ is 1-parameter subgroup.
iii) The group $\mathrm{Ad}_{e^{tp}} = R_p^{2t}$ is the group of rotation w.r.t. the axis $[p] := \mathbb{R}p$ and the Hopf image $\chi(\gamma_{p,m}) = \mathrm{Ad}_{e^{tv}}(-i) = R_v^{2t}(-i)$ is the gaze circle $S_{a,b}^1 = S_{p,m}^1 = S_E^2 \cap \Pi(A, B, -i)$, where $A = \chi(a)$, $B = \chi(b)$ and $\Pi(A, B, -i) = -i + \mathrm{span}\,(i + A, i + B)$.

3 Donder's and Lising's Laws

We consider the sphere $S_E^2 \subset E^3$ in the Euclidean space E^3 as a model of the eye sphere, i.e. the boundary of the eye ball in the primary position, defined by the primary frame $f^1 = (i, j, k)$. Here i indicates the primary position of the gaze (the frontal direction), j is the lateral direction from right to left and k is the vertical direction up.

Donder's Law. If the head is fixed, the configuration space of the eye is 2-dimensional and the position $A \in S_E^2$ of the gaze determines the position of the eye.

The Listing's law describes the configuration space.

Listing's Law. The configuration space of the eye is the Listing's hemisphere. The eye position associated to a gaze $A \in \tilde{S}_E^2$ is given by

$$f(A) = f^{\sigma(A)} := \mathrm{Ad}_{\sigma(A)} f^1.$$

4 Saccades, Three Descriptions

The saccade Sac (A, B) from a gaze direction $A \in \tilde{S}_E^2$ to the gaze direction B is described by the geodesic segment $ab = \sigma(A)\sigma(B)$ and the corresponding gaze curve $AB = \chi(ab)$, which is the segment of the gaze circle $S^1_{a,b}$ with the endpoints A and B.

Let $a, b \in S^+_L$ be two eye positions and $A = \chi(a), B = \chi(b) \in S^2_E$ the associated gaze directions. It defines a unique saccade Sac (A, B) from the eye position a to the position b.

We have three descriptions of the saccade Sac (A, B):

i) As the evolution of the eye position in the configuration space S^+_L that is represented by a segment ab of a geodesic $\gamma_{a,b} \subset S^+_L$ of the Listing hemisphere.

ii) As the evolution of the gaze in the gaze sphere $\tilde{S}^2_E$, described by the gaze curve $AB = \chi(ab)$. It is the segment of the gaze circle $S^1(A, B) = S^2_E \cap \Pi(A, B, -i)$ which is the section of the sphere S^2_E by the plane $\Pi(A, B, -i) = -i + \mathrm{span}\,(A + i, B + i)$.

iii) As the curve $R^{t\varphi}_\Omega$, $t \in [0, 1]$ in the orthogonal group, which is a segment of the one-parameter group of rotations around the axis $[\Omega] = \mathbb{R}\Omega$.

To relate these three characterisations of saccades, we introduce the following definition.

Definition 3. *Let $A \in \tilde{S}^2_E$ be a unit vector. Then the oriented 2-dimensional subspace $L_A = N^\perp_A$ with the positive unit normal $N_A = \frac{i+A}{|i+A|}$ is called A-Listing plane and N_A is called A-Listing's normal.*

Proposition 3. *i) The axis $[\Omega] = \mathbb{R}\Omega$ of rotation of any saccade with initial gaze direction $A \in \tilde{S}^2_E$ is orthogonal to the A-Listing's normal N_A and belongs to the A-Listing's plane $L_A = N^\perp_A$.*

*ii) **Law of half angle:** the (unique) saccade Sac (A, B) from the gaze direction A to the gaze direction B is obtained by the rotation R^φ_Ω around the axis $[\Omega] = \mathbb{R}(N_A \times N_B) = L_A \cap L_B$ by the angle $\varphi = 2\psi$ where $\psi = \angle(N_A, N_B)$ is the angle between the normal vectors N_A, N_B of the A-Listing's plane L_A and the B-Listing's plane L_B.*

iii) Let $\mathrm{Sac}(A_0, A_1)$, $\mathrm{Sac}(A_1, A_2)$ be two consecutive saccades with distinct axes of rotation Ω_1, Ω_2. Then the A_1-Listing's plane is given by

$$L_{A_1} = \mathrm{span}\,(\Omega_1, \Omega_2)$$

and the A_1-Listing's normal is given by:

$$N_{A_1} = \frac{i^*}{|i^*|},$$

where i^* stands for the first dual vector in the dual basis to the basis (i, Ω_1, Ω_2).

Proof. i) Let $B \in \tilde{S}^2_E \setminus \{A\}$. Then during the saccade $\mathrm{Sac}(A, B)$ the gaze belongs to the circle $\Pi(A, B, -i) \cap S^2_E$. Hence the axis of rotation of $\mathrm{Sac}(A, B)$ is orthogonal to the plane $\Pi(A, B, -i)$ and in particular orthogonal to the vector N_A.

ii) Indeed, denote by O' the center of the saccadic circle $S^1_{A,B} = \Pi(A, B, -i) \cap S^2_E$. Then the rotation angle $\varphi = \angle(O'A, O'B)$ is the central angle associated to the inscribed angle

$$\psi = \angle((-i)A, (-i)B) = \angle(N_A, N_B).$$

Hence $\psi = \frac{1}{2}\varphi$.

iii) First note that Ω_1, Ω_2 and i are linear independent. We have

$$i^* = \frac{[\Omega_1, \Omega_2]}{(i, [\Omega_1, \Omega_2])}$$

Thus $\mathbb{R}i^* = \mathbb{R}N_{A_1}$ and $(i^*, i) = 1 > 0$. So $N_{A_1} = \frac{i^*}{|i^*|}$.   □

5   Saccadic n-Cycles

We define a *saccadic n-cycle* as a geodesic polygon $Sac(a_1 a_2 \cdots a_{n+1}) \subset S_L^+, a_1 = a_{n+1}$ and $a_k \neq a_{k+1}$ with the natural parametrization, which is identified with the gaze curve

$$\chi(Sac(a_1 a_2 \cdots a_{n+1})) = Sac(A_1, ..., A_n, A_{n+1}).$$

In terms of the rotations, the saccadic curve is described as the orbit $R^t A_1$ of a piecewise geodesic curve R^t, $t \in [0, T]$ in the orthogonal group $SO(3)$ such that

$$R^{T_j} A_1 = A_j, \; R^{T_j + t} A_1 = R_{\Omega_j}^t A_j, \; 0 \leq t \leq T_{j+1} - T_j,$$

$$0 < T_1 < \cdots < T_{n+1} = T.$$

We call the vectors Ω_j which determine the axes of rotation $[\Omega_j] = \mathbb{R}\Omega_j$ of the saccades $Sac(A_j, A_{j+1})$ the axes of the n-cycle.

5.1   Saccadic n-Cycles via Systems of Axes Expressed at the Primary Frame

Note that the gaze directions A_j, A_{j+1} of the saccade $S_j = Sac(A_j, A_{j+1})$ determine the saccade rotation $R_{\Omega_j}^\varphi$ as follows

$$\Omega_j \equiv N_{A_j} \times N_{A_{j+1}}, \; \varphi_j = 2\angle(N_{A_j}, N_{A_{j+1}})$$

(Where "$\equiv$" means equality up to sign). Now we consider the inverse problem: when a system $(\Omega_1, \cdots, \Omega_n)$ of unit vectors determines a saccadic n-cycle and how to describe the gaze directions $(A_1, \cdots, A_{n+1} = A_1)$ and the angles of rotations of the n-cycle.

We show that it can be done under a mild necessary and sufficient condition and give an explicit formula for the system of gaze directions $(A_1, \cdots, A_{n+1})$.

Theorem 1. *Let $(\Omega_1, \cdots, \Omega_n)$ be a system of $n \geq 3$ unit vectors in E^3. Then there exists a (unique) saccadic n-cycle*

$$Sac(A_1, A_2, \cdots, A_{n+1}), \; A_{n+1} = A_1$$

with axes of rotation

$$[\Omega_1], \cdots, [\Omega_n]$$

if and only if

1. *All triples* $(\Omega_1, \Omega_2, i), \cdots, (\Omega_n, \Omega_1, i)$ *are linear independent.*
2. *All triples* $(\Omega_1, \Omega_2, \Omega_3), \cdots, (\Omega_{n-1}, \Omega_n, \Omega_1), (\Omega_n, \Omega_1, \Omega_2)$ *are linear independent.*

If $span(\Omega_n, \Omega_1) = span(j, k)$ *then* $A_1 = A_{n+1} = i$.

Proof. Necessity follows from Proposition 1 and definition of saccadic cycle. The scheme of proof of sufficiency is as follows. Planes

$$\Pi_1 := span(\Omega_n, \Omega_1), \Pi_2 := span(\Omega_1, \Omega_2), \cdots, \Pi_n := span(\Omega_{n-1}, \Omega_n)$$

define points $a_1, \cdots, a_n$ of configuration space S_L^+ such that Π_k is $\chi(a_k)$-Listing plane. It remains to check that $Sac(a_1, ..., a_n, a_{n+1} = a_1)$ is a desired cycle.   □

5.2   Transition Functions Between Primary Frame and the Moving Frame

In this subsection we show how coordinates of Listing's normal N_A w.r.t the primary frame f^1 and "moving" frame $f^{\sigma(A)}$ are related.

Lemma 1. *Let* $a \in S_L^+$ *and* $A = \chi(a)$ *be the gaze direction. Then coordinates of A-Listing's normal* N_A *w.r.t. the primary frame* f^1 *and "moving" frame* f^a *related as follows:*

$$N_A = N_1 i + N_2 j + N_3 k = N_1 i^a - N_2 j^a - N_3 k^a$$

Proof. We consider the saccade $Sac(1, a) = Sac(i, A)$. By the half-angle law (Proposition 3) we have $N_A = a^{1/2} i a^{-1/2}$. We may write $a = e^{\varphi p}$ for some $p = \cos\theta j + \sin\theta k$ and $\varphi \in [0, \pi/2)$, then w.r.t. the primary frame f^1, the vector N_A is obtained from i by rotation around $[p]$ on $\varphi/2$: $N_A = R_{[p]}^{\varphi/2} i$, whereas w.r.t. the "moving" frame f^a, the vector the vector N_A is obtained from i^a by rotation around $[p^a] = [\cos\theta j^a + \sin\theta k^a]$ on $-\varphi/2$: $N_A = R_{[p^a]}^{-\varphi/2} i^a$.   □

5.3   Saccadic n-Cycles via Systems of Axes Fixed at the Moving Frame

Definition 4. *We associate with a vector* $\Omega \in S_E^2$ *the map*

$$\sigma_\Omega : S_L^+ \to S_E^2, , a \mapsto \Omega^a = Ad_a \Omega = a\Omega a^{-1}$$

We call parameterized by $a \in S_L^+$ *vector* Ω^a *the moving vector.*

The coordinates $\Omega_{f^a}^a$ of the vector Ω^a in the frame f^a does not depend on a and coincide with the coordinates of vector Ω w.r.t. the primary frame. Note that according to the Listing's law the eye can rotate around axis $[\Omega^a]$ if Ω^a belongs to the $\chi(a)$-Listing's plane. For example, $[\Omega^1]$ is the axis of rotation only if it belongs to $\chi(1) = i$-Listing's plane, which is $span (j, k)$.

Given a system $(\Omega_1, \cdots, \Omega_n)$ of vectors from S_E^2, our aim is to describe when there exists a saccadic n-cycle $Sac(a_1, ..., a_n, a_{n+1})$ with axes of rotation

$$[\Omega_1^{a_1}], \cdots, [\Omega_n^{a_n}].$$

We need the following

Proposition 4. *Assume the vectors $\Omega_1, \Omega_2 \in S_E^2$ span the plane $\Pi = span(\Omega_1, \Omega_2)$ which does not contains i. Then there exists unique $a \in S_L^+$, hence gaze direction $A = \chi(a)$, such that $\Pi^a := Ad_a \Pi = span(\Omega_1^a, \Omega_2^a)$ coincides with A-Listing plane L_A.*

Proof. Proposition 3 implies that there is unique $b \in S_L^+$ such that $\Pi = Ad_{b^{1/2}} L_i = L_b$. We have to find $a \in S_L^+$ such that

$$\Pi^a = Ad_{ab^{1/2}} L_i = L_a = Ad_{a^{1/2}} L_i$$

The solution is $a = b^{-1}$. □

Theorem 2. *The system $(\Omega_1, \cdots, \Omega_n)$ of vectors from S_E^2 defines the system of axes of rotation*

$$([\Omega_1^{a_1}], \cdots, [\Omega_n^{a_n}])$$

for a (unique) saccadic n-cycle

$$Sac(a_1, \cdots, a_{n+1}), \; a_1 = a_{n+1}$$

if and only if it satisfies the conditions of Theorem 1. If $span(\Omega_n, \Omega_1) = span(j, k)$ then $a_1 = a_{n+1} = 1$.

Proof. We prove only sufficiency. Let $\Omega_{n+1}^a := \Omega_1^a$. Proposition 4 implies that for each pair $\Omega_k, \Omega_{k+1}, k \in \{1, ..., n\}$ there exists $a_k \in S_L^+$, such that

$$\Pi_k^{a_k} := span(\Omega_k^{a_k}, \Omega_{k+1}^{a_k}) = L_{a_k} \tag{1}$$

Let $a_{n+1} := a_1$. Proof of Proposition 4 implies that $\Pi_k^{a_k} = S_{L_i}(\Pi_k^1)$, where

$$S_{L_i} : \xi_1 i + \xi_2 j + \xi_3 k \mapsto -\xi_1 i + \xi_2 j + \xi_3 k$$

is the symmetry w.r.t.the Listing's plane L_i. Hence we have

$$\Omega_1^{a_n} = \Omega_1^{a_1}, \; \Omega_k^{a_{k-1}} = \Omega_k^{a_k}, k \in \{2, ..., n\} \tag{2}$$

We claim that $Sac(a_1, ..., a_n, a_{n+1})$ is a desired cycle. Indeed $a_k \neq a_{k+1}$ for $k \in \{1, ..., n\}$ due to the assumptions on $\Omega_1, ..., \Omega_n$, the geodesic $a_k a_{k+1}$ corresponds to rotation around $[\Omega_k^{a_k}]$ due to (2) for each $k \in \{1, ..., n\}$; and $[\Omega_k^{a_k}] \subset L_{a_k}$ for each $k \in \{1, ..., n\}$ due to (1). □

Corollary 1. *Let $Sac(a_1, ..., a_{n+1})$ be a saccadic cycle described in terms of Theorem 4 and $[\Omega_m]$, $m = 1, \cdots, n$ be axes of rotation. Then the coordinates of any vector $\Omega := \Omega_m^{a_m}$ w.r.t. the "moving" frame f^{a_m} and the primary frame f^1 are:*

$$\Omega_{f^{a_m}} = (\Omega_1, \Omega_2, \Omega_3)$$
$$\Omega_{f^1} = (-\Omega_1, \Omega_2, \Omega_3)$$

Corollary 1 means that during the saccadic cycle the oculomotor system can easily recalculate coordinates of axis vector w.r.t the primary frame from coordinates of it w.r.t the "moving" frame and vice versa at each switching point. In other words, the way of setting a saccadic cycle in terms of fixed axis vectors in Theorem 1 can be reformulated in terms of moving vectors in Theorem 2 and vice versa.

References

1. Alekseevsky, D.V.: Microsaccades, drifts, hopf bundle and neurogeometry. J. Imaging **8**, 76 (2022)
2. Wikipedia Contributors. Corollary discharge theory. Wikipedia, The Free Encyclopedia (2022). Corollary discharge theory
3. Yarbus, A.: Eye Movements and Vision (B. Haigh, Trans.). Plenum Press, New York (1967)
4. Poletti, M., Rucci, M.: A compact field guide to the study of microsaccades: challenges and functions. Vis. Res. **118**, 83–97.32 (2016)
5. Wurtz, R.H., Joiner, W.M., Berman, R.A.: Neuronal mechanisms for visual stability: progress and problems. Philos. Trans. R. Soc. B **366**(1564), 492–503 (2011)
6. Alexander, R.G., Martinez-Conde, S.: Fixational eye movements. In: Eye Movement Research, pp. 74–108 (2019)
7. Hepp, K.: On Listing's law. Commun. Math. Phys. **132**, 285–292 (1990)

MacKay-Type Visual Illusions via Neural Fields

Cyprien Tamekue, Dario Prandi$^{(\boxtimes)}$ (ID), and Yacine Chitour (ID)

Université Paris-Saclay, CNRS, CentraleSupélec, Laboratoire des Signaux
et Systèmes, 91190 Gif-sur-Yvette, France
{cyprien.tamekue,dario.prandi,yacine.chitour}@centralesupelec.fr

Abstract. To study the interaction between retinal stimulation by
redundant geometrical patterns and the cortical response in the primary
visual cortex (V1), we focus on the MacKay effect (Nature, 1957) and
Billock and Tsou's experiments (PNAS, 2007). We use a controllability
approach to describe these phenomena starting from a classical biological model of neural fields equations with a non-linear response function.
The external input containing a localised control function is interpreted
as a cortical representation of the static visual stimuli used in these
experiments. We prove that while the MacKay effect is essentially a linear phenomenon (i.e., the nonlinear nature of the activation does not
play any role in its reproduction), the phenomena reported by Billock
and Tsou are wholly nonlinear and depend strongly on the shape of the
nonlinearity used to model the response function.

1 Introduction

In many situations, humans perceive an illusory component that is not physically present in a visual stimulus. Helmholtz [5] is probably the first to be
interested in the visual effect induced by the presentation of a pattern consisting of black-and-white zones. In particular, he related the perception of rotating
darker and brighter radial zones after viewing a pattern consisting of black and
white concentric rings to the fluctuation of eye accommodation. In this direction,
[8] reported striking after-effect of visual stimulation by regular geometrical patterns with highly redundant information (see Fig. 3 for the so-called "MacKay
rays") and attributed the phenomena to some part of the visual cortex, which
might profit from such redundancy. In these experiments, an illusory contour
consisting of a pattern of white and black concentric rings (tunnel pattern) is
evoked by all observers as the after-image induced by a pattern consisting of
white and black fan shape (funnel pattern) with high redundant information in
the fovea (the centre of the visual field). Due to the retino-cortical map[1] between

---

[1] The retino-cortical map (see, e.g., [10] and references within) is given by

$$re^{i\theta} \mapsto (x_1, x_2) := (\log r, \theta).$$

© The Author(s), under exclusive license to Springer Nature Switzerland AG 2023
F. Nielsen and F. Barbaresco (Eds.): GSI 2023, LNCS 14072, pp. 501–508, 2023.
https://doi.org/10.1007/978-3-031-38299-4_52

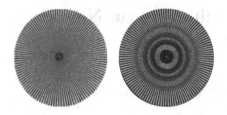

Fig. 1. MacKay effect: the presentation of the stimulus to the *left* ("MacKay rays") induces an illusory perception of the image on the *right*. Adapted from [8,11].

Fig. 2. Billock and Tsou's experiments: the presentation of funnel pattern in the periphery (*left*) induces an illusory perception of tunnel pattern in the centre (*right*) after a flickering. From [2].

the retina and the primary visual cortex (V1, henceforth), the after-images in V1 are superimposed patterns consisting of orthogonal horizontal and vertical stripes. This indicates that neuronal response in V1 tends to favour directions at a right angle to the visual stimulus.

Even more striking visual effects have been obtained in the psychophysical experiments reported by [2], see Fig. 2. As in the case of the MacKay effect, they found that biasing stimuli could induce orthogonal responses in the visual cortex. More precisely, a funnel pattern localised at the fovea (resp. in the periphery) with a background flicker induces the perception of a tunnel pattern in the periphery (resp. at the fovea). The spatial interaction is localized, meaning the hallucination does not extend through the physical stimulus nor into empty non-flickering regions.

This work is concerned with a theoretical description of MacKay and Billock and Tsou's illusory phenomena in V1. This is achieved by studying the properties of the Amari-type neuronal field equation [1] describing the dynamics of the activity $a : \mathbb{R}^2 \to \mathbb{R}$ on V1:

$$\frac{\partial a}{\partial t} = -a + \mu\omega * f(a) + I. \tag{NF}$$

Here, $*$ denotes the spatial convolution operation, $\omega : \mathbb{R}^2 \to \mathbb{R}$ is an interaction kernel modelling cortical connections in V1, f is a sigmoid non-linearity, and $I : \mathbb{R}^2 \to \mathbb{R}$ is the cortical representation of the presented static visual stimulus, that is assumed to be time-independent. Finally, $\mu > 0$ is a parameter measuring the strength of intra-neuron connectivity. In this work, following [10], we assume that the parameter μ is smaller than the threshold parameter μ_c where cortical patterns (e.g., funnels, tunnels, spirals, checkerboards, cobwebs, etc.) spontaneously emerge in V1 (see, e.g. [3,4]). From a neurophysiological point of view, this corresponds to considering an unaltered state where no spontaneous hallucinations emerge in the absence of external inputs.

By an asymptotic analysis of the properties of (NF) we describe why the after-image in the MacKay effect consists of illusory contours in the background of the physical visual stimulus. In particular, the result we provide here implies that a motion in the after-image moves at a right angle to the stimulus pattern. This is because the static physical stimulus and the after-image in V1 are superimposed horizontal and vertical stripes combined with the fact that the inverse retino-cortical map conserves this opponency in the retina.

Our main finding is that while the MacKay effect is essentially a linear phenomenon, Billock and Tsou's experiments are completely non-linear phenomena that strongly depend on the shape of the non-linear function used to model the neuronal response after an activation. Moreover, due to the equivariance of equation (NF) with respect to the plane Euclidean group $\mathbf{E}(2)$, we find that the MacKay effect results from the highly redundant information in the visual stimulus aiming to break its plane Euclidean symmetry. The same is true for Billock and Tsou's phenomenon, where symmetry-breaking arises due to the localization of the visual stimulus in the visual field.

We conclude this section by mentioning that, up to our knowledge, the only other attempt to describe these phenomena theoretically is due to [9]. There, the authors study a different model of neural fields equation containing an adaptation variable and a feedback (state-dependent) external input. Their theoretical result relies on bifurcation and multi-scale analysis, which can be applied only for values of μ near the threshold parameter μ_c and in the presence of fully distributed external inputs. In particular, their analysis does not apply to the range of μ that we consider, nor to localized inputs, such as those used by MacKay and Billock and Tsou's. Nevertheless, they provide numerical results showing the capability of their model to reproduce Billock and Tsou's experiments.

2   Neural Fields Equations

Ermentrout and Cowan, in their seminal paper [4], develop a theory describing (spontaneous) geometric visual hallucinations perceived in the retina. More precisely, using bifurcation techniques near a static Turing-like instability, they found that a simplified biological model of neural fields equation suffices to describe the (spontaneous) formation of cortical patterns (horizontal, vertical and oblique stripes, square, hexagonal and rectangular patterns etc.) in V1. Then, applying the inverse retino-cortical map to these patterns, they obtained some of the geometric visual hallucinations or form constants that [7] had meticulously classified. In their considerations, V1 is treated as a sheet of isotropically interconnected excitatory and inhibitory neurons. A more biologically realistic model of neural fields, including the anisotropic properties of cortical connections in V1 (orientation preference of "simple" cells, see [6]), was done in [3]. The authors were then able to describe all of Klüver's form constants.

Due to the success of the Ermentrout and Cowan model in describing simple patterns, such as funnel patterns, we expect that a similar model (i.e. without orientations preference) should be sufficient to describe sensory hallucinations (visual illusions) induced by these patterns. We, therefore, consider in this work that neuronal activity in V1 evolves according to equation (NF). It models the average membrane potential $a(x, t)$ of a neuron located at $x \in \mathbb{R}^2$ at time $t \geq 0$.

Throughout the following we assume the response function f to be a non-decreasing function of class $C^2(\mathbb{R})$ such that $f(0)$ and $f'(0) = \max_{s \in \mathbb{R}} f'(s) = 1$. Unless explicitly stated otherwise, f is a nonlinear sigmoidal function.

The kernel ω is taken to be a DoG distribution (difference of Gaussians, also called "Mexican hat" distribution). Namely, we let for all $x \in \mathbb{R}^2$

$$\omega(x) = [2\pi\sigma^2]^{-1}e^{-\frac{|x|^2}{2\sigma^2}} - [2\pi\kappa^2\sigma^2]^{-1}e^{-\frac{|x|^2}{2\kappa^2\sigma^2}}, \tag{1}$$

where $\kappa > 1$ and $0 < \sigma < 1$. Clearly, $\omega(x) = \omega(|x|)$ and ω belongs to the Schwartz space $\mathcal{S}(\mathbb{R}^2)$. Moreover, its Fourier transform $\widehat{\omega}$ is maximal at vectors $\xi_c \in \mathbb{R}^2$ such that $|\xi_c|^2 = \frac{\log \kappa}{\pi^2\sigma^2(\kappa^2-1)} =: q_c^2$.

Observe that with this choice, we fall into the framework of [3], i.e., there exists a critical interaction parameter $\mu_c := \widehat{\omega}(\xi_c)^{-1}$ around which spontaneous cortical patterns in V1 emerge.

It is straightforward to show that the r.h.s. of equation (NF) is a Lipschitz continuous map on $L_t^\infty(\mathbb{R}) \times L_x^\infty(\mathbb{R}^2)$. Thus, it is standard to obtain that, for every external input $I \in L^\infty(\mathbb{R}^2)$ and any initial datum $a_0 \in L^\infty(\mathbb{R}^2)$, equation (NF) admits a unique solution $a \in C([0, +\infty); L^\infty(\mathbb{R}^2))$.

Recall that a stationary state $a_I \in L^\infty(\mathbb{R}^2)$ to equation (NF) is a time-invariant solution, viz.

$$a_I = \mu\omega * f(a_I) + I. \tag{2}$$

Via the contraction mapping principle, one obtains the existence of a unique stationary solution whenever $\mu < \mu_0$, see [10, Proposition 1]. Here, we let

$$\mu_0 := \|\omega\|_1^{-1} \le \mu_c. \tag{3}$$

This implies in particular that if $\mu < \mu_0$, the map $\Psi : L^\infty(\mathbb{R}^2) \to L^\infty(\mathbb{R}^2)$ associating to each external input I its corresponding stationary state is well-defined and bi-Lipschitz continuous. Since Ψ is implicitly defined by (2), we deduce that both Ψ and Ψ^{-1} are $\mathbf{E}(2)$-equivariant, see e.g. [4, Appendix A].

Due to the retino-cortical map, funnel and tunnel patterns are respectively given in Cartesian coordinates $x := (x_1, x_2) \in \mathbb{R}^2$ of V1 by

$$P_F(x) = \cos(2\pi\lambda x_2), \quad P_T(x) = \cos(2\pi\lambda x_1), \quad \lambda > 0. \tag{4}$$

This choice is motivated by analogy with the (spontaneous) geometric hallucinatory patterns described in [3,4].

Given the above representation of funnel and tunnel patterns in cortical coordinates, we represent them as contrasting white and black regions to see how they look in terms of images. More precisely, define the binary pattern B_h of a function $h : \mathbb{R}^2 \to \mathbb{R}$ by

$$B_h(x) = \begin{cases} 0, & \text{if } h(x) > 0 \quad \text{(black)} \\ 1, & \text{if } h(x) \le 0 \quad \text{(white)}. \end{cases} \tag{5}$$

It follows that B_h is essentially determined by the zero level-set of h. Since stimuli involved in the MacKay effect and Billock and Tsou experiments are binary patterns, our strategy in describing these phenomena consists in characterising the zero level-set of output patterns. That is, we are mainly devoted to studying the qualitative properties of patterns by viewing them as binary patterns.

3   MacKay Effect

In [10, Theorem 1], we proved that highly redundant information is needed in the funnel and tunnel patterns for equation (NF) to reproduce the MacKay effect if $\mu < \mu_0/2$. More precisely, if the external input $I = P_F$ or $I = P_T$ in equation (NF), I and $\Psi(I)$ have the same binary pattern, then the same geometric shape in terms of images. Moreover, in the case were the response function f is linear, one easily obtains that $\Psi(I)$ is proportional to $I = P_F$ and $I = P_T$. We deduce that the resulting symmetry (the underlying Euclidean symmetry of the interaction kernel) of the system restricts the geometrical shape of visual stimuli that can induce an illusory perception in the after-image. So in describing the MacKay effect with our model of neuronal activity (NF) (provided we are far from the bifurcation point, i.e. $\mu < \mu_0 \leq \mu_c$), it is necessary to break the Euclidean symmetry of the funnel and tunnel patterns by localising redundant information.

Concerning, e.g. the description of the MacKay effect related to funnel pattern, the "MacKay rays" (see the image on the left of Fig 1) is a good candidate. It consists in taking external input

$$I(x) = P_F(x) + \varepsilon H(-x_1) = \cos(2\pi\lambda x_2) + \varepsilon H(-x_1), \tag{6}$$

where $\varepsilon > 0$ and H is the Heaviside step function which would model redundant information in the centre in the funnel pattern.

The following results show that equation (NF) with a linear response function f suffices to describe illusory contours perceived in the after-image induced by "MacKay rays" having (6) as a V1 analytical representation.

Theorem 1. *Assume the response function f is linear and the input I is given by (6). Then, the unique stationary state to equation (NF) is given for all $(x_1, x_2) \in \mathbb{R}^2$ by*

$$a_I(x_1, x_2) = \frac{\cos(2\pi\lambda x_2)}{1 - \widehat{\omega}(\lambda)} + \varepsilon g(x_1). \tag{7}$$

Here $g : \mathbb{R} \to \mathbb{R}$ has a discrete set of zeroes on $(0, +\infty)$.

The hypothesis on f implies that the stationary equation (2) is linear. It follows that the first term in the r.h.s. of (7) is the stationary state associated with the input P_F, and g is the stationary state associated to the second term in the r.h.s. of (6). Therefore, g is the solution of the 1-D stationary equation

$$b(x) = H(-x) + (\omega_1 * b)(x), \qquad x \in \mathbb{R}. \tag{8}$$

where the 1-D kernel ω_1 is readily computed to be a DoG. The proof of Theorem 1 then relies on harmonic and complex analysis techniques. Namely, applying the Fourier transform to both sides of (8), we obtain

$$\widehat{b} = (1 + \widehat{K})\widehat{H(-\cdot)}, \qquad \text{where} \qquad \widehat{K}(\xi) := \frac{\widehat{\omega_1}(\xi)}{1 - \widehat{\omega_1}(\xi)}. \tag{9}$$

The result then follows by estimating the inverse Fourier transform of $\widehat{K}$ via the residue theorem. We refer to a work in preparation for the complete details.

Theorem 1 implies that if the external input is the V1 representation of the "MacKay rays" defined by (6), then the associated stationary state corresponds to the V1 representation of the after-image reported by [8]. Moreover, we have the exponential convergence of $a(\cdot, t)$ on the stationary state when $t \to \infty$. It follows that equation (NF) theoretically describes the MacKay effect associated with the "MacKay rays" at the cortical level. Due to the retino-cortical map, we deduce the theoretical description of the MacKay effect for the "MacKay rays" in the retina.

4   Billock and Tsou's Experiments

In [10], we exhibited numerical results showing the capability of equation (NF) to reproduce Billock and Tsou's experiments. The stimuli used in these experiments are funnel or tunnel patterns localised at the fovea or periphery. Due to the retino-cortical map, this corresponds to taking as external inputs in equation (NF), $I = \varepsilon P_F v$ or $I = \varepsilon P_T v$, where $\varepsilon > 0$ and v is a localised function either in the left or in the right area of the cortex. In particular, the external inputs in these experiments do not fill all of the visual field. The function v can then be thought of as a localised control aiming to break the *global* plane Euclidean symmetry of stimuli patterns.

In this section, we prove that the equation (NF) with a linear response function f cannot describe these phenomena: In contrast to the MacKay effect, the phenomena reported by Billock and Tsou are completely nonlinear.

We will focus on the funnel pattern localised at the fovea. In V1, it corresponds to the following external input.

$$I(x) = \cos(2\pi\lambda x_2)H(-x_1), \qquad \lambda > 0, \quad x \in \mathbb{R}^2, \tag{10}$$

where H is the Heaviside step function. For ease of notation, we assume the kernel ω in (1) is such that $\kappa^2 = 2$. Since the convolution of Gaussians with zero mean remains a Gaussian with zero mean, the following is a direct consequence of Newton binomial formula.

Lemma 1. *Let $I \in \mathcal{S}'(\mathbb{R}^2)$. Assume that the response function f is linear. If $\mu < \mu_0$, the stationary state $a_I \in \mathcal{S}'(\mathbb{R}^2)$ of (NF) is given by*

$$a_I = I + \sum_{n=1}^{\infty} \mu^n \sum_{j=0}^{n} \binom{n}{j} (-1)^j g_{n,j} * I. \tag{11}$$

Here $g_{n,j}$ is the Gaussian defined for $x \in \mathbb{R}^2$, $n \in \mathbb{N}^*$ by

$$g_{n,j}(x) = \frac{1}{2\pi(n+j)\sigma^2} e^{-\frac{|x|^2}{2(n+j)\sigma^2}}, \quad j \in [[0, n]]. \tag{12}$$

Proposition 1. *Assume that the response function f is linear and $I \in L^\infty(\mathbb{R}^2)$ is defined by (10). If $\mu < \mu_0$, the stationary state of (NF) is given by*

$$a_I(x) = [H(-x_1) + R(x_1)] \cos(2\pi\lambda x_2), \quad x \in \mathbb{R}^2, \tag{13}$$

where $R \in L^\infty(\mathbb{R})$.

Proof. By Lemma 1, one computes for all $x \in \mathbb{R}^2$,

$$(g_{n,j} * I)(x) = \frac{\cos(2\pi\lambda x_2)}{\sigma\sqrt{2\pi(n+j)}} e^{-\frac{\lambda^2(n+j)\sigma^2}{2}} \int_{-\infty}^{0} e^{-\frac{(x_1-y_1)^2}{2(n+j)\sigma^2}} dy_1$$

$$= \frac{e^{-\frac{\lambda^2(n+j)\sigma^2}{2}}}{2} \, \mathrm{erfc}\left(\frac{x_1}{\sigma\sqrt{2(n+j)}}\right) \cos(2\pi\lambda x_2),$$

where erfc is the complementary error function.

Proposition 1 shows that the output a_I associated with (10) has a contribution in the right area of the cortex given by

$$a_{I,r}(x) = R(x_1) \cos(2\pi\lambda x_2), \quad x_1 > 0. \tag{14}$$

Since $a_{I,r}$ depends on the factor $\cos(2\pi\lambda x_2)$, the retino-cortical map tells us that a visual stimulus consisting of fan shapes in the centre induces an after-image containing fan shapes in the periphery instead of concentric rings only, as Billock and Tsou reported. Equation (NF) (provided we are far from the bifurcation point, i.e. $\mu < \mu_0 \le \mu_c$) with a linear response function cannot describe Billock and Tsou's experiments. Therefore, these phenomena depend fundamentally on the presence of the nonlinearity f.

5   Numerical Results

The numerical implementation is performed with Julia and is available via the link https://github.com/dprn/MacKay-illusion-2023/. Given an input I, the stationary state a_I is numerically implemented via an iterative fixed-point method.

We exhibit in Fig. 3 the MacKay effect for "MacKay rays" $I(x) = \cos(5\pi x_2) + \varepsilon H(2 - x_1)$, $\varepsilon = 0.025$. Here, we use a linear response function ($f(s) = s$). We stress that the phenomenon can be reproduced with any odd sigmoidal function, see e.g. [10, Fig. 3]. We exhibit in Figs. 4 Billock and Tsou's experiments for a funnel-like stimulus localised at the periphery. Here, we consider the nonlinear response function $f(s) = (1 + \exp(-s + 0.25))^{-1} - (1 + \exp(0.25))^{-1}$.

Fig. 3. Initial stimulus ("MacKay rays", *left*) inducing the MacKay effect (*right*).

Fig. 4. Billock and Tsou's experiments: funnel stimulus localised at the periphery (*left*) and after-image (*right*).

References

1. Amari, S.: Dynamics of pattern formation in lateral-inhibition type neural fields. Biol. Cybern. **27**(2), 77–87 (1977)
2. Billock, V.A., Tsou, B.H.: Neural interactions between flicker-induced self-organized visual hallucinations and physical stimuli. PNAS **104**(20), 8490–8495 (2007)
3. Bressloff, P.C., Cowan, J.D., Golubitsky, M., Thomas, P.J., Wiener, M.C.: Geometric visual hallucinations, euclidean symmetry and the functional architecture of striate cortex. Philos. Trans. R. Soc. Lond., B, Biol. Sci. 356(1407), 299–330 (2001)
4. Ermentrout, G.B., Cowan, J.D.: A mathematical theory of visual hallucination patterns. Biol. Cybern. **34**(3), 137–150 (1979)
5. Helmholtz, H.L.F.: Optic physiologique. Masson (1867)
6. Hubel, D.H., Wiesel, T.N.: Receptive fields of single neurones in the cat's striate cortex. J. Physiol. **148**(3), 574 (1959)
7. Klüver, H.: Mescal and mechanisms of hallucinations. University of Chicago, Chicago (1966)
8. MacKay, D.M.: Moving visual images produced by regular stationary patterns. Nature **180**, 849–850 (1957)
9. Nicks, R., Cocks, A., Avitabile, D., Johnston, A., Coombes, S.: Understanding sensory induced hallucinations: From neural fields to amplitude equations. SIAM J. Appl. Dyn. Syst. **20**(4), 1683–1714 (2021)
10. Tamekue, C., Prandi, D., Chitour, Y.: Reproducing sensory induced hallucinations via neural fields. In: 2022 IEEE-ICIP. pp. 3326–3330 (2022)
11. Zeki, S., Watson, J.D., Frackowiak, R.S.: Going beyond the information given: the relation of illusory visual motion to brain activity. Proc. Royal Soc. B: Biol. Sci. **252**(1335), 215–222 (1993)

Bio-Molecular Structure Determination by Geometric Approaches

Pseudo-dihedral Angles in Proteins Providing a New Description of the Ramachandran Map

Wagner Da Rocha[1,2]([✉]) [iD], Carlile Lavor[3] [iD], Leo Liberti[2] [iD], and Thérèse E. Malliavin[1] [iD]

[1] Laboratoire de Physique et Chimie Théorique, CNRS UMR7019 et Université de Lorraine, Nancy, France
`therese.malliavin@univ-lorraine.fr`
[2] Laboratoire d'Informatique de l'École Polytechnique, CNRS UMR 7161, Palaiseau, France
`wagner.rocha@lix.polytechnique.fr, leo.liberti@polytechnique.edu`
[3] University of Campinas (IMECC - UNICAMP), Campinas, Brazil
`clavor@unicamp.br`

Abstract. Since the first years of structural biology, the Ramachandran map has provided a simple definition of the curvilinear geometry of the protein backbone. Its definition is mainly based on the values of the dihedral angles ϕ and ψ measured between the heavy atoms of the protein backbone. Discontinuities in angle values are observed, particularly in the region of the β-strand secondary structure. We introduce new pseudo-dihedral angles involving hydrogen positions instead of some of the positions of the heavy atoms. We determine simple relationships between the old and new dihedral angles. We show that combining the old and new parameters allows us to overcome the discontinuity problem encountered in the Ramachandran map.

Keywords: interval Branch-and-Prune · pseudo-dihedral angles · Ramachandran map

1 Introduction

The development of structural biology has produced the availability of numerous protein structures. These structures can be considered as geometrical objects provided by nature and are therefore the subject of ongoing interest [6,10,15]. A reasonable quantification of the geometry in proteins is essential for a better definition of the biological function and activity of these molecules. The approaches easing this quantification are thus crucial for application-oriented developments in health and biotechnology.

For a decade, the Branch-and-Prune (BP) and interval Branch-and-Prune (iBP) algorithms have been developed to provide a solution of the Distance

Supported by CNPq, FAPESP, UNICAMP, CNRS, École Polytechnique and Université de Lorraine.

F. Nielsen and F. Barbaresco (Eds.): GSI 2023, LNCS 14072, pp. 511–519, 2023.
https://doi.org/10.1007/978-3-031-38299-4_53

Geometry Problem. These algorithms provide a theoretical framework for the parametrization as well as a systematic enumeration of the atomic coordinates [4,5]. These algorithms are based on the use of known inter-atomic distances as well as dihedral angles [14], and the derivation of relationships between angles would enlarge the possible applications. We present here a geometric relationship between the protein backbone dihedral angles ϕ and ψ to pseudo-dihedral angles defined from backbone atoms, including hydrogens (Sect. 2). Furthermore, we calibrate these formulas experimentally on data from sets of protein structures (Sect. 3). This new way to observe the curvilinear geometry of the backbone angles allows us to solve a discontinuity problem present in the classical definition of the Ramachandran map [11]. Removing the discontinuity helps to improve the systematic enumeration of protein conformations in the iBP algorithm [7] by avoiding the definition of multiple intervals due to the gaps in angle values.

2   Theory

The dihedral (torsion) angles in protein structures are defined for sets of four 3D Cartesian coordinates representing atoms A, B, C and D as the angles between the vectors normal to the planes $\{A, B, C\}$ and $\{B, C, D\}$. Considering two amino acid residues in a protein, consecutive in the primary sequence, R_i and R_{i+1}, the backbone dihedral angles are defined as [9]:

$$
\begin{aligned}
\phi_i &:= C^{i-1} - N^i - C_\alpha^i - C^i, \\
\psi_i &:= N^i - C_\alpha^i - C^i - N^{i+1}, \\
\omega_i &:= C_\alpha^{i-1} - C^{i-1} - N^i - C_\alpha^i,
\end{aligned}
\tag{1}
$$

where the ω_i angle populates two sets of values: $\omega_i \approx 0°$, in the case of the *cis* peptide bond, or $|\omega| \approx 180°$ in the case of the *trans* peptide bond.

By analogy with the definition from (1), and considering the amino acids residues R_j, R_i, and R_{i+1}, we define the following pseudo-dihedral angles:

$$
\begin{aligned}
\nu_{ji}^\kappa &:= C^{i-1} - N^i - C_\alpha^i - H_\kappa^j, & \upsilon_{ji}^\kappa &:= N^i - C_\alpha^i - C^i - H_\kappa^j, \\
\xi_i &:= H_\rho^i - N^i - C_\alpha^i - C^i, & \zeta_i &:= H_\varrho^{i+1} - C_\alpha^i - C^i - N^{i+1}, \\
\mu_{ji}^\kappa &:= H_\kappa^j - N^i - C_\alpha^i - H_\rho^i, & \eta_{ji}^\kappa &:= H_\kappa^j - C_\alpha^i - C^i - H_\varrho^{i+1},
\end{aligned}
\tag{2}
$$

where $\rho = \begin{cases} \alpha_2 \text{ if } R_i = \text{Glycine,} \\ \alpha \text{ otherwise,} \end{cases}$ $\varrho = \begin{cases} \delta_3 \text{ if } R_{i+1} = \text{Proline,} \\ N \text{ otherwise,} \end{cases}$ and κ represents any character that names a hydrogen atom in a protein. We need specific notations for the two amino acid residues Glycine and Proline: indeed, in Glycine residues, two hydrogen atoms are bonded to the C_α atom, H_{α_2} and H_{α_3} [9], and we select H_{α_2} to play the role of H_α; in Proline residues, no H_N atom is present, and we replace the atom H_N by H_{δ_3}.

As the backbone angles are defined in the interval $(-180°, 180°]$, all pseudo-dihedral angles expressed in this paper are also in this interval, and obey the same orientation: the counter-clockwise sense is positive and clockwise sense is negative. Concerning Eqs. (1) and (2), the angles ν_{ji}^κ, ξ_i, and μ_{ji}^κ are associated to the angle ϕ_i and the others to the angle ψ_i. We note that all angle definitions share the same two central atoms.

We present some examples about pseudo-dihedral angles in a protein structure. If R_i is neither a Proline nor a Glycine amino acid residue, assuming $i = j$, then $\kappa = N$ and $\rho = \alpha$ in the pseudo-dihedral angles defined in (2). To simplify the notation in this case, we consider $\nu_i \equiv \nu_{ii}^N$ and $\mu_i \equiv \mu_{ii}^N$. In Fig. 1, the case where $|\nu_i| \approx 180°$ and $\xi_i \approx -120°$ is presented.

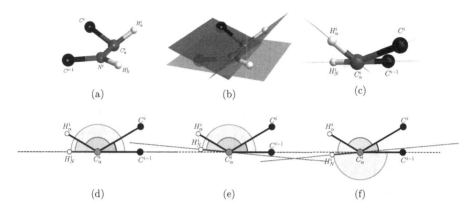

(a)   (b)   (c)

(d)   (e)   (f)

Fig. 1. (a) Spatial positions for the atoms C^{i-1}, N^i, H_N^i, C_α^i, H_α^i, and C^i whether $\nu_i = 180°$, $\xi_i = -120°$, and $\phi_i = 30°$. The numerical values bond length and bond angles are taken from [1]; (b) Definition of the planes $\{C^{i-1}, N^i, C_\alpha^i\}$ (green), $\{N^i, C_\alpha^i, C^i\}$ (blue), and $\{N^i, C_\alpha^i, H_\alpha^i\}$ (red) on the image (a) structure; (c) Rotation of image (b) structure identifying a pertinent point of view; (d) Orthogonal projection of the image (c) structure to the plane perpendicular to the bond $N^i - C_\alpha^i$. The dihedral angle ϕ_i (black) and the pseudo-dihedral angles ν_i (orange), ξ_i (red), and μ_i (light gray) are identified as planar angles and obey the angular relationship: $\phi_i = \nu_i + \xi_i + \mu_i$; (e) Planar approach of image (a) structure concerning $\nu_i = 175°$ with the angular relationship described by $\phi_i = \nu_i + \xi_i + \mu_i$; (f) Analogous case of image (e), though in this case $\nu_i = -175°$ and the angular relationship is expressed by $\phi_i = \nu_i + \xi_i + \mu_i + 360°$. (Color figure online)

The dihedral angles are represented in 2D in Fig. 1(d) to (f) without loss of generality, because we can always define an isometric function to switch the rotation axis defined by atoms N^i and C_α^i to the z-axis of the Cartesian coordinate system [8]. In this case, the angular information required to rotate any point around the z-axis is the same as needed to rotate a point of the plane xy around its origin [8].

The relationship between the dihedral and pseudo-dihedral angles changes depending on the values of the pseudo-dihedral angles involved. Any relationship described from those angles can be considered a composition of rotations in the plane. Indeed, as $\mathbb{R}^2 \cong \mathbb{C}(\mathbb{R})$ [13], we assume without loss of generality that $C_\alpha = 0$ of $\mathbb{C}$ and the atoms C^{i-1}, C^i, H_α^i, and H_N^i are points of $\mathbb{C}$ in a unitary circle. Then, the following rotations in $\mathbb{C}$ allow the definition of the relative atomic positions:

$$C^i = e^{\mathrm{i}\phi_i} C^{i-1}, \ C^i = e^{\mathrm{i}\xi_i} H^i_\alpha, \ H^i_\alpha = e^{\mathrm{i}\mu_i} H^i_N, \ \text{and} \ H^i_N = e^{\mathrm{i}\nu_i} C^{i-1}, \qquad (3)$$

where $\mathbf{i}$ is the imaginary unity. From the rotations of (3), we can write:

$$\begin{aligned} e^{\mathrm{i}\phi_i} C^{i-1} = e^{\mathrm{i}\xi_i} H^i_\alpha = e^{\mathrm{i}\xi_i} e^{\mathrm{i}\mu_i} H^i_N = e^{\mathrm{i}\xi_i} e^{\mathrm{i}\mu_i} e^{\mathrm{i}\nu_i} C^{i-1} = e^{\mathrm{i}(\xi_i + \mu_i + \nu_i)} C^{i-1} \Leftrightarrow \\ \left(e^{\mathrm{i}\phi_i} - e^{\mathrm{i}(\xi_i + \mu_i + \nu_i)} \right) C^{i-1} = 0 \Leftrightarrow e^{\mathrm{i}\phi_i} = e^{\mathrm{i}(\xi_i + \mu_i + \nu_i)}. \end{aligned} \qquad (4)$$

The solution for the complex number equation presented in (4) is given by

$$\phi_i = \nu_i + \xi_i + \mu_i + m \times 360°, \qquad (5)$$

such that $m \in \mathbb{Z}$. As all angles from (5) lie in the interval $(-180°, 180°]$, only $m \in \{-1, 0, 1\}$ must be considered in this solution.

A similar reasoning can be applied to describe the general case solution, which has pseudo-dihedral angles defined in atoms from the residues R_j, R_i and R_{i+1}, providing the Eq. (5):

$$\phi_i = \nu^\kappa_{ji} + \xi_i + \mu^\kappa_{ji} + m \times 360°, \ \text{such that} \ m \in \{-1, 0, 1\}. \qquad (6)$$

For a unique characterization of the result presented in (6), we must specify which value of m has to be considered in each circumstance. Theorem 1 provides a complete analysis.

Theorem 1. *Let R_j and R_i be amino acid residues of a protein, indexed by their positions in the primary sequence; ϕ_i, ν^κ_{ji}, ξ_i, and μ^κ_{ji} be the dihedral and the three pseudo-dihedral angles defined in (1) and (2), respectively. The dihedral angle $\phi_i \in (-180°, 180°]$ can be written in the function of the other three pseudo-dihedral angles by*

$$\phi_i = f_{m(\nu^\kappa_{ji}, \xi_i)}\left(\lambda^\kappa_{ji} + \mu^\kappa_{ji} \right), \qquad (7)$$

where $\lambda^\kappa_{ji} := \nu^\kappa_{ji} + \xi_i$,

$$m(\theta, \vartheta) := \begin{cases} 1 & if -180° < \theta \le -\vartheta, \\ -1 & if \quad -\vartheta < \theta \le 180°, \end{cases} \qquad (8)$$

and

$$f_m(\tau) := \begin{cases} \tau & if \ |\tau| < 180°, \\ \tau + m \times 360° & if \ |\tau| > 180°, \\ 180° & if \ |\tau| = 180°. \end{cases} \qquad (9)$$

Proof. As the pseudo-dihedral angles ν^κ_{ji} and μ^κ_{ji} can assume any value in the interval $(-180°, 180°]$, to analyze all possibilities for these angles, conveniently, we consider the intervals: $ⓘ : -180° < \nu^\kappa_{ji} \le -\xi_i$ and $ⓘⓘ : -\xi_i < \nu^\kappa_{ji} \le 180°$; $① : -180° < \mu^\kappa_{ji} \le 0°$ and $② : 0° < \mu^\kappa_{ji} \le 180°$. As it is mentioned in the examples, $-180° < \xi_i < 0°$; then:

$$ⓘ + ① \Rightarrow -360° < \nu^\kappa_{ji} + \mu^\kappa_{ji} \le -\xi_i \Leftrightarrow \begin{cases} -360° + \xi_i < \lambda^\kappa_{ji} + \mu^\kappa_{ji} \le -180°, \\ -180° < \lambda^\kappa_{ji} + \mu^\kappa_{ji} \le 0°. \end{cases}$$

$$ⓘ + ② \Rightarrow \begin{cases} -180° + \xi_i < \lambda^\kappa_{ji} + \mu^\kappa_{ji} \le -180°, \\ -180° < \lambda^\kappa_{ji} + \mu^\kappa_{ji} \le 180°. \end{cases}$$

$$(10)$$

So, for case ⓘ, we remark:

$$\begin{cases} -360° + \xi_i < \lambda_{ji}^\kappa + \mu_{ji}^\kappa < -180° \\ -180° \le \lambda_{ji}^\kappa + \mu_{ji}^\kappa \le 180° \end{cases} \Leftrightarrow \begin{cases} \xi_i < \lambda_{ji}^\kappa + \mu_{ji}^\kappa + 360° < 180° \\ -180° \le \lambda_{ji}^\kappa + \mu_{ji}^\kappa \qquad \le 180° \end{cases}$$

$$(11)$$

Observing the result from (11) in (6), we can say:

$$\begin{cases} \phi_i = \lambda_{ji}^\kappa + \mu_{ji}^\kappa + 360° & \text{if } \lambda_{ji}^\kappa + \mu_{ji}^\kappa < -180°, \\ \phi_i = \lambda_{ji}^\kappa + \mu_{ji}^\kappa & \text{if } \left|\lambda_{ji}^\kappa + \mu_{ji}^\kappa\right| \le 180°. \end{cases} \qquad (12)$$

With similar arguments from (10), for case ⓘⓘ, we can write:

$$\begin{cases} \phi_i = \lambda_{ji}^\kappa + \mu_{ji}^\kappa & \text{if } \left|\lambda_{ji}^\kappa + \mu_{ji}^\kappa\right| \le 180°, \\ \phi_i = \lambda_{ji}^\kappa + \mu_{ji}^\kappa - 360° & \text{if } \lambda_{ji}^\kappa + \mu_{ji}^\kappa > 180°. \end{cases} \qquad (13)$$

The results from (12) and (13) can be resumed by $\phi_i = f_{m(\nu_{ji}^\kappa, \xi_i)}\left(\lambda_{ji}^\kappa + \mu_{ji}^\kappa\right)$ where f_m and m are defined as the theorem presented them.

To determine the relationship among the angles ψ_i, ν_{ji}^κ, ζ_i, and η_{ji}^κ, we can use the same approach as the one shown to derive (6). As in the previous case, this result is given in Theorem 2, which can be proved similarly to Theorem 1.

Theorem 2. *Let R_j, R_i, and R_{i+1} be the amino acid residues of a protein, indexed by their positions in the primary sequence; ψ_i the dihedral angle defined in (1); ν_{ji}^κ, ζ_i, and η_{ji}^κ the pseudo-dihedral angles defined in (2). The dihedral angle $\psi_i \in (-180°, 180°]$ can be written in the function of the other three pseudo-dihedral by*

$$\psi_i = f_{m\left(\nu_{ji}^\kappa, \zeta_i\right)}\left(\sigma_{ji}^\kappa + \eta_{ji}^\kappa\right), \qquad (14)$$

with $\sigma_{ji}^\kappa := \nu_{ji}^\kappa + \zeta_i$, and m and f_m given by (8) and (9), respectively.

In the next section we experimentally provide the numerical expressions proved by the Theorems 1 and 2. Our experiments yield the association of the dihedral angles ϕ and ψ with the pseudo-dihedral angles μ and η, respectively, defined in two consecutive amino acid residues of a protein primary sequence. This assumption corresponds to the case $j = i$, and in this frame, the results propose an estimation for the angular constants $\lambda_{ii}^{\kappa_1}$ and $\sigma_{ii}^{\kappa_2}$, with $\kappa_1 \in \{N, \delta_3\}$ and $\kappa_2 \in \{\alpha, \alpha_2\}$.

3   Numerical Experiments

A dataset of 226 protein structures was extracted from the list of NMR structures related to the training of the neural network TALOS-N [12], by picking up the first conformer of each structure, as explained in [3].

On the protein structures database, the angles ϕ, ψ, μ, and η are calculated using the python library MDAnalysis [2]. The amino acid residues are sorted into the following types: Glycines, Prolines, and Others. A linear relationship is present by definition between the new angles and the angles ϕ and ψ. In order to explicit these relationship, linear regression was used to determine numerical slopes and intercepts of lines. For each case, at most two lines are observed. Figure 2 shows the results of the plot $\mu \times \phi$; the regression parameters for this case are presented in Table 1. The numerical calculation of $\eta \times \psi$ is given in Table 2. These numerical results allowed us to determine errors on the slopes and intercepts of the linear relationships presented in the theorems.

The distribution plots of the angles calculated on the database of protein structures are plotted (Fig. 3) using couples of dihedral backbone angles: ϕ/ψ (corresponding to the classical view of Ramachandran plots), μ/η, ϕ/η, and μ/ψ. Depending on the choice of the angle variable, the main secondary structure regions (α-Helix, β-Strand, and Loop) are displaced. For each secondary structure type, there is a combination of angles for which this region is connected in the modified Ramachandran map. This connectedness can simplify the application of the iBP algorithm, as the discontinuity due to the periodicity of angle values disappear for the considered secondary structure region.

Table 1. Linear Regression Parameters (L.R.P.), slope and intercept values, of the plot $\phi \times \mu$ regarding the equations $\phi_1(\mu) = a_1\mu + a_0$, $\phi_2(\mu) = b_1\mu + b_0$ analyzed on the protein data set.

L.R.P.	GLY	PRO	Others
a_1	1.0003 ± 0.0015	0.8739 ± 0.0137	0.9974 ± 0.0010
a_0	59.4967 ± 0.1477	16.2654 ± 1.3211	60.3836 ± 0.1375
b_1	0.9937 ± 0.0186	—	0.9767 ± 0.0060
b_0	-299.4858 ± 2.7960	—	-295.3706 ± 0.9490

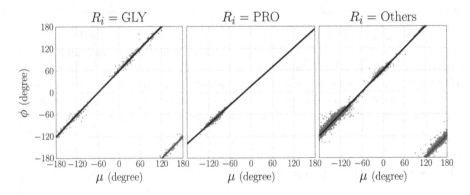

Fig. 2. The data are drawn in blue markers, and its linear regression by continuous lines in red and magenta. (Color figure online)

Table 2. Linear Regression Parameters (L.R.P.), slope and intercept values, of the plot $\psi \times \eta$ regarding the equations $\psi_1(\eta) = a_1\eta + a_0$, $\psi_2(\eta) = b_1\eta + b_0$, analyzed on the protein data set.

L.R.P. R_{i+1} / R_i		cis PRO	trans PRO	Others
a_1	GLY	—	0.9900 ± 0.0189	0.9992 ± 0.0013
a_0		—	139.2018 ± 0.8024	120.4346 ± 0.1393
b_1		—	0.9886 ± 0.0170	0.9999 ± 0.0042
b_0		—	-219.8191 ± 1.6198	-239.6703 ± 0.4781
a_1	Others	0.9163 ± 0.0805	1.0011 ± 0.0063	1.0004 ± 0.0003
a_0		81.5530 ± 5.2555	137.6785 ± 0.2224	119.3149 ± 0.0389
b_1		—	0.9896 ± 0.0187	0.9981 ± 0.0026
b_0		—	-221.6108 ± 2.9543	-240.2788 ± 0.3526

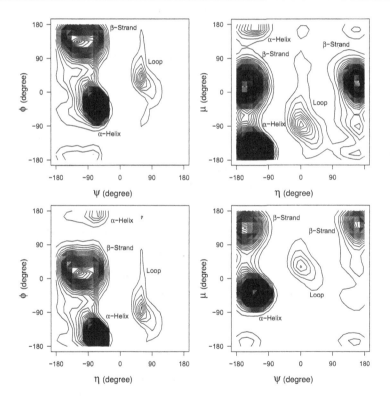

Fig. 3. Distribution of pseudo-dihedral/dihedral $\times$ pseudo-dihedral/dihedral angles. The labels α-Helix, β-Strand, and Loop allow us to follow the displacements of the main secondary structures region on the Ramachandran plot according to the change of variables in angles.

4   Conclusions

The backbone dihedral angles ϕ and ψ were proposed in the 1950s s to describe the curvilinear geometry of the protein backbone in the context of the newly developed X-ray crystallography approach for which the hydrogen positions were not visible in protein structures. Among the three main secondary structure regions, the β-Strand spread on regions in which ψ values display discontinuities jumping from $180°$ to $-180°$.

Extending the dihedral relationships to angles also involving hydrogen atoms, we proposed here new angles μ and η to describe the curvilinear geometry of the backbone angles. We proved precise relationships between ϕ and μ, and ψ and η. Analyzing the database of protein structures presented above, we obtained numerical values and errors for slopes and intercepts describing the relationships between the angles ϕ, ψ, μ, and η. Combining new and old angles, the discontinuities present in the classical Ramachandran map can be suppressed.

During an iBP calculation, the dihedral angle values ϕ and ψ can be used to prune tree branches, or their signed values can be used to create tree branches [14]. In the second case, depending on the considered region of the Ramachandran map, the discontinuities on ϕ and ψ angles can be an obstacle to their straightforward enumeration. In that case, the alternative use of other angle definitions, along with a change of atom ordering, could give the possibility for a straightforward enumeration. A similar approach could be derived, including long-range pseudo-dihedral angles involving atoms located in residues located far apart in the primary protein sequence, and would provide a new point of view on the description of protein structure. Nowadays, most of the long-range are inter-atomic distances, and an angular perspective could bring new insights into protein geometry.

Acknowledgements. Ecole Polytechnique, Lorraine University, ANR PRCI "Multiscale and multi-resolution bio-molecular structure determination by geometric approaches – multiBioStruct" (ANR-19-CE45-0019), CNPq, CNRS, and FAPESP are acknowledged for funding. Wagner Da Rocha thanks the ANR PRCI multiBioStruct (ANR-19-CE45-0019) project for postdoctoral support.

References

1. Engh, R., Huber, R.: Accurate bond and angle parameters for X-ray protein structure refinement. Acta Crystallogr. A **47**, 392–400 (1991)
2. Gowers, R., et al.: MDAnalysis: a Python package for the rapid analysis of molecular dynamics simulations. In: Proceedings of the 15th Python in Science Conference, Austin, TX, 2016 32, pp. 102–109 (2016)
3. Khalife, S., Malliavin, T., Liberti, L.: Secondary structure assignment of proteins in the absence of sequence information. Bioinform. Adv. **1**(1), vbab038 (2021)
4. Lavor, C., Liberti, L., Mucherino, A.: The interval Branch-and-Prune algorithm for the discretizable molecular distance geometry problem with inexact distances. J. Glob. Optim. **56**, 855–871 (2013)

5. Liberti, L., Lavor, C., Maculan, N., Mucherino, A.: Euclidean distance geometry and applications. SIAM Rev. **56**(1), 3–69 (2014)
6. Macari, G., Toti, D., Polticelli, F.: Computational methods and tools for binding site recognition between proteins and small molecules: from classical geometrical approaches to modern machine learning strategies. J. Comput. Aided Mol. Des. **33**(10), 887–903 (2019). https://doi.org/10.1007/s10822-019-00235-7
7. Malliavin, T.E., Mucherino, A., Lavor, C., Liberti, L.: Systematic exploration of protein conformational space using a distance geometry approach. J. Chem. Inf. Model. **59**(10), 4486–4503 (2019)
8. Duncan, M.: Applied Geometry for Computer Graphics and CAD. SUMS, Springer, London (2005). https://doi.org/10.1007/b138823
9. Nelson, D.L., Cox, M.M.: Lehninger Principles of Biochemistry: International Edition. W. H. Freeman & Co Ltd. (2021)
10. Pan, X., et al.: Expanding the space of protein geometries by computational design of de novo fold families. Science **369**(6507), 1132–1136 (2020)
11. Ramachandran, G., Ramakrishnan, C., Sasisekharan, V.: Stereochemistry of polypeptide chain configurations. J. Mol. Biol. **7**(1), 95–99 (1963)
12. Shen, Y., Bax, A.: Protein structural information derived from NMR chemical shift with the neural network program TALOS-N. Methods Mol. Biol. **1260**, 17–32 (2015)
13. Suetin, P.K., Kostrikin, A.I., Manin, Y.I.: Linear Algebra and Geometry. Taylor & Francis (1997)
14. Worley, B., et al.: Tuning interval Branch-and-Prune for protein structure determination. J. Global Optim. **72**(1), 109–127 (2018). https://doi.org/10.1007/s10898-018-0635-0
15. Zhao, J., Cao, Y., Zhang, L.: Exploring the computational methods for protein-ligand binding site prediction. Comput. Struct. Biotechnol. J. **18**, 417–426 (2020)

A Study on the Covalent Geometry of Proteins and Its Impact on Distance Geometry

Simon B. Hengeveld[1]($\boxtimes$), Mathieu Merabti[2], Fabien Pascale[2],
and Thérèse E. Malliavin[2]🄳

[1] IRISA, CNRS UMR 6074, University of Rennes 1, Rennes, France
`simon.hengeveld@irisa.fr`
[2] Laboratoire de Physique et Chimie Théoriques (LPCT), CNRS UMR 7019,
University of Lorraine, Vandoeuvre-lès-Nancy, France
`therese.malliavin@univ-lorraine.fr`

Abstract. The Distance Geometry Problem (DGP) involves determining the positions of a set of points in space based on the distances between them. Due to the importance of measuring spatial proximity's between atoms in structural biology, Distance Geometry finds a natural application to protein structure determination. When an instance of the DGP is discretizable, the search tree can be explored using the (interval) branch-and-prune algorithm (iBP). A statistical study on a database of protein structures shows that slight variations are observed in the covalent geometry of the molecules, depending on the Ramachandran region of the protein residues. These variations are in agreement with observations made on high resolution X-ray crystallographic structures. These slight variations may cause atomic clashes when using iBP algorithms to calculate a folded protein structure. In this contribution, we compare two software implementations (MDJEEP and IB-PNG) that attempt to address this issue.

Keywords: Distance Geometry Problem · interval Branch-and-Prune · local optimisation · protein structure determination · covalent geometry

1 Introduction

The Distance Geometry Problem (DGP) consists of finding the realization of a given set of points in a K-dimensional Euclidean space, where the distances between (some of) the pairs of points are known [20]. Formally, we can define the DGP using a simple weighted undirected graph $G = (V, E, d)$ [14,19], where the vertices V correspond to the points that we want to realize and the existence of an edge $(u, v) \in E$ implies that we know the distance from u to v. The weighting function d assigns each edge to a real interval $\delta(u, v)$. This interval may be degenerate.

CNRS, Lorraine University, University of Rennes.

F. Nielsen and F. Barbaresco (Eds.): GSI 2023, LNCS 14072, pp. 520–530, 2023.
https://doi.org/10.1007/978-3-031-38299-4_54

Definition 1. *Given a simple weighted undirected graph* $G = (V, E, d)$ *and an integer* $K \in \mathbb{Z}_+$ *the (assigned) DGP asks whether a realization*

$$x : v \in V \longrightarrow x_v \in \mathbb{R}^K \tag{1}$$

exists, such that

$$\forall \{u, v\} \in E, \quad ||x_u - x_v|| \in \delta(u, v) \tag{2}$$

where $|| \cdot ||$ *is the Euclidean norm.*

An important application of the DGP is found in the context of structural biology. In this case, the dimension K is equal to 3, the vertices V of G represent atoms, and the edges E (and their weights d) represent distance constraints between these atoms. Generally, these constraints can be divided into four sets. The first set corresponds to *holonomic constraints* [4], derived from the covalent geometry of molecules (using bonds and bond angles). The second set of distances arise from the knowledge of torsion angles within the molecules [10,15]. These torsion angles, combined with the bonds and bond angles lets us compute distances between some of the atoms in the backbone. The third set of distances are measured between atoms not related by a torsion angle and can be obtained using NMR experiments [9] or other structural biology techniques. The fourth set of restraints is imposed by the impossibility of inter-penetrating the van der Waals radii of the atoms. These three last sets of constraints are mostly defined by intervals, except if a torsion angle is considered to be exactly defined, as for the case of the torsion angle ω defining the peptide plane. As the distance constraints may contain a lot of uncertainty, the constraints (2) leave space for the weight $d(u, v)$ to map to a degenerate interval $\delta(u, v)$ (an exact distance):

$$||x_u - x_v|| = \delta(u, v), \tag{3}$$

or to match real a interval $[\underline{\delta}(u, v), \overline{\delta}(u, v)]$:

$$\underline{\delta}(u, v) \leq ||x_u - x_v|| \leq \overline{\delta}(u, v). \tag{4}$$

In the following, we consider the covalent geometry to include, in addition to the holonomic constraints, the torsion ω angle.

The instances of the DGP in the context of protein structure determination have been shown to belong to the DDGP subclass [10,13,15] and may be solved within the branch-and-prune (BP) framework [15], which explores the search tree generated by a discretization process. Early BP algorithms only worked for instances of the DDGP that only contain exact distances. Later, a BP algorithm which could handle both intervals as well as exact distances was proposed, which was referred to as the interval branch-and-prune (iBP) algorithm [13]. This opened the way for calculating protein conformations. Two main implementations of iBP have been introduced over the years: a recursive implementation illustrated by the open-source software MDJEEP [22] and an iterative implementation illustrated by the software IBP-NG [24]. The iBP approach was shown to be effective in various previous works [3,16,18] where NMR data were simulated from known PDB structures [2] or were obtained from experimental restraint files [10].

In the standard iBP approach, as the holonomic distances and torsion angles ω are considered as exact, the covalent geometry is assumed to be static. However, analyses of high resolution X-ray structures have shown [1] that bond angles as well as the torsion angle ω of the peptide plane vary along the Ramachandran regions. Section 2 describes a similar analysis, showing observed variations in the covalent geometry. Studying these tiny variations is of crucial importance to describe the architecture of protein structures. In principle, the variations of the covalent geometry could be tackled in the iBP approach by adding an additional level of branching including discrete variations of the bond angles. From a practical point of view, this would increase the algorithm complexity. To avoid this, one straightforward step is to add a different set of bond length and bond angle values for each residue to the input of the algorithm. In this work, we conducted experiments with two iBP implementations. In Sect. 3 we present more details on the branch-and-prune framework, and describe the two different BP implementations. Next, in Sect. 4 we present experiments with the two implementations to see how they deal with the variations of the covalent geometry of the proteins. Finally, in Sect. 5 we draw conclusions and describe perspectives for the future.

2   Variations in Covalent Geometry

To show that there are in fact variations in the covalent geometry of proteins, we analysed a data set of PDB structures obtained by X-ray crystallography. The data-set was prepared using the PISCES server [23]. The criteria for choosing the structures were: an identity smaller than 20% between protein sequences, an X-ray crystallographic resolution better than 1.6 Å and a *R-factor* better than 0.25. The R-factor is a measure of how well the refined structure predicts the observed data. Protein chains smaller than 100 residues and displaying no *cis* peptide bonds were selected, producing a data set of 391 structures.

To analyze the covalent geometry, we computed the average values for the bond angles between the backbone heavy atoms (N, Cα, C, O) as well as for the torsion angle ω of the peptide plane, using the MDAnalysis package [8]. The results of this analysis are displayed in Fig. 1. Uniform averages (left column) are observed along the amino acid types, except for the residues G, P and L. Next, average values have been calculated on various Ramachandran regions, using the definitions determined from high resolution crystallographic structures [11] (right column). These averages display more variations. In addition, one should notice that the torsion angle ω displays much larger variations than the bond angles, with standard deviations around 15° for ω and around 2° for the bond angles.

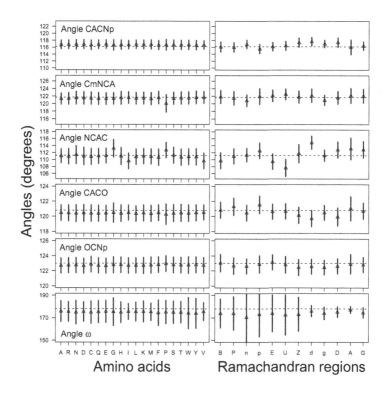

Fig. 1. Average values of the bond angles between the heavy backbone atoms (N, Cα, C, O) as well as on the torsion angle ω of the peptide plane. Np denotes the nitrogen atom N of the next residue, Cm denotes the carboxyl carbon C of the previous residue, and CA denotes the Cα atom. On the left we see the averages according to the amino-acid types, while on the right wee see them according to the Ramachandran region of the residue. The horizontal dashed lines correspond to the angles values from Ref. [5]

3   Branch-and-Prune Implementations

By making certain assumptions on the vertex order, the search space of DGP instance may be discretized [17]. DGP instances related to protein structure determination meet these assumptions [13], making them instances of the Discretizable Distance Geometry Problem (DDGP) [17]. The discrete search space has the structure of a tree, where the nodes of the tree represent a feasible position of an atom $v \in V$. Because the search space is a tree, we can potentially enumerate the complete set of solutions. The search tree is explored by BP algorithms [15]. In the "branching" phase, the BP algorithms will construct the search tree, while in a "pruning" phase branches will be removed if they contain solutions that are infeasible given the input distance constraints. The BP algorithms that can handle the interval distances present in the protein DDGP instances are referred to as interval BP (iBP) algorithms. However, to ensure that the DGP instance relating to proteins can be discretized, some distances must be exact [17]. The inter-atomic distances of which we are most certain are the bond and bond angle related distances, so in the BP-framework these distances are regarded as exact. However, the study in Sect. 2 shows that the covalent geometry contains some variations.

Two error-tolerant iBP-implementations are capable of handling these variations: IBP-NG [24] and MDJEEP [22].

3.1   IBP-NG

IBP-NG[1] makes use of both breadth-first and depth-first iterative tree searches based on a multidimensional index data structure [24]. The iterative order of the tree traversal is based on embedding equations derived from Clifford algebra [12], which makes it ideal for instances with highly repetitive vertex orders. When one converts the torsion backbone angles to a distance, their sign is lost. However, as shown in previous works, the sign of the torsion angles may be very useful during the pruning phase [21]. The aforementioned iterative order lets IBP-NG exploit this sign in the *branching phase* so that it avoids creating unnecessary branches to begin with.

Intervals are handled in a straightforward way. IBP-NG has an input variable B, which is referred to as the *branching factor*. In case an interval is encountered, the associated arc is cut into B pieces uniformly. For each of these pieces, a new branch is created. This means that IBP-NG branches at most $2B$ times for each vertex. In case we have information about the sign of the torsion angle involved in this distance, only B branches are created. A drawback of this approach is that IBP-NG may not be able to identify a solution, even if the instance is feasible.

3.2   MDJEEP

MDJEEP[2] is a recursive implementation of the BP framework [22].

It relies on a depth-first tree search to compute coordinates for each of the vertices. For the coordinate generation step, an efficient method using a rotation matrix is employed [7]. To allow for error in the input distances, MDJEEP makes use of a coarse-grained representation. This allows it to deal with the continuous feasible subsets of potential atomic positions while preserving the general tree structure [18]. A three-dimensional box is assigned to each vertex v, which we suppose contains the *true* position of the atom. After computing an initial position for v within its box, it is then allowed to move in a refinement step. This refinement includes local optimization using a spectral projected gradient method. These boxes also come into play when MDJEEP is handling interval distances. The advantage of this refinement step is that it allows MDJEEP to identify solutions where IBP-NG may not be able to.

The published C-version of MDJEEP only allows for pure instances that only contain distances. To allow for a proper comparison with IBP-NG, the algorithm

---

[1] https://github.com/geekysuavo/ibp-ng.
[2] https://github.com/mucherino/mdjeep.

was reimplemented into Java and extended so that it is capable of using the sign of the dihedral angles to prune. The Java language was chosen because new functionalities may more easily be included in the software tool in the future when a higher-level programming language is used. In the experiments presented below, the Java implementation is used.

4   Computational Experiments

We ran several experiments using both IBP-NG and MDJEEP in order to compare how the iBP-solvers deal with the variations described in Sect. 2.

4.1   Input

The experiments were conducted using a subset of proteins from the PISCES database described in Sect. 2. Then, for each of these proteins we generated distances and angles to create the DDGP instances to be solved by the two iBP methods. The input consists of the following three types of distances.

The first type are holonomic distances, uniform or non-uniform along the sequence. Normally, these are taken from force field parameter files. In these experiments, we only used the bond lengths from the force field PARALLHDG (version 5.3) [5]. For the bond-angles we take a more targeted approach. Usually, in the input of iBP-algorithms, the bond-angle of a triplet atoms is also taken from the force-field parameters, and thus solely depends on the three atom types of the triplet. In this experiment, we vary the bond angles (and thus their associated distances) depending on which residue in the protein sequence the atoms belong to. The bond angles of each residue were taken from the initial PDB structure.

The second type used are distances derived from the backbone dihedral angles ϕ and ψ and ω. The ϕ and ψ angles were taken from the PDB structures, and uncertainty was introduced by creating 10 °C intervals around them. For the ω angles, the procedure differed between the two implementations. The exact ω values of the initial PDB structure were used in the IBP-NG runs, while for MDJEEP 10 °C intervals were used. In order to prune using the torsion angle values directly, we also included the angle intervals directly in the input.

Finally, we have distances arising from van der Waals restraints. The van der Waals radii were scaled by 0.1 in order to allow for some degree of overlap between pairs of var der Waals spheres, as we can observe in real conformations.

Aside from backbone dihedral angles, we also prune using the sign of the improper torsion angles, which defines the stereo-chemistry of L-amino acids. These improper angles were included directly in the input and taken from force-field parameter files.

4.2   Results

The two approaches, IBP-NG and MDJEEP, were applied to the data set of protein structures. Among the 391 structures smaller than 100 residues present in the

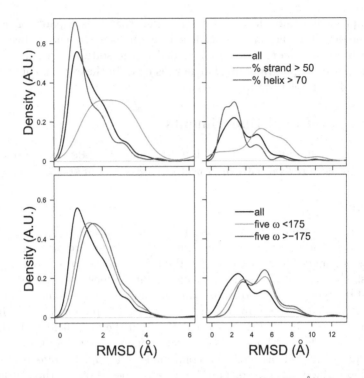

Fig. 2. Distributions of the root-mean-square deviations (RMSD, Å) of atomic coordinates between the initial PDB structure and the structure recalculated by IBP-NG (left) and MDJEEP (right). Upper panel: effect of the percentage of secondary structures: α helices, and β strands, determined by STRIDE [6], on the RMSD distributions. Lower panel: effect of the number of ω outliers outside of the ranges [175,180] or [−180, −175]. (Color figure online)

data set, 32 produced no solution for IBP-NG. MDJEEP produced solutions for all structures in data-set. The quality of the computed structures can be verified by comparing the solutions to the initial PDB structures. In practice, this is done by calculating the root-mean-square deviations (RMSD, Å) between the atomic coordinates. For IBP-NG, for each protein, multiple structures were computed, and the best solution (with the lowest RMSD value) was kept. For MDJEEP, execution was cut after the first found solution, potentially leading to lower quality solutions. This choice was made, because MDJEEP took too long to explore the full search tree.

The distribution of RMSD values (Fig. 2) are different for the two approaches, IBP-NG producing smaller range (0-6 Å) than MDJEEP (0-13 Å). Figure 2 shows the distributions of the RMSD values for IBP-NG (left) and for MDJEEP (right). The RMSD values were computed using the MDAnalysis package [8]. The RMSD values (Fig. 2, upper panels) are also analyzed according to the percentages of α helix and β strand present in the structures, and calculated using STRIDE [6].

The proteins displaying percentage of α helix larger than 70% of the sequence display RMSD distributions shifted towards smaller values (blue curves). At the contrary, the percentages of β strand larger than 50% induce a shift of distributions towards larger values (green curves). Besides, the presence of a larger number of outliers for ω values (Fig. 2, lower panels) induces also a shift of the distributions towards larger RMSD values.

Table 1. Averaged differences of bond and torsion angles ($^\circ$) between the conformations obtained during the runs MDJEEP and IBP-NG, and the PDB structures. The absolute values (above) is the average of the absolute differences. The signed values (below) instead is the average of the signed differences.

Experiment	Averaging	CmNCA	NCAC	CACNp	CACO	OCNp	ϕ	ψ	ω
MDJEEP	absolute	1.336	1.296	1.399	0.087	0.071	1.692	1.701	4.165
	signed	−0.217	−0.264	−0.429	0.016	−0.054	1.655	1.566	4.023
IBP-NG	absolute	1.400	1.253	0.025	0.056	0.029	0.035	0.036	0.024
	signed	−0.162	0.165	−2.05e-4	0.0436	−1.94e-4	2.83e-4	0.66e-4	−0.19e-4

In general, IBP-NG produces solutions that are closer to the X-ray crystallography structures. This difference between the two approaches can be explained by calculating the averaged differences of bond and torsion angles between the obtained conformations and the initial PDB conformations (Table 1). All angles display differences smaller than 5°, which justifies the two strategies used for avoiding atomic clashes: the local optimization approach or the residue-dependent variation of the covalent geometry. Nevertheless, smaller differences for torsion angles ϕ, ψ and ω are observed for IBP-NG compared to MDJEEP. By contrast, the differences between the bond angles are similar for the two runs. For all parameters, the signed difference display smaller values than the absolute ones, which agrees with differences displaying alternating positive and negative signs.

Overall, the analysis of RMSD values (Fig. 2) as well as of the averaged differences of bond and torsion angles, reveals a stronger influence of ϕ, ψ and ω values to the variation of protein conformations. The influence of the ω torsion angle can be put in parallel with the larger variations of this angle among the Ramachandran regions located in the loop and β strand regions (Fig. 2). This supports the conclusion that the conformational drift of conformations is directly related to the variation of the covalent geometry along the protein sequence. The RMSD shifts observed for proteins containing mostly α or β secondary structures can be put in parallel with the standard deviations of the ω angle in the Ramachandran regions corresponding to the α helix and β strand regions (labels A and B respectively in Fig. 1).

Comparing the two implementations, IBP-NG seems to be able to compute structures that are closer to the original X-ray crystallography structures. However, bear in mind that for MDJEEP, only the first found solution for each protein was kept, while for IBP-NG multiple solutions were considered. Furthermore, there was more uncertainty in the input instances of MDJEEP because intervals were used on the ω angles, while for IBP-NG exact values were used. Because MDJEEP takes too long to explore the entire tree, we cannot be sure whether it can potentially give similar quality solutions as ibp-ng. Part of the reason why MDJEEP is slower may be due to the local optimization refinement step. There are many distances in the input, which makes computing the gradient a time-consuming procedure. However, on the other hand, MDJEEP was able to identify solutions for all proteins in the instance because of the local optimization, where as IBP-NG was not.

5   Discussion

The work presented here has confirmed the variability of bond angles and ω torsion angle according to the location of the amino acid residues in the Ramachandran maps. This complicates distance-based approaches, when attempting to determine protein structures in the context of the branch-and-prune framework. We presented two alternatives which may deal with this problem, encapsulated by the MDJEEP and IBP-BG software packages. Although both methods permits to determine solution with bond and torsion angles close to the target values, they did not permit to systematically reach small coordinate deviations. Nevertheless, the drift underlying the calculations may come from patterns of bond and torsion angles along successive residues. This could be investigated by analyzing triplets of amino acids successive in the primary sequence, in place of the analysis of single amino acids performed here.

The approaches presented here to facilitate the generation of protein structures can constitute a first step for the application of the iBP algorithm to the reconstruction of missing regions encountered in X-ray crystallographic or cryo-EM structures. As many of these regions correspond to loop secondary structures, the variability of bond and torsion angles specifically observed for these structures, will be essential to realize an efficient reconstruction. To improve the results in future work, the strong suits of the two implementations should be put together into one.

Acknowledgements. CNRS, Lorraine University and University of Rennes are acknowledged for funding. ANR PRCI "Multi-scale and multi-resolution bio-molecular structure determination by geometric approaches - multiBioStruct" (ANR-19-CE45-0019) was supporting this work.

References

1. Berkholz, D.S., Shapovalov, M.V., Dunbrack, R.L., Karplus, P.A.: Conformation dependence of backbone geometry in proteins. Structure **17**, 1316–1325 (2009)
2. Berman, H., et al.: The protein data bank. Nucleic Acids Res. **28**, 235–242 (2000)
3. Cassioli, A., et al.: An algorithm to enumerate all possible protein conformations verifying a set of distance restraints. BMC Bioinform. **16**, 23 (2015)

4. Crippen, G.M.: Linearized embedding: a new metric matrix algorithm for calculating molecular conformations subject to geometric constraints. J. Comput. Chem. **10**, 896–902 (1989)
5. Engh, R., Huber, R.: Accurate bond and angle parameters for X-ray protein structure refinement. Acta Crystallogr. A **47**, 392–400 (1991)
6. Frishman, D., Argos, P.: Knowledge-based protein secondary structure assignment. Proteins **23**, 566–579 (1995)
7. Gonçalves, D.S., Mucherino, A.: Discretization orders and efficient computation of cartesian coordinates for distance geometry. Optim. Lett. **8**(7), 2111–2125 (2014). https://doi.org/10.1007/s11590-014-0724-z
8. Gowers, R., et al.: MDAnalysis: a python package for the rapid analysis of molecular dynamics simulations. In: Proceedings of the 15th Python in Science Conference, Austin, TX, 2016 32, pp. 102–109 (2016)
9. Harris, R.: Nuclear Magnetic Resonance. Pearson Education Limited (1971)
10. Hengeveld, S.B., Malliavin, T., Lin, J., Liberti, L., Mucherino, A.: A study on the impact of the distance types involved in protein structure determination by NMR. In: Computational Structural Bioinformatics Workshop (CSBW21), IEEE International Conference on Bioinformatics and Biomedicine (BIBM 2021), pp. 2502–2510 (2021)
11. Hollingsworth, S.A., Karplus, P.A.: A fresh look at the Ramachandran plot and the occurrence of standard structures in proteins. Biomol. Concepts **1**, 271–283 (2010)
12. Lavor, C., Alves, R., Figueiredo, W., Petraglia, A., Maculan, N.: Clifford algebra and the discretizable molecular distance geometry problem. Adv. Appl. Clifford Algebras **25**(4), 925–942 (2015). https://doi.org/10.1007/s00006-015-0532-2
13. Lavor, C., Liberti, L., Maculan, N., Mucherino, A.: The discretizable molecular distance geometry problem. Comput. Optim. Appl. **52**, 115–146 (2012)
14. Liberti, L., Lavor, C., Maculan, N., Mucherino, A.: Euclidean distance geometry and applications. SIAM Rev. **56**, 3–69 (2014)
15. Liberti, L., Lavor, C., Maculan, N.: A branch-and-prune algorithm for the molecular distance geometry problem. Int. Trans. Oper. Res. **15**, 1–17 (2008)
16. Malliavin, T.E., Mucherino, A., Lavor, C., Liberti, L.: Systematic exploration of protein conformational space using a distance geometry approach. J. Chem. Inf. Model. **59**, 4486–4503 (2019)
17. Mucherino, A., Lavor, C., Liberti, L.: The discretizable distance geometry problem. Optim. Lett. **6**, 1671–1686 (2012)
18. Mucherino, A., Lin, J.H.: An efficient exhaustive search for the discretizable distance geometry problem with interval data. In: 2019 Federated Conference on Computer Science and Information Systems (FedCSIS19), Workshop on Computational Optimization (WCO19), pp. 135–141 (2019)
19. Mucherino, A.: On the Discretization of Distance Geometry: Theory, Algorithms and Applications. HDR Monograph, University of Rennes 1 (2018)
20. Mucherino, A., Lavor, C., Liberti, L., Maculan, N.: Distance Geometry: Theory. Methods and Applications. Springer, New York (2013). https://doi.org/10.1007/978-1-4614-5128-0
21. Mucherino, A., Lavor, C., Malliavin, T., Liberti, L., Nilges, M., Maculan, N.: Influence of pruning devices on the solution of molecular distance geometry problems. In: Pardalos, P.M., Rebennack, S. (eds.) SEA 2011. LNCS, vol. 6630, pp. 206–217. Springer, Heidelberg (2011). https://doi.org/10.1007/978-3-642-20662-7_18

22. Mucherino, A., Liberti, L., Lavor, C.: MD-jeep: an implementation of a branch and prune algorithm for distance geometry problems. In: Fukuda, K., Hoeven, J., Joswig, M., Takayama, N. (eds.) ICMS 2010. LNCS, vol. 6327, pp. 186–197. Springer, Heidelberg (2010). https://doi.org/10.1007/978-3-642-15582-6_34
23. Wang, G., Dunbrack, R.L.: PISCES: a protein sequence culling server. Bioinformatics **19**, 1589–1591 (2003)
24. Worley, B., et al.: Tuning interval Branch-and-Prune for protein structure determination. J. Global Optim. **72**(1), 109–127 (2018). https://doi.org/10.1007/s10898-018-0635-0

Exploration of Conformations for an Intrinsically Disordered Protein

Shu-Yu Huang[1], Chi-Fon Chang[2], Jung-Hsin Lin[1],
and Thérèse E. Malliavin[3(✉)] (iD)

[1] Research Center for Applied Sciences, Academia Sinica, Taipei, Taiwan
{shuyu,jhlin}@gate.sinica.edu.tw
[2] Genomics Research Center, Academia Sinica, Taipei, Taiwan
chifon@gate.sinica.edu.tw
[3] Laboratoire de Physique et Chimie Théoriques (LPCT), University of Lorraine,
Vandoeuvre-lès-Nancy, France
therese.malliavin@univ-lorraine.fr

Abstract. Intrinsically disordered proteins (IDP) are at the center of numerous biological processes, and attract consequently extreme interest in structural biology. A systematic enumeration of protein conformations, based on distance geometry, was performed on SERF1a, a 62-residue IDP involved in interactions with amyloid peptides. The results obtained with the interval Branch-and-Prune (iBP) approach haven been compared with those produced by *flexible-meccano*, using various predictions for backbone torsion angles ϕ and ψ, provided by TALOS and δ2D. The similarity between profiles of local gyration radii provides to a certain extent a converged view of the SERF1a. A better convergence is observed when using the TALOS inputs than using the δ2D inputs. *flexible-meccano* provides a less converged view of the protein conformational space than TAiBP.

Keywords: intrinsically disordered proteins · interval Branch-and-Prune · conformational space

1 Introduction

The Distance Geometry Problem (DGP) consists of finding the embedding of a given set of points in a dimension K, where the distances between some pairs of points are known. Formally, we can define the DGP using a simple weighted undirected graph G [11].

Definition 1. *Given a simple weighted undirected graph $G = (V, E, d)$ and an integer $K \in \mathbb{Z}_+$ the (assigned) DGP asks whether a realization*

$$x : v \in V \longrightarrow x_v \in \mathbb{R}^K \tag{1}$$

CNRS, Lorraine University, Academia Sinica.

exists, such that

$$\forall \{u, v\} \in E, \quad ||x_u - x_v|| = d(u, v) \tag{2}$$

where $|| \cdot ||$ is the Euclidean norm.

In the context of structural biology, numerous problems of protein structure determination can be rewritten as DGP problems. Indeed, many experimental pieces of information, as in particular Nuclear Magnetic Resonance (NMR) measurements, can be expressed as information on inter-atomic distances. In that frame, the dimension K is equal to 3, the vertices V of G represent atoms, and the edges E (and their weights d) represent distance constraints between these atoms.

In molecular biology, an intrinsically disordered protein (IDP) is a protein that lacks a fixed or ordered three-dimensional structure [3]. IDP are at the center of the attention in the structural biology of proteins, as disordered residues are expected to constitute 35 to 50% of the human proteome [15]. IDPs represent also a challenge for structural biology as most of biophysical techniques concentrate on time-average or space-average data in order to obtain sufficient signal-to-noise ratio. In that frame, the observation of disordered molecules is hampered by the averaging of the observable signal over heterogeneous sets of conformations. On the other hand, the signal produced by individual conformations is in most of the cases not strong enough to be recorded. As the conformational space populated by intrinsically disordered proteins is enormous, an infinite number of conformation sets can be proposed fitting the measurements performed on the IDP sample and most of the approaches used to generate IDP conformations are thus based on Monte Carlo algorithms [16].

During the last years, a new approach for the exploration of the conformational space of IDPs, based on a systematic enumeration of conformations in the frame of DGP, has been proposed [13], based on the interval Branch-and-Prune (iBP) algorithm [10]. This approach of threading-augmented interval Branch-and-Prune (TAiBP) [13,18] provides a framework for the systematic enumeration of protein conformations, while overcoming the combinatorial explosion problem intrinsic to this enumeration. It was recently applied to the analysis of the conformational space of a tandem domain of protein whirlin [12] as well as to Sic1 and pSic1, corresponding to unphosphorylated and phosphorylated states of an IDP [5]. TAiBP is composed by two steps: (i) individual iBP calculations of peptide fragments spanning the studied protein; (ii) enumeration of protein conformations by systematic assembly of fragment conformations. The inputs of the step (i) are predicted ϕ/ψ boxes systematically combined by permutation to prepare individual iBP calculations [13]. Within the iBP step, a tree, in which each node corresponds to one atom position, is built, generating the various possibilities for atom positions (branching step) whereas additional geometric information is used to accept or reject the newly built branch (pruning step). The conformations of peptide fragments are then assembled by superimposing the last and initial residues of the fragments successive in the sequence. The assembled conformations in which Cα atoms closer than 1 Å are observed, are pruned from the calculation.

To scale down the combinatorial nature of the search in the entire conformational space, a clustering approach, the **Self-Organizing Map** (SOM) [1,8,9,14], which is an artificial neural network (ANN) trained using unsupervised learning, was used to reduce the number of conformations. The SOM approach was used after each iBP calculation or assembly step as soon as the number of saved conformations was larger than 100. After each SOM clustering, the set of conformations is replaced by a representative conformation set corresponding to the most homogeneous regions of initial conformations. We studied here the small EDRK-rich factor 1 (SERF1a), a 62-residue protein (NCBI accession number NP_001171558) involved in protein aggregation [7]. Up to now, no conformations have been published for this protein. Its exact function is unknown, but it plays an important role in the diseases such as Alzheimers, Parkinsons, and Huntingtons. A recent study [4] showed that the isolated SERF1a is predominantly disordered and that the interaction of SERF1a with α-synuclein facilitate the conversion of α-synuclein monomers into amyloid fibers. Our NMR study of SERF1a at two close pH values (6.0 and 6.8) allows us to measure the chemical shifts, providing the inputs for TALOS-N [17], and δ2D [2]. The regions of backbone dihedral angles, provided by TALOS-N as well as by δ2D, have been discretized and the individual ϕ and ψ values used for generating tree branches. TAiBP [13] and *flexible-meccano* [16] have then been used to generate protein conformations. Flexible-meccano builds multiple, different copies of the same polypeptide chain by randomly sampling amino acid-specific backbone dihedral angle ϕ/ψ potential wells. The effects of different input predictions for torsion angles ϕ and ψ on the obtained conformations will be investigated here.

2   Materials and Methods

2.1   NMR Experiments

The sample preparation as well as the NMR measurements have been performed by the authors. The NMR spectra were recorded at 298K on Bruker Avance 800MHz equipped with cryoprobe. NMR samples are 0.5mM ^{13}C- ^{15}N- labeled SERF1a in 20mM NaPi (pH6.0 or pH6.8), and 20mM NaCl buffers, in 90% (v/v) H_2O and 10% (v/v) D_2O. NMR spectra of ^1H-^{15}N HSQC, ^1H-^{13}C HSQC, CBCA(CO)NH, HNCACB, HCCH-TOCSY, HCCH-COSY, and HNCO were acquired for backbone and sidechain assignments. NMR spectra were processed using software TopSpin 3.6.5 (Bruker) and peak picked using SPARKY [6].

2.2   Prediction of ϕ and ψ Input Boxes from the Chemical Shifts

The inputs to TAiBP and *flexible-meccano* were determined as boxes of the Ramachandran map, defined from intervals of backbone angles ϕ and ψ. Two sets of boxes were defined starting from the likelihood prediction of the neural network TALOS-N [17], and from the prediction of populated regions of Ramachandran map by δ2D [2].

From the NMR chemical shifts and the protein sequence information, the artificial neural network (ANN) TALOS-N predicts the likelihood that the backbone torsion angles of a given residue n fall in any of the 324 voxels, of $20° \times 20°$

each, that make up the Ramachandran map [17]. The ϕ/ψ distribution produced by TALOS-N for each residue was normalized in order that the sum of all voxels values is equal to one, and voxels for which the normalized value was larger than 0.1, were selected. Boxes were automatically determined starting from maximum value of the probability map and iteratively adding neighbouring pixels up to the situation where all neighboring pixels larger than the threshold have been included in the box.

Starting from the $\delta 2D$ prediction, only the predicted populations larger than 10% were kept for each residue. For these populations, the corresponding Ramachandran regions (helix, strand, polyprolines and coil) were converted to the following regions: α ($-100 < \phi < 0$; $-100 < \psi < 50$), β ($-210 < \phi < -100$; $50 < \psi < 260$), polyproline ($-100 < \phi < 0$; $50 < \psi < 260$), coil ($0 < \phi < 180$; $-50 < \psi < 100$).

The ϕ/ψ boxes are the standard inputs of the TAiBP approach. For *flexible-meccano*, the box information is converted into a Gaussian probability centered to the middle of the box and using widths of 200 or 300°.

2.3   Generation of Protein Conformations

Protein conformations were generated using the TAiBP approach outlined in the introduction and described in more details in [5,12,13,18]. After the assembly step, the sidechains have been added to the conformation backbones, and the conformations were refined by molecular dynamics simulations. It should be noticed that the unsupervised self-organizing map (SOM) approach was shown in the frame of molecular dynamics simulations [14] to sort conformations according to basins of energy. By analogy to the observations done on molecular dynamics trajectories, the use of SOM induces the selection of conformations belonging to energy basins.

Table 1. List of runs performed using torsion angle inputs derived from SERF1a chemical shifts. Each run was performed using chemical shifts measured at two pH values. The duplicated TAiBP runs are labeled 'bis'.

Run name	pH	Input box prediction	Generation of conformers	Number of generated conformations
run1	6	-	*flexible-meccano*	10000
run1	6.8	-	*flexible-meccano*	10000
run2	6	$\delta 2D$ Gaussian 200°	*flexible-meccano*	10000
run2	6.8	$\delta 2D$ Gaussian 200°	*flexible-meccano*	10000
run3	6	$\delta 2D$ Gaussian 300°	*flexible-meccano*	10000
run3	6.8	$\delta 2D$ Gaussian 300°	*flexible-meccano*	10000
run4, run4bis	6	$\delta 2D$	TAiBP	42, 59
run4, run4bis	6.8	$\delta 2D$	TAiBP	125, 33
run5, run5bis	6	TALOS threshold 0.01	TAiBP	191, 93
run5, run5bis	6.8	TALOS threshold 0.01	TAiBP	99, 73
run6, run6bis	6	TALOS threshold 0.02	TAiBP	73, 51
run6, run6bis	6.8	TALOS threshold 0.02	TAiBP	152, 78

The *flexible-meccano* calculations were performed using the resource nmrbox.nmrhub.org and generating 10000 conformations. Two sets of calculations were performed: one set without specific restraints in which the ϕ and ψ values are picked up by Monte Carlo using the generic probability data-set of *flexible-meccano*. In the other set of calculations, specific ϕ and ψ intervals were defined for each residues using a Gaussian shape, as described before.

3   Results

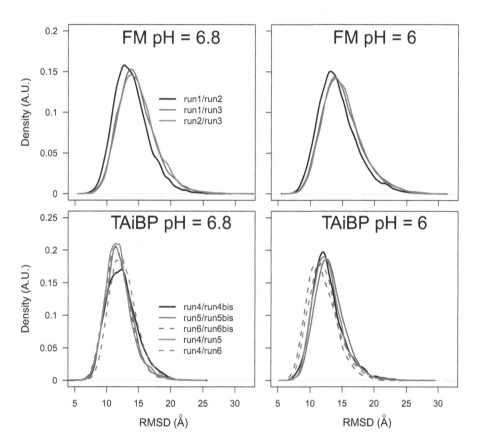

Fig. 1. Root-mean-square deviations (RMSD, Å) between atomic coordinates of conformations obtained using *flexible-meccano* or TAiBP. The run names are those given in Table 1.

The various runs were first analyzed by calculating the root-mean-square deviation (RMSD) between the atomic coordinates of the conformations (Fig. 1). The RMSD values between the TAiBP runs populate 5–20 Å intervals narrower than those of 5–25 Å observed for the *flexible-meccano* calculations. In addition, the RMSD values in the 15–20 Å are less populated in TAiBP than in *flexible-meccano* conformations. For a given type of runs, the variations of ϕ and ψ inputs do not seem to have a strong influence on the distribution of conformations, as the RMSD values are quite superimposed. The narrower RMSD values sampled by TAiBP agree with the selection of conformations by SOM clustering, described in the introduction.

The distribution of gyration radii (R_g) (Fig. 2) displays differences between *flexible-meccano* and TAiBP conformations, similar to those observed for RMSD values. Indeed, the distribution of R_g values is narrower in TAiBP than in *flexible-meccano* conformations. Nevertheless, the maximum of the distributions is around 20 Å for all cases.

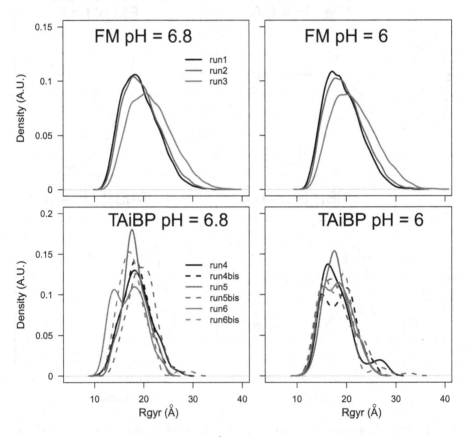

Fig. 2. Distribution of gyration radii (Å) among the sets of conformations obtained using *flexible-meccano* or TAiBP. The run names are those given in Table 1.

Larger gyration radii around 30–35 Å range are observed in *flexible-meccano* than in TAiBP conformations. These values in the 30–35 Å range correspond to a minority of *flexible-meccano* conformations. For TAiBP, the different kind of inputs (TALOS (run5,run6) versus δ2D (run4)) induce variations in the distributions similar to those observed between the duplicated TAiBP runs (Table 1).

For *flexible-meccano*, the input information seems not to have much influence on the distribution of R_g values, as similar distributions are observed for all inputs. Interestingly, small outlier peaks are observed repeatedly around 27 Å at pH 6 for the TAiBP conformations. To describe the local variations in the shape of conformations obtained by *flexible-meccano* and TAiBP, we calculated the cumulative sum of the profiles of local gyration radii. These profiles P_q of local gyration radii, introduced in [5], are calculated along residue number n for each conformation q in the following way:

$$P_q(n) = \sqrt{\frac{1}{N_n} \sum_{i=n-N_{win}}^{n+N_{win}} (\mathbf{X}_i - \mathbf{X}_n^{ave})^2} \tag{3}$$

where $\mathbf{X}_i$ represents the vector of atomic coordinates for the backbone atoms of residue i in the range $n - N_{win}, n + N_{win}$, and $N_{win} = 5$ is the residue window around n on which a local gyration radii is calculated, N_n being the number of backbone atoms located in this window. $\mathbf{X}_n^{ave}$ is the coordinate vector of the centroid of the atomic coordinates of the backbone atoms of residues in the range $n - N_{win}, n + N_{win}$. The profiles give a qualitative description of the conformation separated in extended regions (profile maxima) and in aggregated regions (profile minima).

The profiles of local gyration radii (Fig. 3) calculated from the *flexible-meccano* conformations display variable shape with maxima of profiles located in various places along the protein sequence, depending of the run. At the contrary, for TAiBP conformations, the region of residues 1–25 displays peaks in the profiles, whereas the region 30–50 is more flat. This is specially true for the calculations using TALOS inputs (runs 5, 6), whereas the δ2D inputs (runs 4) produce profiles displaying less variations between the two regions. In addition, the comparison of the profiles obtained for the two pH values shows that more higher profiles are observed in the regions of residues 1–30 for pH 6.8 than pH 6. This result can be put in parallel with the prediction of secondary structures by δ2D and TALOS (Fig. 4) in which larger α helices are predicted in the regions of residues 30–50 at pH 6.8. The α helices being more compact, the corresponding profiles are lower.

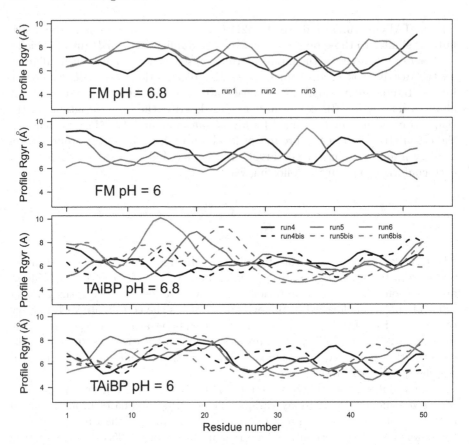

Fig. 3. Profiles of local gyration radii (Å) obtained along the residue numbers of SERF1a for the sets of conformations obtained using *flexible-meccano* or TAiBP. The run names are those given in Table 1.

```
                10        20        30        40        50        60
sequence    MARGNQRELARQKNMKKTQEISKGKRKEDSLTASQRKQRDSEIMQEKQKAANEKKSMQTREK
TALOS pH6   LLLLLLLHHHLLLLLLLLLLLLLLLLLLLLLLLLLHHHHHHHHHHHHHHLLLLLLLLLLLLLLL
TALOS pH6.8 LLLLLLLHHHHLLLLLLLLLLLLLLLLLLLLLLLLLLHHHHHHHHHHHHHHhhHHLLhhLLLLLLLL
d2D pH6     .CCCCCCCCCCCCCCCCCCCCCCCCCCCCCCCCCCCCCCCCCCCCCCCCCCCCCCCCCCCCCC.
d2D pH6.8   .CCCCCCCCCCCCCCCCCCCCCCCCCCCCCCCCCCCHHHCCHHHHHHHHH...CCCC..CCCCCC.
```

Fig. 4. Prediction of secondary structures using TALOS or δ2D on the SERF1a chemical shifts measured at pH values of 6 and 6.8. The predicted secondary structure elements are α-helix (H/h), Loop (L) and coil (C).

4   Discussion-Conclusion

The results obtained by TAiBP and *flexible-meccano* have been compared starting from the chemical shifts measured at two pH values (6 and 6.8) on the protein SERF1a.

Here, we have been comparing the Monte Carlo approach *flexible-meccano*, widely used in the studies of IDP [16] with the approach TAiBP, recently proposed. TAiBP is based on a systematic enumeration of conformations using branch-and-prune approaches on individual atom as well as on protein fragments. This enumeration is coupled to a clustering by the self-organizing maps [1]. This clustering allows to overcome the exponential complexity of the branch-and-prune approach, but, in addition, it induces a convergence of the obtained conformations.

In the comparisons performed here, the induced convergence is visible in the narrower distributions observed for TAiBP conformations for coordinate RMSD between conformations and for gyration radii. These narrower distribution could be seen as a reduction of possible information obtained by TAiBP. But, on the other hand, slightly different distributions of gyration radii are repeatedly observed between the calculations performed at different pH values. Similarly, different profiles of local gyration radii are observed between the different values, and are in agreement with the prediction of secondary structures by TALOS and δ2D.

Different inputs have been tested for the TAiBP calculations. It should be noticed that the ϕ/ψ boxes obtained from the likelihood Ramachandran maps calculated by TALOS, permit to obtain the most converged profiles of local gyration radii. The lack of convergence observed for δ2D inputs could be remediated by dividing the large boxes used for δ2D inputs into smaller boxes and combining the different smaller boxes into iBP steps of the TAiBP calculation.

In the present work, one important aspect is the convergence of the profiles of local gyration radii. Convergence of structural parameters are essential for proving the reliability of structures for folded proteins. Defining structural parameters for which a convergence can be observed could change our point of view on the exploration of IDP conformational space.

Acknowledgements. CNRS, Lorraine University, Academia Sinica and ANR PRCI multiBioStruct (ANR-19-CE45-0019) are acknowledged for funding. High Performance Computing resources were provided by the EXPLOR centre at Lorraine University (2022CPMXX2687). The NMR spectra were acquired at the High-field NMR Center, Academia Sinica (AS-CFII-111-214).

References

1. Bouvier, G., et al.: Functional motions modulating VanA ligand binding unraveled by self-organizing maps. J. Chem. Inf. Model. **54**, 289–301 (2014)
2. Camilloni, C., De Simone, A., Vranken, W.F., Vendruscolo, M.: Determination of secondary structure populations in disordered states of proteins using nuclear magnetic resonance chemical shifts. Biochemistry **51**, 2224–2231 (2012)

3. Dunker, A.K., et al.: What's in a name? Why these proteins are intrinsically disordered. Intrinsically Disord. Proteins **1**, e24157 (2013)
4. Falsone, S.F., et al.: SERF protein is a direct modifier of amyloid fiber assembly. Cell Rep. **2**, 358–371 (2012)
5. Förster, D., Idier, J., Liberti, L., Mucherino, A., Lin, J.H., Malliavin, T.E.: Low-resolution description of the conformational space for intrinsically disordered proteins. Sci. Rep. **12**, 19057 (2022)
6. Goddard, T.D., Kneller, D.G.: SPARKY 3. University of California, San Francisco
7. van Ham, T.J., et al.: Identification of MOAG-4/SERF as a regulator of age-related proteotoxicity. Cell **142**, 601–612 (2010)
8. Kohonen, T.: Self-organized formation of topologically correct feature maps. Biol. Cybern. **43**, 59–69 (1982)
9. Kohonen, T.: Self-Organizing Maps. Springer Series in Information Sciences, Springer, Heidelberg (2001). https://doi.org/10.1007/978-3-642-56927-2
10. Lavor, C., Liberti, L., Mucherino, A.: The interval Branch-and-Prune algorithm for the discretizable molecular distance geometry problem with inexact distances. J. Glob. Optim. **56**, 855–871 (2013)
11. Liberti, L., Lavor, C., Maculan, N., Mucherino, A.: Euclidean distance geometry and applications. SIAM Rev. **56**, 3–69 (2014)
12. Malliavin, T.E.: Tandem domain structure determination based on a systematic enumeration of conformations. Sci. Rep. **11**, 16925 (2021)
13. Malliavin, T.E., Mucherino, A., Lavor, C., Liberti, L.: Systematic exploration of protein conformational space using a distance geometry approach. J. Chem. Inf. Model. **59**, 4486–4503 (2019)
14. Miri, L., et al.: Stabilization of the integrase-DNA complex by Mg^{2+} ions and prediction of key residues for binding HIV-1 integrase inhibitors. Proteins **82**, 466–478 (2014)
15. Oldfield, C.J., Dunker, A.K.: Intrinsically disordered proteins and intrinsically disordered protein regions. Annu. Rev. Biochem. **83**, 553–584 (2014)
16. Ozenne, V., et al.: Flexible-meccano: a tool for the generation of explicit ensemble descriptions of intrinsically disordered proteins and their associated experimental observables. Bioinformatics **28**, 1463–1470 (2012)
17. Shen, Y., Bax, A.: Protein structural information derived from NMR chemical shift with the neural network program TALOS-N. Methods Mol. Biol. **1260**, 17–32 (2015)
18. Worley, B., et al.: Tuning interval Branch-and-Prune for protein structure determination. J. Global Optim. **72**(1), 109–127 (2018). https://doi.org/10.1007/s10898-018-0635-0

Temporal Alignment of Human Motion Data: A Geometric Point of View

Alice Barbora Tumpach[1,2]([✉]) [iD] and Peter Kán[3] [iD]

[1] Institut CNRS Pauli, Oskar-Morgenstern-Platz 1, 1090 Vienna, Austria
[2] University of Lille, Cité scientifique, 59650 Villeneuve d'Ascq, France
alice-barbora.tumpach@univ-lille.fr
[3] Institute of Visual Computing and Human-Centered Technology, TU Wien,
Vienna, Austria
peterkan@peterkan.com
http://math.univ-lille1.fr/~tumpach/Site/home.html

Abstract. Temporal alignment is an inherent task in most applications dealing with videos: action recognition, motion transfer, virtual trainers, rehabilitation, etc. In this paper we dive into the understanding of this task from a geometric point of view: in particular, we show that the basic properties that are expected from a temporal alignment procedure imply that the set of aligned motions to a template form a slice to a principal fiber bundle for the group of temporal reparameterizations. A temporal alignment procedure provides a reparameterization invariant projection onto this particular slice. This geometric presentation allows us to elaborate a consistency check for testing the accuracy of any temporal alignment procedure. We apply this consistency check to some alignment procedures from the literature based on dynamic programming for the task of aligning motions of tennis players. The comparison of the obtained results leads us to propose a version of dynamic programming that incorporates keyframe correspondences. The temporal alignment procedures produced are not only more accurate, but also computationally more efficient.

Keywords: Dynamic Time Warping · Geometric Green Learning · keyframe correspondence

1 Introduction

This work deals with temporal alignment of motions and therefore is connected to motion analysis in general. A state-of-the-art report on motion similarity modeling was presented by Sebernegg et al. [10] and by Senin [9]. We will use a geometric formulation based on group actions and invariances, that can also

Supported by FWF grant I 5015-N, Institut CNRS Pauli, by grant F77 of the Austrian Science Fund FWF (SFB "Advanced Computational Design", SP5), TU Wien and University of Lille.

F. Nielsen and F. Barbaresco (Eds.): GSI 2023, LNCS 14072, pp. 541–550, 2023.
https://doi.org/10.1007/978-3-031-38299-4_56

be used for registration tasks in Shape Analysis. For rate-invariant analysis of motion, see [5]. In the present setting, the relevant features are curves in $\mathbb{R}^3$ or in some Lie groups and homogeneous spaces. Curves in homogeneous spaces were used for animation purposes by Celledoni et al. [1]. Here our main concern is to align motions in order to be able to display them in a synchronized manner. We restrict ourselves to methods using Dynamic Time Warping (DTW). We improve the implementations of temporal alignment procedures introduced in previous research [2–4,7,11]. In these papers dynamic programming is used to compute the best correspondence between two motions relative to a given cost function. This step is computationally relatively expensive, of complexity $O(NM)$ where N and M are the numbers of frames with naive implementation, and of complexity $O(M)$ with a parallel hardware acceleration as explained by Mueen and Keogh [8]. Moreover, most methods use features that are invariant by translations and rotations in $\mathbb{R}^3$, whereas most actions are only invariant by translation along and rotation around the vertical axis, where the vertical axis is aligned with the gravitational field. The discarded information contained in the vertical direction is crucial for accurate synchronization of motions. We propose to incorporate keyframe correspondences into the dynamic programming algorithm based on coarse information extracted from the vertical variations, in our case from the elevation of the arm holding the racket. The temporal alignment procedures produced are not only more accurate, but also computationally more efficient.

Contributions

- we give a geometric formulation of the task consisting of temporal alignment of two motions (Sect. 2);
- this mathematical formulation provides a guide to any temporal alignment procedure, in particular we explain how the consistency of a temporal alignment procedure can be checked (Sect. 3.3);
- we compare different alignment procedures including a coarse alignment procedure that provides a simple and computationally efficient solution to which any other method should be compared in the effort of finding a balance between accuracy and complexity (see Geometric Green Learning [6] and Sect. 3.1);
- we provide a variant of Dynamic Time Warping algorithm that takes into account keyframes correspondences, is computationally more efficient and could be used for different purposes (Sect. 3.4).

2   Temporal Alignment from a Geometric Point of View

2.1   Type of Data Under Consideration

In this paper, we give a geometrical picture of the task consisting of temporal alignment of two motions. As a running example we describe the temporal alignment of motion data of tennis players performing the same action. We use the time evolution of extracted skeletons to characterize the motions. Skeletons extraction is performed using different devices (kinect, simi), but in our examples, the number of joints is fixed, and the extracted features are the same. A priori, the time interval in which a motion is performed could depend on the motion under consideration, but as a pre-processing step we renormalize all motions to the time interval $[0, 1]$, i.e. any action start at time $t = 0$ and ends at time $t = 1$. Nevertheless, the number of frames may differ from actions to actions, depending on the devices used to record the movement. This means that, for a given action, we have a discrete set of times $(t_1, \ldots, t_K)$, where K is the number of frames, at which we know the exact positions of joints.

2.2   What is Temporal Alignment?

The task of aligning two motions $M_1, M_2 : [0, 1] \to \mathbb{R}^{3N}$ can be understood at the theoretical level as the task of finding an optimal time warping (in other words diffeomorphism) $\varphi : [0, 1] \to [0, 1]$ such that $M_2 \circ \varphi$ is visually as close as possible to M_1. On a practical level, the output of time alignment of motions M_1 and M_2 will be a correspondence between the set of frames of M_1, labeled by $\{1, \ldots, K_1\}$ with the set of frames of M_2, labeled by $\{1, \ldots, K_2\}$, where a given frame can be in correspondence with multiple frames. An example of temporal alignment of two motions is given in Fig. 1: the elevation of the arm of a tennis player for two different actions is depicted in Fig. 1 left (blue for the first motion, red for the second). The alignment of the red motion to the blue one results in the green function depicted in the middle picture. The correspondence between the set of frames labeled from 1 to 1400 for the red motion to the set of frames labeled from 1 to 580 for the blue motion is given in the right picture of Fig. 1. It is an increasing function from $[1, 1400]$ to $[1, 580]$. The corresponding time warping $\varphi : [0, 1] \to [0, 1]$ is obtained by rescaling the intervals $[1, 1400]$ and $[1, 580]$ into $[0, 1]$. Conversely, given an optimal time warping $\varphi : [0, 1] \to [0, 1]$ between two motions, the computation of the correspondence between frames presents no difficulty. Therefore, the main challenging task in temporal alignment of motions resides in the computation of an optimal diffeomorphism aligning one motion to the other.

2.3   Temporal Alignment Procedures as Maps from Motions to the Group of Diffeomorphisms

To summarize the setting mathematically at this stage, we consider the manifold $\mathcal{C}$ of all smooth curves parameterized by $[0, 1]$ and with values in $\mathbb{R}^{3N}$:

$$\mathcal{C} = \{c : [0, 1] \to \mathbb{R}^{3N}, c \text{ smooth}\}, \tag{1}$$

and we define the set $\mathcal{M}$ of motions of a skeleton with N joints, fixed set of links $\mathcal{L} \subset \{1, \ldots, N\} \times \{1, \ldots, N\}$ and fixed lengths of bones $c_{jk} > 0$, $\{j, k\} \in \mathcal{L}$ by

$$\mathcal{M} := \{f \in \mathcal{C}, \|f_j(t) - f_k(t)\| = c_{jk}, \forall t \in [0, 1], \forall \{j, k\} \in \mathcal{L}\}. \tag{2}$$

Elements of $\mathcal{M}$ encode positions of body joints, from which one can derive other features like Euler angles and velocities. The lengths of bones are fixed during motion due to physical constraint. The set of time warpings is the group $\mathrm{Diff}^+([0,1])$ of (orientation-preserving) diffeomorphisms $\varphi : [0,1] \to [0,1]$, (i.e. sending 0 to 0 and 1 to 1) acting on curves in $\mathcal{C}$, hence also on motions in $\mathcal{M}$, by $\varphi \cdot f := f \circ \varphi$. Given a reference motion $M_{\mathrm{ref}} \in \mathcal{M}$, a time alignment procedure with respect to M_{ref} is a map which to each motion M in $\mathcal{M}$ associates a unique time warping $\varphi \in \mathrm{Diff}^+([0,1])$ such that M_{ref} and $M \circ \varphi$ are visually as closed as possible.

2.4   Properties of a Temporal Alignment Procedure

Intuitively, a temporal alignment procedure should satisfy:

Property 1. The optimal diffeomorphism aligning M_{ref} with respect to itself should be by the identity map $\mathrm{id} : [0,1] \to [0,1]$, $t \mapsto t$ (reflexivity);

Property 2. if M_2 is visually as close as possible to M_1 than M_1 should be visually as close as possible to M_2 (symmetry);

Property 3. The optimal diffeomorphism aligning $M_{\mathrm{ref}} \circ \varphi$ with respect to M_{ref} should be $\varphi^{-1} \in \mathrm{Diff}^+([0,1])$;

Property 4. If two motions M_1 and M_2 are considered as visually as closed as possible, then for any $\varphi \in \mathrm{Diff}^+([0,1])$, $M_1 \circ \varphi$ and $M_2 \circ \varphi$ should be visually as close as possible ($\mathrm{Diff}^+([0,1])$-equivariance);

Property 5. If M_1 is visually as close as possible to M_2 and M_2 is visually as close as possible to M_3, than M_1 should be visually as close as possible to M_3 (transitivity).

2.5   Temporal Alignment Procedures as Projections on Slices

Properties 1, 2 and 5 imply that "being visually as close as possible" is an equivalence relation $\sim$ on the set of motions $\mathcal{M}$. In this setting, given a reference motion M_{ref}, the set of motions visually as close as possible to M_{ref} is called the equivalence class of M_{ref}. Properties 3 and 4 imply that the group of diffeomorphisms acts equivariently on the set of equivalence classes, i.e. for any $\varphi \in \mathrm{Diff}^+([0,1])$, $M_1 \sim M_2 \Leftrightarrow M_1 \circ \varphi \sim M_2 \circ \varphi$. The uniqueness of the optimal time warping of a motion with respect to a reference motion M_{ref} implies that the orbit of any motion under the action of $\mathrm{Diff}^+([0,1])$ by reparameterizations intersects the equivalence class of M_{ref} at a unique point. Mathematically this means that the equivalence class of a reference motion is a global slice to the set of $\mathrm{Diff}^+([0,1])$-orbits, and that a temporal alignment procedure provides a $\mathrm{Diff}^+([0,1])$-invariant projection on it.

3   Experimental Results

In our experiment all motions are centered and executed facing the camera. We focus in the present paper on temporal alignment. In this section, we first present some alignment procedures that will be used in the paper (Subsect. 3.1). Based on our mathematical formulation, we present a consistency test and compare the accuracy of the temporal alignment procedures under consideration (Subsect. 3.3). We notice that the coarse alignment procedure given by keyframe correspondences has an overall good performance for very low computational cost. For this reason, we incorporate keyframe correspondences into the dynamic programming algorithm in order to improve the other alignment procedures (Subsect. 3.4). This gain can be explained by the fact that the group of invariances of motions under consideration is $\mathbb{R} \times SO(2)$ instead of $SE(3)$, i.e. variations in the elevation of joints contain crucial information for synchronizing two motions. Reincorporating this information, even in a coarse manner, improves the performance of algorithms using $SE(3)$-invariant features.

3.1   Examples of Alignment Procedures

Temporal Alignment Using Keyframes. As a coarse alignment procedure, we have implemented a keyframe correspondence based on the elevation of the arm holding the racket. It allows us to give a temporal bounding box around the movement of interest. For each selected joint of the arm holding the racket, the algorithm detects three keyframes:

1. the first frame with the highest z-coordinate of the joint;
2. the first frame with the lowest z-coordinate of the joint;
3. the second frame with the highest z-coordinate of the joint.

A frame correspondence between two motions is then calculated as the piecewise-linear frame correspondence mapping keyframes to keyframes (see Fig. 1).

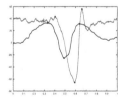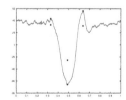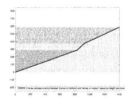

Fig. 1. Elevation of the arm of a tennis player for two different actions before alignment (left) and after alignment (middle), and the frame correspondence aligning keyframes (right).

Temporal Alignment Using SRVT on Trajectories of Joints. We will use a transformation called Square Root Velocity Transform (SRVT) that allows temporal alignment of 3D-curves [11]. For this alignment procedure, Dynamic Programming is used to minimize the L^2-distance between the SRVT-transforms of the trajectories of a selection of joints.

Temporal Alignment Using Gram-Matrices. In this method, we compute the Gram-matrices [4,7] associated to the joints positions of each skeleton and align them in the space of positive semi-definite matrices using Dynamic Programming. In our experiments, we used 10 active joints, namely "*Ankle left*", "*Ankle right*", "*Hip left*", "*Hip right*", "*Knee left*", "*Knee right*", "*Spine low*", "*Spine high*", "*Racket hand*", "*Racket top*".

Temporal Alignment Using Curves on the Group of Rotations SO(3). We construct the moving frame associated to the trajectory of a joint, which is a curve on the group of rotations SO(3) and we reparameterize it in a canonical way. We use Dynamic Programming to align the curvature and torsion functions, which corresponds to a cost function measuring the difference of velocities of the curves on SO(3).

Temporal Alignment Using Curves on the Sphere $\mathbb{S}^2$**.** Given a motion, we create a curve on the sphere $\mathbb{S}^2$ by joining a given joint of the skeleton to the center of the body, and by normalizing the vector obtained. Given two curves on the sphere corresponding to two different motions of a joint, we align them using Dynamic Programming where the cost function uses the SRV transform for homogeneous spaces introduced by Celledoni el al. [2,3].

3.2   Combining Time Warpings for Different Joints

Except the alignment procedure based on Gram-matrices, all the alignment procedures presented in Subsect. 3.1 compute one diffeomorphism per joint. In order to combine the results, we use a weighted average or a median. These methods are more efficient from a computational point of view then the global method using Gram-matrices, because the joints calculations can be parallelized.

3.3   Consistency Check

We designed a consistency check for testing the accuracy of each implemented algorithm. Namely, according to Property 3, each alignment procedure to a reference motion M_{ref} taking $M_{\mathrm{ref}} \circ \varphi$ as input should give as output φ^{-1} for any reparameterization $\varphi \in \mathrm{Diff}^+([0,1])$. To test this, we have implemented a function which takes a skeleton motion and an arbitrary reparameterization as input, and creates a new skeleton motion given by the frame correspondence applied to the initial skeleton. The result is a skeleton motion that performs exactly the same action but with a different rate. In Fig. 2, one can see an example of original motion at the bottom line (joints in green), and the same motion artificially reparameterized at the top line (joints in red) where we can clearly see that the movement starts later than in the initial motion. The reparameterization applied is displayed on the right of the same Figure.

Fig. 2. Bottom left: initial motion, upper line left: artificially reparameterized motion, Right: reparameterization applied.

We have tested the consistency of each algorithm presented in Sect. 3.1. An example of outputs corresponding to the alignment of the upper motion (red joints) of Fig. 2 to the initial motion (green joints) as well as the ground-truth provided by the inverse of the diffeomorphism given in Fig. 2 can be seen in Fig. 3 left. The right picture in Fig. 3 corresponds to an improvement of the algorithms explained in next section. The mean L^1-errors between the output of each alignment procedure and the ground-truth computed over 7 experiments with varying number of frames is recorded in Table 1, as well as average computational times under the same conditions.

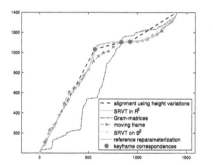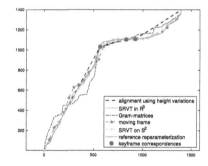

Fig. 3. Frame correspondence obtained with different alignment procedures without keyframe correspondence (left) and with keyframe correspondence (right). Coarse alignment based on elevation of arm (black dashed line), SRVT on $\mathbb{R}^3$ (orange dotted line), Gram-matrices (blue Dash-dotted line), moving frames (magenta dashed line with stars), SRVT on $\mathbb{S}^2$ (cyan dashed line with discs), and reference correspondence (green solid line). (Color figure online)

Table 1. Mean L^1-errors between the frame correspondence provided by each alignment procedure and the ground truth over 7 experiments with number of frames between 50 and 185, and average computational times on a Macbook M1. Left: using dynamic programming (DP), right: using anchored dynamic programming (ADP).

	Error with DP	Time with DP	Error with ADP	Time with ADP
L^1-Error SRVT in $\mathbb{R}^3$	**0.95%**	23 s 227 ms	**0.93%**	7 s 692 ms
L^1-Error Gram-matrices	16.30%	9 m 38 s 706 ms	10.20%	2 m 49 s 257 ms
L^1-Error Moving Frames	6.32%	23 s 771 ms	2.64%	8 s 184 ms
L^1-Error SRVT on $\mathbb{S}^2$	1.15%	23 s 250 ms	1.12%	7 s 640 ms

Baseline: L^1-Error alignment of keyframes = 4.57%, computational time = **5 ms**.

3.4   Incorporating Keyframe Correspondences into Dynamic Programming

In order to take benefit of the stable good performance with low computational cost provided by the coarse alignment procedure based on the elevation of the arm holding the racket, we have modified the dynamic programming algorithm to incorporate keyframe correspondences. Each keyframe correspondence can be thought as a node that should be traversed by the optimal time warping. The resulting anchored dynamic programming finds the path of minimal energy in a landscape that is shaped according to a desired tolerance around each node as in Fig. 4. The modified alignment procedures are computationally more efficient (less nodes to visit) and more accurate (see Table 1). An example of aligned motions by the procedure using moving frames with classical dynamic programming and with anchored dynamic programming is displayed in Fig. 5.

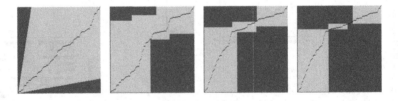

Fig. 4. Energy profile for dynamic programming and example of path of lowest energy. Left: classical dynamic programming, second and third from left: anchored dynamic programming around the nodes provided by the coarse alignment procedure based on keyframes with tolerance of 1/4 and 1/20 of total number of frames respectively, right: anchored dynamic programming with minimal tolerance.

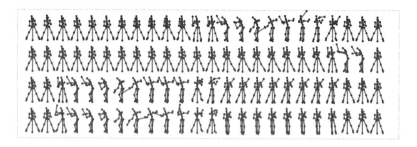

Fig. 5. Bottom: initial motion (ground truth), upper line left: artificially reparameterized motion obtained in Sect. 3.3 given as input, second line: aligned motion using moving frames on 100 frames with classical dynamic programming, third line: aligned motion using moving frames on 100 frames with anchored dynamic programming with zero tolerance.

4   Conclusion

We gave a mathematical formulation of the task consisting of synchronizing human motion data from multiple recordings, which allowed us to elaborate a test to check the consistency of any temporal alignment procedure. In order to include the information gained by a coarse alignment procedure based on keyframes in any method, we implemented a variant of dynamic programming ensuring that associated keyframes are in correspondence. For this algorithm, each keyframe correspondence creates a node by which the optimal time warping has to pass, like a boat that needs to drop anchor in a port. The improvement of the temporal alignment procedures by the anchored dynamic programming can be explained by the fact that the group of symmetries of extracted features mismatched the group of symmetries of the motions under consideration. The lost information was partially recovered by forcing keyframe correspondence. At the same time the complexity of the algorithms decreased significantly.

Acknowledgements. We thank VR Motion Learning GmbH & Co KG for providing us with their dataset of tennis motions. The first author is supported by FWF grant I 5015-N. This work has been funded by grant F77 of the Austrian Science Fund FWF (SFB "Advanced Computational Design", SP5).

References

1. Celledoni, E., Eslitzbichler, M., Schmeding, A.: Shape analysis on Lie groups with applications in computer animation. J. Geom. Mech. **8**(3), 273–304 (2016)
2. Celledoni, E., Eidnes, S., Schmeding, A.: Shape analysis on homogeneous spaces: a generalised SRVT framework. In: Celledoni, E., Di Nunno, G., Ebrahimi-Fard, K., Munthe-Kaas, H.Z. (eds.) Abelsymposium 2016. AS, vol. 13, pp. 187–220. Springer, Cham (2018). https://doi.org/10.1007/978-3-030-01593-0_7
3. Celledoni, E., Eidnes, S., Eslitzbichler, M., Schmeding, A.: Shape analysis on lie groups and homogeneous spaces. In: Nielsen, F., Barbaresco, F. (eds.) GSI 2017.

LNCS, vol. 10589, pp. 49–56. Springer, Cham (2017). https://doi.org/10.1007/978-3-319-68445-1_6

4. Celozzi, E.M., et al.: Modelling the statistics of cyclic activities by trajectory analysis on the manifold of positive-semi-definite matrices. In: 2020 15th IEEE International Conference on Automatic Face and Gesture Recognition (FG 2020) (2020)

5. Drira, H., Tumpach, A.B., Daoudi, M.: Gauge invariant framework for trajectories analysis. In: Proceedings of the 1st International Workshop on Differential Geometry in Computer Vision for Analysis of Shapes, Images and Trajectories (2015)

6. Geometric Green Learning. https://sites.google.com/view/geometric-green-learning

7. Kacem, A., Daoudi, M., Amor, B.B., Berretti, S., Alvarez-Paiva, J.C.: A novel geometric framework on gram matrix trajectories for human behavior understanding. IEEE TPAMI **42**(1), 1–14 (2018)

8. Mueen, A., Keogh, E.: Extracting optimal performance from dynamic time warping. In: KDD 2016, pp. 2129–2130 (2016)

9. Senin, P.: Dynamic time warping algorithm review. Information and Computer Science Department University of Hawaii at Manoa Honolulu, USA, vol. 855, no. 1–23, p. 40 (2008)

10. Sebernegg, A., Kán, P., Kaufmann, H.: Motion Similarity Modeling: A State of the Art Report

11. Srivastava, A., Klassen, E., Joshi, S., Jermyn, I.: Shape analysis of elastic curves in Euclidean spaces. IEEE TPAMI **33**(7), 1415–1428 (2010)

A Linear Program for Points of Interest Relocation in Adaptive Maps

S. B. Hengeveld[1], F. Plastria[2], A. Mucherino[1(✉)], and D. A. Pelta[3]

[1] IRISA, Université de Rennes, Rennes, France
{simon.hengeveld,antonio.mucherino}@irisa.fr
[2] BUTO, Vrije Universiteit Brussel, Brussel, Belgium
Frank.Plastria@vub.be
[3] Department of Computer Science and A.I., Universidad de Granada, Granada, Spain
dpelta@decsai.ugr.es

Abstract. The Point-Of-Interest (POI) relocation problem is a challenge encountered during the construction of personalized maps for given groups of users. This kind of maps was already studied and is known in the scientific literature under the name of "adaptive maps". In this work, we formulate this problem as a subclass of the widely studied Distance Geometry Problem (DGP), where some extra constraints are included for taking into account the local orientation of the POIs in the map. These very same constraints allow us to linearize the problem, and hence to propose a novel linear program for the POI relocation problem. Our initial computational experiments indicate that our approach is promising for further investigations.

1 Introduction

Geographical maps are widely used in our everyday life. In a symbolic fashion, they are designed to depict the main relationships among the various objects represented therein. A classical example is given by city touristic maps, where the main points of touristic attraction are put in evidence in the map, and the connections between them are emphasized, for a potential tourist to easily navigate from one of such points to another. We refer to such *Points-Of-Interest* with the acronym POIs.

In classical geographical maps, the distances between POIs are generally proportional to their actual relative Euclidean distances. In some particular cases, such as in the touristic maps mentioned above, this general rule may be slightly relaxed in order to give more emphasis, for example, to a very important POI. The Euclidean distance, however, remains satisfied in most of the cases, for the final user to more easily interpret the map, and navigate through it.

The maps we are interested in are special maps where the Euclidean distance does not predominate. These maps are meant to give to the user a sense of proximity in some particular conditions impacting the user's mobility conditions. Take for example a person in wheelchair approaching a cathedral from a classical route (the route one can derive from a standard map) and discovering that a few small steps need to be climbed in order to have access to the square where the cathedral stands. While many other people with no mobility restrictions are likely to not even notice the presence of those steps, each step represents instead an important barrier for a wheelchair. This user in wheelchair would have much preferred to have an alternative map where the best route in this particular condition (to seat on a wheelchair) is depicted as the shortest one. We refer to this kind of maps as *adaptive maps*.

The very first reference to this kind of maps seems to have appeared in 1983 in [9]. The idea was then much later employed in [8] for example for the conception of metro maps. More recently, that original idea was extended to the concept of adaptive maps in [11] and [12]. These works indicate that three main steps can be identified for the construction of an adaptive map from a classical geographic map: (*i*) the identification of the POIs and of the main criteria for defining POI proximity; (*ii*) the relocation of the POIs on the basis of the defined proximity criteria; and (*iii*) the modification (basically the *distortion*) of the original map so that the new POI locations are taken into consideration in the final representation of the adaptive map.

We particularly focus our attention in this paper on the step (*ii*) for the construction of an adaptive map. In the following, this step will be referred to as the "POI relocation problem". Since we position ourselves immediately after the step (*i*) above, we can suppose that a set of POIs is already given, and that a position in the two-dimensional Euclidean space (from the original map) is already associated to each POI. In addition, we can suppose that proximity information is associated to pairs of POIs, where such a proximity is appreciated by using a method that does not simply evaluate the classical Euclidean norm between the original locations.

The aim of this work is two-fold. Firstly, we will propose a formulation of the POI relocation problem as a distance geometry problem (see Sect. 2). The interest in such a formulation is given by the large body of work in distance geometry that can as a consequence be exploited in the context of adaptive maps. Secondly, we will focus on some recent research on linearization in distance geometry and we will propose a novel linear program that is particularly tailored to the POI relocation problem (Sect. 3). Finally, computational experiments will be presented in Sect. 4, and Sect. 5 will conclude the paper.

2   Distance Geometry for Adaptive Maps

Given a simple weighted undirected graph $G = (V, E, d)$ and a positive integer K, the Distance Geometry Problem (DGP) [6] asks whether a realization $X : V \to \mathbb{R}^K$ exists such that, every time an edge connects two vertices u and $v \in V$ (i.e. $\{u, v\} \in E$), the Euclidean distance between $X(u)$ and $X(v)$ corresponds to the given weight $d(u, v)$. The DGP belongs to the NP-hard class of problems [10], and several methods and algorithms have been proposed in recent years for its solution [7].

In the context of adaptive maps, we can naturally fix the dimension K to 2. The vertex set V represents the set of selected POIs, while the proximity information is encoded by the edge set E and the corresponding weights d. Since we suppose that an original map containing these POIs is already available, we can notice that a possible realization X_p for our graph G is already available. However, this realization represents the original map, where the distances between POIs reflect the traditional Euclidean distance.

The main interest in the POI relocation problem is to find an alternative location for all these POIs, on the basis of the proximity measure encoded by the edge set E, and of the weight function d.

We can remark that, while the proximity measure is not meant to be Euclidean, its representation in the new realization X is going to be Euclidean. This raises a main issue concerning the compatibility of the proximity measure with the properties of the Euclidean norm. One example is given by the triangular inequality, which needs to be satisfied in Euclidean space, but may not be satisfied by the employed proximity measure. As a consequence, a realization X where *all* weights d are *exactly* satisfied is unlikely to exist, and hence the constraints on the distances $d(u, v)$ need to be relaxed. In other words, the DGP is in this case not considered as a decision problem, but rather as an optimization problem where the deviations from the given weights is minimized [3].

We can additionally remark that the realization related to the original map (the realization we referred to with the symbol X_p) allows us to include some useful additional information to our problem instances. This is explained in the paragraph below: before we do this, we warn the reader that, in order to consider this extra information, it is necessary to enrich our graph representation with an orientation for each of its edges. In other words, our graphs G are *directed* graphs.

Suppose an important POI in the map we wish to adapt is originally placed at its North pole. Since the DGP is only based on distance information, there is no way to predict in advance whether the given distances will force this POI to move to completely different places in the map, and be positioned for example at the South pole, whereas the other POIs basically remain in the same locations. It is evident that this kind of result would be extremely confusing for the final user. Therefore, we include in our model the information about the local orientations of the POIs, and we enforce, in the resulting adaptive maps, that these orientations are preserved.

The local orientations are included as follows. From the original map, we can define two orthogonal axes, $\hat{x}$ indicating the North-South direction, as well as $\hat{y}$ indicating the West-East direction, that we can use to define specific Cartesian systems. For the generic edge $(u, v) \in E$, we can in fact define a Cartesian system centered in a given position $X_p(u)$ for u and having as axes $\hat{x}$ and $\hat{y}$, so that it can be easily verified in which of the four standard quadrants (NW,NE,SW,SE) of this Cartesian system the position $X_p(v)$ of the second POI is contained. Then, we can simply partition the edge set E in four subsets, to which we assign the same names of the Cartesian quadrants: NW, NE, SW and SE. The fact that $(u, v) \in$ NW indicates, for example, that the positions for v need to be in *North-West* quadrant of the Cartesian system defined above and centered in the available position for u. Since our graph contains directed edges, it is necessary to pay attention to the fact that the information about the orientations remains symmetric for each possible realization, i.e. $(u, v) \in$ NW always needs to imply that $(v, u) \in$ SE.

Since we focus on instances in dimension 2, a pair $(x, y) \in \mathbb{R}^2$ will indicate in the following the location of a POI through its x and y components in the Euclidean 2-dimensional space. For example, for the POI $v \in V$, we will write $X(v) = (x_v, y_v)$. On the basis of the comments above, we propose the following definition of POI relocation problem in dimension 2.

Definition 2.1. *Given a simple weighted directed graph $G = (V, E, d)$ and an initial realization X_p of G, where*

- *V is a set of POIs,*
- *E represents the presence of proximity information and it is partitioned in the four subsets NW, NE, SW and SE in order to encode the local orientations in X_p,*
- *the weight function d provides the numerical values for the proximity measures,*

find a realization $X : V \rightarrow \mathbb{R}^2$ that is solution to the following optimization problem:

$$\min_X \sum_{(u,v) \in E} \left(\sqrt{(x_v - x_u)^2 + (y_v - y_u)^2} - d(u,v) \right)^2$$

$$\text{s.t.} : \begin{cases} \forall (u,v) \in NW, \ x_u \geq x_v \text{ and } y_u \leq y_v, \\ \forall (u,v) \in NE, \ x_u \leq x_v \text{ and } y_u \leq y_v, \\ \forall (u,v) \in SW, \ x_u \geq x_v \text{ and } y_u \geq y_v, \\ \forall (u,v) \in SE, \ x_u \leq x_v \text{ and } y_u \geq y_v. \end{cases} \quad (1)$$

Notice the use of the Euclidean norm in dimension 2, and that the function we minimize corresponds to the standard *stress* function already used in the context of the DGP [5], but other penalty functions may also be used. Since the partition of E into NW, NE, SW and SE reflects the relative realworld positions of the POI, these realworld coordinates give a feasible solution to the constraints in (1), which are thereby shown not to be contradictory.

The POI relocation problem can therefore be seen as a particular DGP subclass of instances, where the additional constraints on the relative orientations are enforced. In general, adding constraints to a known problem is likely to increase its complexity. In our case, instead, these new orientation constraints will allow us to formulate the POI relocation problem as a linear program consisting of only real variables. This implies that the DGP subclass consisting of our reformulated POI relocation problem (see next section) is a polynomial case. Moreover, the reduction of the problem to a pure linear problem implies that extremely rapid, well-tested and robust solvers can be used for its solution.

3   A Linear Program for POI Relocation

Let $G = (V, E, d)$ be our simple weighted and directed graph representing an instance of the POI relocation problem. In this section, we propose a "purely" linear programming model for this important problem in the context of adaptive maps.

The optimization problem in Definition 2.1 is based on the Euclidean norm, where the square of the difference between desired and computed distances is summed up to give a global estimate of the error on the distances. In order to avoid the nonlinear terms, we replace the Euclidean norm with the L_1 norm and replace the squared differences by their absolute value.

We point out that we are not the first ones to consider combining these two substitutions: they were already considered for example in [4] (this article moreover, as well

Table 1. The walking distances between 10 pairs of POIs, that we use in our computational experiment. Reproduced from [12].

	P1	P2	P3	P4	P5	P6	P7	P8	P9	P10
P1	0	1553	2783	3614	4240	5674	6972	6064	7408	5387
P2	1553	0	1674	3867	2975	5936	7234	6326	6807	4693
P3	2783	1674	0	3427	2118	5098	6396	5409	5950	3837
P4	3614	3867	3427	0	2471	2571	4609	3745	5222	3292
P5	4240	2975	2118	2471	0	3600	4684	3172	3840	1727
P6	5674	5936	5098	2571	3600	0	2569	3076	5083	3416
P7	6972	7234	6396	4609	4684	2569	0	2692	4966	3903
P8	6064	6326	5409	3745	3172	3076	2692	0	2415	1698
P9	7408	6807	5950	5222	3840	5083	4966	2415	0	2303
P10	5387	4693	3837	3292	1727	3416	3903	1698	2303	0

as [2], also exploited similar substitutions with the L_∞ norm), at the price, however, of adding binary variables and nonlinear constraints, with the consequent loss of polynomiality. It is precisely the additional wish to conserve the relative orientations of the POIs that allows us to use continuous variables only.

Our linear model is a minimization problem:

$$\min \sum_{(u,v)\in E} z_{uv}, \qquad (2)$$

where

$$z_{uv} = \big|\,|x_v - x_u| + |y_v - y_u| - d(u,v)\,\big|, \qquad (3)$$

where the symbol $|\cdot|$ simply represents the absolute value of a real number. The information about the relative orientation for every edge $(u,v) \in E$ (see Eq. (1)) allows us to compute z_{uv} by performing only sums of coordinates and distances (or of their opposite values):

$$\begin{cases} \forall (u,v) \in NW, & z_{uv} \geq y_v - y_u + x_u - x_v - d(u,v), \\ \forall (u,v) \in NW, & z_{uv} \geq d(u,v) - y_v + y_u - x_u + x_v, \\ \forall (u,v) \in NE, & z_{uv} \geq y_v - y_u + x_v - x_u - d(u,v), \\ \forall (u,v) \in NE, & z_{uv} \geq d(u,v) - y_v + y_u - x_v + x_u, \\ \forall (u,v) \in SW, & z_{uv} \geq y_u - y_v + x_u - x_v - d(u,v), \\ \forall (u,v) \in SW, & z_{uv} \geq d(u,v) - y_u + y_v - x_u + x_v, \\ \forall (u,v) \in SE, & z_{uv} \geq y_u - y_v + x_v - x_u - d(u,v), \\ \forall (u,v) \in SE, & z_{uv} \geq d(u,v) - y_u + y_v - x_v + x_u. \end{cases} \qquad (4)$$

Notice that Eq. (3) is therefore only satisfied at optimality. Some computational experiments on our linear model (2)–(1)–(4) are presented in the next section.

We remark that, in our formalism, the graph G allows us to have weights $d(u,v)$ that are different from the weights $d(v,u)$, for some $(u,v) \in E$, because G is a directed

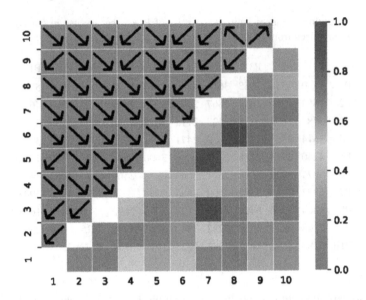

Fig. 1. Some information related to the 10-point instance in [12]. At the upper triangle, a visual representation of the local orientations of the POIs in the original realization, which our linear model is supposed to preserve in the found solution. At the lower triangle, a heatmap where the colors encode the difference in values between the Euclidean distances (in the original realization X_p) and the given proximity distances. The lower the difference, the smaller the value attributed to the corresponding square in the heatmap; all difference values are normalized between 0 and 1.

graph. However, in this work, we will suppose for simplicity that the distance information, even if not Euclidean, is still symmetric, and we will perform our computational experiments for this specific case.

4   Computational Experiments

We present an initial experiment where we use the 10-point POI relocation instance available from [12], reproduced in our Table 1. The graph G is defined so that the vertex set V corresponds to this set of POIs, and its edge set E makes it fully connected. The weight function d associated to the graph maps the edge set to a set of proximity measures where the walking distance between each pair of POIs is estimated. Notice that the walking distance is much likely to differ from the Euclidean distance, while paths along streets resemble more L_1 norm distances. In order to present the data in a visual way, we show in Fig. 1 a colored map depicting the POI distance variations (when comparing the original Euclidean distances and the given proximity distances) in the instance, as well as the local orientations of the POIs.

The IBM's CPLEX solver [1] was used to solve the POI relocation problem for this 10-point instance. The result was obtained in 0.072 s on a MacOS (ventura), 32GB RAM, and with CPU Apple M1 Max (10 cores). In order to superimpose the obtained

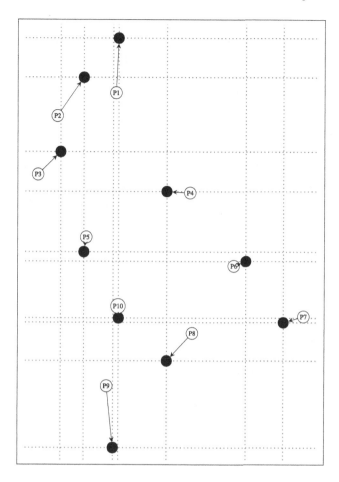

Fig. 2. The solution found by our linear model to the POI relocation problem related to data in Table 1. The points P∗ indicate the original locations of the POI in the initial map. In black, the new positions for the POIs.

coordinates with the original map, we first center both point sets at the origin by subtracting the centroid from each point. Then, we scale the both point sets so that their distances are within the same range. The obtained result, which has objective function value equal to 10273, is shown in Fig. 2.

In order to validate our result, we focus our attention on the "red" edges (appearing in darker gray if in black-and-white) between P6 and P8 in Fig. 1. The distance between these two POIs is supposed to get larger in the adaptive map, and we can notice that this actually happens in our solution (see Fig. 2). We can remark, however, that the locations of some other POIs are subject to even larger modifications, apparently contradicting our input data (because these POIs are mostly related to "green" edges, see for example P9). But a more attentive look at our adaptive map can reveal that these POIs were subject to a change in order to maintain their relative distances to other POIs which were involved in some "red" edges. In the particular case mentioned above, P8 is subject to a substantial relocation because of the edge between itself and P6. As a consequence, P9 is pushed down in the South direction, because otherwise its relative distance from P8 would not be satisfied anymore (Fig. 1 shows that the distance between these two POIs remains approximately the same).

Let us now consider the point P3. Two of the edges related to P7 are "red", indicating a quite important variation on the corresponding relative distances. However, the location of P3 does not seem to have changed much in our adaptive map. This behavior is most likely due to the introduced orientation constraints. In fact, P7 belongs to the SE quadrant of P10, and the little movement it is allowed to take seems to be the consequence of the fact that P10 moved a little in the South direction. Any other movement for P7 in the allowed quadrant would have most likely caused larger distance penalties with its closest neighbors. In general, we can observe that all local orientations are well respected, as imposed by the constraints in Eq. (1) in Definition 2.1, that we use in our linear model.

5   Conclusions

We proposed a new solution method for the POI relocation problem, which is a problem encountered during the construction phase of an adaptive map. We showed that the POI relocation problem represents a particular class of DGP instances, where new constraints on the relative orientations between the POIs can be included. These new constraints allowed us to linearize the POI relocation problem, and thus to propose a linear program to find solutions for it where only real variables are involved.

We presented some initial experiments which show that this new approach is promising. We did not compare against other approaches for distance geometry or adaptive maps because we believe the comparison would have been unfair because we use the orientation constraints, which were, to the best of our knowledge, never used in previous studies.

Future works will be focusing on possible variants of the proposed linear model, to the possibility to use rotated versions of the L_1 norm, and to a large range of experiments aiming at finding the key points for improving our approach.

Acknowledgments. The authors thank the organizers of the 2021 Thematic Program on Geometric Constraint Systems, Framework Rigidity, and Distance Geometry (Fields Institute, Toronto, Canada) because they gave us the opportunity to participate to the Symposium on Sensor Network Localization and Dynamical Distance Geometry, where we set the basis of our collaboration.

AM and SH wish to thank the ANR French funding agency for support (MULTIBIOSTRUCT project ANR-19-CE45-0019). DP acknowledges support from projects PID2020-112754GB-I0, MCIN/AEI /10.13039/501100011033 and FEDER/Junta de Andalucía-Consejería de Transformación Económica, Industria, Conocimiento y Universidades/Proyecto (B-TIC-640-UGR20).

References

1. IBM. ILOG CPLEX 12.6 User's Manual. IBM (2014)
2. Crippen, G.M.: An alternative approach to distance geometry using L^∞ distances. Discret. Appl. Math. **197**, 20–26 (2015)
3. D'Ambrosio, C., Vu, K., Lavor, C., Liberti, L., Maculan, N.: New error measures and methods for realizing protein graphs from distance data. Discrete Comput. Geom. **57**(2), 371–418 (2017). https://doi.org/10.1007/s00454-016-9846-7
4. D'Ambrosio, C., Liberti, L.: Distance geometry in linearizable norms. In: Nielsen, F., Barbaresco, F. (eds.) GSI 2017. LNCS, vol. 10589, pp. 830–837. Springer, Cham (2017). https://doi.org/10.1007/978-3-319-68445-1_95
5. Glunt, W., Hayden, T.L., Raydan, M.: Molecular conformations from distance matrices. J. Comput. Chem. **14**(1), 114–120 (1993)
6. Liberti, L., Lavor, C., Maculan, N., Mucherino, A.: Euclidean distance geometry and applications. SIAM Rev. **56**(1), 3–69 (2014)
7. Mucherino, A., Lavor, C., Liberti, L., Maculan, N. (eds.): Distance Geometry: Theory, Methods and Applications. Springer, New York (2013). https://doi.org/10.1007/978-1-4614-5128-0
8. Raveau, S., Muñoz, J.C., de Grange, L.: A topological route choice model for metro. Transp. Res. Part A: Policy Pract. **45**(2), 138–147 (2011)
9. Rolland-May, C.: A valuation model of subjective spaces. IFAC Proc. Vol. **16**(13), 375–380 (1983)
10. Saxe, J.: Embeddability of weighted graphs in k-space is strongly NP-hard. In: Proceedings of 17th Allerton Conference in Communications, Control and Computing, pp. 480–489 (1979)
11. Torres, M., Pelta, D.A., Verdegay, J.L.: A proposal for adaptive maps. In: Medina, J., Ojeda-Aciego, M., Verdegay, J.L., Perfilieva, I., Bouchon-Meunier, B., Yager, R.R. (eds.) IPMU 2018. CCIS, vol. 855, pp. 657–666. Springer, Cham (2018). https://doi.org/10.1007/978-3-319-91479-4_54
12. Torres, M., Pelta, D.A., Verdegay, J.L., Cruz, C.: Towards adaptive maps. Int. J. Intell. Syst. **34**(3), 400–414 (2019)

Geometric Features Extraction
in Medical Imaging

Diffeomorphic ICP Registration for Single and Multiple Point Sets

Adrien Wohrer[(✉)]

Université Clermont Auvergne, Clermont-Auvergne INP, CNRS, Institut Pascal,
63000 Clermont-Ferrand, France
`adrien.wohrer@uca.fr`

Abstract. We propose a generalization of the iterative closest point (ICP) algorithm for point set registration, in which the registration functions are non-rigid and follow the large deformation diffeomorphic metric mapping (LDDMM) framework. The algorithm is formulated as a well-posed probabilistic inference, and requires to solve a novel variation of LDDMM landmark registration with an additional term involving the Jacobian of the mapping. The algorithm can easily be generalized to construct a diffeomorphic, statistical atlas of multiple point sets. The method is successfully validated on a first set of synthetic data.

Keywords: LDDMM · GMM · Point Set Registration · Medical Atlas

1 Introduction

Registering two, or multiple, sets of points together is a classic task in computer vision, with several applications in computer graphics, medical imaging, pattern recognition, etc. The general problem may be informally defined as follows: a number of point sets (or *point clouds*) $X^{(1)}$, $X^{(2)}$, etc., are considered, each set $X^{(k)}$ consisting of a finite number N_k of points in $\mathbb{R}^d$:

$$X^{(k)} = \left\{ x_n^{(k)} \right\}_{n=1\ldots N_k} \quad \subset \mathbb{R}^d$$

In typical applications, $d = 2$ or 3, and each $X^{(k)}$ may represent: features extracted from a scene, an anatomical structure in a medical image, the surface of a 3d object, etc. (see, e.g., [11,14] for a review).

The *point set registration* problem consists in finding optimal spatial transformations to align the different point sets together. The classic, two-set registration problem seeks a transformation T such that

$$T\big(X^{(1)}\big) \simeq X^{(2)}, \tag{1}$$

whereas the multiple-set registration problem seeks a different transformation T_k associated to each point set $X^{(k)}$, such that

$$T_1\big(X^{(1)}\big) \simeq T_2\big(X^{(2)}\big) \simeq \cdots \simeq T_K\big(X^{(K)}\big) \simeq M \tag{2}$$

F. Nielsen and F. Barbaresco (Eds.): GSI 2023, LNCS 14072, pp. 563–573, 2023.
https://doi.org/10.1007/978-3-031-38299-4_58

with M representing some form of "average" point set. In these informal definitions, symbol $\simeq$ represents a matching of two sets *as a whole*: only the overall spatial overlap of the sets is important, not the identity of the individual points. In particular, the numbers of points N_k in each set need not be equal.

The iterative closest point (ICP) algorithm [2] is a historic method of choice to solve the two-set problem, Eq. (1). Given a starting estimate for the spatial transform T, each point in set $T(X^{(1)})$ is *associated* to its nearest neighbor in set $X^{(2)}$; T is then updated to minimize the distances between associated pairs of points; then the associations are re-computed, etc., until convergence. The ICP algorithm was originally designed with rigid registrations T, and basic ('hard') associations between pairs of points. Rapidly however, "probabilistic" variants of ICP were developed, that allow for smooth associations between points [5,8,13]. In these algorithms, one of the sets (say, $X^{(2)}$) is viewed as the centroids of a Gaussian mixture model (GMM), and the registration T is optimized to maximize the likelihood of the other point set (say, $T(X^{(1)})$) under this GMM distribution. This optimization is generally achieved by an Expectation-maximization (EM) algorithm [3], whose alternation of E step and M step naturally generalizes the two alternating steps of the original ICP algorithm. Generally, the GMM distribution used has a single, isotropic variance parameter σ (see Eq. (3) below) that controls the "smoothness" of the associations between points: when σ is large, each point in $T(X^{(1)})$ is smoothly associated to many points in the second set $X^{(2)}$, depending on their proximity, thereby providing increased stability of the convergence. When $\sigma \to 0$, the original ICP algorithm is recovered.

Recently, the same probabilistic framework has been extended to the multiple point set problem of Eq. (2) [7]. In this algorithm, a single GMM distribution (informally corresponding to M in Eq. (2)) is optimized, in alternation with the registration functions T_k, to maximize the compound likelihood of all registered datasets $T_k(X^{(k)})$. Once convergence is achieved, the resulting, common GMM model summarizes the joint structure of all point sets.

ICP methods have also been extended to incorporate non-rigid transformations T [5,13]. However, (i) this has generally been done through kernel methods that do not guarantee invertibility of the transformations, (ii) the elegant interpretation as a probabilistic inference is lost in the process, (iii) the methods generalize badly to the multiple point set approach of [7]. To ensure invertibility, a better choice would be to use the powerful *large deformation diffeomorphic metric mapping* (LDDMM) framework for diffeomorphic mappings [1,12]. A classic point registration algorithm is already known in this framework, the so-called "landmark registration" [10], however it can only register each point to a predefined target point, so it is not a *point set* registration algorithm as defined above.

In this paper, we propose an ICP-like registration algorithm, based on GMM clustering and LDDMM diffeomorphic mappings, formulated as a well-posed probabilistic inference, and that can readily be extended to multiple point sets. The novel probabilistic formulation requires to solve a variant of LDDMM landmark registration, where the log-Jacobian of the mapping enters the LDDMM energy functional. We solve these equations, and then demonstrate the well-posedness of the resulting algorithm on a first set of synthetic examples.

2   Methods

2.1   Warped GMM Distribution

We consider data points in $\mathbb{R}^d$ ($d = 2$ or 3 typically). We note $\mathcal{M}(\theta)$ the Gaussian mixture model (GMM) distribution, with density

$$f_{\mathcal{M}}(z|\theta) := \frac{1}{(2\pi)^{d/2}\sigma^d} \sum_{c=1}^{C} \pi_c \exp\left(-\frac{|z - \mu_c|^2}{2\sigma^2}\right) \tag{3}$$

$|.|^2$ being the squared Euclidian norm of $\mathbb{R}^d$ (we use single vertical lines, to distinguish it from the RKHS norm over vector fields introduced in Sect. 2.3). This is a mixture of C Gaussian components with centroids μ_c, mixing weights π_c, and uniform isotropic variance σ^2. As in classic probabilistic ICP algorithms, the role of this single variance parameter σ is to control the "smoothness" of associations between points [5,8,13]. We let θ generically denote the subset of parameters $\{\sigma, \mu_c, \pi_c\}$ that should be *optimized* in a given problem, as this will vary depending on the precise application.

The generative model for data points used in this article consists in warping distribution $\mathcal{M}(\theta)$ through a diffeomorphic mapping ψ. That is, each sample point x is assumed to have been generated as $x = \psi^{-1}(z)$, where z is a sample from $\mathcal{M}(\theta)$ in Eq. (3), and ψ is some orientation-preserving diffeomorphic mapping. The resulting probability distribution for x will be noted $\mathcal{M}^*(\psi, \theta)$. It corresponds to the *pullback* by ψ of distribution $\mathcal{M}(\theta)$, with density function

$$f_{\mathcal{M}^*}(x|\psi, \theta) = f_{\mathcal{M}}(\psi(x)|\theta)\det(\mathrm{D}\psi(x)) \tag{4}$$

Here and in the sequel, letter D denotes spatial differentiation. Note that $\mathcal{M}^*(\psi, \theta)$ is indeed a probability distribution, as $\int_{\mathbb{R}^d} f_{\mathcal{M}^*}(x|\psi, \theta)\mathrm{d}x = \int_{\mathbb{R}^d} f_{\mathcal{M}}(z|\theta)\mathrm{d}z = 1$. The fact that the (positive) Jacobian $\det(\mathrm{D}\psi(x))$ is involved in Eq. (4) will be an important specificity of this paper compared to previous work.

2.2   Registration as an Inference Problem

Given an observed point set in $\mathbb{R}^d$, $X = \{x_n\}_{n=1...N}$, we can now reformulate our registration problem as a probabilistic inference problem: *to find GMM parameters θ and diffeomorphism ψ that maximize the likelihood of having generated the points in X*. For this, we make a number of classic assumptions [5,8,13]. First, that the points $\{x_n\}_{n=1...N}$ have been generated as independent samples from distribution $\mathcal{M}^*(\psi, \theta)$. Second, that there is no probabilistic prior over GMM parameters θ (informally, just set $\mathrm{P}(\theta) = 1$). Third, this article considers diffeomorphic mappings ψ belonging to an LDDMM group $\mathcal{G}$, detailed below.

Every mapping $\psi \in \mathcal{G}$ is naturally associated to a number $\mathcal{E}_\mathcal{G}(\psi) > 0$ quantifying the amount of deformation induced by ψ (Eq. (13) below). Hence, we can set a probabilistic prior over mappings $\psi \in \mathcal{G}$ to be

$$P(\psi) \sim \exp(-\lambda \mathcal{E}_\mathcal{G}(\psi)) \tag{5}$$

where $\lambda > 0$ is a model parameter controlling the amount of deformation allowed for ψ, and the normalization is unimportant. Finally, the values of ψ and θ are assumed to be independent, i.e., $P(\psi|\theta) = P(\psi)$. Under all these assumptions, the total likelihood of data and parameters as given by Bayes' law writes $P(X, \psi, \theta) = P(\theta)P(\psi|\theta)P(X|\psi, \theta) = P(\psi) \prod_{n=1}^{N} P(x_n|\psi, \theta)$. We are thus led to solve the following maximum likelihood optimization problem:

$$\max_{\substack{\theta \\ \psi \in \mathcal{G}}} \; P(\psi) \prod_{n=1}^{N} f_{\mathcal{M}^*}(x_n|\psi, \theta) \tag{6}$$

with $f_{\mathcal{M}^*}(x|\psi, \theta)$ given by Eq. (4) and $P(\psi)$ given by Eq. (5). Note that problem (6) can easily be generalized to multiple point sets: see Eq. (19) below. The main exposition will be for a single point set, only to lighten notations.

EM Resolution. The optimization problem Eq. (6) is typically solved with an Expectation-maximization (EM) algorithm [5,7,8,13]. This is the method of choice for maximum-likelihood problems involving mixture distributions; it provides faster and more robust convergence than naive gradient-based methods [3]. Following the classic EM procedure, we introduce responsibility variables $\gamma_{nc} > 0$ between each data point n and GMM component c, constrained by $\forall n, \sum_c \gamma_{nc} = 1$, and derive from Eq. (6) the following EM free energy:

$$\mathcal{F}(\gamma, \theta, \psi) := \sum_{n=1}^{N} \sum_{c=1}^{C} \gamma_{nc} \left(\frac{|\psi(x_n) - \mu_c|^2}{2\sigma^2} + \log \frac{\sigma^d}{\pi_c} + \log \gamma_{nc} \right)$$
$$- \sum_{n=1}^{N} \log \det(D\psi(x_n)) + \lambda \mathcal{E}_\mathcal{G}(\psi) \tag{7}$$

The EM algorithm proceeds to repeated partial minimizations of $\mathcal{F}$ with respect to γ, then θ, then ψ, circularly, until a local minimum is reached. At the minimum, the obtained values for θ (GMM parameters) and ψ (diffeomorphism) also constitute a local maximum of the original likelihood problem, Eq. (6).

The minimizations of $\mathcal{F}$ w.r.t. γ and θ are classic computations from the GMM model [3,5,7,8,13]. In the *E step*, solving $\partial_\gamma \mathcal{F} = 0$ (subject to $\forall n, \sum_c \gamma_{nc} = 1$) yields the following update rule for γ:

$$\gamma_{nc} = \frac{\pi_c e^{-|\psi(x_n) - \mu_c|^2/2\sigma^2}}{\sum_{c'} \pi_{c'} e^{-|\psi(x_n) - \mu_{c'}|^2/2\sigma^2}} \tag{8}$$

In the *M step*, solving $\partial_\sigma \mathcal{F} = 0$, resp. $\partial_{\mu_c} \mathcal{F} = 0$, resp. $\partial_{\pi_c} \mathcal{F} = 0$ subject to $\sum_c \pi_c = 1$, yields the respective update rules for the GMM parameters (of which, only those pertaining to the problem's free parameters θ should be applied):

$$\sigma^2 = \frac{1}{dN} \sum_{n,c} \gamma_{nc} |\psi(x_n) - \mu_c|^2, \quad \mu_c = \frac{\sum_n \gamma_{nc} \psi(x_n)}{\sum_n \gamma_{nc}}, \quad \pi_c = \frac{1}{N} \sum_n \gamma_{nc} \quad (9)$$

In contrast, the minimization of Eq. (7) w.r.t. ψ yields a novel registration problem in the LDDMM framework (Eq. (14) below), which we present now.

2.3 LDDMM Registration with Logdet Term

Due to space constraints, we assume some prior acquaintance with the LDDMM framework [1, 10, 12], only listing rapidly its main elements required in the sequel.

Space $\mathcal{V}$ of Vector Fields. The theory starts by defining a functional space $\mathcal{V}$ of vector fields over $\mathbb{R}^d$, a Reproducing Kernel Hilbert Space (RKHS) whose inner product will be noted $\langle v|w \rangle$, and associated norm $\|v\|^2 := \langle v|v \rangle$. The reproducing kernel of $\mathcal{V}$ is assumed to be translation invariant; we note it as $K_z(y) := K(y - z)$, with $K : \mathbb{R}^d \to \mathbb{R}$ the chosen, radial, kernel function. Evaluation functionals *and their first spatial derivatives* are assumed to be continuous on $\mathcal{V}$; hence we have for every $w \in \mathcal{V}$, indices $i, j \in [\![1; d]\!]$, and $z \in \mathbb{R}^d$,

$$w^i(z) = \langle K_z e_i | w \rangle \quad \text{and} \quad \partial_j w^i(z) = \langle (\partial_j K_z) e_i | w \rangle \quad (10)$$

where w^i denotes the i-th component of vector field w, e_i is the i-th elementary vector of $\mathbb{R}^d$, and ∂_j is spatial derivation w.r.t. to the j-th component.

LDDMM Diffeomorphism Group and Geodesics. We note $\mathcal{G}$ the subgroup of diffeomorphisms on $\mathbb{R}^d$ that can be obtained from the flow of vector fields belonging to $\mathcal{V}$. Precisely, $\psi \in \mathcal{G}$ iif. $\psi = \phi_1$ where $t \mapsto \phi_t$ is the flow (i.e., $\phi_0 = \text{Id}$) associated to the ODE

$$\partial_t \phi_t = v_t \circ \phi_t \quad (11)$$

with $t \mapsto v_t$ a time-evolving vector field with values in $\mathcal{V}$, and sufficient regularity [12]. In the sequel, notation v_t will always represent the time-derivative of ϕ_t according to Eq. (11), without necessarily reminding it.

$\mathcal{G}$ is then equipped with the metric inherited from $\mathcal{V}$: the squared distance along each trajectory $t \mapsto \phi_t$ is defined as $\frac{1}{2} \int_{t=0}^1 \|v_t\|^2 dt$. This allows to define the concept of an LDDMM *geodesic*, i.e., a trajectory $t \mapsto \phi_t$ minimizing the squared distance, given imposed starting point $\phi_0 = \text{Id}$ and endpoint $\phi_1 = \psi$. These curves can be characterized with a classic Euler-Lagrange perturbative approach [1, 12], yielding the following *geodesic equation* on the trajectory:

$$\forall w \in \mathcal{V}, \quad \partial_t \langle v_t | w \rangle = \langle v_t | \mathcal{L}_{v_t} w \rangle \quad (12)$$

with the Lie derivative $\mathcal{L}_v w := \text{D}w.v - \text{D}v.w$. Concretely, Eq. (11)–(12) allow to compute the full geodesic trajectory $t \mapsto \phi_t$ from the initial value $v_0 \in \mathcal{V}$.

Setting $w = v_t$ in Eq. (12), we obtain $\partial_t \|v_t\|^2 = 0$, so $\|v_t\|$ along a geodesic is constant. This leads to introduce, for every $\psi \in \mathcal{G}$, the functional

$$\mathcal{E}_\mathcal{G}(\psi) := \frac{1}{2}\|v_0\|^2 \quad \left(= \frac{1}{2}\|v_t\|^2, \quad \forall t \in [0,1]\right) \tag{13}$$

for the only geodesic $t \mapsto \phi_t$ such that $\phi_0 = \text{Id}$ and $\phi_1 = \psi$. The value of $\mathcal{E}_\mathcal{G}(\psi)$ measures the (minimal) squared distance in $\mathcal{G}$ from Id to ψ. This is the measure that we use as a probabilistic prior on ψ: see Eq. (5) above.

Energy Functional. Returning to our registration problem, and focusing on the dependency of $\mathcal{F}$ in Eq. (7) with respect to ψ, we have to minimize

$$\mathcal{F}(\psi) = C + \sum_{n=1}^{N} \frac{|\phi_1(x_n) - y_n|^2}{2\sigma^2} + \int_{t=0}^{1} \left(\frac{\lambda}{2}\|v_t\|^2 - \sum_{n=1}^{N} \text{div}(v_t)_{\phi_t(x_n)}\right) dt \tag{14}$$

where C denotes terms in $\mathcal{F}$ independent of ψ, and we set $y_n := \sum_c \gamma_{nc}\mu_c$. Notation $t \mapsto (\phi_t, v_t)$ refers to the only geodesic such that $\phi_0 = \text{Id}$ and $\phi_1 = \psi$, and we use the important fact that

$$\partial_t \log \det(D\phi_t) = \text{Tr}((D\phi_t)^{-1}\partial_t D\phi_t) = \text{Tr}(Dv_t) = \text{div}(v_t).$$

If not for the divergence term, Eq. (14) would correspond to the classic "landmark registration" problem in the LDDMM framework [10]. The resolution here will thus be very similar, but with additional terms. Tracing back our equations, the divergence term comes from the Jacobian $\det(D\psi(x))$ in Eq. (4), and thus reflects our modeling of the registration as a *probabilistic inference*.

Finite-Dimensional Geodesic ODE. We now let $\beta := \lambda^{-1}$, and $\varphi_n(t) := \phi_t(x_n)$ the N trajectories of the data points under the flow. It can be shown that if ψ is a local optimum of $\mathcal{F}$ in Eq. (14), there exist N vector-valued weight functions $t \mapsto a_n(t) \in \mathbb{R}^d$, such that the corresponding geodesic's vector field v_t is of the form

$$v_t(z) = \sum_{n=1}^{N} [a_n(t)K(z - \varphi_n(t)) - \beta(\nabla K)(z - \varphi_n(t))] \tag{15}$$

or equivalently, through the RKHS property Eq. (10),

$$\forall w \in \mathcal{V}, \quad \langle v_t | w \rangle = \sum_{n=1}^{N} [a_n(t)^\top w_{\varphi_n(t)} + \beta \text{div}(w)_{\varphi_n(t)}] \tag{16}$$

A rapid, heuristic explanation is that $\langle v_1 |$ has to be of the form Eq. (16) when ψ is an optimum of $\mathcal{F}$ in Eq. (14), and then the form Eq. (15)–(16) is preserved under the geodesic equation (12). Specifically, injecting Eq. (15)–(16) into the geodesic equation (12), and after some simplifications, we find that trajectories

$\varphi_n(t)$ and vector weights $a_n(t)$ along a geodesic must be satisfy the following Hamiltonian ODE:

$$\forall n, \quad \frac{\mathrm{d}}{\mathrm{d}t}\varphi_n = (\partial_{a_n}H)(\varphi, a), \quad \frac{\mathrm{d}}{\mathrm{d}t}a_n = -(\partial_{\varphi_n}H)(\varphi, a) \tag{17}$$

with initial conditions $\varphi_n(0) = x_n$ (data points), $a_n(0) = a_n^0$ (for now, a free parameter), and the Hamiltonian function

$$H(\varphi, a) := \frac{1}{2}\sum_{n,m}\left[(a_m^\top a_n)K + \beta(a_n - a_m)^\top \nabla K - \beta^2(\Delta K)\right](\varphi_m - \varphi_n)$$

Equation (17) provides a concrete embodiment of equations (11)–(12) into a finite dimensional ODE on $2dN$ scalar variables. As in every Hamiltonian system, we recover that the solution verifies $H(\varphi(t), a(t)) = \frac{1}{2}\langle v_t|v_t \rangle = \text{constant} = \mathcal{E}_\mathcal{G}(\psi)$, in accordance with general LDDMM principles.

Geodesic Shooting. Through ODE (17), the final diffeomorphism $\psi = \phi_1$ becomes a function of the initial momentum variables $a^0 = (a_n^0)_{n=1...N}$. Precisely, injecting Eq. (15)–(16) in Eq. (14), we obtain $\mathcal{F}(\psi) = \mathcal{C} + E(a^0)$ for the function

$$E(a^0) := \sum_n \frac{|\varphi_n(1) - y_n|^2}{2\sigma^2} + \frac{\lambda}{2}\int_{t=0}^1 \sum_{m,n}\left[(a_m^\top a_n)K + \beta^2\Delta K\right](\varphi_m - \varphi_n)\mathrm{d}t \tag{18}$$

where $a(t), \varphi(t)$ are implicit functions of a^0 through the geodesic equation (17). The gradient $\nabla_{a^0}E$ can also be estimated: differentiating Eq. (17) w.r.t. a^0 yields a so-called *auxiliary* linear ODE on the quantities $(\partial_{a^0}\varphi)(t)$, $(\partial_{a^0}a)(t)$, and the differentiation of Eq. (18) w.r.t. a^0 involves precisely these quantities $\partial_{a^0}\varphi$, $\partial_{a^0}a$. Furthermore, this computation can be done automatically by numerical libraries such as PyTorch equipped with automatic differentiation.

Hence, a local minimum for $E(a^0)$ can be found with a *geodesic shooting* procedure: start with an initial guess for a^0, numerically implement ODE (17) to estimate $E(a^0)$, and the auxiliary computations to estimate $\nabla_{a^0}E$. This allows to modify a^0 according to some version of gradient descent, and the whole shooting procedure can be repeated, until a local minimum of E is found. This minimum solves the partial minimization of $\mathcal{F}$ in Eq. (7) w.r.t. ψ.

3   Numerical Applications

Implementation. We coded in Python, using libraries PyTorch and KeOps [4]. Some elements of code (LDDMM implementation, visualizations in Figs. 1 and 2) were adapted from KeOps tutorials. RKHS kernel was chosen as $K(z) = \exp(-|z|^2/2\tau^2)$ with $\tau = 0.2$, and LDDMM regularization constant as $\lambda = 500$. The geodesic ODE Eq. (17) was numerically integrated with Ralston's method over 10 discrete time steps. The gradient $\nabla_{a^0}E$ of Eq. (18) was then estimated automatically by back-propagating PyTorch's autograd algorithm through the computations, and input into PyTorch's L-BFGS algorithm to target a local minimum of $E(a^0)$. All code is at https://github.com/AdrienWohrer/diff-icp.

Warping to a Known GMM Distribution. In a first experiment, a known GMM model $\mathcal{M}_g$ (Fig. 1a) and unknown diffeomorphism ψ_g are used to generate a warped point set $(x_n)_{n=1\dots100}$ (Fig. 1b), and the goal is to estimate the unknown diffeomorphism. In our notations, $\theta = \{\}$ (no GMM parameters to optimize), and we seek a mapping ψ to maximize the likelihood in Eq. (6). This is a diffeomorphic generalization of classic "probabilistic ICP" algorithms for two-set registration, Eq. (1), in which the second point set constitutes the centroids of the GMM model [5,8,13] (except that these algorithms also optimize the GMM variance parameter σ, whereas we keep it fixed in this simple illustration).

This optimal ψ is found by looping repeatedly through the E-step update Eq. (8), and the minimization of Eq. (14) w.r.t. ψ (obtained by minimizing Eq. (18) w.r.t. a^0). Before the first loop, variables a^0 are initialized to represent an initial mapping $\psi \simeq \mathrm{Id}$. After a number of loops, convergence is achieved, providing a warping of the point set back to its generative GMM model (Fig. 1c, d).

Registration of Multiple Point Sets. In a second experiment, we extend the model to perform registration of multiple point sets, Eq. (2). The goal is now to register each point set $X^{(k)} = \{x_n^{(k)}\}_{n=1\dots N_k}$ with its dedicated mapping ψ_k, to a common space where all data points can be fitted with a single GMM model, that must also be characterized. That is, in our notations,

$$\max_{\substack{\theta \\ \forall k, \, \psi_k \in \mathcal{G}}} \prod_{k=1}^{K} \mathrm{P}(\psi_k) \left(\prod_{n=1}^{N_k} f_{\mathcal{M}^*}(x_n^{(k)} | \psi_k, \theta) \right) \tag{19}$$

with $\theta = \{\sigma, \mu_c, \pi_c\}$ (all GMM parameters must be optimized). This is a diffeomorphic generalization of the rigid registration algorithm for multiple point sets proposed by [7]. In a typical application, point set k could represent some anatomical features from patient number k, and the GMM $\mathcal{M}(\theta)$ recovered by the algorithm represents a *statistical atlas* of these features across patients, as has been proposed, e.g., in the context of image registration [9].

The resolution algorithm now consists in looping through the two following stages until convergence:

1. Optimize $\mathcal{M}(\theta)$ to the set of *all* warped points $\psi_k(x_n^{(k)})$, given the current mappings ψ_k. We achieve this by looping 10 times through Eq. (8)–(9), with Eq. (9) running over all data points, i.e., $\sum_n$ means $\sum_{k=1}^{K} \sum_{n=1}^{N_k}$.
2. Update each mapping ψ_k to minimize Eq. (14), given the current GMM $\mathcal{M}(\theta)$. This can be done independently for each k.

We tested this method on $K = 10$ point sets generated from the same GMM as Fig. 1a, now considered unknown (Fig. 2a). After a number of loops, all mappings ψ_k and the joint GMM $\mathcal{M}(\theta)$ converge to an equilibrium (Fig. 2b). In particular, the inferred GMM ("statistical atlas") correctly recovers the shape of the generating GMM.

We also tested a modified algorithm, replacing the minimization of Eq. (14) by the classic LDDMM "landmark registration" algorithm [10], which is recovered

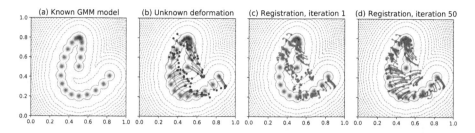

Fig. 1. Warping to a known GMM distribution (see text).

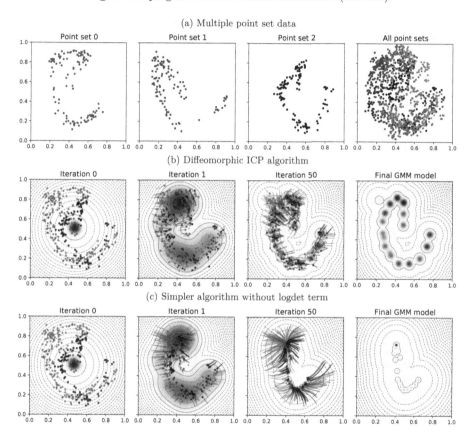

Fig. 2. Multiple point set diffeomorphic registration (see text).

by setting $\beta = 0$ in Eq. (15) and following. In this case, the obtained mappings ψ_k excessively shrink the point sets (Fig. 2c), as this allows to artificially minimize the quadratic error in Eq. (14). This demonstrates the need of the additional divergence term in Eq. (14) to obtain a well-posed algorithm in general.

4   Conclusion

We have proposed a generalization of probabilistic ICP algorithms based on the GMM distribution [5,8,13], and their generalization to multiple point sets [7], to incorporate diffeomorphic registration in the LDDMM framework. First experiments on synthetic data suggest that the algorithm can correctly register two point sets diffeomorphically (Fig. 1), or retrieve a common shape underlying multiple point sets (Fig. 2). Future work will have to confirm these results more quantitatively, and include comparisons with previous algorithms.

The algorithm, presented here in its "raw" form, could easily accomodate the numerous variations that have been proposed to improve performance of previous ICP algorithms: introducing an additional mixture component to handle outliers [5,7], replacing the ICP association rule by one-to-one associations (as in the RPM model with *softassign* [6]), controlling GMM parameter σ through an external annealing program [5,6,8], etc.

We also intend to investigate the roles of the different model meta-parameters: number of GMM classes C, LDDMM regularisation strength λ, RKHS kernel spatial scale τ, and to provide some heuristics for fixing their values. Finally, we will concretely apply the algorithm to the multiple registration of deep brain structures in a cohort of surgically implanted patients.

References

1. Beg, M.F., Miller, M.I., Trouvé, A., Younes, L.: Computing large deformation metric mappings via geodesic flows of diffeomorphisms. Int. J. Comput. Vision **61**, 139–157 (2005)
2. Besl, P.J., McKay, N.D.: Method for registration of 3-D shapes. In: Sensor Fusion IV: Control Paradigms and Data Structures, vol. 1611, pp. 586–606. SPIE (1992)
3. Bishop, C.M.: Pattern Recognition and Machine Learning. Springer, New York (2006)
4. Charlier, B., Feydy, J., Glaunès, J.A., Collin, F.D., Durif, G.: Kernel operations on the gpu, with autodiff, without memory overflows. J. Mach. Learn. Res. **22**(74), 1–6 (2021)
5. Chui, H., Rangarajan, A.: A feature registration framework using mixture models. In: Proceedings IEEE Workshop on Mathematical Methods in Biomedical Image Analysis. MMBIA-2000 (Cat. No. PR00737), pp. 190–197 (2000)
6. Chui, H., Rangarajan, A.: A new point matching algorithm for non-rigid registration. Comput. Vis. Image Underst. **89**(2), 114–141 (2003)
7. Evangelidis, G.D., Horaud, R.: Joint alignment of multiple point sets with batch and incremental expectation-maximization. IEEE Trans. Pattern Anal. Mach. Intell. **40**(6), 1397–1410 (2017)
8. Granger, S., Pennec, X.: Multi-scale EM-ICP: a fast and robust approach for surface registration. In: Heyden, A., Sparr, G., Nielsen, M., Johansen, P. (eds.) ECCV 2002. LNCS, vol. 2353, pp. 418–432. Springer, Heidelberg (2002). https://doi.org/10.1007/3-540-47979-1_28
9. Joshi, S., Davis, B., Jomier, M., Gerig, G.: Unbiased diffeomorphic atlas construction for computational anatomy. Neuroimage **23**, S151–S160 (2004)

10. Joshi, S.C., Miller, M.I.: Landmark matching via large deformation diffeomorphisms. IEEE Trans. Image Process. **9**(8), 1357–1370 (2000)
11. Maiseli, B., Gu, Y., Gao, H.: Recent developments and trends in point set registration methods. J. Vis. Commun. Image Represent. **46**, 95–106 (2017)
12. Miller, M.I., Trouvé, A., Younes, L.: Geodesic shooting for computational anatomy. J. Math. Imaging Vis. **24**, 209–228 (2006)
13. Myronenko, A., Song, X.: Point set registration: coherent point drift. IEEE Trans. Pattern Anal. Mach. Intell. **32**(12), 2262–2275 (2010)
14. Zhu, H., et al.: A review of point set registration: from pairwise registration to groupwise registration. Sensors **19**(5), 1191 (2019)

Chan-Vese Attention U-Net: An Attention Mechanism for Robust Segmentation

Nicolas Makaroff[(✉)] and Laurent D. Cohen

CEREMADE, UMR CNRS 7534, University Paris Dauphine, PSL Research University, 75775 Paris, France
makaroff@ceremade.dauphine.fr

Abstract. When studying the results of a segmentation algorithm using convolutional neural networks, one wonders about the reliability and consistency of the results. This leads to questioning the possibility of using such an algorithm in applications where there is little room for doubt. We propose in this paper a new attention gate based on the use of Chan-Vese energy minimization to control more precisely the segmentation masks given by a standard CNN architecture such as the U-Net model. This mechanism allows to obtain a constraint on the segmentation based on the resolution of a PDE. The study of the results allows us to observe the spatial information retained by the neural network on the region of interest and obtains competitive results on the binary segmentation. We illustrate the efficiency of this approach for medical image segmentation on a database of MRI brain images.

Keywords: Attention Mechanism · Level Set · Chan-Vese · Segmentation

1 Introduction

Medical image segmentation is a crucial task that requires significant time and effort from medical experts. Although various solutions, including convolutional neural networks (CNNs), have been proposed to automate this process, the need for efficient and reliable methods still exists.

While CNNs have shown promising results in medical image segmentation, their lack of transparency and the sensitive nature of medical data raise concerns regarding their applicability in real-world hospital settings, especially for medical staff who are not trained in machine learning.

Convolutional neural networks have revolutionised the study of images for both classification and segmentation, the latter being the objective of interest here. Two architectures stand out among all those studied over the years, the fully convolutional neural network [6] and the famous U-Net [14]. These architectures have been tested many times on various applications such as MRI segmentation of the brain [5] or heart [13], CT scans of organs in the thoracic cavity [3]. Numerous modifications have also been made to improve the efficiency of these

F. Nielsen and F. Barbaresco (Eds.): GSI 2023, LNCS 14072, pp. 574–582, 2023.
https://doi.org/10.1007/978-3-031-38299-4_59

complex structures and in the medical field especially the U-Net has proven to be very versatile for many segmentation tasks. The rest of the paper is organised as follows. In Sect. 2 we introduce our experimental method for Fast Marching Energy CNN. In Sect. 3 we present the main results of our experiments and provide a discussion around our work.

1.1   Related Work

Several attempts have been made to integrate geometric or topological properties in the neural network to incorporate information beyond adjacent pixels for segmentation tasks. In 2019, [4] proposed a lesion segmentation method based on active contours using a U-Net-like neural network. The network predicts a segmentation mask, which is refined using a generalisation of the active contour problem from [2]. Similarly, In 2020, [15] presented a model, where the neural network predicts the parameters for initialising the active contour model and an initial contour. Learning is achieved by combining the error produced by the neural network and that produced during active contour usage. In 2021, [8] proposed a fully integrated geodesic active contour model, where the neural network learns to minimise the energy functional of the model. In this encoder-decoder network, the output is a contour map instead of a probability map for segmentation, based on the active contour method proposed by [1]. Although these methods have shown promise, there is still room for improvement in incorporating geometric and topological properties more effectively and seamlessly into deep learning-based segmentation approaches.

1.2   Contributions

The main contribution of this paper is the development of a novel hybrid segmentation method that combines deep learning with classical functional energy minimization techniques, specifically designed for medical image segmentation applications. Our method features a new attention gate, the *Chan-Vese Attention Gate*, which integrates information from the level sets method of the well-established Chan-Vese functional [2]. Unlike traditional deep learning methods that rely solely on the neural network to improve image segmentation, our approach leverages resolution information to achieve more accurate results.

To demonstrate the effectiveness of our method, we conducted comprehensive experiments on the TCGA_LGG database [10], a repository of brain images for the study of lower grade gliomas. Given the sensitive nature of medical image segmentation, it is crucial to ensure the validity of our results. Our approach achieved at least equivalent results to previous networks while remaining simple to optimise. The training time for our method is only 5% longer than the traditional approach, which is a minor increase considering the benefits. Furthermore, the difference in computation time is imperceptible during inference. Overall, our approach represents a significant advancement in medical image segmentation, offering a more accurate and efficacious solution for this critical field.

2   Methodology

The U-Net Architecture. The U-Net architecture is widely used for medical image segmentation. It maintains the structure of an image during transformation from an image to a vector and back to an image using features extracted during the contraction phase. The architecture takes the shape of a "U" with three parts: contraction, transition, and expansion. The contraction applies several blocks with convolution and pooling layers, doubling the feature maps at each stage. The transition uses convolution layers, while the expansion uses convolution and up-sampling layers. The information is recovered during the contraction to reconstruct the image, with the same number of expansion blocks as contraction blocks. The final output is obtained through the final convolutional layer.

Attention Gate in U-Net Architecture. The authors of [9] presented in their paper a new attention gate especially for the case of the CNN and in particular for the U-Net. The architecture of the U-Net remains the same except for the expansion part. In this part, an attention mechanism is integrated between each block. For each block, the input and the information coming from the connections of the corresponding contraction part block pass through an attention block. The input is up-sampled in parallel and finally the two results are concatenated and sent to the convolution block.

Chan-Vese Energy Minimization. Presented in [2], Chan-Vese's method is used to segment a binary image. Let I be the given grayscale image on a domain Ω to be segmented. The Chan Vese method looks for a piece-wise constant approximation of an image where there are 2 regions separated by an unknown boundary curve C. This is obtained through the minimization of the following energy depending on curve C and the constant values c1 and c2 inside and outside the curve:

$$E(C, c_1, c_2) = \mu \times \text{Length}(C) + \nu \times \text{Area}(inside(C))$$
$$+ \lambda_1 \int_{inside(C)} |I(x,y) - c_1|^2 dxdy + \lambda_2 \int_{outside(C)} |I(x,y) - c_2|^2 dxdy. \quad (1)$$

Energy minimization is simplified by replacing the curve C with a level set function ϕ. The inside region is then the set where $\phi > 0$ and the outside region the set where $\phi < 0$. With the help of the Heavyside function H, the energy becomes:

$$F(c_1, c_2, \phi) = \mu \int_{\Omega} \delta(\phi(x))|\nabla\phi(x)|dx + \nu \int_{\Omega} H(\phi(x))dx$$
$$+ \lambda_1 \int_{\Omega} |I(x) - c_1|^2 H(\phi(x))dx + \lambda_2 \int_{\Omega} |I(x) - c_2|^2(1 - H(\phi(x)))dx, \quad (2)$$

where the term following μ represent the length of the contour, the term following ν the area inside the contour and δ the Dirac mass.

$$(P) : \arg\min_{c_1, c_2, \phi} F(c_1, c_2, \phi) \quad (3)$$

The new variable is ϕ. The energy $F(c_1, c_2, \phi)$ is minimised wit respect to ϕ wit ha gradient descent evolution.

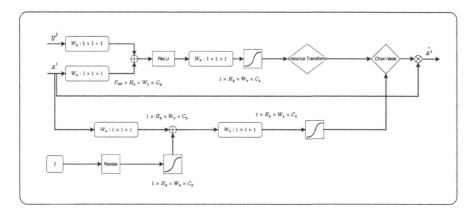

Fig. 1. Scheme of the proposed attention method. The symbols $\oplus$ and $\otimes$ represent respectively the addition and multiplication of the tensors.

Chan-Vese Attention in U-Net Architecture. In this work, we incorporate our novel attention method, the *Chan-Vese Attention Gate*, into the U-Net architecture. The attention method proposed by Oktay et al. [9] allows highlighting areas that provide more information in the skip connections. Our method expands upon this approach by incorporating information from the Chan-Vese method, which performs intermediate segmentation for each block of the expansion phase. As illustrated in Fig. 1, the representation constructed by the network from the skip connection is concatenated with the information from the previous layer before undergoing several transformations. First, let us briefly review the classical attention gate proposed by Oktay et al. [9]. Given a pair of input feature maps, the attention gate computes an attention coefficient that modulates the skip connections. The attention coefficient is obtained through a gating mechanism that involves two 1×1 convolutions followed by an element-wise addition and a non-linear activation function (sigmoid). The attention gate effectively allows the network to learn which regions of the input image are more relevant to the segmentation task. Our method builds upon the classical attention gate by incorporating the Chan-Vese segmentation technique. The first key step is the distance transform applied to the classical attention representation, which serves as the initial contour for the Chan-Vese method. This step is made possible thanks to the use of the distance transform method proposed by [11], which renders the transformation differentiable within a classical automatic differentiation framework. A secondary branch was implemented in the proposed framework to selectively emphasise the tumorous region of the input image. This branch carefully resizes the input image to conform to the dimensions of the current layer and integrates it with the residual feature map via addition. A 1×1 convolution is then applied to the resulting map, which serves as the input for subsequent segmentation. In the second step, the modified Chan-Vese algorithm is used to iteratively segment the image a finite number of times. This segmented image is then used as a control signal to facilitate learning. The proposed approach thus provides a more refined and informative attention for the network.

Let $I \in \mathbb{R}^{N \times H \times W \times K_I}$ represent a batch of input images, where N, H, W, K_I denote the batch size, height, width, and number of channels of the image, respectively. Let $X_i^f \in \mathbb{R}^{F_x \times H_l \times W_l \times K}$ denote the residual feature map f at layer i and $Y_j^f \in \mathbb{R}^{F_x \times H_l \times W_l \times K}$ the previous layer feature map at layer j and feature map f. $W_{1\times1}$ is a 1×1 convolution. Following the additive attention formulation:

$$q_{att}^f = \Psi^T(\sigma_1(\boldsymbol{W}_{1\times1}^T X_i^f + \boldsymbol{W}_{1\times1}^T Y_j^f + b_f) + b_\psi \tag{4}$$

$$\alpha_i^f = \sigma_2(q_{att}^f(X_i^f; Y_j^f; \Theta_{att})) \tag{5}$$

We define D_x as the distance transform that takes a tensor of shape $H_l \times W_l \times K_l$ as input. By applying the distance transform to the former attention gate, we obtain

$$\beta_i^l = D(\alpha_i^l) = -\lambda \log(\alpha_i^l * \exp(-\frac{d(\cdot, 0)}{\lambda})) \tag{6}$$

where $d(\cdot, \cdot)$ is the Euclidean distance and $*$ is the convolution product. This information is passed as an initialization contour.

On the other side, we perform a transformation of the input image, as if using some filter (e.g., CLAHE filter [12]), as follows:

$$\gamma_i^f = \sigma_2(\boldsymbol{W}_{1\times1}^T(\boldsymbol{W}_{1\times1}^T X_i^f + \sigma_2(I)) + b_W) \tag{7}$$

Many areas of the images we wish to segment happen to have the same values as the averages of regions c_1 and c_2 as the region we ultimately want to segment. This sometimes leads the algorithm to add undesirable areas to the segmentation even though the μ and ν hyperparameters have been carefully chosen. The γ_i^f transformation solves this problem by reducing the intensity in areas far from the tumour.

Finally, the resulting mask segmentation and attention coefficient ζ_i^f is given by solving the Chan-Vese problem:

$$\zeta_i^l = CV(\gamma_i^f, \beta_i^f, \mu, \nu) \tag{8}$$

where γ_i^f is the image that supports the initial mask β_i^f. μ and ν are positive parameters.

During the backpropagation procedure, the gradients of the loss function concerning the attention gate's parameters (marked in bold) and intermediate outputs are computed using the chain rule, which can be efficiently executed in modern deep learning frameworks. The differentiable distance transform and the Chan-Vese module enable the backpropagation to update the parameters of the attention gate and the network's other layers. Consequently, the network can learn to focus on the most relevant regions for segmentation, improving its performance in medical image segmentation tasks.

3   Experiments

Evaluation Datasets. In this study, we used the TCGA_LGG database, an openly available online repository [10] containing magnetic resonance imaging (MRI) scans of brain tumour patients. The database consists of 110 patients from The Cancer Genome Atlas (TCGA) lower-grade glioma collection, with genomic cluster data and at least one fluid-attenuated inversion recovery (FLAIR) sequence available. Table 1 provides a summary of the experimental results. The spatial resolution of the images contained in the TCGA LGG dataset is $1mm$ isotropic.

Table 1. Segmentation results (IOU) on the TGCA_LGG brain MRI database. Significant results are highlighted in bold font.

Method	U-Net	Attention U-Net	Chan-Vese U-Net
Dice	**0.832 ± 0.091**	0.830 ± 0.023	0.824 ± 0.019
IOU	0.829 ± 0.075	0.833 ± 0.023	**0.848 ± 0.021**
Hausdorff	2.390 mm ± 0.985	2.416 mm ± 0.775	**2.329 mm ± 0.672**
FPR	0.010 ± 0.003	**0.009 ± 0.002**	0.012 ± 0.004
FNR	0.013 ± 0.004	0.015 ± 0.005	**0.013 ± 0.003**

Implementation Details. We used a large batch of 32 for gradient update and the model parameters are optimised using an adamW optimiser [7] with learning rate 5×10^{-4} and batch normalisation. We applied standard data augmentation (resize, horizontal flip, vertical flip, random rotate, transpose, shift and scale, normalise). The Chan-Vese parameters μ and ν are set respectively to 0.1 and 1.0. The loss is computed using the addition of Dice loss and Binary Cross Entropy. The added attention layer slows down the training by an average of 1 sec out of 6 sec per batch. The code is written in Jax using Haiku framework and will soon be available.

Segmentation Results. In this study, we conducted a comparative evaluation of our proposed model with the classical U-Net and the original Attention U-Net. Table 1 provides a summary of the experimental results. Dice, IOU (Intersection Over Union), Hausdorff, FPR (False Positive Rate), and FNR (False Negative Rate) values are reported, along with their respective standard deviations (denoted as sd_i). The image resolution should be considered when interpreting the Hausdorff distance values. Our proposed model demonstrated superior IOU scores and improved false negative performance. This can be attributed to the model's ability to focus on a smaller area of interest and the integration of the Chan-Vese method, which enables more effective capture of relevant information and reduces the risk of information loss.

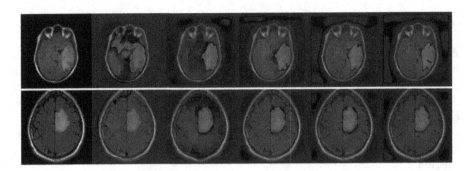

Fig. 2. Output of the Chan-Vese Attention Layer at different stage of learning (epochs: 1, 50, 100, 200, 300) from left to right.

Chan-Vese Attention Masks Analysis. We can observe from the results of the attention layer (see Fig. 2) that with the use of Chan-Vese the attention mask quickly converges to an apparently tumour-like segmentation. It takes advantage of the minimization of the Chan-Vese energy from the initialization of the mask thanks to the part inspired by the attention method of [9] but also of the use of the initial image to be segmented. Gradually the contours of the tumour become more precise and the active intensity on the tumour is the confidence on the energy to be minimised. The 0 level set was used to enable the neural network to selectively prioritise the tumour area during segmentation. The upper level set was subsequently employed to refine the tumour segmentation.

Comparison with Attention U-Net. Figure 3 shows the attention output of the Chan-Vese Attention Module and the classical Attention Module. Both methods allow the neural network to focus on the tumour area. It should be noted that the method proposed by [9] obtains a finer mask on certain details of the tumour but does not manage in the framework of our study to rank the confidence of the presence of the tumour. In many places outside the tumour area we observe artifacts that do not correspond to the object of interest in the image. In contrast

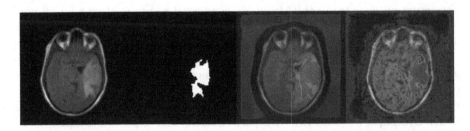

Fig. 3. Comparison of the Attention Mask between a Chan-Vese Attention and the Original Attention. (From left to right: the input MRI, the tumour to be segmented, Chan-Vese Attention mask, original Attention mask).

to these observations the method supported by Chan-Vese focuses only on the tumour area inside the skull.

4   Conclusion

In this paper, we have presented a novel segmentation approach that effectively combines classical energy minimization techniques with Deep Learning. Our proposed model, which integrates the Chan-Vese algorithm into the attention mechanism of a U-Net, demonstrates the value of incorporating non-deep learning sources of shape and structure information, particularly when dealing with sensitive medical data.

Our experimental analysis reveals that the proposed model achieves close results, with some improvements compared to both the classical U-Net and Attention U-Net in key performance metrics such as IOU scores and false negative rates. The Chan-Vese Attention Module successfully narrows down the model's focus to the tumour area, contributing to enhanced segmentation accuracy and precision. This work highlights the potential of combining diverse segmentation methods, paving the way for future research in blending various sources of shape and structure information with Deep Learning models. Ultimately, our approach aims to contribute to the development of more effective and versatile segmentation solutions.

Acknowledgements. This work is in part supported by the French government under management of Agence Nationale de la Recherche as part of the "Investissements d'avenir" program, reference ANR-19-P3IA-0001 (PRAIRIE 3IA Institute).

References

1. Caselles, V., Kimmel, R., Sapiro, G.: Geodesic active contours. Int. J. Comput. Vision **22**, 61–79 (1995)
2. Chan, T., Vese, L.: Active contours without edges. IEEE Trans. Image Process. A Publication of the IEEE Signal Processing Society **10**(2), 266–77 (2001)
3. Gerard, S.E., Reinhardt, J.M.: Pulmonary lobe segmentation using a sequence of convolutional neural networks for marginal learning. In: 2019 IEEE 16th International Symposium on Biomedical Imaging (ISBI 2019), pp. 1207–1211 (2019). https://doi.org/10.1109/ISBI.2019.8759212
4. Hatamizadeh, A., et al.: Deep active lesion segmentation. In: Suk, H.-I., Liu, M., Yan, P., Lian, C. (eds.) MLMI 2019. LNCS, vol. 11861, pp. 98–105. Springer, Cham (2019). https://doi.org/10.1007/978-3-030-32692-0_12
5. Kleesiek, J., et al.: Deep MRI brain extraction: a 3D convolutional neural network for skull stripping. Neuroimage **129**, 460–469 (2016)
6. Long, J., Shelhamer, E., Darrell, T.: Fully convolutional networks for semantic segmentation. In: The IEEE Conference on Computer Vision and Pattern Recognition (CVPR), June 2015
7. Loshchilov, I., Hutter, F.: Fixing weight decay regularization in adam. CoRR abs/1711.05101 (2017). arxiv.org/abs/1711.05101

8. Ma, J., He, J., Yang, X.: Learning geodesic active contours for embedding object global information in segmentation CNNs. IEEE Trans. Med. Imaging **40**(1), 93–104 (2020)

9. Oktay, O., et al.: Attention U-Net: Learning Where to Look for the Pancreas (2018). https://doi.org/10.48550/ARXIV.1804.03999. arxiv.org/abs/1804.03999

10. Pedano, N., et al.: The cancer genome atlas low grade glioma collection (TCGA-LGG) (version 3) [data set]. Cancer Imaging Archive (2016). https://doi.org/10.7937/K9/TCIA.2016.L4LTD3TK

11. Pham, D.D., Dovletov, G., Pauli, J.: A differentiable convolutional distance transform layer for improved image segmentation. In: Akata, Z., Geiger, A., Sattler, T. (eds.) DAGM GCPR 2020. LNCS, vol. 12544, pp. 432–444. Springer, Cham (2021). https://doi.org/10.1007/978-3-030-71278-5_31

12. Pizer, S.M., et al.: Adaptive histogram equalization and its variations. Comput. Vision Graph. Image Process. **39**(3), 355–368 (1987). https://doi.org/10.1016/S0734-189X(87)80186-X

13. Pop, M., Sermesant, M., Jodoin, P.-M., Lalande, A., Zhuang, X., Yang, G., Young, A., Bernard, O. (eds.): STACOM 2017. LNCS, vol. 10663. Springer, Cham (2018). https://doi.org/10.1007/978-3-319-75541-0

14. Ronneberger, O., Fischer, P., Brox, T.: U-Net: convolutional networks for biomedical image segmentation. In: Navab, N., Hornegger, J., Wells, W.M., Frangi, A.F. (eds.) MICCAI 2015, Part III. LNCS, vol. 9351, pp. 234–241. Springer, Cham (2015). https://doi.org/10.1007/978-3-319-24574-4_28

15. Zhang, M., Dong, B., Li, Q.: Deep active contour network for medical image segmentation. In: Martel, A.L., et al. (eds.) MICCAI 2020, Part IV. LNCS, vol. 12264, pp. 321–331. Springer, Cham (2020). https://doi.org/10.1007/978-3-030-59719-1_32

Using a Riemannian Elastic Metric for Statistical Analysis of Tumor Cell Shape Heterogeneity

Wanxin Li[1], Ashok Prasad[2] (ID), Nina Miolane[3(✉)] (ID), and Khanh Dao Duc[4(✉)] (ID)

[1] Department of Computer Science, University of British Columbia,
Vancouver, BC V6T 1Z2, Canada
wanxinli@cs.ubc.ca

[2] Department of Chemical and Biological Engineering and School of Biomedical Engineering, Colorado State University, Fort Collins, CO 80523, USA
ashok.prasad@colostate.edu

[3] Department of Electrical and Computer Engineering,
University of California Santa Barbara, Santa Barbara, CA 93106, USA
ninamiolane@ucsb.edu

[4] Department of Mathematics, University of British Columbia,
Vancouver, BC V6T 1Z4, Canada
kdd@math.ubc.ca

Abstract. We examine how a specific instance of the elastic metric, the Square Root Velocity (SRV) metric, can be used to study and compare cellular morphologies from the contours they form on planar surfaces. We process a dataset of images from osteocarcoma (bone cancer) cells that includes different treatments known to affect the cell morphology, and perform a comparative statistical analysis between the linear and SRV metrics. Our study indicates superior performance of the SRV at capturing the cell shape heterogeneity, with a better separation between different cell groups when comparing their distance to their mean shape, as well as a better low dimensional representation when comparing stress statistics. Therefore, our study suggests the use of a Riemannian metric, such as the SRV as a potential tool to enhance morphological discrimination for large datasets of cancer cell images.

Keywords: elastic metric · shape analysis · cell morphology · dimensionality reduction

1 Introduction

Cells cultured on planar surfaces adopt a variety of morphological shapes, that are tightly coupled with molecular processes acting on the cellular membrane and cytoskeleton [19]. This tight coupling suggests various potential applications of quantitative measurements of cellular morphology, e.g. in morphogenesis [26], morphological screening [12] and image guided medical diagnosis, for example to

© The Author(s), under exclusive license to Springer Nature Switzerland AG 2023
F. Nielsen and F. Barbaresco (Eds.): GSI 2023, LNCS 14072, pp. 583–592, 2023.
https://doi.org/10.1007/978-3-031-38299-4_60

improve the accuracy of cancer tumor grading by cytologists [22]. Using textural or boundary image information, a vast number of features can possibly be used in this context, ranging from human-crafted features to basis function expansions [2,21]. Interestingly, biophysical measurements of cell elasticity in cancer cell have also shown some heterogeneity across cell types, as well as cultures and measurement conditions, which makes their use for diagnosis and treatment promising, yet challenging [9].

In this paper, we consider the so-called *elastic metric*, a Riemannian metric that aims to quantify local deformations of curves by evaluating how they bend and stretch. As a natural geometric tool to investigate the elasticity of cell shape, we apply it to a dataset of tumor cell images that include different cell lines and conditions. Upon processing this dataset to extract cell shapes and using a specific instance of the elastic metric, namely the square root velocity metric (SRV), we find that in comparison with the linear metric, the SRV is superior at capturing and representing the cell shape heterogeneity. Therefore, our study suggests that the elastic metric can be a potential tool to enhance morphological discrimination for heterogeneous datasets of cancer cell images.

2   Background

2.1   Elastic Metric and Square Root Velocity Metric for Planar Curve Comparison

Definition: The family of *elastic metrics*, introduced by Mio *et al.* [13], can be defined over the space $\mathcal{C}$ of smooth parametrized curves $c : [0,1] \mapsto \mathbb{R}^2$ with nowhere-vanishing derivative. With $a, b > 0$ denoting the parameters of the family, one associates with every curve $c \in \mathcal{C}$ an inner product $g_c^{a,b}$ over the tangent space $T_c\mathcal{C}$, given by [3,18]

$$g_c^{a,b}(h,k) = a^2 \int_{[0,1]} \langle D_s h, N\rangle\langle D_s k, N\rangle ds + b^2 \int_{[0,1]} \langle D_s h, T\rangle\langle D_s k, T\rangle ds, \quad (1)$$

where h, k are two curve deformations in the tangent space $T_c\mathcal{C}$, that can also be considered as planar curves [13]; $<, >$ is the Euclidean inner-product in $\mathbb{R}^2$, $D_s = \frac{1}{\|c'(s)\|}\frac{d}{ds}$, is a differential operator with respect to the arc length s, and N and T respectively are the local unit normal and tangent from a moving frame associated with c. Intuitively, elements in $T_c\mathcal{C}$ represent infinitesimal deformations of c, and $g_c^{a,b}$ quantifies the magnitude of these deformations, with the two factors a and b that can be interpreted as weights penalizing the cost of bending (for a) and stretching (for b) the curve c. In this paper, we specifically consider the case $(a, b) = (1, 1/2)$ that defines the so-called *Square Root Velocity metric*, as it allows in practice for an efficient evaluation [10,23]. In Fig. 1.A, we illustrate how the metric can be interpreted for a local deformation h of c: As we project the derivative of h (with respect to its arc length) along the tangent and normal vectors of the reference frame associated with c, increasing the bending in h results in a relatively higher contribution from the normal component, and thus the integral weighted by a^2, according to Eq. (1). Similarly, stretching increases the contribution from the tangent component, and the integral weighted by b^2.

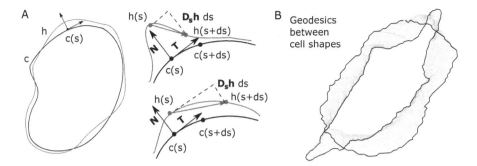

Fig. 1. Elastic metric on cell shapes: (**A**) We illustrate how the elastic metric applies to a given shape c (shown in left) and a local deformation h. According to Eq. (1), this metric is given by the sum of two components, which integrate the projection of the derivative of h with respect to the arc length ($\mathbf{D_s}h \, ds$), on $\mathbf{N}$ and $\mathbf{T}$ respectively, which are the local normal and tangent vectors of c (shown in right). The projection on $\mathbf{N}$ ($\mathbf{T}$) emphasizes bending (stretching) deformations, as shown in top (bottom) right. (**B**) Upon implementing the metric in Geomstats, we can construct a geodesic path between two cell shapes, as a continuous deformation (with intermediate cells in grey) that minimizes the path length (see Eq. (3)) and yields a geodesic distance (see Material and Methods).

Geodesic Distance: As a Riemaniann metric [13,23], the elastic metric yields a geodesic distance over $\mathcal{C}$: For two curves c_0 and c_1 and a regular parameterized path $\alpha : [0,1] \mapsto \mathcal{C}$ such that $\alpha(0) = c_0$ and $\alpha(1) = c_1$, the length of α, associated with the elastic metric $g^{a,b}$ is given by

$$L^{a,b}[\alpha] = \int_0^1 g_{\alpha(t)}^{a,b}(\alpha'(t), \alpha'(t))^{1/2} dt, \qquad (2)$$

and the *geodesic distance* between c_0 and c_1 is

$$d^{a,b}(c_0, c_1) = \inf_{\alpha:[0,1] \mapsto \mathcal{C} \ | \ \alpha(0)=c_0 \ ; \ \alpha(1)=c_1} L^{a,b}[\alpha]. \qquad (3)$$

Figure 1.B illustrates the shortest path joining two cell shapes using the elastic metric.

Fréchet Mean: With the space of curves equipped with this distance, the so-called Fréchet mean of n curves $(c_1, \ldots, c_n)$ [16] is defined as

$$\bar{c} = \underset{c \in \mathcal{C}}{\mathrm{argmin}} \sum_{i=1}^n (d^{a,b}(c, c_i))^2. \qquad (4)$$

Note that the existence and uniqueness of the Fréchet mean is a priori not guaranteed, but requires that the data is sufficiently concentrated (which we will assume for our datasets) [8].

Comparison with Euclidean Linear Metric: We compare the performance of the elastic metric on cell shapes with the Euclidean linear metric, associated with the $\mathcal{L}_2$ distance in $\mathbb{R}^2$. For two cell contours c_0 and c_1 defined as above, this linear distance is given by $d_E(c_0, c_1) = \left(\int_{[0,1]} \|c_0(s) - c_1(s)\|_2^2 \, ds \right)^{1/2}$, and the *linear mean* of n curves $(c_1, \ldots, c_n)$ is $\bar{c}_E = \frac{1}{n} \sum_{i=1}^{n} c_i$. In practice, we compare the use of both metrics in the space of curves quotiented by the action of the group of rigid transformations (i.e. scaling, translation, rotation), via optimal alignment of curve elements [10], using the recent implementation from the *Geomstats* Python library [14] (see Sect. 3.1).

Implementation: An approximation of the geodesic distance associated with the elastic metric $g^{a,b}$ can be computed as a pull-back of the linear metric: Upon applying a transformation that maps the geodesic associated with $g^{a,b}$ into a straight line, the geodesic distance is equal to the $\mathcal{L}^2$ distance between the two transformed curves [18]. While the procedure to construct the mapping can be numerically unstable [3,18], it is simple for the SRV, with the geodesic distance being the $\mathcal{L}^2$ distance obtained upon representing the curve by its speed, renormalized by the square root of its norm as $q(c) = \dot{c}/\sqrt{|\dot{c}|}$ [4].

2.2 Related Works

The elastic metric has been the object of several theoretical and computational developments that primarily focused on curves (for a recent overview, see [4]), with applications to various kinds of biological shapes, including tumor images from MRI (but no cell shapes), plant leafs, or protein backbones [5,6,23]. More recent studies have applied the elastic metric to cell shapes [7,11,15,17] for classification [7,15], dimensionality reduction [11] and regression with metric learning [17]. To our knowledge, the present study is the first to perform a comparative statistical analysis of the elastic metric for tumor cells across different conditions. More generally, there has been a vast number of approaches and features that have been used to model and study cell and/or nuclear shape, as reviewed in [21]; however none of them uses the elastic metric. Overall, the interdependence and relative complexity of the parameters that are classically used to describe biological cells [1,19] make the search of simple interpretable features, such as the elastic metric, useful. For example, a model with two independent nondimensional parameters, that respectively capture flatness and scale, was recently proposed to analyze nucleus shapes of populations for multiple cell lines [2].

3   Methods

3.1   Datasets

Our dataset consists of images of two murine osteosarcoma cell lines, DUNN and DLM8, which were also previously used for data analysis of cell images [1]. These two cell lines are closely related except for their degree of cancer invasiveness, as the DLM8 line is derived from the DUNN cell line with selection for metastasis. We here consider these two lines separately and present the main results of our statistical analysis for the DUNN line, with results for the DLM8 line being shown at this link. Both lines have been either treated with DMSO (control) or by cytochalasin D (cytD) or jasplakinolide (jasp). cytD and jasp are two cancer drugs that differently affect the cellular morphology, as cytD leads to actin depolymerization, while jasp enhances it [1]. More details about the experimental methods are available in [1]. We remove outliers from artefacts due to bad segmentation [11], and discretize the cell contour into 100 2D points. After processing, the DUNN cell lines contains 392 cells, including 203 cells in the control group, 96 cells treated by jasp and 93 cells treated by cytD. Similarly, the DLM8 cell line contains 258 cells, including 114 cells in the control group, 82 cells treated by cytD and 62 cells treated by jasp. The datasets are publicly available at this link) and have been added to Geomstats [14]. We preprocessed the datasets by scaling, translation and rotation of the curves: For scaling, we simply normalized the length of each curve to one. For translation, we centered the curves around the plane's origin. For rotation, we aligned every curve to a reference by finding an optimal rotation to minimize the L^2 distance between two curves. Note that the reparameterization is approximately invariant to the starting points as we selected 200 candidate starting points for each cell when computing the optimal alignment.

3.2   Experiments

Our study compares the performance of the SRV metric with the linear metric on analyzing the datasets of tumor cell images, illustrated in Fig. 2.A). To do so, we evaluated the associated pairwise distances between cells from Eq. (3), as well as their distance to the mean shape from Eq. (4), upon removing the action of translations and rotations by optimally aligning the curve elements for both the linear and SRV metrics. The analysis is done in Python 3.8 and using the implementation of the SRV in Geomstats [14] with the version from Aug 23, 2022, with scripts reproducing our results available in this github repository.

4   Results

4.1   Visualization of Shape Quantiles

Upon considering the linear and SRV metrics, we evaluate for each cell of the dataset the distance to the mean shape associated with each metric (the mean shapes can be visualized in Figs. 2.B and C. For the three conditions that reflect different treatments applied to the cells (see Sect. 3.1), we visualize in Figs. 2.D and E the quantiles associated with the 0th, 30th, 60th and 90th percentiles of the resulting distance distribution. The results suggest that while the linear and SRV mean shapes are quite similar, the SRV metric captures irregularities in shape, as increasingly irregular shapes appear as the quantile increases in

Fig. 2.E, with farthest cells having large invaginations and overhangs in the cell perimeter, which are captured in the "bending" term of the SRV metric (Eq. (1)). In comparison, the linear metric detects less regular cells on the 0th percentile for cells in all control and treatment groups. We also obtained similar findings with the DLM8 cell line (data not shown here, available at this link).

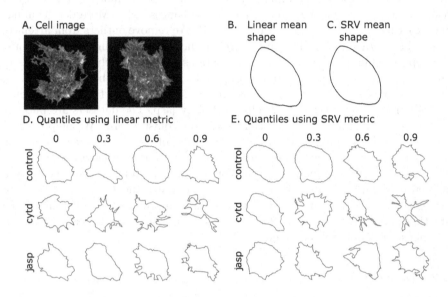

Fig. 2.   **A:** Two examples of osteocarcoma cell image obtained from fluorescence microscopy. **B:** Mean cell shapes computed over the cells from the DUNN cell line using the linear metric. **C:** Same as **(B)** with the SRV metric. **D:** Quantiles of distance to the mean shape for different conditions using the linear metric (dataset of DUNN cells). **E:** Same as (D) for the SRV metric.

4.2   Histograms of Distances to the Mean

To confirm the visual impression from Fig. 2, we plot in Fig. 3 the histograms of distances from the global mean for the different conditions (control and treatments). Upon comparing the histograms produced by the linear (Fig. 3.A and SRV (Fig. 3.B) metrics (defined in Sect. 2.1), we find that while the linear metric does not indicate significant differences between the jasp and control conditions, the SRV metric globally yields a better separation across the conditions (with similar results for the DLM8 cell line at this link). Using the SRV metric, the distribution of cytD cells is shifted to the right, while jasp treated cells are shifted to a less extent, yielding a narrower distribution. Interestingly, these observations are in line with the biological mechanisms of action of cytD and jasp: As cytD disassembles actin filaments and causes cytosolic aggregation of actin masses [25], the resulting loss of internal tension forces is expected to give rise to shapes with many invaginations, and increasing distance from the smooth ellipsoid. In comparison, the effects of jasp should be less dramatic since it leads to the formation of a more diffuse actin network, but without affecting stress fibers [20]. The SRV metric also remarkably yields a bimodal distribution for the control cells, which possibly suggests the existence of multiple states within the cell line. To visualize the difference between these the two modes identified by the SRV metric, we show the mean cell shapes over the subsets of cells in the bins 0.06-0.08 (Fig. 3.C) and 0.18-0.20 (Fig. 3.D) of the histogram, which correspond to the two observed peaks. The two mean cells are remarkably different, with the edge of mean cell in bin 0.06-0.08 being smooth and the edge of the mean cell in bin 0.18-0.20 being rough. This observation is expected since we previously found that cells that are farther away from the mean are more irregular. Using the DLM8 cell line, we also find a better (but less significant) separation across the conditions with the SRV metric (link here).

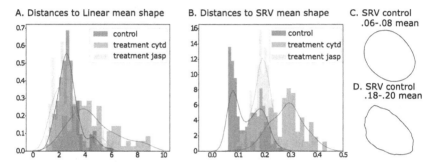

Fig. 3. Histograms of distances to linear mean (**A**), and SRV Fréchet mean (**B**) for the DUNN cell line. We observe that the distances of cells in the control group are closer than those in the treatment groups using SRV metric. The curves present kernel-density estimate for each group using Gaussian kernels. Mean cell shapes of control cells with SRV Fréchet mean in 0.06-0.08 (**C**), and with SRV Fréchet mean in 0.18-0.20 (**D**) using the SRV metric.

4.3   Visualization in Lower Dimension Space

As dimensionality reduction methods are usually employed to visualize and interpret large datasets of cell images [1], we finally compare the results obtained from projecting the data into a lower dimension space using both metrics. To do so, we perform a multidimensional scaling (MDS) on the pairwise distance matrix, with the results for the DUNN line shown in 2D (Fig. 4) and 3D (results available here). Our results suggest that the SRV metric tends to better capture the cell shape heterogeneity, as cells are overall more spread in both 2D and 3D, with the control group mostly being centered in the middle, surrounded by jasp treated cells and cytD treated cells located further. We also examine the stress statistics obtained for the different dimensions tested in MDS, as an indicator of the goodness-of-fit [24]. The MDS always achieves a better stress statistic for the SRV and all dimension tested (results available here), suggesting that the metric captures more informative patterns when it comes to dimensionality reduction. We obtain similar results for the DLM8 line (link here).

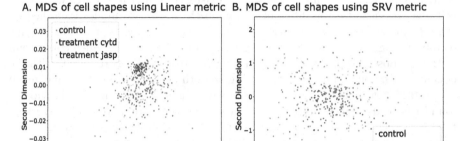

Fig. 4. Projections for MDS in 2D using the linear metric **(A)**, and the SRV metric **(B)**. We observe that cells in different control and treatment groups are more uniformly spread using the SRV metric in 2D.

5   Conclusion

We used the SRV metric to statistically analyze 2D shapes from tumor cell images, and showed some superior performance over the linear metric in all the tasks tested. The results of our comparative analysis were consistent across different cell lines of our datasets, suggesting that the elastic metric is better suited for interpreting morphological changes in the cell shape. While our study suggests the potential use of this Riemaniann metric for cancer cell detection, it is still limited to a relatively small dataset. The development of the elastic metric as a tool for practitioners requires some more thorough investigation of the morphological changes observed across different cell types and conditions, jointly with other features of the literature. The evaluation of the SRV distance in our paper could also be refined, by notably quotienting out the action of reparametrization [4,10], and it would be interesting to generalize our results to the whole family of elastic metrics, and compare their performance with some alternative metrics. In particular, considering the H_1 metric would be useful to determine if the performance of the elastic metric comes from its non-linear nature, or from the derivative information it contains. We are currently pursuing these directions.

References

1. Alizadeh, E., Xu, W., Castle, J., Foss, J., Prasad, A.: TISMorph: a tool to quantify texture, irregularity and spreading of single cells. PLoS ONE **14**(6), e0217346 (2019)

2. Balakrishnan, S., Raju, S.R., Barua, A., Pradeep, R.P., Ananthasuresh, G.K.: Two nondimensional parameters for characterizing the nuclear morphology. Biophys. J . **120**(21), 4698–4709 (2021)
3. Bauer, M., Bruveris, M., Marsland, S., Michor, P.W.: Constructing reparameterization invariant metrics on spaces of plane curves. Differential Geom. Appl. **34**, 139–165 (2014)
4. Bauer, M., Charon, N., Klassen, E., Kurtek, S., Needham, T., Pierron, T.: Elastic metrics on spaces of euclidean curves: theory and algorithms. arXiv preprint arXiv:2209.09862 (2022)
5. Bharath, K., Kurtek, S.: Analysis of shape data: from landmarks to elastic curves. Wiley Interdiscip. Rev. Comput. Stat. **12**(3), e1495 (2020)
6. Cho, M.H., Asiaee, A., Kurtek, S.: Elastic statistical shape analysis of biological structures with case studies: a tutorial. Bull. Math. Biol. **81**, 2052–2073 (2019)
7. Epifanio, I., Gual-Arnau, X., Herold-Garcia, S.: Morphological analysis of cells by means of an elastic metric in the shape space. Image Anal. Stereol. **39**(1) (2020)
8. Hartman, E., Sukurdeep, Y., Klassen, E., Charon, N., Bauer, M.: Elastic shape analysis of surfaces with second-order Sobolev metrics: a comprehensive numerical framework. Int. J. Comput. Vis. **131**, 1183–1209 (2023)
9. Kwon, S., Yang, W., Moon, D., Kim, K.S.: Comparison of cancer cell elasticity by cell type. J. Cancer **11**(18), 5403 (2020)
10. Le Brigant, A.: A discrete framework to find the optimal matching between manifold-valued curves. J. Math. Imaging Vis. **61**(1), 40–70 (2019)
11. Li, W., Mirone, J., Prasad, A., Miolane, N., Legrand, C., Dao Duc, K.: Orthogonal outlier detection and dimension estimation for improved MDS embedding of biological datasets. bioRxiv https://doi.org/10.1101/2023.02.13.528380 (2023)
12. Marklein, R.A., Lam, J., Guvendiren, M., Sung, K.E., Bauer, S.R.: Functionally-relevant morphological profiling: a tool to assess cellular heterogeneity. Trends Biotechnol. **36**(1), 105–118 (2018)
13. Mio, W., Srivastava, A., Joshi, S.: On shape of plane elastic curves. Int. J. Comput. Vision **73**(3), 307–324 (2007)
14. Miolane, N., et al.: Geomstats: a python package for Riemannian geometry in machine learning (2020)
15. Miolane, N., et al.: ICLR 2021 challenge for computational geometry & topology: Design and results. arXiv preprint arXiv:2108.09810 (2021)
16. Miolane, N., et al.: Introduction to geometric learning in Python with geomstats. In: SciPy 2020–19th Python in Science Conference, pp. 48–57 (2020)
17. Myers, A., Miolane, N.: Regression-based elastic metric learning on shape spaces of elastic curves. arXiv preprint arXiv:2210.01932 (2022)
18. Needham, T., Kurtek, S.: Simplifying transforms for general elastic metrics on the space of plane curves. SIAM J. Imag. Sci. **13**(1), 445–473 (2020)
19. Prasad, A., Alizadeh, E.: Cell form and function: interpreting and controlling the shape of adherent cells. Trends Biotechnol. **37**(4), 347–357 (2019)
20. Rotsch, C., Radmacher, M.: Drug-induced changes of cytoskeletal structure and mechanics in fibroblasts: an atomic force microscopy study. Biophys. J. **78**(1), 520–535 (2000)
21. Ruan, X., Murphy, R.F.: Evaluation of methods for generative modeling of cell and nuclear shape. Bioinformatics **35**(14), 2475–2485 (2019)
22. Sailem, H.Z., Bakal, C.: Identification of clinically predictive metagenes that encode components of a network coupling cell shape to transcription by image-omics. Genome Res. **27**(2), 196–207 (2017)

23. Srivastava, A., Klassen, E., Joshi, S.H., Jermyn, I.H.: Shape analysis of elastic curves in euclidean spaces. IEEE Trans. Pattern Anal. Mach. Intell. **33**(7), 1415–1428 (2010)
24. Sturrock, K., Rocha, J.: A multidimensional scaling stress evaluation table. Field Methods **12**(1), 49–60 (2000)
25. Tsakiridis, T., Vranic, M., Klip, A.: Disassembly of the actin network inhibits insulin-dependent stimulation of glucose transport and prevents recruitment of glucose transporters to the plasma membrane. J. Biol. Chem. **269**(47), 29934–29942 (1994)
26. Yin, Z., Sailem, H., Sero, J., Ardy, R., Wong, S.T., Bakal, C.: How cells explore shape space: a quantitative statistical perspective of cellular morphogenesis. BioEssays **36**(12), 1195–1203 (2014)

Perturbation of Fiedler Vector: Interest for Graph Measures and Shape Analysis

Julien Lefevre[1]([✉]), Justine Fraize[2,3], and David Germanaud[2,3]

[1] Institut de Neurosciences de la Timone, Aix Marseille Univ, CNRS 7289,
Marseille, France
`julien.lefevre@univ-amu.fr`
[2] CEA Paris-Saclay, Joliot Institute, NeuroSpin, UNIACT, Gif-sur-Yvette, France
[3] Université Paris Cité, Inserm, U1141 NeuroDiderot, inDEV, Paris, France

Abstract. In this paper we investigate some properties of the Fiedler vector, the so-called first non-trivial eigenvector of the Laplacian matrix of a graph. There are important results about the Fiedler vector to identify spectral cuts in graphs but far less is known about its extreme values and points. We propose a few results and conjectures in this direction. We also bring two concrete contributions, i) by defining a new measure for graphs that can be interpreted in terms of extremality (inverse of centrality), ii) by applying a small perturbation to the Fiedler vector of cerebral shapes such as the corpus callosum to robustify their parameterization.

Keywords: Graph Laplacian · Fiedler vector · Shape analysis

1 Introduction

Let $G = (V, E)$ be an undirected graph where $V = \{v_i\}_{i=1...n}$ are the vertices and $E = \{(v_i, v_j)\}$ the edges. The adjacency matrix A is defined by $A_{i,j} = 1$ if $i \neq j$ and $e_{i,j} \in E$. $A_{i,j} = 0$ otherwise. The degree matrix D is a diagonal matrix where $D_{i,i} = \deg(v_i) := \sum_{j=1...n} A_{i,j}$. The (unnormalized) graph Laplacian is the matrix $L := D - A$. L is a symmetric, semi-definite positive matrix. It has n eigenvalues $0 = \lambda_1 \leq \lambda_2 \leq ... \leq \lambda_n$. The multiplicity of 0 equals the number of connected components of G. In our case we will consider connected graphs and in that case the associated eigenfunction is constant. In this article we will focus more precisely on the second smallest eigenvalue (*algebraic connectivity*) and associated eigenvector that is called Fiedler vector, denoted by Φ. In the following we will assume that $||\Phi|| = 1$ and the eigenvalue λ_2 is simple, so Φ is uniquely defined up to a sign. We have first a very classical and useful result that is obtained from the Courant min-max principle:

$$\lambda_2 = \min_{||X||_2=1} {}^\top XLX = \sum_{i,j} A_{i,j}\big(\Phi(i) - \Phi(j)\big)^2 \tag{1}$$

Given that eigenvectors are orthogonal and the eigenvector associated to eigenvalue 0 is constant we have $\sum_i \Phi(i) = 0$ and Φ has sign changes. Since

F. Nielsen and F. Barbaresco (Eds.): GSI 2023, LNCS 14072, pp. 593–601, 2023.
https://doi.org/10.1007/978-3-031-38299-4_61

seminal works by Fiedler [4] there has been a considerable amount of theoretical results on spectral properties of graph Laplacian. Even if not of interest in our case, the subgraph of G induced on the vertices v with $\Phi(v) \geq 0$ is connected. This property allows to decompose a graph in two sub-domains, according to the sign of Φ.

In the past ten years stimulating connections have been made between the Fiedler vector and the first non trivial Laplacian eigenfunction u with Neumann boundary conditions on ∂D. In this continuous setting it has been postulated since the 70's that the maximum and minimum of u are located on ∂D. The underlying *hot-spots conjecture* turns out to be false but it has raised new interests regarding the extreme points of the Fiedler vector. In [1] it has been conjectured that such points, for a closed surface with no holes, maximized the geodesic distance. The conjecture has been proved to be false on a specific class of trees called *Rose graphs* [3,9]. A few recent works have generalized the previous examples to offer better characterization of extreme points of the Fiedler vector for trees [7,8]. To state it very rapidly, the most general and simple result we have comes from a theorem by Fiedler [5] (see also [8]) stating that the Fiedler vector is monotonic along branches of a tree which implies that the maximal and minimal values are attained in vertices with degree 1 (*pendant vertices*).

Given this rapid state of the art, it is possible to present our contributions:

- We are interested in understanding more the properties of extreme values of Fiedler vector and for that we will use perturbations of graph Laplacian. A natural question is to know whether extreme points of Fiedler vectors are *stable under perturbations* of the graph.
- The perturbation we will consider first consists in adding a pendant vertex to any vertex of the graph. Besides we will vary the weight x on the new edge, *not only for small values* but also by looking at the limit $x \to +\infty$.
- In the previous process we can wonder for which value of x the Fiedler vector of the new graph has an extrema on the new vertex. This will allow us to define a new measure of graphs that can be interpreted in terms of centrality/periphery.
- Last we apply the (small) perturbation procedure to characteristic points of medical shapes and demonstrate that it allows to robustify the description of a longitudinal structure such as the corpus callosum.

2   Perturbation of Fiedler Vector

2.1   Intuitions

First we can do a basic representation of our situation of interest involving a graph G and a perturbation consisting in adding a weighted edge between a vertex v and a new vertex $n + 1$.

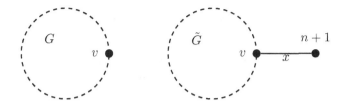

The weight x can be interpreted as the inverse of a distance. Namely considering for G the line graph with n vertices, $L(G)$ can be seen as a finite difference approximation of the second derivative operator on a segment sampled by n regularly spaced points t_i, since $f''(t_i) \approx f(t_{i+1}) + f(t_{i-1}) - 2f(t_i)$. By adding a perturbation at one end n we obtain the following matrix:

$$L(\tilde{G}) = \begin{pmatrix} 1 & -1 & 0 & \ldots & 0 & 0 \\ -1 & 2 & -1 & \ldots & 0 & 0 \\ \ldots & \ldots & \ldots & \ldots & & 0 \\ 0 & \ldots & -1 & 2 & -1 & 0 \\ 0 & \ldots & 0 & -1 & 1+x & -x \\ 0 & \ldots & 0 & 0 & -x & x \end{pmatrix}$$

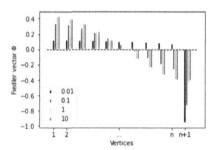

Fig. 1. Evolution of the Fiedler vector when x varies in $[0.01, 10]$.

We can see that it corresponds approximately to an irregular sampling of the segment $[0, n - 1 + 1/x]$ with n intervals of length 1 and the last interval of length $1/x$. When x is large we expect the Fiedler vector to be close to a cosine function with only one oscillation on the segment. Besides the Fiedler vector on n and $n+1$ converges to a same value. Conversely, if x is small, the n first points will tend to share the same value of the Fiedler vector and the last point to have the largest magnitude, of opposite sign. It is confirmed by numerical simulations on Fig 1.

In the previous case, the graph has been perturbed at a very specific position - one of the two extremities. In the following we will investigate first the more general situation of a perturbation at any vertex.

2.2 Classical Results

We recall classical results when considering eigenvalues and eigenvectors of a symmetric matrix M and its perturbation by another symmetric matrix P [12].

Theorem 1 (Weyl's inequalities). *Let* $\alpha_1 \geq \ldots \geq \alpha_n$, $\delta_1 \geq \ldots \geq \delta_n$, $\gamma_1 \geq \ldots \geq \gamma_n$ *be the spectra of* M, P *and* $M + P$ *respectively. Then we have:*

$$\gamma_{i+j-1} \leq \alpha_i + \delta_j \leq \gamma_{i+j-n} \tag{2}$$

We have also a local counterpart when the perturbation is small.

Theorem 2. *Let $\lambda_1 \leq ... \leq \lambda_n$ and $\tilde{\lambda}_1 \leq ... \leq \tilde{\lambda}_n$ be the spectra of matrices M and $\tilde{M}$; $\Phi_1,...\Phi_n$ and $\tilde{\Phi}_1,...\tilde{\Phi}_n$ corresponding eigenvectors. Then if the eigenvalue λ_i is simple, we have the two following approximations:*

$$\tilde{\lambda}_i = \lambda_i + {}^\top\Phi_i(\tilde{M} - M)\Phi_i + o(||\tilde{M} - M||) \tag{3}$$

$$\tilde{\Phi}_i = \Phi_i + \sum_{j \neq i} \frac{{}^\top\Phi_j(\tilde{M} - M)\Phi_i}{\lambda_i - \lambda_j}\Phi_j + o(||\tilde{M} - M||) \tag{4}$$

In practice, the formula are of little use in our case because the perturbed matrix has the eigenvalue 0 with multiplicity 2 and of course when the perturbation is large.

2.3   A New Result for Small Perturbations

Proposition 1. *We consider an undirected connected graph $G = (V, E)$ and a Fiedler vector Φ. Given a vertex v, we look at a weighted graph $\tilde{G} = (\tilde{V}, \tilde{E}, \tilde{W})$ where $\tilde{V} = V \cup \{n + 1\}$, $\tilde{E} = E \cup \{(v, n + 1)\}$. The weights $\tilde{W}_{i,j}$ are 0 or 1 depending on the adjacency between i and j except for $\tilde{W}_{v,n+1} = x > 0$. Calling $\Phi(x, \cdot) \in \mathbb{R}^{n+1}$ a Fiedler vector of the graph Laplacian of $\tilde{G}$. Then, there exist $a(v) > 0$ that satisfies:*

$$a(v) = \max\{a/\forall x \ 0 \leq x \leq a, \ \Phi(x, n + 1) = \arg\max_i \Phi(x, i)\} \tag{5}$$

Proof. First we obtain an upper bound on $\lambda_2(x)$, the algebraic connectivity of $\tilde{G}$. Indeed the graph Laplacian of $\tilde{G}$ can be expressed as:

$$\begin{pmatrix} L & 0 \\ {}^\top 0 & 0 \end{pmatrix} + x \begin{pmatrix} E_{v,v} & -\mathbf{e_v} \\ -{}^\top\mathbf{e_v} & 1 \end{pmatrix}$$

where $\mathbf{e_v} \in \mathbb{R}^n$ is 0 everywhere except 1 on the row v and $E_{v,v} = \mathbf{e_v}{}^\top\mathbf{e_v}$. Eigenvalues of those two matrices are respectively $0, 0, \lambda_2, ..., \lambda_n$ and 0 (multiplicity n), 2ϵ. Then by the left inequality in Theorem 1, with $i = n$ and $j = 1$ we get $\lambda_2(x) \leq 0 + 2x$. Since algebraic connectivity is positive then $\lambda_2(x) \rightarrow 0$ when $x \rightarrow 0$.

Next we use Courant's theorem:

$$\lambda_2(x) = x(\Phi(x, n + 1) - \Phi(x, v))^2 + \sum_{1 \leq i < j < n+1} A_{i,j}(\Phi(x, i) - \Phi(x, j))^2$$

The sum on the right tends to 0 and since $A_{i,j} \geq 0$ we obtain that as soon as i and j are neighbors and different from $n+1$, $\Phi(x, i) - \Phi(x, j) \rightarrow 0$. But for $i \neq n+1$ $(L\Phi(x, 1 : n))(i) = \sum_j A_{i,j}(\Phi(x, i) - \Phi(x, j))$ and so $(L\Phi(x, 1 : n))(i) \rightarrow 0$. We conclude that $\Phi(x, 1 : n)$ converges to the eigenspace of L associated to eigenvalue 0, i.e. the span of the constant vector.

By using the fact that $||\Phi(x,\cdot)|| = 1$ and $\sum_{i=1}^{n+1} \Phi(x,i) = 0$ we get that:

$$\Phi(x,\cdot) \rightarrow \frac{1}{\sqrt{n(n+1)}} \begin{pmatrix} -\mathbf{1} \\ n \end{pmatrix}$$

By continuity of $\Phi(x,\cdot)$, it is possible to find an interval $[0, a(v)[$ where $n+1$ remains a maximum point. □

Since our previous result is independent from the choice of v one can be naturally interested to know what is the maximum value of $a(v)$.

2.4 The Case of Large Perturbations

Here we consider the situation where x is large. It is interesting to examine first the case of complete graph with n vertices. As before we add a vertex $n+1$ to the vertex n with weight x large. By arguments of symmetry it is reasonable to look at a perturbed Fiedler vector of the form $^\top(-1, ..., -1, a, b)$ (up to a constant). The $n-1$ first lines of the eigenvalue problem are the same: $\lambda_2(x) = a+1$. The last lines yields $-x(a-b) = \lambda_2(x)b$. Rearranging all those terms and considering that $n-1-a-b = 0$ we obtain that a should be one of the root of the polynomial $a^2 - a(2x + n - 2) + (n-1)(x-1)$. Asymptotically one is like $2x$ and the other one tends to $(n-1)/2$. So $\lambda_2(x) \rightarrow (n+1)/2$ with associated Fiedler vector $^\top(-1, ..., -1, (n-1)/2, (n-1)/2)$. Given that $b-a > 0$ we conclude that $n+1$ is an extremum of the Fiedler vector for x sufficiently large.

The empirical result observed for the line graph is preserved here and we can propose the following conjecture:

Conjecture 1. Given an undirected connected graph G. We consider v an extremum of the Fiedler vector of the graph G. $\tilde{G}$ is the graph obtained from G and v as in Proposition 1. Then for all $x > 0$ the Fiedler vector $\Phi(x,\cdot)$ of $\tilde{G}$ has an extremum at $n+1$.

3 Applications

3.1 A New Measure for Graphs

Definition 1. *We consider an undirected connected graph $G = (V, E)$. Given a vertex v we consider $a(v) > 0$ as defined in Eq. 5 of Proposition 1. We will denote $Fcd(v) := 1/a(v)$ the Fiedler centrality distance (Fcd), which is a finite and positive number.*

Following the previous conjecture we expect that $d(v) = 0$ if v is an extremum of the Fiedler vector of G. This measure is supposed to reflect a distance to what could be a boundary.

On Fig. 2 we illustrate the evolution of $\Phi(x,\cdot)$ on an Erdös-Renyi random graph $G(n,m)$ with $n = 20$ vertices and $m = 45$ edges. Note that when x exceeds the threshold $a(v)$, $\Phi(x, 1:n)$ is very similar to the Fiedler vector of the unperturbed graph.

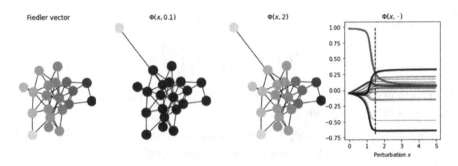

Fig. 2. From left to right: Fiedler vector of a Erdös-Renyi random graph, values are encoded from blue (−) to yellow (+) ; Fiedler vector of the perturbed graphs with $x = 0.1$ and $x = 2$; Plot where each curve corresponds to $\Phi(x, i)$ with i fixed. In blue $i = v$, in red $i = n+1$ and in black the vertices of the two extrema of the initial Fiedler vector. The dotted line corresponds to the value $a(v)$ after which $n + 1$ is no more an extremum of $\Phi(x, \cdot)$ (Color figure online)

Implementation Aspects. We can propose a variation on the previous definition by considering the quantity $\bar{a}(v) = \max\{x > 0 /\ \Phi(x, n + 1) = \arg\max_i \Phi(x, i)\}$ Clearly $a(v) \leq \bar{a}(v)$ and we conjecture that $a(v) = \bar{a}(v)$ based on empirical observations. If this conjecture is true, it allows a fast computation of $a(v)$ based on a dichotomous search in an interval $[x_{min} = 10^\alpha, x_{max} = 10^\beta]$ then iterating by computing the geometric mean $\bar{x}$ of the extremities and choosing the good side depending if v is an extrema of $\Phi(\bar{x}, \cdot)$. It requires at most $K = \log_2(\beta/\alpha)$ steps.

Thus, computing $a(v)$ needs K steps where a symmetric eigenproblem is solved. In practice we have used the function `eigh` of `scipy` to obtain the two first eigenpairs. Experiments on random graphs $G(n, m)$ with n varying in $[10, 200]$ and $m = pn$ with p varying in $[0.2, 0.9]$ have revealed that the time complexity is between $O(n^2)$ and $O(n^3)$ which is consistent with existing results[1]. All the codes used for the article are available on github[2].

Comparison with Other Centrality Measures. Next we can compare our new centrality measure with existing ones such as betweenness centrality, closeness centrality and eigenvector centrality. Figure 3 left illustrates visually the similarity between those 4 measures. We also generate $N = 100$ random graphs from the $G(n, m)$ model and compute all the correlations between the 4 measures. For $n = 20$ we observe on Fig. 3 right the evolution of the correlations with m in the same spirit as in [10]. There are differences between our measure and the 3 others when the graphs are fully or weakly connected and a good correlation in between.

---

[1] For instance on https://stackoverflow.com/questions/50358310/how-does-numpy-linalg-eigh-vs-numpy-linalg-svd.

[2] https://github.com/JulienLefevreMars/GSI_2023.

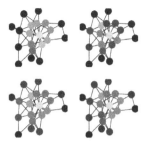 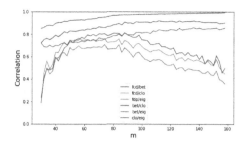

Fig. 3. Left: Same graph as in Fig. 2 with Fiedler centrality distance, betweenness centrality, closeness centrality and eigenvector centrality. Right: Correlations between 3 centrality measures and ours (fcd) for the $G(n, m)$ model with $n = 20$ and m varying in $[30, 160]$.

Those preliminary results suggest to test the Fiedler distance centrality on real networks to see whether complementary information can be obtained with respect to classical centralities.

Finally, we would like to draw the reader's attention to an important point. Worst case complexity of centrality algorithms is $O(n^3)$ which makes them not scalable on very large graphs. Our method is no exception to this situation and it is tempting to follow the general trend to use deep learning methods to approximate the centrality metrics [13]. At this stage, an essential question is to know the interest and benefits of this choice, especially with regard to the risks linked to the massification of deep learning and its environmental and societal impacts [11].

3.2   Longitudinal Parameterization of the Corpus Callosum

Finally, we show a very practical and useful application of the previous theoretical framework in the context of shape morphometry. The corpus callosum is a cerebral structure composed of axons of the two hemispheres joining in the center of the brain. The corpus callosum is easily visualized in MRI brain imaging, on medial sagittal slices. This structure can be affected in some neurological disorders, as in the fetal alcohol syndrome. In their study, the authors [6] measured manually the thickness of the corpus callosum. They needed to replicate their results by making fully automated measurements of this geometrical 2D shape.

From the MRI acquisitions a segmentation of the corpus callosum is obtained and it is possible to build a planar graph modelling this 2D shape. Given the elongated shape of the corpus callosum (see Fig. 4) we use the Fiedler vector of this graph to compute a quasi-isometric parameterization [2]. Figure 4 illustrates how we can obtain at the end a map of the corpus callosum thickness on regularly spaced slices following the longitudinal orientation of the shape given by the level sets of the Fiedler vector.

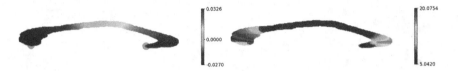

Fig. 4. Left: Fiedler vector of the corpus callosum S_1 and the two extrema in yellow. Right: thickness map on each isoline. (Color figure online)

The anatomical main axis of the corpus callosum is well described by the Fiedler vector but among the 125 processed shapes, 38 showed a discordance between the maximum of the Fiedler vector and the tip of the corpus callosum as defined by the expert neurologist. This situation is illustrated on Fig. 5. Then it is possible to inject the information of the correct position v of this maximum by perturbating the graph Laplacian by adding a pendant vertex at v with a weight close to $a(v)$. Then the new Fiedler vector follows the correct elongation of the corpus callosum as shown on Fig. 5 left. Eventually the parameterization procedure can be applied without any adaptation. On Fig. 5 right we can observe that the unperturbed thickness of S_2 has a value at the tip much more comparable to the one of S_1. The thickness remains almost the same on the rest of the shape for S_2 with and without perturbations.

Our approach allows a more realistic evaluation of the thickness which will benefit group studies of corpus callosum shapes in a future work.

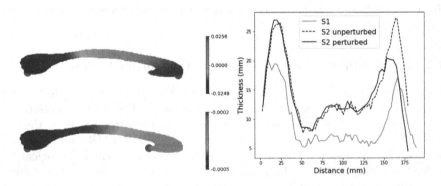

Fig. 5. Top Left: example of corpus callosum S_2 where the maximum of Fiedler vector (yellow dot on the right) is not correctly located. Bottom left: perturbed Fiedler vector from the correct position of the maximum. Right: Thickness profiles for the two corpus callosum S_1 and S_2. In black the same shape with the perturbed and unperturbed parameterization. (Color figure online)

Acknowledgements. This project is funded by the French National Agency for Research (ANR-19-CE17-0028-01) and the French National Institute for Public Health research (IRESP-19-ADDICTIONS-08).

References

1. Chung, M.K., Seo, S., Adluru, N., Vorperian, H.K.: Hot spots conjecture and its application to modeling tubular structures. In: Suzuki, K., Wang, F., Shen, D., Yan, P. (eds.) MLMI 2011. LNCS, vol. 7009, pp. 225–232. Springer, Heidelberg (2011). https://doi.org/10.1007/978-3-642-24319-6_28
2. Coulon, O., Lefevre, J., Klöppel, S., Siebner, H., Mangin, J.F.: Quasi-isometric length parameterization of cortical sulci: application to handedness and the central sulcus morphology. In: 2015 IEEE 12th International Symposium on Biomedical Imaging (ISBI), pp. 1268–1271. IEEE (2015)
3. Evans, L.: The fiedler rose: On the extreme points of the fiedler vector. Arxiv preprint arXiv:1112.6323 (2011)
4. Fiedler, M.: Algebraic connectivity of graphs. Czech. Math. J. **23**(2), 298–305 (1973)
5. Fiedler, M.: A property of eigenvectors of nonnegative symmetric matrices and its application to graph theory. Czech. Math. J. **25**(100), 619–633 (1975)
6. Fraize, J., Garzón, P., Ntorkou, A., Kerdreux, E., Boespflug-Tanguy, O., Beggiato, A., Delorme, R., Hertz-Pannier, L., Elmaleh-Berges, M., Germanaud, D.: Combining neuroanatomical features to support diagnosis of fetal alcohol spectrum disorders. Developmental Medicine & Child Neurology (2022)
7. Gernandt, H., Pade, J.P.: Schur reduction of trees and extremal entries of the fiedler vector. Linear Algebra Appl. **570**, 93–122 (2019)
8. Lederman, R.R., Steinerberger, S.: Extreme values of the fiedler vector on trees. arXiv preprint arXiv:1912.08327 (2019)
9. Lefèvre, J.: Fiedler vectors and elongation of graphs: a threshold phenomenon on a particular class of trees. arXiv preprint arXiv:1302.1266 (2013)
10. Li, C., Li, Q., Van Mieghem, P., Stanley, H.E., Wang, H.: Correlation between centrality metrics and their application to the opinion model. Eur. Phys. J. B **88**(3), 1–13 (2015). https://doi.org/10.1140/epjb/e2015-50671-y
11. Ligozat, A.L., Lefevre, J., Bugeau, A., Combaz, J.: Unraveling the hidden environmental impacts of AI solutions for environment life cycle assessment of AI solutions. Sustainability **14**(9), 5172 (2022)
12. Stewart, G.W., Sun, J.G.: Matrix perturbation theory (1990)
13. Wandelt, S., Shi, X., Sun, X.: Complex network metrics: Can deep learning keep up with tailor-made reference algorithms? IEEE Access **8**, 68114–68123 (2020)

Applied Geometric Learning

Generative OrnsteinUhlenbeck Markets via Geometric Deep Learning

Anastasis Kratsios[1]([✉]) [ID] and Cody Hyndman[2] [ID]

[1] Department of Mathematics and Statistics, McMaster University,
Hamilton, Canada
kratsioa@mcmaster.ca
[2] Department of Mathematics and Statistics, Concordia University,
Montréal, Canada
cody.hyndman@concordia.ca
https://anastasiskratsios.github.io/,
http://mypage.concordia.ca/alcor/chyndman/

Abstract. We consider the problem of simultaneously approximating the conditional distribution of market prices and their log returns with a single machine learning model. We show that an instance of the GDN model of [13] solves this problem without having prior assumptions on the market's "clipped" log returns, other than that they follow a generalized Ornstein-Uhlenbeck process with a priori unknown dynamics. We provide universal approximation guarantees for these conditional distributions and contingent claims with a Lipschitz payoff function.

Keywords: Geometric Deep Learning · Market Generation · Optimal Transport · Mathematical Finance · Gaussian Measures

Mathematics Subject Classification (2020): 68T07 · 91G20 · 91G60

1 Introduction

In classical portfolio theory, one considers a portfolio comprised of D predetermined risky assets and a riskless asset. The objective is to identify the most "efficient portfolios" by which we mean portfolios exhibiting the greatest gains while not exceeding a fixed level of risk or variability. Here, a portfolio's gains are quantified by its expected (log) returns, and its risk is quantified by the variance of its (log) returns. Thus, efficient portfolios are defined by optimizers of the following problem

$$\hat{w}(\gamma, \mu, \Sigma) \triangleq \operatorname*{argmin}_{\substack{w \in \mathbb{R}^D \\ \mathbb{1}^\star w = 1}} \left(-\gamma \mu^\star w + \frac{w^\star \Sigma w}{2} \right). \tag{1}$$

In (1), w is the vector of portfolio weights expressed as the proportion of wealth invested in each risky asset, $\mu \in \mathbb{R}$ is the vector of the expected (log) returns of

© The Author(s), under exclusive license to Springer Nature Switzerland AG 2023
F. Nielsen and F. Barbaresco (Eds.): GSI 2023, LNCS 14072, pp. 605–614, 2023.
https://doi.org/10.1007/978-3-031-38299-4_62

the risky assets, Σ is the covariance matrix of that portfolio's (log) returns, γ is a parameter balancing the objectives of maximizing the portfolio return versus minimizing the portfolio variance, $\bar{1}$ is the vector with all its components equal to 1, and $*$ denotes matrix transpose operator. If Σ is non-singular, the unique optimal solution to Eq. (1) is given in closed-form by

$$\hat{w}(\gamma, \mu, \Sigma) = \frac{\Sigma^{-1}\bar{1}}{\bar{1}^\star \Sigma^{-1}\bar{1}} + \gamma\left(\Sigma^{-1}\mu - \frac{\bar{1}^\star \Sigma^{-1}\mu}{\bar{1}^\star \Sigma^{-1}\bar{1}}\Sigma^{-1}\bar{1}\right). \tag{2}$$

The particular case where γ is set to 0 is the minimum variance portfolio of [15]. The minimum-variance portfolio $\hat{w}(0, \mu, \Sigma)$ may also be derived by minimizing the portfolio variance subject to the budget constraint $\bar{1}^\star w = 1$. Accordingly, we consider the case where Σ is *non-singular*. The optimality of (1) is contingent on the *normality* of the asset's (log) returns in this *static* picture of the market.

In reality, any financial market is continually and randomly evolving. Therefore, one must actively update the risky asset's mean μ and covariance matrix Σ in (2) to maintain an efficient portfolio. Since future market prices are unknown, so are the efficient portfolios in (1). Thus, our *objective* will be to forecast *both* the conditional evolution of market prices and the distribution of their log returns up to a regularizing factor.

Encoding Market Dynamics via Clipped Log Returns. In stochastic finance, the market's continual random evolution is typically formalized by a $(0, \infty)^D$-valued stochastic process $S. \overset{\text{def.}}{=} (S_t)_{t \geq 0}$ defined on a complete filtered probability space $(\Omega, \mathcal{F}, \mathbb{F} \overset{\text{def.}}{=} (\mathcal{F}_t)_{t \geq 0}, \mathcal{P})$. The components of $S.$ describe the evolving market prices. For simplicity, we omit the riskless asset, assuming that the continuously compounded interest rate is a constant, $r \geq 0$.

Since problem (1), concerns the log returns of the market's assets, i.e. one often models a latent Gaussian process $X.$ driving the market prices where $S_t \approx e^{X_t}$ (where the exponential map is applied component-wise). This is primarily due to three reasons: 1) stock prices cannot be non-positive, 2) most stock returns are somewhat log-normally distributed on an appropriate time-scale, and 3) the distribution of a stock's log returns are mathematically convenient.

We note that any asset's (log) returns can be substantial, either in the negative or positive directions, but realistically they cannot be arbitrarily large. With this in mind, it will be analytically convenient to work with "clipped (or regularized) log returns" which also satisfy the heuristics (1)–(3). By "clipped log returns" we encode the evolution of the market's prices $S.$ as

$$S_t \overset{\text{def.}}{=} \mathcal{E}(X_t) \text{and} \mathcal{E}(x) \overset{\text{def.}}{=} \exp\left(\frac{1}{\min\{1, \|x/M\|\}} \cdot x\right) \tag{3}$$

for all $t \geq 0$, where the exponential map exp is *applied component-wise* to any vector in $\mathbb{R}^D$, and the "clipping threshhold" $M > 0$ is a fixed and large. From a practical standpoint, both ways of encoding the evolution of market prices e^{X_t} and $\mathcal{E}(X_t)$, into the latent "log returns-like" Gaussian process $X.$, are virtually indistinguishable for M large enough. The main technical advantage of $\mathcal{E}$ over exp is that it Lipschitz; thus, it is compatible with the optimal-transport toolbox.

The transformation $\mathcal{E}$ is also appealing from the *stochastic analytic* vantage point. This is because it is the composition of a convex function with a smooth function; whence, if $X.$ is a semi-martingale, then we can directly compute the dynamics of $S.$ from those of $X.$ using a non-smooth Itô formula (e.g. [3,7]).

An Interpretable but Model-Agnostic Approach. We operate in the interpretable scenario where the clipped log returns process $X.$'s are not only conditionally Gaussian, but they are a strong solution to a simple and interpretable *stochastic differential equation (SDE)*. We consider the generalized Ornstein-Uhlenbeck (OU) process

$$X_t^x = x + \int_0^t (\mu_s + M_s\,X_s^x)\,ds + \int_0^t \sigma_s\,dW_s, \tag{4}$$

where $W. \overset{\text{def.}}{=} (W_t)_{t \geq 0}$ is a D-dimensional $\mathbb{F}$-Brownian motion, $\alpha : \mathbb{R} \to \mathbb{R}^D$ and $\beta : \mathbb{R} \to \mathbb{R}^{D \times D}$ are a-priori *unknown* continuously differentiable Lipschitz functions, and $\sigma : \mathbb{R} \to \mathbb{R}^{D \times D}$ is an a-priori *unknown* Lipschitz functions; further each σ_t a symmetric positive definite matrix (for $t \geq 0$). We drop the superscript emphasizing the dependence of $X_.^x$ on the initial condition x whenever clear from the context.

The first appeal of (4) is that, given any $\mu.$ and any $\sigma.$, the dynamics of X_t and $S_t \approx X_t$ are readily *interpretable*. The second appeal of (4), after a simple/classical computation, shows that each X_t follows a D-dimensional Gaussian distribution with mean $\int_0^t \mu_s\,ds$ and *non-singular* covariance $\int_0^t \sigma_s\sigma_s^\top\,ds$; which we denote $\mathcal{N}_D\big(\bar{\mu}_t, \int_0^t \sigma_s\sigma_s^\top\,ds\big)$ where $\bar{\mu}_t$ solves the ODE $\partial_t\bar{\mu}_t = \mu_t + M_t\bar{\mu}_t$ for the initial condition $\mu_0 = x$. Note that if $M_t = 0$ then $\bar{\mu}_t = \int_0^t \mu_s\,ds$.

As an informal illustration, suppose that $\mu_t = \mu_0 - \sigma_0^2/2$, $M_t = 0$, and $\sigma_t = \sigma_0$ in (4). Then, as M tends to infinity, we see S_t tends to the classical Geometric Brownian Motion (GBM) model used to derive the classical Black-Scholes formula and used to derive tractable optimal investment strategies [8,17].

We remain *agnostic* to specifications of μ and of σ and instead, we adopt a machine learning approach. Our first main objective is to implicitly infer the dynamics of $X.$by explicitly approximating its *regular conditional distribution function* $x \mapsto \mathbb{P}(X_t \in \cdot | X_0 = x)$. Then, our second goal is to deduce the same for $S.$. Thus, we instead only postulate minimal regularity of the functions $\mu.$ and $\sigma.$, just enough for a *deep neural network approximation* to the conditional probability distribution function of $X.$ to be viable.

Contributions. We will show that the *geometric deep network* modelling framework of [13], as specified in [13, Corollary 39], provides a universal solution to the problem of simultaneously predicting the regular conditional distributions of $X.$ and of $S.$, conditioned on the current state of the market x for any given future time t. In this case, the GDN implements a principled extension of the so-called *deep Kalman filter* of [14], which has recently also entered the mathematical finance literature in [11].

Relation to Other Deep Learning Models

There have recently been several other probability-measure-valued deep learning models proposed in the literature. For instance, [1] proposes a deep learning framework for approximating any regular conditional distribution function when the target space of probability measures is equipped with the 1-Wasserstein or adapted p-Wasserstein distances. In the case of the simple market dynamics (4), we will find that the GDN model is more economical in its theoretically guaranteed parameter count. Unlike those models, its approximation-theoretic guarantees are necessarily limited to markets evolving according to generalized OU dynamics such as (4), with non-singular volatility/diffusion. Gaussian-measure-valued deep learning models were experimentally considered in [14].

Additional results can be found in the arXiv version, while experimental support is provided at [12].

2 Preliminaries

We review the necessary background required to formulate our main results.

2.1 2-Wasserstein Riemannian Geometry

We equip the set of D-dimensional Gaussian distributions with non-singular covariance, denoted by $\mathcal{N}_D$, with a smooth structure induced by the global chart

$$\varphi : \mathbb{R}^D \times \mathbb{R}^{D(D+1)/2} \to \mathcal{D}$$
$$(\mu, \sigma) \mapsto \mathcal{N}_D\Big(\mu, \exp \circ \mathrm{sym}(\sigma)\Big), \tag{Chart}$$

where exp is the matrix exponential and sym is the linear map sending any vector $X \in \mathbb{R}^{D(D+1)/2}$ to $D \times D$ symmetric matrix

$$\mathrm{sym}(X) \overset{\text{def.}}{=} \begin{pmatrix} X_1 & X_2 & \dots & X_D \\ X_2 & X_{D+1} & \dots & X_{2D-1} \\ \vdots & & \ddots & \vdots \\ X_D & & \dots & X_{D(D+1)/2}. \end{pmatrix} \tag{5}$$

Following [16], we equip $\mathcal{N}_D$ with a Riemannian metric $g_{\mathcal{W}_2}$ defined at any D-dimensional Gaussian distribution $\mathcal{N}_D(\mu, \Sigma)$ with non-singular covariance matrix (i.e. any point in $\mathcal{N}_D$)) by

$$g_{\mathcal{W}_2,(\mu,\Sigma)}(u,v) \overset{\text{def.}}{=} \langle u_1, v_1 \rangle + \mathrm{tr}\big(\mathrm{sym}(u_2)\Sigma\,\mathrm{sym}(u_2)\big),$$

where we have identified the tangent vectors u, v at $\mathcal{N}_D(\mu, \Sigma)$ with Euclidean vectors via $u = (u_1, u_2), v = (v_1, v_2) \in \mathbb{R}^D \times \mathbb{R}^{D(D+1)/2}$. Together $(\mathcal{N}_D, g_{\mathcal{W}_2})$ is a well-defined simply connected Riemannian manifold (whence it has a well-defined geodesic distance between any two points). In [16, Proposition A], the authors show that the geodesic distance on $(\mathcal{N}_D, g_{\mathcal{W}_2})$ coincides with the 2-Wasserstein distance $\mathcal{W}_2$ on $\mathcal{N}_D$. By [9], $\mathcal{W}_2$ admits the following closed-form for any $\mathcal{N}_D(\mu_1, \Sigma_1), \mathcal{N}_D(\mu_2, \Sigma_2) \in \mathcal{N}_D$

$$\mathcal{W}_2^2\big(\mathcal{N}_D(\mu_1, \Sigma_1), \mathcal{N}_D(\mu_2, \Sigma_2)\big) = \|\mu_1 - \mu_2\|^2 + \operatorname{tr}\big(\Sigma_1 + \Sigma_2 - 2(\Sigma_2^{1/2}\Sigma_1\Sigma_2^{1/2})^{1/2}\big),$$

where $\Sigma_i^{1/2}$ denotes the square-root of the positive-definite matrices Σ_1 and Σ_2.

2.2 The GDN Model

Figure 1 illustrates the GDN implements the top arrow between the (non-Euclidean) Riemannian manifolds $(\mathcal{N}_D, g_{\mathcal{W}_2})$ in two phases. First, it transforms that vector via a deep feedforward neural network with ReLU activation function; then, it decodes the deep feedforward neural network output by interpreting them as the parameters defining D-dimensional Gaussian distribution with a non-degenerate covariance matrix, thus generating a $\mathcal{N}_D$-valued prediction.

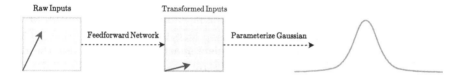

Fig. 1. *Summary of the GDN Model Processing:* First, it transforms the vectorial data using a deep feedforward neural network with a suitable activation function; next, the neural network output vectors are decoded as the parameters defining a Gaussian mean and covariance. This Gaussian distribution is the GDN generated prediction.

Definition 1 (Geometric Deep Network (GDN)). *Fix a non-polynomial smooth "activation function"* $\sigma : \mathbb{R} \to \mathbb{R}$, *and* $D, W, J \in \mathbb{N}_+$, *a geometric deep network (GDN) on* $\mathcal{N}_D$ *of width* W *and depth* J *is a map* $\hat{f} : \mathbb{R}^{1+d} \to \mathcal{N}_D$ *with representation: for every* $\mathcal{N}_D(\mu, \Sigma) \in \mathcal{N}_D$

$$\hat{f}(x) \stackrel{\text{def.}}{=} \varphi(A^{(J)} x^{(d)} + b^{(J)})$$
$$x^{(k+1)} \stackrel{\text{def.}}{=} \sigma \bullet (A^{(k)} x^{(k)} + b^{(k)}) \qquad \text{for } k = 0, \ldots, J-1$$
$$x^{(0)} \stackrel{\text{def.}}{=} (x, t),$$

where $(x, t) \in \mathbb{R}^{1+d} \cong \mathbb{R}^d \times \mathbb{R}$, *each* $A^{(k)} \in \mathbb{R}^{d_k \times d_{k+1}}$, $b^{(k)} \in \mathbb{R}^{d_{k+1}}$, $D = d_0$, $d_J = D(D+1)/2$, $d_k \leq W$ *for every* $k = 1, \ldots, J-1$, *and* φ *as in (Chart).*

3 Main Results

Our first result guarantees that the GDN model can approximate the distribution of X_t^x at any future time t, given any initial state x, log returns imposing any modelling assumptions for the "asset's drift" μ. nor for its "volatility" σ..

Theorem 1 (GDNs can Approximately Implement the Distribution of (4)). *Fix a parameter* $\delta > 0$. *Let* $K \subseteq \mathbb{R}^D$ *be (non-empty) and compact and consider a "time-horizon"* $T > \delta > 0$. *For every "approximation error"* $\epsilon > 0$, *there is a GDN* $\hat{f} : \mathbb{R}^{1+D} \to \mathcal{N}_D$ *satisfying the uniform estimate*

$$\max_{x \in K, \, \delta \leq t \leq T} \mathcal{W}_2\big(\hat{f}(x,t), \mathbb{P}(X_t^x \in \cdot)\big) < \epsilon.$$

Moreover, if t *is fixed then* $\hat{f}$ *has width* $D(6 + 2D + D^2)/2$ *and depth* $\mathcal{O}\big(\frac{1}{\epsilon^{2D}}\big)$.

The power of the GDN model is that it can *simultaneously approximate* the regular conditional distribution (RCD) of the clipped log returns process and the market prices.

Theorem 2 (Simultaneous Approximation of the Market RCD). *Consider the setting of Theorem 1, and let* $\hat{f}$ *the GDN obtained from that result. For every* $x \in K$ *and each* $t \in [\delta, T]$

$$\mathcal{W}_2\big(\mathbb{P}(S_t^x \in \cdot), \mathcal{E}_\#(\hat{f}(t,x))\big) < \sqrt{D} e^M \, \varepsilon.$$

The Fundamental Theorem of Asset Pricing [5], implies that risk-neutral prices for contingent claims exist and are expressed as conditional expectations of the claim payoffs, computed under an equivalent martingale measure for the discounted market prices $(e^{-rt} S_t^x)_{t \geq 0}$. For illustrative simplicity, suppose that $r = 0$ and that S^x is a $\mathbb{P}$-martingale. Then contingent claims are computed as conditional expectations under $\mathbb{P}$.

Theorem 3 (Automatic Contingent Claim Pricing). *Consider the setting and conclusion of Theorem 2, let* $r = 0$, *and suppose that* $S.$ *is a* $\mathbb{P}$-*martingale. For every Lipschitz payoff function* $V : \mathbb{R}^D \to \mathbb{R}$, *and every* $(x,t) \in K \times [\delta, T]$

$$\Big| \mathbb{E}_{U \sim \hat{f}(x,t)} \big[V\big(\mathcal{E}(U)\big) \big] - \mathbb{E}_{\mathbb{P}} \big[V(S_t^x) \big] \Big| < C\epsilon,$$

for some constant $C \geq 0$ *depending only on* V, D *and on* M.

Theorem 3 implies that once $\hat{f}$ is trained, then we can directly approximate any contingent claim on $S.$ by simply sampling $V(\mathcal{E}(U))$ where U is distributed according to $\hat{f}(x,t)$. We conclude by proving our guarantees for the GDN model.

4 Proofs

Lemma 1 (Gaussianity of the SDE (4) Solutions). *For any* $x \in \mathbb{R}^D$ *and any* $t > 0$, *the random vector* S_t^x *is distributed according to a* D-*dimensional Gaussian distribution with non-singular covariance; more precisely*

$$S_t^x \sim \mathcal{N}_D\Big(\bar{\mu}_t, \int_0^t \sigma_s \sigma_s^\top \, dt\Big);$$

where $\bar{\mu}$ *is continuous and solves* $\partial_t \bar{\mu}_t = \mu_t + M_t \bar{\mu}_t$ *with initial condition* $\bar{\mu}_0 = x$.

Lemma 1 implies that the map $(x,t) \mapsto \mathbb{P}(S_t^x \in \cdot)$ takes values in $(\mathcal{N}_D, g_{\mathcal{W}_1})$ so that we can apply the universal approximation theorem of [13]. However, need to verify that the target function is continuous. The next Lemma implies that $(x,t) \mapsto \mathbb{P}(S_t^x \in \cdot)$ has the required regularity to apply the results of [13].

Lemma 2 (Stability Estimate for $(t,x) \mapsto X_t^x$). *Fix a compact subset $K \subseteq \mathbb{R}^D$ and a positive "time–horizon" $T > 0$. Then the map*

$$\mathbb{R}^D \times [0,\infty) \to (\mathcal{N}_D, g_{\mathcal{W}_2})$$
$$(x,t) \mapsto X_t^x$$

is Lipschitz in x and $1/2$-Hölder in t, over $K \times [0,T]$.

Proof. For each $t, s > 0$ and every $x, \tilde{x} \in \mathbb{R}^D$, we have

$$\mathcal{W}_1\big(\mathbb{P}(X_t^x \in \cdot), \mathbb{P}(S_s^{\tilde{x}} \in \cdot)\big) \leq \mathbb{E}\big[\|X_t^x - X_s^{\tilde{x}}\|^2\big]^{1/2}. \tag{6}$$

Applying [4, Propositions 8.15 and 8.16] to the right-hand side of (6) yields

$$\mathcal{W}_2\big(\mathbb{P}(X_t^x \in \cdot), \mathbb{P}(X_s^{\tilde{x}} \in \cdot)\big) \leq \mathbb{E}\big[\|X_t^x - X_s^{\tilde{x}}\|^2\big]^{1/2} \leq C(\|t - s\|^{1/2} + \|x - \tilde{x}\|), \tag{7}$$

for some constant $C \geq 0$ depending on $K \times [0,T]$. The result then follows from [16, Proposition A], which states that the geodesic distance on $(\mathcal{N}_D, g_{\mathcal{W}_2})$ coincides with the restriction of the 2-Wasserstein distance thereto.

Proof (Proof of Theorem 1). By Lemma 1 for every $(x,t) \in K \times [\delta, T]$ the random vector X_t^x is distributed according to a Gaussian distribution with a non-singular covariance matrix. Thus, the map $f(x,t) \mapsto \mathbb{P}(X_t^x \in \cdot)$ takes values in $\mathcal{N}_D$. By our stability estimate, namely Lemma 2, f is a Lipschitz function; in particular, it is continuous. Therefore, [13, Corollary 40] applies; whence, for every given $\epsilon > 0$ there is a GDN satisfying $\max_{(x,t) \in K \times [\delta, T]} \mathcal{W}_2\big(f(x,t), \hat{f}(x,t)\big) < \epsilon$. Furthermore, if t is fixed, then the depth and width of $\hat{f}$ are given in the first row of [13, Table 1]; since $x \mapsto \mathbb{P}(X_t^x)$ is Lipschitz.

Lemma 3. *The push-forward $\mathcal{E}_{\#}$ is a well-defined map from $(\mathcal{P}_2(\mathbb{R}^D), \mathcal{W}_2)$ of $\sqrt{D}e^M$-Lipschitz continuity. In particular, $\mathcal{E}_{\#}$ is a Lipschitz map to $(\mathcal{P}_2(\mathbb{R}^D), \mathcal{W}_1)$.*

Proof (Proof of Lemma 3). We first observe that $\mathcal{E}$ is Lipschitz. To see this, note that $x \mapsto \exp\big((\min\{1, \|x/M\|\})^{-1} \cdot x\big)$ is precisely the orthogonal projection P of $\mathbb{R}^d$ onto the closed Euclidean ball $\overline{B}_{\mathbb{R}^D, \|\cdot\|_2}(0, M)$ of radius $M > 0$ of about $0 \in \mathbb{R}^D$. Since $\overline{B}_{\mathbb{R}^D, \|\cdot\|_2}(0, M)$ is a closed convex set, then this projection is well-defined and 1-Lipschitz (see [2, Example 12.25 and Proposition 12.27]). Since $\mathcal{E}$ is given by the composition $\mathcal{E} = \exp \circ P$ (here exp is "composed" component-wise), P is 1-Lipschitz, and since the composition of Lipschitz functions is again Lipschitz, then $\mathcal{E}$ if exp is Lipschitz on the range of P. By Rademacher's theorem (see [6, Theorem 3.16]), if exp were to be L-Lipschitz on the range of P, then

$\sup_{x \in P(\mathbb{R}^D)} \|\nabla \exp(x)\|$ must be finite; in which case this quantity is equal to its Lipschitz constant L. This is indeed the case, since

$$\sup_{x \in p(\mathbb{R}^D)} \|\nabla \exp(u)\| \leq \sqrt{D} \max_{-M \leq v \leq M} \sqrt{D} e^M < \infty. \tag{8}$$

Thus, $\mathcal{E}$ is -Lipschitz, with Lipschitz constant bounded-above by $L \overset{\text{def.}}{=} \sqrt{D} e^M$.

It is straight-forward to see that $\mathcal{E}_\#$ is well-defined and maps $\mathcal{P}_2(\mathbb{R}^D)$ to itself, since we have just seen that $\mathcal{E}$ is Lipschitz. To see that $\mathcal{E}_\#$ is $\sqrt{D} e^M$-Lipschitz, fix any two $\mu, \nu \in \mathcal{P}_f(\mathbb{R}^m)$ a transport plan π between them. Define the "induced diagonal transport plan" $\tilde{\pi} := (\mathcal{E}, \mathcal{E})_\# \pi$ and simply note that $\tilde{\pi}$ is indeed a transport plan between $\mathcal{E}_\# \mu$ and $\mathcal{E}_\# \nu$. We then compute

$$\begin{aligned}
\mathcal{W}_2^2(\mathcal{E}_\# \mu, \mathcal{E}_\# \nu) &\leq \mathbb{E}_{(U_1, U_2) \sim \tilde{\pi}}[\|U_1 - U_2\|^2] \\
&= \mathbb{E}_{(V_1, V_2) \sim \pi}[\|\mathcal{E}(V_1) - \mathcal{E}(V_2)\|^2] \\
&\leq \mathbb{E}_{(V_1, V_2) \sim \pi}[D e^{2M} \|V_1 - V_2\|^2] \\
&= D e^{2M} \mathbb{E}_{(V_1, V_2) \sim \pi}[\|V_1 - V_2\|^2].
\end{aligned}$$

We complete the proof by first square-rooting both sides of the inequality and then taking the infimum over all transport plans π between μ and ν; thus

$$\mathcal{W}_2(\mathcal{E}_\# \mu, \mathcal{E}_\# \nu) \leq \sqrt{D} e^M \inf_\pi \mathbb{E}_{(V_1, V_2) \sim \pi}[\|V_1 - V_2\|^2]^{1/2} = \sqrt{D} e^M \mathcal{W}_2(\mu, \nu).$$

Since $\mathcal{W}_1 \leq \mathcal{W}_2$ for any probability measures, the second claim follows.

Proof (Proof of Theorem 2). By (3), we have that $\mathbb{P}(S_t^x \in \cdot) = \mathcal{E}_\# \mathbb{P}(X_t^x \in \cdot)$. By Lemma 1 $\mathbb{P}(X_t^x \in \cdot)$ belongs is a D-Dimensional Gaussian measure and therefore it belongs to $\mathcal{P}_2(\mathbb{R}^D)$. Likewise, by construction $\hat{f}$ is also D-Dimensional Gaussian measure; thus, it also belongs to $\mathcal{P}_2(\mathbb{R}^D)$. Therefore, Lemma 3 applies, allowing us to deduce that

$$\mathcal{W}_2\big(\mathbb{P}(S_t^x \in \cdot), \mathcal{E}_\#(\hat{f}(t, x))\big) \leq \sqrt{D} e^M \mathcal{W}_2\big(\mathbb{P}(X_t^x \in \cdot), \hat{f}(t, x)\big). \tag{9}$$

Since $\hat{f}$ is as in Theorem 1 then, the right-hand side of (9) is less than $\sqrt{D} e^M \varepsilon$.

Proof (Proof of Theorem 3). If V is constant, then the result is clear. Therefore, assume that V is non-constant. Since $\mathcal{W}_1 \leq \mathcal{W}_2$, then Theorem 2 implies that, for every $(x, t) \in K \times [\delta, T]$

$$\mathcal{W}_1\big(\mathcal{E}_\# \hat{f}(x, t), \mathbb{Q}(\tilde{S}_t^x \in \cdot)\big) \leq C_1 \varepsilon, \tag{10}$$

were the for the constant $C_1 \overset{\text{def.}}{=} \sqrt{D} e^M$. By the Kantorovich-Rubinstein duality (see [10, Theorem 9.6]) the left-hand side of (10) can be rewritten as

$$\sup_{\tilde{V} \in \text{Lip}(\mathbb{R}^D, \mathbb{R}; 1)} \left| \mathbb{E}_{U \sim \mathcal{E}_\# \hat{f}(x, t)}[\tilde{V}(U)] - \mathbb{E}_\mathbb{P}[\tilde{V}(\tilde{S}_t^x)] \right| = \mathcal{W}_1\big(\hat{f}(x, t), \mathbb{Q}(\tilde{S}_t^x \in \cdot)\big) \tag{11}$$

for every $x \in K$ and each $t \in [\delta, T]$; where $\mathrm{Lip}(\mathbb{R}^D, \mathbb{R}; 1)$ is the set of real-valued Lipschitz maps on $\mathbb{R}^D$ with Lipschitz norm $\|\tilde{V}\|_{Lip} \stackrel{\text{def.}}{=} \sup_{x \in \mathbb{R}^D} |\tilde{V}(x)| + \mathrm{Lip}(\tilde{V})$ at-most 1 *(note, we the Lipschitz norm as ∞ if the map $\tilde{V}$ is not Lipschitz since its "Lipschitz constant" $\mathrm{Lip}(\tilde{V})$ would be infinite)*. Since V is non-constant then $\mathrm{Lip}(V) > 0$ is positive. Thus, $\tilde{V} \stackrel{\text{def.}}{=} [\|V\|_{Lip}]^{-1} \cdot V$ is well-defined and (11) implies

$$\frac{1}{\|V\|_{Lip}} \left| \mathbb{E}_{U \sim \mathcal{E}_\# \hat{f}(x,t)}[V(U)] - \mathbb{E}_{\mathbb{P}}[V(\tilde{S}_t^x)] \right| = \mathcal{W}_1 \left(\mathcal{E}_\# \hat{f}(x,t), \mathbb{Q}(\tilde{S}_t^x \in \cdot) \right), \quad (12)$$

for $(x, t) \in K \times \{T\}$. We conclude by multiplying (12) by $\|V\|_{Lip}$, setting $C \stackrel{\text{def.}}{=} C_1 \|V\|_{\mathrm{Lip}}$, and using the change-of-variable formula for push-forward measures.

References

1. Acciaio, B., Kratsios, A., Pammer, G.: Designing universal causal deep learning models: the geometric (hyper)transformer. Math. Finan. 1–65 (2023). https://onlinelibrary.wiley.com/doi/10.1111/mafi.12389, (to appear) Special Issue: Machine Learning in Finance
2. Bauschke, H.H., Combettes, P.L.: Convex Analysis and Monotone Operator Theory in Hilbert Spaces. CBM, Springer, Cham (2017). https://doi.org/10.1007/978-3-319-48311-5
3. Carlen, E., Protter, P.: On semimartingale decompositions of convex functions of semimartingales. Illinois J. Math. **36**(3), 420–427 (1992)
4. Prato, G.: Introduction to Stochastic Analysis and Malliavin Calculus. PSNS, vol. 13. Scuola Normale Superiore, Pisa (2014). https://doi.org/10.1007/978-88-7642-499-1
5. Delbaen, F., Schachermayer, W.: A general version of the fundamental theorem of asset pricing. Mathematische Annalen **300**(1), 463–520 (1994)
6. Federer, H.: Geometric measure theory. Die Grundlehren der mathematischen Wissenschaften, Band 153, Springer-Verlag, New York Inc, New York (1969)
7. Föllmer, H., Protter, P.: On Itô's formula for multidimensional Brownian motion. Probab. Theor. Related Fields **116**(1), 1–20 (2000)
8. Gatheral, J., Schied, A.: Optimal trade execution under geometric Brownian motion in the Almgren and Chriss framework. Int. J. Theor. Appl. Finance **14**(3), 353–368 (2011)
9. Gelbrich, M.: On a formula for the l2 Wasserstein metric between measures on euclidean and hilbert spaces. Mathematische Nachrichten **147**(1), 185–203 (1990)
10. Gozlan, N., Roberto, C., Samson, P.M., Tetali, P.: Kantorovich duality for general transport costs and applications. J. Funct. Anal. **273**(11), 3327–3405 (2017)
11. Jaimungal, S.: Reinforcement learning and stochastic optimisation. Finance Stochast. **26**(1), 103–129 (2022)
12. Kratsios, A.: Universal pricing in markets driven by Itô processes - Code. https://github.com/AnastasisKratsios/GSI_2023.git (2023)
13. Kratsios, A., Papon, L.: Universal approximation theorems for differentiable geometric deep learning. J. Mach. Learn. Res. **23**(196), 1–73 (2022)
14. Krishnan, R.G., Shalit, U., Sontag, D.: Deep Kalman filters. arXiv preprint arXiv:1511.05121 (2015)

15. Markowitz, H.M.: Portfolio selection. J. Finan. (1968)
16. Takatsu, A.: Wasserstein geometry of Gaussian measures. Osaka J. Math. **48**(4), 1005–1026 (2011)
17. Yu, F., Ching, W.K., Wu, C., Gu, J.W.: Optimal pairs trading strategies: a stochastic mean-variance approach. J. Optim. Theor. Appl. **196**(1), 36–55 (2023)

SL(2, ℤ)-Equivariant Machine Learning with Modular Forms Theory and Applications

Pierre-Yves Lagrave$^{(\boxtimes)}$ ⓘD

Morgan Stanley, Paris R&D Center, 61 rue de Monceau, 75008 Paris, France
pierre-yves.lagrave@morganstanley.com

Abstract. This paper introduces an approach for building Machine Learning (ML) algorithms embedding equivariance mechanisms to the Lie group $SL(2,\mathbb{Z})$ by leveraging on modular forms theory. More precisely, we propose using Eisenstein series to build parametric equivariant operators which can then be combined within usual ML architectures to solve both supervised and unsupervised tasks. We substantiate the interest of using $SL(2,\mathbb{Z})$-equivariance on simulated Toeplitz Hermitian Positive Definite matrices datasets built to reproduce some of the challenges associated with financial time series analysis.

Keywords: Equivariant Machine Learning · Modular Forms Theory · Quantitative Finance

1 Introduction and Related Work

The statistical analysis of covariance matrices has a wide range of applications in several domains, including for example classification and clustering of Magnetic Resonance Imaging (MRI) images [8], radar signals [1,5,6] and financial time series [2,3,21]. Recently, simulation tasks have also been considered in [20] where generative models were used to produce realistic correlation matrices for financial assets.

Designing algorithms exploiting the native geometry of the space of covariance matrices and the corresponding symmetries is a very active field of research belonging to that of Geometric Deep Learning [4,14]. In [18], it was in particular shown that the use of equivariance mechanisms to the group $SU(1,1)$ within Neural Networks increases both accuracy and robustness in the context of Toeplitz Hermitian Positive Definite (THPD) matrices classification and that Helgason-Fourier analysis on the Poincaré disk $\mathbb{D}$ could be used to compute convolution operators that are equivariant to the action of $SU(1,1)$ when regularly represented. Linear operators equivariant to more general representations of $SU(1,1)$ on the Fock-Bargmann spaces were considered in [15], where a parameterization of $SU(1,1)$ from its Bloch-Messiah decomposition was suggested for Monte-Carlo computations.

F. Nielsen and F. Barbaresco (Eds.): GSI 2023, LNCS 14072, pp. 615–623, 2023.
https://doi.org/10.1007/978-3-031-38299-4_63

More generally, efficiently estimating group-based equivariant convolutions remains an open challenge, in particular for non-compact groups. Significant progress has been made with this respect in [13] where a scalable approach is introduced to solve the kernel constraints in steerable neural networks [9] when working with Lie groups with finite dimensional representations. Variance reduction methods for Monte-Carlo estimators of group-based convolutions were considered in [17]. While the former does not apply to groups such as $SU(1,1)$ and $SL(2,\mathbb{Z})$ that will be of interest in the following, the latter remains still very costly in practice.

By leveraging on modular forms theory, this paper introduces a new approach for building parametric operators equivariant to the group $SL(2,\mathbb{Z})$, which can be seen as a discretization of $SU(1,1)$ modulo its isomorphism with $SL(2,\mathbb{R})$. By anchoring in [18] and on the one-to-one correspondence between $\mathbb{D}$ and the Poincaré half-plane $\mathbb{H}_2$, we show that such operators can be used to process THPD matrices and substantiate their interest through numerical experiments with simulated datasets.

2   Hyperbolic Space, Group Action and Equivariance

There are several models that are commonly used in practice for representing the 2-dimensional hyperbolic space, including in particular, the Poincaré Half-Plane $\mathbb{H}_2$ and the Poincaré Disk $\mathbb{D}$ models which are defined as it follows:

$$\mathbb{D} = \{z = x + iy \in \mathbb{C},\ x^2 + y^2 < 1\} \qquad \mathbb{H}_2 = \{z = x + iy \in \mathbb{C},\ y > 0\} \quad (1)$$

The Poincaré Half-Plane and Disk are actually isomorphic [7] and are both homogeneous spaces in the sense that they can be written as a quotient space G/H between a given group G and one of its stabilizer subgroup H. More precisely, we have

$$\mathbb{H}_2 = SL(2,\mathbb{R})/SO(2) \quad \text{and} \quad \mathbb{D} = SU(1,1)/U(1) \qquad (2)$$

where we have made use of the following Lie groups, with the matrix multiplication as internal composition law:

$$SU(1,1) = \left\{ g_{\alpha,\beta} = \begin{bmatrix} \alpha & \beta \\ \bar{\beta} & \bar{\alpha} \end{bmatrix},\ |\alpha|^2 - |\beta|^2 = 1,\ \alpha, \beta \in \mathbb{C} \right\} \qquad (3)$$

$$SL(2,\mathcal{R}) = \left\{ g_{a,b,c,d} = \begin{bmatrix} a & b \\ c & d \end{bmatrix},\ ad - bc = 1,\ a,b,c,d \in \mathcal{R} \right\} \qquad (4)$$

$$U(1) = \left\{ \begin{bmatrix} \frac{\alpha}{|\alpha|} & 0 \\ 0 & \frac{\bar{\alpha}}{|\alpha|} \end{bmatrix},\ \alpha \in \mathbb{C} \right\} \qquad (5)$$

where $\mathcal{R} \in \{\mathbb{C}, \mathbb{R}, \mathbb{Z}\}$. The groups $SU(1,1)$ and $SL(2,\mathbb{R})$ are isomorphic and conjugate in $SL(2,\mathbb{C})$ and act respectively on $\mathbb{D}$ and $\mathbb{H}_2$ with their transitive actions $\circ_{\mathbb{D}}$ and $\circ_{\mathbb{H}_2}$ given by

$$\forall x \in \mathbb{D},\ \forall g_{\alpha,\beta} \in SU(1,1),\ g_{\alpha,\beta} \circ_{\mathbb{D}} x = \frac{\alpha x + \beta}{\bar{\beta} x + \bar{\alpha}} \qquad (6)$$

and

$$\forall y \in \mathbb{H}_2, \ \forall g_{a,b,c,d} \in \mathrm{SL}(2, \mathbb{R}), \ g_{a,b,c,d} \circ_{\mathbb{H}_2} y = \frac{ay + b}{cy + d} \quad (7)$$

As a subgroup of $\mathrm{SL}(2, \mathbb{R})$, $\mathrm{SL}(2, \mathbb{Z})$ also acts on $\mathbb{H}_2$. Finally, an operator Φ from a space $\mathcal{X}$ to $\mathcal{Y}$ is said to be equivariant with respect to a group G acting on $\mathcal{X}$ with $\circ_{\mathcal{X}}$ and on $\mathcal{Y}$ with $\circ_{\mathcal{Y}}$ if $\forall g \in G$ and $\forall x \in \mathcal{X}$, we have $\Phi(g \circ_{\mathcal{X}} x) = g \circ_{\mathcal{Y}} \Phi(x)$. When there is no confusion about the considered space and group, we simply denote the action $\circ$ to ease the notations.

2.1 Equivariant Modular Forms

We start with some background about modular forms theory and refer to [10,23] for more details. We then highlight how parametric operators can be built from the basis of Eisenstein series in order to achieve equivariance with respect to the action of $\mathrm{SL}(2, \mathbb{Z})$ on elements of $\mathbb{H}_2$.

General Theory. For $\mathbb{P}^1(\mathbb{Q}) = \mathbb{Q} \cup \{\infty\}$ denoting the set of cusps, a modular form of weight $k \in \mathbb{Z}$ is a meromorphic function f on the extended half-plane $\bar{\mathbb{H}}_2 = \mathbb{H}_2 \cup \mathbb{P}^1(\mathbb{Q})$ such that $\forall g_{a,b,c,d} \in \mathrm{SL}(2, \mathbb{Z})$ and $\forall z \in \bar{\mathbb{H}}_2$, the following equation holds true:

$$f(z) = (cz + d)^{-k} f(g_{a,b,c,d} \circ z) \quad (8)$$

A particular example of modular forms of weight k for $k \in 2\mathbb{Z}$ and $k \geq 4$ are the Eisenstein series G_k, which are defined by the following equation for $z \in \bar{\mathbb{H}}_2$

$$G_k(z) = \sum_{\substack{m, n \in \mathbb{Z} \\ mz + n \neq 0}} \frac{1}{(mz + n)^k} \quad (9)$$

Moreover, the complex vector space M_k of modular forms of weight k is generated by the modular forms $G_4^a G_6^b$ for $a, b \in \mathbb{N}$ such that $4a + 6b = k$. Hence, any $f \in M_k$ can be written as it follows

$$f(z) = f_\alpha(z) = \sum_{\substack{a, b \in \mathbb{N} \\ 4a + 6b = k}} \alpha_{a,b}^f G_4^a(z) G_6^b(z) \quad (10)$$

with $z \in \bar{\mathbb{H}}_2$, $\alpha_{a,b}^f \in \mathbb{C}$ and $\alpha = \left(\alpha_{a,b}^f \right)_{a,b \in \mathbb{N}}^{4a+6b=k}$.

Equivariance and Modular Forms. Equivariant modular forms were investigated in [11,12,22]. More precisely, a meromorphic function $h : \mathbb{H}_2 \to \mathbb{C}$ is said to be an equivariant form if $z \to h(z) - z$ is meromorphic at the cusps and if the following equation holds true

$$h(g_{a,b,c,d} \circ z) = g_{a,b,c,d} \star h(z) \quad \forall z \in \mathbb{H}_2 \text{ and } \forall g_{a,b,c,d} \in \mathrm{SL}(2, \mathbb{Z}) \quad (11)$$

where $\star$ refers to the natural extension of $\circ$ to $\mathbb{C}$. Rational equivariant forms [11] will be of particular interest in the following and are parameterized by modular forms. More precisely, given a modular form $f \in M_k$ of weight k, a corresponding rational equivariant form h_f can be defined according to

$$h_f(z) = z + k\frac{f(z)}{f'(z)} \tag{12}$$

Leveraging on the previous paragraph, we can therefore obtain a set $\mathcal{E}_k$ of parametric representations of equivariant modular forms by leveraging on the basis M_k, with more precisely

$$\mathcal{E}_k = \left\{ h_\alpha^k : h_\alpha^k(z) = z + k\frac{f_\alpha(z)}{f_\alpha'(z)}, \ \alpha = \left(\alpha_{a,b}^f\right)_{a,b\in\mathbb{N}}^{4a+6b=k}, \ \alpha_{a,b}^f \in \mathbb{C} \right\} \tag{13}$$

3   SL(2, ℤ)-Equivariant Machine Learning

3.1   Approach

We consider in the following that our input samples x are points in a product space of $\mathbb{H}_2$ spaces. Modulo isomorphism, it was shown in [18] that generalized convolutions can be used to build parameterized operators equivariant to $SL(2, \mathbb{R})$. These convolutions can then be embedded into the structure of a Group-Convolutional Neural Network (G-CNN) [9] to provide end-to-end differentiability for training within the usual Machine Learning development frameworks.

Although promising results were obtained for several applications using this approach [16,19], the computational cost associated with $SL(2, \mathbb{R})$-convolutions currently prevents from scaling to industrial applications. We here propose approximating the action of $SL(2, \mathbb{R})$ by that of $SL(2, \mathbb{Z})$ and to leverage on the equivariant forms introduced in 2.1 in order to build a $SL(2, \mathbb{Z})$-equivariant architecture for which the outputs can be efficiently computed.

More precisely, for a given input $x = (x_1, ..., x_p) \in \mathbb{H}_2^p$, we can parameterize the building of m $SL(2, \mathbb{Z})$-equivariant features by leveraging on the set $\mathcal{E}_k$ and by computing $h_{\alpha_j}^{k_j}(x) = \left(h_{\alpha_j}^{k_j}(x_1), ..., h_{\alpha_j}^{k_j}(x_p)\right) \in \mathbb{C}^p$, for $j = 1, ..., m$. These features are then processed by appropriate layers (e.g., fully connected, softmax, etc.) to produce adequate outputs for the considered learning task. The trainable parameters of our architecture are therefore the parameters $\alpha_j = \left(\alpha_{a,b}^{f_j}\right)_{a,b\in\mathbb{N}}^{4a+6b=k_j}$, $j = 1, ..., m$, and the weights of the downstream layers.

3.2   Scalability and Expressivity

Computing the $SL(2, \mathbb{Z})$-equivariant operators can be done very efficiently as only requiring to evaluate Eisenstein series and their derivatives. To do so, our implementation relies on the following formula

$$G_k(z) = 2\zeta(k) + 2\frac{(2i\pi)^k}{(k-1)!} \sum_{n=1}^{\infty} \sigma_{k-1}(n) q(z)^n \tag{14}$$

where $q(z) = \exp(2i\pi z)$, ζ is the Riemann zeta function and for any $t \geq 0$, we have

$$\sigma_t(n) = \sum_{1 \leq d|n} d^t \tag{15}$$

With this respect, there is therefore no practical issue with the scaling to wide neural architectures (large values of m). However, our approach relies on the assumption that the parameterized sets of rational equivariant forms $\mathcal{E}_k$ are rich enough to approximate the considered target function. Although this paper provides some preliminary evidence for supporting the expressivity of our approach, obtaining corresponding rigorous approximation theorems is deferred to some further work.

4   Application to THPD Matrices Classification

We give here an application of our SL(2, ℤ)-equivariant architecture introduced in Sect. 3 to the problem of THPD matrices classification. Before highlighting results from some of our experiments, we remind ourselves about the existing relationship between such matrices and hyperbolic product spaces.

4.1   Hyperbolic Embedding and Equivariant ML

As described in [6], the Trench-Verblunsky theorem allows to build an isomorphism between the space of THDP matrices $\mathcal{T}_n^+$ and the product space of $\mathbb{R}_+^*$ with $n-1$ copies of the Poincaré disk $\mathbb{D}$. Leveraging on the isomorphism between $\mathbb{D}$ and $\mathbb{H}_2$ component-wise, we can therefore represent an element $\Gamma \in \mathcal{T}_n^+$ by a power coefficient and $n-1$ elements $x_i \in \mathbb{H}_2$ which are the $\mathbb{H}_2$ representation of the usual reflection coefficients $\mu_i \in \mathbb{D}$. We will consider in the following only rescaled THDP matrices in $\mathcal{T}_n^+$, which can therefore be represented by a vector $x = (x_1, ..., x_{n-1}) \in \mathbb{H}_2^{n-1}$. The SL(2, ℤ)-equivariant machine learning approach introduced in Sect. 3 can therefore be leveraged on to perform THPD matrices classification tasks, as illustrated below.

4.2   Motivations from Quantitative Finance

Detecting regime changes in time series has many applications in the financial industry, including in particular the design of efficient trading strategies. A classical approach for analyzing the time series of a given asset prices is to use an Auto-Regressive (AR) model on its daily returns, assuming in particular stationarity. Under this paradigm, the corresponding auto-correlation matrix of order H is THPD of dimension H and, following 4.1, can therefore be represented

$H - 1$ coefficients in $\mathbb{H}_2$. Detecting changes between two known regimes (e.g., usual/distressed market conditions) in the time series can then be seen as a classification task on $\mathcal{T}_H^+$. However, the empirical estimation of the auto-correlation matrices on rolling time windows is subject to measurement noise [2], which can affect the accuracy of the classification. This motivates in particular the need for achieving robustness to noise when classifying THPD matrices with Machine Learning algorithms.

4.3   Numerical Experiments

We present here some experiments that we have run to substantiate the interest of our approach for quantitative finance applications by anchoring in particular in the paradigm described in 4.2.

More precisely, we have followed the approach introduced in [6] and considered that the returns of a given time series of interest in the ith regime can be represented by the random vector $Z_{(i)}$ defined by the following equation

$$Z_{(i)} = \Gamma_{(i)}^{1/2} X + \epsilon \tag{16}$$

where $\Gamma_{(i)}$ is a simulated THPD matrix, $X \sim \mathcal{N}_{\mathbb{C}}(0, \sigma_X)$, $\epsilon \sim \mathcal{N}_{\mathbb{C}}(0, \sigma_\epsilon)$, and $\mathcal{N}_{\mathbb{C}}(0, \sigma)$ refers to the multivariate complex Gaussian distribution with zero-mean vector and covariance matrix σId. We consider in the following a two-classes classification problem ($i = 2$) for which we have computed a training set of 1000 matrices (500 of each class) from simulated returns distributed according to (16), for $\epsilon \sim \mathcal{N}_{\mathbb{C}}(0, 1)$. We are then interested in evaluating our approach on similar test sets T_{σ_ϵ} that are obtained by empirical measurement from new returns, still distributed according to (16), but with an error term $\epsilon \sim \mathcal{N}_{\mathbb{C}}(0, \sigma_\epsilon)$. This formalism in particular allows us to study the robustness of the approach with respect to the measurement noise ϵ by varying its variance σ_ϵ.

More precisely, we have instantiated the algorithm described in Sect. 3 by building a layer of 16 equivariant features, which are then flattened to a 2 dimensional logit layer. To appreciate the improvement provided by our approach, we have compared it with a Fully Connected Neural Network (FCNN) of 2 layers with ReLu activation functions. Both algorithms share roughly the same number of parameters and we will denote them $\mathcal{A}_{SL(2,\mathbb{Z})}$ and $\mathcal{A}_{FCNN}$ when trained on the previously introduced training data.

As illustrated on Fig. 1, the algorithm $\mathcal{A}_{SL(2,\mathbb{Z})}$ is more robust than $\mathcal{A}_{FCNN}$ with respect to the increase in the noise variance σ_ϵ, while achieving a slightly higher accuracy in the baseline scenario where $\sigma_\epsilon = 1$. Our results were obtained by averaging over 10 trainings with random weights initialization. We can further observe that the accuracy of $\mathcal{A}_{SL(2,\mathbb{Z})}$ has a higher standard deviation than that of $\mathcal{A}_{FCNN}$, which we explain by some numerical noise due to small values occurring in the denominator of Equation (12) when computing the equivariant features. While not impacting the performance ranking of the two algorithms, improving this behavior will be subject to some further work.

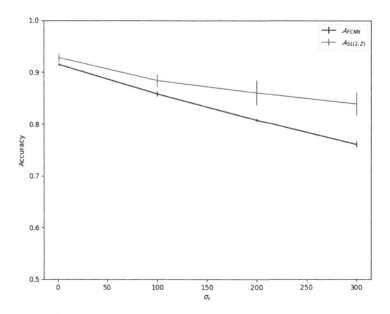

Fig. 1. Average accuracy results for the algorithms $\mathcal{A}_{SL(2,\mathbb{Z})}$ and $\mathcal{A}_{FCNN}$ on the testing sets T_{σ_ϵ} shown as a function of the variance σ_ϵ of the measurement noise, together with the corresponding standard deviations represented as error bars.

5   Conclusion and Further Work

In this paper, we have introduced a Machine Learning approach allowing to process inputs lying in the 2D hyperbolic space $\mathbb{H}_2$ while providing equivariance to the action of SL $(2, \mathbb{Z})$. Thanks to the use of modular forms theory, our proposed approach relies on closed form-formulas evaluations and does not require computing group-based convolution operators, which makes it scalable. Finally, we have substantiated its interest on simulated data in the context of THDP matrices classification, a task which has a wide range of applications in Quantitative Finance. Further work will include analysing the expressivity of our approach through the lens of approximation theorems, studying the numerical behavior of our equivariant layers and investigating their usage for other applications, including for instance generative tasks by following the spirit of [20]. Benchmarking with other approaches such as [3] which leverages on Random Matrix Theory will be also of interest, in particular when working with non-Gaussian stationary returns.

References

1. Barbaresco, F.: Radar micro-doppler signal encoding in Siegel unit poly-disk for machine learning in Fisher metric space. In: 2018 19th International Radar Symposium (IRS), pp. 1–10 (2018). https://doi.org/10.23919/IRS.2018.8448021

2. Bouchaud, J.P., Potters, M.: Financial applications of random matrix theory: a short review (2009). https://doi.org/10.48550/ARXIV.0910.1205, https://arxiv.org/abs/0910.1205

3. Bouchaud, J.P., Mastromatteo, I., Potters, M., Tikhonov, K.: Excess out-of-sample risk and fleeting modes (2022). https://doi.org/10.48550/ARXIV.2205.01012, https://arxiv.org/abs/2205.01012

4. Bronstein, M.M., Bruna, J., Cohen, T., Veličković, P.: Geometric deep learning: Grids, groups, graphs, geodesics, and gauges (2021)

5. Brooks, D., Schwander, O., Barbaresco, F., Schneider, J., Cord, M.: A hermitian positive definite neural network for micro-doppler complex covariance processing. In: 2019 International Radar Conference (RADAR), pp. 1–6 (2019). https://doi.org/10.1109/RADAR41533.2019.171277

6. Cabanes, Y., Barbaresco, F., Arnaudon, M., Bigot, J.: Non-supervised machine learning algorithms for radar clutter high-resolution doppler segmentation and pathological clutter analysis. In: International Radar Symposium. Ulm, Germany, June 2019. https://doi.org/10.23919/IRS.2019.8768140, http://hal.archives-ouvertes.fr/hal-02875365

7. Cannon, J.W., Floyd, W., Kenyon, R., Parry, W.R.: Hyperbolic geometry. Flavors Geom. **31**(59-115), 2 (1997)

8. Chakraborty, R., Bouza, J., Manton, J.H., Vemuri, B.C.: Manifoldnet: a deep neural network for manifold-valued data with applications. IEEE Trans. Pattern Anal. Mach. Intell. **44**(2), 799–810 (2022). https://doi.org/10.1109/TPAMI.2020.3003846

9. Cohen, T.S., Geiger, M., Weiler, M.: A general theory of equivariant CNNs on homogeneous spaces. In: Wallach, H., Larochelle, H., Beygelzimer, A., Alché-Buc, F., Fox, E., Garnett, R. (eds.) Advances in Neural Information Processing Systems, vol. 32, pp. 9145–9156. Curran Associates, Inc. (2019). http://proceedings.neurips.cc/paper/2019/file/b9cfe8b6042cf759dc4c0cccb27a6737-Paper.pdf'

10. Diamond, F., Shurman, J.: A First Course in Modular Forms. GTM, vol. 228. Springer, New York (2005). https://doi.org/10.1007/978-0-387-27226-9

11. El Basraoui, A., Sebbar, A.: Rational equivariant forms. Int. J. Number Theor. **08**(04), 963–981 (2012). https://doi.org/10.1142/S1793042112500571

12. Elbasraoui, A., Sebbar, A.: Equivariant forms: Structure and geometry. Can. Math. Bull. **56**(3), 520–533 (2013). https://doi.org/10.4153/CMB-2011-195-2

13. Finzi, M., Welling, M., Wilson, A.G.: A practical method for constructing equivariant multilayer perceptrons for arbitrary matrix groups (2021). https://doi.org/10.48550/ARXIV.2104.09459, https://arxiv.org/abs/2104.09459

14. Gerken, J.E., et al.: Geometric deep learning and equivariant neural networks (2021)

15. Lagrave, P.Y., Barbaresco, F.: Generalized SU(1,1) Equivariant Convolution on Fock-Bargmann spaces for robust radar doppler signal classification, July 2021. http://hal.science/hal-03309817, working paper or preprint

16. Lagrave, P.Y., Barbaresco, F.: Hyperbolic equivariant convolutional neural networks for fish-eye image processing, February 2022. http://hal.science/hal-03553274, working paper or preprint

17. Lagrave, P.Y., Barbaresco, F.: Adaptive importance sampling for equivariant group-convolution computation. Phys. Sci. Forum **5**(1) (2022). https://doi.org/10.3390/psf2022005017, www.mdpi.com/2673-9984/5/1/17

18. Lagrave, P.Y., Cabanes, Y., Barbaresco, F.: su(1,1) equivariant neural networks and application to robust toeplitz hermitian positive definite matrix classification. In: Nielsen, F., Barbaresco, F. (eds.) Geometric Science of Information, pp. 577–584. Springer International Publishing, Cham (2021)

19. Lagrave, P.Y., Cabanes, Y., Barbaresco, F.: An equivariant neural network with hyperbolic embedding for robust doppler signal classification. In: 2021 21st International Radar Symposium (IRS), pp. 1–9 (2021). https://doi.org/10.23919/IRS51887.2021.9466226

20. Marti, G.: Corrgan: sampling realistic financial correlation matrices using generative adversarial networks. In: ICASSP 2020–2020 IEEE International Conference on Acoustics, Speech and Signal Processing (ICASSP), pp. 8459–8463 (2020). https://doi.org/10.1109/ICASSP40776.2020.9053276

21. Marti, G., Nielsen, F., Bińkowski, M., Donnat, P.: A review of two decades of correlations, hierarchies, networks and clustering in financial markets, pp. 245–274, March 2021. https://doi.org/10.1007/978-3-030-65459-7_10

22. Sebbar, A., Sebbar, A.: Equivariant functions and integrals of elliptic functions. Geometriae Dedicata **160**(1), 373–414 (2012). https://doi.org/10.1007/s10711-011-9688-7

23. Stein, W.: Modular Forms, a ComputationalAapproach. Graduate Studies in Mathematics, American Mathematical Society (2007). http://books.google.fr/books?id=blaZAwAAQBAJ'

K-Splines on SPD Manifolds

Margarida Camarinha[1]([✉])(iD), Luís Machado[2,3](iD), and Fátima Silva Leite[2,4](iD)

[1] CMUC, University of Coimbra, Department of Mathematics,
3000-143 Coimbra, Portugal
mmlsc@mat.uc.pt

[2] Institute of Systems and Robotics - University of Coimbra, Rua Silvio Lima - Polo
II, 3030-290 Coimbra, Portugal
fleite@mat.uc.pt

[3] Department of Mathematics, University of Trás-os-Montes e Alto Douro (UTAD),
Quinta de Prados, 5000-801 Vila Real, Portugal
lmiguel@utad.pt

[4] Department of Mathematics, University of Coimbra, 3000-143 Coimbra, Portugal

1 Introduction

Symmetric positive definite (SPD) matrices are widely used in data science applications. In computer vision, for instance, image and video information is encoded by SPD matrices. Identified as diffusion tensors, the SPD matrices are fundamental tools in medical imaging for many neuroscientific studies, including schizophrenia, multiple sclerosis, autism, depression, hypoxia-ischemia, trauma, Alzheimer's disease and other dementias. More details about these applications can be found, for instance, in [11] and references therein. An essential task in image processing requires to interpolate known data to obtain new data. In this context, developing interpolation schemes for SPD matrices is clearly very important. But working on the interpolation of SPD matrices can be quite demanding, since the geometry of the SPD space has to be chosen to comply with the specific area of application and properties of the data.

The most natural geometric framework is given by a Riemannian metric, and in this context Lie groups and symmetric spaces also play a role. Different Riemannian and Lie group structures on the SPD space have been considered in the literature (see for instance the recent work of Lin [4], Lin et al. [5], Pennec [12], Thanwerdas and Pennec [15], and Arsigny et al. [1]). The choice of the metric should be guided by the invariance and computational properties that are important for each specific application. This analysis can be given by observing different effects that can occur (swelling, fattening and shrinking effects). One inconvenient feature of Frobenius and other metrics is the swelling effect in the geodesics connecting two SPD matrices. The swelling can hinder the correct

The work of Margarida Camarinha was partially supported by the Centre for Mathematics of the University of Coimbra - UIDB/00324/2020, funded by the Portuguese Government through FCT/MCTES. The work of Luís Machado and Fátima Silva Leite has been supported by Fundação para a Ciência e Tecnologia (FCT) under the project UIDP/00048/2020.

F. Nielsen and F. Barbaresco (Eds.): GSI 2023, LNCS 14072, pp. 624–633, 2023.
https://doi.org/10.1007/978-3-031-38299-4_64

interpretation of the information carried by the SPD matrices, with evident consequences in data interpolation problems. The Log-Cholesky and Log-Euclidean metrics were designed to eliminate this setback. A comparison of the properties of the most commonly used metrics in the SPD space can be found in [12,15].

The main objective of this paper is to study high-order interpolation on SPD manifolds. A high-order interpolation method in Riemannian manifolds based on the optimization properties of the Euclidean splines was introduced in Camarinha et al. [2]. This method gave rise to the so-called geometric splines. The generalization of Euclidean splines to Riemannian manifolds, initiated with the work of Noakes et al. [7] and Crouch and Silva Leite [3], was motivated by trajectory planning problems for rigid body motion, but quickly became quite relevant in many other areas of science and technology. In SPD manifolds, geometric splines were studied in Zhang and Noakes [16] and Machado and Silva Leite [6].

In this paper, we first review the theory of high-order geometric splines for general Riemannian manifolds and its specialization to Lie groups, based on the work of Camarinha et al. [2] and Popiel [14], respectively. We then study geometric splines in SPD manifolds, by considering the Log-Cholesky metric and the Lie group structure introduced in [4]. Using that geometric structure, we derive a necessary and sufficient condition for a curve in SPD to be a geometric spline. We also present a closed form expression for cubic polynomials satisfying boundary conditions on position and velocity. The choice of the Log-Cholesky metric enables to obtain easy-to-compute expressions for higher order interpolation curves. The Cholesky factor representation of SPD matrices reduces substantially the computational costs in comparison with other metrics, which makes possible to work with larger dimensional input data.

2   Riemannian Splines

Let $(M, \langle \cdot, \cdot \rangle)$ be a n-dimensional Riemannian manifold. Denote by $\frac{D}{dt}$ the covariant derivative along a curve associated with the Levi-Civita connection on M, and by R the curvature tensor. For a curve x in M, the notation $\frac{D^{i+1}x}{dt^{i+1}}$ will be used to represent $\frac{D^i}{dt^i}\left(\frac{dx}{dt}\right)$, $i \geq 0$.

We consider the following natural generalization of the variational problem that gave rise to the Euclidean splines of odd degree.

Problem ($\mathcal{P}$):

$$\min_{x \in \Gamma} \frac{1}{2} \int_0^1 \left\langle \frac{D^m x}{dt^m}, \frac{D^m x}{dt^m} \right\rangle dt, \tag{1}$$

over the class Γ of $\mathcal{C}^{2m-3}$ paths x on M satisfying $x|_{[t_i, t_{i+1}]}$ is smooth,

$$x(t_i) = x_i, \ 0 \leq i \leq N, \tag{2}$$

for a distinct set of points $x_i \in M$ and fixed times t_i, $0 \le i \le N$, where $0 = t_0 < t_1 < \cdots < t_{N-1} < t_N = 1$, and, in addition,

$$\frac{D^j x}{dt^j}(0) = v_0^j, \quad \frac{D^j x}{dt^j}(1) = v_1^j, \quad 1 \le j \le m-1, \tag{3}$$

where v_i^j, with $i = 0, 1$ and $1 \le j \le m-1$, are fixed tangent vectors.

Proposition 1. ([2]) *A necessary condition for x to be a minimizer of the functional (1) is that x is C^{2m-2} and, for $0 \le i \le N-1$,*

$$\frac{D^{2m} x(t)}{dt^{2m}} + \sum_{j=2}^{m} (-1)^j R\left(\frac{D^{2m-j} x(t)}{dt^{2m-j}}, \frac{D^{j-1} x(t)}{dt^{j-1}}\right) \frac{dx(t)}{dt} = 0, \; \forall t \in [t_i, t_{i+1}]. \tag{4}$$

Definition 1. *We say that a curve $x \in \Gamma$ is a* geometric spline of degree $2m-1$ *on M if x is C^{2m-2} and each curve segment $x|_{[t_i, t_{i+1}]}$ satisfies equation (4).*

In the absence of interpolating points, these curves are smooth and we naturally call them *geometric polynomials of degree $2m-1$*. Hence, geometric splines of degree $2m - 1$ can be described as C^{2m-2} curves obtained by concatenation of geometric polynomials at the interpolating points.

Geometric polynomials of degree one are geodesics in a Riemannian manifold and geometric polynomials of degree three are given through the equation

$$\frac{D^4 x}{dt^4} + R\left(\frac{D^2 x}{dt^2}, \frac{dx}{dt}\right)\frac{dx}{dt} = 0. \tag{5}$$

These curves were studied in Noakes et al. in [7] and Crouch and Silva Leite [3], to develop dynamical interpolation schemes on Lie groups and symmetric spaces.

Suppose now that the Riemannian manifold M is a connected Lie group G endowed with a bi-invariant metric and denote by $\mathfrak{g}$ its Lie algebra. Equation (4) can be reduced to a $2m - 1$ order differential equation in $\mathfrak{g}$ using the so-called Lie reduction of a vector field along a curve. Given a curve x in G, we define the curve in $\mathfrak{g}$ by $V = d\ell_x^{-1} \circ \frac{dx}{dt}$, where ℓ_x denotes the left translation in G. We denote by $V^{(s)}$ the usual s-order derivative of V. In order to write the reduced equation in V for all values of m, we define $V_0 := V$ and introduce the following auxiliary variables V_k, $k = 1, \ldots, 2m - 2$, and Z_m.

$$V_k = V_{k-1}^{(1)} + \frac{1}{2}[V, V_{k-1}], \quad k = 1, \ldots, 2m-2, \tag{6}$$

$$Z_m = V_{2m-2} + Y_m, \tag{7}$$

with $Y_m = \frac{1}{2} \sum_{j=2}^{m} (-1)^j [V_{2m-j-1}, V_{j-2}]$. This method was proposed in [14] and permits to express the Eq. (4) in terms of a Lax equation.

Proposition 2. *([14]) A curve x is a geometric polynomial of degree $2m - 1$ iff*

$$\frac{dx}{dt} = d\ell_x \circ V \tag{8}$$

$$Z_m^{(1)} = [Z_m, V], \tag{9}$$

with Z_m given by (6–7).

The reduced equation in $\mathfrak{g}$ is highly complex, but the auxiliary variable Z_m enables to rewrite it in Lax form and easily identify conserved quantities. There are very few examples where a closed form expression for geometric polynomials is known, even for the lowest values of m. For the case $m = 2$, it is important to mention the extensive work done by Noakes and its collaborators [8–10]. In this case, the Eq. (9) is

$$Z_2^{(1)} = [Z_2, V], \tag{10}$$

with $Z_2 = V_2 + Y_2$ and $V_2 = V_1^{(1)} - Y_2$. Then $Z_2 = V^{(2)}$ and we have

$$V^{(3)} = [V^{(2)}, V]. \tag{11}$$

Equation (11) was first obtained for $G = SO(3)$ in [7] and for general Lie groups in [3].

When the Lie group G is Abelian, the Eqs. (6)–(9) simplify substantially. Geometric polynomials are then obtained through Euclidean polynomials in the Lie algebra. Moreover, they give rise to the solution of Problem $(\mathcal{P})$, since the necessary condition in Proposition 1 is also sufficient.

In the next section, we obtain geometric splines in SPD manifolds endowed with the so-called Log-Cholesky metric introduced by Lin in [4].

3   Splines on SPD Manifolds

3.1   Geometry of SPD with Respect to the Log-Cholesky Metric

In this subsection, we review the results introduced in [4] that are most relevant to establish our main result.

Given a $n \times n$ real matrix $A = [a_{i,j}]$, we use $\mathbb{L}(A)$ to represent the $n \times n$ matrix whose (i, j) element is a_{ij} whenever $i > j$ and 0 otherwise. We also use $\mathbb{D}(A)$ to denote the diagonal matrix whose (i, i) element is a_{ii}.

Denote by $\mathfrak{t}(n)$ the set of $n \times n$ lower triangular matrices. A matrix $X \in \mathfrak{t}(n)$ can be written as $X = \mathbb{L}(X) + \mathbb{D}(X)$, where $\mathbb{L}(X)$ is the strictly lower triangular part and $\mathbb{D}(X)$ is the diagonal part of X.

Now, denote by $\mathfrak{t}^+(n)$ the subset of $\mathfrak{t}(n)$ with positive diagonal elements. A matrix $L \in \mathfrak{t}^+(n)$ can be parametrized by a lower triangular matrix via the diffeomorphism

$$\begin{aligned}
\mathscr{D} : \mathfrak{t}^+(n) &\longrightarrow \mathfrak{t}(n) \\
L &\longmapsto \mathscr{D}(L) = \mathbb{L}(L) + \log(\mathbb{D}(L)).
\end{aligned} \tag{12}$$

We can use the map $\mathscr{D}$ to pull back to $\mathfrak{t}^+(n)$ the additive product in $\mathfrak{t}(n)$, obtaining the following product $\odot$ in $\mathfrak{t}^+(n)$

$$L \odot K = \mathbb{L}(L) + \mathbb{L}(K) + \mathbb{D}(L)\mathbb{D}(K), \tag{13}$$

which gives an Abelian Lie group structure to $\mathfrak{t}^+(n)$. Then the diffeomorphism $\mathscr{D}$ becomes an isomorphism between the Lie groups $(\mathfrak{t}^+(n), \odot)$ and $(\mathfrak{t}(n), +)$.

The tangent map of the left translation ℓ_L at $K \in \mathfrak{t}^+(n)$ is given by

$$d(\ell_L)_K(Y) = \mathbb{L}(Y) + \mathbb{D}(L)\mathbb{D}(Y), \quad \text{for } Y \in T_K \mathfrak{t}^+(n), \tag{14}$$

where the tangent space of $\mathfrak{t}^+(n)$ at a point K is identified with $\mathfrak{t}(n)$.

Let $\langle \,.\,,\,.\,\rangle_F$ denote the Frobenius inner product on $\mathfrak{t}(n)$, given by $\langle A, B \rangle_F = \mathrm{Tr}(A^\top B)$. The following defines an inner product on each tangent space of $\mathfrak{t}^+(n)$, obtained throughout the diffeomorphism $\mathscr{D}$,

$$\ll X, Y \gg_L = \langle \mathbb{L}(X), \mathbb{L}(Y) \rangle_F + \langle \mathbb{D}(L)^{-1}\mathbb{D}(X), \mathbb{D}(L)^{-1}\mathbb{D}(Y) \rangle_F, \; L \in \mathfrak{t}^+(n). \tag{15}$$

Consequently, $\mathfrak{t}^+(n)$ is a Riemannian manifold with bi-invariant metric $\ll \cdot,\cdot \gg$.

Now, let $\mathfrak{s}(n)$ be the set of $n \times n$ real symmetric matrices and $\mathfrak{s}^+(n)$ the open convex half cone of symmetric and positive definite matrices. Given $P \in \mathfrak{s}^+(n)$, there exists a unique $L \in \mathfrak{t}^+(n)$, such that $P = LL^\top$. The matrix L is called the *Cholesky factor* of P. The *Cholesky map* is the following diffeomorphism,

$$\begin{aligned} \mathscr{L} : \mathfrak{s}^+(n) &\longrightarrow \mathfrak{t}^+(n) \\ P &\longmapsto \mathscr{L}(P) = L, \end{aligned} \tag{16}$$

where $P = LL^\top$. Its inverse is defined by

$$\begin{aligned} \mathscr{S} : \mathfrak{t}^+(n) &\longrightarrow \mathfrak{s}^+(n) \\ L &\longmapsto \mathscr{S}(L) = LL^\top. \end{aligned} \tag{17}$$

The composition $\mathscr{D} \circ \mathscr{L}$ gives the Log-Cholesky parametrization presented in [13]. Using this single chart, a matrix $P = LL^\top \in \mathfrak{s}^+(n)$ can be represented by the lower triangular matrix $\mathbb{L}(L) + \log(\mathbb{D}(L))$.

In order to equip the SPD manifold $\mathfrak{s}^+(n)$ with a Riemannian structure, define

$$S_{\frac{1}{2}} := \mathbb{L}(S) + \frac{1}{2}\mathbb{D}(S), \quad \text{for a square matrix } S. \tag{18}$$

The tangent map of $\mathscr{S}$ at $L \in \mathfrak{t}^+(n)$ is given by

$$\begin{aligned} d\mathscr{S}_L : T_L \mathfrak{t}^+(n) &\longrightarrow T_{LL^\top}\mathfrak{s}^+(n) \\ X &\longmapsto XL^\top + LX^\top, \end{aligned} \tag{19}$$

with inverse

$$\begin{aligned} d\mathscr{L}_{LL^\top} : T_{LL^\top}\mathfrak{s}^+(n) &\longrightarrow T_L \mathfrak{t}^+(n) \\ W &\longmapsto L(L^{-1}WL^{-\top})_{\frac{1}{2}}. \end{aligned} \tag{20}$$

Now, given $P = LL^\top \in \mathfrak{s}^+(n)$ and $V, W \in T_P\mathfrak{s}^+(n)$, the diffeomorphism $\mathscr{S}$ induces a Riemannian metric in $\mathfrak{s}^+(n)$, called *Log-Cholesky metric*, defined by

$$\langle V, W \rangle_P = \ll L(L^{-1}VL^{-\top})_{\frac{1}{2}}, L(L^{-1}WL^{-\top})_{\frac{1}{2}} \gg_L. \tag{21}$$

Introducing the multiplication $\otimes$ on $\mathfrak{s}^+(n)$, such that the Log-Cholesky map $\mathscr{L}$ is a homomorphism, $(\mathfrak{s}^+(n), \otimes)$ becomes an Abelian Lie group and the Log-Cholesky metric is bi-invariant.

These Riemannian and Lie structures are clearly based on the Log-Cholesky parametrization mentioned above.

3.2 K-Splines on SPD Manifolds with the Log-Cholesky Metric

The main goal of this section is to solve Problem $(\mathcal{P})$ when the SPD manifold $\mathfrak{s}^+(n)$ is equipped with the Lie group structure $\otimes$.

Theorem 1. *A necessary and sufficient condition for x to be a minimizer of the functional (1) over the class Γ of C^{2m-3} paths x on $\mathfrak{s}^+(n)$, such that $x|_{[t_i, t_{i+1}]}$ is smooth, satisfies $x(t_i) = P_i$, $0 \leq i \leq N$, and also*

$$\frac{D^j x}{dt^j}(0) = V_0^j, \quad \frac{D^j x}{dt^j}(1) = V_1^j, \quad 1 \leq j \leq m - 1,$$

is that, x is C^{2m-2}, and, $\forall t \in [t_i, t_{i+1}]$ and $0 \leq i \leq N - 1$, the following holds

$$x(t) = \mathscr{S}\Big(\mathbb{L}(L_i) + \sum_{j=1}^{2m-1} \frac{(t - t_i)^j}{j!} \mathbb{L}(Y_j) + \mathbb{D}(L_i) \exp\Big(\sum_{j=1}^{2m-1} \frac{(t - t_i)^j}{j!} \mathbb{D}(Y_j)\Big)\Big), \tag{22}$$

where $L_i = \mathscr{L}(P_i)$ and $Y_j \in \mathfrak{t}(n)$, $j = 1, \ldots, 2m - 1$, are determined by the interpolation and boundary conditions.

Proof. Let us consider a curve $x \in \Gamma$ satisfying the necessary conditions of Proposition 1. Since the Log-Cholesky map $\mathscr{L}$ in (16) defines an isometry between $\mathfrak{s}^+(n)$ and $\mathfrak{t}^+(n)$, the curve $\widetilde{x} = \mathscr{L}(x)$ in $\mathfrak{t}^+(n)$ also satisfies equation (4). But, being $(\mathfrak{t}^+(n), \odot)$ an Abelian Lie group, the Lie bracket vanishes identically and equations (6–9) simply reduce to

$$\widetilde{V}^{(2m-1)}(t) = 0, \ t \in [t_i, t_{i+1}], \ (0 \leq i \leq N - 1), \tag{23}$$

where $\widetilde{V}$ is the curve in $\mathfrak{t}(n)$ defined by $\widetilde{V} = d\ell_{\widetilde{x}}^{-1} \circ \frac{d\widetilde{x}}{dt}$. Therefore,

$$\widetilde{V}(t) = \sum_{j=0}^{2m-2} \frac{(t - t_i)^j}{j!} Y_{j+1}, \quad \text{with } Y_j \in \mathfrak{t}(n), j = 1, \ldots, 2m - 1.$$

Writing $\widetilde{V} = \mathbb{L}(\widetilde{V}) + \mathbb{D}(\widetilde{V})$, and using the expression (14) for the differential of left translation, Eq. (23) can be decomposed in the following two equations.

$$\frac{d\mathbb{L}(\widetilde{x})}{dt} = \mathbb{L}(\widetilde{V}), \quad \frac{d\mathbb{D}(\widetilde{x})}{dt} = \mathbb{D}(\widetilde{x})\mathbb{D}(\widetilde{V}). \tag{24}$$

On the other hand, the decomposition $Y_j = \mathbb{L}(Y_j) + \mathbb{D}(Y_j)$, $j = 1, \ldots, 2m - 1$, allows us to obtain

$$\mathbb{L}(\widetilde{V}(t)) = \sum_{j=0}^{2m-2} \frac{(t - t_i)^j}{j!} \mathbb{L}(Y_{j+1}), \quad \mathbb{D}(\widetilde{V}(t)) = \sum_{j=0}^{2m-2} \frac{(t - t_i)^j}{j!} \mathbb{D}(Y_{j+1}).$$

Now, integrating Eqs. (24) in the interval $[t_i, t_{i+1}]$, it is immediate to conclude that $\widetilde{x}$ is given explicitly by $\widetilde{x}(t) = \mathbb{L}(\widetilde{x}(t)) + \mathbb{D}(\widetilde{x}(t))$, where

$$\mathbb{L}(\widetilde{x}(t)) = \mathbb{L}(L_i) + \sum_{j=1}^{2m-1} \frac{(t - t_i)^j}{j!} \mathbb{L}(Y_j),$$

$$\mathbb{D}(\widetilde{x}(t)) = \mathbb{D}(L_i) \exp\left(\sum_{j=1}^{2m-1} \frac{(t - t_i)^j}{j!} \mathbb{D}(Y_j) \right). \tag{25}$$

Moreover, taking into account that diagonal matrices commute with each other, one can also write the analytical expression of $\widetilde{x}$, in the interval $[t_i, t_{i+1}]$, $0 \leq i \leq N - 1$, as

$$\widetilde{x}(t) = \mathbb{L}(L_i) + \sum_{j=1}^{2m-1} \frac{(t - t_i)^j}{j!} \mathbb{L}(Y_j) + \mathbb{D}(L_i) \prod_{j=1}^{2m-1} \exp\left(\frac{(t-t_i)^j}{j!} \mathbb{D}(Y_j) \right).$$

Since the necessary conditions satisfied by $\widetilde{x}$ are also sufficient and the matrices Y_j, $j = 1, \ldots, 2m - 1$, are uniquely determined from the regularity conditions and from the following interpolation and boundary conditions,

$$\widetilde{x}(t_i) = L_i, \quad \frac{D^j \widetilde{x}}{dt^j}(0) = L_0(L_0^{-1} V_0^j L_0^{-\top})_{\frac{1}{2}}, \quad \frac{D^j \widetilde{x}}{dt^j}(1) = L_1(L_1^{-1} V_1^j L_1^{-\top})_{\frac{1}{2}},$$

the result follows.

$\square$

The problem of finding the geodesic connecting two SPD matrices P_0 and P_1 is the case $m = 1$ of Problem $(\mathcal{P})$. Using the boundary conditions, we obtain

$$\mathbb{L}(Y_1) = \frac{\mathbb{L}(L_1) - \mathbb{L}(L_0)}{t_1 - t_0}, \quad \mathbb{D}(Y_1) = \frac{\log(\mathbb{D}(L_0)^{-1} \mathbb{D}(L_1))}{t_1 - t_0},$$

and the following holds.

Corollary 1. *The geodesic in $\mathfrak{s}^+(n)$ connecting the point P_0 (at $t = t_0$) to the point P_1 (at $t = t_1$) is given explicitly by*

$$x(t) = \mathscr{S}\left(\mathbb{L}(L_0) + \frac{t - t_0}{t_1 - t_0}(\mathbb{L}(L_1) - \mathbb{L}(L_0)) + \mathbb{D}(L_0) \exp\left(\frac{t - t_0}{t_1 - t_0} \log(\mathbb{D}(L_0)^{-1} \mathbb{D}(L_1)) \right) \right),$$

where $L_i \in \mathfrak{t}^+(n)$ is the Cholesky factor of P_i, $i = 0, 1$, $t \in [t_0, t_1]$.

Corollary 2. *The cubic polynomial x in $\mathfrak{s}^+(n)$ satisfying the boundary conditions*

$$x(t_0) = P_0, \quad \frac{Dx}{dt}(t_0) = V_0, \quad x(t_1) = P_1, \quad \frac{Dx}{dt}(t_1) = V_1,$$

is given explicitly by

$$x(t) = \mathscr{S}\Big(\mathbb{L}(L_0) + \sum_{j=1}^{3}\frac{(t-t_0)^j}{j!}\mathbb{L}(Y_j) + \mathbb{D}(L_0)\exp\Big(\sum_{j=1}^{3}\frac{(t-t_0)^j}{j!}\mathbb{D}(Y_j)\Big)\Big), \quad t \in [t_0,t_1],$$

where Y_i, $i = 1,2,3$, are given by

$$\mathbb{L}(Y_1) = \mathbb{L}(X_0),$$
$$\mathbb{L}(Y_2) = \frac{2}{(t_1-t_0)^2}\Big(3\big(\mathbb{L}(L_1) - \mathbb{L}(L_0)\big) - (t_1-t_0)\big(2\mathbb{L}(X_0) + \mathbb{L}(X_1)\big)\Big),$$
$$\mathbb{L}(Y_3) = \frac{6}{(t_1-t_0)^3}\Big(2(\mathbb{L}(L_0) - \mathbb{L}(L_1)) + (t_1-t_0)\big(\mathbb{L}(X_0) + \mathbb{L}(X_1)\big)\Big),$$

$$\mathbb{D}(Y_1) = \mathbb{D}(X_0)\mathbb{D}(L_0)^{-1},$$
$$\mathbb{D}(Y_2) = \frac{2}{(t_1-t_0)^2}\Big((t_0-t_1)(2\mathbb{D}(L_0)^{-1}\mathbb{D}(X_0) + \mathbb{D}(L_1)^{-1}\mathbb{D}(X_1)) + 3\log(\mathbb{D}(L_1)\mathbb{D}(L_0)^{-1})\Big),$$
$$\mathbb{D}(Y_3) = \frac{6}{(t_0-t_1)^3}\Big((t_0-t_1)(\mathbb{D}(L_0)^{-1}\mathbb{D}(X_0) + \mathbb{D}(L_1)^{-1}\mathbb{D}(X_1)) + 2\log(\mathbb{D}(L_1)\mathbb{D}(L_0)^{-1})\Big),$$

where L_i is the Cholesky factor of P_i and $X_i = L_i(L_i^{-1}V_iL_i^{-\top})_{\frac{1}{2}}$, $i = 1,2$.

Figure 1 bellow illustrates geometric polynomials of degree 1 and 3 using the Log-Cholesky metric and the ones obtained in [6] for the Log-Euclidean metric. In Table 1, we register the corresponding determinants. With respect to swelling effect, we don't observe significant changes in the value of those determinants when the degree of the polynomial increases.

Fig. 1. Interpolation through geodesics and geometric cubic polynomials joining the same elements. First row: Log-Cholesky geodesic interpolation. Second row: Log-Cholesky cubic interpolation. Third row: Log-Euclidean geodesic interpolation. Fourth row: Log-Euclidean cubic interpolation.

Table 1. Values of the determinant of each iteration of the curve joining P_0 to P_1.

Geodesic	36.3214	34.8781	33.4923	32.1615	30.8836	29.6564	28.4780	27.3465
Cubic	36.3214	32.7492	23.3877	15.4079	10.9067	9.6617	12.4752	27.3465

4   Conclusion

In this article, we constructed high-order polynomial spline curves on the SPD Riemannian manifold equipped with the Log-Cholesky metric. These smooth curves minimize a certain energy functional, interpolate a given set of data points and satisfy some boundary conditions.

The Abelian Lie group structure on the SPD manifold introduced in [4] enables considerable simplifications. In particular, the variational problem could be solved efficiently and closed form expressions for polynomial splines were obtained.

With the chosen structure, the interpolation problem was reduced to the Lie algebra, as shown in the proof of Theorem 1. This easily follows from the fact that the Riemannian exponential map is a global diffeomorphism that coincides at the identity with the Lie group exponential.

References

1. Arsigny, V., Fillard, P., Pennec, X., Ayache, N.: Geometric means in a novel vector space structure on symmetric positive-definite matrices. SIAM J. Matrix Anal. Appl. **29**(1), 328–347 (2007)
2. Camarinha, M., Silva Leite, F., Crouch, P.: Splines of class C^k on non-Euclidean spaces. IMA J. Math. Control Inf. **12**, 399–410 (1995)
3. Crouch, P., Silva Leite, F.: The dynamic interpolation problem on Riemannian manifolds, Lie groups and symmetric spaces. J. Dyn. Control Syst. **1**(2), 177–202 (1995)
4. Lin, Z.: Riemannian geometry of symmetric positive definite matrices via Cholesky decomposition. SIAM J. Matrix Anal. Appl. **40**(4), 1353–1370 (2019)
5. Lin, Z., Müller, H.G., Park B.U.: Additive models for symmetric positive-definite matrices and Lie groups. Biometrika. (2022). asac055. https://doi.org/10.1093/biomet/asac055
6. Machado, L., Silva Leite, F.: Interpolation and polynomial fitting in the SPD manifold. In: Proceedings of 52nd IEEE Conference on Decision and Control, Firenze, Italy, pp. 1150–1155 (2013)
7. Noakes, L., Heinzinger, G., Paden, B.: Cubic splines on curved spaces. IMA J. Math. Control Inform. **6**, 465–473 (1989)
8. Noakes, L.: Null cubics and Lie quadratics. J. Math. Phys. **44**(3), 1436–1448 (2003)
9. Noakes, L.: Duality and Riemannian cubics. Adv. Comput. Math. **25**(1–3), 195–209 (2006)
10. Noakes, L., Popiel, T.: Quadratures and cubics in $SO(3)$ and $SO(1, 2)$. IMA J. Math. Control Inform. **23**(4), 463–473 (2006)
11. O'Donnell, L.J., Westin, C.F.: An introduction to diffusion tensor image analysis. Neurosurg Clin N Am. **22**(2), 185–196 (2011)

12. Pennec, X.: Manifold-valued image processing with SPD matrices. In: Riemannian Geometric Statistics in Medical Image Analysis, pp. 75–134, Academic Press (2020)
13. Pinheiro, J.C., Bates, D.M.: Unconstrained parametrizations for variance-covariance matrices. Stat. Comput. **6**, 289–296 (1996)
14. Popiel, T.: Higher order geodesics in Lie groups. Math. Control Sig. Syst. **19**, 235–253 (2007)
15. Thanwerdas, Y., Pennec, X.: Theoretically and computationally convenient geometries on full-rank correlation matrices. SIAM J. Matrix Anal. Appl. **43**(4), 1851–1872 (2022)
16. Zhang, E., Noakes, L.: Riemannian cubics and elastica in the manifold $SPD(n)$ of all $n \times n$ symmetric positive-definite matrices. J. Geom. Mech. **11**(2), 235–253 (2019)

Geometric Deep Learning: A Temperature Based Analysis of Graph Neural Networks

M. Lapenna[1]([✉]) [ID], F. Faglioni[2] [ID], F. Zanchetta[1] [ID], and R. Fioresi[1] [ID]

[1] University of Bologna, Bologna, Italy
{michela.lapenna4,ferdinando.zanchetta2,rita.fioresi}@unibo.it
[2] University of Modena, Modena, Italy
francesco.faglioni@unimore.it

Abstract. We examine a Geometric Deep Learning model as a thermodynamic system treating the weights as non-quantum and non-relativistic particles. We employ the notion of temperature previously defined in [5] and study it in the various layers for GCN and GAT models. Potential future applications of our findings are discussed.

Keywords: Geometric Deep Learning · Statistical Mechanics · Machine Learning

1 Introduction

Machine learning and statistical mechanics share a common root; starting from the pioneering works by Jaynes [7] and Hopfield [6], up to the visionary theory of (deep) Boltzmann machines [1,10], it is clear there is a common ground and the understanding of statistically inspired machine learning models can bring a new impulse to the field. The powerful language of statistical mechanics, connecting the elusive microscopic and measurable macroscopic physical quantities seems the perfect framework to tackle the difficult interpretation questions that the successful deep neural networks present. Indeed many researchers (see [3,4] and refs. therein) have conducted a thermodynamic study, through analogies, of various actors in the most popular algorithms, Deep Learning above all, and yet such analogies were not able to fully elucidate the mechanisms of generalization and representability that still elude our understanding. Along the same vein, new mathematical modeling, inspired by thermodynamics, brought along new interesting mathematics, see [2,9], and in particular [11], that seems especially suitable to model the dissipative phenomenon we observe in the SGD experiments.

The purpose for our present paper is to initiate this thermodynamic analysis for the Geometric Deep Learning algorithm along the same line of our previous works [5,8] on more traditional Convolutional Neural Networks (CNN). We shall treat the parameters of the model as a thermodynamic system of particles and we exploit the sound notion of temperature we have previously given in [5,8]. Then, we study the temperature of the system across layers in Graph Convolutional Networks (GCN) [18] and Graph Attention Networks (GAT) [20].

F. Nielsen and F. Barbaresco (Eds.): GSI 2023, LNCS 14072, pp. 634–643, 2023.
https://doi.org/10.1007/978-3-031-38299-4_65

Our paper is organized as follows. In Sect. 2 we briefly recall the correspondence between thermodynamics concepts and neural networks ones [5]. In Sect. 3 we present a Geometric Deep Learning model on the MNIST Superpixels dataset [13] and we study the temperature of layers comparing with the behaviour found for the CNN architecture in [5,8]. In particular, we study the dependence of temperature from the two hyperparameters learning rate and batch size at the end of the training, when loss and accuracy have reached their equilibrium values. We also analyse the dynamics of the weights inside a single Graph Convolutional layer. We consider both a GCN and a GAT models and compare the results. In Sect. 4 we draw our conclusions and we lay foundations for future work.

2 Thermodynamics and Stochastic Gradient Descent

We briefly summarize the thermodynamic analysis and modelling appearing in [3,5,8].

Stochastic Gradient Descent (SGD) and its variations (e.g. Adam) are common choices, when performing optimization in Deep Learning algorithms.

Let $\Sigma = \{z_i \mid 1 \le i \le N\} \subset \mathbb{R}^D$ denote a dataset of size N, i.e., $|\Sigma| = N$, $L = (1/N)\sum L_i$ the loss function, with L_i the loss of the i-th datum z_i and $\mathcal{B}$ the minibatch. The update of the weights $w = (w_k) \in \mathbb{R}^d$ of the chosen model (e.g., Geometric Deep Learning model), with SGD occurs as follows:

$$w(t+1) = w(t) - \eta\nabla_{\mathcal{B}}L(w), \quad \text{with} \quad \nabla_{\mathcal{B}}L := \frac{1}{|\mathcal{B}|}\sum_{i\in\mathcal{B}}\nabla L_i \qquad (1)$$

where η denotes the learning rate. Equation (1) is modelled in [3] by the stochastic ODE (Ito formalism [19]) expressed in its continuous version as:

$$dw(t) = -\eta\nabla L(w)dt + \sqrt{2\zeta^{-1}D(w)}dW(t) \qquad (2)$$

where $W(t)$ is the *Brownian motion* term modelling the stochasticity of the descent, while $D(w)$ is the *diffusion matrix*, controlling the anisotropy of the diffusivity in the process. The quantity $\zeta = \eta/(2|\mathcal{B}|)$ in [3], is called the *temperature*. It accounts for the "noise" due to SGD: small minibatch sizes or a high learning rate will increase the noise in the trajectories of the weights during training.

In [5], the time evolution of the parameters is written in continuous and discrete version as:

$$dw(t) = -\eta\nabla_{\mathcal{B}}L(w)dt, \qquad w(t+1) = w(t) - \eta\nabla_{\mathcal{B}}L(w) \qquad (3)$$

The stochastic behaviour modelled by (3) is then accounted for introducing a microscopic definition of temperature mimicking Boltzmann statistical mechanics. We first define the *instantaneous temperature* $\mathcal{T}(t)$ of the system as its kinetic energy $\mathcal{K}(t)$ divided by the number of degrees of freedom d (in our case the dimension of the weight space) and a constant $k_B > 0$ to obtain the desired units:

$$T(t) = \frac{\mathcal{K}(t)}{k_B d} = \frac{1}{k_B d} \sum_{k=1}^{d} \frac{1}{2} m_k v_k(t)^2 \qquad (4)$$

where $v_k(t)$ is the instantaneous velocity of one parameter, computed as the difference between the value of the kth parameter at one step of training and its value at the previous step (the shift in time Δt is unitary since we are computing the instantaneous velocity between consecutive steps or epochs):

$$v_k(t) = \frac{w_k(t) - w_k(t-1)}{\Delta t} \qquad (5)$$

In the formula for the kinetic energy, m_k is the mass of parameter w_k and we set it to 1. This is because we have do not know if the different role of the parameters can be modelled through a parallelism with the concept of mass.

The *thermodynamic temperature* is then the time average of $\mathcal{T}(t)$:

$$T = \frac{1}{\tau} \int_0^\tau \mathcal{T}(t) \, dt = \frac{1}{\tau k_B d} \int_0^\tau \mathcal{K}(t) = \frac{K}{k_B d} \qquad (6)$$

where K is the average kinetic energy and τ is an interval of time long enough to account for small variation in temperature. In [5] we interpret this system as evolving at constant temperature: at each step the temperature is reset (analogy with system in contact with heat reservoir). Hence we do not have a *constant energy* dynamics, as it is commonly referred to in atomic simulations, but we are faced with a dissipative effect occurring at each step.

The thermodynamic analysis performed in [5] and summarized here implies that with SGD we have a residual velocity for each particle even after equilibrium is reached. Our system does not evolve according to Newton dynamics and in particular the mechanical energy is not constant. The fact we maintain a residual temperature at equilibrium with a constant temperature evolution means that we achieve a minimum of *free energy*, not of the potential energy i.e. our loss function. This fact is stated in [3] and is well known among the machine learning and information geometry community (see also [2]).

Let us summarize the key points of the system dynamics. We have:

– No costant mechanical energy $K + V$, where $K = \sum_{k=1}^{d}(1/2)m_k v_k^2$ is the kinetic energy and $V = L$ (L the loss function) is the potential energy;
– No maximization of entropy,

All of this is due to the stochasticity of SGD which is enhanced by small sizes of minibatch and high learning rate, as we shall elucidate more in our experimental section together with an analysis of the temperature in the layers.

3   Experiments

In this section we perform experiments with Geometric Deep Learning models on MNIST Superpixels PyTorch dataset [13], in order to test the dependence of the thermodynamic temperature from the hyperparameters learning rate and batch size. We examine the temperature of different layers and we look at the mean squared velocities of the weights of a single layer. We also analyze some key differences with the findings in [5] and in [8], where we proposed pruning techniques based on the notion of temperature.

We choose to investigate the behaviour of two separate and important Graph Neural Network (GNN) architectures. First, we implement a model using Graph Convolutional Network (GCN) layers from [18]. Then, we make a comparison with a Graph Attention Network (GAT) model employing attention mechanism during the convolution [20]. We stress that, in the literature, GAT models have outperformed GCN's in classification problems on superpixel images. We consider both models in order to compare the weights' dynamics and reason on the thermodynamic modelling we propose.

3.1   GCN Architecture

The architecture we use consists of four GCNConv layers followed by a concatenation of a mean and a max pooling layers and at the end a dense layer (Fig. 1). We use tanh as activation function.

```
GCN(
  (initial_conv): GCNConv(1, 32)
  (conv1): GCNConv(32, 64)
  (conv2): GCNConv(64, 64)
  (conv3): GCNConv(64, 64)
  (out): Linear(in_features=128, out_features=10, bias=True)
)
```

Fig. 1. Architecture of our GCN model in PyTorch framework. In the parenthesis next to each layer, the first and second number indicate the embedding dimension of the input and output respectively. The final linear layer takes as input an embedding which is twice the size of the output from the previous layer, due to concatenation of the two pooling operations.

We do not apply batch normalization [12] to the convolutional layers, since normalizing the weights could bias our experiments (compare with [5]). We optimize the network with SGD with a Cross Entropy loss and Adam optimizer, without any form of weight regularization.

We train this model on the MNIST Superpixels dataset obtained in [13], where the 70.000 images from the original MNIST dataset were transformed into graphs with 75 nodes, each node corresponding to a superpixel. We train the model for 600 epochs, starting with a learning rate of 10^{-3} and decaying it of a 1/10 factor every 200 epochs. After training, the model reaches an accuracy of 64%, which is worse than the performance obtained on the same dataset in [13]. We think this is due to the fact that our architecture is much simpler than MoNET [14] used in [13]. Once equilibrium of accuracy and loss is reached, we further train the model for 100 epochs and we focus our thermodynamic analysis on these last epochs at equilibrium. In particular, to investigate the dependence of temperature from learning rate and batch size, we further train the same equilibrium model by changing either the learning rate or the batch size.

In Fig. 2 and 3, we show the behaviour of the temperature T, as defined in our previous section, depending on the learning rate η and the inverse of the batch size $1/\beta$. We try values of learning rate in the range from $7 \cdot 10^{-4}$ to $3 \cdot 10^{-3}$ (batch size fixed to 32) and values of batch size in the range from 8 to 128 (learning rate fixed). We stress that, for each layer of the architecture, the temperature was computed as the mean kinetic energy of the weights averaged over the 100 equilibrium epochs.

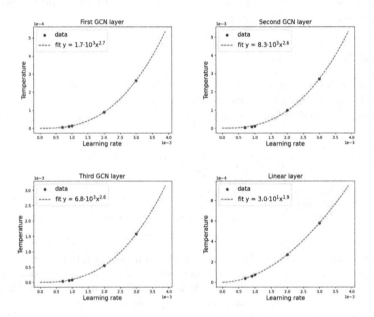

Fig. 2. Temperature dependence from the learning rate for selected layers of the GCN architecture (the other behaving similarly). The equation resulting from the fit is shown in the top left.

Despite in the literature ([3] and refs therein) the temperature ζ (Eq. 2) is commonly believed to behave proportionally to such parameters, we observe quite a different behaviour. As in [5], the dependence of temperature from the learning rate is parabolic for the linear layer, whereas it is almost parabolic for the GCNConv layers, where the exponent of x in the fit is greater than 2 (see Fig. 2). As far as the dependence on the batch size, we notice that all layers of the

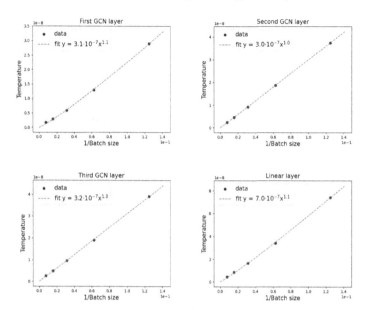

Fig. 3. Temperature dependence from the inverse of the batch size for selected layers of the GCN architecture.

architecture exhibit a linear dependence of temperature from $1/\beta$. This is in line with the results obtained by [3], but differs from the ones in [5]. Indeed, in [5], the linear dependence of the temperature from $1/\beta$ appears only for the output linear layer, while the other layers exhibit an essential non linearity. However, overall, as for the CNN architecture previously studied [5,8], we find that, given our thermodynamic definition of temperature, T is not proportional to η/β.

Furthermore, if we look at the mean squared velocities of the weights over the epochs, without averaging on the number of weights, we discover quite an interesting behaviour. Inside the same layer, the weights do not show all the same mean squared velocity at equilibrium, but different rows of the weight matrix show completely different thermal agitation (Fig. 4). This is similar to what happens in [8], where we use the concept of temperature to distinguish between "hot" and "cold" filters in a CNN layer and we discover that high temperature filters can be removed from the model without affecting the overall performance. We believe the same reasoning can be applied here for GNN layers and it could imply that some rows of the weight matrix are redundant to the learning. We expect to use this analysis to eliminate useless features or to reduce the dimensions of the feature embedding and speed up the optimization.

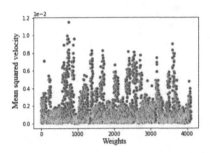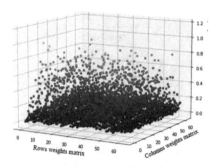

Fig. 4. Mean squared velocity for the weights of the second GCNConv layer. The weight matrix has dimensions 64 × 64 and different rows of the matrix show different temperature. The plot on the left is obtained by flattening the matrix.

3.2 GAT Architecture

The architecture we use is inspired by [15] and consists of three GATConv layers followed by a final mean pooling layer and three dense layers (Fig. 5). We take ReLU as activation function [16].

```
GAT(
    (initial_conv): GATConv(1, 32, heads=2)
    (conv1): GATConv(64, 64, heads=2)
    (conv2): GATConv(128, 64, heads=2)
    (out): Linear(in_features=128, out_features=32, bias=True)
    (out1): Linear(in_features=32, out_features=32, bias=True)
    (out2): Linear(in_features=32, out_features=10, bias=True)
)
```

Fig. 5. Architecture of our GAT model in PyTorch framework. In the parenthesis next to each layer, the first number and second number indicate the embedding dimension of the input and output respectively. The third parameter indicates the number of heads. Since each GAT in the architecture has 2 heads, each layer following a GAT has input dimension twice the size of the ouput from the previous GAT layer.

As for the previous GCN model, we do not apply either batch normalization [12] or dropout [17] or other forms of regularization to the weights. We optimize the network as described for GCN and we obtain a test set accuracy of 74%. To inspect the dependence of temperature on the two hyperparameters η and β, again we train the equilibrium model for other 100 epochs and we restrict the analysis to these final epochs. In Fig. 6 and 7, we report the behaviour of the temperature dependence on the learning rate η and the inverse of the batch size $1/\beta$ (range of values of the hyperparameters as for the GCN model). Similarly to the model without attention, the dependence of T from η is parabolic for the final linear layer and almost parabolic for the other GATConv and linear layers, since for these layers the exponent of x in the fit is greater than 2 (Fig. 6).

Furthermore, the dependence of T from $1/\beta$ is again almost linear for every layer (only the final linear layer shows a more parabolic behaviour).

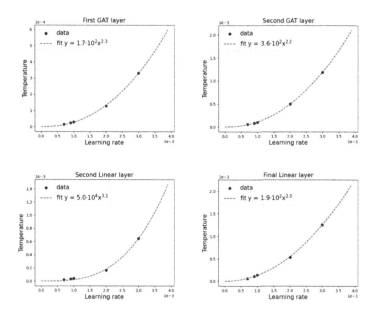

Fig. 6. Dependence of temperature from the learning rate for some layers of the GAT architecture.

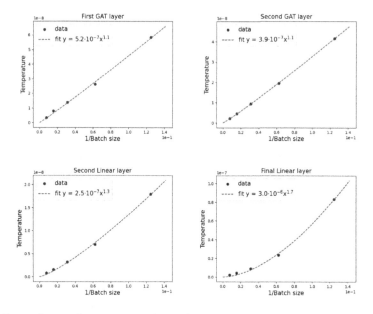

Fig. 7. Dependence of temperature from the inverse of the batch size for some layers of the GAT architecture.

4 Conclusions

We investigate the parallelism between SGD dynamics and thermodynamic systems extending the study in [5,8] to Geometric Deep Learning algorithms. Experiments show that similarly to the Deep Learning setting of CNN architectures [5,8], also for Geometric Deep Learning the temperatures of each layer behave independently. The temperature of the linear layers in the Geometric Deep Learning models considered behaves similarly to that of the linear layers in the CNN models studied in [5] and does not exhibit any simple dependence from η and β as originally assumed in [3]. On the contrary, the temperature of GCN and GAT "convolutional" layers behaves differently from the one of CNN layers considered in [5]: while the dependence of temperature from η is again almost parabolic, the one from β becomes linear. Furthermore, we find that different areas of GCNConv and GATConv layers have different temperature (see Fig. 4) as observed for the filters of a CNN. This suggests a future technique of parameter pruning, based on temperature, as in [8], that may help speed up the optimization and make it more effective. More mathematical and physical modelling is needed to further advance in this direction.

References

1. Ackley, D.H., Hinton, G.E., Sejnowski, T.J.: A learning algorithm for Boltzmann machines. Cogn. Sci. **9**(1), 147–169 (1985)
2. Barbaresco, F.: Lie group statistics and Lie group machine learning based on Souriau Lie groups thermodynamics and Koszul-Souriau-Fisher metric: new entropy definition as generalized Casimir invariant function in coadjoint representation. Entropy **22**(6), 642 (2020)
3. Chaudhari, P., Soatto, S.: Stochastic gradient descent performs variational inference, converges to limit cycles for deep networks. In: 2018 Information Theory and Applications Workshop (ITA), pp. 1–10 (2018)
4. Chaudhari, P., et al.: Entropy-sgd: biasing gradient descent into wide valleys. In: International Conference on Learning Representations ICLR 2017, 1611.01838 (2017)
5. Fioresi, R., Faglioni, F., Morri, F., Squadrani, L.: On the thermodynamic interpretation of deep learning systems. In: Geometric Science of Information: 5th International Conference (2021)
6. Hopfield, J.J.: Neurons with graded response have collective computational properties like those of two-state neurons. PNAS **81**(10), 3088–3092 (1984)
7. Jaynes, E.T.: Information theory and statistical mechanics. Phys. Rev. **106**(4), 620 (1957)
8. Lapenna, M., Faglioni, F., Fioresi, R.: Thermodynamics modeling of deep learning systems. Front. Phys. **11** (2023). https://doi.org/10.3389/fphy.2023.1145156
9. Marle, C.-M.: From tools in symplectic and Poisson geometry to J. M. Souriau's theories of statistical mechanics and thermodynamics. Entropy **18**(10), 370 (2016)
10. Salakhutdinov, R., Hinton, G.: Deep Boltzmann machines. In: Artificial Intelligence and Statistics, pp. 440–455. PMLR (2009)

11. Anahory Simoes, A., De Leon, M., Lainz Valcazar, M., De Diego, D.M.: Contact geometry for simple the rmodynamical systems with friction. Proc. R. Soci. A **476**(2241), 20200244 (2020)
12. Ioffe, S., Szegedy, C.: Batch normalization: accelerating deep network training by reducing internal covariate shift. In: Proceedings of the 32nd International Conference on Machine Learning, PMLR 37, pp. 448–456 (2015)
13. Monti, F., Boscaini, D., Masci, J., Rodolà, E., et al.: Geometric deep learning on graphs and manifolds using mixture model CNNs. In: 2017 IEEE Conference on Computer Vision and Pattern Recognition (CVPR) (2017). https://doi.org/10.1109/CVPR.2018.00335
14. Gou, M., Xiong, F., Camps, O., Sznaier, M.: MoNet: moments embedding network. In: 2018 IEEE/CVF Conference on Computer Vision and Pattern Recognition (CVPR), 1611.08402 (2018)
15. Avelar, P.H.C., Tavares, A.R., da Silveira, T.L.T., et al.: Superpixel image classification with graph attention networks. In: 33rd SIBGRAPI Conference on Graphics, Patterns and Images (SIBGRAPI), pp. 203–209 (2020). https://doi.org/10.1109/SIBGRAPI51738.2020.00035
16. Nair, V., Hinton, G.E.: Rectified linear units improve restricted Boltzmann machines. In: Proceedings of the 27th International Conference on Machine Learning (ICML-10), 21–24 June (2010)
17. Srivastava, N., Hinton, G., Krizhevsky, A., et al.: Dropout: a simple way to prevent neural networks from overfitting. J. Mach. Learn. Res. **15**(1), 1929–1958 (2014)
18. Kipf, T.N., Welling, M.: Semi-supervised classification with graph convolutional networks. In: International Conference on Learning Representations ICLR 2017 (2017). https://openreview.net/forum?id=SJU4ayYgl
19. Risken, H.: The Fokker-Planck Equation. In: Methods of Solution and Applications, Springer, Berlin (1996). https://doi.org/10.1007/978-3-642-61544-3
20. Velikovi, B.Y., et al.: Graph attention networks. In: International Conference on Learning Representations ICLR 2018 (2018). https://doi.org/10.40550/ARXIV.1710.10903

Correction to: Geometric Science of Information

Frank Nielsen⬭ and Frédéric Barbaresco⬭

Correction to:
F. Nielsen and F. Barbaresco (Eds.): *Geometric Science of Information*, **LNCS 14072,**
https://doi.org/10.1007/978-3-031-38299-4

The original version of the book was inadvertently published with a typo in the frontmatter. In the headline of page xiii it should read "GSI'23 Keynote Speakers" instead of "GSI'21 Keynote Speakers" and a typo sponsor's section. Instead of "European COST CaLISTA" it should read "European Horizon CaLIGOLA". This has been corrected.

The updated original version of the book can be found at
https://doi.org/10.1007/978-3-031-38299-4

Author Index

F. Nielsen and F. Barbaresco (Eds.): GSI 2023, LNCS 14072, pp. 645–648, 2023.
https://doi.org/10.1007/978-3-031-38299-4

Printed in the United States
by Baker & Taylor Publisher Services